Advances in Turbulence 3

A.V. Johansson P.H. Alfredsson (Eds.)

Advances in Turbulence 3

Proceedings of the
Third European Turbulence Conference
Stockholm, July 3-6, 1990

With 366 Figures

Springer-Verlag
Berlin Heidelberg New York
London Paris Tokyo
Hong Kong Barcelona
Budapest

Professor Arne V. Johansson
Department of Mechanics,
Royal Institute of Technology,
S-100 44 Stockholm, Sweden

Professor P. Henrik Alfredsson
Department of Gasdynamics,
Royal Institute of Technology,
S-100 44 Stockholm, Sweden

ISBN-13: 978-3-642-84401-0 e-ISBN-13: 978-3-642-84399-0
DOI: 10.1007/978-3-642-84399-0

Softcover reprint of the hardcover 1st edition 1991

55/3140 – 543210 – Printed on acid-free paper

Preface

The Third European Turbulence Conference was held at the Royal Institute of Technology, Sweden, from the 3rd to the 6th of July 1990 under the auspices of the European Mechanics Committee. This series of conferences is primarily devoted to fundamental aspects of turbulence and aimed at bringing together engineers, physicists and mathematicians.

The scientific committee – serving also as the European Turbulence Conference Committee – consisted of the following members: G. Comte-Bellot (Lyon), H.H. Fernholz (Berlin), H. Fiedler (Berlin), U. Frisch (Nice), J.C.R. Hunt (Cambridge), J. Jiménez (Madrid), A.V. Johansson (Stockholm), E. Krause (Aachen), G. Ooms (Amsterdam), and I. Rhyming (Lausanne). The local organizing committee consisted of A.V. Johansson and P.H. Alfredsson.

The conference programme comprised six invited lectures, given by D. Bechert, Y. Gagne, J. Jiménez, M.T. Landahl, C.H. Priddin and G.M. Zaslavsky, and 96 contributions given as either oral or poster presentations.There were 172 participants from 17 countries. The members of this scientific committee acted as referees for the selection and scrutiny of the papers published in these proceedings.

The main topics of this meeting were turbulence structure; transition and dynamical systems; turbulent combustion and mixing; turbulence affected by body forces; turbulence modelling; drag reduction and turbulence control; and novel experimental techniques. During the first evening of the conference the new MTL low-speed, low-turbulence wind-tunnel at KTH was inaugurated. The wind-tunnel outline, also acting as decoration for the "official conference cakes", is shown below.

We wish to thank all participants, in particular the survey lecturers and the session chairmen, for their contributions to the conference. Our special thanks go to the conference secretary M. Olsson, to the technical staff E. Liljeholm, L.G. Sundström, N. Tillmark and I. Ullberg, and to all others who in one way or another contributed to the success of the conference.

Generous financial support by AA Systems, ABB, ABB Stal, Alliant Computer Systems, Fläkt AB, Göran Gustafssons Stiftelse, IBM, SAAB, STEV, STU, Vattenfall and Volvo Flygmotor is gratefully acknowledged. Finally, we thank Springer-Verlag for their cooperation in publishing these proceedings.

Stockholm, December 1990

A.V. Johansson
P.H. Alfredsson

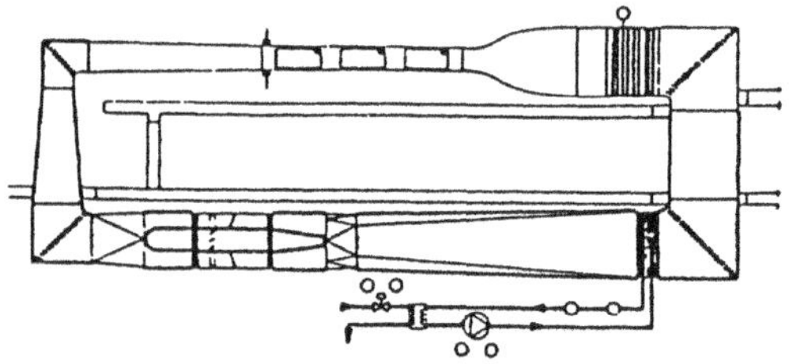

Contents

Part I Turbulence Structure

Part II Transition and Dynamical Systems

Part V Turbulence Modelling

Part I

Turbulence Structure

Structure and Statistics of Intermittency in Homogeneous Turbulent Shear Flow

James G. Brasseur and Wen-Quei Lin

Department of Mechanical Engineering, Pennsylvania State University, University Park, PA. 16802

1. Introduction

Fully developed turbulence, by nature, is inherently intermittent and topologically complex - particularly in the smaller scales. The influence of intermittency in stochastic models of global spectral dynamics, first elucidated by Kolmogorov (1962), remains a major point of difficulty (Chen *et al.* 1989). More recently, there is growing interest in the role of intermittency in scale similarity as described by multifractal geometry (Sreenivasan & Meneveau 1988, Novikov 1990). The dynamic role of local pockets of intense activity in vorticity, dissipation, Reynolds stress, etc., is not understood. It is exceedingly difficult to address this issue because the regions of interest are inherently three-dimensional, topologically and structurally complex, highly time-dependent, and exist over a wide range of length and time scales.

We describe in this paper a methodology for the extraction of intermittent regions of concentrated activity, or 'events,' in arbitrary fluctuating variables from numerically generated data, ordered by the intensity within each region. The method is applied to the study of homogeneous turbulent shear flow structure at moderate nondimensional shear-rate using direct numerical simulations at relatively low Reynolds numbers. Although the Reynolds numbers are low, the simulations yield the full three-dimensional time-dependent complexity of a large degree-of-freedom dynamical system, here-to-fore impossible experimentally. Our approach is to combine quantitative measuresJwith graphical imaging to study the topology and kinematic characteristics of the more intermittent regions within different fluctuating variables, and the temporal and dynamical interrelationships during flow evolution. From this methodology quantitative measures may be designed for application to joint numerical/experimental studies. We concentrate here on the vorticity field and the fluctuating uv component of the Reynolds stress tensor, these being variables with particular dynamic significance, yet very different in topology and scale. To what extent the highest intensity events dominate the local and global dynamics, the *significance* of these localized regions in kinematic and dynamic evolution, and the *interrelationships* among events within different fluctuating variables are primary issues of interest.

We analyze direct numerical simulations of homogeneous turbulent shear flow carried out by Rogers (Rogers et al. 1986, Roger & Moin 1987) using Rogallo's code (1981) at total mean shear $St = 8$, where S is the mean shear-rate. The pseudo-spectral calculations were carried out on a $128 \times 128 \times 128$ grid, with periodic boundary conditions applied on the boundaries of a distorting computational domain. Statistical averages are taken over the computational domain at fixed time. The initially Gaussian statistics settle down to experimentally measured values (Harris *et al.* 1977, Tavoularis & Corrsin 1981) by roughly $St = 2$. Since all scales grow with time, kinetic energy sat-

Advances in Turbulence 3
Editors: A.V. Johansson · P.H. Alfredsson

urates around $St \sim 14 - 16$. At the time step chosen for this study ($St = 8$) the flow is well-formed, but far from numerical saturation. Reynolds number based on longitudinal Taylor microscale is 74. The turbulence-to-mean time scale, $S^* = Sq^2/\varepsilon = 9$ (q^2 is twice the turbulent kinetic energy, ε is dissipation-rate). This value for S^* matches that outside the buffer layer of a flat-plate turbulent boundary layer; it is interesting that, both statistically and structurally, homogeneous turbulent shear flow appears to be similar to wall-bounded shear flows outside the buffer layer at the same S^* (Rogers & Moin 1987, Kim & Lee, 1989).

2. Definition of an Intermittent 'Event'

In the design of a method for extracting structural elements from a flow, one should take into account the *subjectivity* which is inherent in any definition of structure. A useful technique for 3D structural studies is the graphical visualization of iso-contours of the three-dimensional fluctuating fields at fixed points in time (e.g., Robinson *et al.* 1989). The approach, however, is extremely subjective, offering the viewer little idea of the compactness or temporal integrity of the structures under study. The contour level, for example, is arbitrary, and rather high contour levels are usually chosen to reduce visual clutter. The approach can easily leave false impressions, since one observes only the peaks of what are three-dimensional 'mountains,' with no information on the diffuseness or compactness of the 'mountain', the spatial extent of the 'mountain,' or the degree to which neighboring 'mountains' overlap. For these same reasons, the interpretation of quantified kinematic or dynamic characteristics based on global iso-contours is unclear. One cannot quantify, for example, the space-filling lower-magnitude fluctuations which surround the more intermittent regions.

We approach this issue with great care, recognizing that any quantitative definition of structure has some element of subjectivity involved. In our methodology, we attempt not only to quantify the *definition* of 'event,' but also the *subjective nature* of the definition. Secondly, the scheme allows for both visual and quantifiable information to be obtained concurrently. Only a simplified description of the methodology can be given here. Further details and additional analysis will be forthcoming.

The methodology is based on the over-simplified illustration given in figure 1, where a one-dimensional 'mountain' of high-intensity fluctuations is shown rising from a lower-lying sea of low-level fluctuations (both major peaks are included in the definition of 'mountain'). 1D 'mountains' correspond to 3D volumes in the data set, and a line from one side of the 'mountain' to the other at fixed height corresponds to a volumetric subset of a complete 3D 'mountain' at fixed threshold. One must choose between the whole three- dimensional 'mountain,' or some subset of the 'mountain,' in defining an 'event.' The process begins by identifying an event as the peak value in the data set. The volume occupied by the 'mountain' underlying the peak (figure 1a) is mapped by incrementally lowering the threshold from the peak value, level by level, while at each successive level searching for *adjacent* nodes in the data-set with magnitudes above the current threshold level; the event 'node list,' given as a function of threshold level, is illustrated in figure 1b. At threshold levels where a small peak within the 'mountain' is uncovered, event volume increases suddenly. These jumps in volume occur at 'stages,' as shown in figure 1c, where a sudden growth the the ratio of 'new volume' to the 'total volume' just before the jump occurs. Usually these 'stages' of growth add small amounts of volume to the event. Occasionally a large new peak is uncovered over

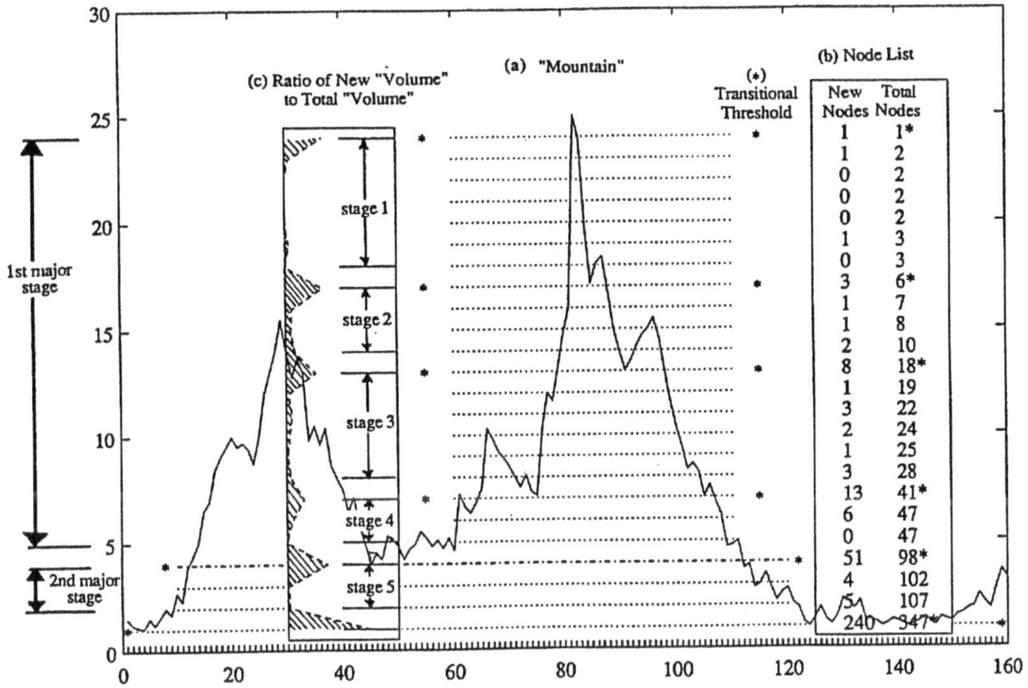

Figure 1. An over-simplified one-dimensional illustration of the process by which volumetric regions of concentrated activity are extracted from the data set for subsequent analysis. In 1D, local concentrations of fluctuating variables are shown as 'mountains' with multiple peaks. (a) A single 'mountain' is illustrated here with two major peaks, and many minor peaks. At each 'threshold level' the volumetric extent of the mountain is 'uncovered' as a list of nodes occupying the interior of the 'mountain.'(b) An example node list; i.e., the nodes within the 'mountain' at each threshold level. (c) The ratio of new volume uncovered at each lower threshold level divided by the total volume within the event up to the previous threshold. Peaks in this ratio identify 'transitional threshold' levels ($*$) where significant relative growth in event volume occurs during the extraction process.

what we call 'major stages.' We have found through extensive graphical analysis that major stages correspond to the uncovering of well-defined 'sub-structures' contiguous to previously uncovered structures. The 'final threshold' occurs when the lowest magnitude fluctuations are uncovered. Because an exceptionally large relative growth in event volume occurs here, this level is easily determined.

The advantage of the technique is that the intermittent structure is mapped out in a way which conforms to what are typically highly convoluted three-dimensional shapes defining the volumes in which the fluctuations are most highly concentrated. Subjectivity enters in the choice of the 'cutoff threshold' that defines the event boundary. We have explored this issue in great detail (Brasseur & Lin, 1990), and have developed an automated technique which separates the 'mountain' into sub-structures, allowing the user to decide how many substructures to include in each intermittent region for subsequent analysis. Once this decision is made, the first intermittent region is extracted from the data set, and the process is repeated to identify and extract events 2, 3, 4, etc.. As they are extracted, the intermittent regions are naturally ordered by peak value within the event. We use peak value as a measure of intensity in this paper. Other definitions of event intensity have been explored, with essentially the same conclusions as given here (Brasseur & Lin, 1990).

3. Some Results

3.1 Contributions to the Global Mean

Two different data sets of several thousand events each, both in enstrophy (ω^2) and Reynolds shear stress (uv), were created. In the first data set, the entire 'mountain' was defined as the event (i.e., both major peaks in figure 1), whereas in the second data set the first substructure encountered while uncovering the 'mountain' was defined as the event (the first, tallest, peak in figure 1). uv events were selected with no regard to sign ($| uv |$). In figure 2 the accumulated contributions to $< \omega^2 >$ and $< (uv)^2 >$ are plotted against the accumulated contributions to the global volume. The solid line is for the first data set (the 'entire mountain') and the dotted lines are for the second data set (the first substructure). Two important observations can be made. First, the contributions of the intermittent regions to the global mean are large relative to the contributions to the global volume. This is especially true with the second data set, where only 11% of the more intense intermittent regions within ω^2, and only 7% within uv, contribute 50% to the global mean. Clearly the intermittent regions are

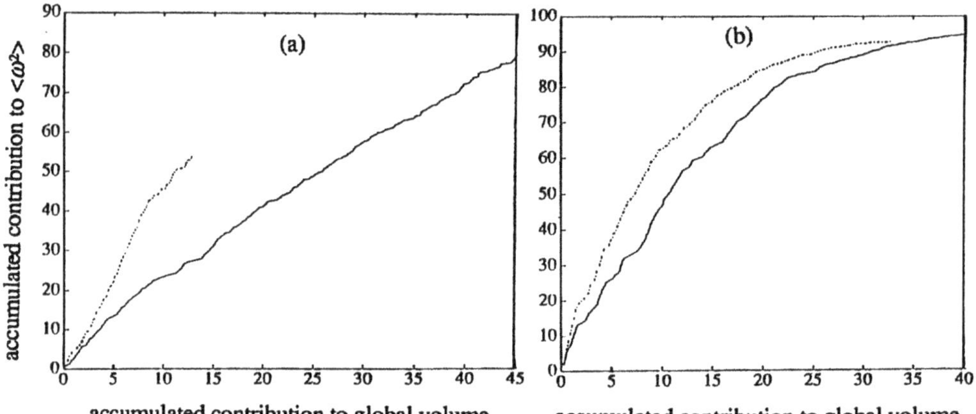

Figure 2. Contributions of intermittent regions to the global mean relative to their contributions to the global volume. (a). Accumulated mean of enstrophy from the most intense enstrophy events to the least intense events relative to the global mean, $< \omega^2 >$; plotted against accumulated event volume relative to the global volume of the data set. (b) Accumulated mean of $| uv |^2$ from the most intense $| uv |$ events to the least intense events relative to the global mean, $<| uv |^2>$; plotted against accumulated event volume relative to the global volume of the data set.

—— : events extracted as whole 'mountains,' including substructures (see text)

$\cdots$: events extracted as the first encountered substructure (see text).

major, if not dominant, contributors to the global statistics. Consequently, one might expect these regions to play an important role in global dynamical evolution as well. A second point follows partially from the relative difference between the two curves in figure 2a is very much larger than the relative difference between the two curves in figure 2b. This difference arises from the fact that, on average, intermittent ω^2 structures are topologically more complex, with more contiguous substructures, than intermittent regions in $| uv |$. Vortical structures are overwhelmingly tube-like and disk-like, stretched in the direction of mean principle strain-rate; uv structures are larger, more 'blob-like,' and somewhat less stretched.

The discussion which follows is restricted to the first data set - intermittent regions defined as whole 'mountains.'

3.2 Intermittency in Vorticity

An intriguing discovery is illustrated in figure 3; intermittent vortical structures appear to be self-similar. To see this we separate the intermittent regions into intensity groups, where each group contains approximately 4% of the total volume of the data set. The intensity within each group is given in terms of the global rms of ω^2. Group 1 contains all events with a peak value in the range 18-9 rms, group 2 between 9-7 rms, group 3 between 7-6 rms, etc., with the final group in the range 2-1.5 rms. We then calculate the local pdf within each group; that is, each pdf is normalized by the sample and rms within that group. The local pdfs of each group, weighted by ω^2, are plotted in figure 3. The 'weighted' pdfs of the first three groups collapse nicely onto a single curve (figure 3a), as do the pdf's of the last 4 groups (figure 3b), but the collapse occurs on two significantly different pdfs. Groups 4 and 5 are transitions between these two pdfs. The collapse is even more dramatic within local *joint* pdfs between ω_x, ω_y, and ω_z, indicating that the internal structure within the more intense intermittent regions (18 to 6 rms) is similar, but different from the internal structure within less intense intermittent regions (4.0 to 1.5 rms). By contrast, structural similarity was not found within uv events, suggesting that the property of structural similarity might be associated both with *scale* and *intensity* of flow structures.

A second interesting observation is that the non-Gaussian nature of the vorticity field in this shear flow is directly associated with the intermittent regions in vorticity magnitude. The source of the non- Gaussianity is primarily in the spanwise (ω_z) vorticity fluctuations (ω_x and ω_y have symmetrical pdf's, whereas ω_z is highly negatively skewed). The pdfs of ω_x and ω_z, weighted by ω_x and ω_z, are plotted in figure 4. The area under the outermost curves is the variance of the individual components. The interior curves give the relative contributions of groups of intermittent structures. In fluctuating ω_x and ω_y (not shown) events at all intensity levels contribute equally to the positive and negative sides of the pdf's. The pdf of ω_z, however, is negatively skewed. The intermittent regions contribute *twice* as much to the negative side of the pdf as to the positive side. Thus, the intermittent regions are dominant contributors to the skewness of ω_z, and to the non- Gaussian character of the vorticity field. On average, higher-intensity enstrophy events are dominated by negative spanwise vorticity fluctuations - ω_x and ω_y are evenly balanced in sign. Is it possible that lower intensity fluctuations are more isotropic than the intermittent regions in a turbulent flow, and therefore more amenable to classical statistically-based closures? In homogeneous turbulent shear flow the answer appears to be 'yes,' in that the more intense intermittent regions are the dominant contributors to the anisotropy of the vorticity- vorticity

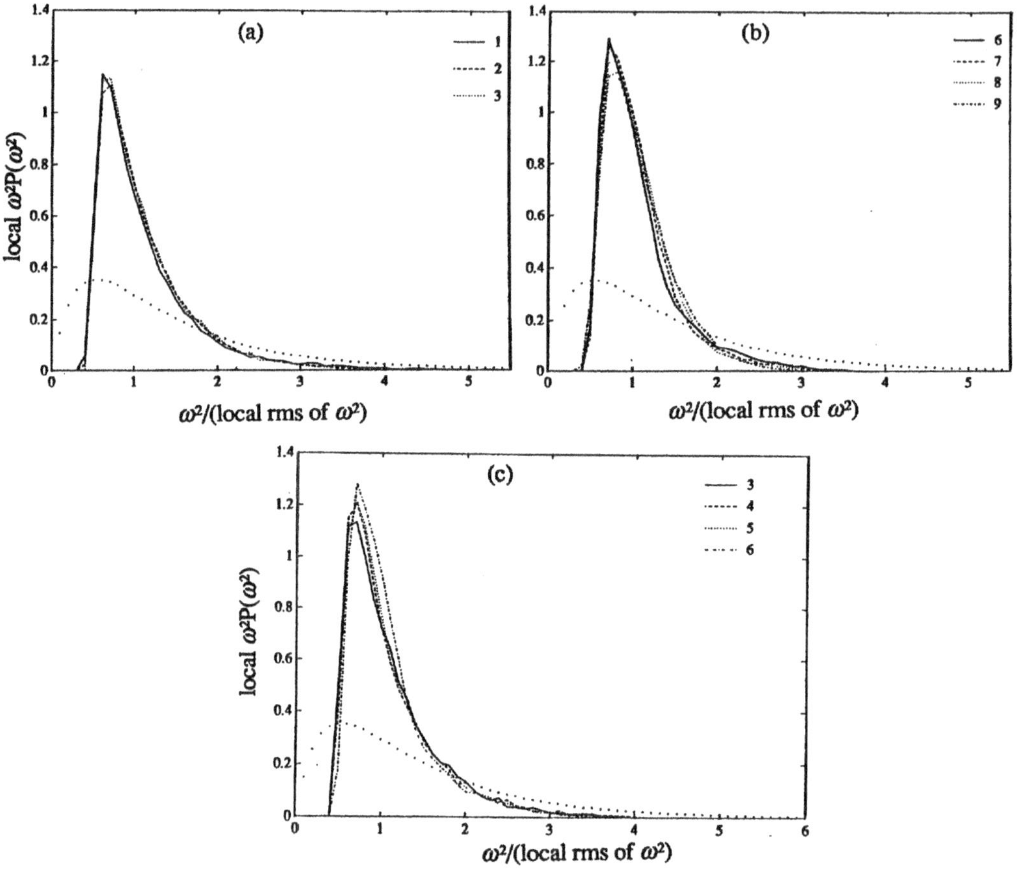

Figure 3. Local pdfs of ω^2, weighted by ω^2, within event intensity groups 1, 2, 3, ..., 10, where each group contains approximately 4% of the data set volume. 'Local' implies that each pdf is defined and normalized by the sample and rms within that group. The intensity levels (peak value) within each group (*grp*) are (*grp*1): 18 $\rightarrow$ 9 rms, (*grp* 2): 9 $\rightarrow$ 7 rms, (*grp* 3): 7 $\rightarrow$ 6 rms, (*grp* 4): 6 $\rightarrow$ 5 rms, (*grp* 5): 5 $\rightarrow$ 4 rms, (*grp* 6): 4 $\rightarrow$ 3 rms, (*grp* 7): 3 $\rightarrow$ 2.5 rms, (*grp* 8): 2.5 $\rightarrow$ 2 rms, (*grp* 9): 2 $\rightarrow$ 1.5 rms, (*grp* 10): $\sim$ 1.5. (a) Local pdfs for groups 1-3. (b) Local pdfs for groups 6-9. (c) Local pdfs for groups 3-6. For comparison, the lowest dotted line is the global pdf of ω^2, weighted by ω^2.

correlation tensor, $\omega_i\omega_j$. The isotropic state is reached as the second and third principle invariants of the anisotropy tensor, II_v and III_v vanish. The 'anisotropy invariant map' is given in figure 5. The filled symbol gives the global anisotropy of the entire data set. Each subsequently lower symbol gives the anisotropy of subsets of the data set, where each subset is created by *excluding* groups 1, 1-2, 1-3, 1- 4,..., 1-10, respectively, from the data set. We find that the level of anisotropy decreases as each subsequent group of intermittent structures is excluded from consideration, indicating that vorticity fluctuations within the low magnitude fluctuations are significantly more isotropic than the vorticity fluctuations within the intermittent structures. There is reason to believe, however, that the smallest magnitude fluctuations retain a certain level of anisotropy due to persistent distortion by the mean shear.

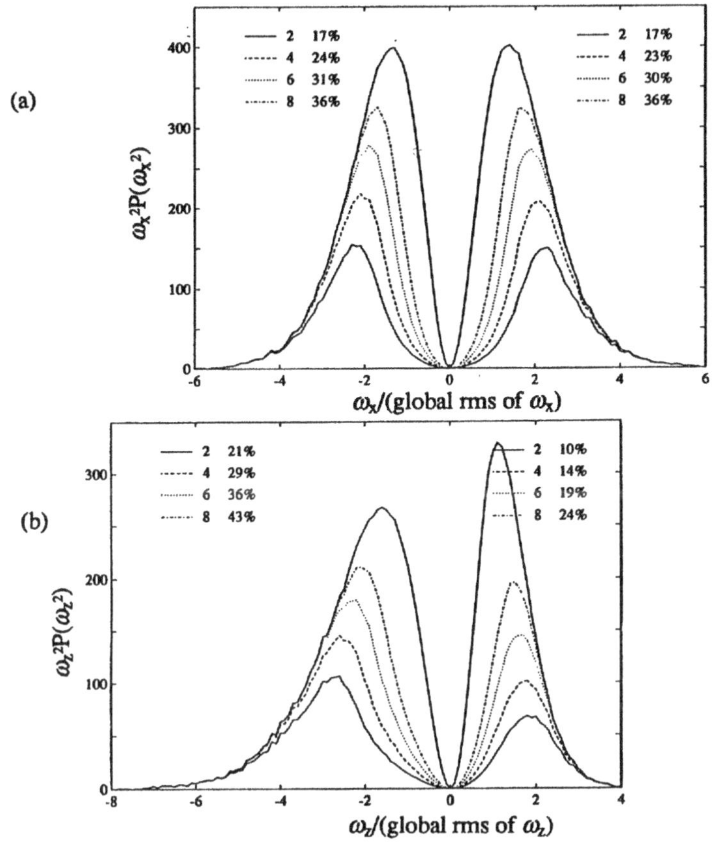

Figure 4. The outer curves give the global pdfs of the square of a vorticity component, weighted by the vorticity component: (a) $\omega_x P(\omega_x^2)$, (b) $\omega_z P(\omega_z^2)$. The inner curves give the contributions to the global pdfs from enstrophy event groups 1-2, 1-4, 1-6, and 1-8, respectively, from the innermost to the outermost curves. The % figures shown on the positive and negative sides of the 'weighted' pdfs are the relative contributions to the positive or negative side of the 'weighted' pdf of event groups 1 to the event group indicated.

3.3 Intermittency in Reynolds Shear Stress

Two brief comments will be made about the intermittent regions in Reynolds shear stress. Firstly, extensive analysis has shown that the more intermittent enstrophy events occupy different volumetric regions in space than intermittent regions in enstrophy. Furthermore, contrary to analyses of near-wall structure in the turbulent boundary layer (Robinson *et al.* 1979), there appears *not* to be a one-to-one correspondence between regions of concentrated Reynolds shear stress and hairpin-like vortices. The second point is made in figure 6, where peak intensity within each $\mid uv \mid$ event is plotted against event number. Note that event intensity decreases rather rapidly at first, then slowly beyond event 200-300. Those events which are dominant *negative uv* are plotted above the horizontal axis, and those events which are dominant positive

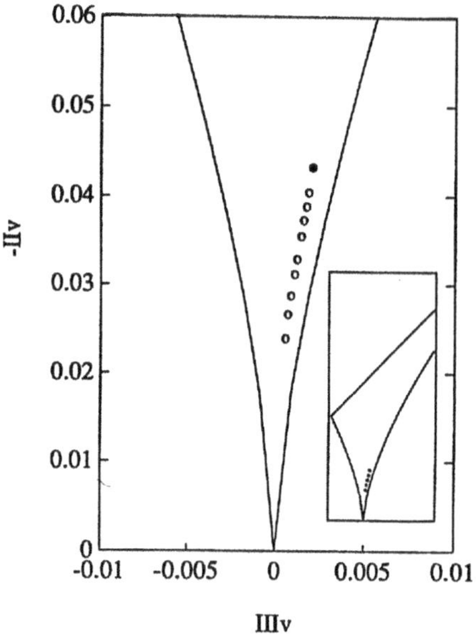

Figure 5. 'Anisotropy invariant map' of the vorticity-vorticity correlation tensor, $\omega_i\omega_j$, given by $< v_{ij} > \equiv < \omega_i\omega_j > / < \omega_i\omega_j > -1/3\delta_{ij}$. Here the second principle invariant, II_v, is plotted against the third principle invariant, III_v. The isotropic state is at the origin. The filled symbol defines the global anisotropic level over the entire data set. Each open symbol defines the anisotropic state over a subset of the computational domain, where each subset is formed by excluding high intensity event groups. Beginning from the closed symbol, each subse-quently lower open symbol excludes group 1, groups 1-2, groups 1-3, etc., to groups 1-9. The insert gives the overview of the anisotropy invariant map.

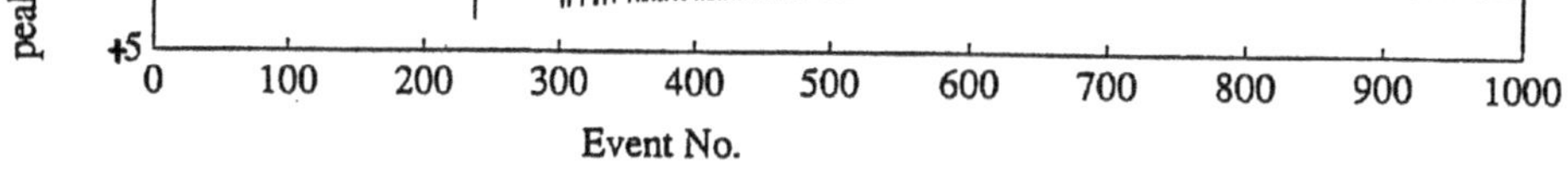

Figure 6. Intensity of $\mid uv \mid$ events, defined by peak magnitude within the event and normalized by the global rms, plotted against event number. Events with dominant negative uv fluctuations are plotted above the axis, and events with dominant positive uv fluctuations are plotted below the axis. The nearly solid black region above the axis up to event 300 indicates a total dominance of negative uv fluctuations within the most intense intermittent regions.

uv are plotted below the horizontal axis. The most intense $| uv |$ events (18 to 2.5 rms) are entirely dominated by negative uv, or energy-producing fluctuations. At lower intensities, a more even mix of positive and negative uv fluctuations exist (the joint pdf of uv suggests that the lowest magnitude fluctuations are an even mix of positive and negative uv fluctuations exist (the joint pdf of uv suggests that the lowest magnitude fluctuations are an even mix of positive and negative uv fluctuations). Regrouping the intense negative uv events by volume shows that the smaller negative uv events are dominantly quadrant 2 (Q2) events, whereas the larger events are dominantly Q4.

4. Conclusions

We have developed a methodology for the extraction of intermittent volumetric structures from a three-dimensional data set at fixed time. The technique works exceptionally well, and has the advantages of allowing for interactive quantitative and graphical analysis of topological, structural, and other kinematic attributes of the more intermittent regions in a turbulence field, and holds considerable promise for local analyses of dynamical processes underlying global evolution. This paper provides a brief overview of the method, and a few of the many results which have been obtained using conditional statistical analyses coupled with graphical imaging. Relative to the analysis of homogeneous turbulent shear flow, we have made the following observations:

- Contrary to what has been reported elsewhere in low Reynolds number turbulence simulations (e.g., Reutsch & Maxey 1990), we find that the more intense intermittent structures in both the vorticity and Reynolds shear stress fields are *dominant* contributors to the global statistics. The apparent contradiction in conclusions is explainable, and important. Based on a pdf approach, one finds that the tails of the pdf are relatively small contributors to the global statistics - volumetrically, there are considerably more low magnitude fluctuations than high-magnitude fluctuations. However, intermittent *structures*, volumetric regions with identifiable form and which contain both high and low magnitude fluctuations, provide large contributions to the global statistics relative to their volumetric contribution. We argue that the latter approach is, physically, more relevant. We presume that these intermittent structures retain some level of coherence over a few eddy turnover times. Preliminary analyses suggest that this is the case.
- We have discovered the existence of structural similarity within intermittent vortical structures. The more intermittent events in the vorticity field appear to share common structural characteristics, as do relatively lower intensity intermittent structures. However, the apparent self-similar structure is different within these two intensity groups. Although these conclusions are based on events being extracted as whole 'mountains,' structural similarity is observed also within vortical substructures. Similarity was *not* found within uv events.
- Intermittent vortical structures are dominant contributors both to the negative skewness of the vorticity field, and to the anisotropy of the vorticity-vorticity correlation tensor. The implication is that, whereas regions of concentrated vorticity appear to dominate the global character of the vorticity field, the statistics of the low magnitude fluctuations surrounding the vortical structures may be well- approximated using classical isotropic models.
- The more intense regions of concentrated $| uv |$ activity are wholly dominated by negative uv fluctuations which, statistically, contribute significantly to the production of turbulent kinetic energy. Among these structures, the smaller more concentrated ones

are dominant Q2 events, moving turbulent fluid in the direction of the mean gradient. Combined with 3D graphical imaging, we hope, in the future, to add considerably to our understanding of the local dynamics which underlies the evolution of turbulent shear flows.

Acknowledgements

We gratefully acknowledge support from the United States Air Force Office of Scientific Research, and past support from the Center for Turbulence Research. Many useful discussions took place with M.M. Rogers, K.R. Sreenivasan, and A.J. Smits. We especially thank Mike Rogers for his help in obtaining the data and in using Rogallo's code.

References

Brasseur, J.G.and Lee. M.J. 1989 *Advances in Turbulence 2*, Springer-Verlag, 306.

Brasseur, J.G. and Lin, W-Q. 1990 manuscript in preparation.

Chen, H., Herring, J.R., Kerr, R.M., and Kraichnan, R.H. 1989 *Phys. Fluids A* **1**, 1844.

Harris, V.G., Graham, J.A.H., and Corrsin, S. 1977 *J. Fluid Mech.* **81**, 657.

Kim, J. and Lee, M.J. 1989 *Proc. Seventh Symp. Turb. Shear Flows*, 1.1.1.

Kolmogorov, A.N. 1962 *J. Fluid Mech.* **13**, 82.

Novikov, E.A. 1990 *Phys. Fluids A* **2**, 814.

Ruetsch, G.R., and Maxey, M.R. 1990 *Bull. Amer. Phys. Soc.* **35**(10), 2268.

Robinson, S., Kline, S.J. and Spalart, P.R. 1989 *NASA Tech. Memo.* 102191.

Rogallo, R.S. 1981 *NASA Tech. Memo.* 81315.

Rogers, M.M., Moin, P. and Reynolds, W.C. 1986 TF-26, Department of Mechanical Engineering, Stanford University.

Rogers, M.M. and Moin, P. 1987 *J. Fluid Mech.* **176**, 33.

Sreenivasan, K.R. and Meneveau, C. 1988 *Phys. Rev. A* **38**, 6287.

Tavoularis, S. and Corrsin, S. 1981 *J. Fluid Mech.* **104**, 349.

Vortex-Shedding and the Maintenance of Thick Turbulent Boundary Layers on Long Cylinders in Axial Flow

M.K. Bull and R.E. Luxton

Dept. of Mechanical Engineering, University of Adelaide,
Adelaide S.A. 5001, Australia

It has been found that vortices with cellular structure can be shed, over a range of Reynolds numbers, from circular cylinders in near-axial flow, even at yaw angles as small as 1°. Experimental data for shedding frequencies and vortex orientation are presented. It is suggested that piece-wise vortex shedding, resulting from cross-flows within the turbulent boundary layer, may play a crucial role in the mechanism of turbulence production necessary to sustain a very thick axisymmetric turbulent boundary layer.

1. Introduction

The axisymmetric constant-pressure turbulent boundary layer on a cylinder is subject to the same governing parameters as a two-dimensional flat-plate turbulent boundary layer, and, in addition, to the effect of transverse curvature. In principle, the curvature, with a length scale characterised by the cylinder radius a, introduces only one additional independent parameter; but its effects can be expected to depend on its length scale in relation to both the inner flow (wall) length scale ν/U_τ and the outer flow scale δ, i.e. on the two parameters $\nu/aU_\tau = 1/a_+$ and δ/a. (Here $U_\tau = \sqrt{\tau_w/\rho}$, τ_w is the surface shear stress, ν and ρ are the kinematic viscosity and density of the fluid and δ is the boundary layer thickness.) When, in a flow with free-stream velocity U_1, the cylinder Reynolds number $Re_a = U_1 a/\nu$ is small and the Reynolds number $Re_x = U_1 x/\nu$ (x a streamwise coordinate) is large, a_+ becomes relatively small and the boundary layer thickness becomes very much greater than the cylinder radius. In these circumstances both $1/a_+$ and δ/a are large; the effects of transverse curvature can then be expected to be significant in both the inner and outer regions of the boundary layer. The present work is concerned with turbulent boundary layers of this type.

In such layers the volume of turbulence maintained by unit area of the boundary surface is larger than for a two-dimensional flat-plate layer by a factor of $\delta/2a$, which may have a value in the order of twenty. This raises the question of whether the mechanism of turbulence production in the axisymmetric boundary layer is the same as that in a two-dimensional layer or whether the implied increased rate of turbulence production is the result of some other or a more effective form of the same generation process.

In a previous investigation of the thick axisymmetric turbulent boundary layer on a cylinder at low cylinder Reynolds numbers, Luxton, Bull & Rajagopalan (1984) found that a feature of the time-histories of streamwise (u) velocity fluctuations was the repeated occurrence, at all locations in the boundary layer, of large negative-going spikes. It was recognised that, when $\delta/a \gg 1$, the cylinder will present little inhibition to the passage over it of fluid associated with large scale (of order δ) outer-flow structures. It was then postulated that the large structures, sweeping over the cylinder, remove

Advances in Turbulence 3
Editors: A.V. Johansson · P.H. Alfredsson

low-speed fluid from the inner layer and convect it to all parts of the boundary layer, thereby giving rise to the low-velocity spikes in the u-signals. Each time the inner layer is disrupted in this way a new inner layer must be established, and at the beginning of the re-establishment process the mean-velocity gradient near the cylinder will be very high. This would constitute a very efficient process for maintaining the vorticity in the boundary layer.

The preceding postulate was tested by experiments with marked (heated) fluid (Van de Loo, 1985; Bull, Luxton, Rajagopalan & Van de Loo 1986). The cylinder was locally heated over a short axial length and the resulting temperature fluctuations in the fluid examined at various distances from the cylinder at downstream stations. Temperature spikes similar in character to the u-velocity spikes were detected and were almost always accompanied by velocity spikes, indicating that the u-velocity spikes are due to low-speed fluid which has originated at points upstream very close to the cylinder surface.

Subsequent examination of radial (v) velocity fluctuations (Bull et al. 1986, 1988) by sampling conditioned on the occurrence of axial-velocity spikes has shown that the u-velocity spikes are accompanied by similar positive-going v-velocity spikes at all radial positions in the boundary layer. The (kinematic) Reynolds shear-stress, $-\overline{uv}$, is also characterised by positive-going spikes at all radial positions. Quadrant resolution of the Reynolds stress shows that the u-velocity and v-velocity spikes make the dominant contribution to it. Close to the cylinder the Reynolds-stress spikes are associated with large values of mean shear and consequently with a high rate of production of turbulence energy. Cross-flows in the axisymmetric boundary layer and their relation to turbulence production and Reynolds stresses have also been investigated by Lueptow and Haritonidis (1987). They interpret their measurements in terms of the familiar burst-sweep mechanisms of turbulence generation identified in planar boundary layers, allowing modification of that process by curvature. However, here, as a result of the experimental work cited above, the process of spike generation is seen as central to the ability of the cylinder to maintain such a large volume of turbulence in relation to its surface area.

Measurements with rakes of hot-wires (Bull et al. 1986, 1988) indicate that the low-speed fluid stripped from the cylinder forms a front, inclined at an angle in the order of 45° to the cylinder axis, which advances in the streamwise direction at a velocity of roughly $0.6U_1$. From consideration of the average frequency of generation of the velocity spikes and the low-frequency spectral levels of the radial velocity fluctuations, it was also suggested that the cross-flow velocities involved in the process by which inner-layer fluid is stripped from the cylinder may well be produced by large-scale undulation of the boundary layer.

Apart from the general characteristics outlined above, details of the flow associated with the postulated peeling process have not been closely investigated. The possibility that the process might involve piece-wise vortex-shedding from the cylinder under the influence of cross-flows was originally discounted on the basis of the belief that vortex-shedding does not occur in uniform flows over cylinders at very small angles of yaw. However subsequent tests have shown that vortices can be shed at yaw angles very much smaller than previously believed. Results of these tests will be presented and the possible role of vortex-shedding in the peeling process will then be discussed.

2. Experimental Apparatus and Procedures

Experiments on yawed cylinders were made in the same closed circuit wind-tunnel as that used for the experiments reported by Luxton et al. (1984) and Bull et al. (1986, 1988) on long cylinders in axial flow. The working section of the tunnel is rectangular with horizontal dimension 650 mm and vertical dimension 300 mm, and is 3 m long. Air speed is continuously variable from 1 to 27 m/s .Over the length of the test section, boundary layer growth on the walls causes an increase in free stream velocity by about 13% at 2 m/s reducing to about 5% at 23 m/s. Test cylinders less than 3 mm in diameter were of nylon and those of greater than 3-mm-diameter were of steel. The nylon cylinders were held at their upstream ends in the wind-tunnel contraction by wire stays; they were tensioned by means of a weight and a pulley at their downstream ends and meassures were taken to ensure that vortex shedding was not influenced by cylinder vibration. The angle of yaw could be varied in the horizontal plane, up to a maximum of $\beta = 10^o$. The steel cylinders were mounted in the vertical plane, within the test section, and parallel to its 300 mm dimension. The smaller diameter cylinders were used for vortex-shedding frequency measurements and the larger ones for frequency measurements and flow-visualisation. Most flow-visualisation experiments made use of cylinders with $d = 6.0\,mm$.

Velocity fluctuations were detected by constant-temperature hot-wires 5 μm in diameter, with a sensing length of 1 mm, located 2700 mm downstream of the start of the test section and positioned so that vortex-shedding from only one side of the cylinder was detected. Frequency spectra of the hot-wire signals were obtained by means of a Hewlett-Packard HP 3582A analyser. Vortex shedding frequency was determined from the large dominant spike in the spectrum. At the limits of the regular shedding regime the dominant spectral spike rapidly decreases in strength and increases in width until it fades into the spectrum of background turbulence.

Flow visualisation was effected by means of a plane array of fine smoke streaks generated by evaporating light oil from an electrically-heated 0.1-mm-diameter stainless steel wire spanning the wind-tunnel. Flow patterns were recorded by a National F-10N CCD video camera. Vortex angles were determined from the video pictures (Circelli, 1988).

For one particular yaw angle, $\beta = 5^o$, of a cylinder with $d = 0.9\,mm$, vortex angles were also determined from relative phase measurements between two hot-wires whose separation could be varied in a plane parallel to the plane of cylinder yaw. By measuring vortex spacing in two different directions (normal to the flow and at 45° to the cylinder axis) the orientation of the vortex axes could be determined. Such measurements were made in planes $1.89\,d$ and $2.56\,d$ above and below the plane of cylinder yaw (Mackenzie, 1989).

3. Experimental Results for Yawed Cylinders

3.1 Limits of regular vortex shedding

It has been found (Bull & Dekkers, 1989) that vortices can be shed from yawed cylinders at much smaller angles of yaw than previously believed, and that the maximum Reynolds number $Re_d = U_1 d/\nu$ (d=cylinder diameter) for which regular shedding occurs varies with yaw angle. Data for cylinders of diameter 0.90, 1.93 and 2.95 mm are shown in figure 1.

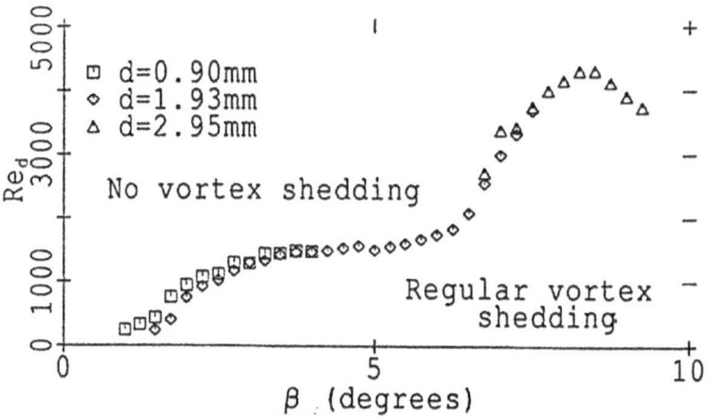

Figure 1. Vortex-shedding regimes for circular cylinders at small angles of yaw.

The boundary between regular shedding and no shedding is not as abrupt as figure 1 might suggest, but, as indicated in the previous section, the spectral spike corresponding to regular shedding fades out over a range of Reynolds number which is broadly of the order of 10% of the demarcation value shown. Figure 1 shows results for $\beta \geq 1^o$. At still smaller angles of yaw, down to about 0.5^o, continuous but rather irregular vortex shedding could still be observed. No lower limiting Reynolds number for vortex shedding was detected down to the lowest flow speeds investigated.

In the previous work of Luxton et al. (1984) on axisymmetric cylinder boundary layers, the layers at large downstream distances ($x/a \sim 3000$) were found to be still turbulent at Re_d as low as 500. But there may be some doubt that the layers for which shedding is observed are turbulent for $\beta \leq 1.5^o$. For $\beta > 1.5^o$ the boundary layer can be expected to be turbulent over the greater part of the shedding range.

3.2 Vortex shedding frequencies

Measurements for a 0.90-mm-diameter cylinder for yaw angles in the range $0 \leq \beta \leq 10^o$ and Reynolds number in the range $120 \leq Re_d \leq 1560$ (Bull & Dekkers, 1989) show that the shedding frequencies can be represented by an empirical relation of the form

$$F = A\beta^n, \tag{1}$$

where F is the frequency parameter fd^2/ν and A and n are Reynolds-number-dependent "constants". For Reynolds number $Re_d \geq 600$ the relation for the corresponding Strouhal number $St = fd/U$ can be approximated by

$$St/Re_d^{3/2} = (25.6 \times 10^{-6})\beta^n. \tag{2}$$

The index n varies very nearly linearly with Reynolds number from about 0.74 at $Re_d = 600$ to about 1.2 at $Re_d = 1500$. Frequency data plotted in the form of equation (1) are shown in figure 2. Bull & Dekkers also showed that these results form a consistent extension to smaller values of β of the data obtained by Van Atta (1968) for yaw angles $20^o < \beta < 90^o$.

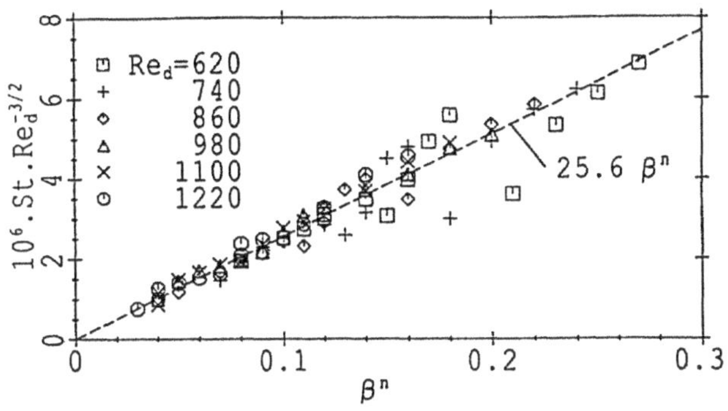

Figure 2. Dependence of Strouhal number on Reynolds number and yaw angle for circular cylinders at small angles of yaw.

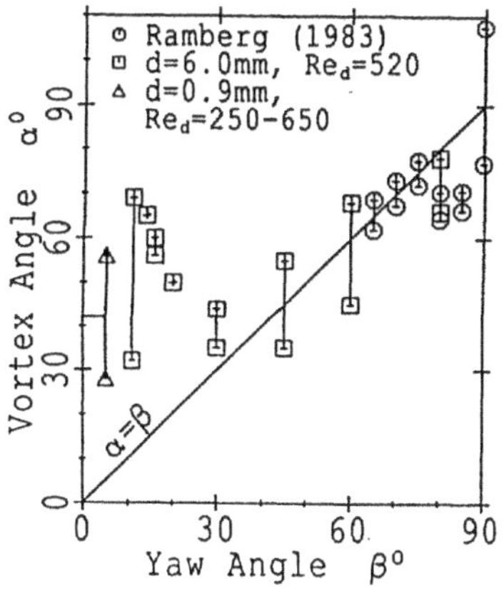

Figure 3. Variation of vortex angle for shedding from yawed circular cylinders.

3.3 Vortex angles

The variation of the inclination α of shed vortices to the flow direction, as determined from video recordings of the flow-visualisation and in the case of $\beta = 5^o$ from phase variations determined by hot-wire traverses, is shown in figure 3. Data obtained by Ramberg (1983) for yaw angles greater than 65° are also shown. Ramberg measured vortex angles in the flow over a yawed cylinder with one end set in the wind-tunnel wall and the other free. For this configuration two modes of vortex shedding were present simultaneously, one influenced by end conditions and another over the portion of the cylinder remote from the free end. The results shown in figure 3 are those for shedding remote from the free end.

The data obtained in the present work for $\beta < 45^o$ are in accord with Ramberg's conclusion that, because the boundary layer flow on a yawed cylinder has both axial and normal components of vorticity, the resultant vorticity vector will be inclined to the flow direction at an angle greater than β, with the result that the vortex angle α should be greater than the yaw angle β.

It will be noted that in many cases there is a considerable range of values of α shown for a given yaw angle; the range shown results from both variations in vortex angle from frame to frame in the video recordings and the fact that the vortex lines are not always exactly straight in any given frame. The results for $\beta = 5^o$ were obtained at various Reynolds numbers in the range $250 \leq Re_d \leq 650$ and include differences at different Reynolds numbers and different distances of the traverse plane from the plane of the cylinder.

3.4 Vortex cells

As indicated in section 2, there is a significant increase in velocity between upstream and downstream ends of the wind-tunnel test section. This variation of velocity along the span of the yawed cylinder has an exactly analogous effect to that of the gradual variation of diameter of a slender cone with its axis normal to a uniform stream, as investigated by Gaster (1969, 1971). Gaster observed that in such a case a series of adjacent spanwise vortex cells is formed with a change in shedding frequency from one cell to the next.

For the 6-mm-diameter yawed cylinders investigated here, video records for $\beta = 60^o$ and 45° show straight vortices continuous over the whole span of the cylinder except where they encounter the tunnel-wall boundary layers. A typical flow pattern, for $\beta = 45^o$ is shown in figure 4. When β is reduced to 30° there is evidence of formation of spanwise vortex cells, which become more pronounced as β is further reduced. They can be clearly seen in figure 5 for $\beta = 11^o$.

The shedding frequency from 0.9-mm-diameter cylinders at small angles of yaw, $0 \leq \beta \leq 10^o$, measured at the downstream end of the working section is markedly higher than at the upstream end – consistent with an increase in flow velocity. This difference can be seen in figure 6(a) for $\beta = 8^o$. However, when the data are plotted against local flow velocity they fall on a common, non-linear, curve (figure 6(b)), implying that the Strouhal number in a particular cell depends only on Reynolds number. The change in dimensional shedding frequency along the span of the cylinder and the consequent phase

Figure 4. Vortex shedding from yawed circular cylinder at $\beta = 45^o$, $Re_d = 520$ (flow from left to right).

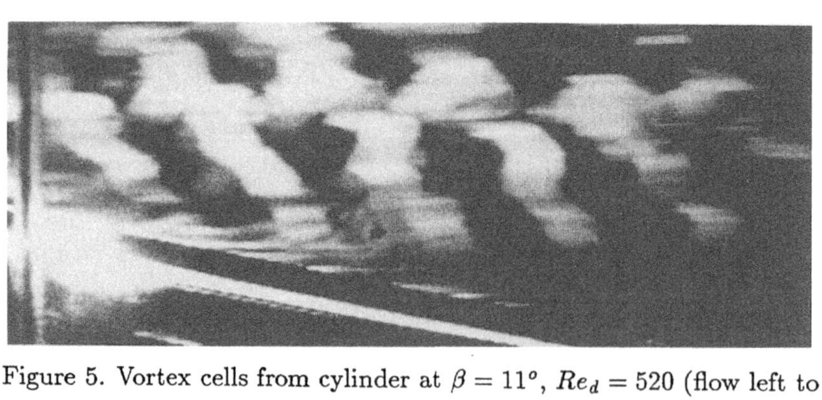

Figure 5. Vortex cells from cylinder at $\beta = 11^o$, $Re_d = 520$ (flow left to right).

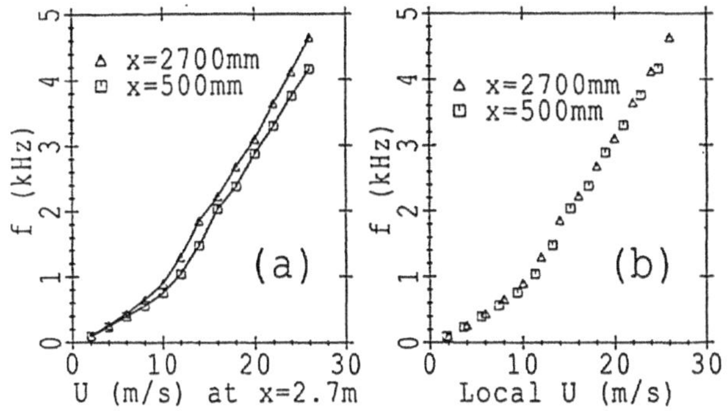

Figure 6. Dependence of shedding frequency on (a) free-stream velocity at downstream end of yawed cylinder and (b) local free-stream velocity, for $\beta = 8^o$.

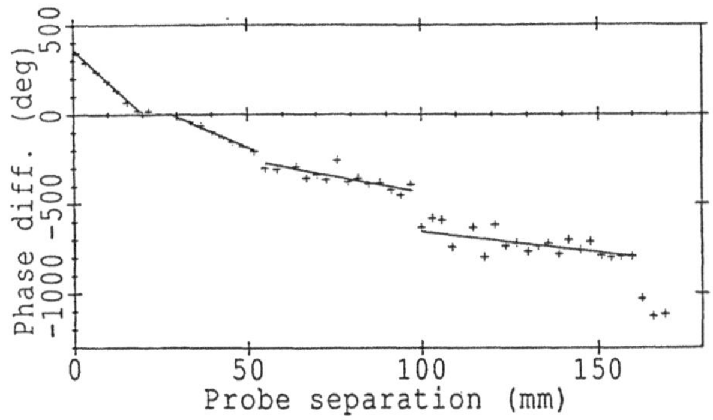

Figure 7. Phase variations in vortex cells from yawed cylinder $\beta = 5^o$, $Re_d = 350$.

variations must imply the formation of spanwise vortex cells to preserve continuity of vortex lines.

Evidence of vortex cells was also provided by relative phase data obtained in the measurement, by hot-wire traverses, of the inclination of vortices shed by a 0.9-*mm*-diameter cylinder at an angle of yaw of 5°. Figure 7 shows the variation of phase angle between two hot-wire probes as a function of separation distance along a line normal to the flow at $Re_d = 350$. The data indicate four vortex cells over a lateral distance of about 160 *mm* which corresponds to a streamwise distance of about 1830 *mm*. Over the latter distance the increase in free-stream velocity at this flow Reynolds number would be about 6%.

4. Role of Vortex Shedding in the Axisymmetric Turbulent Boundary Layer

The process of formation of propagating fronts of low-speed fluid identified in the axisymmetric turbulent boundary layer on a cylinder in the investigation of Bull et al. (1986, 1988) is shown diagrammatically in figure 8. The axial and transverse velocities associated with the low-speed inner-layer fluid peeled from the cylinder were found to be typically $U_l = 0.67\, U_1$ and $V_l = 0.06\, U_1$ respectively; a typical inclination of the resultant flow to the cylinder axis is therefore about 5°. If this is interpreted as the

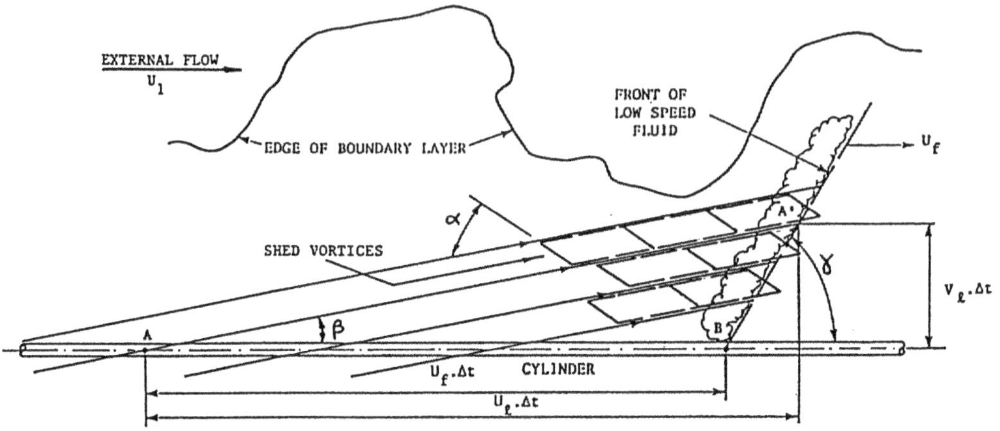

Figure 8. Diagrammatic illustration of formation and progression of fronts of low-speed fluid and associated vortex cells.

yaw angle of a cylinder in a uniform flow, figure 1 shows that vortex shedding could be expected for Reynolds numbers up to $Re_d \simeq 1500$. Figure 3 would indicate in this case an average vortex inclination angle to the resultant flow of $\alpha \simeq 43^o$.

It should be noted immediately that this does not indicate similar orientations of the vortex axes and the front of low-speed fluid: referred to an axis along the cylinder centre-line in the flow direction the low-speed front has an inclination of $\gamma \simeq 45^o$ while the vortex axes have an inclination of $(\pi + \beta - \alpha)$ or about 142°. Thus, if the process of peeling of low speed fluid from the cylinder does involve vortex formation, then for this to be compatible with the aggregation of low-speed fluid as a front inclined to the cylinder axis at an angle $\gamma < 90^o$ the formation of a series of adjacent vortex cells would be necessary. Within these cells, fluid particle paths would still be along lines inclined at angle $\beta \simeq 5^o$ to the cylinder axis. Because of the limited duration of the peeling process, the vortex cells would consist of closed vortex loops of finite length, such as those shown in figure 8.

There is little systematic information on which to base an estimate of the width of the vortex cells. Gaster's (1971) experiments with a uniform flow over a slender cone suggest that the width of a cell corresponds to about a 16% change in cylinder diameter i.e. to a 16% change in Reynolds number. In contrast, in the present experiments with 0.9-mm-diameter cylinders at $\beta = 5^o$ a variation of Reynolds number along the length of the cylinder of only about 6% gives rise to four cells (figure 7); and the flow over the 6-mm-diameter cylinder at $\beta = 11^o$ shows eight cells for a Reynolds number change of about 13% over the length of the cylinder (figure 5). These two yawed-cylinder results are broadly consistent and suggest that the maximum Reynolds number change which can be sustained across a vortex cell from a cylinder at a small angle of yaw is only of the order of one or two per cent. Streamwise variations in the cross-flow velocities involved in any given peeling event, at least of the order of those obtaining for $\beta = 5^o$ and 11°, are to be expected, and it can probably be said that for a front of low-speed fluid extending over the greater part of a boundary layer thickness the number of vortex cells might be expected to be in the order of ten. This would certainly seem to be sufficient to allow vortex cells with appropriate vortex inclinations to generate a low-speed fluid front of the form observed.

These ideas are clearly speculative and experimental work is required to test their validity. Such experiments involving flow visualisation in conjunction with hot-wire arrays and u-velocity spike detection are being undertaken.

Acknowledgement

This work is supported by the Australian Research Council.

References

Bull, M.K. & Dekkers, W.A. 1989 Vortex shedding from cylinders in near axial flow. Tenth Australasian Fluid Mechanics Conference, Melbourne, 11-15 Dec. 1989.

Bull, M.K., Luxton, R.E. & Rajagopalan, S. 1988 Flow mechanisms in the thick axisymmetric turbulent boundary layer on a circular cylinder. Univ. of Adelaide, Dept. Mech. Eng., Tech. Report 11/88.

Bull, M.K., Luxton, R.E., Rajagopalan, S. & Van de Loo, P. 1986 On the maintenance of turbulence in the thick axisymmetric turbulent boundary layer on a slender cylinder. First European Turbulence Conference, Lyon.

Circelli, T. 1988 Investigation of the vortex shedding from a yawed circular cylinder. Univ. of Adelaide, Dept. Mech. Eng., Final-year Honours Report.

Gaster, M. 1969 J. Fluid Mech. **38**, 565-576.

Gaster, M. 1971 J. Fluid Mech. **46**, 749-756.

Lueptow, R.M. & Haritonidis, J.H. 1987 Phys. Fluids **30**, 2993-3005.

Luxton, R.E., Bull, M.K. & Rajagopalan, S. 1984 Aeron. J. **88**, 186-199.

Mackenzie, N.C. 1989 Investigation of the vortex shedding structure produced by flow over a yawed cylinder. Univ. of Adelaide, Dept. Mech. Eng., Final-year Honours Report.

Ramberg, S.E. 1983 J. Fluid Mech. **128**, 81-107.

Van Atta, C.W. 1968 AIAA Jl. **6**, 931-933.

Van de Loo, P. 1985 The turbulent boundary layer on a slender cylinder aligned with the flow. Univ. of Adelaide, Dept. Mech. Eng., Final-year Honours Report.

Properties of Fine Scales in High Reynolds Number Turbulence

Y. Gagne

Institut de Mécanique de Grenoble,
C.N.R.S.umr101, Université J. Fourier and Institut National Polytechnique
B.P. 53X, F-38041 Grenoble Cédex, France.

Scaling properties are one of the basic concepts used to study fully developed turbulence. We review here some recent experimental, numerical and theoretical results on the fine scale intermittency in high Reynolds number turbulence. In particular, we show that *local scaling* is consistent with experimental data. The scaling exponents take on a whole range of values, the most frequent ones being close to the value 1/3 predicted by Kolmogorov in 1941. Negative exponents are also obtained.

1. Introduction

A central problem of fully developed turbulence is generally viewed to be the energy-cascading process. It has resisted all attempts at a full physical understanding or mathematical formulation. The main reasons for this failure are related to the wide hierarchy of scales involved, the highly nonlinear character inherent in the Navier-Stokes equations, and the spatial intermittency in the dynamically regions (Monin & Yaglom (1975), Frisch & Orszag (1990)). The only systematic theory of fully developed turbulence is the well-known scaling theory of Kolmogorov (1941) which is based on the *global scaling* invariance of the Navier-Stokes equations. Indeed, in the limit of Reynolds number tending to infinity, the equations are invariant by simultaneous rescaling of distances by λ, velocities by λ^h, and time by λ^{1-h}, where h is an arbitrary real number, called the scaling exponent. In his 1941 theory, Kolmogorov assumed that, average nondimensionalized inertial quantities are scale-invariant, whereas detailed structures need not be. Furthermore, the mean energy transfer rate, when expressed in terms of the characteristic velocity v_ℓ of eddies of size ℓ and their scale ℓ, namely $\epsilon_\ell \sim v^3{}_\ell/\ell$, should be independent of ℓ in the scaling region. This singles out the scaling exponent $h = 1/3$. The well-known power law $k^{-5/3}$ of the energy spectrum follows. Another immediate consequence of Kolmogorov's 1941 theory is that the velocity structure function of order p, defined as the average of the pth order of velocity increments $\delta v(\ell) = v(x+\ell) - v(x)$, measured over inertial separations ℓ, scales as $< [\delta v(\ell)]^p > \sim \ell^{p/3}$. Obviously, such a derivation is essentially global and cannot give scaling exponents which depend on the position as may be needed to describe intermittency.

Indeed, for forty years (Batchelor & Townsend (1949)), it has been known that the energy dissipation rate at small scales is intermittent both in space and time. In particular, measurements done by Kuo & Corrsin (1972) have clearly shown that the spottiness of dynamical regions is increasing with the wavenumber.

During the last thirty years, considerable experimental work has been done to study intermittency of fine structures both at inertial and dissipation scales (Van Atta & Chen

Advances in Turbulence 3
Editors: A.V. Johansson · P.H. Alfredsson

(1970); Van Atta & Park (1972); Frenkiel & Klebanoff (1975); Antonia & Chambers (1980); Anselmet, Gagne, Hopfinger & Antonia (1984); Gagne, Hopfinger & Frisch (1989)).

In the same time, statistical models have been proposed to take into an account this fundamental property. All of them are mainly based on the energy cascading (or energy dissipation) process, into which intermittency is put in by hand, using "had hoc" probability distributions of the energy transfer (or energy dissipation) rate (Kolmogorv (1962); Novikov & Stewart (1964); Frisch, Sulem & Nelkin (1978); Fujisaka & Mori (1979); Nakano & Nelkin (1985); Meneveau & Sreenivasan (1987)). Mandelbrot (1974) was the first to give a fractal description of intermittency of the fine structures, and he noticed that the intermittency models developed by the Russian school were consistent with fractal behaviour. He proposed a classification of intermittency models including either homogeneous fractal models or non homogeneous ones (see Mandelbrot (1982)). The latter were later recognized by Parisi & Frisch (1984) as having a multifractal structure.

In this paper, we emphasize scaling properties of three-dimensional turbulence by highligthing recent experimental, numerical and theoretical results. In section 2, we give some new experimental data (mostly obtained in the large S1 wind tunnel of ONERA at Modane), and also recent numerical simulation results concerning the statistical properties of velocity (and temperature) at small scales. In section 3, these data are put into theoretical context of intermittency models, and are in particular compared with predictions of the multifractal model. Finally, in section 4 we examine how a data processing method using *wavelet decomposition* can give us local information about very strong and rare events which characterize the dynamics of high Reynolds number turbulence (Argoul, Arnéodo, Grasseau, Gagne, Hopfinger & Frisch (1989); Bacry, Arnéodo, Frisch, Gagne & Hopfinger (1990)).

2. Statistical Experimental Data

2.1 Probability Density Functions

Fully developed turbulent flows are characterized by a wide hierachy of scales. The relevant experimental data to study the velocity statistics of fine structures are the velocity increments already defined as $\delta v(\ell) = v(x+\ell) - v(x)$ (Monin & Yaglom (1975)). This difference of velocity measured at two points separated by a distance ℓ gives roughly the instantaneous local velocity v_ℓ of an eddy of size ℓ. By varying this separation ℓ from the integral scale ℓ_0 to the dissipation scale ℓ_d, we can study the inertial range. Here, these velocity increments $\delta v(\ell)$ will be denoted $\alpha = \delta v(\ell)/[< (\delta v)^2 >]^{1/2}$.

Figure 1 shows the probability density function (hereafter p.d.f.) of the difference of streamwise velocity components α measured in the wind tunnel S1 of ONERA at Modane. Measurements have been done in the turbulent flow of the return channel (with a diameter of $24m$), corresponding to a kind of "chunk" turbulence (Gagne (1987)). The Taylor-scale based Reynolds number is $R_\lambda = 2720$ and the extent of the inertial scales is almost three decades from 25η until $10^4\eta$,(η is the Kolmogorov scale, approximatively $0.35mm$). In figure 1, the separation ℓ (experimentally obtained by using the Taylor hypothesis) of the normalized increment α is $\ell \simeq 2900\eta$, a typical size of the largest inertial scale of this flow.

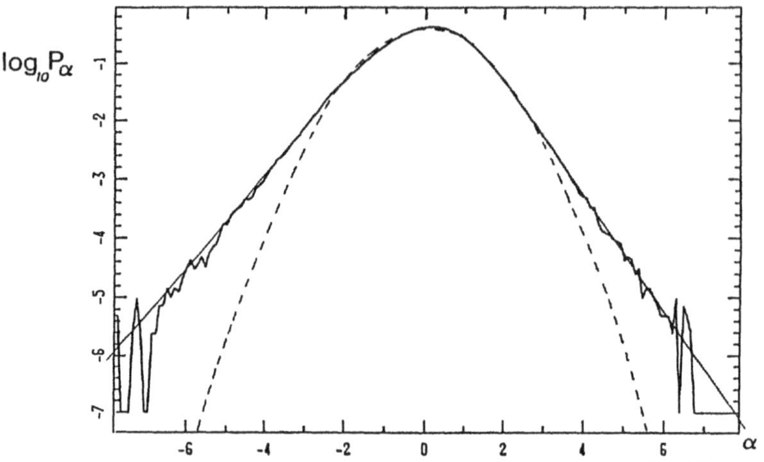

Fig. 1. P.d.f. of the normalized velocity increment $\alpha = \delta v(\ell)/[< (\delta v)^2 >]^{1/2}$ in the ONERA wind-tunnel. $\ell \simeq 2900\eta$. Dashed line corresponds to a Gaussian distribution.

In the linear-logarithmic coordinates used in figure 1, a large discrepancy is observed for large fluctuations of α between the experimental p.d.f. and a Gaussian distribution plotted in dashed line (the Gaussian has a parabolic shape in such coordinates). If this discrepancy is a statistical signature of small-scale intermittency, then figure 1 shows that this intermittency appears not only at dissipative scales but also at large inertial scales. Nevertheless, the most important feature of this p.d.f. is the quasi-exponential shape of tails for large α's. Furthermore, the negative tail corresponding to strong negative fluctuations of velocity increments has a higher probability than the positive tail. This is consistent with the skewness (and higher odd order statistical moment) of α being negative. It is well known that the negative sign of skewness of velocity derivatives is connected with the streching of vortex filaments (Tennekes (1968)). So, figure 1 suggests that this dynamical process characteristic of three dimensional turbulence is linked to very strong negative velocity events with rare occurence.

These results have recently received some confirmation from moderate R_λ numerical simulations (Siggia (1981); Métais & Herring (1989); Yamamoto & Hosokawa (1988); Kida & Murakami (1989); She, Jackson & Orszag (1990)). Figure 2 shows the p.d.f. of velocity gradient $\partial u/\partial x$ (adapted from Meneguzzi & Vincent (1990)), obtained in a numerical simulation of isotropic turbulence with $R_\lambda \simeq 150$. Although we find again the same features than those observed in figure 1, we note that the tails display a weak curvature. This tendency has been also observed in experiments at the smallest scales (Gagne (1987); Castaing, Gagne & Hopfinger (1990)). Experimental data coming from the wind tunnel of ONERA clearly indicates indeed that the tails of the p.d.f. of $\delta v(\ell)$ are increasingly upward bent as the size ℓ decreases and reaches the dissipation scales.

The comparison between figure 1 and figure 2 suggests that this typical shape of p.d.f.'s tails depends neither on Reynolds number, nor on boundary conditions (which are very different in the numerical simulations and the S1 experiment).

All these results lead up to the following question: is the velocity distribution observed in figures 1 and 2 universal and characteristic of the nonlinearities of the Navier Stokes equations? In this context, we notice that similar features have been observed in

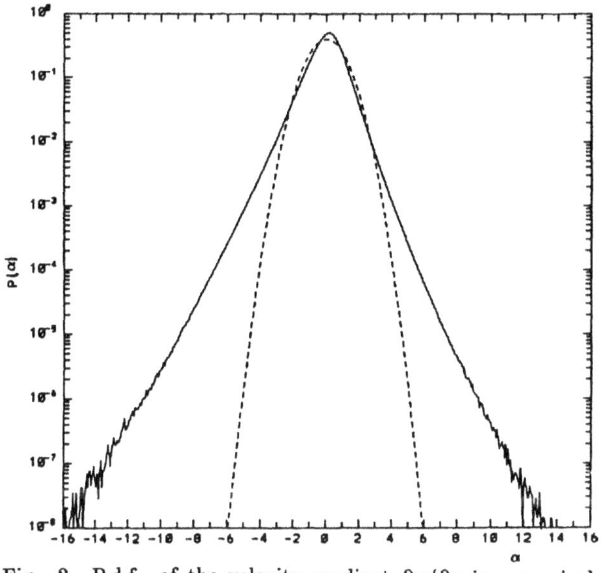

Fig. 2. P.d.f. of the velocity gradient $\partial u/\partial x$ in numerical simulation (from Meneguzzi & Vincent (1990)). Dashed line: Gaussian distribution.

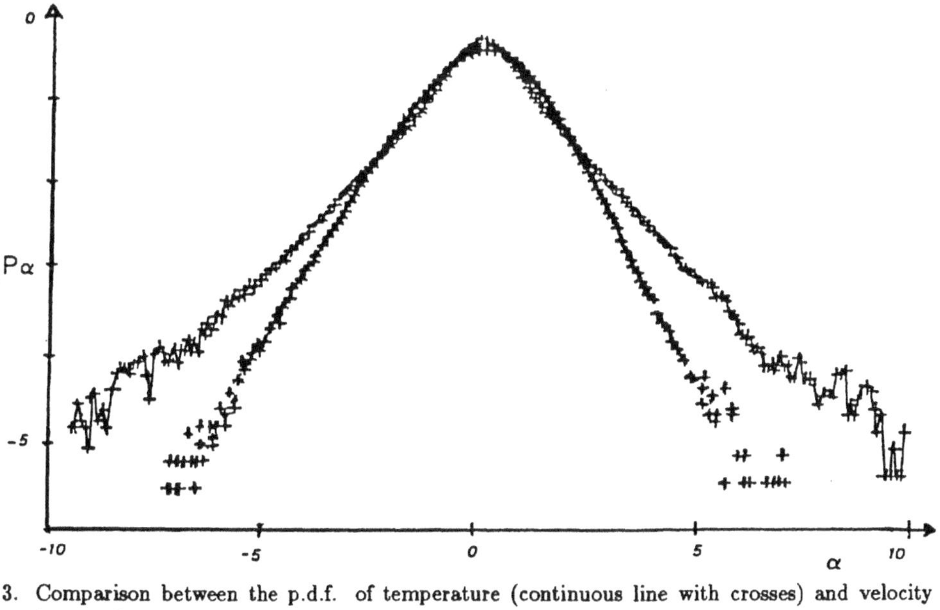

Fig. 3. Comparison between the p.d.f. of temperature (continuous line with crosses) and velocity (separated crosses) increments measured in ONERA wind-tunnel. ($\ell \simeq 1300\eta$).

thermal convection experiments at high Rayleigh number done by the Chicago group. Measurements have shown exponential-like tails of temperature p.d.f. characteristic of the so-called "hard" turbulence régime (Castaing, Gunaratne, Heslot, Kadanoff, Libchaber, Thomae, Wu, Zaleski & Zanetti (1989)).

Passive scalar structures in fully developed turbulent flows are also intermittent at small scale. Figure 3 gives a comparison between p.d.f. of velocity and temperature

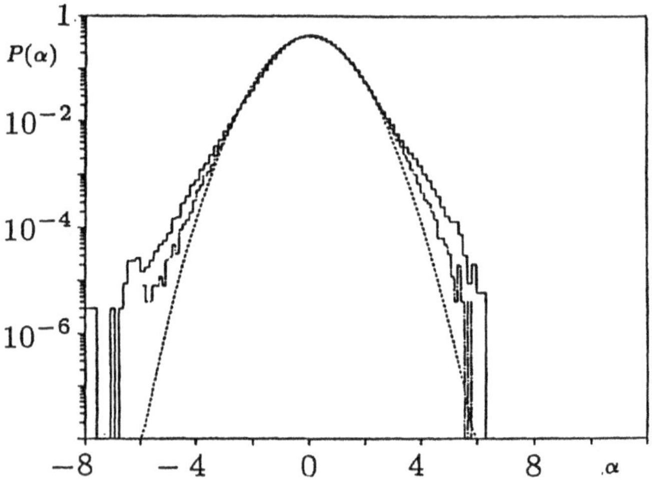

Fig. 4. P.d.f. of temperature fluctuations on numerical simulation done by Métais and Lesieur (1990). Continuous line: direct simulation; chain-dotted line: large eddy simulation; dashed line: Gaussian distribution.

increments simultaneously measured in the same flow than that of figure 1. Here, the common separation ℓ of $\delta v(\ell)$ and $\delta\theta(\ell)$ is $\ell \simeq 1300\eta$, a typical size of large inertial scales. The p.d.f. of the temperature deviates from a Gaussian distribution even at small temperature excursions. We can observe that an upward curvature appears at small α's. In contrast, the tails of the velocity p.d.f. are barely bent, since the size of the analyzed scale ℓ is here rather large. However, measurements suggest that p.d.f. of velocity and temperature increments are roughly the same for separations ℓ of the order of the dissipation scale, at least when the Prandlt number is near unity (Gagne (1987)).

The tendency of the temperature to deviate from Gaussianity even at large inertial scales is also observed in numerical simulations. As shown in Métais & Lesieur (1990), figure 4 gives the p.d.f. of temperature θ itself in numerical calculations, for freely decaying isotropic turbulence at a moderate Reynolds number $R_\lambda \simeq 80$. We observe that this p.d.f. strongly deviates from a Gaussian distribution (in dashed line).

2.2 High Order Structure Functions.

Experimental data gathered during the last twenty years indicate that the statistical moments of the velocity increments $\delta v(\ell)$ (also called velocity structure functions) have scaling properties: for ℓ lying in inertial range

$$< [\delta v(\ell)]^p > \sim \ell^{\zeta_p}.$$

Van Atta & Chen (1970) and Van Atta & Park (1972) were the firsts to give the experimental behaviour of the ζ_p scaling exponents for large values of the order p. As shown in figure 5 from Anselmet et al. (1984), the prediction $\zeta_p = p/3$ of the 1941 Kolmogorov theory (chain-dotted line) is only marginally supported by the experiment for small p and not at all for large p's. In others words, figure 5 indicates that *global statistical scaling* invariance is broken by small scale intermittency of fully developed turbulence.

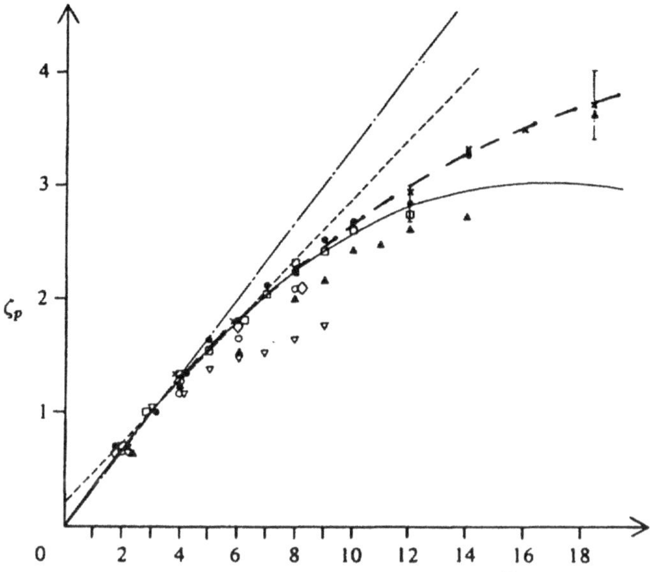

Fig. 5. Variation of exponent ζ_p as a function of the order p. (figure 14 of Anselmet *et al.*). Experimental data: o(Mestayer); $\triangle$(Vasilenko *et al.*); ∇(Van Atta and Park); $\diamond$(Antonia *et al.*); •, □, ×(Anselmet *et al.*). Intermittency models: solid curve (lognormal model); dotted line (β-model); chain-dotted line (1941 Kolmogorov theory); dashed curve (multifractal fit of Benzi*et al.*).

3. Multifractal Models of Inertial-scale Intermittency

The experimental nonlinear dependence of the exponents ζ_p of structure functions shown in figure 5 has been a challenge for theoritical models of intermittency. These models are based either on intermittency of inertial quantities such as the energy transfer rate, or on intermittency of the local dissipation rate ϵ_ℓ. Among models of the former class, there are so-called "homogeneous fractal" turbulent models in which the energy cascading process is assumed to occur on a single fractal set. The first homogeneous model has been proposed by Novikov & Stewart (1964); it uses the local dissipation. A more dynamical version called the "β-model" has been constructed by Frisch et al. (1978), in which the flux of energy is transfered to only a fixed fraction β of the eddies half the size. All of these homogeneous fractal models lead to the same prediction, namely, a linear plus constant function ζ_p which can try to fit experimental ζ_p's (the dotted line in figure 5 corresponds to the β-model prediction).

Actually, the data appear not to fall on a straight line, but display curvature, i.e. *nonlinear* dependence on p. In order to explain this nonlinear behaviour of ζ_p scaling exponents, Parisi & Frisch (1984) introduced a *multifractal* model of the inertial-scale intermittency. This model is based on the following remark. As we saw in the previous section, the concept of global scaling invariance is not consistent with experiments. Therefore, if we still want to preserve the concept of scaling, because it is such an important physical concept, one is led to assume *local scaling* invariance of the Navier Stokes equations. In the multifractal model of Parisi & Frisch (1984) it is indeed assumed that velocity increments $\delta v(\ell)$ scale as ℓ^h with a variety of possible local scaling exponents $h(x)$ depending on the position x at a given instant. For each value of h there is a set in physical space of Hausdorff dimension $D(h)$ near which the velocity behaves as l^h. More precisely the velocity behaves locally as a self-similar process of scaling exponent h (Vergassola & Frisch (1990)). It follows that, for a given h and a inertial scale ℓ, the contribution to the velocity structure function of order p, $< [\delta v(\ell)]^p >$, is proportional to ℓ^{ph} times the probability of intersecting the fractal set of dimension $D(h)$ with a ball of diameter ℓ. The latter is proportional to $\ell^{3-D(h)}$. (If not familiar with fractal dimension concepts, the reader may wish to check this for integer-dimension objects such as a curve, a surface and a volume.) Hence, the velocity events with scaling exponent h contribute a term proportional to $\ell^{ph+3-D(h)}$. For $\ell \to 0$, the contributions with the smallest exponent dominate. It follows that the structure function has an exponent ζ_p given by

$$\zeta_p = min_h(ph + 3 - D(h)).$$

This exponent is thus the Legendre transform of the codimension $3 - D(h)$ of the fractal set. It is important to stress that in the multifractal model, it is the *local scaling* assumption which permits to obtain a nonlinear dependence of ζ_p on p. However, the multifractal model contains an arbitrary function which specifies the distribution of dimensions $D(h)$. To day, the determination of this distribution has not been yet derived theoretically, so it is necessary to try to extract it from experimental data. Benzi, Paladin, Parisi & Vulpiani (1984) have shown that a simple form of a multifractal model (random β-model) depending only on two parameters can give a good fit for the ζ_p's (dashed line in figure 5).

Concerning multifractal measurements, there is an alternative strategy, recently used by Meneveau & Sreenivasan (1987). It consits in apply the multifractal analysis not to velocity increments in the inertial range, but to the local energy dissipation ϵ_ℓ, a spatial average of the dissipation over a distance ℓ. In fact, this idea introduced by Mandelbrot (1974), generalizes the well-known lognormal model proposed by Kolmogorov (1962) and eliminates inconsistencies noted by Mandelbrot (1974) and Kraichnan (1974). (In spite of its inconsistencies, the lognormal model gives a good quadratic fit for the ζ_p's as it shown in figure 5 (solid curve)). The attempt of Meneveau & Sreenivasan (1987) to test multifractality by measuring the local energy dissipation ϵ_ℓ is theoretically very interesting since they work with a positive quantity. Hence, methods developed for multifractality of the invariant measures of attractors become applicable. Still, such measurements involve difficulties because squared time-derivatives of the velocity are not obtained with a good accuracy by existing probes and may not be fully representative of the local dissipation. This dissipation-scale multifractal model also accounts for a nonlinear dependance of ζ_p on p for inertial size ℓ if it is assumed that ϵ_ℓ and δv_ℓ are related locally by $\epsilon_\ell \sim (\delta v)^3/\ell$. There are however difficulties for negative moments as explained in Bacry et al. (1990).

4. Wavelet Analysis of Turbulence

The multifractal concept is a good theoretical framework to give a simple interpretation of fine scale intermittency in fully developed turbulence. However, these statistical models are insufficient because they cannot give any local information about the intermittent (and occasionally strong) events in physical space. In this section, we shall show that the *wavelet transform* can be used for direct measurements of local scaling exponents h of the velocity increments.

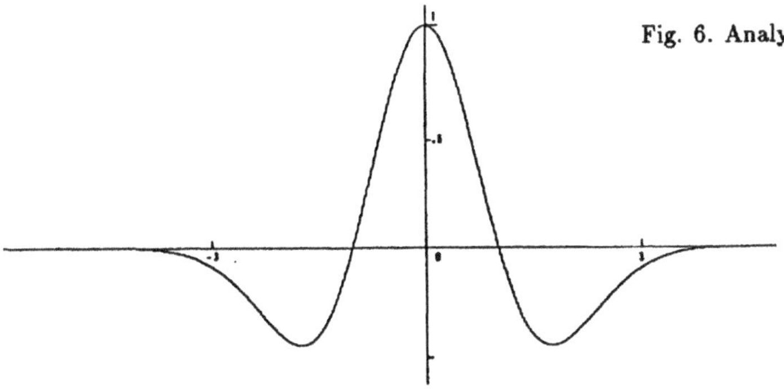

Fig. 6. Analysing wavelet: The Mexican hat.

The wavelet transform of a signal, as explained in Grossman & Morlet (1987), permits an analysis both in physical space and in scale space (scales are roughly inverse of wavenumbers). The wavelet transform consists in expanding functions in terms of "wavelets" which are constructed from one single function g by means of dilatations and translations. Figure 6 gives an exemple of such a function g, which is called the analysing wavelet.

It is necessary, in order to perform wavelet transforms, that the analysing wavelet should be localized around $x = 0$ and have a vanishing integral. Further conditions may be needed. From the single analysing function g, we construct a set of wavelet functions

$$g_{a,x_0}(x) = \frac{1}{a} g(\frac{x - x_0}{a})$$

by translations and dilatations. The wavelets $g_{a,x_0}(x)$ are then used as "mathematical microscopes", in which g characterizes the optics, while x_0 is the position analysed and $1/a$ the magnification. The parameter a which has dimension of a length, will henceforth be called the (analysing) scale. The wavelet transform of a signal $s(x)$ is defined as

$$T_g(a, x_0) = \frac{1}{a} \int_{-\infty}^{+\infty} s(x) g(\frac{x_0 - x}{a}) dx \qquad a > 0.$$

In others words, $T_g(a, x_0)$ is simply the convolution product of the signal and the scaled analysing function $\frac{1}{a} g(\frac{x}{a})$. The wavelet transform may also be regarded as a local Fourier transform: for a given position x_0, the amplitude of $T_g(a, x_0)$ will be largest when the scale a is roughly equal to the inverse local wavenumber of the signal. Here, we choose as analysing wavelet g, the second derivative of the Gaussian function called "Mexican hat", shown in figure 6. Occasionally, we also used a stepwise constant approximation of the Mexican hat, called "French top hat" which is piecewise. (For details, see Bacry et al. (1990)).

The turbulent velocity signal analysed comes from the wind tunnel S1 of ONERA, as in section 1 . Figure 7 shows the wavelet transform obtained with the French top hat wavelet applied to this velocity signal. The coordinates are linear (abcissa: position x, ordinate: scale a, grey levels for the amplitude $T(a, x)$). It is important to note that only positive values of $T(a, x)$ are coded; thus white regions correspond to both zero and negative values of $T(a, x)$. The grey level is defined for each value of the scale a by using 32 levels; so, for a given scale a_0 the grey level changes from white to black when the amplitude $T(a, x)$ increases from zero to its maximum value. It is seen that

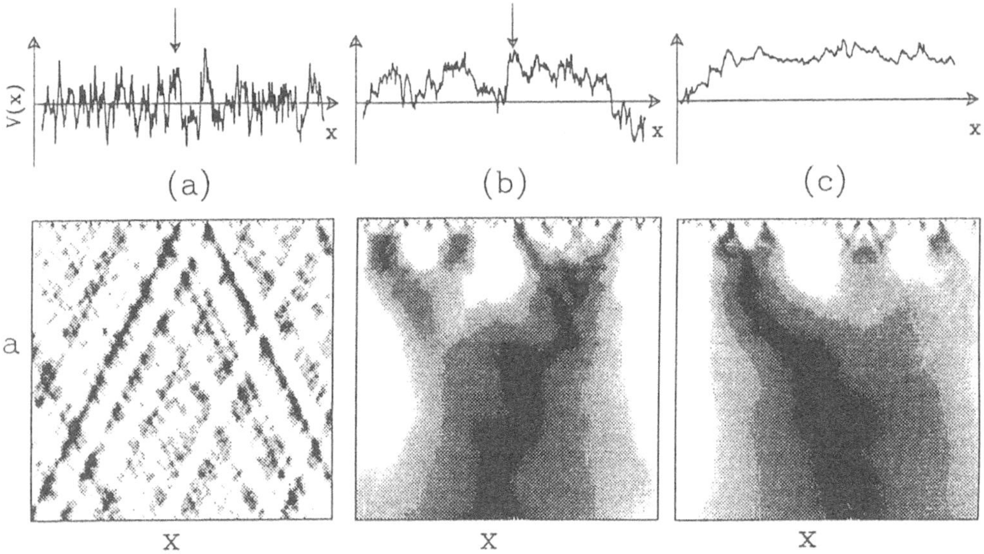

Fig. 7. (a) Wavelet transform of turbulent signal shown in upper part (french top hat): large scales; (b) Same as 7a at intermediate scales (near arrow on 7a); Same as 7b at small scales (near arrow on 7b).

patterns revealed by figure 7a are drastically different from those displayed on the others figures 7b and 7c. Figure 7a displays only large scales, ranging from $\ell_0/10$ to $28\ell_0$. As we expect, it reveals structures of size of order ℓ_0, randomly distributed in space. The cone-like patterns seen point towards the strongest structures and are generated by the edges of the French top hat. Figures 7b and 7c are obtained by analysing much smaller scales a, reduced by a factor 20 (7b) and 20^2 (7c). The analysis shown in these figures is made near the sharpest velocity gradient noted by arrows on the velocity sample. The full range of scales displayed in figures 7b and 7c covers the inertial range from the integral scale ℓ_0 down to about one quarter of the Taylor scale. These figures show a hierarchical branching process for which the physical interpretation is not completely yet clear to day. In order to test the sensivity of this branching process to the particular representation chosen, others representations have been used and are reported in Argoul et al. (1989) and in Bacry et al.(1990).

An immediate consequence of the definition of the wavelet transform is the following: if at a given position x_0, the signal behaves as $s(x_0+\ell)-s(x_0)\sim \ell^{h(x_0)}$ for $\ell\to 0$, then $T_g(a,x_0)\sim a^{h(x_0)}$ for $a\to 0$ (Holschneider (1988)). Thus, we can measure the local scaling exponent by analysing the behaviour of the wavelet coefficient $T_g(a,x_0)$ as a function of the scale a, for a fixed position x_0.

A log-log plot of such wavelet amplitude versus the scale a, for given position x_0, which reveals the local exponent $h(x_0)$ is given in figure 8. We see that at scales smaller than the integral scale, the amplitude of the wavelet coefficients $T(a,x_0)$ has a power law behaviour with superimposed oscillations. Such oscillations are not completely understood; they may reflect lacunarity of fractal (Smith, Fournier & Spiegel (1986)) or intrinsic fluctuations already observed for Gaussian self-similar processes (Vergassola & Frisch (1990)). Using a least square fit to a power law over inertial-range scales, we

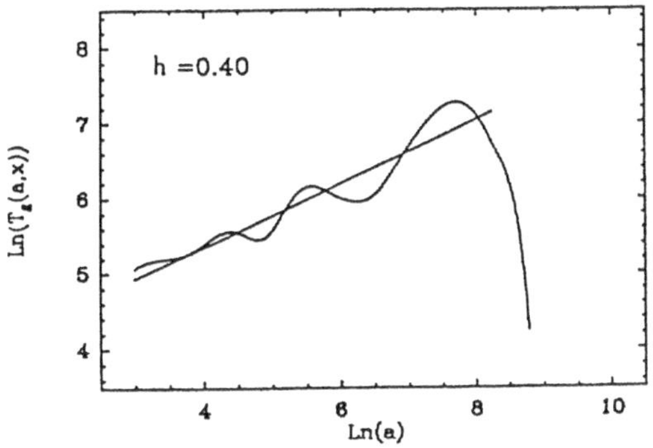

Fig. 8. Local scaling exponent measured in log-log plot of wavelet amplitude $T(a, x)$ vs. scale a.

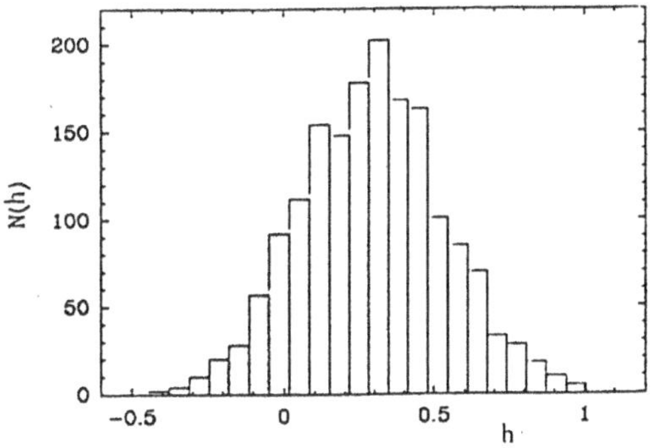

Fig. 9. Histogram of scaling exponents $h(x_0)$ (from 2000 data points x_0 out of 4 million).

can extract the local scaling exponents. For the case shown in figure 8, this exponent is $h(x_0) \simeq 0.4$ not very far the Kolmogorov 1941 value of 1/3.

The same kind of local analysis has been performed on a systematic investigation involving more than 2000 points x_0 randomly selected among the 4.10^6 points in our experimental sample. The histogram of the measured scaling exponents is given in figure 9. We can observe that the most probable value of the local scaling exponents is $h = 1/3$ corresponding to the Kolmogorov 1941 value. The values of the reliably measured exponents $h(x_0)$ range between -0.2 and 0.8, thereby suggesting that the multifractal picture proposed by Parisi and Frisch (1984) is appropriate. We believe that the data are statistically significant for negative exponents down to at least -0.1. One tentative interpretation of such negative local scaling exponents could be the occasional passage near the probe of slender vortex filaments of the sort observed in numerical simulations of Siggia (1981), Yamamoto & Hosokawa (1988) and She et al. (1990).

5. Concluding Remark

We expect, in the future, even more synergy between computational fluid dynamicists and experimental approaches of high Reynolds number turbulence. Among the challenging questions to be answered: (i) Is the small-scale intermittency of high Reynolds number flows universal?, (ii) Do small-scale vortex filaments observed in moderate Reynolds number simulations subsist at high Reynolds number and do they form some kind of fractal tangle?

Acknowledgements

The author has greatly benefitted from discussions with U. Frisch.

References

Anselmet, F., Gagne, Y., Hopfinger, E.J. & Antonia, R.A. 1984 *J. Fluid Mech.*, **140**, 63-89.

Antonia, R.A. & Chambers, A.J. 1980 *Bound. Layer Meteo.*, **18**, 399-410.

Antonia, R.A., Satyaprakash, B.R. & Chambers, A.J. 1982 *Phys. Fluids*, **25**, 29-37.

Argoul, F., Arnédo, A., Grasseau, G., Gagne, Y., Hopfinger, E.J. & Frisch, U. 1989 *Nature*, **338**,no 6210, 51-53.

Bacry, E., Arnéodo, A., Frisch, U., Gagne, Y. & Hopfinger, E.J. 1990 *in Turbulence 89: organized structures and turbulence in fluid mechanics*, Kluwer Academic Publishers.

Batchelor, G.K. & Townsend, A.A. 1949 *Proc. Roy. Soc.*, **A 199**, 238.

Benzi, R., Paladin, G., Parisi, G. & Vulpiani, A. 1984 *J. Phys. A*, **17**, 3521-3531.

Castaing, B., Gagne, Y. & Hopfinger, E.J. 1990 *Physica D* in press.

Castaing, B., Gunaratne, G., Heslot, F., Kadanoff, L., Libchaber, A., Thomae, S., Wu, X-Z., Zaleski, S. & Zanetti, G 1989 *J. Fluid Mech.*, **204**, 1-27.

Frenkiel, F.N. & Klebanoff, P.S. 1975 *Bound. Layer Meteo.*, **8**, 173-200.

Frisch, U. & Orszag S.A. 1990 in *Physics Today*, January, pp. 24-32.

Frisch, U., Sulem, P.L. & Nelkin, M. 1978 *J. Fluid Mech.*, **87**, 719-736.

Fujisaka, H. & Mori, H. 1979 *Prog. Theor. Phys.*, **62**, 54.

Gagne, Y. 1987 Thèse. Université de Grenoble.

Gagne, Y., Hopfinger, E.J. & Frisch, U. 1989 in *New Trends in Nonlinear Dynamics and Pattern Forming Phenomena: The Geometry of Non Equilibrium*, Plenum Press.

Grossman, A. & Morlet, J. 1987 in *Mathematics and Physics, Lecture on Recent Results*, World Scientific.

Holschneider, M. 1988 *J. Stat. Phys.*, **50**, 963.

Kida, S. & Murakami, Y. 1989 *Fluid Dyn. Res.*, **4**, 347-370.

Kolmogorov, A.N. 1941 *Dokl. Akad. Nauk SSSR*, **30**, 299-305.

Kolmogorov, A.N. 1962 *J. Fluid Mech.*, **13**, 82-85.

Kraichnan, R.H. 1974 *J. Fluid Mech*, **62**, 305-330.

Kuo, A.Y-S. & Corrsin, S. 1972 *J. Fluid Mech.*, **56**, 447-479.

Mandelbrot, B. 1974 *J. Fluid Mech.*, **62**, 331-358.

Mandelbrot, B. 1982 in *The Fractal Geometry of Nature*, Freeman and Co.

Meneguzzi, M. & Vincent, A. 1990 to appear in *J. Fluid Mech.*.

Meneveau, C. & Sreenivasan, K.R. 1987 *Nuclear Physics* **B2** (Suppl.), 49-67.North-Holland.

Mestayer, P. 1980 *J. Fluid Mech.*, **125**, 475-503.

Métais, O. & Herring, J. 1989 *J. Fluid Mech.*, **202**, 117-148.

Métais, O. & Lesieur, M. 1990 submitted to *J. Fluid Mech.*.

Monin, A.S. & Yaglom, A.M. 1975 in *Statistical Fluid Mechanics*, vol. 2, MIT Press.

Nakano, T. & Nelkin, M. 1985 *Phys. Rev. A*, **31**, 1980-1982.

Novikov, E.A. & Stewart, R.W. 1964 *Izv. Akad. Nauk. SSSR*, Ser. Geofiz. **3**, 408-413.

Parisi, G. & Frisch, U. 1984 in *Turbulence and Predictability in Geophysical Fluid Dynamics and Climate Dynamics*, 84. North-Holland.

She, Z.S., Jackson, E. & Orszag, S.A. 1990 in *Proc. Newport Conference on Turbulence*, Springer-Verlag, (in press).

Siggia, E.D. 1981 *J. Fluid Mech.*, **107**, 375-406.

Smith, L.A., Fournier, J.D. & Spiegel, E.A. 1986 *Phys. Lett.*, **A114**, 465-473.

Tennekes, H. 1968 *Phys. Fluids*, **11**, 669-671.

Van Atta, C.W. & Chen, W.Y. 1970 *J. Fluid Mech.*, **44**, 145-159.

Van Atta, C.W. & Park, J. 1972 in *Statistical Models and Turbulence*, Lecture Notes in Physics, vol. 12, 402-426. Springer-Verlag.

Vasilenko, V., Lyubimtsev, M.N. & Ozmidov, R.V. 1975 *Izv. Atmos. Ocean.Phys.*, **11**, 926-932

Vergassola, M. & Frisch, U. preprint 1990. Observatoire de Nice.

Yamamoto, K. & Hosokawa, I. 1988 *J. Phys. Soc. Japan*, **57**, 1532-1535.

Dynamics and Topology of Coherent Structures in a Plane Wake

F.F. Grinstein, F. Hussain and J.P. Boris*

Laboratory for Computational Physics & Fluid Dynamics
US Naval Research Laboratory
Washington, D.C. 20375, USA

We report results of the study of coherent structures (CS) using numerical simulations of the transition region of a compressible, (subsonic), spatially evolving wake behind a thick splitter plate at moderately high *Re*. The numerical model solves the time-dependent, gasdynamic, conservation equations using the Flux-Corrected Transport (FCT) algorithm and directional timestep-splitting techniques. The three-dimensional flow dynamics in the regions of vortex formation and near-wake are investigated by examining their response to cross-stream sinusoidal spanwise excitation at the inflow. The results show evidence of vortex reconnection processes leading to the formation of staggered vortex loops in the saddle regions between spanwise rollers of opposite sign.

1. Introduction

Characterizing the underlying fluid-dynamic processes in bluff body near-wake flows is necessary to achieve control of lift and drag in many practical external flows. Recent theoretical studies using linear stability analysis have shed new light on the local absolute/convective nature of wake-flow instabilities (Triantafyllou, Kupfer & Bers 1987), providing a new framework for understanding the dynamics of vortex-street formation and evolution. Numerical simulation of wake flows evolving in space and time extend these theoretical studies to practical regimes by accounting for the nonparallel nature of the flows, for global instabilities due to feedback, and for effects induced by the presence of stationary bodies on the near-wake flow. Numerical simulations of bluff body wake flows can provide information leading to the understanding of the dynamics and topology of the dominant CS, and thus point to mechanisms for practical flow control through exercise of appropriate influence on the formation, evolution and interaction of CS (Hussain 1986).

Experimental investigations over a wide range of Reynolds numbers have shown that the flow dynamics in the near wake of circular cylinders and other bluff bodies is dominated by large-scale two-dimensional (2D) spanwise vortex rollers arranged in von Karman vortex streets (e.g., Unal & Rockwell 1988, and references therein). Significant three-dimensional (3D) effects have been observed in experimental investigations of wakes behind circular cylinders (Hayakawa & Hussain 1985; Meiburg & Lasheras 1988), where the dynamics of the 3D CS appears to have distinct features due to the presence and interaction of spanwise rollers of opposite vorticity sign. As in the case of the

* *Department of Mechanical Engineering, University of Houston*

Advances in Turbulence 3
Editors: A.V. Johansson · P.H. Alfredsson

mixing layer – where the presence of 3D CS can significantly enhance the entrainment and mixing of the coflowing streams (Grinstein, Hussain & Oran 1987, 1988) the detailed CS dynamics in the near wake and its dependence on initial conditions have not yet been fully established.

Here we report results of the study of CS using numerical simulations of the transition region of a compressible (subsonic), wake behind a thick splitter plate with a rectangular trailing edge. The main focus of this work is the flow dynamics in the regions of vortex formation and near wake, where we examine the dynamics and topology of 3D CS by studying the flow evolution and its response to spanwise excitation.

2. Numerical Model

The basic flow configuration used in the simulations is shown schematically in Fig. 1, where streams of air are initially separated by a thick splitter plate. The identical laminar streams have free-stream velocity, U_∞=2.0 × 10^4 cm/s, and free-stream Mach number, $M_\infty = 0.6$, at standard initial temperature and pressure conditions.

The numerical model solves the time-dependent inviscid conservation equations for mass, momentum, and energy density for an ideal gas to examine the evolution of the large-scale CS. A passive scalar, convected separately with the velocity field, is used to evaluate the flow entrainment. The model uses the FCT algorithm with directional and timestep-splitting techniques on structured grids (Boris et al. 1985; Grinstein et al. 1989). FCT is an explicit, conservative, monotone, finite-difference algorithm, requiring no explicit artificial viscosity for stabilization. No explicit subgrid modeling other than the natural FCT high-frequency filtering has been included. The nonlinear properties of the FCT algorithm effectively act as a minimal subgrid model by maintaining the large-scale structures without aliasing while numerically smoothing the scales with wavelengths smaller than a few computational cells. The approach has been shown to be suitable for simulating the large-scale features of high-Reynolds-number vorticity dynamics in transitional free shear flows (Grinstein et al. 1989, 1990). A fourth-order phase-accurate FCT algorithm is used in the calculations.

Reflecting free-slip wall conditions are imposed at the edge of the splitter plate, inflow and outflow boundary conditions are imposed where appropriate in the streamwise direction (x), free-stream boundary conditions in the cross-stream direction (y), and periodic boundary conditions are imposed in the spanwise direction (z). The density and velocities are specified at the inflow guard cells, as $\rho = \rho_\infty$, $u = U_\infty$, and $v = 0$, respectively. A zero-slope condition on the energy allows the inflow pressure to vary in response to acoustic waves generated by events downstream. At the outflow, linear advection equations are imposed using the local velocity near the boundary, and the pressure is relaxed towards a specified ambient value P_∞. The inflow and outflow

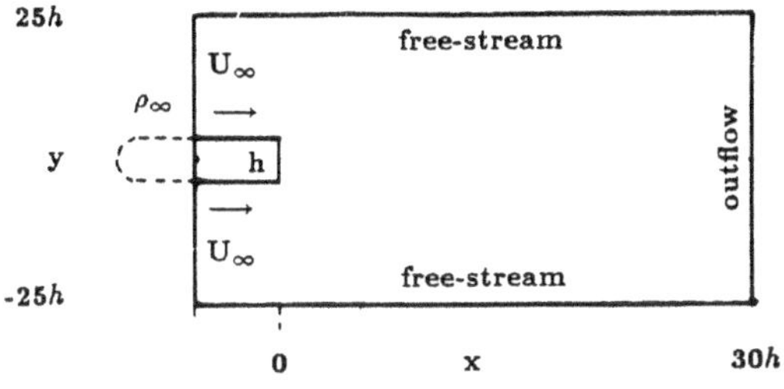

Fig. 1. Schematic flow configuration.

boundary conditions used in the model were developed and tested in previous shear-flow simulations (Grinstein et al., 1989) Initial conditions for the fully 3D simulations are based on developed vortex streets obtained in 2D simulations.

An initial square-well velocity profile is imposed at the inflow. The sharp edges of this profile are smoothed within a few computational cells by the small residual numerical viscosity of the algorithm into a doubly inflectional wake profile. For low Mach numbers ($M \lesssim 0.5$) and when the gridding involves evenly spaced computational cells, an effective viscosity, ν_{eff}, can be defined, behaving as $\nu_{eff} \sim O\ (\Delta r)^{1.5}$ for fixed Courant numbers, where Δr is the mesh spacing (Grinstein & Guirguis 1990). Characteristic Reynolds numbers (based on U_∞, h, and ν_{eff}), for the simulated flows in this work are in the range $10^3 \lesssim Re \lesssim 4 \times 10^4$.

The gridding is uniform in the regions where entrainment takes place. The computational cells are stretched in the cross-stream direction as the free-stream boundaries in the transverse direction are approached, but are otherwise uniformly spaced in the flow region of interest. The number of computational cells varies in the ranges 116 – 232 in the cross-stream direction, 173 – 600 in the streamwise direction, and 60–120 in the spanwise direction. The grid is held fixed in time. Courant numbers in the range 0.3–0.5 are used in the numerical simulations. Resolution studies doubling and coarsening the mesh indicated the convergence of calculated quantities, such as base-pressure coefficients and shedding frequencies, within 3–4%. The codes are fully vectorized and optimized for processing on CRAY-YMP computers, where typical calculations involve about 17 μs of CPU time per timestep per grid point.

3. Results and Discussion

We focus on the temporal and spatial evolution of the 3D CS for the plane wake. Previous numerical simulations of 3D plane wakes have been performed for the temporal, spatially periodic configuration using inviscid vortex methods in the incompressible regime (Meiburg & Lasheras 1988) and direct numerical simulation with pseudo-spectral methods in the compressible regime (Chen, Mansour & Cantwell 1989). The onset of 3D instabilities in the plane wake studied here was triggered by sinusoidal spanwise excitation of the transverse inflow velocity. Such spanwise excitation simulates the excitation obtained by putting a spanwise indentation in the edge of the thick splitter plate, as in the laboratory experiments of Meiburg and Lasheras (1988). In the latter experiments, no differences in the growth rate of the 3D instabilities were found for ratios $\lambda_{span}/\lambda_{st}$, in the range $\frac{1}{3} - \frac{5}{3}$, where λ_{st} is the natural wake wavelength and λ_{span} is the wavelength of the spanwise excitation. For the sake of computational convenience, $\lambda_{span}/\lambda_{st} = \frac{5}{3}$ was chosen, one spanwise excitation wavelength was included along the span, and the base of the thick splitter plate was located at the inflow boundary. The spanwise excitation was gradually added to the free streams at the inflow, over a time equal to a small fraction of a shedding period, during which the rms perturbation level grew from zero to its final value of of $0.02U_\infty$.

Figure 2 shows the calculated evolution of the 2D flow dynamics of the unforced wake visualized through contours of a passive mixing scalar, ϕ, defined to be of unit magnitude but of opposite signs for the two initial free streams. The figure shows the alternate shedding of spanwise rollers of opposite sign which give rise to the vortex street characteristic of wake flows. Typical shedding frequencies for the wakes simulated in this work are $St_h \approx 0.2$. Validation of the numerically simulated wake based on comparisons

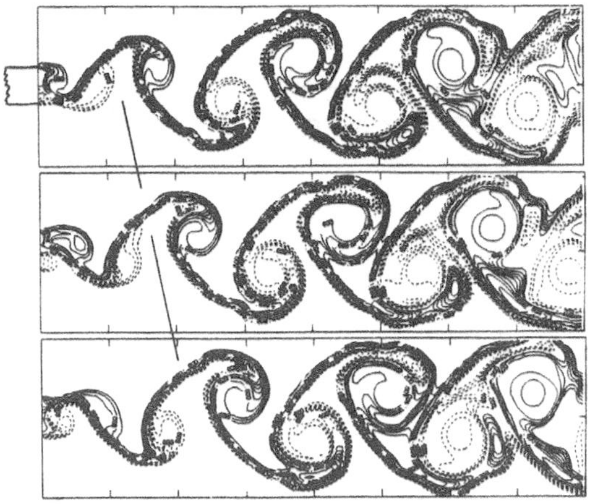

Fig. 2. Evolution of 2D wake flow.
Passive mixing scalar. —— :> 0; – – – :< 0.

of shedding frequencies and base-pressure coefficients with those of experiments has been reported elsewhere (Grinstein, Boris & Griffin 1990).

The spanwise perturbation of the transverse velocity at the inflow is a symmetric function of the spanwise variable z of the form $cos(2\pi z/\lambda_{span})$. In response to this excitation, the centers and ends of the rollers are advanced in opposite streamwise directions, in such a way that the undulation of the rollers on one side of the wake has a 180° phase shift relative to that of the rollers on the other side.

The vorticity introduced by the spanwise excitation in the transverse and streamwise directions, is stretched and concentrated into longitudinal vortices as a result of the high strain-rates between spanwise rollers. The longitudinal structures appear in counterrotating pairs distinctly superimposed on the spanwise rollers, and can be initially interpreted consistently as legs of horseshoe vortices, as in the case of the mixing layer (Bernal & Roshko 1986; Grinstein et al. 1988). The temporal sequence in Fig. 3 shows the formation of these horseshoe vortices in response to the spanwise excitation. The time of the visualizations is measured after 3D initialization in terms of the sheddding period ($t = \tau$). The initial perturbation first approaches the outflow boundary at $t \approx 4\tau$. As modeled schematically, the horseshoe structures initially develop together with the undulation of a newly shed roller '3' in the saddle region 'I' between this roller and the next one of the same sign being shed upstream ('1'). Wrapping the roller '2' of opposite sign shed on the other side of the wake, a horseshoe vortex is likewise formed with the opposite circulation in the saddle region III. As it becomes possible for fluid from one side of the wake to be entrained on the other side of the wake, the high-strain saddle region I rotates into II – in a reference frame moving with the rollers – and aligns itself with the divergent separatrix between the rollers of opposite sign '2' and '3'. As this topological change takes place, there is vorticity cancellation and reconnection of vortex lines in roller '2', as indicated schematically in Fig. 3. Through this 'cut and connect' process (Takaki & Hussain 1985) the vorticity redistributes to form closed vortex loops such as that in the top frame of Fig. 3.

The larger longitudinal structures on the spanwise (periodic) ends in Fig. 3 are also part of closed vortex loops generated through the same mechanism discussed above;

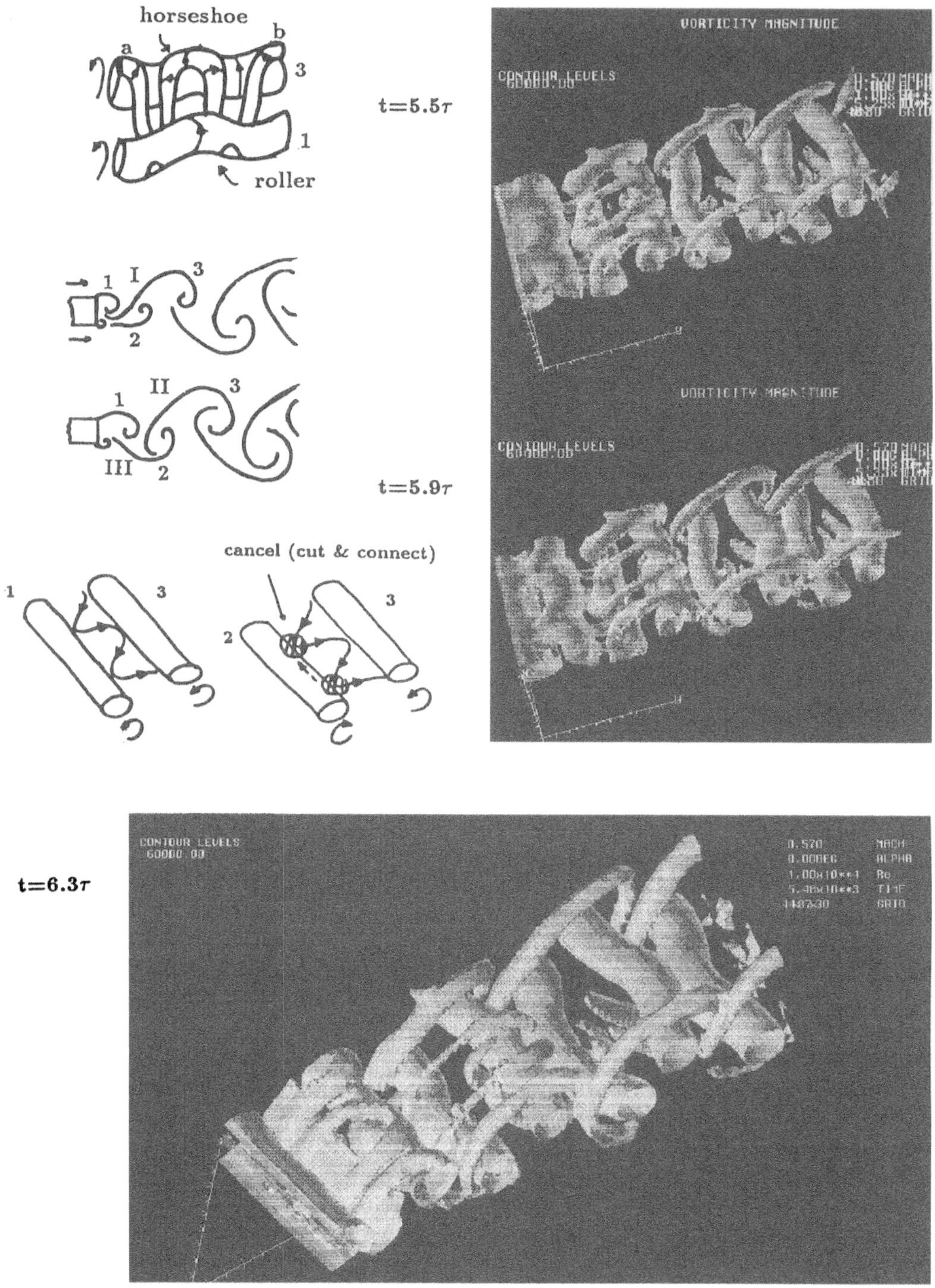

Fig. 3 Horseshoe vortices and vortex loops in the near wake. Perspective view from $y > 0$.

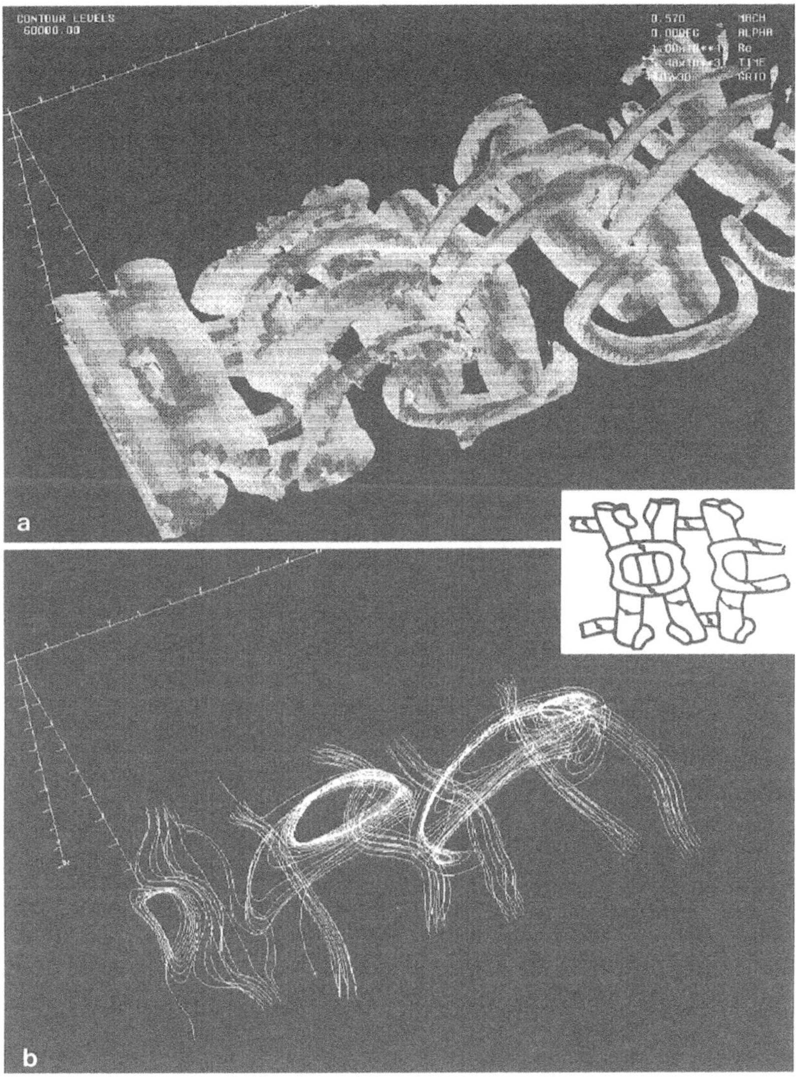

Fig. 4 Closed vortex loops in the near wake. Perspective view from $y < 0$ at $t = 6.3\tau$. a) constant enstrophy, b) vortex lines.

they originate on the end legs 'a' and 'b' which have opposite signs and initially form horseshoe vortices by closing into each other through periodicity while wrapping the rollers. The vorticity surfaces and vortex lines in Fig. 4 clearly show similar vortex loops formed in the same way on the other side of the wake in the center of the spanwise section studied.

The vortex loops are distributed following a preferred staggered pattern as observed in Figs. 3 and 4; they fit in the regions of the saddles where either centers or ends of rollers are closer, and where the strain field due to the latter is higher. The vortex loops *on each side* of the wake can partially overlap in the fully developed near wake, as the region where a vortex loop wraps a roller becomes very close to that where the next loop downstream wraps a roller of opposite sign (e.g., possible overlap between the two last loops in the center in Fig. 4a). Before the region of developed near-wake flow is attained, there is a transitional region where weaker nonfavored loops are also observed; the vorticity in these loops redistributes as the distortion of the rollers develops, in such a way that the saddle portions of the loops become progressively thinner as they are convected downstream. An example of this process is indicated by the decay of the loop initially located between rollers '2' and '3' in the top frame of Fig. 3. Farther on downstream, such weaker decaying loops are also observed at later times when the flow becomes more developed in time.

The presence of closed vortex loops in plane wakes has been previously suggested (Roshko 1976) and modeled (Meiburg & Lasheras 1988) by assuming single continuous vortex lines going from one roller to the other, as in some of the proposed 3D topological vorticity models of the mixing layer (Bernal & Roshko 1986). In particular, Meiburg & Lasheras (1988) found a staggered distribution pattern of the vortex loops in the case with transverse spanwise excitation, which is consistent with that reported here. The present simulations give direct evidence of CS interactions leading to the formation and development of *unconnected* vortex loops in the 3D-excited plane wake.

An effect of the distortion of the rollers induced by the spanwise excitation is to modify the pressure field near the base, in such a way that the resulting pressure minimum there is somewhat less focused. The 3D runs were not long enough to perform adequate statistics, to conclusively assess the effects of spanwise excitation on the base pressure. The instantaneous base-pressure coefficient, C_{pb}, was found to be somewhat larger in the 3D case, thus implying reduced pressure-drag. However, the differences found between 2D and 3D instantaneous calculations of C_{pb} were not significant, (within 15%), thus suggesting that for the flow regimes and spanwise excitation levels considered, the base pressure is essentially determined by the 2D features of the spanwise rollers.

4. Summary and Conclusions

We have presented results from finite-difference numerical simulations of a spatially evolving plane wake. The focus has been on the flow dynamics in the region of vortex formation and in the near-wake as relevant for the control of lift and drag. We have examined the dynamics and topology of the 3D CS in the plane wake by examining its response to the presence of transverse spanwise perturbations on the inflowing free-stream flow. The simulations have shown the presence of a staggered distribution of unconnected closed vortex loops. The vortex loops consist of counterrotating pairs of longitudinal vortices closing the loops as they wrap the rollers; they lie on the high strain saddle regions between consecutive rollers of opposite sign and can partially overlap on each side of the wake.

Acknowledgements

We thank O.M. Griffin, K. Kailasanath, and E.S. Oran for useful discussions. This work is supported by the Office of Naval Research, the Naval Research Laboratory, and the DARPA Applied & Computational Mathematics Program. The calculations were performed at the computing facilities of NAS at NASA Ames Research Center.

References

Bernal, L.P. & Roshko, A. 1986 Streamwise Vortex Structures in Plane Mixing layers *J. Fluid Mech.* **170**, 499, and references therein.

Boris, J.P., Oran, E.S., Gardner, J.H., Grinstein, F.F., and Oswald, C.E. 1985 Direct Simulations of Spatially Evolving Compressible Turbulence, *9th Int. Conf. Num. Meth. Fluid Mech.* (eds. Soubbaramayer & J.P. Boujot), p. 98, Springer Verlag, New York.

Chen, J., Cantwell, B., and Mansour, N. 1989 The Effect of Mach Number in the Stability of a Plane Supersonic Wake, AIAA Paper 89-0285, Reno.

Grinstein, F.F., Oran, E.S., & Hussain, F. 1987 Numerical Simulation of Three-Dimensional Coherent Structures in a Spatially Evolving Mixing Layer *Proc. 6th Symp. Turb. Shear Flows*, Toulouse.

Grinstein, F.F., Oran, E.S., & Hussain, F. 1988 Three-Dimensional Numerical Simulation of a Compressible, Spatially Evolving Mixing Layer, AIAA Paper 88-0042, Reno.

Grinstein, F.F., Guirguis, R.H., Dahlburg, J.P., & Oran, E.S., 1989 Three-Dimensional Numerical Simulation of Compressible Spatially Evolving Shear Flows, *11th Int. Conf. Num. Meth. Fluid Mech.* (eds. D.L. Dwoyer, M.Y. Hussaini, & R.G. Voigt), p. 283, Springer Verlag, New York.

Grinstein, F.F., Hussain, F. and Oran, E.S. 1990 Vortex-Ring Dynamics in a Transitional Subsonic Free Jet. A Numerical Study, *Europ. J. Mech. B / Fluids*, in press.

Grinstein, F.F., Boris, J.P., and Griffin, O.M. 1990 Passive Pressure-Drag Control in a Plane Wake, to appear in *AIAA J.*

Grinstein, F.F., and Guirguis, R.H. 1990 Effective Viscosity in the Simulation of Spatially Evolving Shear Flows with Monotonic FCT models, submitted to *J. Comp. Phys.*.

Hussain, F. 1986 Coherent Structures and Turbulence *J. Fluid Mech.* **173**, 303.

Hayakawa, M. & Hussain, F. 1985 Eduction of Coherent Structures in the Turbulent Plane Wake, *Proc. 5th Symp. Turb. Shear Flows*, Cornell.

Meiburg, E. & Lasheras, J.C. 1988 Experimental and Numerical Investigation of the Three-Dimensional Transition in Plane Wakes, *J. Fluid Mech.* **190**, 1

Roshko, A. 1976 Structure of Turbulent Shear Flows: A New Look, *AIAA J.* **14**, 1349.

Takaki, R. and Hussain, F. 1985 Recombination of Vortex Filaments and its Role in Aerodynamic Noise, *Proc. 5th Symp. Turb. Shear Flows*, Cornell.

Triantafyllou, G.S., Kupfer, K. & Bers, A. 1987 Absolute Instabilities and Self-Sustained Oscillations in the Wakes of Circular Cylinders, *Phys. Rev. Lett.* **59**,1914.

Unal, M.F. & Rockwell, D. 1988 On Vortex Formation From a Cylinder. Part 1. The Initial Instability, *J. Fluid Mech.* **190**, 491.

The Mechanisms of Formation and Breakdown of Soliton-like Coherent Structures in Boundary Layers

Y.S. Kachanov

Institute of Theoretical and Applied Mechanics, Novosibirsk 630090, USSR

The paper is devoted to a review of recent studies (mainly experimental ones) where a set of principle results on the problem of the generation, development and breakdown of specific nonlinear wave packets (coherent structures) in a transitional boundary layer were obtained. These results are generalized and applied to the description of the developed turbulent boundary layer. A new resonant-soliton concept of the wall turbulence is proposed and experimentally substantiated.

1. Introduction

The idea about the analogy between the transitional and developed turbulent flows regarding the mechanisms of the turbulence generation was advanced long ago and it is often discussed now. New experimental and theoretical results on the nonlinear stages of the laminar-turbulent transition obtained recently permit us to look at this analogy from a new viewpoint and to develop some new notions about the processes of the turbulence generation both in transitional and turbulent flows.

It is well known now that there are two main types of the laminar-turbulent transition in the boundary layer. The classical regime found by Klebanoff with coauthors [1] was termed later as the "K-regime" of the breakdown. A new type of the breakdown was found in 1976 in Novosibirsk [2] and was termed as the "N-regime" of the transition (or the "subharmonic" one). The nature of the N-regime of the transition had been revealed both experimentally and theoretically soon after its first detection (see for review work [3]). In contrast, the progress in understanding the K-breakdown was very slow till recently. In 1980 in Novosibirsk a new detailed experimental study of the K-regime of the breakdown was conducted [4]. The results obtained provide for the first time a systematic information about the frequency-wavelength spectral structure of the disturbances and radically changed the existing notions about the mechanisms of the K-breakdown. The present paper is devoted to the following investigations of the K-regime of the laminar boundary-layer breakdown and to the application of their results to the description of the developed turbulent flow with coherent structures.

2. Detection of solitons in the transitional boundary layer

One of the most typical property of the K-regime of the transition is the amplification of a strong spanwise modulation of the flow and disturbances, and the formation at "peak" spanwise positions (where the disturbance amplitudes have their maxima) of the flashes of high-frequency secondary fluctuations - so-called "spikes". Starting with the pioneer work [1] the generation of the spikes in the K-regime of the transition and following doubling, tripling etc. was usually associated with the start of a stormy breakdown

Advances in Turbulence 3
Editors: A.V. Johansson · P.H. Alfredsson

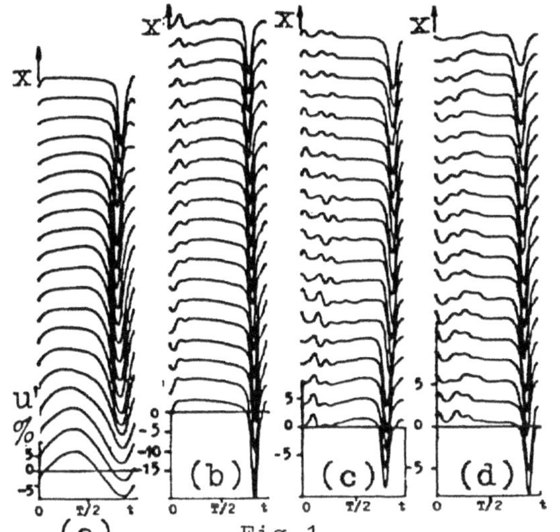

Fig.1

and randomization of the laminar flow. However, it was clearly shown in works [3, 4] that the multiplication of the spikes is of deterministic character and observed only at downstream displacement along the line $y = const.$ At the same time the spike by itself quickly moves away from the wall during its formation and then acquire a rather conservative form which almost does not change further downstream. This property is visually shown in Fig. 1 obtained by the author together with V.I. Borodulin and reported in work [5]. The traces are displayed for the downstream positions from $x = 400$ to 600 mm. The last position corresponds to almost a turbulent flow (initiated by the spike) inside the boundary layer.

The authors of work [6] payed their attention to such a stable, conservative behaviour of the spike and noted its resemblance with solitons observed, for example, in thin films of liquid flowing along an inclined plane. The behaviour of such solitons was described in work [7] from the viewpoint of the theory of nonlinear dynamic systems as a movement of a phase point along a homoclinic trajactory near the saddle-focus irregular point. The resemblance of the form of the spikes (Fig. 2a) and solitons form [7] (Fig. 2b) as well as of their phase trajectories (shown in Fig. 2c, d and 2e, f) proved to be so close that the authors of [6] surmise that the spikes are also solitons. Further investigations corroborate the correctness of this hypothesis. Some of the results obtained by the author together with V.I. Borodulin [5] are shown in Fig. 3. The soliton moves quickly towards the external boundary of the boundary layer (Fig. 3) and then travels downstream almost along the line $y_s = \delta$. The amplitude of the spike-soliton A_a is changing very slowly downstream after its formation has been completed ($x \approx 500$ mm) despite the fact that the spike consists of high-frequency harmonics which are rapidly decreased according to a linear stability theory (Fig. 3). The typical spanwise width of the soliton h_z is also practically unchanged (Fig. 3). A typical angle of the packet dispersion is about 0.4° and this value is 45 times less than that of the linear wave packet [8].

Recently it was shown theoretically in [9, 10] that the equations for the description of the nonlinear wave packets in the boundary layer can be reduced to the Benjamin-

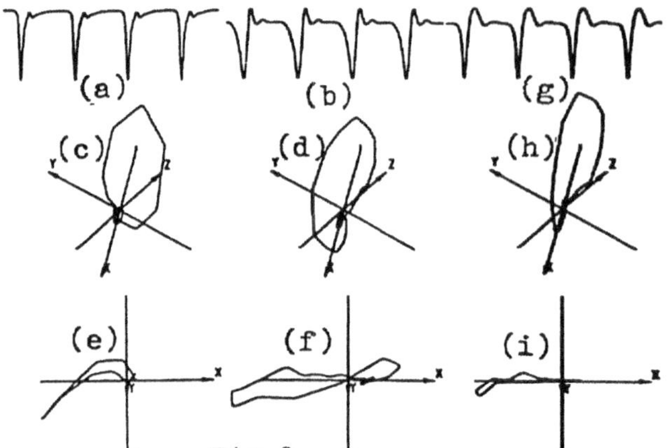

Fig.2

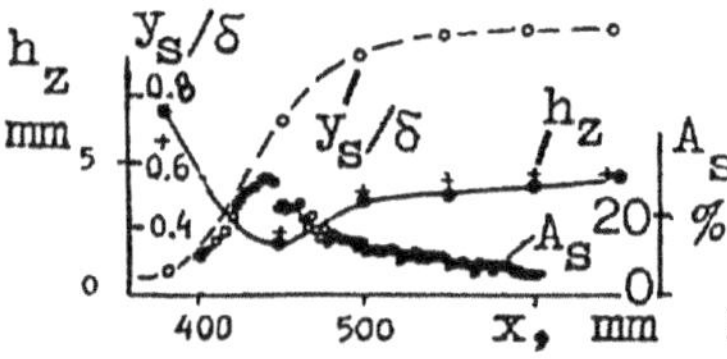

Fig.3

Ono equation which has soliton solutions. One of these solutions obtained in [10] is shown in Fig. 2g. It also closely resembles the spikes observed in the experiments (Figs. 1, 2a). The same can be said about the phase trajectories (Fig. 2h, i).

3. Two concepts of the spikes' generation

During almost thirty years the concept of local high-frequency secondary (LHS) instability of the flow was the most widespread explanation of the causes of spikes' generation in the K-regime of the breakdown (see for example [1, 11]). The essence of this mechanism consists in amplification of a packet of high-frequency fluctuations under the influence of an unstable inflexional instantaneous velocity profile which is formed (locally in time and space) by a low-frequency primary wave. The first direct experiment devoted to the corroboration of the LHS instability concept was carried out in a flat channal flow in [12]. It confirmed the possibility of the realization of this mechanism. However, the question whether this mechanism is the immediate cause of the spikes' generation, was not answered. It was found later (in [6]) that the answer to this question is negative. The results obtained in work [6] led its authors to the two main conclusions: (i) the mechanism of the LHS instability of the flow actually takes place in the K-regime of the boundary-layer transition (the same as in work [12] for the channel flow); however (ii) the generation of the flashes-spikes is not connected immediately with this mechanism.

On the basis of a detailed analysis of the experimental and theoretical data a physical model of the development of the deterministic stages of the K-regime of the breakdown has been developed in [3]. It was termed the "wave-resonant" (WR) concept of the breakdown. The essence of the WR concept is explained in brief by Fig. 4. An idea about the four-wave resonances proposed in [13] and developed in [14] (and in the following works) was taken as the basis of the WR concept. It consists in the following.

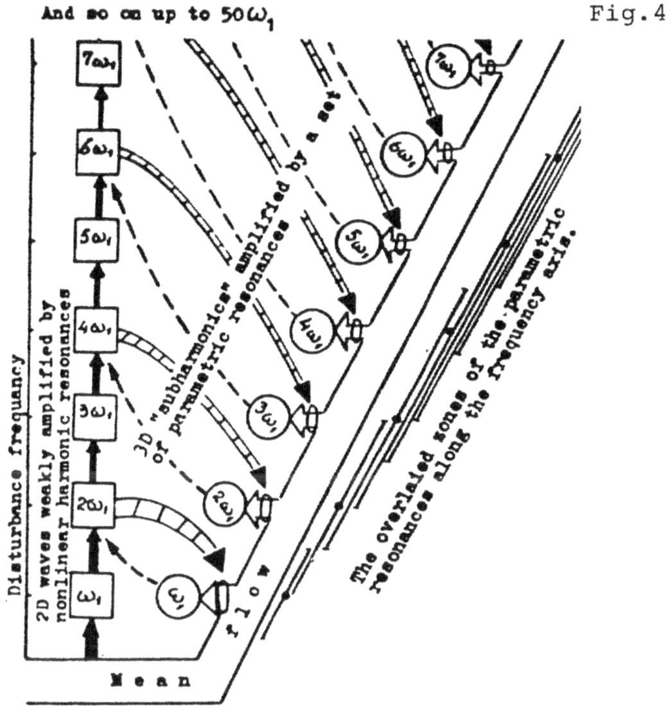

The amplification of the 3D oscillations with the frequency of the fundamental wave ω_1, of type $(\omega_1, \pm\beta_r)$ could be initiated by a four-wave resonant interaction of waves $(\omega_1, 0)$, $(2\omega_1, 0)$ and $(\omega_1, \pm\beta_r)$. The WR concept proposes the generalization of this idea for the case of a great number of waves of the type $(n\omega_1, 0)$ and $(m\omega_1, \pm\ell\beta_1)$, where $n = 2, 3, 4, 5, \ldots\ldots, m \sim n/2$ and $\ell\beta_1 \approx \beta_r$ are the spanwise wavenumbers of the 3D priming disturbances which turned out within the ranges of the resonant amplification; β_1 is the characteristic wavenumber of an initial spatial inhomogeneity. It is supposed that these priming subharmonic disturbances, which are necessary for the parametric amplification, arise mainly as the result of a very slow (nonresonant) combination interaction of initial disturbances: primary wave $(\omega_1, 0)$ and small primary modulation of the mean flow $(0, \beta_1)$. A more detailed description of the WR concept is presented in [3].

It should be noted that the nonlinearity in subharmonic resonances [15,16] results in an increase both of the fundamental wave and of the subharmonics; their phase synchronism is maintained [15,17] providing an additional mechanism of the generation of synphase flat higher harmonics of the fundamental wave. In Fig. 4 these nonlinear connections are shown by dotted arrows. As a result, when the amplitudes of the subharmonics $(m\omega_1, \pm\ell\beta_1)$ are growing a set of high-frequency waves can be detached ("taken off") from its base - disturbances with the fundamental frequency ω_1 - and probably it could exist separately.

It was shown in work [3] that if the cascade of resonances mentioned above had been realized, it would have inevitably resulted in the appearance of the spikes typical for the K-breakdown located just in those places of time and space where they are observed in the experiments. Most of the experimental results obtained in [1,4] and unexplained before were qualitatively or even quantitatively explained in the framework of the WR concept. This good correlation with the observations was an indirect substantiation of the concept. However, the direct evidence of the existance of the cascade of the resonances, postulated within the framework of the WR concept, was not obtained at that time (1986-87). It appeared later.

The corresponding evidence was obtained recently in work [18]. The existance of the phase synchronism between flat harmonics of the fundamental wave ($n\omega_1$) and their "subharmonics" (with frequences close to $n\omega_{1/2}$) was found directly for a large number of resonant groups of the instability modes. The results of work [18] showed that the system (cascade) of the parametric subharmonic resonances which was postulated in the framework of the WR concept [3]), is actually realized in the K-regime of the breakdown on the stage of the formation of the spikes-solitons and plays the role of the mechanism of their amplification.

4. Amplification of background turbulence by the spikes-solitons

It was usually supposed that the amplification of random fluctuations in the K-regime of the breakdown and the flow randomization start with a "sudden", "explosion" appearance of flashes-spikes, that is usually associated with the mechanism of a local high-frequency secondary instability. In the light of the results presented above this explanation cannot be regarded as satisfactory now. How does the randomization really occur? In Fig. 5 (taken from [19]) families of realizations of the velocity fluctuations, observed in a boundary layer near the wall (y=1 mm) are shown for different x-positions. The periodical flow (which appears at the spikes' stage) is seen to become gradually distorted by nonperiodic, random additions in a form of "antinodes". What is the cause of the amplification of these random fluctuations? It was supposed in [3] that the cause is the same as in the N-regime of the breakdown - the subharmonic resonances with background disturbances. The following analyses confirms in the main this supposition.

Families of the oscilloscope traces which are analogous to those in Fig. 5, but obtained after the filtration of the subharmonic frequency, are shown in Fig. 6a. It is seen that within the range where the random antinodes appear (x=470-520 mm) the subharmonic fluctuations are amplified (see also Fig. 7) and acquire a characteristic form with nodes and loops. Points of the phase trajectory of the subharmonic fluctuations, corresponding to these traces, demonstrate a typical evolution of the subharmonic amplitude and phase in a slow time scale (Fig. 6b). Such a behaviour of the subharmonic

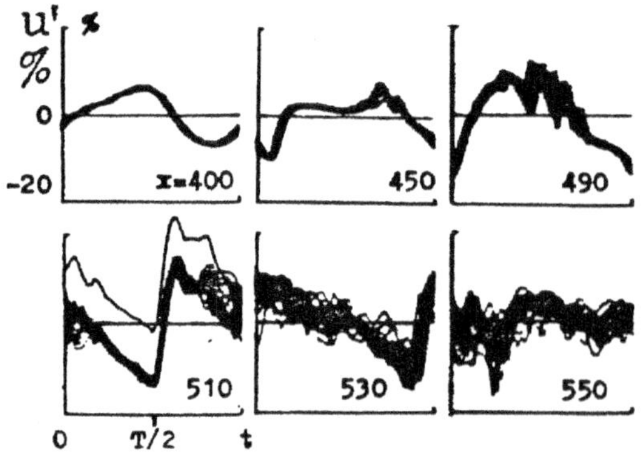

Fig.5

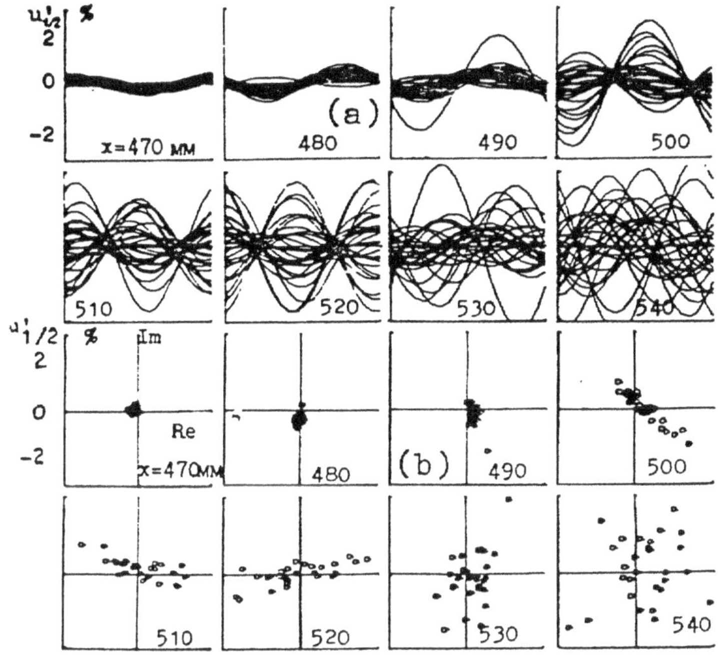

Fig.6

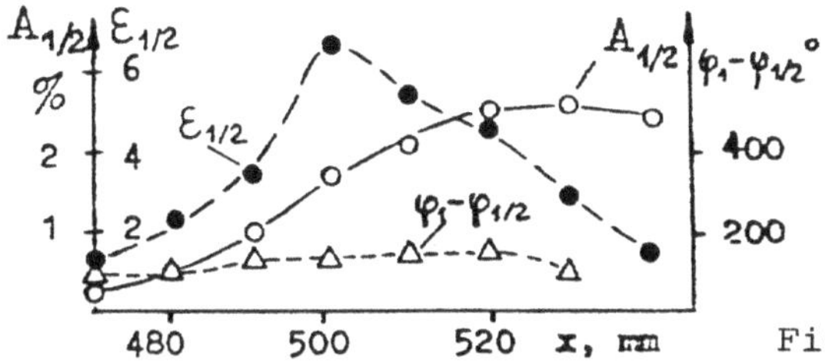

Fig.7

disturbances has been already observed in the N-regime of the breakdown in [17]. It corresponds to an amplification of subharmonic random background disturbances under the influence of a parametric resonant interaction with the fundamental wave. An additional confirmation of this conclusion is shown in Fig. 7, where the connection between the amplitude $A_{1/2}$ of the amplified subharmonic, lengthening of the cloud of the points of its phase trajectory $\varepsilon_{1/2} = a/b$ (where a, b are the dimensions of the cloud along its main axis) and the difference between the phase of the fundamental wave and that of the subharmonic $\varphi_1 - \varphi_{1/2}$ are demonstrated. It is seen that the greatest amplification rates of $A_{1/2}$ correspond to the largest lengthening of the cloud that is connected with the resonance. The phase synchronism between the subharmonic and fundamental waves is also observed. However, in the present case, in contrast to the N-breakdown, a great number of higher random harmonics of the subharmonic wave ($m\omega_1/2$ with odd and even m) are observed (Fig. 8). Their appearance could be initiated either by a set of the subharmonic resonances with the participation of the random priming subharmonic disturbances and the deterministic forcing waves $n\omega_1$ (which have been amplified earlier on the stage of the formation of spikes-solitons) [3], or by a combination interaction of the random subharmonic with the higher harmonics of the fundamental wave (i.e.

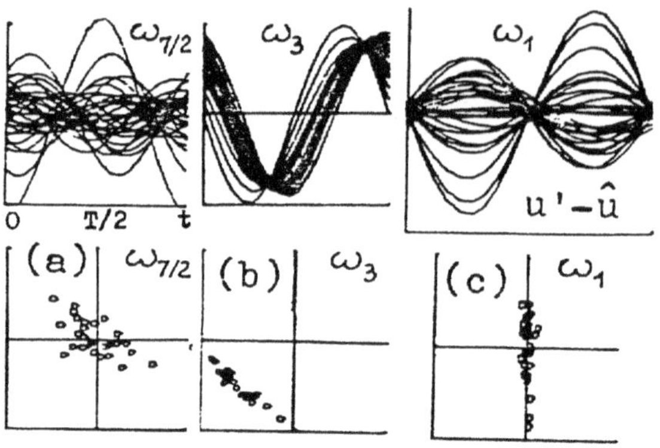

Fig.8

with the soliton) [2]. This question needs a more precise examination. Maybe both mechanisms are realized.

5. Hypothesis on the structure of the developed turbulence.

5.1. Initial promises.

The idea about the existence of the mechanism of the turbulence generation through the development of the coherent structures was advanced long ago [20], The notions about the "continuous transition" to turbulence carried out in a viscous (laminar) sublayer are also well known (see for example [21]). A formal resemblance of the coherent structures of the turbulence, observed near the wall by means of visualization and called very often as horseshoe vortices, with Λ-vortices observed in transitional boundary layers were pointed out repeatedly, the same as the resemblance of the accompanying inflexional instantaneous velocity profiles [22]. However, all these ideas have been developed on a comparatively superficial, qualitative base and they have not resulted in a theory on the turbulence generation. At present a rational foundation appears for the development not only of a basis of a theory on the coherent structures but also of a physical basis for a new theory of turbulence which could be termed as the "resonant-soliton" theory of the turbulence.

5.2 The main ideas of the resonant-soliton concept of the turbulence.

An amplification of packets of relatively low-frequency fluctuations, leading to the phenomena which are analogous to the N- and K-regimes of the breakdown, takes place sporadically in the wall region (near the boundary of the viscous sublayer) of the turbulent boundary layer under the influence of the turbulent fluctuations initiated upstream. The K-regime of the breakdown is of a paramount importance. It appears when the wave packet has a sufficiently representative frequency-wavelength spectrum and when the amplitudes of the packets' disturbances reach some critical, threshold values (see [17, 23]). The take-over to the K-breakdown leads to a start (near the wall) of the cascade of the resonant amplification of the harmonics described by the wave-resonant concept (see Fig. 1 and paragraph 3) and the spike-soliton corresponding to the coherent structure of the turbulence starts to form. The following destiny of the coherent structure-soliton (CS-soliton) is analogous to the one in the transitional boundary layer. It moves away

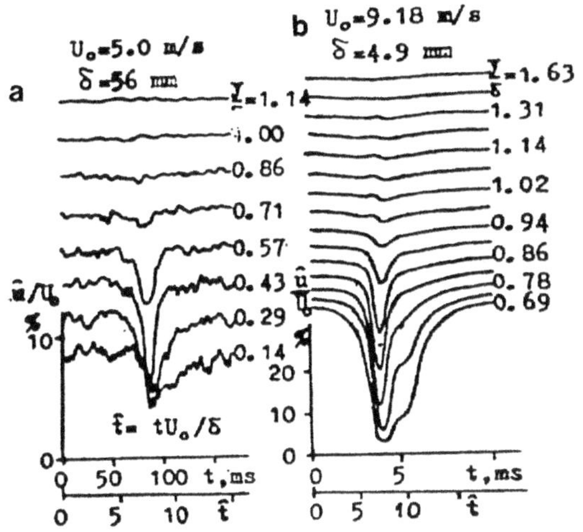

Fig.9

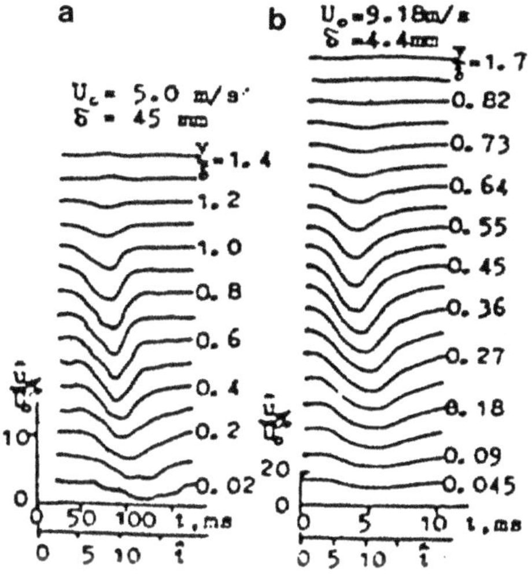

Fig.10

from the wall piercing the boundary layer and propagates downstream along the external boundary of the boundary layer almost without changing its form. Near the wall, the CS-solitons are displayed as a characteristic tooth-shaped jump of the flow velocity from smaller to greater values (see Fig. 9). The interaction of the CS-soliton with the turbulent noise leads (the same as in the K-breakdown) to the parametric resonant amplification of random background fluctuations, i.e. to the refuelling of the turbulence energy from the mean flow owing to the "catalytic" influence of the CS-soliton (see paragraph 3). At the same time the soliton is gradually destroyed probably due to its nonlinear interaction with the turbulence (this problem is practically not examined for the present).

5.3 Some experimental evidence of the correctness of the resonant-soliton concept of the turbulence.

There exist now much evidence of this kind. Some examples are given below. Traces corresponding to the coherent structures of the turbulence (experimentally observed in [24-26] are shown in Figs. 9a, 10a, 11b. The corresponding traces for the spikes-solitons, observed in the K-breakdown in [6, 4] are shown in Figs. 9b, 10b, 11a. The graphs testify to a striking resemblance of these phenomena both near the wall and near the external boundary of the boundary layer. In particular, despite a considerable difference (about 1-2 orders) of the dimensional characteristic time and spatial scales, the nondimensional time and y-coordinates almost coincide. The downstream velocity of the "teeth" is equal to $0.67U_0$ for the coherent structures and to $0.57U_0$ for the K-breakdown. A number of other characteristics of the disturbances also coincide (or almost coincide).

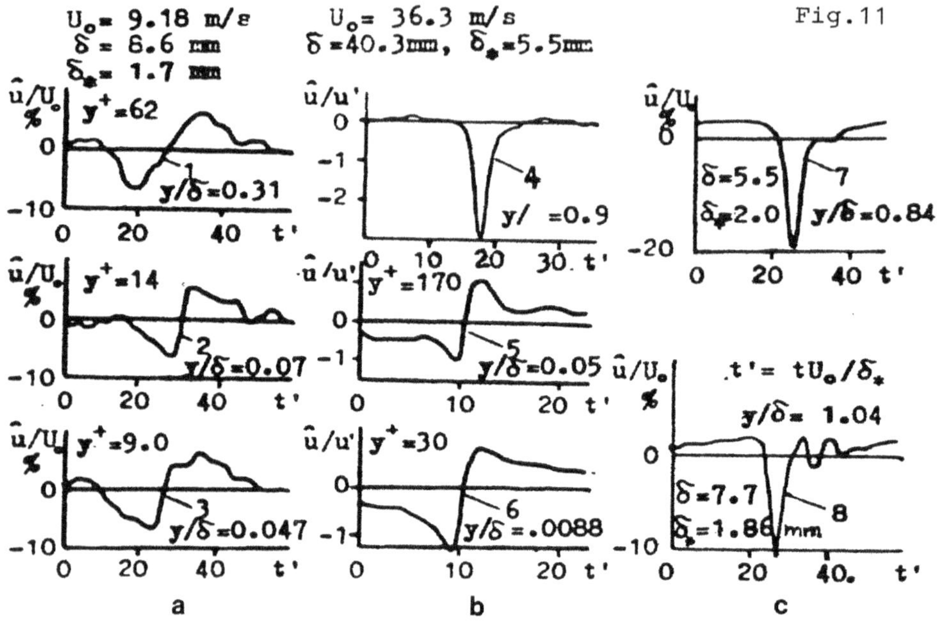

Fig.11

6. Conclusions

A number of very important conclusions follow from the results mentioned above.

1. The flashes-spikes (typical for the K-breakdown) possess the properties of the solitons [5].

2. The mechanism of the local high-frequency secondary instability is actually realized in the K-breakdown of the boundary layer but it is not responsible for the appearance of the spikes-solitons [6]. Their amplification is explained within the framework of the wave-resonant concept proposed earlier [3].

3. A set (cascade) of the subharmonic parametric resonances postulated by the WR-concept is detected experimentally [18].

4. The spikes-solitons amplify background turbulence through the parametric resonant interaction with random quasi-subharmonic fluctuations and result in the randomization of the flow in the K-regime of the breakdown [19].

5. A hypothesis on the soliton nature of the turbulent coherent structures formed (as in the K-breakdown) through the cascade of the resonances, described by the WR-concept, is proposed.

6. A new physical mechanism of the wall turbulence - resonant-soliton concept of turbulence - is proposed.

References

1. Klebanoff P.S., Tidstrom K.D. & Sargent L.M. 1962 *J. Fluid Mech.* **12**, 1-41.

2. Kachanov Y.S., Kozlov V.V. & Levchenko V.Y. 1977 *Proc. Acad. Sci. USSR, Fluid Mech.* **3**, 49-58 (in Russian).

3. Kachanov Y.S. 1987 *J. Fluid Mech.* **184**, 43-74.

4. Kachanov Y.S., Kozlov V.V., Levchenko V.Y. & Ramazanov M.P. 1985 In: *Laminar-Turbulent Transition.* Springer 61-73.

5. Kachanov Y.S. 1990 To be published in: *Proc. Third IUTAM Symp. Laminar-Turbulent Transition.* Toulouse 1989 Springer.

6. Borodulin V.L. & Kachanov Y.S. 1988 *Proc. Siberian Div. Acad. Sci. USSR.* **5**, 65-77 (in Russian).

7. Demyohin E.A. & Shkadov V.Y. 1986 *Proc. Acad. Sci. USSR. Fluid Mech.* **3**, 91-97 (in Russian).

8. Kachanov Y.S. 1985 In: *Laminar-Turbulent Transition.* Springer, 115-123.

9. Zhuk V.I. & Ryzhov O.S. 1982 *Dokl. Akad. Nauk SSSR.* **263**, N° 1 (in Russian).

10. Rothmayer A.P. & Smith F.T. 1987 *Trans. ASME*, Cincinnatti, Ohio, June 1987.

11. Landahl M.T. 1972 *J. Fluid Mech.* **56**, 755-802.

12. Nishioka M., Asai M. & Iida S. 1980 In: *Laminar-Turbulent Transition.* Springer 37-46.

13. Craik A.D.D. 1971 *J. Fluid Mech.* **50**, 393-413.

14. Nayfeh A.H. & Bozatli A.N. 1979 *AIAA-79-1456.*

15. Zelman M.B. & Maslennikova I.I. 1982 In: *Instability sub- and supersonic flows.* I.T.A.M., Novosibirsk 5-15 (in Russian).

16. Herbert T. 1988 *Ann. Rev. Fluid Mech.* **20**.

17. Kachanov Y.S. & Levchenko V.Y. 1984 *J. Fluid Mech.***138**, 209-247.

18. Borodulin V.I. & Kachanov Y.S. 1989 In: *Simulation in Mechanics.* I.T.A.M., Novosibirsk **3**(20) N° 2, 38-45 (in Russian).

19. Dryganets S.V., Kachanov Y.S, Levchenko V. Y. & Ramazanov M.P. 1990 *J. Appl. Math. Tech. Phys.* **2** (in Russian).

20. Kline S.J. & Rundstadler P.W. 1959 *Trans. ASME E: J. Appl. Mech.* **26**.

21. Repik E.U. & Sosedko Y.P. 1974 In: *Turbulent Flows.* Moscow, Nauka.

22. Blackwelder R.F. 1983 *Phys. Fluids* **26**, 2807-2815.

23. Lifshits A.M., Rahamtullaev R.D. & Shtern V.N. 1989 *Proc.Acad. Sci. USSR.* 4 (in Russian).

24. Fukinishi Y., Sato H. & Inous O. 1987 *AIAA-87-1253.*

25. Makita H., Sassa K., Abe M. & Itabashi, A. 1987 *AIAA-87-1232.*

26. Thomas A.S.W. & Bull M.K. 1983 *J. Fluid Mech.* **128**, 283-322.

Deformation and Splitting of Pseudo-elliptical Vortex Rings

M. Kiya and H. Ishii

Department of Mechanical Engineering, Hokkadido University, Sapporo, 060 Japan

A vortex-dynamics simulation has been made on the deformation of a pseudo-elliptical vortex ring which consists of two parallel line segments connected by two semicircles of the same radius, the major-to-minor axis ratio being between 2 and 20. The deformation has four fundamental patterns depending on the aspect ratio: the periodic axis switching; the splitting into two vortex rings after the axis switching; the splitting into two vortex rings without the axis switching; the splitting into three vortex rings without the axis switching.

1. INTRODUCTION

Turbulent jets issued from non-circular nozzles appear to have a greater rate of the growth and the mixing with surrounding fluid than those issued from a circular nozzle of the same cross-sectional area. Immediately downstream of a nozzle a vortex ring is formed whose circumference is approximately the same as that of the nozzle. The vortex ring is deformed both by the non-uniform self-induced velocity along the circumference and by the velocity induced by neighbouring vortex rings.

The behaviour of a single vortex ring whose shape is the same as that of the nozzle provides useful information on the vortex structure and turbulence property in the initial region of the jets. A few examples are as follows: The cross section of a square jet experiences the axis switching, which is also observed in single square vortex rings (Toyoda et al. 1989). An elliptical vortex ring with the major-to-minor axis ratio of 4 deforms to split into two vortex rings; this corresponds to the bifurcation of an elliptical jet with the same value of the axis ratio (Hussain et al. 1989). The splitting into two or three vortex rings is also observed for a rectangular vortex ring with an aspect ratio of 4.5 (Kambe et al. 1971). Thus it is useful to examine the behaviour of non-circular vortex rings by systematically changing representative aspect ratios.

This paper describes results of a simulation on the deformation of a vortex ring defined in Fig. 1. The simulation was made by the three-dimensional vortex-blob method. The centreline of the vortex ring consists of two parallel line segments connected by two semicircles of the same radius R, the length of the major axis being denoted by L. This vortex ring, which will hereinafter be referred to as the pseudo-elliptical vortex ring, has zero curvature along the parallel parts and a constant curvature along the semi-circular parts with a discontinuous change of

Advances in Turbulence 3
Editors: A.V. Johansson · P.H. Alfredsson

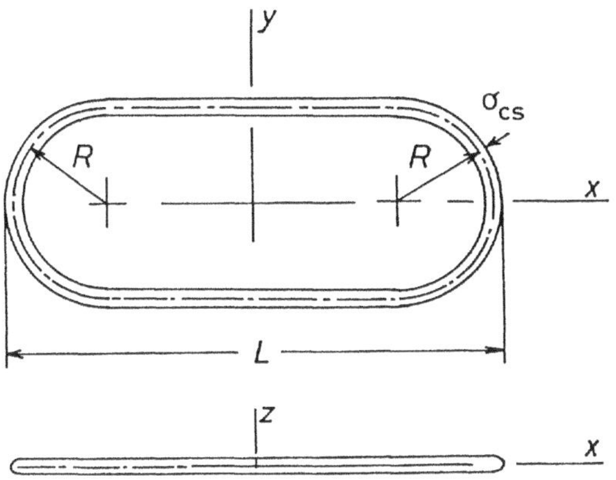

Figure 1. Configuration of pseudo-elliptical vortex ring.

curvature at the junctions. On the other hand, an elliptical vortex ring has a continuous distribution of curvature along the circumference but the curvature near the ends of the major axis becomes excessively large for a large value of the major-to-minor radius ratio. This is unfavourable for the numerical simulation. This is the main reason why the pseudo-elliptical vortex ring was employed in the present study. In passing it may be noted that the pseudo-elliptical nozzles can more easily be constructed than the elliptical nozzles. The simulation showed that the pseudo-elliptical vortex rings had four fundamental patterns of deformation depending on the aspect ratios $L/(2R) = 2$ - 20.

2. METHOD OF NUMERICAL SIMULATION

The circular cross section of the pseudo-elliptical vortex ring shown in Fig. 1 has a distribution of vorticity which attains a maximum value at the centre, decaying smoothly to zero with increasing radial distance. The radial length scale of the decay of vorticity is defined as the core radius σ. The region where the vorticity is non-zero will be referred to as the vortex core.

We consider a uniform-density inviscid fluid. The pseudo-elliptical vortex ring was represented by a single chain of vortex blobs arranged along its circumference. In other words, the cross section of the vortex ring was represented by one vortex blob. A vortex blob α is characterized by the position of its centre $\mathbf{x}^{\alpha}$, the vorticity $\boldsymbol{\omega}^{\alpha}(\mathbf{x}^{\alpha})$, the volume $d^3\mathbf{x}^{\alpha}$, and the core radius σ_{α}; the strength $\boldsymbol{\gamma}^{\alpha}$ is defined by $\boldsymbol{\omega}^{\alpha}(\mathbf{x}^{\alpha})d^3\mathbf{x}^{\alpha}$. In the present study the vortex blob was taken as a cylindrical vortex stick with radius δ_{α} and length l_{α}. Thus the volume is given by $d^3\mathbf{x}^{\alpha} = \pi(\delta_{\alpha})^2 l_{\alpha}$ and the circulation of the blob by $\Gamma_{\alpha} = \pi(\delta_{\alpha})^2|\boldsymbol{\omega}^{\alpha}(\mathbf{x}^{\alpha})|$, which is equal to the circulation of the vortex ring Γ. The radius of the vortex blob δ_α need not necessarily be equal to the core radius σ_α although they should be of the same order of magnitude. These radii were treated differently in the following two schemes of the calculations:

(i) The radius of the vortex blob δ_α was taken equal to the core radius σ_α. Thus both δ_α and σ_α were varied with time to meet the conservation of mass and circulation. Most of the present simulations were made by this scheme.

(ii) The core radius σ_α was maintained constant while the radius δ_α was varied with time to meet the conservation laws. This scheme was more stable than the first one in the sense that the divergence of vorticity of the vortex blobs occurred at later times for the same initial conditions. Thus this scheme was employed when the first scheme failed to calculate the evolution of the vortex ring for a sufficiently long time.

The vorticity field near the vortex blob α is written as

$$\omega^\alpha(\mathbf{x}) = \frac{1}{\sigma_\alpha^3} p\left(\frac{|\mathbf{x} - \mathbf{x}^\alpha|}{\sigma_\alpha}\right)\gamma^\alpha \tag{1}$$

where, for the Rosenhead-Moore core structure, the smoothing function has the form $p(\xi) = (3a/4\pi)(\xi^2 + a)^{-5/2}$, the value of a depending on the assumed distributions of vorticity around the position $\mathbf{x}^\alpha$. For the second-order Gaussian core, we have a = 0.4131 (Leonard 1985). The velocity induced at the centre $\mathbf{x}^\alpha$ of the vortex blob α is given by the modified Biot-Savart law

$$\frac{D\mathbf{x}^\alpha}{Dt} = \mathbf{u}^\alpha = -\frac{1}{4\pi}\sum_\beta \frac{\mathbf{r}^{\alpha\beta} \times \gamma^\beta}{r_{\alpha\beta}^3} q\left(\frac{r_{\alpha\beta}}{\sigma_\beta}\right) \tag{2}$$

where $\mathbf{r}^{\alpha\beta} = \mathbf{x}^\alpha - \mathbf{x}^\beta$, $r_{\alpha\beta} = |\mathbf{r}^{\alpha\beta}|$ and $q(\xi) = \xi^3(\xi^2 + a)^{-3/2}$ (Leonard 1985). The evolution of the strength γ^α is governed by the Helmholtz vorticity equation

$$\frac{D\gamma^\alpha}{Dt} = \frac{1}{4\pi}\sum_\beta \left\{\frac{\gamma^\beta \times \gamma^\alpha}{(r_{\alpha\beta}^2 + a\sigma_\beta^2)^{3/2}} - \frac{3(\gamma^\beta \times \mathbf{r}^{\alpha\beta})(\gamma^\alpha \cdot \mathbf{r}^{\alpha\beta})}{(r_{\alpha\beta}^2 + a\sigma_\beta^2)^{5/2}}\right\} \tag{3}$$

The position $\mathbf{x}^\alpha$ and the strength γ^α were updated by the first-order Euler forward scheme. The second-order Adams-Bashforth scheme was also employed for several cases to find insignificant changes in the pattern of deformation of the vortex rings.

For a finite region of vorticity in an unbounded uniform-density, inviscid fluid which is at rest at infinity, we have five invariants (Saffman et al. 1986), namely, the total vorticity, the total linear momentum, the total angular momentum, the total helicity and the total kinetic energy . By the use of Eqs. (1) and (2), these invariants can be expressed in terms of the position, strength and core radius of the individual vortex sticks (Kiya et al. 1988).

If a smooth field of vorticity is represented by a collection of a large number of the vortex blobs with overlapping cores, the vortex method is expected to yield a solution which converges to an exact smooth solution of the Euler equations as the density of the blobs is increased (Beale et al. 1982a,b; Anderson et al. 1985). This is the case despite the fact that the vorticity field of Eq. (1) is not divergence-free (Greengard et al. 1988). As mentioned before, the cross section of the vortex ring is represented in this study by one vortex blob with the core radius $\delta = \sigma$, so that the convergence theorem may not necessarily be applied. Thus this thin-tube model should be interpreted as a method of discretization of the vorticity field for numerical computations. The applicability of the thin-tube model should be judged by examining the extent to which the five invariants are really conserved.

3. RESULTS AND DISCUSSION

The numerical calculations were performed for the pseudo-elliptical vortex rings with the core radius $\delta/R = 0.10 - 0.20$ and the aspect ratio $L/(2R) = 2 - 20$. All the vortex blobs had the initial core radius $\sigma = \delta$ and the initial length $l/R = 0.1571$. The overlap between neighbouring vortex blobs was $\sigma/l = 1.27$ for all the vortex rings. The time step of advancing the vortex blobs was chosen as $\Gamma\Delta t/R^2 = 0.00749$; halving the time step produced insignificant changes in the deformation of the vortex rings. The length of the vortex blobs generally increased with time owing to the vortex stretching associated with the deformation. If the length of a vortex blob l was increased beyond twice its initial value, the vortex blob was divided into two vortex blobs with the same vorticity and the length $l/2$.

The calculations showed the following four fundamental patterns of deformation of the elliptical vortex rings. These patterns (referred to as patterns I-IV) are presented in Fig. 2.

(a) In pattern I (Fig. 2(a)) the vortex ring experienced the axis switching. This pattern appeared in a range of the aspect ratios $2 < L/(2R) < 4$ for the core radius $\sigma/R = 0.20$. For the vortex ring of Fig. 2(a) three periods of the axis switching were observed although the exact initial shape of the vortex ring was not repeated. At a non-dimensional time t^* ($= \Gamma t/R^2$) $= 50.6$ a significant local irregularity appeared in the arrangement of the vortex blobs. In this irregular region the number of the vortex blobs rapidly increased with time owing to the large vortex stretching to give rise to the divergence of vorticity. This calculation was made by the scheme

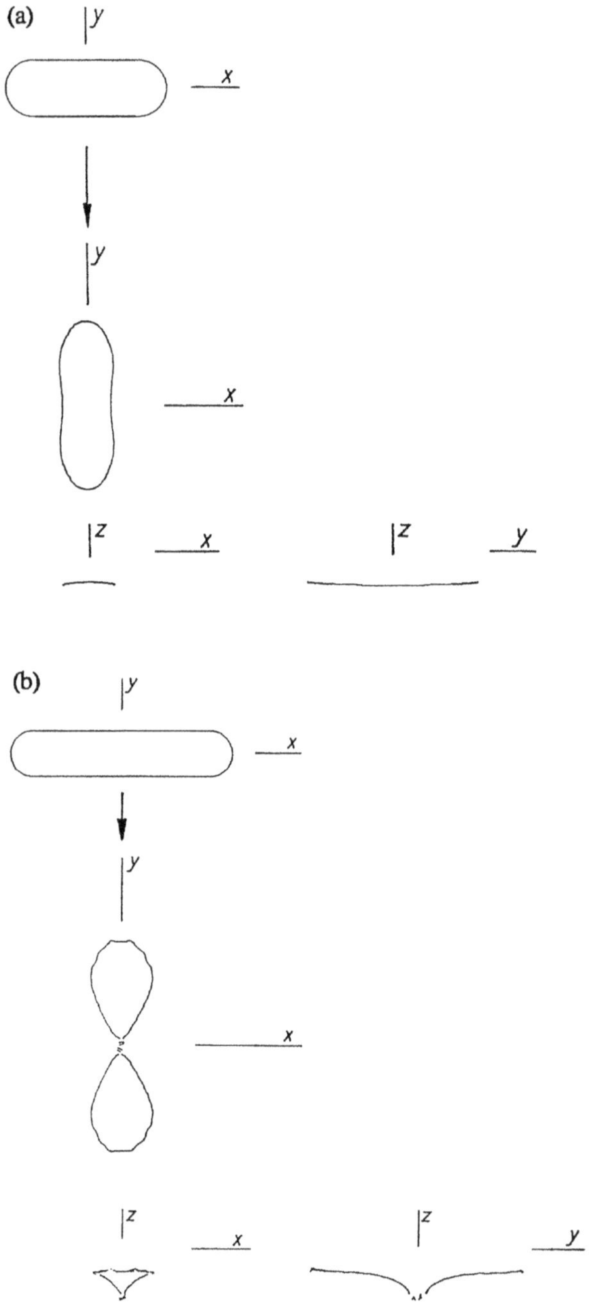

Figure 2. Four fundamental patterns of deformation of pseudo-elliptical vortex ring for $\sigma/R = 0.20$. (a) Pattern I: periodic axis switching ($L/(2R) = 2.89$); (b) Pattern II: splitting into two distorted vortex rings after axis switching ($L/(2R) = 4.93$); (c) Pattern III: splitting into two distorted vortex rings without axis switching ($L/(2R) = 8.85$); (d) Pattern IV: splitting into three distorted vortex rings without axis switching ($L/(2R) = 12.0$). Plan at root of arrow indicates initial configuration. Note that origin of z-axis is arbitrarily shifted.

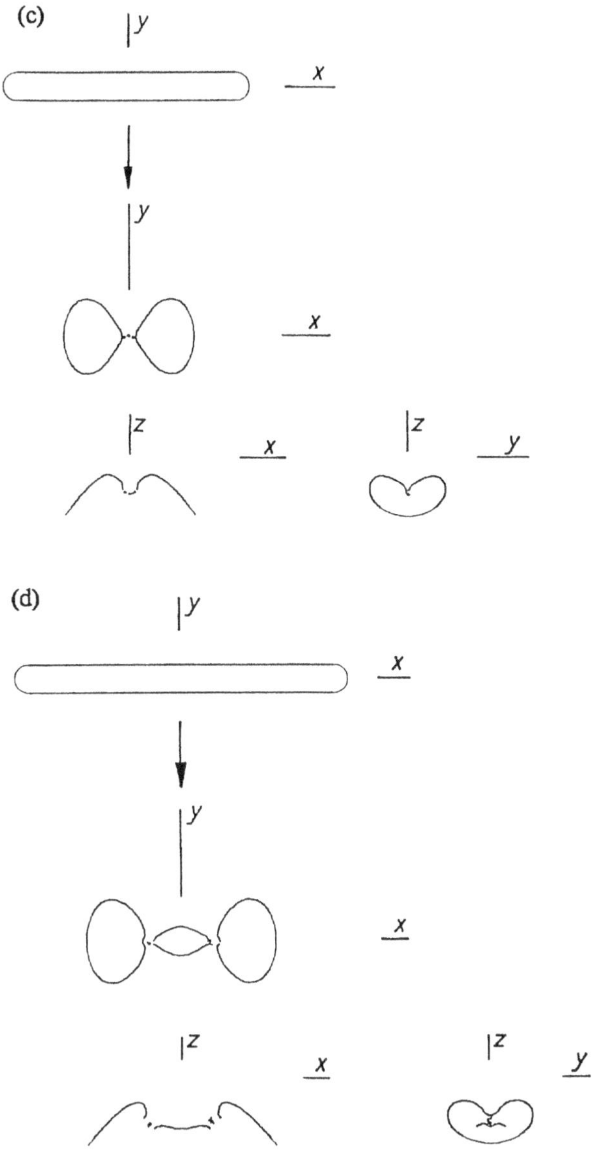

Figure 2(c,d). For caption see page 56.

(i). When the scheme (ii) was employed, the divergence of vorticity could be avoided much later than the time $t_* = 50.6$.

Figure 3 shows the first period of the axis switching τ_{sw} plotted against the aspect ratio. The period was approximately the same for the second and the third axis switching. The deformation of the vortex ring initially started along the semi-circular parts, propagating towards the parallel

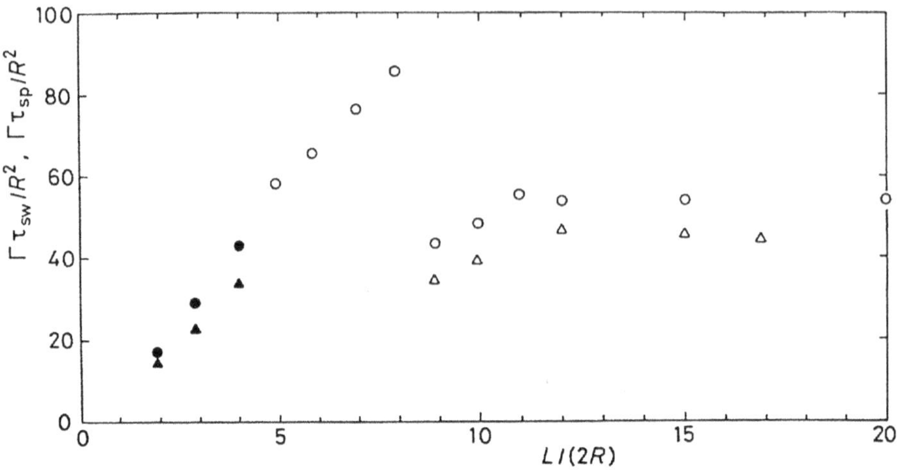

Figure 3. Period of axis switching τ_{sw} and time of splitting τ_{sp} versus aspect ratio L/(2R). o, •, σ/R = 0.20; Δ, ▲, σ/R = 0.10. Solid symbols for τ_{sw} and open symbols for τ_{sp}.

sides. Thus, the larger the aspect ratio, the larger was the time required for the axis switching to be completed. This is the reason why the switching period increases with the aspect ratio. The switching period is seen to be smaller for a smaller core radius σ/R = 0.10 than for a larger one σ/R = 0.20. This is because the induced velocity and thus the rate of the distortion are larger for the smaller core radius.

(b) In pattern II (Fig. 2(b)) the vortex ring experienced the axis switching once and then the splitting into two distorted vortex rings. This pattern was obtained in a range of the aspect ratios 5 < L/(2R) < 8. The distorted vortex rings appear to have the velocity of translation such that they move in the opposite direction. The time at which the splitting occurred, τ_{sp}, is shown in Fig. 3 for the core radius σ/R = 0.20. The splitting time is seen to increase with increasing aspect ratio.

(c) In pattern III (Fig. 2(c)) the vortex ring split into two distorted vortex rings without experiencing the axis switching. This pattern appeared for the aspect ratios approximately in a range 8.9 < L/(2R) < 11. The resulting vortex rings have the velocity of translation such that they collide each other at some time later. The collision is expected to produce the reconnection of the two vortex rings to lead to the emergence of either two new rings which will attempt to separate (Aref et al. 1990) or a single deformed vortex ring. The present calculation could not be extended to this stage because the divergence of vorticity occurred immediately after the first splitting. In order to proceed the calculation further, the surgery is perhaps required to remove the vortex blobs whose vorticity tends to diverge. The time of the first splitting is shown in Fig. 3. The splitting time decreases in a step-like manner at L/(2R) ≃ 8.

(d) In pattern IV (Fig. 2(d)) the vortex ring split into three distorted vortex rings without experiencing the axis switching. This pattern appeared for the aspect ratios $11 < L/(2R) < 20$. Note that the value 20 is the maximum aspect ratio in the present simulation. An exact value of $L/(2R)$ demarcating patterns III and IV was difficult to obtain because for values of $L/(2R)$ near 11 the central vortex ring formed by the splitting was so slender that it looked like anti-parallel vortex filaments which are seen in the region of contact in pattern III, see Fig. 2(c). The splitting time τ_{sp}, which is included in Fig. 3, remains approximately constant in this pattern IV. The splitting time becomes smaller for a smaller core radius $\sigma/R = 0.10$.

The arrangement of the three vortex rings formed by the splitting (Fig. 2(d)) is such that they probably collide at some time later. What will emerge after the collision is interesting but could not be obtained in the present study owing to the divergence of vorticity.

The distance which the pseudo-elliptical vortex ring moved from the initial position until the splitting was completed was approximately $z_{sp}/(2R) = 3.0$, 2.0 and 2.5 for the aspect ratio $L/(2R) = 4.93$ (pattern II), 9.95(pattern III) and 14.98 (pattern IV). Thus the splitting of the vortex ring is expected to occur within a few minor diameters downstream of the exit of the nozzle.

Among the five invariants of the vortical motion, the total energy and the longitudinal z-component of the total linear momentum varied within 10% of their initial values until the splitting took place. The variation of the total helicity was less than $0.016\Gamma^2$, its initial value being zero. The total vorticity, the total angular momentum and the normal and transverse components of the linear momentum maintained their initial value of zero. A catastrophic increase in the invariants was observed when the vorticity tended to diverge immediately after the splitting. Thus the results of the present calculations appear to be reliable until the divergence of vorticity occurs.

4. CONCLUSION

The vortex blob simulation has shown that the pseudo-elliptical vortex rings with aspect ratios less than 20 have four fundamental patterns of deformation due to the self-induced velocity: the periodic axis switching; the splitting into two distorted vortex rings after the axis switching; the splitting into two vortex rings without the axis switching; the splitting into three vortex rings without the axis switching. The period of the axis switching and the time of splitting are presented as functions of the aspect ratio and core radius. The invariants of the three-dimensional vortical motion remained constant within approximately 10% of the corresponding initial values. The deformation of vortex core can be included into the simulation by representing the cross section of the vortex by a number of vortex blobs.

REFERENCES

Anderson, C. & Greengard, C. 1985 *SIAM J. Numer. Anal.* **22**, 413.

Aref, H. & Zawadzki, I. 1990 *Topological Fluid Mechanics* (eds. H. K. Moffatt and A. Tsinober), Camb.Univ. Press 535.

Beale, J. T. & Majda, A. 1982a *Math. Comp.* **39**, 1.

Beale, J. T. & Majda, A. 1982b *Math. Comp.* **39**, 29.

Greengard, C. & Thomann, E. 1988 *Phys. Fluids* **31**, 2810.

Hussain, F. & Husain, H. S. 1989 *J. Fluid Mech.* **208**, 257.

Kambe, T. & Takao, T 1971 *J. Phys. Soc. Japan* **31**, 591.

Leonard, A 1985 *Ann. Rev. Fluid Mech.* **17**, 523.

Kiya, M. & Ishii, H. 1988 *Fluid Dyn. Res.* **3**, 197.

Saffman, P.G. & Merion, D. I. 1986 *Phys. Fluids* **28**, 2373.

Toyoda, K.& Hussain, F 1989 *Trans. JSME* **B55**, 1542 (in Japanese).

The Structure of Round Turbulent Jets in Counterflow: A Flow Visualization Study

O. König[1] *and H.E. Fiedler*[2]

1. Institute für Luft und Raumfahrt, Technische Universität Berlin.

2. Hermann-Föttinger-Institut für Thermo- und Fluiddynamik, Technische Universität Berlin.

The round free jet in a counterflowing external stream is an interesting configuration both from the practical point of view of mixing enhancement as well as scientifically, pertaining e.g. to stability. Only little is found about this flow in the literature, presumably owing to its inherent experimental difficulties. Two different states of flow are observed: a stable case at low velocity ratios $U_{jet}/|U| \lesssim 1.4$, and an unstable situation at high velocity ratios $U_{jet}/|U| > 1.4$, which is dominated by random motions of high amplitude.

1. Experimental setup

The non-buoyant, swirl-free jet is discharged from a nozzle of 25 mm exit diameter into a wind tunnel flow of opposite direction. The jet has a top-hat profile at the nozzle exit. The velocity ratio $R = U_{jet}/|U|$ is varied via the jet velocity, while the ambient velocity is constant throughout, except for the case $U_{jet}/|U| = 1$, where the ambient stream is increased to maintain turbulence in the jet flow. Loudspeakers at the entrance of the pipe upstream of the nozzle are used for axial excitation. Orbital excitation of the jet is achieved by mechanically moving the nozzle exit around a small radius. All experiments were done in a Reynolds number range of $5000 < U_{jet}D/\nu < 20000$.

For the purpose of visualization oil vapour is added to the jet flow. The photographs were taken with an ordinary (video) camera.

2. Results and observations

The jet in the counterflowing configuration is a combination of jet and wake - where the jet generates its own wake. Therefore, we may distinguish between a jet region, a deflection region and a wake region. Two characteristic situations, depending on the velocity ratio $R = U_{jet}/|U|$, are identified:

The stable case at low velocity ratios $R \lesssim 1.4$ (figure 1)

In this situation the jet region is not developed. The flow is bent over shortly downstream of the pipe exit, showing no oscillations about its axis. Instead, the flow wraps itself stably around the pipe, changing rapidly into a wake flow. In this situation axial and/or orbital excitations produce significant influence:

Axial excitation at amplitudes < 15% of the jet velocity leads to distinct formation of vortex rings (0th mode), which are carried downstream by the ambient flow. Orbital excitation or a combination of axial and orbital excitation generates a well defined

Advances in Turbulence 3
Editors: A.V. Johansson · P.H. Alfredsson

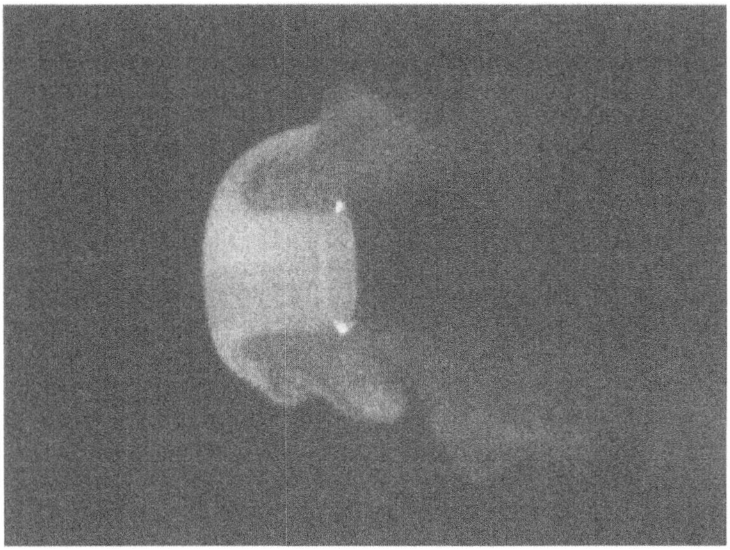

Figure 1. Stable case for $U_{jet}/|U| = 1$. Laser sheet visualization.

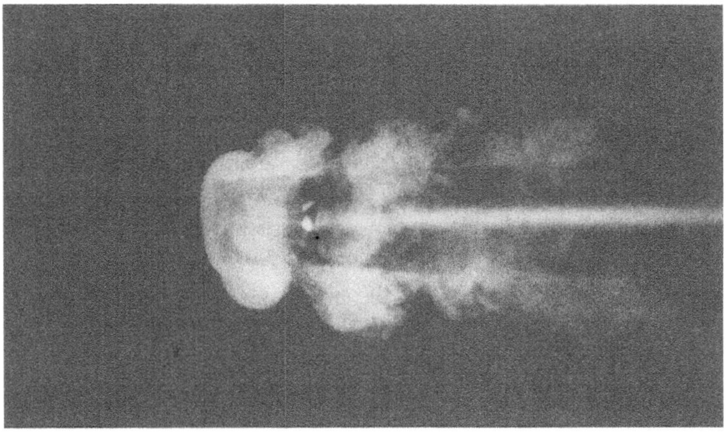

Figure 2. Helical structure of the orbitally excited jet. $U_{jet}/|U| = 1 . f_{orb} = 25Hz$. Flash light illumination.

helix winding downstream around the inlet nozzle and the attached pipe (fig. 2). Axial excitation of large amplitude (> 20%) causes the jet to break off from the nozzle, thereby increasing its penetration depth at unchanged volume flux considerably. The behaviour is then similar to the unstable case.

The unstable case at velocity ratios $R > 1.4$ (fig. 3)

In the mean the immediate near-field of the inner jet shortly after the inlet orifice behaves like an ordinary jet, except for the fact that the half-velocity-line has a convex shape as was reported by Beltaos and Rajaratnam (1973). This picture is, however, misleading. A short time exposure (fig. 4) shows the jet region to be always deflected

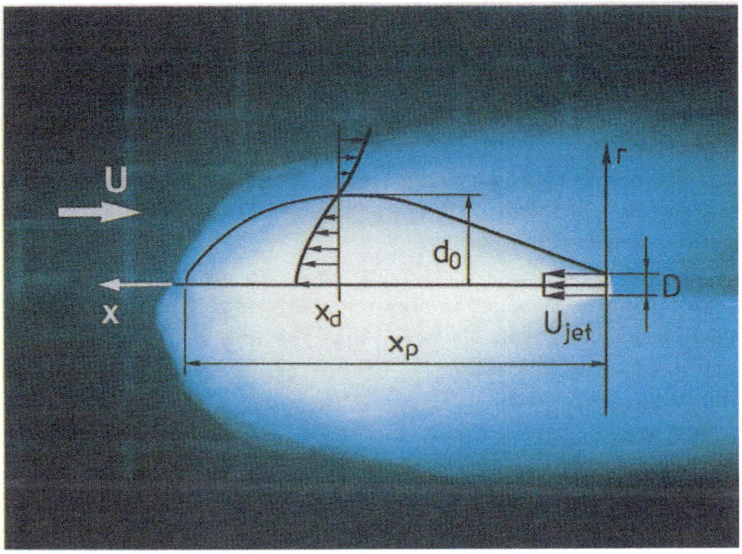

Figure 3. Mean flow field for $U_{jet}/|U| > 1.4$ with definition sketch. Laser sheet visualization.

Figure 4. Instantaneous flow field for $U_{jet}/|U| = 7$. Incandescent light.

as a whole to one side. The temporal developement analysed from a video film reveals an oscillation of the jet axis both around and through the mean (x-)axis. This may explain the convex shape of the half-velocity-line as found from time averaged measurements..

An image processing procedure locating the point of maximum smoke concentration of a cross section from a video, hence the location of the instantaneous jet axis, shows a random oscillation of large amplitude through and around the x-axis, switching, as it were, from one quasi-stable angular position to another on the opposite side. The characteristic frequency of this switching is very low, owing to the low convection velocities of the major structures, which are related to the velocity difference between jet and ambient stream. This frequency depends weakly on the momentum flux ratio or, for equal density of jet and external fluid, on the velocity ratio.

The oscillation is also observed for jets with small inclination angles up to 30 degrees between counterflow and primary jet axis. Its amplitude decreases with increasing inclination angle (König and Fiedler (1990)).

Axial excitation produces vortex rings in the jet region of short lifetime, when compared to those of a co-flowing jet or a jet in stagnant surrounding. The excitation seems to equalise the oscillating motion in the way, that the amplitudes are slightly smaller than in the unexcited case, an at the same time more regular. Still, the stochastic character of the oscillation remains.

In the deflection zone the high turbulence level causes a rapid mixing of jet and external fluid, and thus a rapid growth. In this region, nor in the subsequent wake zone, we found virtually no influence of axial and/or orbital excitation. This must be understood in view of the exceedingly large and stochastic velocity fluctuations particularly in the stagnation region, where stabilization seems to be conceivable only by unrealistically large amplitudes of low frequency.

Taking this fluctuation into account, we define the actual (fluctuating) penetration depth, being in itself a function of the velocity ratio only, as the distance on the jet centerline between the orifice and the point where the jet is bent over to a 90 degree angle against the external flow direction. This is identical with the penetration length x_p as obtained from the location of the the average stagnation point in long time exposures.

In fig. 3 we have sketched the general flow situation, the major parameters and dimensions, and the coordinate system used.

The normalized penetration depth x_p/D versus the velocity ratio R as obtained from long-exposure smoke photographs is shown in fig. 5 for the natural ($S = 0$) and two cases of axial excitation. The value of $S = 0.37$ corresponds to the preferred mode in the usual definition $S = fD/U_{jet}$ and, for comparison - in view of the counterflow situation - based on the velocity difference, i. e. $S = fD/(U_{jet} - |U|)$. We found no evidence

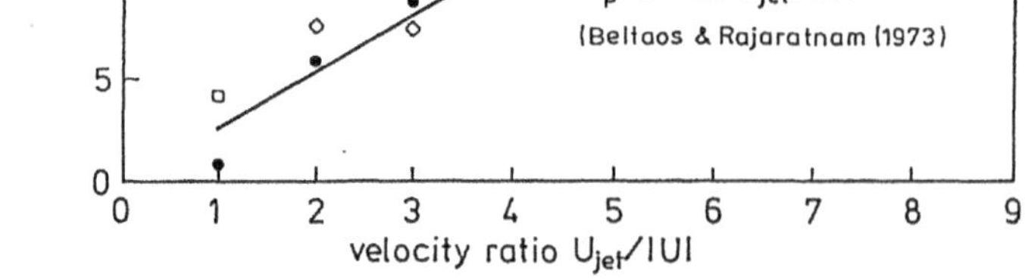

Figure 5. Penetration depth as obtained from long time exposures.

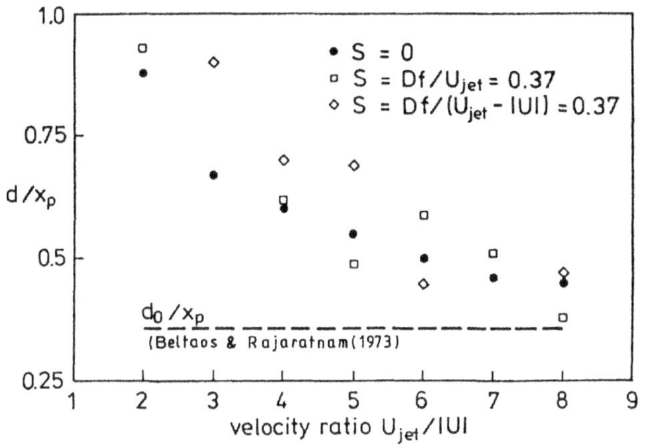

Figure 6. Maximum diameter of mean forward flow region.

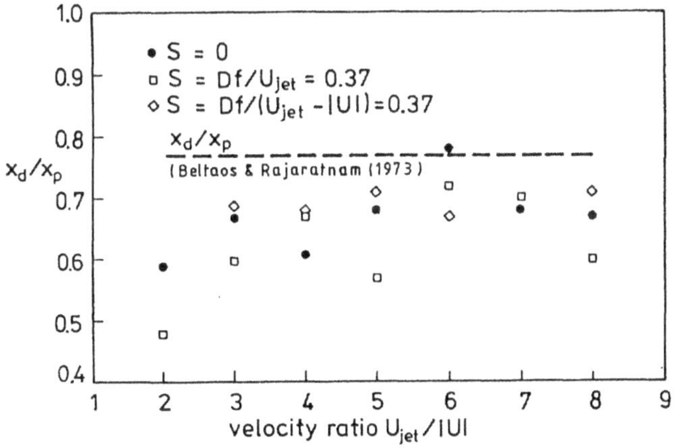

Figure 7. Location of maximum diameter of mean forward flow region.

for any influence of this excitation, nor at any other frequency examined. The relation $x_p/D = 2.7R$ for the penetration length as given by Beltaos and Rajaratnam(1973) was found to hold for all cases considered.

The maximum diameter d_0 of the forward flow region (see fig. 3), identified as the field of high smoke concentration on long time exposures, depends on the velocity ratio $U_{jet}/|U|$ as is shown in fig. 6. The value given by Beltaos and Rajaratnam (1973) (the dotted line in fig. 6) is therefore only asymptotically valid for high velocity ratios.

The location x_d/x_p of the maximum diameter of the foreward flow region remains approximately constant for velocity ratios $3 < R < 8$ (fig. 7). The values found in the present work are, however, lower than those by Beltaos and Rajaratnam (1973).

3. Summary

The jet in a counterflow has two characteristic states, depending on the velocity ratio R: a stable situation at low velocity ratios $R < 1.4$, and an unstable situation at $R > 1.4$. The structures in the stable case may be controlled by axial and/or orbital excitation, leading to either ring vortices or helices. Excitation at axial mode with amplitudes $> 20\%$ of the jet exit velocity leads to a significant increase of the jet's penetration depth into the counterflow.

In the unstable case the jet oscillates stochastically at a low frequency around and through its mean axis. An axial or orbital excitation with amplitudes $\leq 15\%$ of the jet velocity has no discernible influence on the behaviour of the deflection zone or the return flow.

The distinction between mean and instantaneous behaviour introduces a new physical concept for the jet in counterflow. It may provide an explanation for the failure of the theoretical models based on the mean flow behaviour as found in long-time visualizations.

References

Beltaos S., Rajaratnam N.: Circular Jets in an Opposing Stream, *Proc. 1st Can. Hydraul. Conf.*, Edmonton, 1973

König O., Fiedler H.E.: On the asymptotic behaviour of axisymmetric turbulent jets in an ambient stream under arbitrary inclination angles, *Zeitschr. Flugwiss. Weltraumf.*. Accepted for publication in 1990.

Waves in Turbulence

Mårten T. Landahl

Massachusetts Institute of Technology, Cambridge, Ma 02139, USA

The paper discusses the role of wave motion in fully developed turbulent shear flows. The equations of motion are written so as to single out the terms describing the linear interaction with the mean shear flow (approximated as a parallel one) and the remaining nonlinear terms treated as spectrally broad-band source terms (Landahl, 1967). The resulting fluctuations may be expressed in terms of travelling wave modes. For a mean shear flow with a small damping rate one finds from this model that the fluctuations will have frequency-wave number spectrum with a peak at the frequency near that for the eigenfrequency of the free wave, at the wave number considered, with a magnitude inversely proportional to the square of the damping rate. For an unstable mean field, as with a free shear layer, the exponential growth of the excited instability waves leads to a nonhomogeneous random field that is a function of the distance downstream from the beginning of the shear layer. Comparisons of the predictions of the wave model with experimental results (Bark, 1975, Gaster, Kit & Wygnanski, 1983) confirm the presence of the shear waves and the importance of the linear interaction with the mean shear.

The nature of the nonlinear excitation of the waves is discussed next. Strong nonlinear interactions are expected during sublayer bursting, which is observed to be a highly intermittent event that may be the result of local shear layer instability. The mean-shear linear interaction model predicts that thin internal shear layers with unstable inflectional velocity profiles will form from a convected eddy. The excitation due to the Reynolds stresses produced by the local instability may initiate growing streamwise streaks which may be analyzed with a simple theory.

1. Introduction

Much of the recent research efforts in turbulence has been devoted to the search for "coherent structures", i.e., recurring flow structures with recognizable temporal-spatial connectivity. In view of the accepted classical picture of the initiation of turbulence as being due to the growth of instability waves, it would be natural to consider the possibility that waves or wave-like motion would be present in such coherent structures. Although many different kinds of wave motion may be present in turbulent flow fields (e.g., acoustical waves, surface waves, inertial waves, Alfven waves, etc.) we shall here consider only one kind, namely shear waves. With a wave one usually means a motion which is periodic in both space and time and which has a well defined relation between wave length and frequency, a dispersion relation. Waves propagate, they do not simply convect with the background flow. This means that in a flow with shear the disturbance created by a wave (a modal wave) would be expected to propagate with the same velocity at all levels and thus could have an important influence on the whole cross-section of the flow.

Advances in Turbulence 3
Editors: A.V. Johansson · P.H. Alfredsson

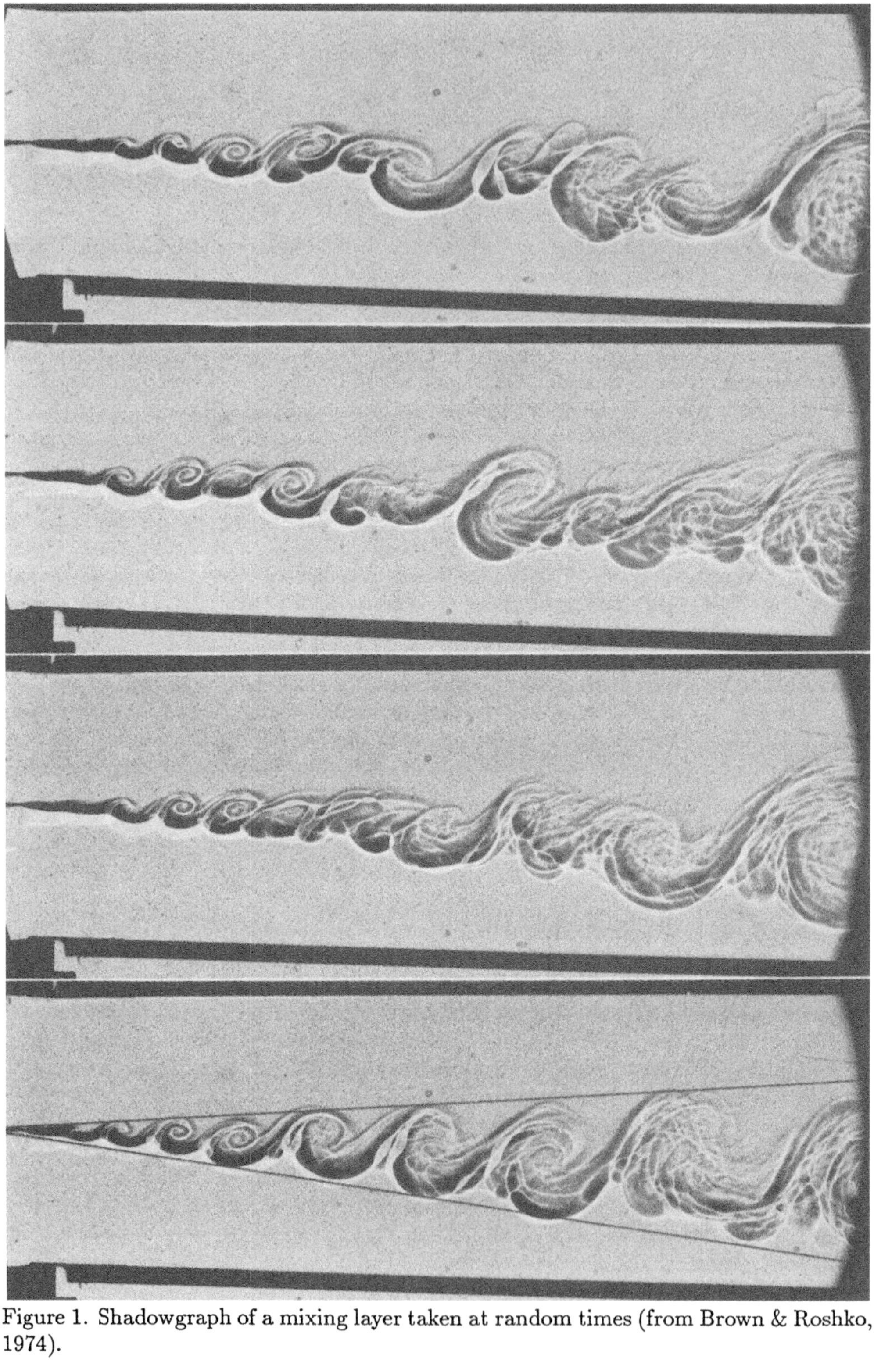

Figure 1. Shadowgraph of a mixing layer taken at random times (from Brown & Roshko, 1974).

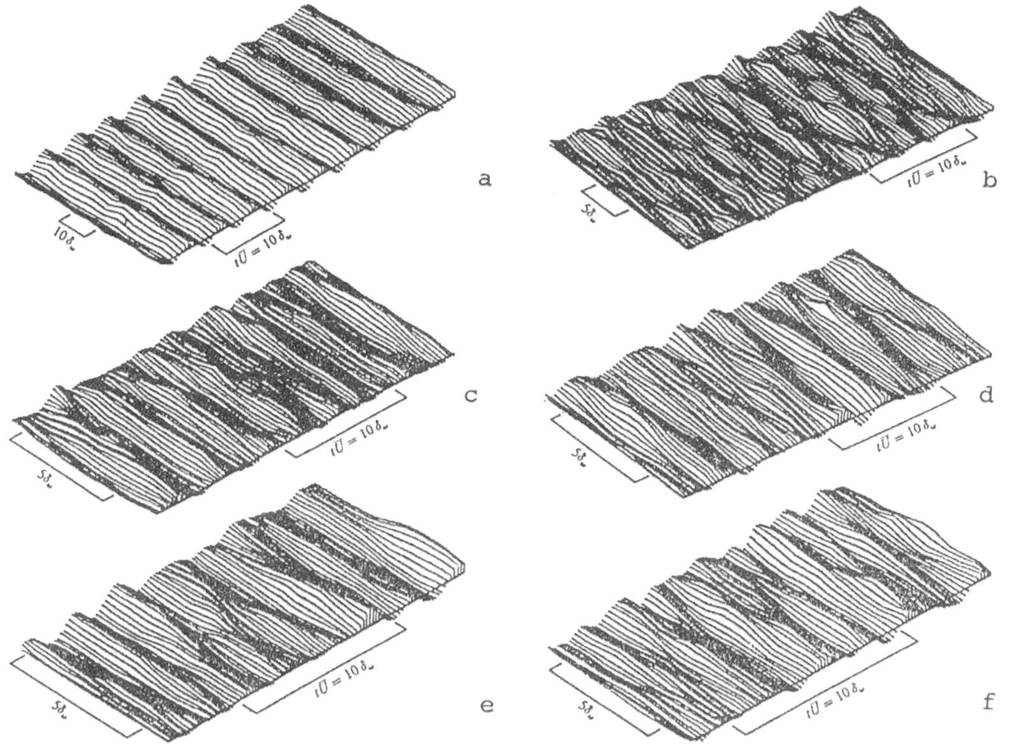

Figure 2. Visualizations of the longitudinal velocity fluctuating field in a turbulent mixing layer (from Browand & Troutt, 1985).

What experimental evidence is there for wave-like motion? The clearest one is found in free shear layers and mixing layers. The early visualization experiments of Brown & Roshko (1974) on flows with a sharp density gradient showed the appearance of Kelvin-Helmholtz instability waves producing rollup into a row of more-or-less concentrated vortices (Figure 1). Detailed investigations of a free shear layer carried out by Browand & Troutt (1985) showed that the vortices produced by the initially two-dimensional and regular wave field tends to merge in a somewhat irregular three-dimensional fashion (see Figure 2) and to give rise to a frequency spectrum with a fairly broad spectrum with a peak at a frequency nearly twice that for the most amplified waves. Similar results have been reported by Hussain & Zaman (1985). By studying forced oscillatory disturbances in a free shear layer Gaster *et al.* (1985) were able to demonstrate that the ensuing wave motion was governed primarily by the linear interaction with the mean shear flow, thereby finding strong support for the simple instability wave concept.

A free shear layer has mean velocity profile possessing an inflection point and would therefore from the linear and inviscid point of view be expected to promote wave-like instability. For boundary layers, however, the situation is quite different since the mean velocity profile is linearly stable for all frequencies (Landahl, 1967). As will be further explained below, however, the presence of lightly damped waves show up as ridges in the wave-number frequency spectrum, as demonstrated by Bark (1975) (see Figure 3a). Later detailed flow investigations have also shown that high-frequency wave packets

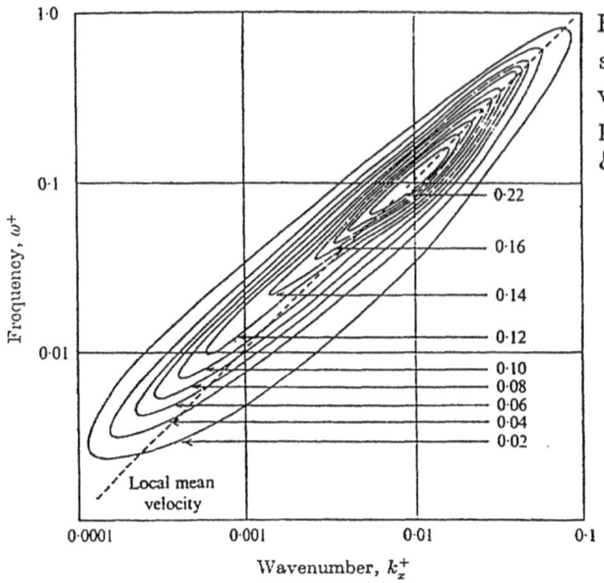

Figure 3a. Wave number-frequency spectrum for the longitudinal velocity fluctuations in a turbulent pipe flow (from Morrison, Bullock & Kronauer, 1971).

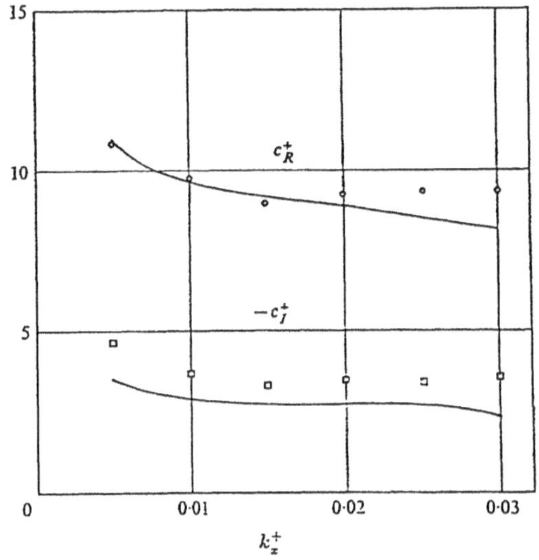

Figure 3b. Calculated real and imaginary part of phase velocity with values obtained from the spectrum (from Bark, 1975).

appear intermittently in the near-wall flow region. From his measurements of wall pressures using pressure pickups of very small size Schewe (1983) inferred the presence of such packets with a wave length of approximately 150 in viscous wall units and a propagation speed of approximately 11 times the wall friction velocity (see Figure 4). Employing a special conditional sampling technique, Haritonidis, Gresko & Breuer (1988) could show that wave packets of wave lengths and frequencies consistent with those discovered by Schewe (1983) show up in the velocity fluctuations as well.

Numerical simulation of turbulence based on the Navier-Stokes equations may be regarded as a new experimental tool for the study of turbulence in which the supercom-

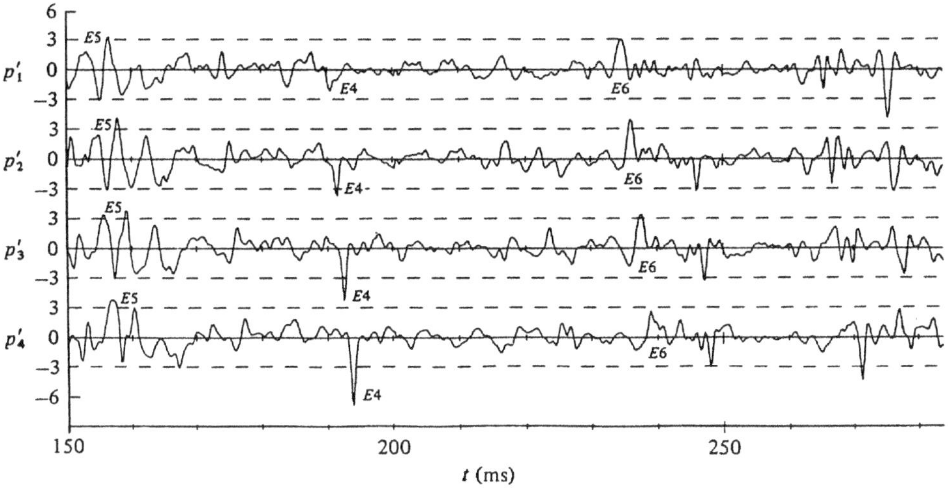

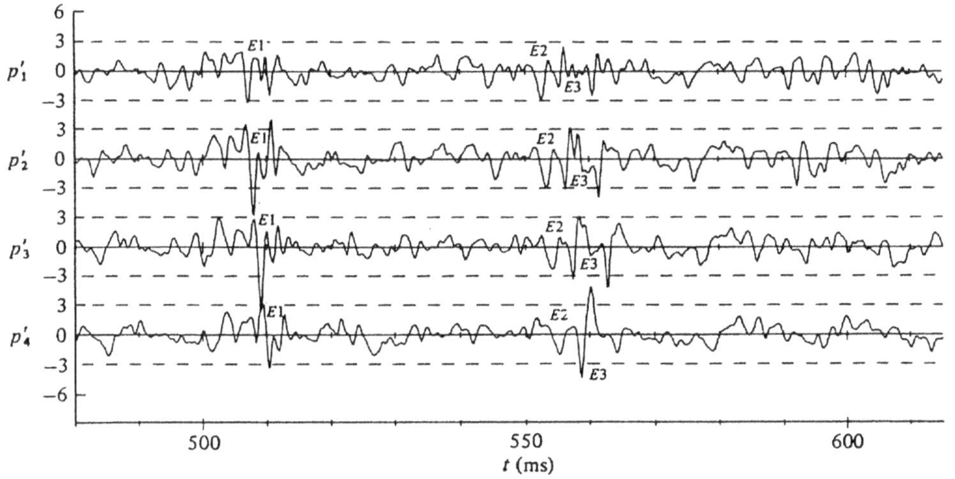

Figure 4. Instantaneous wall pressures measured in four consecutive pressure transducers (from Schewe, 1983).

puter takes over the role of the wind tunnel. The advantage of numerical experiments is that one can make more detailed studies of the flow field with much less severe limitations in the spatial resolution than what is usually experienced in laboratory experiments because of the finite sizes of the probes used. As a remarkable example of what can be accomplished with this new tool we show in Figures 5 and 6 results from recent simulations of the flow in a square duct obtained by Gavrilakis (1990). In the figures are shown instantaneous distributions of the spanwise component of the vorticity at different instants of time (all numbers are given in nondimensional viscous wall units). Figure 5 shows the near-wall shear layer at one particular instant. For clarity of presentation the vertical coordinate has been greatly expanded (by a factor of 2.5 in Figure 5 and by 3 in Figure 6) so that the thinness and flatness of the appearing shear layers becomes even more accentuated when this is taken into account. Figure 6 shows

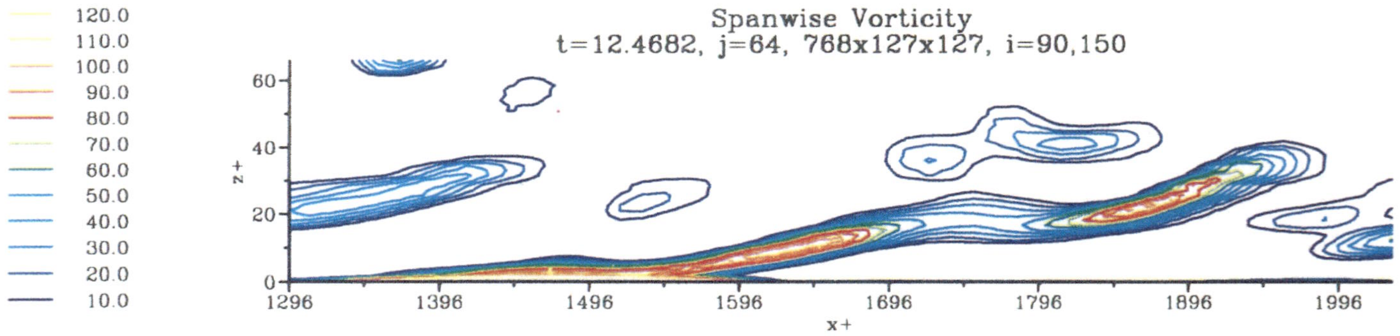

Figure 5. Instantaneous spanwise vorticity near the wall in turbulent duct flow (from Gavrilakis, 1990).

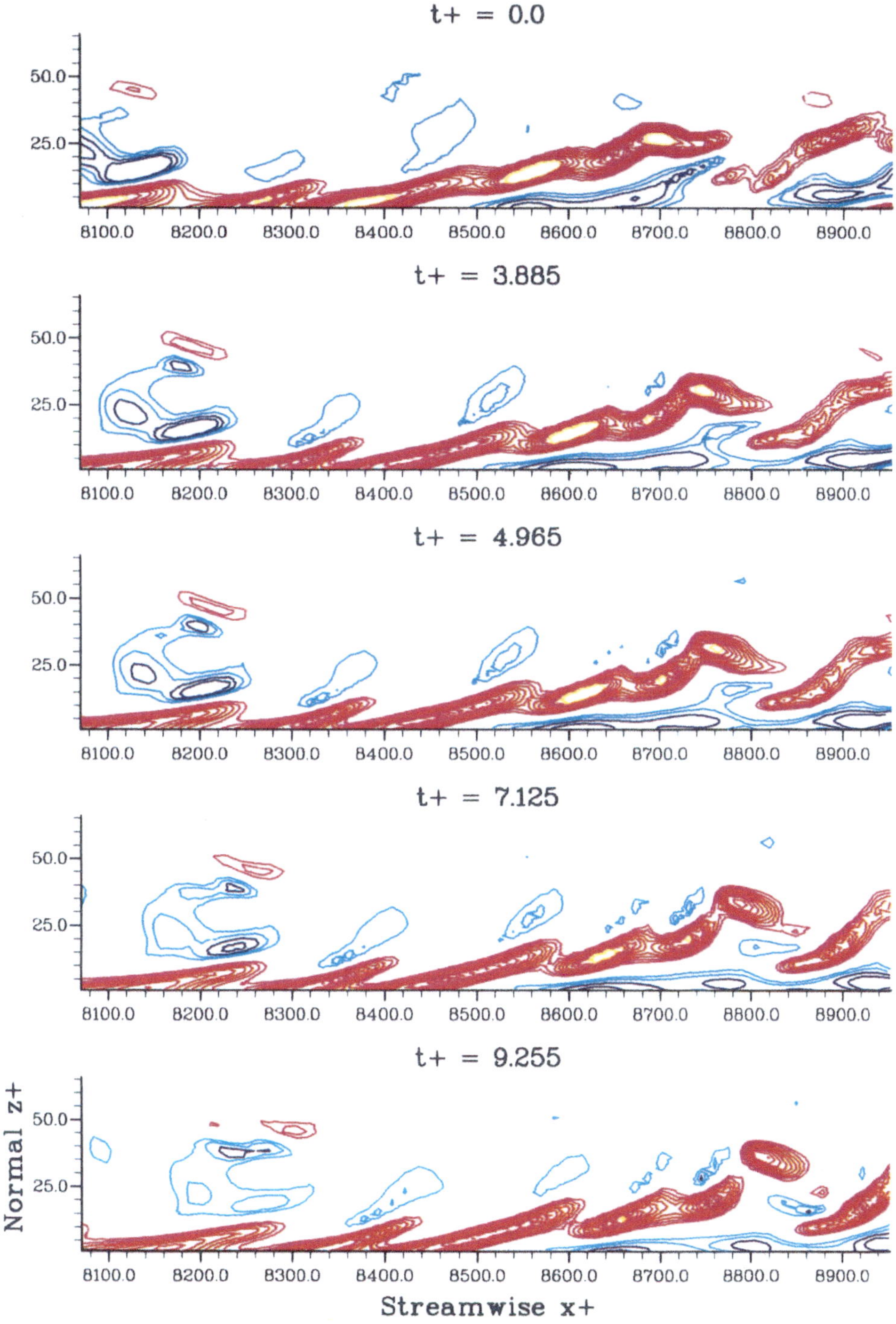

Figure 6. Instability and breakdown of shear layer in turbulent duct flow (from Gavrilakis, 1990).

a time sequence of the evolution of the shear layer. One can clearly see the evidence of an instability of the layer with a wave length of about 100 - 150 in viscous wall units and the beginnings of the breakup of the layer. The shear layer instability is a crucial component in the theoretical model presented below.

2. Equations of motion

Consider the Navier-Stokes equations for turbulent velocity fluctuations $u_i(x_j,t)$ where $(x_1 = x, x_2 = y, x_3 = z)$ and fluctuating pressures $p(x_j,t)$ in an incompressible shear flow having a mean velocity distribution $\overline{U}_i(x_j,t)$ which is approximated as a parallel one, $\overline{U}(y)\delta_{i1}$. One finds

$$U_i(x_j,t) = \overline{U}(y)\delta_{i1} + u_i(x_j,t) \tag{1}$$

$$\frac{\overline{D}u_i}{Dt} + \overline{U}'(y)u_2\delta_{1i} = -\frac{1}{\rho}\frac{\partial p}{\partial x_i} + \frac{\partial}{\partial x_j}(\mu\frac{\partial u_i}{\partial x_j} + \tau_{ij}) \tag{2}$$

$$\frac{\partial u_i}{\partial x_i} = 0 \tag{3}$$

where

$$\frac{\overline{D}}{Dt} = \frac{\partial}{\partial t} + \overline{U}\frac{\partial}{\partial x} \tag{4}$$

$$\tau_{ij} = -\rho(u_i u_j - \overline{u_i u_j}) \tag{5}$$

A single equation for $u_2 = v$ may be obtained by taking the divergence of the system (1) to obtain a Poisson type equation for the pressure in terms of the velocities and then use this to eliminate the pressure from the y-component of (1). This yields the following nonhomogeneous Orr-Sommerfeld type equation (Landahl, 1967):

$$\frac{\overline{D}}{Dt}\nabla^2 v - v_x\overline{U}'' - \nu\nabla^4 v = q \tag{6}$$

where

$$q = \nabla^2 T_2 - \frac{\partial^2 T_i}{\partial x_i \partial x_2} \approx \frac{\partial^2}{\partial y^2}(\frac{\partial uv}{\partial x} + \frac{\partial vw}{\partial z}) + ... \tag{7}$$

$$T_i = \frac{1}{\rho}\frac{\partial \tau_{ij}}{\partial x_j} \tag{8}$$

On the anticipation that the ensuing structures are highly flattened in the vertical direction (see Figures 5 and 6) only the terms involving the highest y-derivatives are shown in the expression for the fluctuating quadratic terms comprising the nonhomogeneous "source term" q.

3. Shear flow turbulence regarded as the response of a driven linear system

The fundamental reason for setting down the equation (6) is of course to treat it as if q were known, obtain the formal solution for v, and then employ some suitable model for q for use in the calculations. This is the idea underlying the wave-guide model proposed in Landahl (1967) and further elaborated upon by Bark (1975).

For the case of the boundary layer one may regard the fluctuation field to be statistically homogeneous in the horizontal direction x, z. Then, take Fourier transform in x, z, and t, which gives

$$L_{O-S}(\hat{v}) = \frac{\hat{q}}{i\alpha} \tag{9}$$

$$L_{O-S}(\hat{v}) = (\overline{U} - c)(D^2 - k^2)\hat{v} - \overline{U}''\hat{v} - \frac{1}{i\alpha Re}(D^2 - k^2)^2\hat{v} \tag{10}$$

where $D = d/dy, c = \omega/a$. Here the 'hat' denotes transformed quantities with $\mathbf{k} = (\alpha, \beta)$ and ω denoting the space and time transform variables, respectively. The formal solution may be obtained by expansion in terms of the eigensolutions, $v^{(n)}$, of the Orr-Sommerfeld problem (see Landahl, 1967, Bark, 1975). By including only the lowest order, least-damped, mode, $v^{(0)}$, one finds for the wave number-frequency spectrum Ψ_{vv} for v (which is proportional to $|v|^2$) an expression of the form:

$$\Psi_{vv} \approx \frac{\Psi_{qq}}{|\omega - \omega^{(0)}|^2} = \frac{\Psi_{qq}}{(\omega - \omega_r^{(0)})^2 + (w_i^{(0)})^2} \tag{11}$$

where

$$\omega^{(0)} = \omega_r^{(0)} + i\omega_i^{(0)} = \alpha(c_r^{(0)} + ic_i^{(0)}) \tag{12}$$

is the complex eigenvalue for the temporal Orr-Sommerfeld problem and Ψ_{qq} is the inner product of the absolute square of q and the adjoint eigenfunction.

For small decay rates $\omega_i^{(0)} = \alpha c_i^{(0)}$ the Ψ_{vv}-spectrum, and hence the spectra for the pressure and the other velocity components, as well, will thus have a peak, for each wave number $\mathbf{k} = (k_x, k_z)$, at $\omega = \omega_r^{(0)}(\mathbf{k})$ which shows up in the contour curves of the spectra in the $(\omega, \mathbf{k})$- plane as a ridge of a width proportional to $\omega_i^{(0)}$. This is illustrated by the results obtained by Morrison, Bullock & Kronauer (1971) for the longitudinal fluctuations in a turbulent pipe flow reproduced in Figure 3a.

As demonstrated by Bark (1975), by fitting curves of the form (11) to the measured spectra, one can determine $c_r^{(0)}$ and $c_i^{(0)}$ directly from the experimental data. In Figure 3b are shown the results he obtained through this procedure together with the theoretical values obtained from a numerical solution of the Orr-Sommerfeld problem for the mean velocity profile. As seen, the agreement between the experimental and theoretical values are remarkably good, even though the experimental decay rates are somewhat greater than the theoretical ones.

For a free shear layer the model has to be modified somewhat. The thickness of the free shear layer grows downstream, therefore one cannot regard it as a linear system that is homogeneous in the streamwise direction. Also, the free shear layer has a mean velocity distribution which is inflectional, therefore it is unstable to small disturbances. Each wave component grows exponentially downstream, therefore from spatial instability theory

$$\text{amplitude} \sim \exp(-\alpha_i x) \tag{13}$$

The nondimensional spatial eigenvalue $\theta\alpha^{(0)} = \theta(\alpha_R + i\alpha_I)$ (θ = momentum thickness) is a function of the nondimensional frequency,

$$\overline{\omega} = \omega\theta/U_0 \tag{14}$$

θ increases approximately linearly with x, hence the total exponential growth becomes

$$\exp[\int_0^{\overline{\omega}} -\frac{\overline{\alpha}_I}{\theta'}\frac{d\overline{\omega}}{\overline{\omega}}] \tag{15}$$

with $\overline{\alpha}_I = \theta\alpha_I$ and $\theta' = d\theta/dx$. Thus for the frequency spectrum Φ_{vv}

$$\Phi_{vv} \approx \Phi_{qq}\exp[\int_0^{\overline{\omega}} -\frac{\overline{\alpha}_I}{\theta'}\frac{d\overline{\omega}}{\overline{\omega}}] \tag{16}$$

where Φ_{qq} is the frequency spectrum of q. The integral will receive positive contributions for all frequencies lower than that for neutral stability, which, according to the inviscid stability theory for a free shear layer, is approximately twice that for which the maximum amplification rate occurs. The peak of Φ_{vv} could thus be expected to be located at a frequency higher than that for maximum amplification. Michalke (1964) gives the Strouhal number $St = \overline{\omega}/2\pi$ for max amplification rate as 0.017. This is to be compared to the Strouhal number for passage of coherent structures 0.024, obtained in the experiments by Hussain & Zaman (1985).

4. The formation of thin shear layers

Perhaps the most significant observation in the pioneering visualization experiments by Kline *et al.* (1967) was that the turbulence in the near-wall region was found to be highly intermittent in that it shows short periods of strong activity, "bursts", separated by longer periods of unsteady but laminar-like activity. A useful idealized model for the fluctuation field is therefore to assume that the nonlinear "source term" q in (9) is active only during short periods in which the eddies forming the coherent structures are created. Away from such a formation period the eddy interacts with the mean shear flow in a passive and linear manner.

The mean shear flow is free from inflection and is stable to small wave-like disturbances. Therefore, after a transient initiation period the vertical perturbation velocity, and also the perturbation pressure, (but not necessarily the horizontal components, see Landahl, 1975, 1984) will decay and tend towards zero for long times after the eddy formation. With the nonlinear terms, and the pressure, tending to zero, the streamwise fluctuation velocity component is thus governed in the inviscid long-time limit,

$$tl_1\overline{U}' \to \infty \tag{17}$$

l_1 being the streamwise scale of the eddy, by (see eq. 2)

$$\frac{Du}{Dt} = 0, \tag{18}$$

the solution of which may be written as

$$u = u_\infty(\xi, y, z), \xi = x - \overline{U}(y)t \tag{19}$$

(c. f. Taylor's "frozen field" hypothesis).

In (19) y and z enter only as parameters. One finds for the perturbation shear

$$\frac{\partial u}{\partial y} = -t\overline{U}'(y)u_{\infty\xi} + u_{\infty y} \tag{20}$$

(partial derivatives denoted by subscripts). Thus, the shear intensifies linearly with time, the mechanism involved being spanwise stretching of the mean shear by the spanwise component $w_z = -u_{\infty\xi}$, as follows from continuity, with $v = 0$. The interaction with the mean shear will thus lead to the formation of internal shear layers in regions where $u_{\infty\xi} < 0$. However, for long times viscous diffusion will limit the intensity of the induced shear and hence set a lower limit of the thickness of the shear layer formed.

To include the effects of viscosity one needs to solve the diffusion equation for u, which results from setting p and v equal to zero in the linearized equation (2) for the u-component,

$$\frac{\overline{D}u}{Dt} - \nu\nabla^2 u = 0 \tag{21}$$

An approximate solution valid for large times may be found by setting in (21)

$$u = F(\xi, y, z, t) \tag{22}$$

which yields

$$\nabla^2 u = t^2(\overline{U}')^2 F_{\xi\xi} + O(t) + O(1) \tag{23}$$

Here, only the term involving the highest power of t has been retained as being the dominating one for large times. Equation (21) with (23) may be solved by the introduction of the viscous time

$$T = \nu(\overline{U}')^2 t^3/3 \tag{24}$$

which gives

$$F_T - F_{\xi\xi} \cong 0 \tag{25}$$

The initial condition for F may be found from considering the inviscid limit $T \to 0$, which, in the sense of matched asymptotic expansions, may be regarded as the inner limit of F. Thus

$$F(\xi, y, z, 0) = u_\infty(\xi) \tag{26}$$

y and z again entering only as parameters. Application of the standard method of solution for the diffusion equation yields

$$F(\xi, T) = \frac{1}{\sqrt{2\pi T}} \int_{-\infty}^{+\infty} u_\infty(\xi_1)\exp[-\frac{(\xi-\xi_1)^2}{4T}]d\xi_1 \tag{27}$$

The viscous long-time limit is found by letting T tend to infinity holding $\xi/\sqrt{T}$ constant. This gives that the disturbance for long times decays as $t^{-3/2}$ and has a streamwise amplitude variation in form of a "Gaussian hat"

$$u \sim t^{-3/2} \exp[-\xi^2/(4\overline{U}' \sqrt{t^3\nu/3})] \tag{28}$$

Such a disturbance has, in a sense, a self-preserving shape, since an inviscid long-time limit of the form

$$u_\infty \sim \exp[-(\frac{\xi}{l_1})^2] \tag{29}$$

yields a streamwise velocity distribution of

$$\frac{u}{u_\infty} = \frac{l_1}{L} \exp[-(\frac{\xi}{L})^2] \tag{30}$$

where

$$L = [l_1^2 + 4\nu(\overline{U}')^2 t^3/3]^{1/2} \tag{31}$$

The viscous diffusion thus lengthens the disturbed region and its amplitude diminishes such as to retain the original total streamwise momentum.

From (30), (31) one can make an estimate of the maximum shear intensity in the shear layer and hence its smallest thickness. The maximum shear (setting $\partial u/\partial y \approx -t\overline{U}'(y) u_{\infty\xi}$) is found to occur at $\xi/L = 1/\sqrt{2}$ and $t = [3l_1^2/8\nu(\overline{U}')^2]^{1/3}$ and is given by

$$(\frac{\partial u}{\partial y})_{max} = 2\frac{u_\infty}{l_1}(\frac{Re_1}{9})^{1/3} \tag{32}$$

where Re_1 is the Reynolds number based on the initial length of the structure and $\overline{U}'$,

$$Re_1 = \frac{\overline{U}' l_1^2}{\nu} \tag{33}$$

From this one may estimate the smallest thickness of the shear layer as

$$\frac{\delta_{min}}{l_1} = \frac{u}{l_1}(\frac{\partial u}{\partial y})_{max} = C(Re_1)^{-1/3} \tag{34}$$

where the numerical constant $C = O(1)$ is

$$C = 3^{1/6} 2^{-1/2} \approx 0.85$$

For example, an initial length of the structure of $l_1^+ = 150$ gives $\delta_{min} = 16$ in viscous units, occurring at a time when the effective streamwise length of the shear layer is $L^+ = \sqrt{2},\ l_1^+ = 184$. Thus, the shear layer is thin compared to its length, and an approximate localized treatment of the stability of this layer taking it to behave as in a parallel flow seems reasonable. The inviscid theory shows that a free shear layer is unstable for wave numbers smaller than unity corresponding to a wave length greater than 2π times the thickness of the shear layer. The wave length of the observed wave packets of about 150 in viscous units is thus not inconsistent with the above estimate of the smallest shear layer thickness.

5. Reynolds stresses due to a local instability

The thin shear layer formed in the convected eddy in the manner analyzed above could be expected to be susceptible to local inflectional instability. In view of the slow development of the eddy in the viscous long-time limit, and the large length to thickness ratio of the shear layer (see the estimates above), an approximate analysis of this instability could be carried out on basis of parallel-flow theory, supplemented as needed by the application of kinematic-wave theory to account for the effects of slow temporal and spatial variations (Landahl, 1982).

First, we subdivide the disturbance flow field in a large-scale and small-scale portion by setting

$$u = \tilde{u} + u', v = \tilde{v} + v', w = \tilde{w} + w' \tag{35}$$

where "tilde" denotes the large-scale flow field due to the convected eddy and where u', v', w' are the contributions that arise from a small-scale wave-like instability of the form (real part implied)

$$u' = \hat{u}(y)E, v' = \hat{v}(y)E, w' = \hat{w}(y)E \tag{36}$$

$$E = \exp[i\alpha(x - ct) + i\beta z], c = c_r + ic_i \tag{37}$$

the instability occurring on the parallel flow field

$$\mathbf{U}(y) = (U(y), W(y)), U = \overline{U} + \tilde{u}, W = \tilde{w} \tag{38}$$

Note that the spanwise component is included to allow for the possibility of cross-flow instability. Inflectional instability is governed by the inviscid equations which, in the linearized approximation, give

$$-\frac{d}{dy}(\hat{u}\hat{v}^*) = A \sin\Theta_w + B\cos\Theta_w \tag{39}$$

$$-\frac{d}{dy}(\hat{v}\hat{w}^*) = -A\cos\Theta_w + B\sin\Theta_w \tag{40}$$

where * denote complex conjugate and

$$\Theta_u = \tan^{-1}\frac{W}{U}, \Theta_w = \tan^{-1}\frac{\beta}{\alpha} \tag{41}$$

Θ_u and Θ_w being the local flow and wave angles, respectively, and where

$$A = \tan(\Theta_w - \Theta_u)\frac{d}{dy}[\tilde{U}'(y)\frac{d}{dt}(\frac{1}{2} \mid \hat{l} \mid^2)] \tag{42}$$

$$B = \tilde{U}''\frac{d}{dt}(\frac{1}{2} \mid \hat{l} \mid^2) = \frac{\tilde{U}'' \mid \hat{v}^2 \mid c_i}{(\tilde{U} - c_r)^2 + c_i^2}, \alpha\tilde{U} = \alpha U + \beta W \tag{43}$$

and where $l' = \hat{l}E$ is the liftup of the fluid element due to the instability

The highest growth rate is found for waves with fronts normal to the local flow direction. Hence, for the most rapidly growing waves

$$\Theta_w \cong \Theta_u, A \cong 0 \tag{44}$$

or

$$\frac{d}{dy} < u'v' >= -\frac{1}{2} Re \frac{d}{dy}(\hat{u}\hat{v}^*) \cong B \cos \Theta_w \tag{45}$$

$$\frac{d}{dy} < v'w' >= -\frac{1}{2} Re \frac{d}{dy}(\hat{v}\hat{w}^*) \cong B \sin \Theta_w \tag{46}$$

where <> denotes ensemble or conditional average.

6. Moment of momentum due to a localized disturbance

The flow field due to a localized disturbance will have an overall effect on the large-scale flow field which may be assessed by considering the Reynolds stresses associated with the disturbance. By integration of (2) over the volume, δV, that the disturbance occupies, it is found that the net momentum transport associated with it is zero, as is to be expected since the Reynolds stresses serve only to redistribute the momentum between different parts of the flow field. The nature of this redistribution may be clarified by determining the moment of momentum impulse (M_x, M_y, M_z) due to a localized disturbance, active during some finite time δT, namely

$$M_x = \iiint_{\delta V} \rho dV \int_{\delta T} (y\frac{\partial w}{\partial t} - z\frac{\partial v}{\partial t})dt = -\int_{\delta T} dt \int_{-\infty}^{+\infty}\int_{-\infty}^{+\infty} z p_{wall} dx dz \tag{47}$$

$$M_y = 0 \tag{48}$$

$$M_z = \iiint_{\delta V} \rho dV \int_{\delta T} (x\frac{\partial v}{\partial t} - y\frac{\partial u}{\partial t})dt = \int_{\delta T} dt \int_{-\infty}^{+\infty}\int_{-\infty}^{+\infty} x p_{wall} dx dz \tag{49}$$

With the use of the relation for the disturbance pressure,

$$\frac{\nabla^2 p}{\rho} = -\frac{\partial^2 (u_i u_j)}{\partial x_i \partial x_j} - 2U'\frac{\partial v}{\partial x} \tag{50}$$

one finds, after some manipulations, taking advantage of the requirement from continuity that the integral of the component v over each plane y=constant must be zero, that

$$M_x = -\iiint_{\delta V} \rho dV \int_{\delta T} vw dt \tag{51}$$

$$M_z = -\iiint_{\delta V} \rho dV \int_{\delta T} uv dt \tag{52}$$

Thus, the moment of momentum impulse transmitted to the flow is directly related to the Reynolds stress pulse produced in the flow by the localized disturbance. These results were found originally by Landahl (1977) from the Euler equations, but the restriction to inviscid flow made in that analysis is not necessary since (50) holds also for a viscous flow.

A localized region of positive Reynolds shear stress, $-\rho < uv >$, will thus induce a rotation of a sense opposite to that of the mean shear, as would be manifested through a region of ejection downstream of a sweep. Of particular interest is the spanwise shear stress component, $-\rho < vw >$, which will induce streamwise rotation and thus would be the one responsible for the formation of streamwise vortices and streak formation. This process will be further discussed below.

7. The formation of streamwise streaks

Waves or wave-like disturbances are not the only, or perhaps not even the most interesting, coherent structural component to be found in shear-flow turbulence. The early visualization experiments of Kline *et al.* (1967) showed that the most prominent flow structures found in the near-wall flow region are long streamwise streaks (of a streamwise length of 1000 viscous wall units or more) of alternating low- or high-speed flow having an average spanwise scale of approximately 100 viscous wall units. The Navier-Stokes numerical flow simulations of Moin & Kim (1982) confirm this picture.

A theoretical model for the streak formation has been proposed recently by Landahl (1990). He assumes that the initiation of a streak is due to a localized and highly intermittent nonlinear activity at the bursting instant $t = t_n$ of the form

$$q = Q(x, y, z)\delta(t - t_n) \tag{53}$$

where $\delta(t)$ denotes the Dirac delta function. The spatial variation of the source term may be found from the application of conditional sampling, making use of, e.g., the VITA technique. To be consistent with the local instability idea as expressed in relations like (42) – (46) above he chose for the conditionally sampled driving term $< Q >$ the following model:

$$< Q >= \frac{\partial^2}{\partial y^2}\left(\frac{\partial}{\partial x} < uv > + \frac{\partial}{\partial z} < vw >\right) = \frac{\partial^2}{\partial y^2}\left(\cos\Theta_w \frac{\partial B}{\partial x} + \sin\Theta_w \frac{\partial B}{\partial z}\right) \tag{54}$$

taking for B a modified Gaussian "hat"

$$B = Cy^3[1 - 2(\frac{z}{l_3})^2]\exp[-(\frac{x}{l_1})^2 - (\frac{y}{l_2})^2 - (\frac{z}{l_3})^2] \tag{55}$$

C being an arbitrary constant. In the examples presented the scale factors l_1, l_2, l_3 were chosen as

$$l_1^+ = 100, l_2^+ = 15, l_3^+ = 30$$

(The factor $1 - 2(z/l_3)^2$ was included to make $\int_{-\infty}^{+\infty} B dz = 0$). Some of the results of Landahl's (1990) calculations are reproduced in Figure 7. Clearly, the theoretical results resemble quite closely in character the numerical simulations of Moin & Kim (1982), shown in Figure 8.

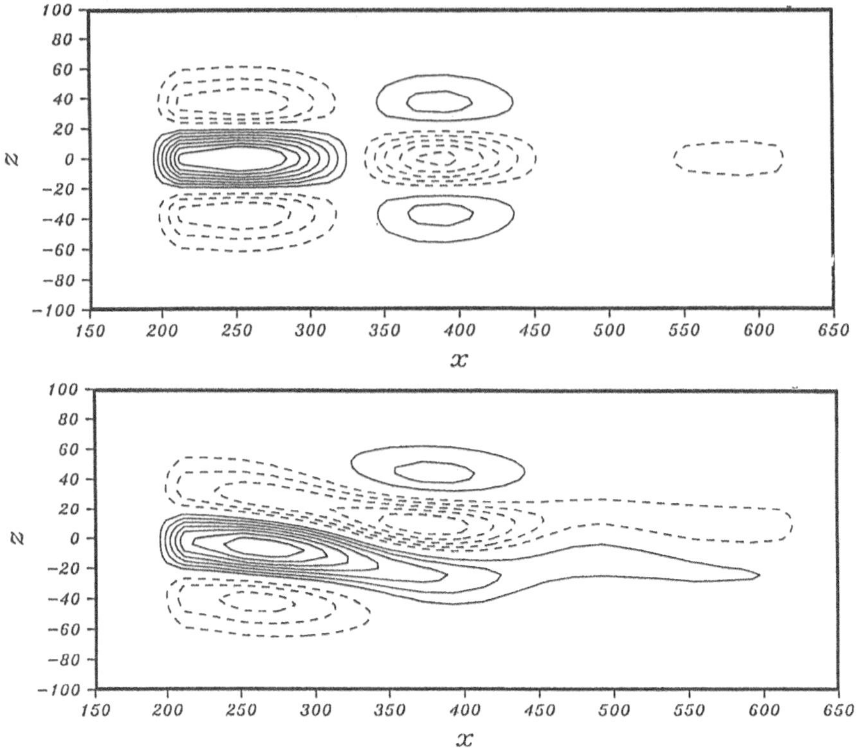

Figure 7. Contours of constant u in the plane $y^+ = 15$, $t^+ = 40$. (a) symmetrical structure; (b) asymmetrical structure, $\theta = 5^o$ (from Landahl, 1990).

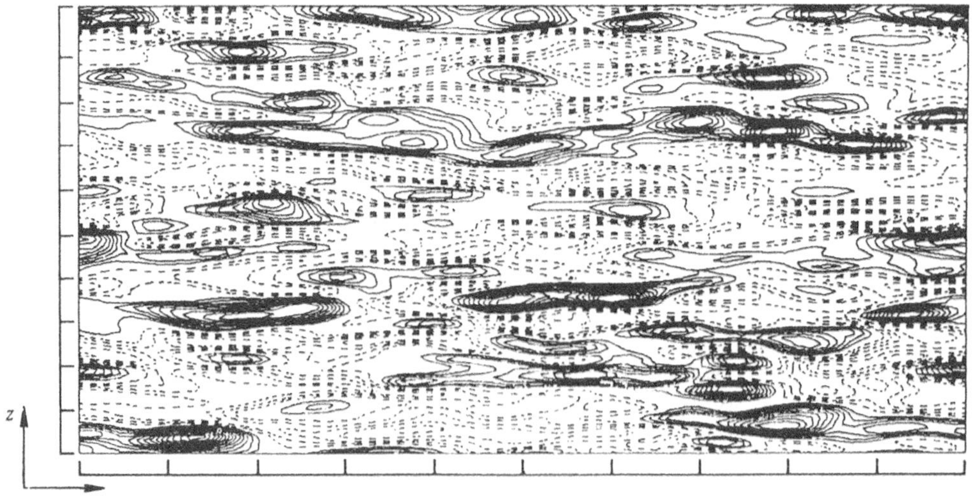

Figure 8. Contours of fluctuating streamwise velocities in the plane $y^+ = 6.14$ obtained by numerical simulation for a channel flow (from Moin & Kim, 1990).

8. Conclusions

The experimental and theoretical results reviewed demonstrate that turbulent flows with inflectional mean velocity profiles (free shear layers, mixing layers) show the presence of wave-like components in their fluctuations. The response of such flows to forced excitation, as well as to excitation by intermittent nonlinear activity, is well described by linear theory. Evidence of damped Orr-Sommerfeld type waves is seen in the wave-number-frequency spectra of wall-bounded turbulent shear flows.

In wall-bounded turbulence short wave packets with a wave length of ≈ 150 in viscous wall units appear intermittently. They appear to be the result of local instability of the thin shear layers that are formed near the wall by convected disturbances through spanwise stretching of the mean vorticity. The Reynolds stresses from these unstable wave packets can act as random sources driving the larger-scale fluctuations. The spanwise component $< vw >$ from an oblique instability wave packet induces longitudinal vorticity and also can initiate streak growth.

The further study of wave motion in a shear flow subjected to random fluctuations could provide deepened insight into the dynamics of turbulent flows (e.g., cross-flow local instability waves). Further study of non-wave-like disturbance components are also warranted for the understanding of streak formation, and the behaviour of convected eddies.

8. Acknowledgements

This research was supported by the Aeronautical Research Institute of Sweden.

References

Bark, F.H. 1975 On the wave structure of the wall region of a turbulent boundary layer. *J. Fluid Mech.* **70**, 229.

Brown, G.L. & Roshko, A. 1974 On density effects and large structures in turbulent mixing layers. *J. Fluid Mech.* **64**, 775.

Browand, F.K. & Troutt, T.R. 1985 The turbulent mixing layer: geometry of large vortices. *J. Fluid Mech.* **158**, 489.

Gaster, M., Kit, E., & Wygnanski, I. 1983 Large-scale structures in a forced turbulent mixing layer. *J. Fluid Mech.* **150**, 23.

Gavrilakis, 1990 (to be published).

Haritonidis, J.H., Gresko ,L.S. & Breuer, K.S. 1988 Wall pressure peaks and waves. In: *Proc. of Zoran Zaric Memorial International Seminar on Near-Wall Turbulence* (ed. S.J. Kline), Dubrovnik, Yugoslavia, Hemisphere Publ. Corp.

Hussain, A.K.M.F. & Zaman, K.B.M.Q. 1985 An experimental study of organized motion in the turbulent plane mixing layer. *J. Fluid Mech.* **159**, 85.

Kline, S.J., Reynolds, W.C., Schraub, F.A. & Runstadler, P.W. 1967 The structure of turbulent boundary layers. *J. Fluid Mech.* **30**, 741.

Landahl, M.T. 1967 A wave-guide model for turbulent shear flow. *J. Fluid Mech.* **27**, 443.

Landahl, M.T. 1975 Wave breakdown and turbulence. *SIAM J. Applied Math.* **28**, 775.

Landahl, M.T. 1977 Dynamics of boundary layer turbulence and the mechanism of drag reduction. *Phys. Fluids* **20**, S55.

Landahl, M.T. 1982 The application of kinematic wave theory to wave trains and packets with small dissipation. *Phys. Fluids* **25**, 1512.

Landahl, M.T. 1984 On the dynamics of large eddies in the wall region of a turbulent boundary layer. In: *Turbulence and Chaotic Phenomena in Fluids* (ed. T. Tatsumi), Elsevier, p. 467.

Landahl, M.T. 1990 On sub-layer streaks. *J. Fluid Mech.* **212**, 593.

Michalke, A. 1964 On the inviscid instability of the hyperbolic-tangent velocity profile. *J. Fluid Mech.* **19**, 543.

Moin, P. & Kim, J. 1982 Numerical investigation of turbulent channel flow. *J. Fluid Mech.* **118**, 341.

Morrison, W.R.B., Bullock, K.J. & Kronauer, R.E. 1971 Experimental evidence of waves in the sublayer. *J. Fluid Mech.* **47**, 639.

Schewe, G. 1983 On the structure and resolution of wall-pressure fluctuations associated with turbulent boundary-layer flow. *J. Fluid Mech.* **134**, 311.

The Evolution of Helical Waves on Axisymmetric Jets

Eckart Meiburg and James E. Martin

Center for Fluid Mechanics, Division of Applied Mathematics,
Brown University, Providence, RI 02912 USA

We consider the three-dimensional evolution of a nominally axisymmetric transitional jet for the case of helical perturbations. Inviscid vortex filament calculations provide insight into the mechanisms leading to the concentration, reorientation, and stretching of the vorticity. For the case of a helical perturbation only, the streamwise vorticity forming in the braid is of the same sign everywhere, with the vortex helix representing streamwise vorticity of the opposite sign. Due to the helical symmetry, concentrated structures do not form in the braid. If this symmetry is broken by introducing an additional periodic perturbation in the azimuthal direction, we observe the emergence of concentrated streamwise braid vortices that are amplified in the strain field set up by the vortex helix. In contrast to jet flows dominated by vortex rings, the streamwise vortical structures are all of the same sign. It appears that a Kelvin- Helmholtz-like instability of the braid vorticity layer plays a significant role in their generation.

1. Introduction

It is well established that laminar jets are unstable with respect to axisymmetric and helical waves, as, for example, the stability analyses by Batchelor and Gill (1962), Michalke and Hermann (1982) as well as Cohen and Wygnanski (1987) show. Furthermore, some recent experimental investigations, e.g. Dimotakis, Miake-Lye and Papantoniou (1983), Tso and Hussain (1989) as well as Mungal and Hollingsworth (1989), to name just a few, indicate that even fully turbulent jets are dominated by ringlike and helical structures whose dynamics become largely independent of the Reynolds number when this dimensionless parameter is large. Hence a more complete understanding of the dynamics of these structures is a prerequisite for a successful manipulation and control of jets. In a recent paper (Martin and Meiburg 1990), we numerically investigated the three-dimensional evolution of jets that emerge from a corrugated nozzle and are perturbed by axisymmetric waves. Under these circumstances, vortex rings form that set up a strain field with a free stagnation point in the braid region. Small perturbations in the braid vorticity due to the corrugation are subsequently amplified, so that pairs of concentrated streamwise vortices form in between the vortex rings. The mechanism for their generation is similar to the model problem investigated by Lin and Corcos (1984) as well as by Neu (1984) for the plane mixing layer. If, on the other hand, the axisymmetric jet is perturbed by a helical wave, only one sign of streamwise vorticity forms in the braid region (Martin, Meiburg and Lasheras 1990). As long as no further perturbations are present, this streamwise braid vorticity has the form of a smooth layer, due to the helical symmetry. In the present paper, we intend to investigate the evolution of the helically perturbed jet under the influence of a periodic perturbation in the azimuthal direction which breaks the helical symmetry. The main questions of interest concern the possibility of the formation of concentrated streamwise vortices as

Advances in Turbulence 3
Editors: A.V. Johansson · P.H. Alfredsson

well as the mechanism for their generation. We will apply an inviscid vortex filament technique for the numerical investigation, which will briefly be described in section 2. In section 3, we will compare the evolution of helical waves on axisymmetric jets without and with azimuthal perturbations and discuss the differences between the two cases.

2. Numerical Technique

The non-divergent nature of the velocity field in incompressible flows, along with the definition of vorticity, allows for a complete description of the kinematics of the flow in the form of the Biot-Savart law (Batchelor 1967). Using the theorems of Kelvin and Helmholtz for inviscid dynamics and following the general concepts reviewed by Leonard (1985), vortex filaments are used for the representation of the vorticity field. Each filament is represented by a number of node points along its centerline, through which a cubic spline is fitted to give it a smooth shape. The three-dimensional Biot-Savart integral is reduced to a line integral by assuming an invariant algebraic vorticity distribution around the filament centerline. We can then obtain the velocity at any position by integrating over the arclength of all filaments in the flowfield. For the numerical simulation, we limit ourselves to the temporally growing problem, i.e., our flow is periodic in the streamwise direction. Consequently, the Biot-Savart integration has to be carried out over the periodic images of the vortex filaments as well. Based on a series of test calculations, we set the number of periodic images carried along equal to three. We take the velocity difference between the centerline and infinity as our characteristic velocity. The thickness of the axisymmetric shear layer, defined as the velocity jump divided by the maximum slope of the velocity, serves as the characteristic length scale, which results in the filament radius of 0.5. In these units, the radius of the jet considered here is 5. Hence, the important ratio of jet radius to momentum thickness of the jet shear layer is 22.6. The Biot-Savart integration is carried out with second order accuracy both in space and in time by employing the predictor-corrector time-stepping scheme and the trapezoidal rule for spatial integration, respectively. We typically discretize the axisymmetric shear layer into 59 filaments per streamwise wavelength of initially 120 nodes each. As the flow develops a three-dimensional structure, the vortex filaments undergo considerable stretching. To maintain an adequate resolution, additional nodes are introduced, based on a criterion involving distance and curvature. Furthermore, the time-step is repeatedly reduced as local acceleration effects increase. The filament core radius decreases as its arclength increases to conserve its total volume. A more detailed discussion of the numerical method can be found in Ashurst and Meiburg (1988).

3. Results

3.1 The evolution of a helical wave only

In this section, we focus on the evolution of a nominally axisymmetric jet perturbed by a helical wave of azimuthal wavenumber 1. The axisymmetric shear layer is represented by vortex filaments that initially have the form of vortex rings. Numerically, the helical perturbation wave is introduced by slightly displacing the vortex filament centerlines in the streamwise direction. The form of this displacement is the same for all vortex filaments. However, there is an azimuthal phase shift between the displacements of successive vortex filaments, which results in the helical nature of the overall perturbation. Figure 1 shows the flow field at times $t = 0.31, 2.81$ and 8.75. For each time, a side

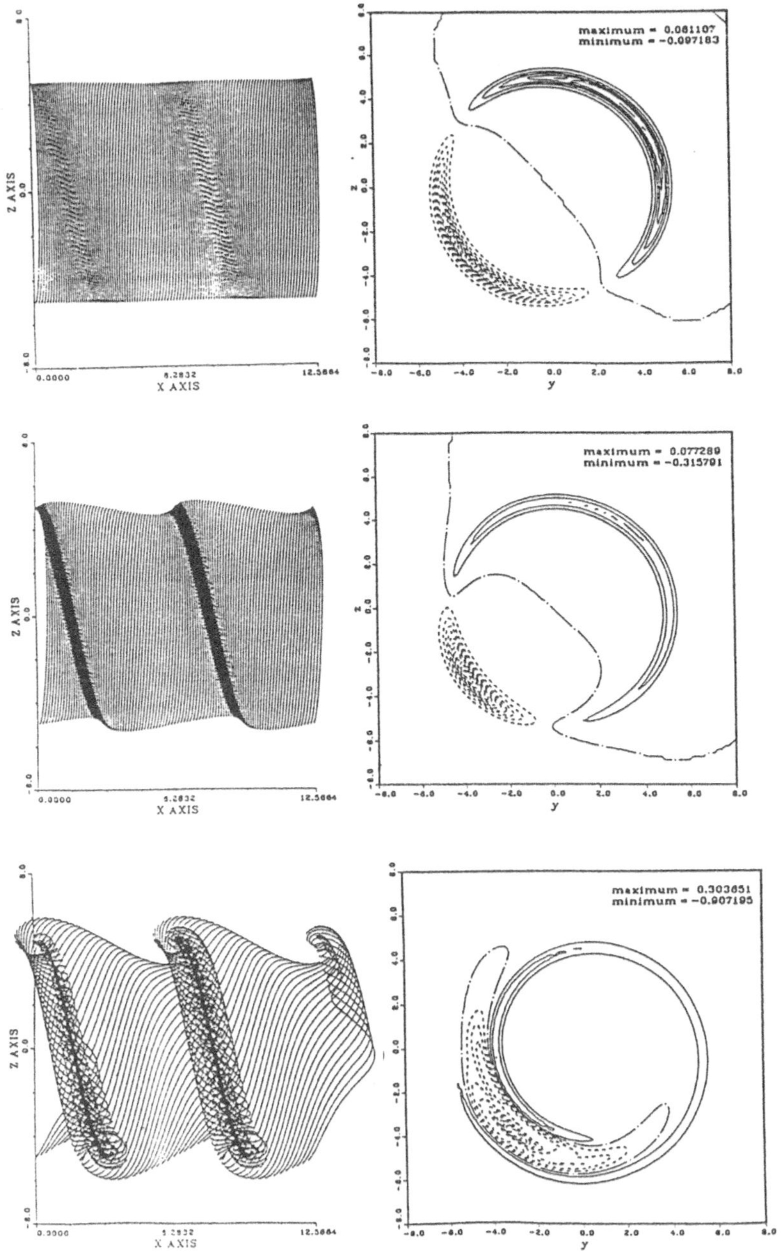

Fig. 1: The growth of a helical wave on an axisymmetric jet. Shown are views of the vortex filaments (left, $y > 0$ only) and contours of constant streamwise vorticity (right) for the times 0.31, 2.81 and 8.75, respectively. Observe the growth of a strong vortex helix, corresponding in this case to regions of negative streamwise vorticity. The strain field set up by this helix leads to a reorientation of the braid vorticity, which alignes itself with the direction of extensional strain, thus resulting in positive streamwise braid vorticity.

view and a streamwise view of the vortex filaments are shown, along with a contour plot of the streamwise vorticity at $x = 4$. At $t = 0.31$, the side view, which for clarity only shows the filament sections at $y > 0$, already indicates the emergence of a helical structure representing more concentrated vorticity. The nearly symmetric form of the streamwise vorticity plot still reflects the initial perturbation. It should be pointed out that the streamwise vorticity contours look identical at all x-locations except for a rotation around the jet axis.

At $t = 2.81$, the growth of the vortex helix is clearly visible. It causes a strain field with a free stagnation point to form in the braid region between successive helix segments. The axis of extensional strain becomes aligned with the direction normal to the emerging helix. Consequently, the braid region between segments of the helix becomes increasingly depleted of vorticity, as the vorticity layer continues to wrap around the helical structure, thus reinforcing it. It is important to notice that, in contrast to a jet dominated by vortex rings, the principal axis of extensional strain now has a circumferential component as well. Furthermore, the stagnation line in the braid region is a helix itself. As a result, streamwise vorticity is formed as the braid vorticity becomes reoriented in the strain field generated by the growing helix. This evolution also leads to the loss of the initial symmetry observed in the streamwise vorticity contour plots. The concentrated region of negative streamwise vorticity represents the helix intersecting a plane $x =$const., whereas weak positive streamwise vorticity indicates the presence of the braid vorticity layer.

At time $t = 8.75$, which is the final time of our simulation, the vortex helix has grown in strength, while the braid vorticity keeps acquiring a larger streamwise component by reorienting itself along the direction of extensional strain. Notice in the streamwise vorticity contour plots how the braid region of positive streamwise vorticity has acquired a more sheet-like structure and begun to "spiral" around the region of concentrated negative streamwise helix vorticity, thus leading to significant stretching of the vortex filaments. This reflects the fact that each filament, as it has to remain closed, is part of the helix in one region and part of the braid in some other region. Where it is part of the helix, the filament is oriented in the upstream direction, while it has a downstream orientation over most of the circumference. In summary, perturbing the jet by a helical wave leads to the evolution of a strong helical vortex, whose successive segments are connected by a layer of streamwise braid vorticity. In the contour plots of the streamwise vorticity, the vortex helix corresponds to a concentrated region of, in this case, negative vorticity, while the braid has the form of a layer of positive streamwise vorticity. This is in contrast to a ring-dominated jet, where under the effect of azimuthal perturbations both signs of streamwise vorticity occur in both the ring as well as the braid region. Due to the helical symmetry of the above flow, concentrated streamwise vortices cannot form in the braid region. In the following section, we will study the evolution of the flow when an additional periodic azimuthal perturbation is introduced.

3.2 The evolution of a helically perturbed jet emanating from a corrugated nozzle

As in the previous section, we perturb the initially axisymmetric jet with a helical wave by displacing the vortex filament centerlines correspondingly. In addition, we also simulate the effect of a corrugated nozzle by displacing each vortex filament centerline periodically in the radial direction, as can be seen in the streamwise view of fig. 2. This

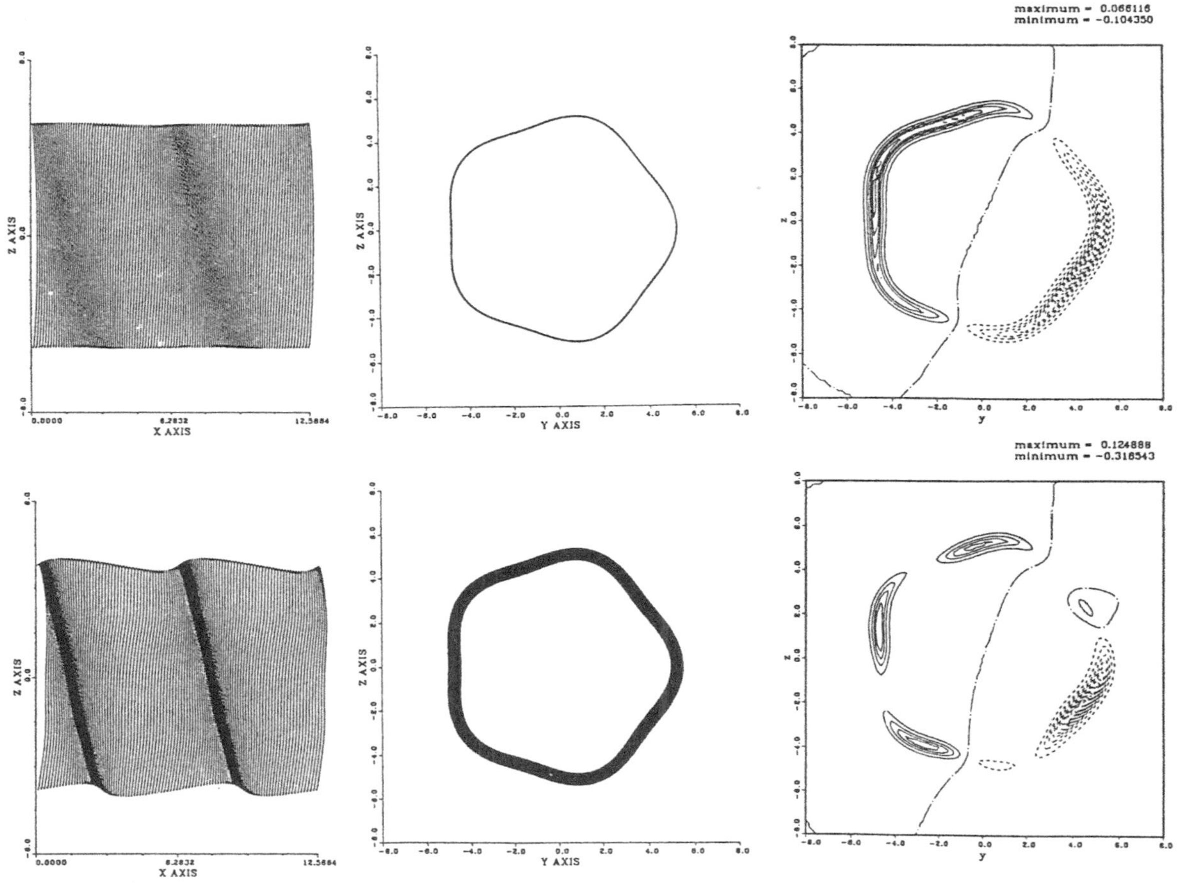

Fig. 2: The growth of a helically perturbed jet under the additional influence of a periodic perturbation in the azimuthal direction. Displayed are side and streamwise views of the vortex filaments along with contours of constant streamwise vorticity for times 0.31, 2.65, 9.45 and 15.70. The azimuthal perturbation breaks the helical symmetry, and as a result concentrated streamwise vortices form in the braids. In contrast to jets dominated by vortex rings, the streamwise braid vortices are all of the same sign.

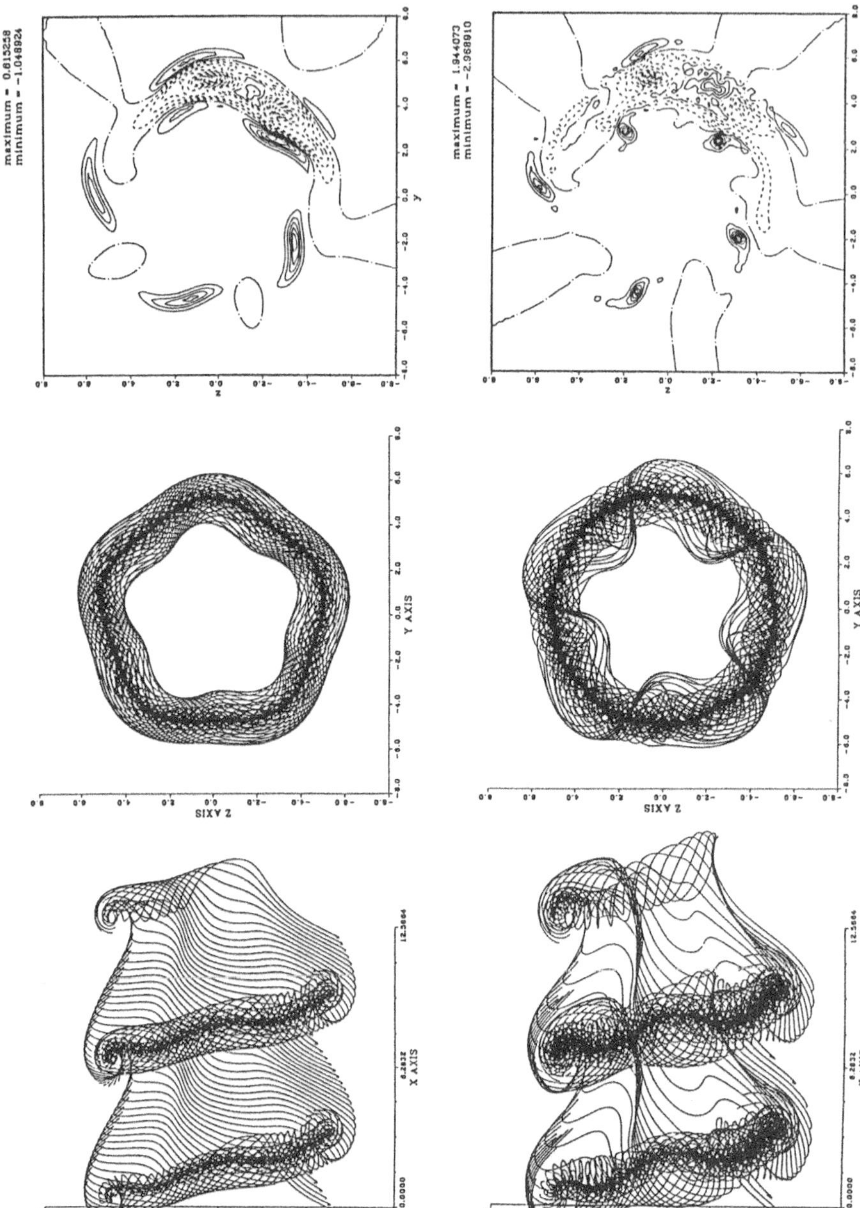
maximum = 0.615258
minimum = -1.048924
maximum = 1.944073
minimum = -2.968910
y
z
Y AXIS
Z AXIS
X AXIS

Fig. 2 cont'd

figure shows the resulting flow field at times $t = 0.31, 2.65, 9.45$, and 15.70. Again, side views and streamwise views of the vortex filaments are provided along with contour plots of the streamwise vorticity at $x = 2$.

Soon after the start of the calculation, the symmetry-breaking effect of the azimuthal perturbation becomes visible. It results in a wavy modulation of the emerging layer of positive braid vorticity, whose subsequent evolution represents a combined effect of several distinct mechanisms. As in the case of a ring-dominated jet emerging from a corrugated nozzle, the radial displacement brings the local and global induction effects into play, cf. Martin and Meiburg (1990). But in contrast to the ring-dominated flow, the braid vorticity has the form of a smooth layer and is of the same sign everywhere in the present case. Consequently, the braid vorticity layer has the possibility to undergo a Kelvin-Helmholtz-like instability, an indication of which can already be seen in the streamwise vorticity contour plot at time $t = 2.65$. This is in marked contrast to the three-dimensional evolution of an axisymmetric jet with ring-like structures, which results in the formation of counterrotating streamwise vortex pairs. At the same time, the emerging structures in the braid region are being intensified by the extensional strain field generated by the emerging vortex helix. At $t = 9.45$, one can clearly recognize in the streamwise vorticity contour plots how the Kelvin-Helmholtz-unstable braid vorticity layer spirals around the vortex helix, which is also acquiring a wavy modulation, as can be seen in the side view. The streamwise view of the filaments shows that the corrugation amplitude itself is not undergoing rapid growth.

At time $t = 15.70$, the concentrated streamwise braid vortices connecting successive segments of the vortex helix have formed. Simultaneously, the waviness of the vortex helix has grown, and small regions have formed in which the streamwise vorticity component of the vortex helix changes its sign. In summary, it is important to notice that the loss of symmetry resulting from the periodic perturbation in the azimuthal direction results in the formation of concentrated streamwise vortices in the braid region. These concentrated streamwise braid vortices are all of the same sign, in contrast to jets dominated by ring-like structures. One important aspect in their generation is a Kelvin-Helmholtz-type instability of the braid vorticity layer. In addition, they are amplified by the extensional strain field set up by the vortex helix. We are currently investigating how the above three-dimensional evolution depends on the amplitudes of the initial perturbations as well as on the ratio of jet radius to momentum thickness of the jet shear layer. A more detailed analysis will be published elsewhere.

Acknowledgements

This work is supported by the National Science Foundation under grant #MSM-8809438 and by DARPA under URI contract #N00014-86-K0754 to Brown University. The San Diego Supercomputer Center is providing computing time on its CRAY-Y/MP.

References

Ashurst, W.T. and Meiburg, E. 1988 Three-dimensional shear layers via vortex dynamics. *J. Fluid Mech.* **189**, 87.

Batchelor, G.K. 1967 *An introduction to fluid mechanics.* Cambridge University Press.

Batchelor, G.K. and Gill, A.E. 1962 Analysis of the stability of axisymmetric jets. *J. Fluid Mech.* **14**, 529.

Cohen, J. and Wygnanski, I. 1987 The evolution of instabilities in the axisymmetric jet. Part1. The linear growth of disturbances near the nozzle. *J. Fluid Mech.* **176**, 191.

Dimotakis, P.E., Miake-Lye, R.C., and Papantoniou, D.A. 1983 Structure and dynamics of round turbulent jets. *Phys. Fluids* **26**, 3185.

Leonard, A. 1985 Computing three-dimensional incompressible flows with vortex elements. *Ann. Rev. Fluid Mech.* **17**, 523.

Lin, S.J. and Corcos, G.M. 1984 The mixing layer: deterministic models of a turbulent flow. Part 3: The effect of plane strain on the dynamics of streamwise vortices. *J. Fluid Mech.* **141**, 139.

Martin, J.E. and Meiburg, E. 1990 The three-dimensional evolution of an axisymmetric jet: a numerical investigation. Submitted to *J. Fluid Mech.*

Martin, J.E. and Meiburg, E., and Lasheras, J.C. 1990 Three-dimensional evolution of axisymmetric jets: A comparison between computations and experiments. Submitted to the Proc. of the IUTAM Symp. on Separated Flows and Jets, Novosibirsk, USSR.

Michalke., A. and Hermann, G. 1982 On the inviscid instability of a circular jet with external flow. *J. Fluid Mech.* **114**, 343.

Mungal, M.G. and Hollingworth, D.K. 1989 Organized motion in a very high Reynolds number jet. *Phys. Fluids A* **1**, 1615.

Neu, J.C. The dynamics of stretched vortices. 1984 *J. Fluid Mech.* **143**, 253.

Tso, J. and Hussain, F. 1989 Organized motions in a fully developed turbulent axisymmetric jet. *J. Fluid Mech.* **203**, 425.

The Interaction Between Particles and Coherent Structures in a Plane Turbulent Jet

R.J. Perkins, S. Ghosh and J.C. Phillips

Department of Applied Mathematics and Theoretical Physics, University of Cambridge, Silver Street, Cambridge CB3 9EW

In a series of experiments dense particles were released above a horizontal plane turbulent air jet and collected in trays beneath the jet. The experiments were simulated using a Random Flight model, and the predicted distributions compared with those measured in the experiments. Good agreement was obtained for the large particles, but the agreement was very poor for the smaller particles which seemed to leave the jet much sooner than the model predicted. This may be due to interaction between the particles and the large scale coherent structures in the jet; the relevant parameters are of the right order for this to occur. Measurements of particle trajectories using a particle-tracking technique provide some direct experimental evidence of this phenomenon.

1. Introduction

Many environmental and industrial processes involve the interaction between coherent structures and particles, bubbles or droplets. Typical examples include droplet dispersion in agricultural sprays and fuel injectors, the entrainment and deposition of dust by air jets and the transport of sediment in rivers and coastal regions.

Recent investigations have shown how large scale structures such as vortices can 'trap' bubbles (Auton 1984) and either trap or disperse dense particles (Perkins and Hunt 1986). These ideas have been used to explain the entrainment of bubbles into a mixing layer (Hunt *et al* 1988) and the dispersion of dense particles in an axisymmetric air jet (Yule 1980, Chung and Troutt 1988). In particular, Chung and Troutt found that when the particle response time was of the same order as the eddy turnover time, the particles appeared to be 'thrown out' of the jet by the eddies and the particle diffusivity could then bé as much as 1.6 times the fluid diffusivity. To study this, we are conducting an experimental and numerical investigation of the entrainment and dispersion of dense particles in a plane turbulent air jet.

2. Experimental arrangement

The general experimental arrangement is illustrated in figure 1, and the experimental methods are described in detail in Perkins and Hunt (1990). A fan blew air through a horizontal nozzle (width 10 mm) with an exit velocity which could be varied between 1 m/s and 25 m/s. Particles were released from a vibrating hopper 289 mm above the jet centreline, and collected in trays (width 25 mm) 235 mm beneath the jet centreline. The downstream release distance varied between 80 mm and 500 mm. The particles were discharged fairly slowly ($\sim$ 10 gm/min) to prevent them simply falling through the jet in a narrow stream; the particle concentrations in the jet were correspondingly low

Advances in Turbulence 3
Editors: A.V. Johansson · P.H. Alfredsson

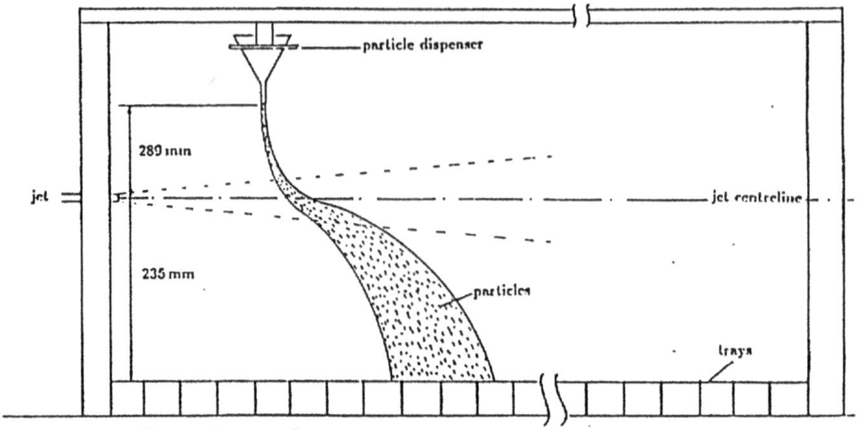

Figure 1. Experimental arrangement

so it seems reasonable to neglect particle-particle interactions and the effect of particles on the flow. At the end of each experiment the trays were weighed to determine the deposition concentration distribution along the centreline of the jet.

The experiments were carried out with two different particle sizes – 244 μm and 86 μm; both had a density of 2.5×10^3 kg/m^3. The particle terminal velocities (U_T) were about 1.8 m/s and 0.4 m/s respectively, with corresponding aerodynamic response times (τ_A) of about 0.05 s and 0.48 s. The exit velocity for the 244 μm particles was 12.1 m/s and that for the 86 μm particles was 6.9 m/s, giving jet exit Reynolds numbers ($U_o b_o/\nu$) of 8 300 and 4 800. Experiments with other exit velocities have been carried out, and are reported in Perkins and Hunt (1990).

3. Modelling particle motion

We consider a small spherical particle, diameter ϕ, density ρ_p, moving with velocity $\mathbf{u}_p$ in a fluid of density ρ_f and kinematic viscosity ν. The fluid velocity at the location of the particle ($\mathbf{x}_p$) and in the absence of the particle, is denoted by $\mathbf{u}$. The relative density of the particle is given by $\gamma = \rho_p/\rho_f$ and the particle Reynolds number by $\mathcal{R}_p = u_r\phi/\nu$ where $u_r = |\mathbf{u} - \mathbf{u}_p|$. In these experiments, $\gamma \sim 2.5 \times 10^3$ and $1 < \mathcal{R}_p < 50$.

The particle is subject to buoyancy, drag, pressure gradient, lift and history forces and the general equation of motion is complicated (*e.g.* Auton, Hunt and Prud'homme 1988), but for these conditions it can be simplified considerably. History effects can be neglected for $\gamma > 10^3$ (Clift, Grace and Weber 1978), and the pressure gradient term will be small compared with the drag and buoyancy terms. Several lift forces have been suggested (*e.g.* Rubinow and Keller (1961), Saffman (1965), Auton (1987)) and there is experimental evidence (*e.g.* White and Schulz (1977), White (1982)) that some of these can be important in shear flows. In this model, however, we have neglected all lift forces because they are likely to be small compared with the drag and buoyancy terms. (Perkins and Hunt (1990) gives a full discussion of the relative importance of the different terms.) The equation of motion for the particle therefore becomes:

$$\frac{d\mathbf{u}_p}{dt} = \mathbf{g} + \frac{3C_D}{4\phi\gamma}|\mathbf{u} - \mathbf{u}_p|(\mathbf{u} - \mathbf{u}_p)\,. \tag{1}$$

The drag coefficient, C_D, is a function of $\mathcal{R}_p$, but there are empirical correlations (*e.g.* Morsi and Alexander 1972) for shapes such as spheres.

4. Random Flight Model

The fluctuating velocities sampled by a fluid element in a turbulent velocity field can be modelled as a stochastic process derived from the Langevin equation:

$$\mathbf{u}(t+\Delta t) = \left(1 - \frac{\Delta t}{T_L}\right)\mathbf{u}(t) + \sigma_u \sqrt{\frac{2\Delta t}{T_L}}\, \boldsymbol{\xi} \tag{2}$$

where T_L is the Lagrangian integral timescale of the turbulence, σ_u is the rms velocity and $\boldsymbol{\xi}$ is a random variable drawn from a Gaussian distribution, with $\overline{\xi_i}=0$ and $\overline{\xi_i^2}=1$. This type of model has been used to compute dispersion in a range of homogeneous and inhomogeneous turbulent flows (van Dop *et al* 1985, Thomson 1986, Luhar and Britter 1989).

4.1 *Inhomogeneous, anisotropic turbulence*

For a flow such as a jet, the equation for the fluid velocities has to be modified to include correlations between the streamwise and transverse fluctuating velocities, and the effects of an inhomogeneous turbulent velocity field; following Legg and Raupach (1982) and Legg (1983), (2) becomes:

$$u(t+\Delta t) = r_u u(t) + \sigma_u \sqrt{1-r_u^2}\, \xi_i + (1-r_u) T_{Lu} \frac{\partial \overline{u'v'}}{\partial y} \quad , r_u = \exp\left\{-\frac{\Delta t}{T_{Lu}}\right\}, \tag{3a}$$

$$v(t+\Delta t) = r_v v(t) + \sigma_v \sqrt{1-r_v^2}\, \xi_j + (1-r_v) T_{Lv} \frac{\partial \sigma_v^2}{\partial y} \quad , r_v = \exp\left\{-\frac{\Delta t}{T_{Lv}}\right\}, \tag{3b}$$

$$\xi_i = b\xi_j + \sqrt{1-b^2}\,\xi' \ , \ b = \frac{\overline{u'v'}(1-r_u r_v)}{\sigma_u \sigma_v \sqrt{1-r_u^2}\sqrt{1-r_v^2}} \tag{3c}$$

4.2 *The effect of particle inertia*

Hunt and Nalpanis (1985) used a similar approach to model the fluid velocity *sampled by a dense particle* in a turbulent flow. Because of particle inertia, the fluid velocities sampled by the particle will not be the same as those sampled by a fluid element – the particle will tend to 'cut through' the eddies, so that the integral timescale (T_L^*) of the velocities sampled by a dense particle will be less than T_L. A particle and a fluid element which are coincident at some instant will have separated by $u_r \Delta t$ at time Δt later, so T_L^* depends on both the space and the time correlations of the fluid turbulence (figure 2).

Hunt and Nalpanis (1985) used an inertial subrange scaling for the spatial velocity correlations (*e.g.* Batchelor 1953) to derive a relationship between T_L^* and T_L:

$$T_L^* = \frac{T_L}{1 + \alpha\left(\frac{u_r}{\sigma_u}\right)^{\frac{2}{3}}\left(\frac{T_L}{\Delta t}\right)^{\frac{1}{3}}} \tag{4}$$

where α is an empirical coefficient which also depends on Δt. Calibration against the experimental results of Snyder and Lumley (1971) yielded $\alpha = (7.8\Delta t^{1/3})^{1.43}$. However this form is not entirely satisfactory, because of the dependence on Δt, and it is the subject of current research (*e.g.* Fung and Perkins 1989).

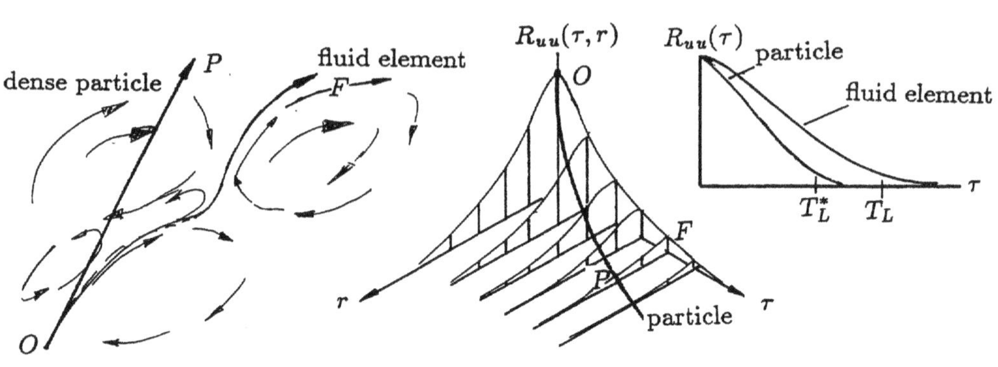

Figure 2. The relationship between T_L and T_L^*

4.3 *The mean velocity field*

The mean velocities $\overline{u}$ and $\overline{v}$ were obtained from the theoretical solution for a plane jet (Goertler 1942):

$$\frac{U_c}{U_o} = \sqrt{\frac{3\beta}{4}\frac{b_o}{x}} \tag{5a}$$

$$\frac{\overline{u}}{U_c} = 1 - \tanh^2\left\{\frac{\beta y}{x}\right\} \tag{5b}$$

$$\frac{\overline{v}}{U_c} = \frac{1}{\beta}\left[\frac{\beta y}{x} - \frac{\beta y}{x}\tanh^2\left\{\frac{\beta y}{x}\right\} - \frac{1}{2}\tanh\left\{\frac{\beta y}{x}\right\}\right] \tag{5c}$$

where U_o is the jet exit velocity, b_o is the width of the nozzle and $\beta \sim 7.67$, from experimental measurements (Rajaratnam 1976).

Turbulence profiles (σ_u, σ_v and $\overline{u'v'}$) were obtained from the measurements of Bradbury (1965) by least squares fitting of a function of the form:

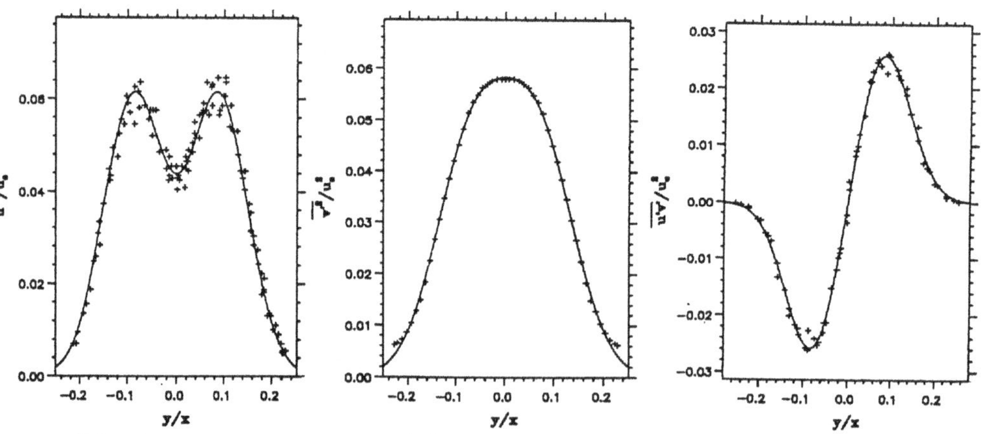

(a) $\overline{u'^2}/U_c^2$ vs. y/x (b) $\overline{v'^2}/U_c^2$ vs. y/x (c) $\overline{u'v'}/U_c^2$ vs. y/x

Figure 3. Turbulence profiles. (Data from Bradbury 1965.)

$$\left(\frac{\sigma}{U_c}\right)^2 = a_1\left[\exp\left\{a_2\left(\frac{z}{x}-a_3\right)^2\right\}+\exp\left\{a_2\left(\frac{z}{x}+a_3\right)^2\right\}\right]. \tag{6}$$

For σ_u we obtained a_1=0.0604, a_2=-131 and a_3=0.088, and 0.0445, -91.4 and 0.0680 respectively for σ_v. An anti-symmetric form of equation (6) was used for the cross correlation $\overline{u'v'}$, with a_1=0.0266, a_2=-136 and a_3=0.080; all three curves are plotted together with the data in figures 3*a-c*.

Estimates for the integral timescales T_{Lu} and T_{Lw} were obtained using the approximation $T_L \sim L/\sigma$ (Corrsin 1963), in which L was taken as the half width of the jet at the axial location of the particle.

4.4 *Results*

Particles were released from the same location as in the experiments, and their trajectories were calculated by integrating (1) using a simple Euler scheme, with a velocity field generated as described in §4.1-§4.3. As in the experiments, the particles were collected in 'trays', from which the concentration distribution was then calculated. Figure 4a shows the predicted and measured distributions for the 244 μm particles. The mean and variance of the predicted distribution agree very closely with those actually measured in the experiments. The spread in the distribution must be due to the interaction between the particles and the turbulence, so it seems that the RFM simulates this process quite adequately. Figure 4b shows the corresponding distributions for the 86 μm particles; the agreement between the two is very poor. In the RFM the particles are carried much farther by the jet, and the spread is much greater than that measured in the experiments.

A possible explanation for this is that the particles are being 'flung out' of the jet by large scale eddies. Chung and Troutt (1988) found this in their Discrete Vortex simulation of particle dispersion in an axisymmetric jet, when the particle response time τ_A was of the same order as the eddy turnover time $\mathcal{T}_e$. To estimate $\mathcal{T}_e$ we assume that a typical eddy radius is of the order of the jet half width, δ, and that a typical eddy velocity is about half the centreline velocity U_c, from which $\mathcal{T}_e \sim 4\pi\delta/U_c$. In the 86 μm

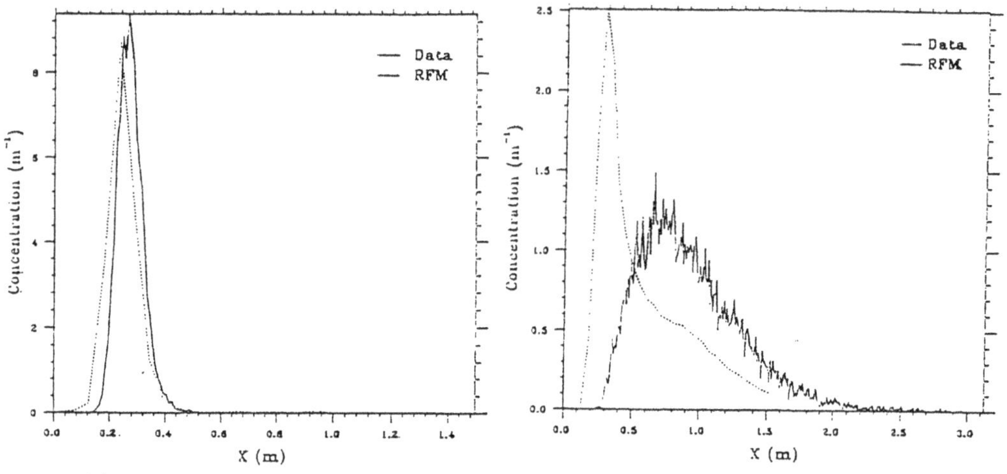

(a) 244 μm particles (b) 86 μm particles

Figure 4. Comparison of measured and predicted deposition distributions

experiments the particles were released at $x = 150\,\text{mm}$, and at this point $\mathcal{T}_e \sim 0.05\,\text{s}$. Since this is of the same order as τ_A for the $86\,\mu\text{m}$ particles, this sort of interaction could be occurring. (We do not expect it to occur in the $244\,\mu\text{m}$ experiments, since $\mathcal{T}_e \sim 0.03$ and $\tau_A \sim 0.5$.) The RFM does not model the spatial structure of the velocity field so it cannot reproduce this sort of interaction; this would therefore explain the difference between the predicted and measured distributions, and the difference between the comparisons for the $86\,\mu\text{m}$ and the $244\,\mu\text{m}$ particles.

5. Flow visualisation and particle tracking

It was possible to watch the motion of the $86\,\mu\text{m}$ particles using laser sheet illumination, and this seemed to show them being 'ejected' from the jet, on curved trajectories, as though they were indeed being 'flung out' by the eddies. We wanted to film this so that we could analyse individual particle trajectories using the particle tracking techniques we have developed (Perkins and Hunt 1989). Unfortunately, it was not possible to capture this on film, and we had to use air-filled soap bubbles produced by an Armfield F11-10 Helium bubble flow visualisation system. By adjusting the supply rates of the soap solution and the air we were able to create bubbles with a diameter of the order of $3\,\text{mm}$ a fall velocity of about $0.4\,\text{m/s}$ and τ_A of the order of $0.1\,\text{s}$.

The bubbles were discharged upwards from a nozzle, at the same location as the hopper, so that they fell under gravity and had reached terminal velocity when they were entrained into the jet. The flow was illuminated by a vertical light plane (thickness $\sim 20\,\text{mm}$) at the centre of the jet, and filmed from the side at 500 fps, using a Hadland Hyspeed-2 camera. The image plane was calibrated at the beginning of each film by placing a grid in the light plane, and stationary reference points around the edges of the image were used to position each frame during analysis.

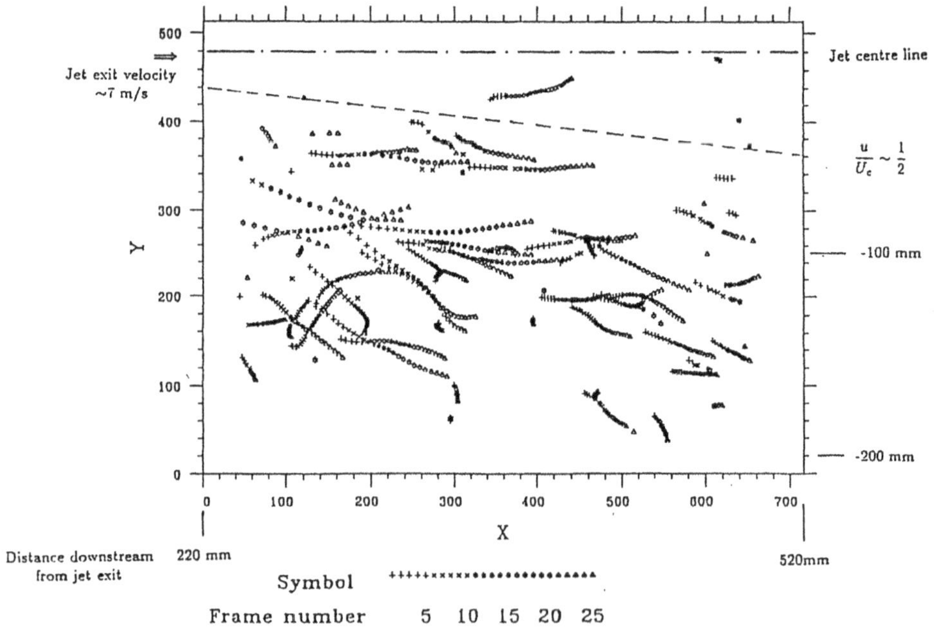

Figure 5. Soap bubble trajectories in a plane air jet

Figure 5 shows the results of processing 25 frames from that film; the field of view is somewhat below the jet centreline, between 220 mm and 520 mm downstream of the jet exit. Although τ_A is a bit large for interaction between the bubbles and the eddies there are some trajectories which show this abrupt ejection from the jet. We are now processing these films to obtain Lagrangian statistics and to extract the trajectories which show this sort of behaviour.

6. Acknowledgements

RJP was supported by the DTI (Warren Spring Laboratory) and the Wolfson Foundation; SG and JCP are funded by the AFRC. The experimental work was made possible by grants from Statoil and from the Hasselblad Foundation.

7. References

AUTON, T.R. 1984 The Dynamics of Bubbles, Drops and Particles in Motion in Liquids. Ph.D. thesis, University of Cambridge.

AUTON, T.R. 1987 The lift force on a spherical particle in a rotational flow. *J. Fluid Mech.* **183**, 199-218.

AUTON, T.R., HUNT, J.C.R. & PRUD'HOMME, M. 1988 The force exerted on a body in inviscid unsteady non-uniform rotational flow. *J. Fluid Mech.* **197**, 241-257.

BATCHELOR, G.K. 1953 *The Theory of Homogeneous Turbulence.* Cambridge University Press.

BRADBURY, L.J.S. 1965 The structure of a self-preserving turbulent plane jet. *J. Fluid Mech.* **23**, 31-64.

CHUNG, J.N. & TROUTT, T.R. 1988 Simulations of particle dispersion in an axisymmetric jet. *J. Fluid Mech.* **186**, 199-222.

CLIFT, R., GRACE, J.R. & WEBER, M.E. 1978 *Bubbles, Drops, and Particles.* Academic Press.

CORRSIN, S. 1963 Estimation of the relation between Eulerian and Lagrangian scales in large Reynolds number turbulence. *J. Atmos. Sci.* **20**, 115-119.

FUNG, J.C.H. & PERKINS, R.J. 1989 Particle trajectories in turbulent flow generated by time-varying random Fourier modes. In *Advances in Turbulence 2* (ed. H.-H. Fernholz & H.E. Fiedler), pp. 322-328. Springer-Verlag.

GOERTLER, H. 1942 Berechnung von Aufgaben der freien Turbulenz auf Grund eines neuen Naherungsansatzes. *ZAMM* **22**, 244-254.

HUNT, J.C.R., AUTON, T.R., SENE, K., THOMAS, N.H. & KOWE, R. 1988 Bubble motions in large eddies and turbulent flows. In *Transient Phenomena in Multiphase Flow* (ed. N.H. Afgan), p. 103. Hemisphere.

HUNT, J.C.R & NALPANIS, P. 1985 Saltating and suspended particles over flat and sloping surfaces. I. Modelling concepts. *Proceedings of* International Workshop on the Physics of Blown Sand, Aarhus, Denmark.

LEGG, B.J. 1983 Movement of plant pathogens in the crop canopy. *Phil. Trans. R. Soc. Lond. B* **302**, 559-574.

LEGG, B.J. & RAUPACH, M.R. 1982 Markov-chain simulation of particle dispersion in inhomogeneous flows : mean drift induced by a gradient in Eulerian velocity variances. *Bound.-Layer Met.* **24**, 3-13.

LUHAR, A.K. & BRITTER, R.E. 1989 A random walk model for dispersion in inhomogeneous turbulence in a convective boundary layer. *Atmos. Env.* **23**, No. 9, 1911-1924.

MORSI, S.A. & ALEXANDER, A.J. 1972 An investigation of particle trajectories in two-phase flow systems. *J. Fluid Mech.* **55**, 193-208.

PERKINS, R.J. & HUNT, J.C.R. 1986 Particle Transport in Turbulent Shear Flows. *Report to the SERC.*

PERKINS, R.J. & HUNT, J.C.R. 1989 Particle tracking in turbulent flows. In *Advances in Turbulence 2* (ed. H.-H. Fernholz & H.E. Fiedler), pp. 286-291. Springer-Verlag.

PERKINS, R.J. & HUNT, J.C.R. 1990 The Design of Air Curtains for Dust Control. *Report to Warren Spring Laboratory.*

RAJARATNAM, N. 1976 *Turbulent Jets.* Elsevier.

RUBINOW, S.I. & KELLER, J.B. 1961 The transverse force on a spinning sphere moving in a viscous fluid. *J. Fluid Mech.* **11**, 447-459.

SAFFMAN, P.G. 1965 The lift force on a sphere in a slow shear flow. *J. Fluid Mech.* **22**, 385-400.

SNYDER, W.H. & LUMLEY, J.L. 1971 Some measurements of particle velocity autocorrelation functions in a turbulent flow. *J. Fluid Mech.* **48**, 41-71.

THOMSON, D.J. 1986 A random walk model of dispersion in turbulent flows and its application to dispersion in a valley. *Q. Jl. R. met. Soc.* **112**, 511-530.

VAN DOP, H., NIEUWSTADT, F.T.M. & HUNT, J.C.R. 1985 Random walk models for particle displacements in inhomogeneous unsteady turbulent flows. *Phys. Fluids* **28**, No. 6, 1639-1653.

WHITE, B.R. 1982 Two-phase measurements of saltating boundary layer flow. *Int. J. Multiphase Flow* **8**, No. 5, 459-473.

WHITE, B.R. & SCHULZ, J.C. 1977 Magnus effect in saltation. *J. Fluid Mech.* **81**, 497-512.

YULE, A.J. 1980 Investigations of eddy coherence in jet flows. In *The Role of Coherent Structures in Modelling Turbulence and Mixing* (ed. J. Jimenez), pp. 188-207. Springer.

Large Eddy Simulation of the Interaction Between Two Distinct Turbulent Velocity Scales

L. Shao, J.P. Bertoglio and M. Michard

Laboratoire de Mécanique des Fluides et d'Acoustique, U.R.A. C.N.R.S. 263 Ecole Centrale de Lyon 36, av. Guy de Collongue, 69130 Ecully, France

A Large Eddy Simulation of a shearless turbulence mixing layer is proposed with a view to understand the transport mechanisms in a simple situation of inhomogeneous turbulence. The numerical technique is shortly presented. Results are found to be in qualitative agreement with experimental data. Special attention is paid to transport terms by triple velocity correlations and by pressure velocity correlations, in order to permit comparisons with existing models.

1. Introduction

The purpose of the present work is to use Large Eddy Simulation to study the properties of a turbulence mixing layer without mean shear, i.e. to study the interaction of two distinct turbulences, situated side by side, and having different energy and length scales. As recently pointed out by Veeravalli and Warhaft (1989) (hereinafter referred to as V.W.), the shearless turbulence mixing layer is a subject which has received little attention in the past. It is however a case which is, in some respect, more fundamental than the traditional turbulent mixing layer since the interaction between two different turbulent fields can be studied independently of instability mechanisms. Since the work of Gilbert (1980) and the recent paper of V.W. (1989), experimental data are providing new insights into the complex interaction between diffusion, pressure transport and dissipation, and in particular on the transport mechanism due to inhomogeneity which is still not well understood.

Experimentally, the shearless turbulence mixing layer is formed in two decaying grid turbulences with a larger mesh spacing on one side of the grid than on the other (see V.W. 1989), as appears on the sketch in figure 1. The mean flow is nearly uniform in the experiments, hence no significant effect of shear is present. In the sketch in figure 1, a grid with square mesh is represented for simplicity, in fact in the V.W. experiment two different grids were used; one was formed of parallel bars, the other was a perforated plate with circular holes.

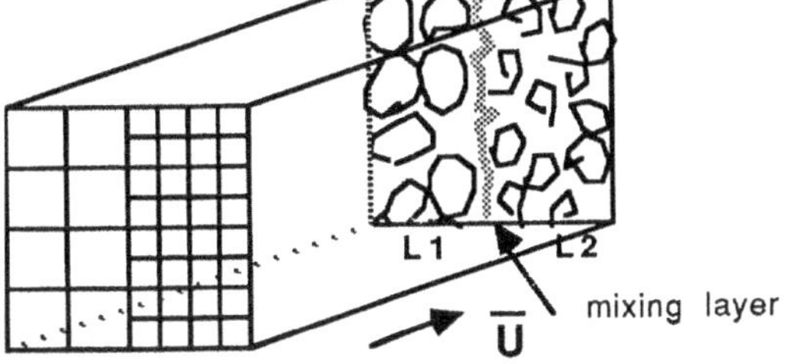

Fig. 1

Advances in Turbulence 3
Editors: A.V. Johansson · P.H. Alfredsson

The shearless turbulence mixing layer can be considered as a test case particularly convenient to check the validity of closure models for inhomogeneous turbulence, since it is the simplest situation in which inhomogeneity plays a significant role. It is for example much simpler than channel flow, in which wall effects and mean shear can strongly interact with the transport mechanism. It is also simpler than the case of the usual mixing layer where the presence of mean velocity gradients plays an important role. Consequently, it is a convenient flow to test the validity of the modelling of transport by triple correlations (and by pressure velocity correlations) in Reynolds stress models or k-e models. One can also consider that the shearless mixing layer is one of the few inhomogeneous flows that could be predicted in the framework of two-point closures, like the Direct Interaction Approximation or an extented Eddy Damped Quasi-Normal Markovian model, since the degree of symmetry of the statistical field remains high (axisymmetry). Furthermore Lagrangian types of predicting methods have already been applied to this flow (Pope and Haworth, 1987).

In order to provide comparisons with experiments, and to study in more detail effects like those of pressure that are difficult to measure, a Large Eddy Simulation of the shearless turbulence mixing layer is presented in the present paper. In section 2, the assumptions introduced in the simulation are discussed (use of periodic boundary conditions) and the method of generation of the initial velocity field is described. Different cases corresponding to different ratios of length scales are investigated. To study the behaviour of third-order and fourth-order correlations, the results are averaged over several realizations. One-dimensional spectra and probability densities are also evaluated, as well as pressure velocity correlations. Results are presented in section 3. In section 4, the problem of the influence of the subgrid modelling is discussed. In the conclusion, some results of comparisons with one-point closure models are shortly presented.

2. Large Eddy Simulation of the shearless mixing layer

In order to use a pseudo-spectral Fourier code, periodicity of the field is assumed, and computations are made on a numerical cube (see sketch in figure 2). Due to the periodic boundary conditions, two mixing layers are actually simulated instead of one: one in the middle of the box, and an other one at the edges of the computational domain. The numerical code used is a code written for homogeneous turbulence simulations (Bertoglio, 1985) which is well adapted to the use of different subgrid models. The inhomogeneity of the turbulent field is only due to the inhomogeneity of the initial conditions. The numerical problem is therefore very simple, the difficulty being the generation of initial conditions that must satisfy the incompressibility condition.

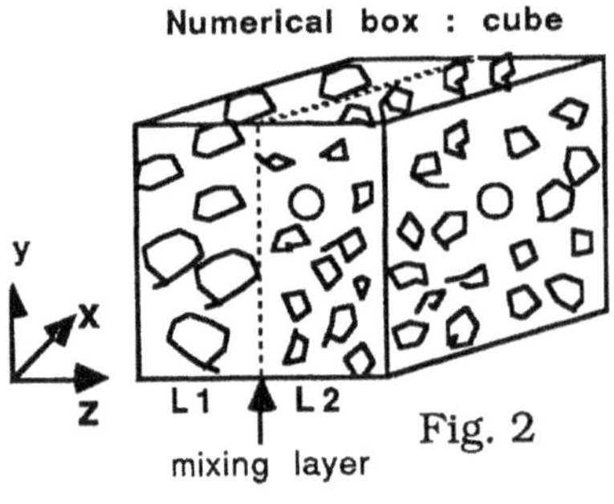

Fig. 2

To generate the initial field, a simple method is retained. Two homogeneous fields are first created in two separated numerical boxes corresponding to the whole domain. They are generated using a classical random Fourier mode technique (Rogallo, 1981), which allows specification of the spectra of the two fields, and consequently permits chosing the length scales, as well as the energy levels. The two fields are thereafter truncated by putting to zero the velocity components in one half of each domain. The two resulting fields are then put side by side. The incompressibility condition is satisfied by eliminating the velocity components parallel to the wave-vector $\mathbf{K}$, in Fourier space: the velocity fluctuation $u_i(\mathbf{K},0)$ is multiplied by the solenoidal projector: $P_{ij}(\mathbf{K}) = \delta_{ij} - K_i K_j / K^2$.

The method was tested on Taylor vortices as well as on random fields, in two-dimensional and in three-dimensional situations. As a result, the influence of the application of the solenoidal projector appears to be localized in space; it induces a perturbation on the velocity field whose size is typically the length scale of a turbulent eddy. In the case of Taylor's vortices, the size of the perturbation clearly appears to be the size of the vortices. In the case of a random field, if average is made over a sufficient number of realizations, results show that reasonably flat profiles of turbulent kinetic energy are obtained.

Before presenting the results of the present simulations, it is necessary to discuss the influence of the periodicity of the boundary conditions that was imposed in the computations. Two consequences of the periodicity have been identified. To discuss the first effect, it is necessary to write the equation for the mean velocity in the inhomogeneity direction (Z direction). In this equation

$$\frac{\partial < W >}{\partial t} + \ldots = -\frac{1}{\rho}\frac{\partial < p >}{\partial Z} - \frac{\partial < w^2 >}{\partial Z}$$

the presence of the last term shows that the inhomogeneity of turbulence tends to create a mean velocity parallel to Z which, in turn, could affect the turbulent field by advection. In the periodic case, due to the continuity equation, the mean velocity is not generated, and the effect of $\partial < w^2 > / \partial Z$ is counterbalanced by the presence of a mean pressure gradient. In the V.W. experiment, it is not clear whether the presence of the wind tunnel walls is sufficient to generate a mean pressure gradient, or whether a small transverse component of mean velocity does exist. One can for example imagine that the continuity equation can be satisfied through the presence of secondary flows at the walls, or by a slight modification of the longitudinal mean velocity. One must therefore be careful when comparing the present simulations to the experimental results.

The second effect introduced by periodicity is that two mixing layers are simulated instead of one. For large values of time, an interaction between the two mixing layers can therefore develop. The criterion adopted to detect this interaction is based on the flatness factor of the velocity fluctuation. In each mixing layer, a bump in the flatness factor is detected. When time increases, the positions of the bumps tend to shift toward the small scale region of the flow. When the two bumps start to merge, it is considered that the two layers begin to interact and the computations have to be stopped. For most of the results presented here, computations are run up to values of time slightly larger than the limits deduced from the flatness factor criterion.

Due to the symmetries of the problem, the equations for the averaged turbulent quantities are drastically simplified, and the important terms clearly appear. The equations for the double velocity correlations at one-point, or Reynolds stress tensor, are:

$$\frac{D < w^2 >}{Dt} = -\frac{\partial < w^3 >}{\partial Z} - \frac{2}{\rho}\frac{\partial < pw >}{\partial Z} + \frac{2}{\rho} < p\frac{\partial w}{\partial Z} > + \text{ Viscous term}$$

$$\frac{D < u^2 >}{Dt} = -\frac{\partial < wu^2 >}{\partial Z} - \quad 0 \quad + \frac{2}{\rho} < p\frac{\partial u}{\partial X} > + \text{ Viscous term}$$

The first two terms of the right hand sides are respectively transport terms by triple correlations and by pressure velocity correlations. The third terms are the pressure redistribution terms, whose effect is the interchange of energy between components of the velocity fluctuation and which do not affect energy repartition in space, since the trace of the tensor $< p\partial u_i/\partial X_j >$ is zero. The viscous terms essentially account for molecular dissipation effects. Viscous effects also contribute to turbulent transport, but their contribution is not discussed here since it vanishes for high Reynolds number flows.

From symmetry considerations, $< v^2 >$ is equal to $< u^2 >$ and the transport term in the $< v^2 >$ equation, $\partial < wv^2 > /\partial Z$ is equal to the transport term in the $< u^2 >$ equation, since $< wu^2 >=< wv^2 >$. Extra-diagonal components of the double velocity correlation tensor must also be equal to zero due to symmetries, as well as the triple correlations $< v^3 >$ and $< u^3 >$. Consequently, the important statistical quantities for turbulent transport are: $< w^3 >, < wv^2 >=< wu^2 >$, and $< pw >$.

In the present computations, only the large scales of the fluctuating field are resolved. The small scale effects are parameterized. Two different subgrid models were used: the spectral eddy viscosity model proposed by Chollet and Lesieur (1981), and the turbulent viscosity introduced by Smagorinsky (1963). In the case of the Smagorinsky model, the viscosity is formulated in physical space with $C_{smag} = 0.25$. The influence of the subgrid parameterization is discussed in section 4. The Chollet-Lesieur model is used without the "cusp", and the value of the constant is 0.267.

3. Results

Different runs were made corresponding to different initial conditions. They are summarized in table 1, in which subscripts 1 and 2 indicate quantities on the large scale side and the small scale side of the box respectively. Results are averaged over (X, Y) planes, for each value of the coordinate Z in the inhomogeneity direction. It needs to be pointed out that, in the present simulation, it is not possible to average the results over the three directions, or to introduce a time average as it is the case for inhomogeneous fields which are statistically stationnary. Consequently there is less sampling in the present simulation than is usually the case. This is the reason why computations were performed on independent realizations, which differed one from the other by the random phases of the initial Fourier modes.

Table 1 : characteristics of the different runs and initial conditions

Test case	Number of grid points	Nb. of realiza-tions	$<q_1^2>$ $(cm/s)^2$	$<q_2^2>$ $(cm/s)^2$	L_1 (cm)	L_2 (cm)	$\frac{<q_1^2>}{<q_2^2>}$	$\frac{L_1}{L_2}$	$R_{e_{L_1}} = \frac{(q^2/3)^{1/2} L_1}{\nu}$
32-01	32^3	100	400	200	3.23	0.517	2	6.32	~248.6
64-01	64^3	1	~1200	~600	0.37	0.112	2	3	∞
64-02	64^3	3+9	~2000	~1000	0.71	0.33	2	2	∞
64-03	64^3	1	~1800	~1000	0.91	0.45	1.8	2	∞

The first test case (32-01) was run on a 32^3 grid, and up to one hundred realizations with independent initial conditions were computed in order to perform statistical averages. The three other test cases were run on 64^3 grids. Only one realization was run in test cases 64-01 and 64-03. For test case 64-02, three realizations were performed with the Chollet Lesieur (1981) spectral subgrid model, and nine realizations were run using Smagorinsky's model. The other test cases were run using only the Chollet Lesieur model. Due to the fact that a smaller number of runs was used in the case of 64^3 computations, smaller integral length scales were introduced than in the case of the 32^3 runs where a better statitical sampling was available.

The ratio of the energies contained in both sides of the computational box is approximatively 2 in all cases. The ratio of the length scales varies between 2 and 6, covering the range of ratios corresponding to the experiments of Gilbert (1980) and of V.W.(1989). The shapes and characteristics of the spectra introduced as initial conditions are given in table 2. In the case of 32^3 computations a top hat spectrum was used, whereas in the case of 64^3 runs, spectra with a $K^{-5/3}$ range were adopted in order to facilitate comparisons with one-point closures. The computational domain is a cubic box of size $L = 32$ cm. No mean flow is present in the computation.

Table 2 : spectral characterization of initial conditions

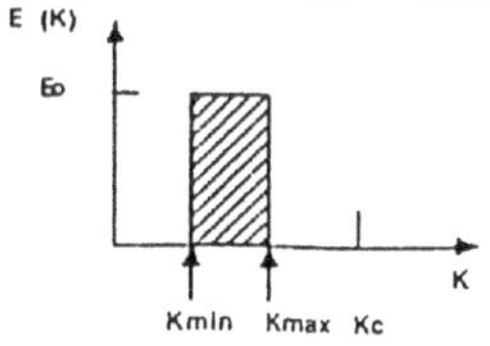

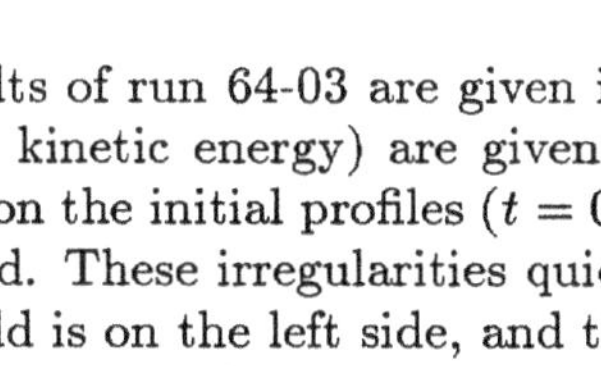

Test case		K_c (cm^{-1})	K_{min} (cm^{-1})	K_{max} (cm^{-1})	E_0 (cm^3 s^{-2})
Test case 32-01	Large scales	2.81	0.2	0.6	~500
	Small scales	2.81	1.9	2.7	~125
Test case 64-01	Large scales	5.9	1.1	1.9	~250
	Small scales	5.9	4.1	4.9	~200
Test case 64-02	Large scales	5.9	0.85	1.65	500
	Small scales	5.9	1.94	2.74	250
Test case 64-03	Large scales	5.9	0.3	1.1	500
	Small scales	5.9	0.85	1.65	250

The results of run 64-03 are given in figure 3, where the profiles of $< q^2 >$ (twice the turbulent kinetic energy) are given for different values of time. The presence of irregularities on the initial profiles ($t = 0$) is an artefact of the method used to generate the initial field. These irregularities quickly disappear as time evolves. In figure 3, the large scale field is on the left side, and the small scales are on the right. The turbulent mixing layer is situated in the middle of the figure (near $Z = 16.5$ cm). Due to the periodic conditions, another mixing layer is situated at the edges of the numerical box ($Z = 32.5$ cm or $Z = 0.5$ cm). Smaller integral length scales having been introduced as initial conditions for test cases 64-01 and 64-02, a lower effect of diffusion is found in the $< q^2 >$ profiles than in case 64-03.

In figure 4, the time evolution of the turbulent kinetic energy is given at four different values of the Z coordinate, respectively corresponding to the middle of the

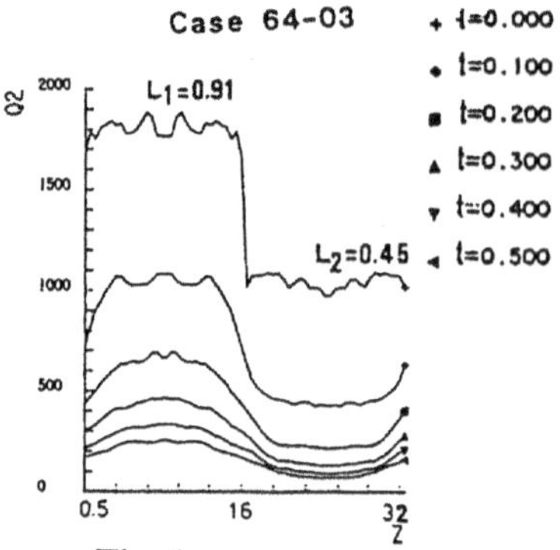

Fig. 3

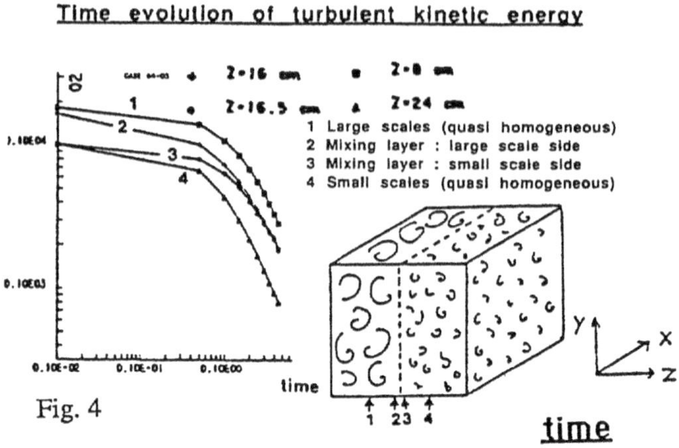

Fig. 4

EVOLUTION Q2

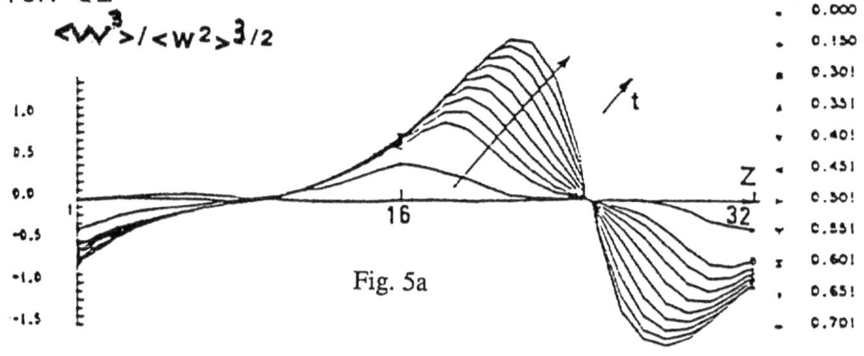

Fig. 5a

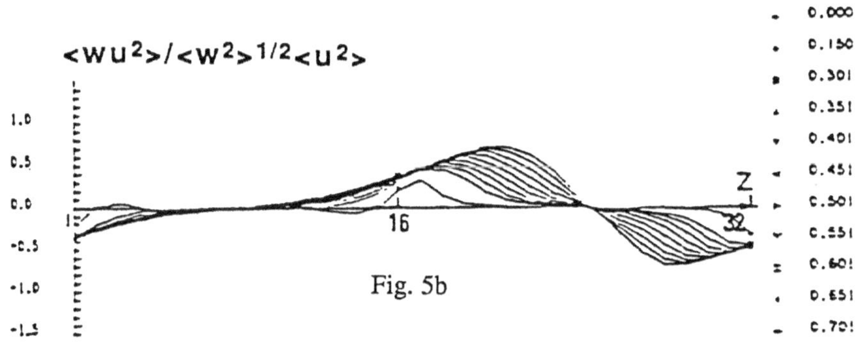

Fig. 5b

large scale region, to a position in the mixing layer close to the edge on the large scale side ($Z = 16$ cm), to a position in the mixing layer in the small scale region ($Z = 16.5$ cm), and to the middle of the small scale field. In the two nearly homogeneous parts of the flow ($Z = 8$ cm and $Z = 24$ cm) usual decay laws are obtained, whereas near the edge of the mixing layer a faster decay is obtained on the large scale side ($Z = 16$ cm) and a slower decay on the small scale side ($Z = 16.5$ cm) due to the exchange of energy across the layer.

Profiles of non-dimensional triple correlations, or skewness factors, are given in figure 5 where $< w^3 > / < w^2 >^{3/2}$ and $< wu^2 > /(< w^2 >^{1/2} < u^2 >)$ are plotted as functions of the coordinate in the inhomogeneity direction. Results are given for test case

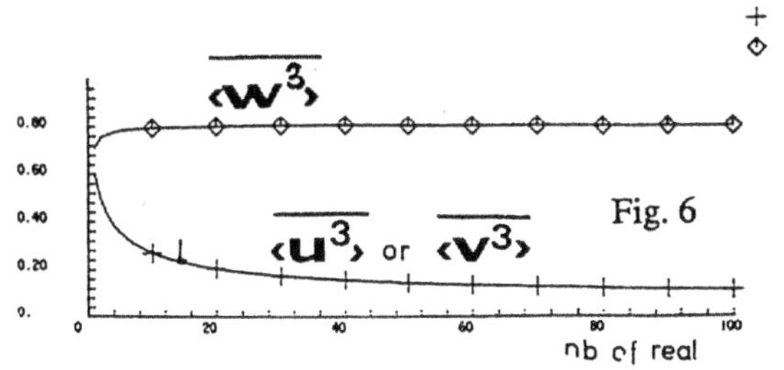

Fig. 6

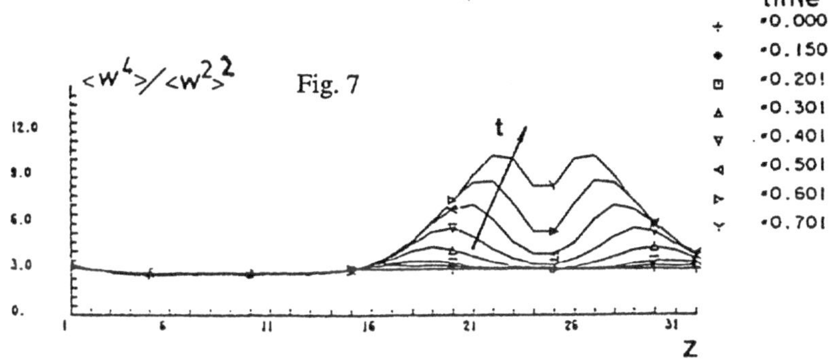

Fig. 7

32-01, in which the best sampling is available. As could be expected from symmetry considerations, $< wu^2 >$ is found to be equal to $< wv^2 >$. Non dimensional triple correlations are found to increase with time, and the position of the extremum values is found to move towards the small scale side which is consistent with the experimental trends. The correlation $< w^3 >$ is found to be larger than $< wv^2 >$ or $< wu^2 >$.

Some comments should be made concerning triple correlations in the directions perpendicular to the inhomogeneity direction: $< u^3 >$ and $< v^3 >$. As stated above, these quantities should identically vanish due to symmetry considerations. However when results of L.E.S. were averaged over a small number of realizations, they were found to reach values of about 0.3. This was rather puzzling result since the same departure from zero was also found in the experiments (see Veeravalli and Warhaft, 1987). The behaviour of $< u^3 >$ and $< v^3 >$ is now well understood. The departure from zero of these quantities in the simulation is only due to the lack of statistical sampling. This clearly appears in figure 6, where the quantity:

$$\overline{< u^3 >} = \frac{1}{N}\sum^{N} \int |< u^3 > (Z)| \, dZ$$

is plotted as a function of the number of realizations N. The two quantities $< u^3 >$ and $< v^3 >$ tend to zero as a power of N which is close to $-1/2$. The reason for the non zero values obtained in the experiment is also well understood now and the explanation is given in V.W. 1989 ; it is attributed to the existence of small gradients of mean velocity.

The flatness factor (figure 7) indicates strong departure from gaussianity of the w component. As stated above, for large values of time ($t \approx 0.5$ s for the 32-01 test case) the two bumps in the flatness factor merge, indicating that the interactions between the two simulated mixing layers begins to occur.

The probability density function of w is found not to be symmetrical. The histogram of w at $Z = 16.5$ cm, i.e. on the small scale side of the field and very close to

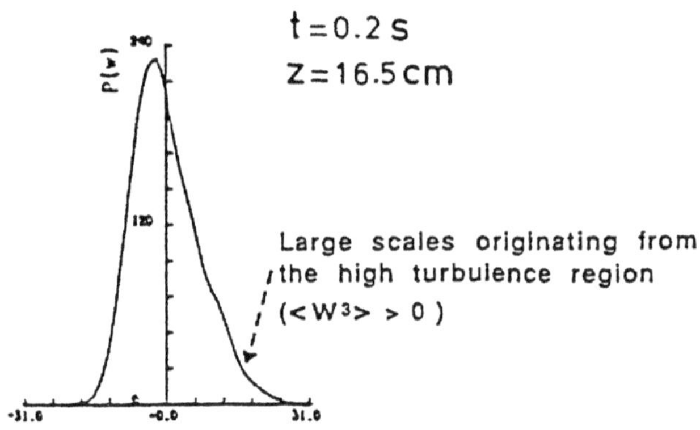

Fig. 8 : Probability of w in the mixing layer (case 64-02)

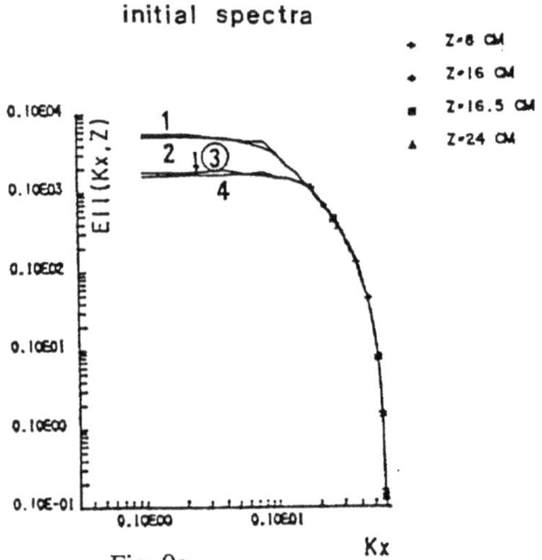

Fig. 9a

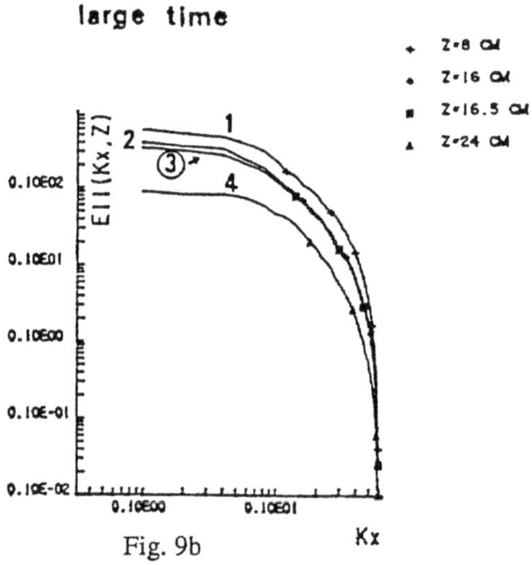

Fig. 9b

the centre of the mixing layer, indicates that large positive values of w are obtained, corresponding to large scales originating from the high turbulent region (figure 8).

One dimensional spectra are given in figure 9. The longitudinal spectrum $E_{11}(K_x)$ (i.e. the spectrum of the u component in the X direction) is plotted for different values of Z ($Z = 8$ cm: nearly homogeneous large scale field; $Z = 16$ cm: mixing layer on the large scale side; $Z = 16.5$ cm: mixing layer on the small scale side ; and $Z = 24$ cm: nearly homogeneous small scale field). The increase of the spectrum at $Z = 16.5$ cm clearly indicates transport processes. The same behaviour is found for the transverse spectrum $E_{33}(K_x)$ (spectrum of the w component). Figure 10 shows the plots of $K_x E_{33}(K_x)/ < w^2 >$ and $K_x E_{11}(K_x)/ < u^2 >$ for different values of Z, and can be compared to figure 18 of the paper of V.W. (1989). The results are obtained by averaging over 9 realizations for case 64-02. The same behaviours are found: on the one hand, for the longitudinal spectra, the peak is found to regularly shifts into larger wave numbers when proceeding from the large scale turbulent region to the small scale

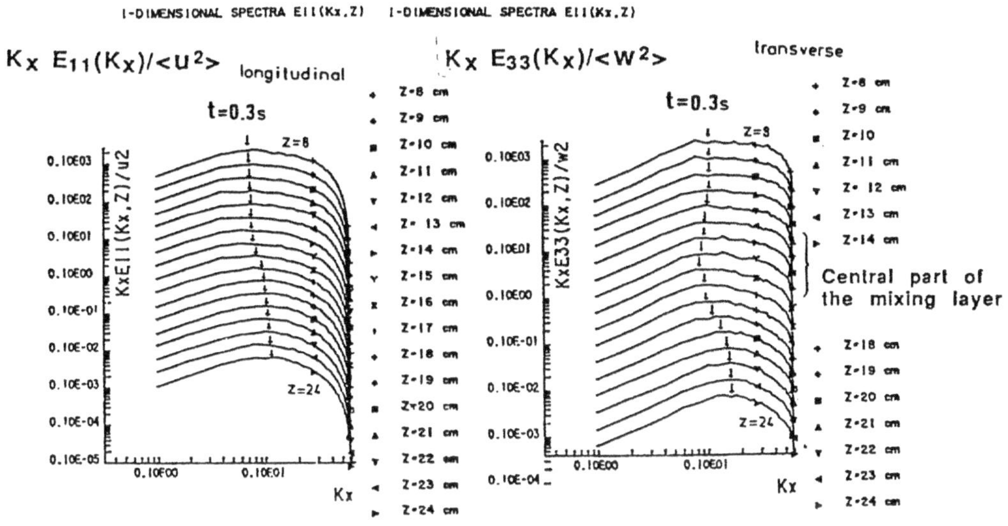

Fig. 10a ↓ extrapolated positions of the maxima Fig. 10b

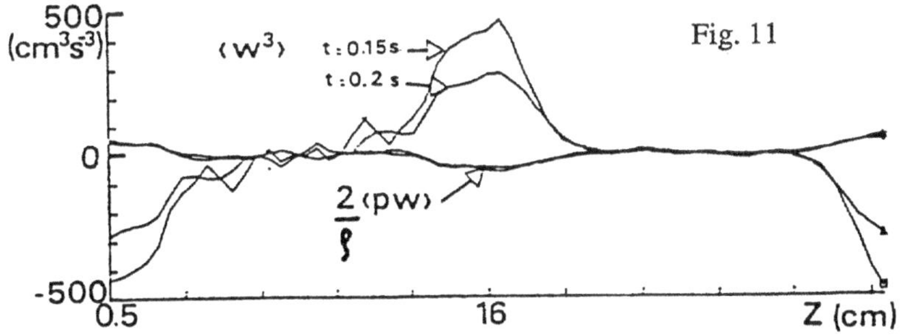

Fig. 11

Transport by pressure velocity correlation compared with transport by triple correlations

region; on the other hand, the peak of $K_x E_{33}(K_x)/ < w^2 >$, for the transverse spectra, is seen first to shift slightly towards lower wave numbers before increasing to the value attained in the central region of the small scale field.

Pressure-velocity correlations were also evaluated. In figure 11, the term responsible for the transport of $< w^2 >$ by the pressure effect is compared to the contribution due to the triple correlations. It is found that the pressure velocity correlations tend to counterbalance triple correlations transport, but their quantitative influence remains small.

4. Influence of the subgrid model

Two different models were used in the computation. The Chollet-Lesieur (1981) spectral model, and the Smagorinsky (1963) model (in physical space), both based on the

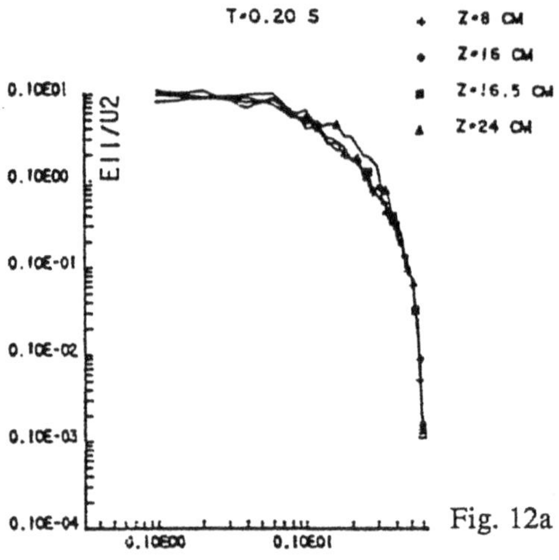

Fig. 12a

Normalized 1D Spectra obtained with Chollet-Lesieur Subgrid model (1981)

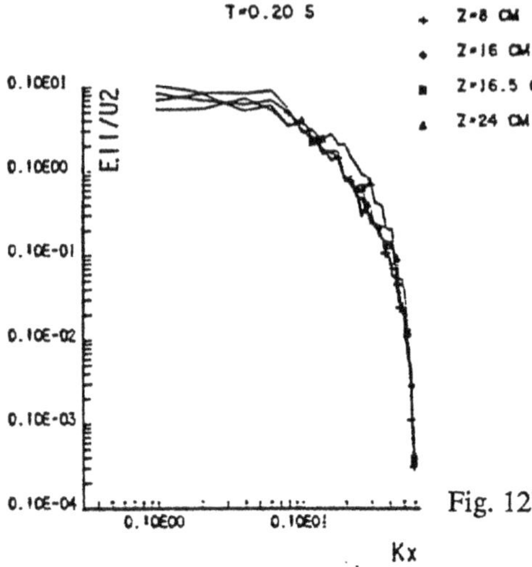

Fig. 12b

Normalized 1D Spectra obtained with Smagorinsky Subgrid model (same realization)

introduction of the eddy viscosity concept. In the case of Chollet-Lesieur model, the viscosity is specified in spectral space: $\nu_t(K,t) = C(E(K_c)/K_c)^{1/2}$, in which Kc is the cutoff wave-number and $E(K_c)$ is the value of the kinetic energy spectrum at the cutoff. Such a model is typically local in **K**-space, but not in physical space. As for the Smagorinsky model, it is local in physical space, but not in **K**-space. In the case of inhomogeneous turbulence, it can be suspected that the spectral subgrid models, like the Chollet-Lesieur model, will not be appropriate since the viscosity in each region of the flow, will depend on the energy contained in the small scales present in the entire domain of computation. One can therefore infer that the spectral eddy viscosity model will somehow be misled by inhomogeneities in the small scale field. The problem is of pratical importance and the shearless mixing layer is a good test case to address this point.

Figure 12 shows the comparison between results obtained using the model of Chollet-Lesieur (1981), and the model of Smagorinsky (1963) in the case of test case 64-02 with exactly the same initial conditions. There are slight differences between the spectra obtained with the two subgrid models. The model recently proposed by Métais and Lesieur (1990), based on a local eddy viscosity defined by introducing the second order velocity structure function, will be implemented in the present computation in a near future.

5. Conclusion

Most of the results found in the present Large Eddy Simulation study of the shearless mixing layer are in agreement with the experimental results of V.W. (1989). Some data, not directly available in the experiment, were also evaluated, like pressure velocity correlations. Comparisons between the results of the present Large Eddy Simulation and the results obtained with one-point closure models will be presented in an other paper (Shao, Le Penven and Bertoglio, 1990). The main conclusion of the comparison with $k - \varepsilon$ and Reynolds stress models is that triple velocity correlations are found to be underestimated by the diffusion models when the classical values for the constants are retained. Concerning the modelling of the pressure transport term, the model of Lumley (1978) is found to give the appropriate tendancy, but the constant usually employed ($C = 1/5$) appears to be overestimated.

Directions for future works include the study of length scale transport, more sampling being necessary before integral scale profiles could be evaluated. The effect of the subgrid scale parameterization will also be studied in more details. Comparisons with the simplified spectral closure proposed by Bertoglio and Jeandel (1987) could also be proposed, as well as comparisons with the results obtained by Pope and Haworth (1987) using a Lagrangian model. In order to study the influence of an external force on turbulent transport, the case of a shearless mixing layer submitted to solid body rotation will be investigated. Finally, the study could be extended to the case of passive scalar fields. The case of particle dispersion is also of interest.

Acknowledgements

This work was supported by D.R.E.T., Contract no. 88351

References

Bertoglio J.P., 1985, A Stochastic Subgrid Model for Sheared Turbulence, *Lecture Notes in Phys., Macrosc. Mod. of Turbulent Flows*, Sophia-Antipolis, Dec. 1984, Springer-Verlag.

Bertoglio J.P. and Jeandel D., 1987, A Simplified Spectral Closure for Inhomogeneous Turbulence: Application to the Boundary Layer, *Fifth Turb. Shear Flows*, Springer-Verlag.

Chollet J.P. and Lesieur M., 1981, Parametrization of Small Scales of Three-dimensional Isotropic Turbulence Utilizing Spectral Closures, *J. Atm. Sci.* **38**, pp.2747-2757.

Gilbert B., 1980, Diffusion Mixing Layer in Grid Turbulence without Mean Shear, *J. Fluid Mech.* **100**, pp.349-365.

Lumley J.L., 1978, Computational Modeling of Turbulent Flow, *Adv. Appl. Mech.* **18**, pp.123-176.

Métais O. and Lesieur M., 1990, Spectral Large-Eddy Simulation of Isotropic and Stably-Stratified Turbulence, Submitted to *J. Fluid Mech.*

Pope S.B. and Haworth D.C., 1987, The Mixing Layer between Turbulent Fields of Different Scales, in *Turb. Shear Flows 5*, pp.44-53, Springer-Verlag.

Rogallo R.S., 1981, Numerical Experiments in Homogeneous Turbulence, *Nasa Techn. Mem.* no. 81315, Sept.81.

Shao L., Le Penven L. andBertoglio J.P., 1990, Study of the Modelling of the Transport Terms in One-point Closures using Large Eddy Simulation of Turbulence, *Int. Symp. on Eng. Turbulence Modelling and Measurements*, Dubrovnik, Yu., Sept. 90.

Smagorinsky J., 1963, General Circulation Experiments with the Primitive Equations, *Monthly Weather Review* **91**.

Veeravalli S. and Warhaft Z., 1987, The Interaction of Two Distinct Turbulence Velocity Scales in the Absence of Mean Shear, ***Fifth Turb. Shear Flow Symp.***, Springer-Verlag.

Veeravalli S. and Warhaft Z., 1989, The Shearless Turbulence Mixing Layer, *J. Fluid Mech.* **207**, pp.191- 229.

Coherent Structures in Favourable and Adverse Pressure Gradient Boundary Layers

F. Unger and R. Friedrich

Lehrstuhl für Strömungsmechanik
Technische Universität München
8000 München 2, Federal Republic of Germany

Statistically two-dimensional and steady, high Reynolds number, flat-plate boundary layers with a sudden change in the freestream longitudinal pressure gradient are simulated. Three different flow situations are considered, corresponding to zero-, negative and positive pressure gradients, respectively. In all three cases the same inflow boundary conditions are used, taken from a separate LES of a zero-pressure gradient boundary layer. Thus, a direct comparison of the effects of acceleration and deceleration on the turbulence structure is possible. This comparison is made on the basis of instantaneous velocity contour lines, contour surfaces and the time development of velocity correlations at a fixed position. The instantaneous vorticity field is analysed in terms of vector plots, vorticity lines and inclination angles of the vorticity vectors.

1. Introduction

The existence of organized, so-called coherent structures in turbulent shear layers is generally accepted in the turbulence community. However, their precise definition is still a point of discussion. In view of the variety of such structures and their complexity, the search for an exact definition on the other hand, may be as fatal as the earlier assumption that randomness is the characterizing feature of turbulence. This idea was responsible for the late discovery of coherent structures. Hussain (1986) has given a useful description of such structures as open volumes with distinctive large-scale vorticity distributions. Fiedler (1987) defines them as being recurrent, appearing at best quasi-periodically and having a characteristic life-span. They are comparable in size with the global dimension of the flow field, are dynamically significant and therefore amenable to prediction methods such as the large eddy simulation (LES) technique.

2. Mathematical model and numerical solution

The Navier-Stokes equations for an incompressible fluid are low-pass filtered via integration over the mesh volume of a Cartesian, equidistant staggered grid leading to a second-order accurate finite-difference scheme in space which conserves mass exactly, Schumann (1975):

$$\delta_j\, {}^j\overline{u_j} = 0, \tag{1}$$

$$\delta_t\, {}^i\overline{u_i} = -\delta_j({}^j\overline{u_i}\,{}^j\overline{u_j} + {}^j\overline{u_i' u_j'}) - \delta_i\, {}^v\bar{p}/\rho + \nu\delta_j\, {}^j\overline{D_{ij}}. \tag{2}$$

Equations (1),(2) contain grid-surface and grid-volume averaged flow variables, viz:

Advances in Turbulence 3
Editors: A.V. Johansson · P.H. Alfredsson

$$^j\overline{\phi_i} = \frac{1}{\Delta S_j} \int_{\Delta S_j} \phi_i dS, \qquad ^v\overline{\phi} = \frac{1}{\Delta V} \int_{\Delta V} \phi dV. \tag{3a, b}$$

δ_j stands for the central finite difference operator

$$\delta_j \phi = \frac{1}{\Delta x_j}(\phi(x_j + \frac{\Delta x_j}{2}) - \phi(x_j - \frac{\Delta x_j}{2})) \tag{4}$$

and D_{ij} describes the deformation tensor

$$D_{ij} = \delta_j u_i + \delta_i u_j. \tag{5}$$

The subgrid scale (SGS) stress is related to the filtered deformation tensor following Schumann's (1975) suggestion:

$$^j\overline{u_i' u_j'} = -\mu_{iso}(^j\overline{D_{ij}} - \langle ^j\overline{D_{ij}} \rangle) - \mu_{inh} \langle ^j\overline{D_{ij}} \rangle + \frac{1}{3}\delta_{ij} \, ^k\overline{u_k' u_k'}. \tag{6}$$

The right-hand side consists of a fluctuating part (in Reynolds' sense) and a statistical part which accounts for inhomogeneities in the SGS turbulence. The eddy viscosities μ_{iso}, μ_{inh} contain length scales which are related to the grid spacing (in the case of μ_{iso}) or are empirical. The corresponding velocity scales are computed from the fluctuating and from the statistically averaged deformation tensor. For more details see Schmitt (1988).

We use an explicit scheme for *integration* of (2) *in time*. It is a variant of Chorin's (1968) projection method. The integration proceeds in cycles of 49 leapfrog-steps and one averaging step in order to avoid $2\Delta t$-oscillations which may result from the leapfrog scheme. The initial integration step is an Euler step. The use of the projection method leads to a Poisson equation for the pressure which is evaluated with a fast elliptic solver. It uses a fast fourier transformation in the homogeneous direction in which there is no mean flow and the cyclic reduction technique, Schumann & Sweet (1976), in a vectorized version. The time step Δt is chosen such that the Courant number is well below one.

3. Boundary conditions

Crucial points in the LES of spatially developing flows are the specification of boundary conditions. This applies especially to *inflow planes* where the instantaneous velocity vector and the sub-grid-scale (SGS) energy have to be specified at each time step in the simulation. As described in the next section we take these data from a corresponding simulation of a boundary layer flow with weak streamwise development. This is the zero-pressure gradient boundary layer CASE 0, shown in figure 1.

A special treatment of the velocity vector in the *outflow plane* has proved successful. It uses linear extrapolation of the mean streamwise component, constant extrapolation of the remaining two velocity components and a convection equation for the velocity fluctuations, with the time-averaged longitudinal velocity as the proper transport velocity. The conditions are discussed in more detail in Friedrich & Unger (1990).

Periodic boundary conditions are used in the *lateral direction*.

It follows from the filtered equations (1,2) that *wall boundary conditions* can be specified for the normal velocity and the shear stress components. According to Schumann (1975) the wall shear stress is in phase with the tangential velocity component.

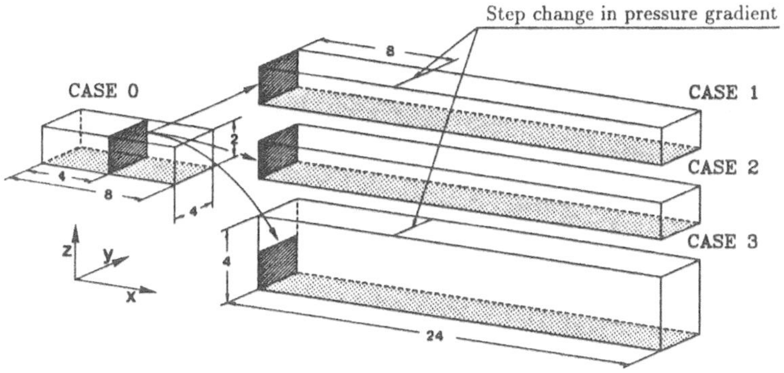

Figure 1. Geometry of the simulated flow cases.

In the *upper plane* of the computational domain (z=2(4)) the longitudinal velocity component is constant in the range $0 \le x \le 8$ and varies downstream of x=8 according to the pressure gradient. The spanwise velocity component is set to zero whereas the vertical velocity follows from continuity.

4. Specification of the simulated flow cases

An overview of the simualated flow cases is given in figure 1. CASES 1,2 and 3 correspond to the three flow situations to be compared each with a different pressure gradient downstream of x=8. In CASE 0 a zero-pressure gradient boundary layer is simulated in order to produce proper and equal inflow boundary conditions for CASES 1,2 and 3.

The length and velocity scales used for normalization are the boundary layer thickness δ_0 and the friction velocity $u_{\tau,0}$ in the entrance plane of CASE 0. In this case the computational domain measures 8×4×2 units in the (x, y, z)-direction and is resolved by an equidistant 64×32×32 grid. Based on δ_0 and $u_{\tau,0}$ the Reynolds number is 3240, which corresponds to a value of approximately 90000 based on the freestream velocity $U_{\infty,0}$. The distance from the wall to the wall-next grid point in units of $\nu/u_{\tau,0}$ is 100. Storing the velocity vectors from plane x=4 on magnetic tape over 12000 time steps provides the Dirichlet inflow conditions for CASES 1,2 and 3.

Three different constant values of dp/dx downstream of x=8 are investigated in the CASES 1,2 and 3, namely $-10, 0, +10$ (nondimensionalized with δ_0 and $u^2_{\tau,0}$). They are obtained by specifying the freestream velocity accordingly. The computational domains have the sizes 24×4×2(4) and are resolved by equidistant 192×32×32(64) grids. The values in brackets refer to the retarded boundary layer which grows much faster than that in the other cases. The Reynolds numbers correspond to that of CASE 0.

5. Discussion of results

It is found that among the three velocity components the one in the spanwise direction gives the best impression of the typical pressure gradient effect. Figure 2 shows contour lines of the instantaneous v-component in a (x, z)-plane for the three different flow cases. Since the contour levels and increments are the same in the three plots we are able to follow the history of a large scale structure under the different external conditions. In the region $6 \le x \le 8$ the flow fields are nearly identical because

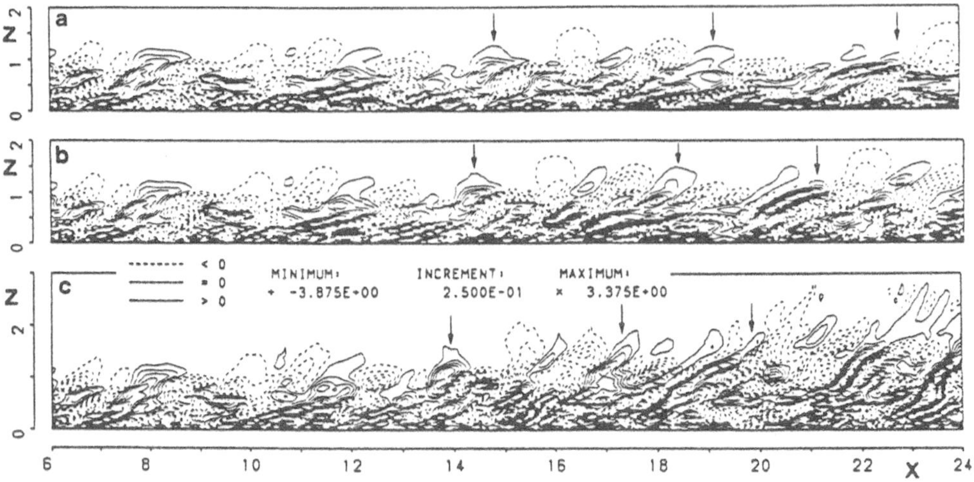

Figure 2. Contour plot of the instantaneous spanwise velocity. a,b,c : CASE 1,2,3.

the step change in pressure gradient at x=8 has only a weak upstream influence. Further downstream the pressure-gradient effect can be seen from the three structures marked with arrows.

The effect is to increase/decrease their spacing due to acceleration/deceleration (figure 2a,c) and is visable in a pronounced change in their inclination angle, which reaches values of $\approx 50°$ in the adverse pressure-gradient case. The flattening (figure 2a) or steepening (figure 2c) of the inclination angle is a result of the strong negative/positive vertical velocities leading to a reduced or increased boundary layer thickness.

A perspective view of the constant value surfaces of the spanwise velocity v confirms these observations (figure 3). Especially the decrease in number density of large-scale structures in CASE 1 and the occurrence of a greater number of steeper and intenser structures in CASE 3 are evident.

In figure 4 we have isolated two structures of the v-field which correspond to each other, but have undergone different histories. Acceleration of the flow (figure 4a) reduces the volume of the instantaneous structures and also their intensity. It makes them travel faster. This is why their positions downstream differ at a certain instant.

The side view of these two v-structures plotted in one diagram (figure 5) proves the reduction in volume and in inclination angle in the outer region of the boundary layer as a consequence of acceleration (shaded structure). The decrease in intensity is a secondary effect. Since the v-field obtains its energy primarily from the u-fluctuations via the pressure-strain correlation, it reflects the reduced production rate of these fluctuations due to acceleration.

The change in number density of large scale structures does not explain the inhibition or enhancement of turbulence intensities due to favorable or adverse pressure gradient. The variation in spacing between the energy-carrying structures must be interpreted together with the change in their convection speed. A time series can give the proper answer. Figure 6 shows time series of the stress-producing term $-u''w''$ over 3000 timesteps for CASE 1 (figure 6a) and CASE 2 (figure 6b). To compare the structures directly, the time series are shifted on the time axis according to the different

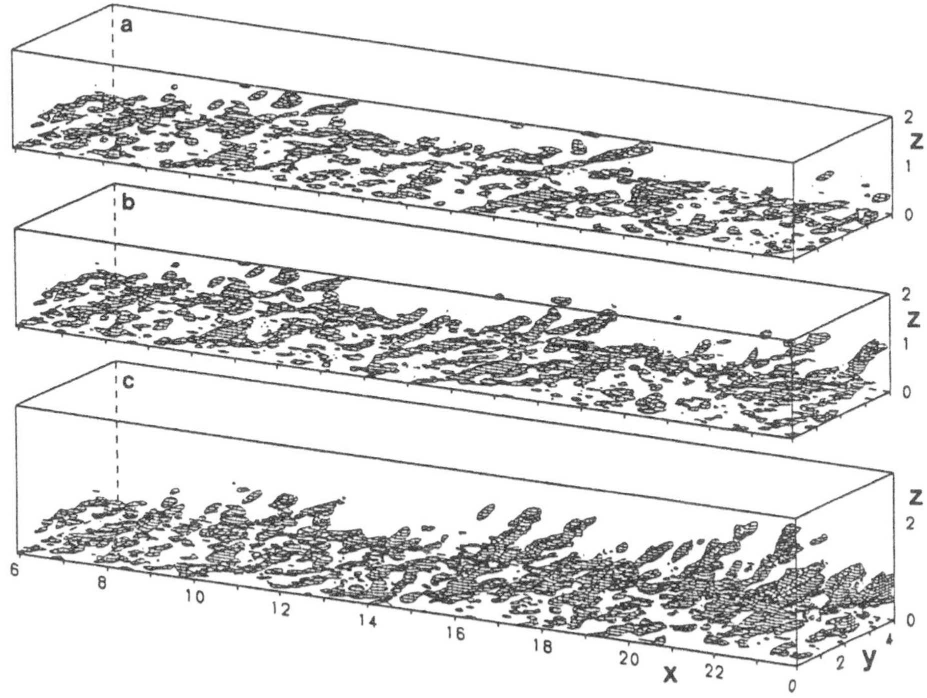

Figure 3. Contour surfaces of the spanwise velocity, v=1.5. a,b,c : CASE 1,2,3.

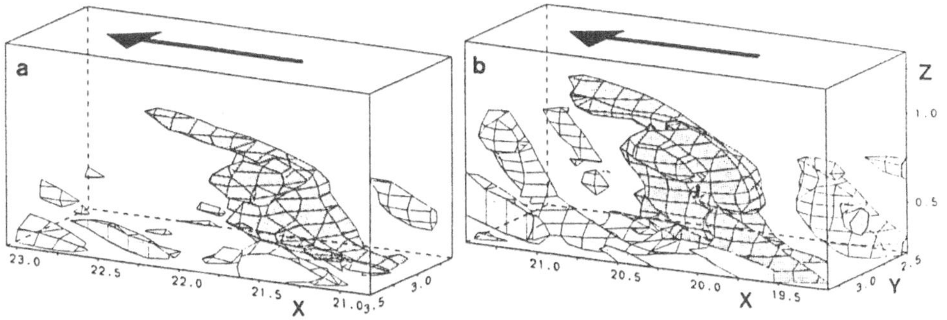

Figure 4. Contour surfaces of two corresponding v-structures, v=1.5. a,b : CASE 1,2.

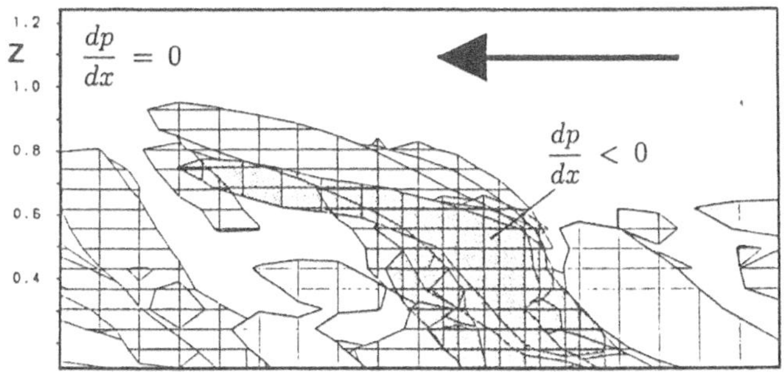

Figure 5. Side view comparison of the v-structures in figure 4a,b.

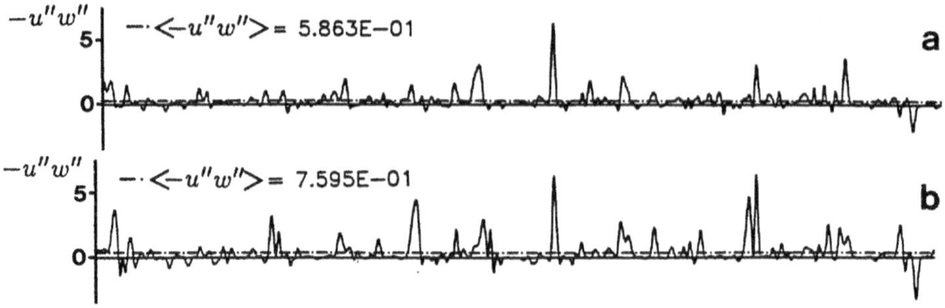

Figure 6. Time series of $-u''w''$ over 3000 timesteps. a,b : CASE 1,2.

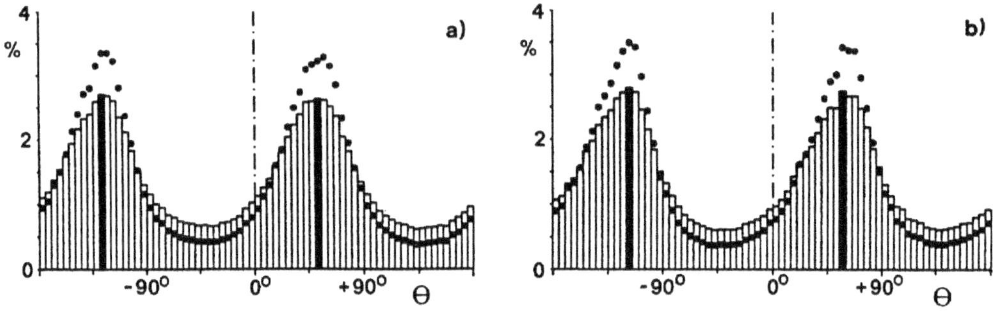

Figure 7. Probability distribution of the inclination angle Θ at x=20.0, z=0.1875.

convection velocities in both flow cases. It can be seen that the distance between strong structures is nearly unaltered on the time axis. However, the signal amplitudes have on the average decreased in the accelerated flow.

The inclination angles of the vorticity vectors with respect to the wall are defined as

$$\Theta = \arctan \frac{\omega_z}{\omega_x} \qquad \text{and} \qquad \Psi = \arctan \frac{\omega_z}{\omega_y} \tag{7}$$

and have been calculated for several grid points. Probability distributions were determined from these by taking 32 samples in the homogeneous direction at each of 15000 time steps. Additional distributions are calculated by weighting the original distributions with the magnitude of the (x, z)-or (y, z)-components of the vorticity vectors.

The value of the most probable angle is $\approx 60°$ for the zero-pressure gradient boundary layer (CASE 2). The positive pressure gradient (CASE 3) increases this angle by roughly 5°. The negative pressure gradient (CASE 1) has the opposite effect, reducing the angle by approximately 5° (see figure 7). This small change can be explained by inspection of the inviscid vorticity equation:

$$\frac{D\omega_i}{Dt} = \omega_j \frac{\partial u_i}{\partial x_j} = \omega_j s_{ij}. \tag{8}$$

Equation (8) does not contain the mean pressure explicitly. The pressure term enters implicitly through its effect on the velocity field as can be seen from the Reynolds equation. We split the velocities into their time averaged part $\langle u_i \rangle$ and their fluctuating

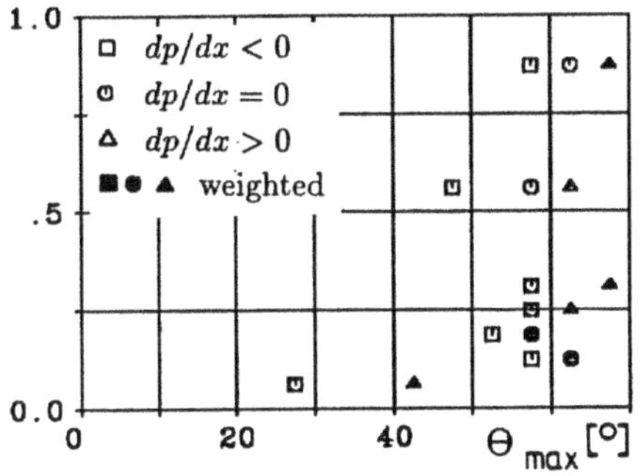

Figure 8a.Variation of the most probable Θ_{max}-value at x=20.0.

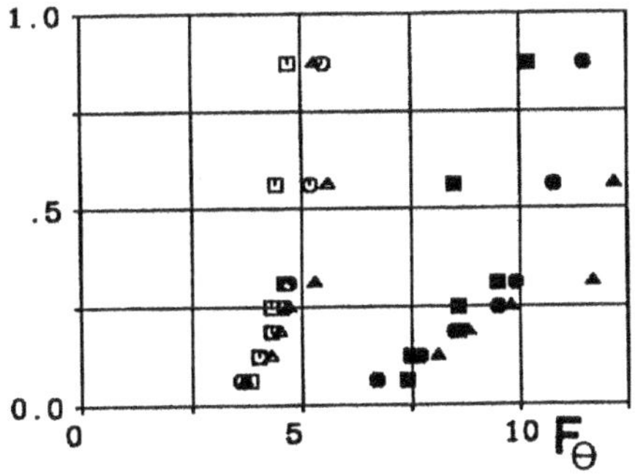

Figure 8b.Variation of the unweighted and weighted ratio F_Θ.

part u_i'' keeping in mind that there is no mean flow in the spanwise direction. From the continuity equation $\partial\langle w\rangle/\partial z$ can be replaced by $-\partial\langle u\rangle/\partial x$. Then the transport of the ω_x- and ω_z-components is governed by:

$$\frac{D\omega_x}{Dt} = \omega_x \frac{\partial\langle u\rangle}{\partial x} + \omega_z \frac{\partial\langle u\rangle}{\partial z} + \omega_j s_{ij}'', \tag{9}$$

$$\frac{D\omega_z}{Dt} = \omega_x \frac{\partial\langle w\rangle}{\partial x} - \omega_z \frac{\partial\langle u\rangle}{\partial x} + \omega_j s_{ij}''. \tag{10}$$

Considering equations (9) and (10) we are able to give a qualitative argument for the decrease in Θ in the acceleration case ($\partial\langle u\rangle/\partial x > 0$, $\partial\langle u\rangle/\partial z > 0$, $\partial\langle w\rangle/\partial x \approx 0$). In this case the first and second terms on the right-hand side of (9) are positive. Assuming that these terms are dominant and provided that both vorticity components are positive, the ω_x value will increase. At the same time ω_z will decrease because the corresponding terms in (10) are negative. Similar arguments hold in the case of retarded flow.

The variation of the most probable Θ-values is plotted in figure 8a as a function of the position in the boundary layer. It can be seen, that the most probable value remains nearly constant throughout most of the boundary layer. However in the near wall region there is a noticeable decrease. Figure 8b shows the maximum-to-minimum ratio

$$F_\Theta = \frac{\text{Probability of } \Theta_{max}}{\text{Probability of } \Theta_{min}} \tag{11}$$

for the three flow cases plotted against z at x=20.0. The definition of F_Θ is that given by Moin & Rogers (1987). It is a measure of the extent to which the flow field consists of organized structures. An examination of figure 8b shows that the flow field in CASE 3 is more organized than that in CASES 1 or 2. One notices a weak increase in F_Θ as one moves away from the wall in all three cases. In addition, the weighted values are higher and increase more quickly with z than the unweighted values.

This can be understood as follows: High intensity vorticity structures show a more typical behavior than the weaker ones. Only these structures have a life history which enables them to reach the outer region of the boundary layer.

Looking at the weighted Ψ-histograms one can see a great difference between the distributions near the wall and these in the outer region of the boundary layer. Figure 11 gives an impression of CASE 3 only. Near the wall the histogram shows a sharp peak

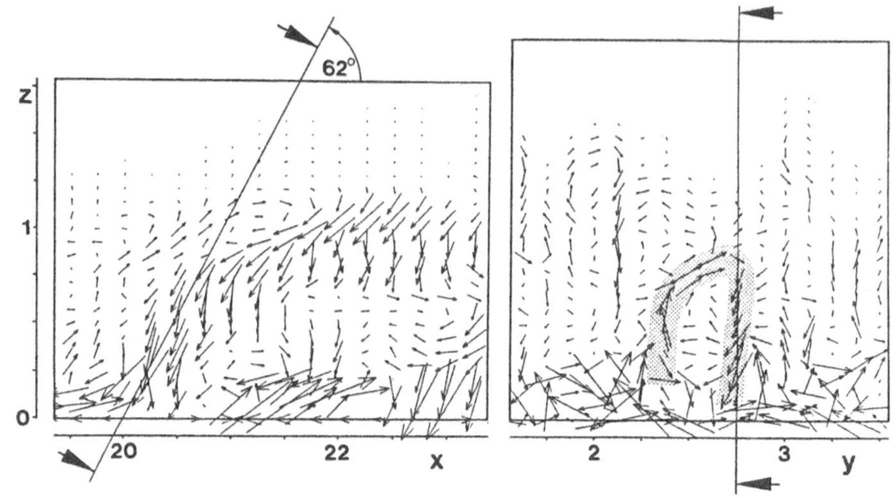

Figure 9. Inst. vorticity vectors. a: (x, z)-plane, b: projection on an inclined plane.

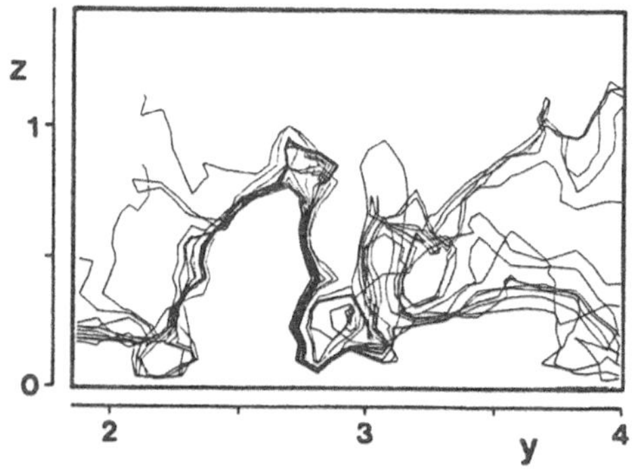

Figure 10. Front view of three-dimensional vorticity lines.

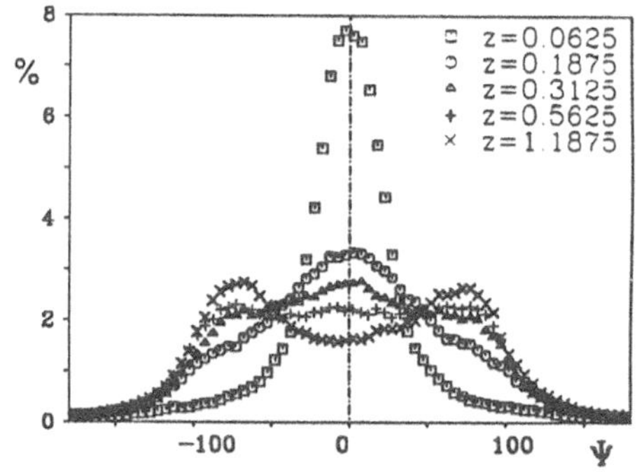

Figure 11. Histograms of Ψ at different vertical positions, CASE 3.

at 0°. In the outer region this peak has been replaced by a nearly constant plateau extending from approximately −90° to +90°. Beyond the edge of the boundary layer this plateau is, in turn, replaced by a symmetric double peak profile. The information drawn from figures 7, 8 and 11 can be explained by the presence of hairpin-like vortex structures.

To verify this assertion, figure 9a,b shows (x, z)-vorticity vectors and projections of the total vorticity vector on a plane inclined at 62° with respect to the (x, y)-plane. In this projection we observe a hairpin-like vortex structure. Figure 10 shows a front view of three dimensional vorticity lines which were initiated near x=20.3, y=2.75 and z=0.3. These figures illustrate the existence of organized vortex structures embedded in a chaotic surrounding.

Acknowledgement

This work was supported by the German Research Society (DFG). We also gratefully acknowledge the support of the Leibniz-Computing-Center of the Bavarian Academy of Sciences.

References

Chorin, A.J. (1968) Numerical solution of the Navier-Stokes equations. *Mathematics of Computations* **22**, 745-762.

Fiedler, H.E. (1987) Coherent structures. In *Advances in Turbulence* (ed. G. Comte-Bellot & J.Mathieu), Proc. 1st Europ. Turbulence Conference, 320-336.

Friedrich, R. & Unger, F. (1990) Large eddy simulation of boundary layers with a step change in pressure gradient. In *Turbulence and Coherent Structures*, Selected papers from "Turbulence 89", M. Lesieur and O. Metais, Eds., Kluwer academic publisher.

Hussain, A.K.M.F. (1986) Coherent structures and turbulence. *J. Fluid Mech.* **179**, 303-356.

Rogers, M.M. & Moin, P. (1987) The structure of the vorticity field in homogeneous turbulent flow. *J. Fluid Mech.* **176**, 33-66.

Schmitt, L. (1988) Grobstruktursimulation turbulenter Grenzschicht-, Kanal- und Stufenströmung. Dissertation, Lehrstuhl für Strömungsmechanik, TU München.

Schumann, U. (1975), Subgrid scale model for finite difference simulations of turbulent flows in plane channels and annuli. *J. Comp. Phys.* **18**, 376-404.

Schumann, U. & Sweet, R.A. (1976) A direct method for the solution of Poisson's equation with Neumann boundary conditions on a staggered grid of arbitrary size. *J. Comp. Phys.* **20**, 171-182.

Part II

Transition and Dynamical Systems

The Multifractal Nature of Turbulent Dispersion

M.S. Borgas

CSIRO Division of Atmospheric Research, Private bag #1, Mordialloc. VIC. 3195. AUSTRALIA

Lagrangian properties of particle trajectories in turbulent flows are determined in the form of multifractal scaling results. These results are then used to examine the effects of intermittency upon dispersion.

1. Introduction

Characterising properties of a turbulent flow by multifractal measures is a recent development which shows considerable promise as a universal description of intrinsic high-Reynolds-number ($Re \gg 1$) properties (Meneveau & Sreenivasan, 1987a; Chhabra *et al*, 1989). Here we consider isotropic, homogeneous and stationary turbulence to illustrate the concepts involved.

Usually these measures are defined as spatial distributions of local scaling behaviour (inertial-range properties) and may be regarded essentially as probability distributions. If we have a one-dimensional instantaneous measurement record, of say local rate of energy dissipation, $\{\epsilon(\underline{x}) : \underline{x} \in [0, \mathcal{L}]\}$, where $\mathcal{L}$ is "sufficiently" large, the scaling behaviour may be analyzed as follows. Divide $[0, \mathcal{L}]$ into $\mathcal{L}/\Delta$ $(= N)$ segments of length Δ and locally average over each segment to give N random variables $\{\epsilon_i\}$. Although the mean dissipation, $\bar{\epsilon} = \frac{1}{N}\sum \epsilon_i$, is a well defined parameter which together with the typical velocity fluctuation, σ, set the scales of the turbulence ($L = \sigma^3/\bar{\epsilon}, Re = \sigma L/\nu$ where ν is the kinematic viscosity), the values of ϵ_i are typically either much larger or much smaller than $\bar{\epsilon}$ with relatively few $O(\bar{\epsilon})$. This is the well-known characteristic of intermittency. In order to quantify intermittency, the set of values $\{\epsilon_i\}$ is divided into equi-scale subsets S_α such that in each S_α we have $\epsilon_i \approx \bar{\epsilon}\left(\frac{\Delta}{L}\right)^{\gamma-1}$ for $\alpha \leq \gamma \leq \alpha + \mathrm{d}\alpha$. The number of Δ-segments in each S_α is then supposed to be $\mathcal{N}_\alpha \approx \left(\frac{\Delta}{L}\right)^{-f(\alpha)}\mathrm{d}\alpha$ for some function $f(\alpha)$. The sense of these *scaling* results is readily understood when a number of subdivisions (differing Δ lengths) are used and say $\mathcal{N}_\alpha$ is plotted against Δ on $\log - \log$ axes. The straight line (when it is evident) has slope $-\ f(\alpha)$.

The interpretation of the multifractal representation as a probability distribution is clear when we consider the moments generated by $\{\epsilon_i\}$:

$$\langle \epsilon_i^q \rangle \approx \bar{\epsilon}^q \int_{\alpha_{min}}^{\alpha_{max}} \rho(\alpha) \left(\frac{\Delta}{L}\right)^{(\alpha-1)q+1-f(\alpha)} \mathrm{d}\alpha \approx \bar{\epsilon}^q \left(\frac{\Delta}{L}\right)^{(q-1)(D_q-1)} \tag{1}$$

where $\rho(\alpha)$ is an "$O(1)$" function necessary for the correct normalisations; henceforth such non-scaling factors will be ignored. Equation (1) defines D_q in terms of $f(\alpha)$ via a steepest-descent calculation, and provides an alternative means of data analysis to quantify the multifractal behaviour. The Legendre transforms (Feder, 1988) between D_q and $f(\alpha)$ are

Advances in Turbulence 3
Editors: A.V. Johansson · P.H. Alfredsson

$$\alpha = \frac{\mathrm{d}}{\mathrm{d}q}(q-1)D_q \quad \text{and} \quad f(\alpha) = \alpha q - (q-1)D_q. \tag{2}$$

Both the $f(\alpha)$ and the D_q representation of scaling properties have been independently verified experimentally (Meneveau & Sreenivasan, 1987a; Chhabra *et al*, 1989); furthermore, relationship (2) has also been established from the experimental data.

Given that the multifractal structure is supported by measurement, it is worthwhile to ask whether the structure can be used to make useful predictions of other properties. It is proposed that spatial characteristics of other physical quantities (besides dissipation) are also given and that the *Lagrangian* scaling characteristics, which are considered later, are specified. The former predictions are not entirely new, and the scaling properties of velocity structure functions are proposed by Meneveau & Sreenivasan (1987a). The reasoning, essentially due to Obukhov (1962), is the following: for the collection of Δ-segments where $\epsilon_i \approx \bar{\epsilon}\left(\frac{\Delta}{L}\right)^{\alpha-1}$ (i.e. S_α) the average magnitude of the velocity difference across the segments must be

$$u_\alpha \approx (\epsilon_i \Delta)^{1/3} \approx (\bar{\epsilon} L)^{1/3} \left(\frac{\Delta}{L}\right)^{\alpha/3} \tag{3}$$

provided viscous effects can be ignored, which, for the large *Re* situation that we consider, is for most Δ-segments. More precisely, this latter requirement is that $\Delta \gg \eta_\alpha$, where η_α is the local length scale of a dissipative event with the local dissipation rate $\epsilon_i \approx \bar{\epsilon}\left(\frac{\eta_\alpha}{L}\right)^{\alpha-1}$. Therefore, as ν parameterises energy loss to heat,

$$\epsilon_i \approx \bar{\epsilon}\left(\frac{\eta_\alpha}{L}\right)^{\alpha-1} \approx \nu^3 \eta_\alpha^{-4} \quad \Rightarrow \quad \eta_\alpha = L Re^{-\frac{3}{3+\alpha}} \tag{4}$$

which differs from the Kolmogorov scale for dissipation, $\eta_{Kol} = LRe^{-3/4}$, because α takes values from α_{min} (≈ 0.51) to α_{max} (≈ 1.74). However, the bulk of space in a turbulent flow is occupied by structures with α near unity so that η_{Kol} is a good representative value. Now, assuming that $\Delta \gg \eta_\alpha$ for all relevant α, the q-th order velocity-difference structure function (i.e. we now average u_α^q over α) is given by

$$\left\langle |\underline{u} - \underline{u}'|^q \right\rangle \approx (\bar{\epsilon}L)^{\frac{q}{3}} \int_{\alpha_{min}}^{\alpha_{max}} \left(\frac{\Delta}{L}\right)^{q\alpha/3+1-f(\alpha)} \mathrm{d}\alpha \approx (\bar{\epsilon}L)^{\frac{q}{3}} \left(\frac{\Delta}{L}\right)^{q/3+(q/3-1)\left(D_{q/3}-1\right)} \tag{5}$$

which differs from the non-intermittent statistical theories by the exponent corrections $c_q = (q/3-1)\left(D_{q/3}-1\right)$. Meneveau & Sreenivasan (1987a) have established that this prediction is consistent (and superior to other intermittency models) with the measurements of the structure functions given by Anselmet *et al* (1985). With similar reasoning many other physical scaling predictions can be made, but none (other than those for velocity structure functions) have been tested experimentally.

2. Pointwise averages

An extension of multifractal scaling is to consider its implications on pointwise averages. Here η_α plays a crucial role. We suppose that at leading order the number of events with length scale $O(\eta_\alpha)$ is given by $\left(\frac{\eta_\alpha}{L}\right)^{-f(\alpha)} \mathrm{d}\alpha$, which is consistent with the multifractal scaling. However, when we subdivide into Δ boxes much smaller that η_α, each event with local scale η_α subdivides into $\left(\frac{\eta_\alpha}{\Delta}\right)$ *like* events with equivalent local dissipation rate (in a scaling sense). Thus

$$\langle \epsilon(\underline{x})^q \rangle = \lim_{\Delta \to 0} \langle \epsilon_i^q \rangle \approx \overline{\epsilon}^q \int_{\alpha_{min}}^{\alpha_{max}} \left(\frac{\eta_\alpha}{L}\right)^{(\alpha-1)q+1-f(\alpha)} d\alpha \approx \overline{\epsilon}^q Re^{-3(q-\tilde{q})} \tag{6}$$

where the equation

$$4(q-\tilde{q}) = (\tilde{q}-1)(D_{\tilde{q}}-1) \tag{7}$$

determines $\tilde{q}(q)$ as a function of the parameter q and hence the Reynolds-number exponent in (6). Equation (6) is a useful result because it contains the scaling information of the probability distribution for pointwise dissipation, say $\mathcal{P}(\epsilon)$. Similar reasoning allows the calculation of other sub-dissipation range properties, for example velocity structure functions for $\Delta \ll \eta_{\alpha_{min}}$:

$$\langle |\underline{u} - \underline{u}'|^q \rangle \approx (\overline{\epsilon}L)^{q/3} \left(\frac{\Delta}{L}\right)^q Re^{-3\left(\frac{q}{3} - \tilde{q}\left(\frac{2q-1}{4}\right)\right)}.$$

With this information it is possible to show that multifractal scaling properties are consistent with pointwise moments of the Navier-Stokes equations, in particular the Karman-Howarth equation and the equation for the mean-square acceleration (Borgas, 1990).

3. Lagrangian Multifractals

Here we outline a multifractal theory for the *history* of dissipation events encountered by a single fluid particle meandering freely through our turbulent flow, which is given in more detail by Borgas (1990). Thus, we suppose that $\mathcal{H} \equiv \{\epsilon(t) : t \in [0, \mathcal{T}]\}$ is such a record of dissipation and that the set of $M = \frac{\mathcal{T}}{\tau}$ dissipations, $\{\epsilon_j\}$, each averaged over a short τ-duration sub-interval of the history, has multifractal scaling properties. In particular, for each equi-scale subset of $\mathcal{H}$, $\tilde{\mathcal{S}}_\kappa$ (defined similarly to $\mathcal{S}_\alpha$ above) there are approximately $\left(\frac{\tau}{T}\right)^{-\tilde{f}(\kappa)} d\kappa$ τ-segments with each scaling as $\epsilon_j \approx \overline{\epsilon}\left(\frac{\tau}{T}\right)^{\kappa-1}$. Here $T = \frac{L}{\sigma}$.

While the actual experimental construction of $\mathcal{H}$ is not technically feasible at present, there will be a strong connection between the spatial and temporal ($\equiv$ Lagrangian) multifractals because of the ergodic property of turbulence (Tennekes & Lumley, 1981). This is because we may infer the instant-wise scaling properties (in terms of Re) from the knowledge of $\tilde{f}(\kappa)$, i.e. obtain $\mathcal{P}(\epsilon)$ in terms of $\tilde{f}(\kappa)$ (as well as $\mathcal{P}(\epsilon)$ in terms of $f(\alpha)$).

Let the local *Lagrangian* time scale for dissipation events with the local dissipation rate ϵ_j be t_κ; then t_κ is determined by

$$\epsilon_j \approx \overline{\epsilon}\left(\frac{t_\kappa}{T}\right)^{\kappa-1} \approx \frac{\nu}{t_\kappa^2} \Rightarrow t_\kappa \approx T Re^{-\frac{1}{1+\kappa}}. \tag{8}$$

Similarly as for η_α in (4), t_κ differs in general from the traditionally assumed Kolmogorov microscale. However, the preponderance of dissipation events correspond to κ near unity so that again $t_{Kol} \approx T Re^{-1/2}$ is a good estimate for dissipative time scales.

With (8) and the reasoning above, the instant-wise moments follow as

$$\langle \epsilon(t)^q \rangle = \lim_{\tau \to 0} \langle \epsilon_j^q \rangle \approx \bar{\epsilon}^q \int_{\kappa_{min}}^{\kappa_{max}} \left(\frac{t_\kappa}{T} \right)^{(\kappa-1)q+1-\tilde{f}(\kappa)} d\kappa \approx \bar{\epsilon}^q Re^{-(q-\hat{q})} \tag{9}$$

where the equation

$$2(q-\hat{q}) = (\hat{q}-1)(\mathcal{D}_{\hat{q}}-1) \tag{10}$$

determines $\hat{q}(q)$ (and $\mathcal{D}_q$ is a function of q defined analogously as D_q in (1), but here for the temporal multifractal). However, comparison between (6) and (9), together with the ergodic hypothesis, evidently requires

$$q - \hat{q} = 3\,(q - \tilde{q})\,. \tag{11}$$

Equations (7), (10) and (11) represent a closed system for implicitly determining $\mathcal{D}_q$ in terms of D_q. This is a well defined problem and can be solved numerically (Borgas, 1990). However, when recast in terms of $f(\alpha)$ and $\tilde{f}(\kappa)$, an explicit result is obtained:

$$\tilde{f}(\kappa) = -\frac{1}{2}\kappa + \left(1 + \frac{1}{2}\kappa\right) f\left(3\frac{\kappa}{2+\kappa}\right) \tag{12}$$

which is remarkable for its simplicity and also demonstrates a one-to-one correspondence of the temporal and spatial multifractals. There is also a simple interpretation of the argument of f in (12); this is that the dissipation at a point ϵ_α (say) must be equivalent to some instantaneous value ϵ_κ in the Lagrangian record, whence (4) and (8) give

$$\bar{\epsilon}\left(\frac{\eta_\alpha}{L}\right)^{\alpha-1} \approx \bar{\epsilon}\left(\frac{t_\kappa}{T}\right)^{\kappa-1} \Leftrightarrow \alpha = 3\frac{\kappa}{2+\kappa}\ .$$

Figures 1 and 2 show our results, with both the temporal and spatial representations illustrated. The spatial multifractals are represented by an idealised model representa-

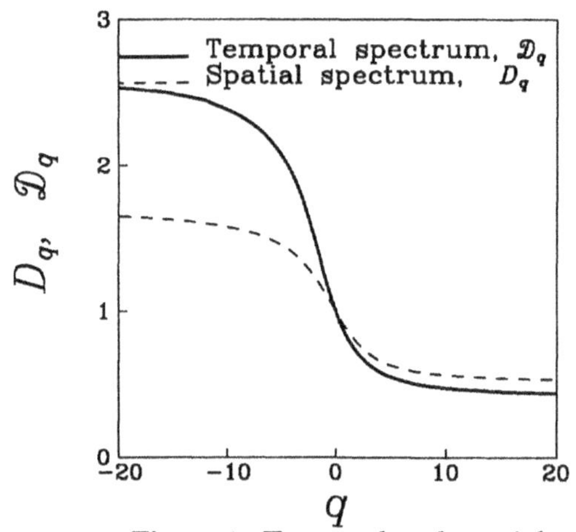

Figure 1. Temporal and spatial moment scaling exponents.

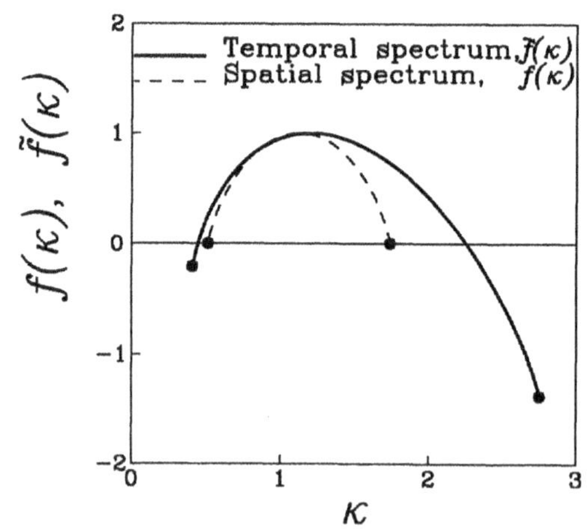

Figure 2. Temporal and spatial equi-scale fractal dimensions.

tion, as a binomial multiplicative process (Feder, 1988), giving a simple form for D_q, which Meneveau & Sreenivasan (1987b) show is a good fit to their experimental data. However, it is in principle possible to use any empirical form for D_q (or $f(\alpha)$).

Important observations are that the Lagrangian "spectrum" $\tilde{f}(\kappa)$ is both much broader than $f(\alpha)$ (i.e. κ values span a greater range than α values) and is negative for κ near either κ_{min} or κ_{max}. Both these results imply that the Lagrangian signal, $\mathcal{H}$, is more intermittent than the spatial record. In practice, enormously long signals are needed to experimentally resolve portions of the spectrum with $\tilde{f}(\kappa) < 0$ because on average there is much less than one such scaling event in a signal of duration T. In fact it is likely that the spatial record has similar behaviour near α_{min} or α_{max} (Chhabra & Sreenivasan, 1990) but this has been neither adequately measured nor modelled yet.

4. Turbulent Dispersion

Having obtained the small-scale scaling characteristics for a particle moving in a turbulent flow we may consider the implications of these on the important practical problem of dispersion. The success with which the spatial multifractal representation of dissipation *implicitly* captured the scaling properties of velocity structure functions, essentially due to local similarity arguments, encourages the application of the same principles to Lagrangian statistics. Thus, moments of temporal velocity *differences* are predicted as

$$\langle|\underline{u}(t+\tau)-\underline{u}(t)|^q\rangle \approx (\bar{\epsilon}T)^{q/2}\left(\frac{\tau}{T}\right)^{q/2+(q/2-1)(\mathcal{D}_{q/2}-1)} \tag{13}$$

where $\tau \gg$ dissipative time scales. Note that, when $q = 2$, the result indicates *no* intermittency correction to the mean-square velocity increment. This is equivalent to the third-order (spatial) velocity structure function being free of intermittency corrections; both are a consequence of linear proportionality to the dissipation-rate parameter in the inertial range, as noticed by Novikov (1989, 1990).

It is possible, following Borgas (1990), to construct an *approximate* sequence of velocity increments, $\delta\underline{u}_k$, which constitute a model history of the small-scale properties of a particle's motion. We write

$$\delta\underline{u}_k = \sqrt{C_0\epsilon_k\tau}\underline{\theta}$$

where $\underline{\theta}$ is a random direction ($|\underline{\theta}| = 1$; each component of $\underline{\theta}$ is a random number drawn from a Gaussian distribution, then the vector length is normalised); ϵ_k is the dissipation in the $k-th$ τ-duration subinterval of a binomial multiplicative process with multifractal "spectra" that approximate both $\mathcal{D}_q$ and $\tilde{f}(\kappa)$; and C_0 is some universal constant. Previously, such sequences have been taken to be Gaussian, with mean zero and variance $\approx 3C_0\bar{\epsilon}\tau$ (Thomson ,1987), i.e.

$$\delta\underline{u}_k = \sqrt{C_0\bar{\epsilon}\tau}\underline{v}$$

where $\underline{v}$ is a Gaussian random vector with $<\underline{v}>= 0$ and $<\underline{v}.\underline{v}>=3$ and C_0 has a value in the range 4 ± 2. The multifractal sequence is a generalisation which models intermittent velocity changes, but with the same mean and variance!

The total velocity change, from say $\underline{u}_0$ initially, to say $\underline{u}(t)$ at time t cannot be obtained simply by using the naive assumption that

$$\underline{u}(t) = \underline{u}_0 + \sum_{k=0}^{K} \delta \underline{u}_k \qquad (t = K\tau)$$

as this gives velocity fluctuations, $< \underline{u}.\underline{u} >$, which grow linearly with time, indefinitely, whereas $< \underline{u}.\underline{u} > = 3\sigma^2$ is actually a fundamental constant. This problem occurs for both parameterisations of $\delta \underline{u}_k$ above. The simple remedy is to suppose that the velocity evolves according to

$$\underline{u}(t+\tau) - \underline{u}(t) = -\frac{1}{2} C_0 \underline{u}(t) \left(\frac{\tau}{T}\right) + \delta \underline{u}_k \qquad (t = k\tau) \tag{14}$$

which is equivalent to assuming that the velocity process is Markovian † when the $\delta \underline{u}_k$ increments are Gaussian (white noise). Note that $\delta \underline{u}_k$ is formally the larger of the contributions to the right-hand-side of (14), however the cumulative effect for the sum of many $\delta \underline{u}_k$ means that the "drift" term is significant and consequently that $< \underline{u}.\underline{u} > \approx 3\sigma^2$ for all time. Despite this success in modelling the Lagrangian velocity variance, even this model is partly deficient and its limitations are discussed by Borgas (1990).

With the model for velocity history it is trivial to integrate once to obtain the particle displacement history, i.e. $\underline{x}(t+\tau) \approx \underline{x}(t) + \underline{u}(t)\tau$. Furthermore, it is possible to calculate large numbers of trajectories (each corresponding to a *different* turbulent flow) so that ensemble average statistics may be calculated.

Firstly, we consider properties of a typical trajectory; figure 3 shows a model history for the local dissipation encountered by a meandering fluid particle. This model dissipation record has approximate scaling properties as required by (9) and (12) with D_q and $f(\alpha)$ provided by Meneveau & Sreenivasan's (1987b) parameterisation of experimental data. Next, in figure 4, we have used (14), and a subsequent integration, to

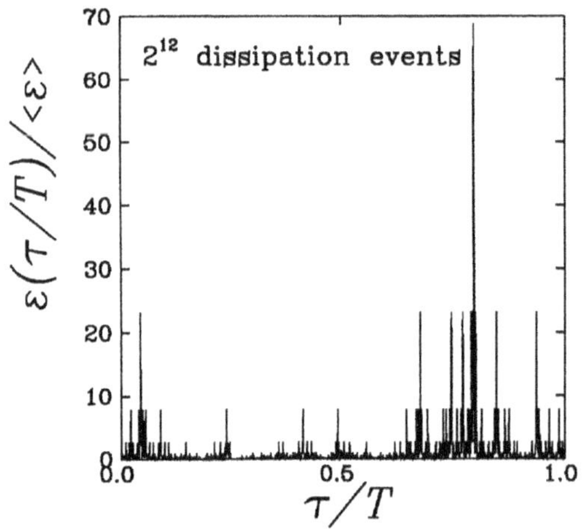

Figure 3. Simulated dissipation history. $\bar{\epsilon} = \frac{\sigma^2}{T}$. $Re \approx 10^5$.

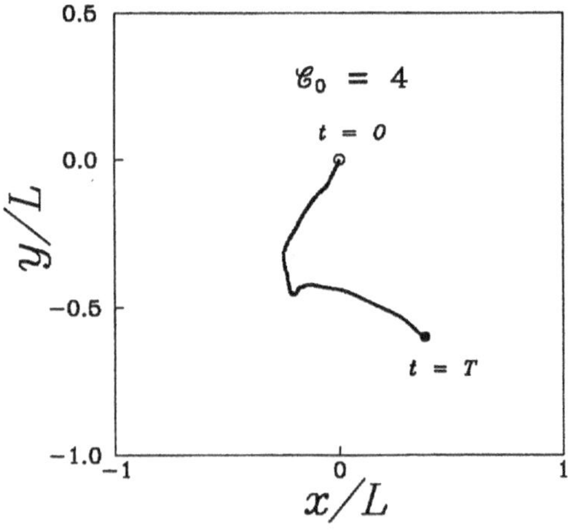

Figure 4. Simulated path using equation (14) and ϵ from fig. 3.

† Thus particle velocities have in the past often been modelled using Langevin equations (Thomson, 1987).

generate a typical particle path which is shown in projection. This path begins at the origin and is shown for an elapsed time T. At first sight the lack of path contortions seems remarkable, and in the past it has been assumed (Mandelbrot, 1983; and Viecelli 1989), without much evidence, that such paths are fractal, with ***significant*** convolutions and deformations all the way down to Kolmogorov microscales. This cannot be true, however, since this assumption is equivalent to the path length being an increasing and unbounded function of Re, i.e. effectively infinite for the flows we consider. To follow such a tortuous path it is clear that the average speed of the particle must be similarly large with Re; a fact that contradicts the basic parameterisation of the turbulence.

To clarify the context of the present multifractal discussion, consider a particle that has been meandering for time $\mathcal{T} \gg T$. The path will almost certainly be of $O(\sigma\mathcal{T})$ in length but be contained within a sphere of radius $O(\sigma(\mathcal{T}T)^{1/2})$ (Taylor, 1921) and therefore is convoluted and highly complex. It may be considered to be a fractal with dimension $\frac{3}{2}$, which is equivalent to a Brownian process. Thus, on time and length scales large compared to the energy containing scales (L and T), the path is a fractal and the dispersion is effectively a (non-anomalous) diffusion process. However, on smaller scales the path becomes sufficiently smooth so that it is no longer sensible to think of it in a fractal sense, despite the underlying multifractal character of the velocity-differences and energy dissipation.

For an ensemble of paths the principle interest lies in the prediction for mean-square displacements. We simply note that intermittency has only a weak effect here since in the inertial-range, where this quantity is linearly dependent upon dissipation rate, we have Novikov's (1989) result

$$\left\langle |\underline{x}(t) - \underline{u}_0 t - \underline{x}_0|^2 \right\rangle \approx C_0 \bar{\epsilon} t^3 \qquad (t_{\kappa_{max}} \ll t \ll T) \tag{15}$$

and in the "diffusion" range

$$\left\langle |\underline{x}(t) - \underline{x}_0|^2 \right\rangle \approx \sigma^2 \frac{12}{C_0} T t \qquad (T \ll t), \tag{16}$$

which is Taylor's (1921) result, and neither of which is affected by intermittency.

5. Conclusions

Here we have considered recent developments in the representation of inertial-range properties of turbulence as multifractals. Elsewhere (Meneveau & Sreenivasan, 1987a; Chhabra *et al*, 1989), these are shown to be good representations of the spatial energy-dissipation field, and subsequently successful in predicting velocity structure functions. We have extrapolated to the dissipation subrange for which there are no comprehensive measurements to verify our results. However, the predictions can be shown to be at least self-consistent with moments of the Navier-Stokes equations (Borgas, 1990).

A parallel development is made for a temporal multifractal representation of Lagrangian inertial-range properties, and a one-to-one relationship between the spatial and temporal multifractal spectra is obtained.

Turbulent dispersion is discussed with reference to these new results; in particular new stochastic equations are proposed which model particle trajectories and account for intermittency. However, the effect of intermittency is not substantial so that the traditional dispersion results (Taylor, 1921; Monin & Yaglom, 1975) for mean-square displacements are unchanged. Furthermore, particle paths in a turbulent flow are not fractals on integral time (O(T)) and length (O(L)) scales, and typically traverse an O(L) distance during time T. On much larger scales the path is essentially Brownian and the traditional notion of a diffusion process is correct.

References

Anselmet, F., Gagne, Y., Hopfinger, E. & Antonia, R.A. (1984) *J. Fluid Mech.* **140**, 63-89.

Borgas, M.S. (1990) *to be submitted to Proc. Roy. Soc. London.*

Chhabra, A.B., Meneveau, C., Jensen, R.V. & Sreenivasan, K.R. (1989) *Phys. Rev. A*, **40**(9), 5284-5294.

Chhabra, A.B. & Sreenivasan, K.R. (1990) *New Perspectives in Turbulence (ed. L. Sirovich)*, Springer.

Feder, J. (1988) *Fractals.* Plenum.

Mandelbrot, B.B. (1983) *The Fractal Geometry of Nature.* W.H. Freeman.

Meneveau, C. and Sreenivasan, K.R. (1987a) *Nuclear Physics B (Proc. Suppl.)* **2**, 49-76.

Meneveau, C. and Sreenivasan, K.R. (1987b) *Phys. Rev. Letters*, **59**(13), 1424-1427.

Monin, A.S. & Yaglom, A.M. (1975) *Statistical Fluid Mechanics* **II**. MIT Press (p. 534).

Novikov, E.A. (1989) *Phys. Fluids* A**1**, 326.

Novikov, E.A. (1990) *Phys. Fluids* A**2** (5), 814-820.

Obukhov, A.M. (1962) *J. Fluid Mech.* **13**, 77

Taylor, G.I. (1921) *Proc. Lond. Math. Soc.* **20**, 196.

Tennekes, H. & Lumley, J.L. (1981) *A first Course in Turbulence.* MIT Press.

Thomson, D.J. (1987) *J. Fluid Mech.* **180**, 529-556.

Viecelli, J.A. (1989) *Phys. Fluids* A**1**(11), 1836-1843.

The Structure and Splitting of Puffs in Transitional Pipe Flow

Kenneth S. Breuer and Joseph H. Haritonidis
Department of Aeronautics and Astronautics
M.I.T. Cambridge, MA 02139, USA

The structure of the turbulent puff in a laminar pipe flow is investigated experimentally. It is found that the rear interface of the puff appears to generate toroidal vortices by means of an inflectional instability. These vortices propagate through the puff and break up near the leading interface. The number of such vortices in the puff at any one time is typically two or three at the equilibrium Reynolds number of 2200. At $Re = 2300$, the puff elongates and supports three to four vortices at any one time. At this Reynolds number, the puff also undergoes a splitting process in which a vortex establishes itself independently from the rear interface, becoming a new puff which accelerates away from the parent. It is also found that the the puff is uniformly coherent around the pipe with a correlation coefficient close to 0.55.

1. Introduction

The puff is a turbulent structure that exists in an otherwise laminar pipe flow. Although turbulent spots have been observed in several flow geometries, the puff is unique because, at the equilibrium Reynolds number of 2200 (based on the bulk velocity and the pipe diameter), it neither grows nor decays in length, traveling down the pipe at the bulk velocity of the flow ($U_B = 0.5U_{\max}$). This was first documented by Wygnanski & Champagne (1973) and later studied in some detail by Wygnanski, Sokolov & Friedman (1975), who showed that close to the trailing interface, the velocity profile and fluctuation levels of the puff resemble that of turbulent pipe flow, while the velocities slowly relax to the laminar parabolic profile towards the downstream end. The puff was originally thought to consist of a large toroidal vortex at the upstream end of the puff (Wygnanski *et al.*, 1975) and later (Rubin, Wygnanski & Haritonidis, 1979) of a number of toroidal vortices fixed with respect to the trailing interface. However, as shown by Breuer (1984), the puff consists of a variable number of vortices that are not fixed with respect to the trailing interface but, rather, define it by their intermittent formation. Once formed, these vortices appear to propagate through the puff and breakup as they approach the downstream end. This was also shown in the flow visualization studies of Bandyopadhyay & Hussain (1983) and, in addition, Bandyopadhyay (1986) argued that the vortices are not axisymmetric (toroidal) but, rather, helical in nature.

At Reynolds numbers lower than critical, puffs generated through a strong disturbance eventually decays and the flow reverts to laminar flow everywhere, while at higher Reynolds numbers puff splitting occurs, whereby two puffs are formed in tandem from a single puff. Wygnanski & Champagne (1973) speculated that the splitting process originates at the rear (upstream) end of the puff while Breuer's (1984) work suggests that

Advances in Turbulence 3
Editors: A.V. Johansson · P.H. Alfredsson

the splitting process originates towards the middle of the puff. At even higher Reynolds number, the puffs join to form slugs - long regions of fully developed turbulent flow.

The present work is an experimental study of both the nature of the large scale structures formed at the upstream end of the puff and the splitting process of puffs.

2. Results and Discussion

The experiments were carried out in a chrome-plated steel pipe, 2.703cm in diameter and 450 diameters long. The inlet flow was conditioned using a settling chamber, honeycomb, three screens and a 49:1 contraction before entering the pipe. Natural transition was found to occur at $Re_d > 50000$. Measurements were taken using hot wire probes, arranged both radially (eight wires extending from the centerline to $r/R = 0.8$), and azimuthally (ten wires, each spaced 36 degrees apart at $r/R = 0.55$).

2.1 The Equilibrium Puff

Figure 1 shows contour plots of the perturbation streamwise velocity for three different puffs measured at the critical Reynolds number of $Re = 2200$. The puffs were measured with a radial rake of hot-wires so that the entire structure from the centerline to the wall could be captured "instantaneously" (assuming a frozen-flow hypothesis). The three puffs were chosen at random and are examples of the puff's typical appearance. The horizontal axis is non-dimensional time ($\tau = tU_B/D$), shifted so that the rear interface is at $\tau = 0$,

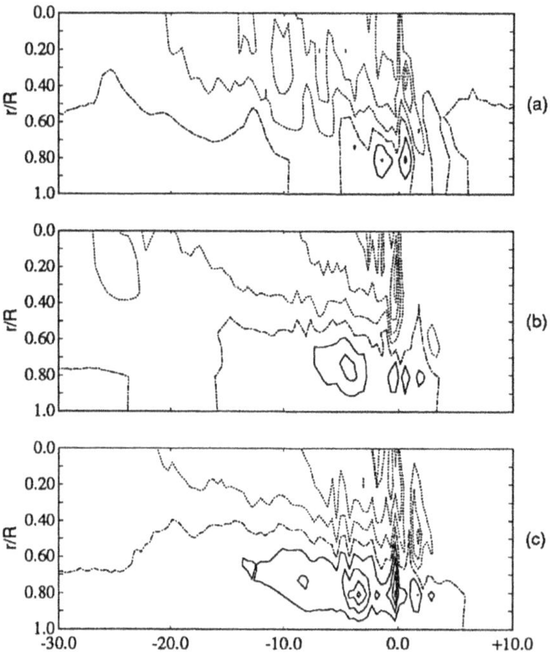

Figure 1: Contours of perturbation streamwise velocity illustrating the radial structure of the puff at $Re = 2200$. Contour levels: $0.2U_B$. The time axis is shifted such that 0.0 corresponds to the sharp trailing interface. Solid line represent positive contours; dotted lines represent negative contours while the dot-dash line indicates the zero contour line.

while the vertical axis is the radial coordinate r/R, with the pipe centerline at the top of each plot and the wall at the bottom. The rear interface is defined as the point at which the centerline velocity falls to 65% of its undisturbed value. Since the rear interface is so sharp, the drop in velocity from its undisturbed value to this threshold typically takes place within 0.3 diameters.

The first observation is that while the general features of the puff are reproduced in all three cases, the details vary from one event to the next. For example, the sharp deceleration at the centerline (indicated by the closely-spaced negative contour lines) and the toroidal vortices located at about $r/R = 0.8$ are present in all three cases. However, the number, strength and axial position of of each vortex varies considerably between puffs. We should note that from these measurements it is not proven that the ring-like structures that are observed are in fact vortices. However, from previous measurements of both the streamwise and radial velocity by Breuer (1984), and from the flow visualization studies of Bandyopadhyay (1986), this conclusion appears fully justified. The observed variation of the features between individual puffs is in support of the results of Bandyopadhyay (1986) and Breuer (1984) who found that the vortical structure of the puff was not fixed, but a dynamic and evolutionary process. Vortices appear to be generated at the rear interface and then propagate through the puff, breaking down as they move to the downstream end of the structure. Evidence of this process is given in figure 1c. The vortex structure suggests that a vortex is forming at $\tau = 4$, and that a vortex appears to be dissipating at $\tau = -10$. This is repeated in the other two frames, consistent with the proposed scenario for the vortex dynamics inside the puff.

Bandyopadhyay (1986) argued that the vortex generation at the rear is due to an inflectional instability mechanism caused by the high speed fluid at the center of the pipe (from the laminar part of the flow) intruding into the rear of the puff which is characterized by a flatter turbulent-like profile. This results in a "jetting" at the rear of the puff which is unstable and allows the toroidal vortices to grow on the locally inflectional profiles. This mechanism is strongly supported by these results and, although it is not shown here, the inflectional profiles as well as the detection of what appear to be instability waves have been observed at the rear of the puff.

If, as proposed, the vortices are generated at the rear interface and then propagate through the puff, then the number of vortices present in the puff at any one time, and their positions relative to the rear interface will not be fixed as the puff travels down the pipe. In addition, these quantities will also vary from puff to puff. Since we cannot follow a single puff as it travels down the pipe, we have computed the probability distributions for the number of vortices present in the puff, and the positions of each vortex relative to the rear interface for a large ensemble of puffs (200 events). These are shown in figure 2 (Note that the vertical axis of this histogram, and all histograms presented in this paper, is normalized such that the integrated area under the histogram is equal to one). The algorithm for counting the vortices in each puff was as follows: the signal at $r/R = 0.8$ was low-pass filtered to remove fluctuations on time scales faster than $\tau = 1.1$. Then, the number of "sufficiently large" maxima in the signal was counted. This was taken to be a measure of the number of vortices in the puff. The term "sufficiently large" denotes a selection procedure designed to eliminate small fluctuations and to isolate only the large scale features. The algorithm was checked against several puffs (approximately 50) to ensure that the number of vortices counted by the algorithm agreed with the number of vortices counted visually. As a check, the same algorithm was repeated, checking for *minima* on the *centerline* velocity signal. The results were essentially identical.

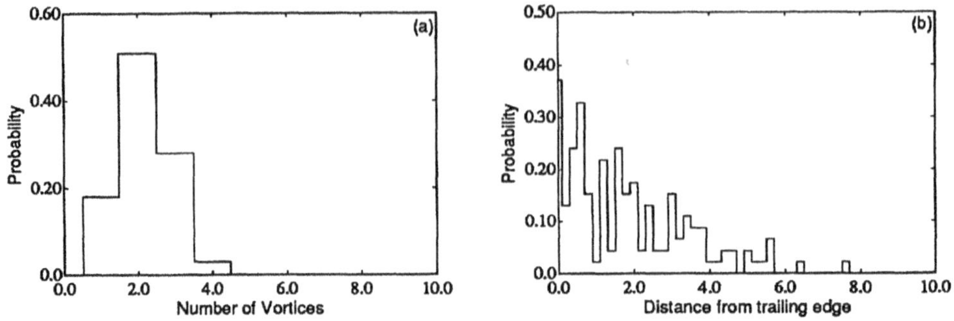

Figure 2: histograms showing (a) the number of vortices located in each puff at $Re = 2200$, and (b) The distance of each vortex from the rear interface.

As figure 2 indicates, the number of vortices in the puff is generally two or three (Rubin *et al.* (1979) arrived at the same conclusion), and their positions are widely distributed in the puff (although the number of vortices does decrease as the distance from the rear interface increases). Thus, while the structures are strongly associated with the rear interface, their positions are not fixed with respect to the interface, but vary as they propagate through the puff.

2.2 Puff Splitting

As the Reynolds number increases above the critical value of $Re = 2200$, some fraction of the puffs that emerge at the measurement station are observed to be in the process of splitting, that is to say, a second puff is emerging, or has emerged from the original single puff. At this higher Reynolds number, the turbulent production mechanism (the generation and activity of the toroidal vortices) and the diffusion process (the breakdown of the vortices in the forward part of the pipe and their dissipation) are no longer in balance. This is initially manifested in an increased activity inside the puff, illustrated in Figure 3a and b which shows histograms of the puff lengths both at $Re = 2200$ and $Re = 2300$. At the higher Reynolds number, the puffs are approximately ten diameters longer, illustrating the increased turbulent activity that the higher Reynolds number

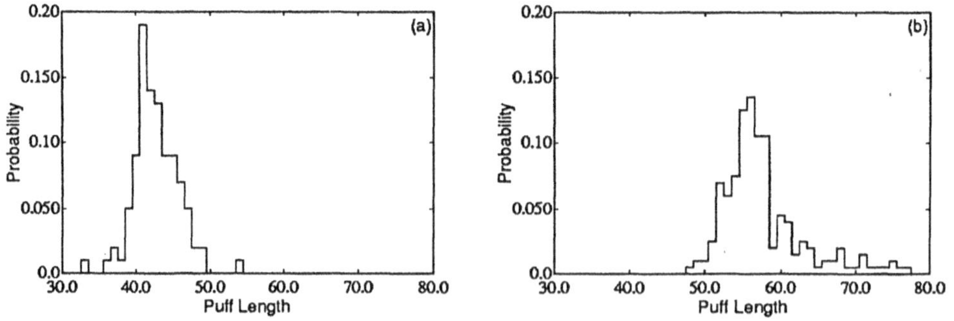

Figure 3: Histograms showing the length of the puff at (a) $Re = 2200$ and (b) at $Re = 2300$ illustrating the variation, resulting from the dynamic nature of the puff structure, and the elongation that occurs as the Reynolds number rises above its equilibrium value

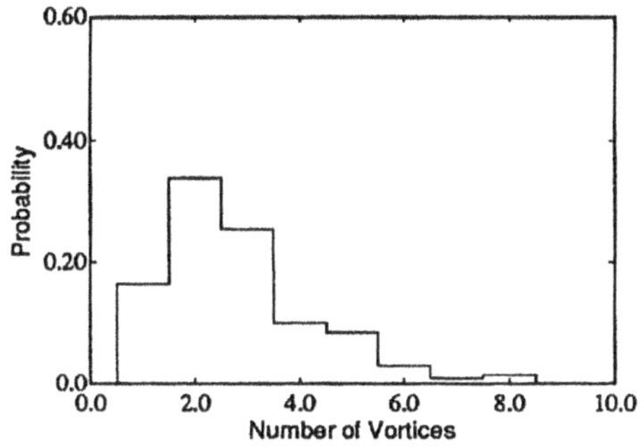

Figure 4: Histogram of the number of vortices located in each puff at $Re = 2300$. illustrating the increased number of vortices that the puff supports at the higher Reynolds number. The few events with several vortices (6, 7, 8) are probably puffs that are in the late stages of the splitting process

supports. Similarly, figure 4 shows the probability distribution of the number of vortices counted in puffs at $Re = 2300$. When compared with figure 2b, we see that although the most probable number of vortices is still two, the average number of puffs in the pipe has increased from two to three, and that there are a significant number of puffs which support several more vortices (we will show that these are puffs which are undergoing the splitting process).

In earlier studies (cf Wygnanski & Champagne, 1973) it was believed that the puff splitting originated at the rear of the puff. However, Breuer (1984) determined that the splitting first appeared at the center of the puff. This result is confirmed in figure 5 which shows contours of the streamwise perturbation velocity for three *different* puffs at $Re = 2300$. The three frames were chosen because they illustrate the puff at what appear to be three different stages of the splitting process. In the first frame, (a), the puff appears very similar in structure to the puffs at $Re = 2200$. However, one's attention is drawn towards the persistence of the vortex in the forward region of the puff, and additionally, what appears to be some oscillations immediately above the high velocity region.

In the second frame, (b), we see a strong vortex in the middle of the puff. In addition, the velocity at the centerline shows significant activity at $\tau = -15$, unlike the centerline behavior of puffs at $Re = 2200$ which all show a slow and gentle relaxation to the laminar profile. There are two possible explanations for the presence of this vortex far from the rear interface. It is possible that the observed vortex is one which was originally generated at the rear interface, has propagated through the puff, but has not broken down. Instead, it is still sufficiently strong and coherent to distort the local mean profile and allow for new vortex generation, similar to, but independent from the mechanism already in place at the rear interface. A second possibility is that it is the *breakdown* of vortices generated at the rear interface which creates a local mean flow distortion sufficiently large to allow the generation of new vortices, and it is a newly generated vortex which we observe here (rather than an old vortex which has survived from the rear interface). In the last frame, (c), we see that the downstream vortex has established itself, completely independently of the main structure and has become a new puff, accelerating away from its parent. In addition, it appears that a second splitting is underway (at $\tau = -11$).

Thus we propose the following series of events that lead to puff splitting: At the higher Reynolds number, the puff can sustain more turbulent activity and consequently

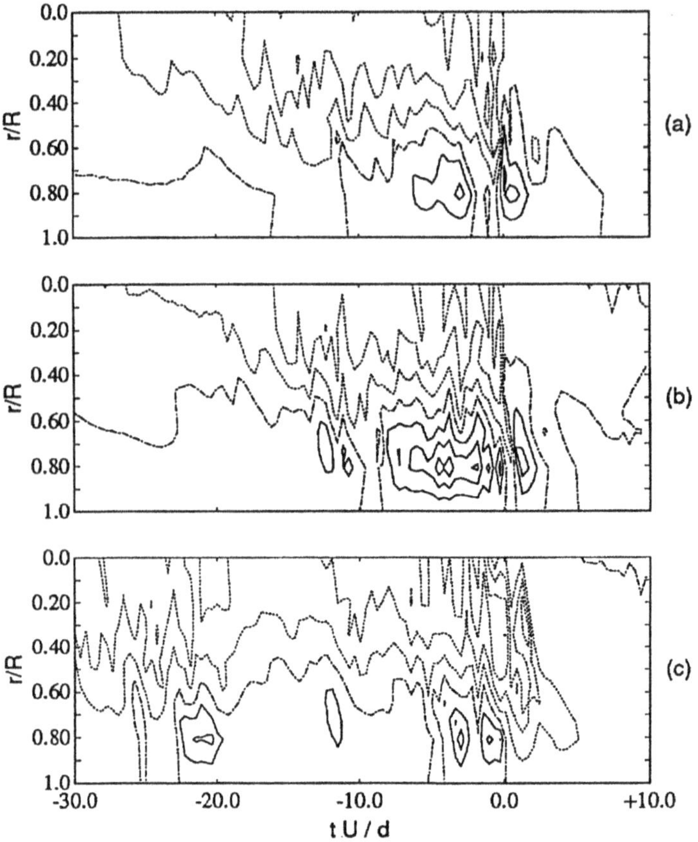

Figure 5: Contours of perturbation streamwise velocity illustrating the radial structure of the puff at $Re = 2300$. Contour levels: $0.2U_B$. Although each picture corresponds to *different* realizations, it does show puffs at several stages during the splitting process, illustrating how the process appears to develop. Solid lines indicate positive contours; dotted lines, negative contours, while the dot-dash line indicates the zero contour level.

it elongates. Each vortex, generated at the rear interface, can survive longer before it breaks down and thus the number of vortices in the "typical" puff increases. For some fraction of puffs, a local mean-flow distortion in the middle of the puff (either caused by a "renegade" vortex, or by the redistribution of momentum from vortex breakdown) is strong enough to initiate additional instabilities and vortex generation. This becomes a new rear interface for the emerging puff, which accelerates from its parent, completing the splitting process.

Although it is difficult to precisely determine the splitting process from snapshots of different structures, the three events shown here are indicative of several similar events, and the proposed splitting mechanism was supported by all the measurements thus far examined. At no time did we observe an event which would contradict this scenario.

2.3 Azimuthal Structure of the Equilibrium Puff

The azimuthal structure of the puff at $Re = 2200$ has been a matter of discussion in the literature. Unpublished data by Rubin *et al.* suggests that the puff appears to be well correlated around the azimuth of the pipe, and that the vortices are axisymmetric. However, Bandyopadhyay (1986), based on flow visualizations, reported that the puff appears to have a helical nature. The cross-pipe correlation of the puff was assessed by measuring the azimuthal structure of the puff with ten hot-wires arranged in a circle. The wires were located evenly around the pipe at $r/R = 0.55$ and the puff's passage was recorded by the simultaneous sampling of the ten wires, each at 16.67 KHz - fast enough to resolve the puff to within 0.005 diameters (a total of 300,000 data points/puff). For each puff in an ensemble of 100 puffs, the signal from each wire was cross-correlated with every other wire, yielding 1000 cross-correlations for each value of the angle separating two wires (which ranges from 36 to 180 degrees). The average correlation coefficient as a function of the wire separation is shown in figure 6. The correlation coefficient is approximately 0.55, independent of the wire separation, indicating that the puff is quite well correlated around its azimuth, and that the puff maintains its coherence across the pipe.

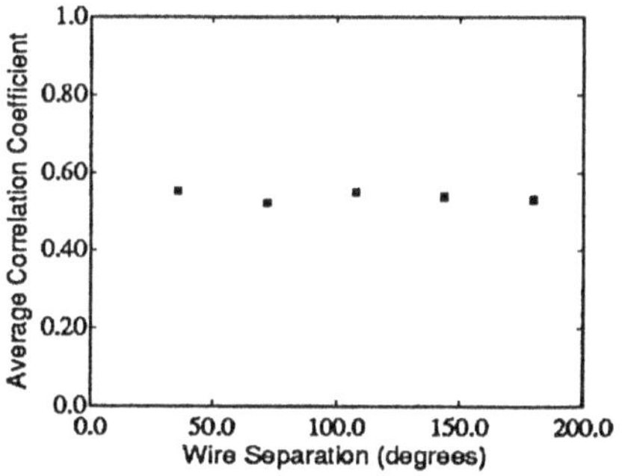

Figure 6: The variation of the average correlation coefficient with wire separation indicating the uniform coherence of the puff around the azimuth.

By measuring the phase shift between two signals from adjacent wires, one can determine whether there is any helical structure to the puff, or whether it is axisymmetric. In addition, by comparing the measured phase shift between two hot wires at the same streamwise location, with measured shifts between two hot wires which are at slightly different streamwise locations (i.e. one slightly downstream from its neighbor), one can resolve both the advection speed of the large scale structures and their angle to the streamwise axis of the pipe. This work in currently underway and at this time, the results indicate that the most probable phase shift between two adjacent wires is zero, suggesting that the puff is largely axisymmetric. However, at this stage, more analysis of the results is necessary and we regard this issue as still unresolved.

3. Concluding Remarks

The study has revealed the structure of the puff both at its equilibrium state, in which it neither grows nor decays, and at a higher Reynolds number in which splitting is observed to take place. Based on the examination of several separate realizations, we believe that we have qualitatively explained the splitting process as being one in which an independent vortex generation mechanism establishes itself in the middle of the puff, either from an old vortex which has not broken down, or simply by mean-flow distortion resulting from vortex breakdown. Once this independent generation mechanism is in place, a new puff grows out of the debris of the old one, accelerating away with an increased propagation velocity. It is worth noting that although we do not observe splitting in all of the puffs, it seems likely that at a Reynolds number above 2200, and in a sufficiently long pipe, all puffs will eventually split and continue to split as they travel downstream forming a ever-increasing chain of puffs. At some stage, they will merge together to form a continuous slug of turbulence, as has been found by Rubin *et al.* (1979) and Wygnanski *et al.* (1973).

Lastly, the similarities between puffs in pipe flow and turbulent spots in plane Poiseuille flow deserve some mention. Although the detailed features of the two structures are quite different, their production mechanisms are in some ways very similar. In both cases, the energy production mechanism derives from the interaction between the disturbance and the surrounding laminar flow. In the case of the Poiseuille flow, the results of Henningson & Alfredsson (1987) indicate that the spot generates its energy from the growth and breakdown of waves which grow due to an instability in the distorted mean flow surrounding the wing-tips of the spot. In a similar manner, the production mechanism in the puff is derived from the distortion of the laminar profiles by the puff itself. These distorted profiles allow for vortex rollup and the generation of turbulence. In both cases, the traditional wall-turbulence production mechanisms (bursts, sweeps, etc) are unable to establish themselves (because of the low Reynolds numbers and the constraints of the geometry) and consequently alternative mechanisms, incorporating the localized nature of the disturbed portion of the flow, exist to sustain the disturbance's structure.

The support of NSF through Grant MEA-8204972, and the Center for Fluid Mechanics at Brown University, supported by DARPA-URI N00014-86-K0754 (KSB) are gratefully acknowledged. JHH also enjoyed the support of ONR Grant N00014-90-J-1302 during the preparation of this article.

References

Bandyopadhyay, P. R. 1986. Aspects of the equilibrium puff in transitional pipe flow. *J. Fluid Mech.* **163**, 439–458.

Bandyopadhyay, P. R. & Hussain, A. K. M. F. 1983. The organized motions in "puffs" in transitional pipe flow. In *Proc. Third Int. Symp. Flow Visualization*, Univ. of Michigan, Ann Arbor MI.

Breuer, K. S. 1984. An experimental investigation into transitional pipe flow. Master's thesis, Department of Aeronautics and Astronautics, Massachusetts Institute of Technology, Cambridge, MA.

Henningson, D. S. & Alfredsson, P. H. 1987. The wave structure of turbulent spots in plane Poiseuille flow. *J. Fluid Mech.* **178**, 405.

Rubin, Y., Wygnanski, I. & Haritonidis, J. H. 1979. Furthur observations on transition in a pipe. In *Proc. IUTAM Symp. on Laminar-Turbulent Transition*, Springer-Verlag.

Wygnanski, I. & Champagne, F. H. 1973. On transition in a pipe. Part 1. The origins of puffs and slugs and the flow in a turbulent slug. *J. Fluid Mech.* **59**, 281–335.

Wygnanski, I., Sokolov, M. & Friedman, D. 1975. On transition in a pipe. Part 2. The equilibrum puff. *J. Fluid Mech.* **69**, 283–304.

The Subharmonic Growth of a Wave-Packet in a Laminar Boundary Layer

Jacob Cohen†, *Kenneth S. Breuer*⋆ *and Joseph H. Haritonidis*⋆

† Faculty of Aerospace Engineering
Technion - Israel Institute of Technology, Haifa 32000, Israel

⋆ Department of Aeronautics and Astronautics
M.I.T. Cambridge, MA 02139, USA

The evolution of a wave packet is investigated experimentally. Two regimes are studied: a linear regime in which the wave packet has small amplitude and may be described as a linear superposition of least damped Orr-Sommerfeld modes, and a subharmonic regime in which spanwise modes with half the fundamental frequency gain energy through nonlinear interactions. It is argued that the subharmonic stage is governed by a Craik-type resonance mechanism. The role of the vertical vorticity on the horizontal velocities inside the boundary layer is also assessed.

1. Introduction

In the study of transition to turbulence in boundary layers, much effort has been focused on the evolution of localized disturbances such as wave packets. The structure and growth of a linear wave packet was measured by Gaster & Grant (1975), and their results showed excellent agreement with the theoretical results of Gaster (1975) who modeled the wave packet as a summation of least-damped Orr-Sommerfeld modes.

The present theories for the nonlinear evolution of wave systems in boundary layers include several types of nonlinear interactions (See Herbert (1988) for a recent review of the literature), although the available experimental data (for example, Kachanov & Levchenko, 1984; Saric & Thomas, 1984; Corke & Mangano, 1989) are all in the context of the three-dimensional secondary instability of two-dimensional, single-frequency wave trains in which the frequency and wave angle of the three-dimensional modes are externally forced. The unforced nonlinear evolution of a broad band of frequencies, such as those present in the wave packet, is less well understood.

The aim of the present work is twofold. Firstly, to determine the structure of the wave packet throughout the entire boundary layer, extending the work of Gaster & Grant (1975) who only considered the amplitude of the wave packet above the boundary layer at $y/\delta^* = 3.2$. Secondly, to observe the evolution of the wave packet into the nonlinear regime and to determine the nature of the nonlinear interactions that ensue.

2. Results and Discussion

The experiments were conducted in the low-turbulence wind tunnel in the Turbulence Research Laboratory at MIT. The wave packet was generated by a sinusoidal air pulse introduced into the boundary layer at $x_0 = 81$cm through a small circular perforated disk mounted flush with the flat plate (For a more complete description of the facilities and

Advances in Turbulence 3
Editors: A.V. Johansson · P.H. Alfredsson

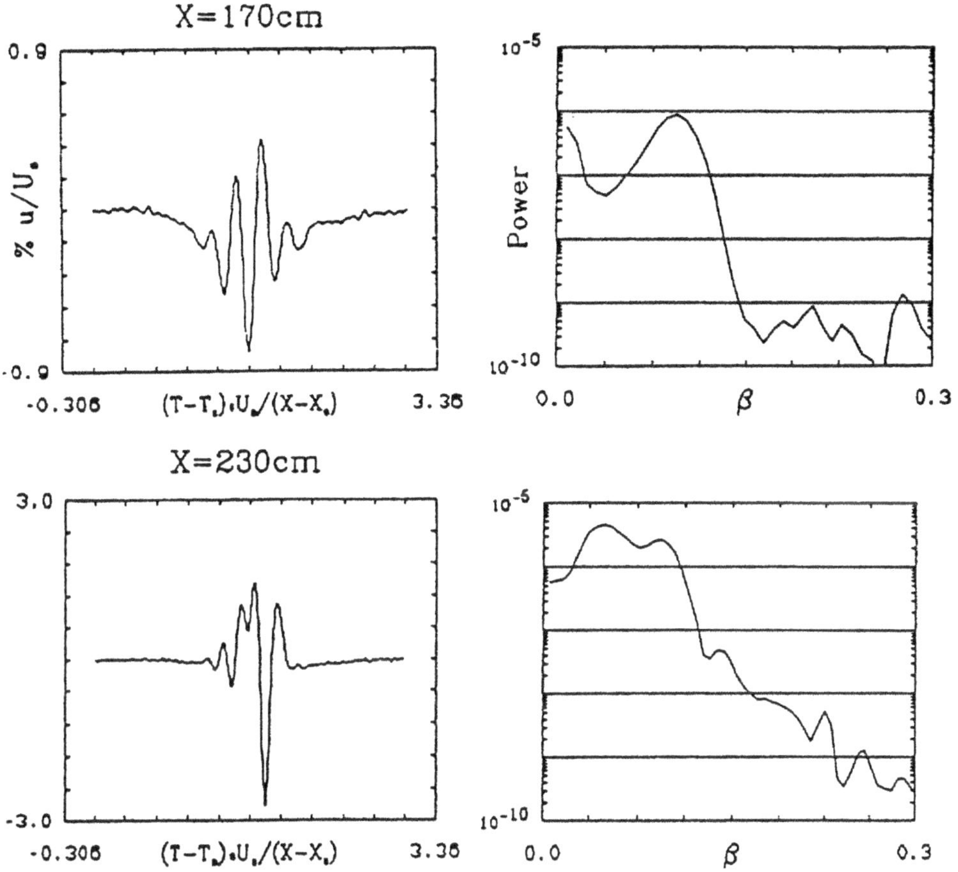

Figure 1: Ensemble averaged u-component of velocity at $z = 0$, $y/\delta^* = 0.6$ (plotted versus nondimensional time τ), and its corresponding power spectrum (plotted versus nondimensional frequency, β) at two downstream locations, $x = 170$cm and $x = 230$cm.

of the experimental techniques, see Cohen, Breuer & Haritonidis, 1990). The free-stream velocity, U_0, was 6.65 m/s, so that the local Reynolds number, R, (based on displacement thickness, δ^*) at the location of the disturbance generator was approximately 1000. The flow measurements were made using constant temperature hot-wire anemometry. All three velocity components (u, v, w) were measured at several (x, y, z)-locations so that a complete map of the entire disturbance was obtained (x, y and z denote the streamwise, wall-normal and spanwise directions respectively).

Figure 1 shows the streamwise component of velocity at the centerline and inside the boundary layer at $y/\delta^* = 0.6$. The results are plotted in two forms: on the left of each pair, the velocity is graphed versus a non-dimensional time, $\tau = U_0(t - t_0)/(x - x_0)$ where $t_0 = (x - x_0)/0.935U_0$. The right side shows plots of the corresponding power spectrum at each station versus nondimensional frequency, $\beta = 2\pi f\delta^*/U_0$. The two x-stations are chosen because they represent two characteristic stages of the wave packet's development. In the first stage, at $x = 170$cm, the wave packet has a smooth and even appearance. At this stage it has a small amplitude (0.8% of U_0) and its spectrum indicates that

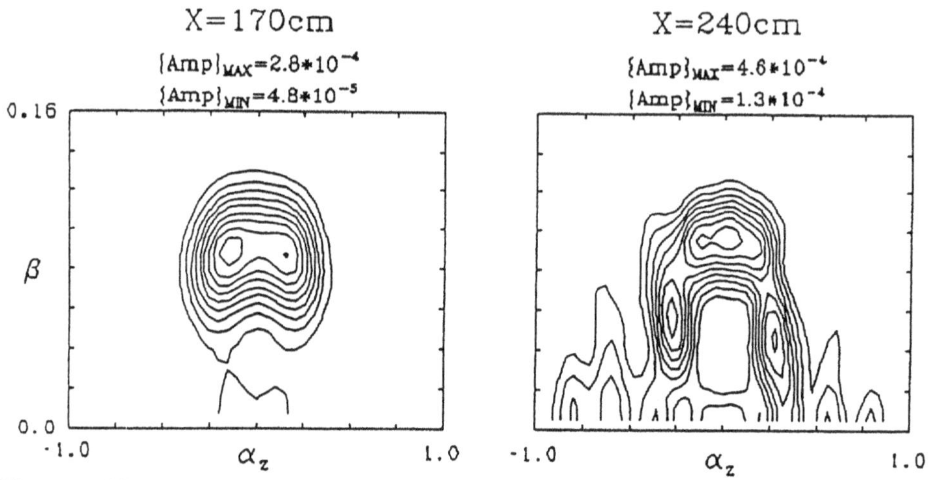

Figure 2: Wavenumber-frequency spectra of the u-component of velocity at $y/\delta^* = 0.6$ at the linear ($x = 170$cm) and subharmonic stages ($x = 240$cm).

the disturbance is comprised of a smooth, broad peak of frequencies centered about a nondimensional frequency of $\beta = 0.1$. This is characteristic of the linear stage and the results here are in very good agreement with the measurements of Gaster & Grant (1975) and the calculations of Gaster (1975).

At $x = 230$cm, however, the appearance of the disturbance has changed. At this stage it has grown to an amplitude of approximately 3% of U_0, and has almost completely lost one of its cycles. The power spectrum reveals that this is due to the presence of a subharmonic peak at $\beta = 0.05$ which, at this x-location, has become the most dominant spectral peak. Although not shown here, the gradual growth of the subharmonic peak was observed by several measurements at intervening x-locations. The growth of the subharmonic peak is not predicted by the linear evolution equations, indicating that, at this stage, nonlinear interactions have become significant in the disturbance's growth.

The linear and nonlinear development of the wave packet is more clearly seen in Figure 2 which shows the wavenumber-frequency spectra of the streamwise component of velocity computed at two downstream stations characteristic of the linear and subharmonic regimes. The two-dimensional spectra were computed by measuring the spanwise structure of the wave packet (at $y/\delta^* = 0.6$) and then performing a double Fourier transform on the two-dimensional velocity field. In the linear stage, most of the energy is concentrated about frequencies close to the fundamental frequency of $\beta = 0.1$ and "near-two-dimensional" modes. The actual maxima in the energy spectrum are for oblique modes with $\alpha_z = 0.15$. This is in contrast to the results of Gaster & Grant (1975) who show (for the linear stage) spectra with their maxima at $\alpha_z = 0$. However, their measurements were taken above the boundary layer at $y/\delta^* = 3.2$ while these results are for $y/\delta^* = 0.6$. The significance of this difference with regard to the spectra of u is discussed in the next section.

In the nonlinear stage, the spectrum is completely different. Three distinct centers of energy are now apparent. The first is the two-dimensional fundamental mode, centered at $\beta = 0.1$ and $\alpha_z = 0.0$, while the two others are at the subharmonic frequency of $\beta = 0.05$ and at $\alpha_z = \pm 0.25$. This clearly shows the subharmonic nature of the nonlinear stage.

2.1 Theoretical Considerations

Following Gaster (1975), we consider the wave packet as being a superposition of the least stable two- and three-dimensional Orr-Sommerfeld modes. For given values of the frequency, β, and spanwise wavenumber, α_z, the mode structure is governed by the linearized equations of motion (cf. Benney & Gustavsson):

$$\left(\frac{d^4}{dy^4} - 2k^2\frac{d^2}{dy^2} + k^4 - iR\alpha_x\left[(\bar{U} - c)(\frac{d^2}{dy^2} - k^2) - \frac{d^2\bar{U}}{dy^2}\right]\right)\hat{v} = 0 \tag{1}$$

$$\left(\frac{d^2}{dy^2} - k^2 - iR\alpha_x(\bar{U} - c)\right)\hat{\eta} = i\alpha_z R\frac{d\bar{U}}{dy}\hat{v} \tag{2}$$

where $k^2 = \alpha_x^2 + \alpha_z^2$ and $c = \beta/\alpha_x$. R is the local Reynolds number based on the local displacement thickness and free-stream velocity.

Equation (1) is the familiar Orr-Sommerfeld equation for the vertical velocity amplitude, $\hat{v}$, while equation (2) is the equation for the vertical vorticity amplitude, $\hat{\eta}$. The amplitudes of the streamwise and spanwise fluctuating velocities are given by:

$$\hat{u} = \frac{i}{k^2}\left(\alpha_x\frac{d\hat{v}}{dy} - \alpha_z\hat{\eta}\right) \quad \text{and} \quad \hat{w} = \frac{i}{k^2}\left(\alpha_z\frac{d\hat{v}}{dy} + \alpha_x\hat{\eta}\right) \quad . \tag{3}$$

One consequence of these equations is shown in Figure 3 which shows the wavenumber-frequency spectra for the u-, v- and w-components of velocity in the linear regime, at $x = 170$cm and $y/\delta^* = 0.9$. While the v spectrum shows the maximum energy located in a two-dimensional mode, the horizontal velocities indicate that the spanwise waves are the most energetic modes. This was observed above in Figure 2 and its explanation is found in the structure of the linearized equations, (1) and (2). The amplitude of the vertical velocity is governed by the Orr-Sommerfeld equation which, for this value of R, dictates that the two-dimensional modes are the most amplified and should be the most energetic modes in a linear disturbance. However, for $\alpha_z \neq 0$, the horizontal velocities also include a contribution from the vertical vorticity, η. Thus, while the two-dimensional mode may be the most amplified mode, the influence of the vertical vorticity results in a three-dimensional mode being the most energetic mode (for u and w). This point will be discussed further in a later section in connection with the structure of the disturbance inside and outside the boundary layer.

2.2 Subharmonic Resonance

Based on the power spectra of the disturbance (Figure 2), we believe that the nonlinear stage observed in the previous results is due to a subharmonic resonance mechanism of the type proposed by Craik (1971). This nonlinear interaction is characterized by the interaction between a two-dimensional wave and two oblique waves, both with frequency one half that of the fundamental. Inspection of Figure 2 confirms that this frequency relationship exists. The resonance conditions also require that the three waves have the same phase velocity. This was confirmed by solving equation (1) and determining the eigenvalue, α_x, for both the most energetic two-dimensional mode and the most energetic three-dimensional mode. The phase velocity c_r is given by $\beta/\Re(\alpha_x)$. For Figure 2, the phase velocities of the two- and three-dimensional waves were $0.34U_0$ and $0.36U_0$ respectively, satisfying the resonance conditions to within experimental accuracy.

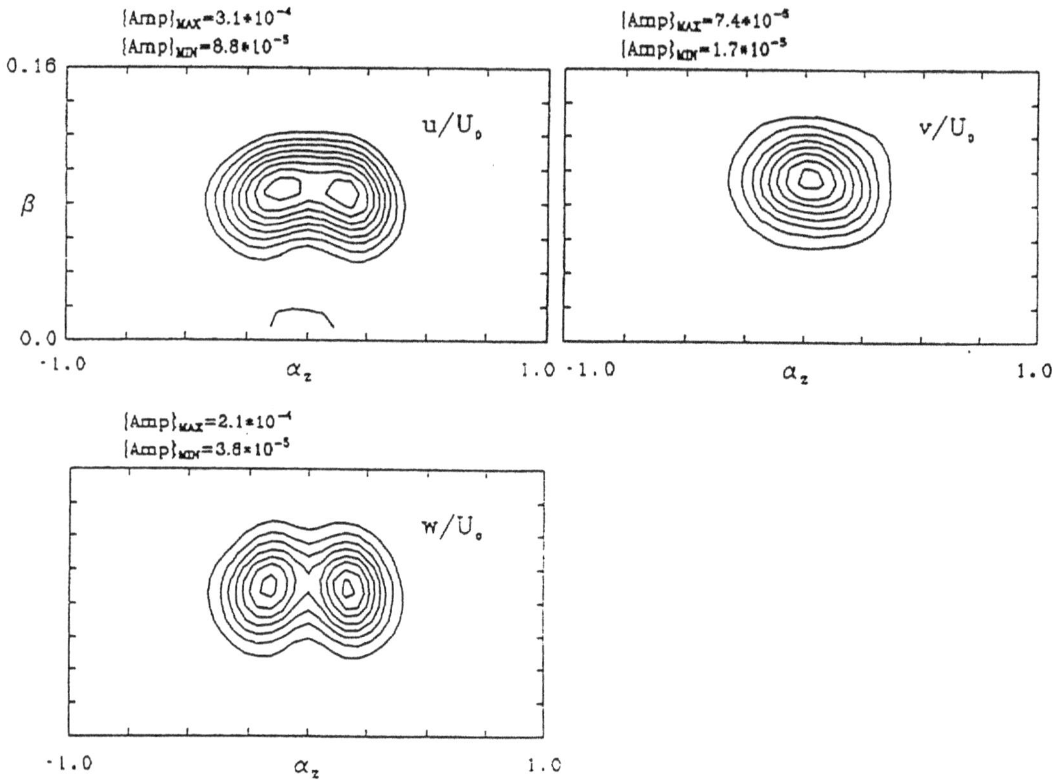

Figure 3: Wavenumber-frequency spectra of the $u-$, v- and w-components of velocity at x = 170cm, $y/\delta^* = 0.6$ showing the differences between the vertical and horizontal components of velocity due to vertical vorticity.

Further support for the belief that the observed nonlinear interactions are Craik-type resonances is given by the fact that the amplitude of the wave packet remains small during the stage of subharmonic growth. Herbert's (1988) results indicate that the Craik resonance is the only subharmonic secondary instability mechanism that remains viable as the amplitude of the fundamental wave decreases. In the present case, growth of the subharmonic is observed to begin when the wave packet has an amplitude of less than $0.01U_0$. However, in a companion series of experiments, the subharmonic was observed to grow when the peak-to-peak amplitude of the fundamental wave packet was less than $0.002U_0$. These amplitudes are too small to be accounted for by the higher amplitude broadband instabilities, again pointing towards the parametric resonance mechanism as the primary nonlinear interaction.

One should note that in this flow, the entire evolution of the wave packet, including the nonlinear interactions are *unforced*, and all of the observed frequencies and spatial scales arise naturally. In order to assure that there were not any preferential scales imposed by the experimental setup, the experiment was repeated several times, varying both the Reynolds number and the initial excitation amplitude. In all cases, the results were consistent with those presented here.

2.3 Vertical mode structure

At the two x-stations characterizing the linear and subharmonic regimes, the disturbance was mapped in $y - z - t$ space and the wavenumber-frequency power spectrum was computed for each y-location. From this, the vertical mode structure could be examined for any given values of β and α_z. Figure 4 shows the mode structure for the most energetic two- and three-dimensional modes at both the linear and subharmonic stages.

As might be expected, the correspondence between the computed and measured eigenfunctions in the linear stage is very strong. Both the measured amplitude and phase of all three components (two in the case of the streamwise mode) agree very well with the computed distributions, both in terms of shape and relative amplitudes. It should be noted that only one free constant is available in matching the amplitudes of the measured and computed eigenfunctions for all three velocity components. Similarly, one constant is available for matching the measured and computed phases of all three components.

In the subharmonic stage, the agreement is also excellent and indicates that, although nonlinear interactions are important in predicting the rise of the subharmonic modes, the individual mode shapes and phases are still well represented by linear theory. This suggests the use of weakly nonlinear theory, in which the vertical mode structure is assumed to be given by linear theory, while the the amplitudes of each mode are predicted by a system of nonlinear evolution equations. Computations based on this approach are presently underway.

2.4 The role of vertical vorticity

As an illustration of the importance of the effect of vertical vorticity on the horizontal velocities, Figure 5 shows the computed vertical distribution of the u-component of perturbation velocity for the same three-dimensional subharmonic mode as in Figure 4. The solid line shows the amplitude of the complete eigenfunction (the same as shown in Figure 4), while the dotted line shows the contribution to u from the Orr-Sommerfeld portion of the eigenmode (i.e. the term $i\alpha_x v'/k^2$). It is clear that within the boundary layer, the contribution of η is dominant, while outside the boundary layer, η rapidly goes to zero (because U' goes to zero) and the Orr-Sommerfeld portion becomes the sole contribution to the total amplitude. This has implications with regard to the contrast between the two-dimensional spectra for different velocity components (Figures 2 and 3) and we now see that the twin-peaked spectra of u and w will approach the spectra of v as y increases and the influence of η decreases.

The distinction between the behavior of the modes inside and outside the boundary layer has a significant effect on the spatial structure of the disturbance, as is illustrated in Figure 6 which shows contour plots of the streamwise perturbation velocity in the wave packet at two heights: $y/\delta^* = 0.7$ and $y/\delta^* = 4.0$, and in both the linear and the subharmonic stages. In the linear stage, in which the wave packet is dominated by modes centered around the fundamental frequency with $\alpha_z \approx 0$, the wave packet has the bowed shape, familiar from the results of Gaster (1975), and Gaster & Grant (1975). The only visible difference between the interior and exterior of the boundary layer is a change in sign, a result in accordance with the linear mode shape which changes phase as y passes through the edge of the boundary layer. However, in the subharmonic regime, the pattern inside the boundary layer is dramatically different and the smooth wave crests have broken up into peaks on either side of the centerline. In contrast to this, the pattern above the boundary layer still retains much of its "linear" characteristics. The

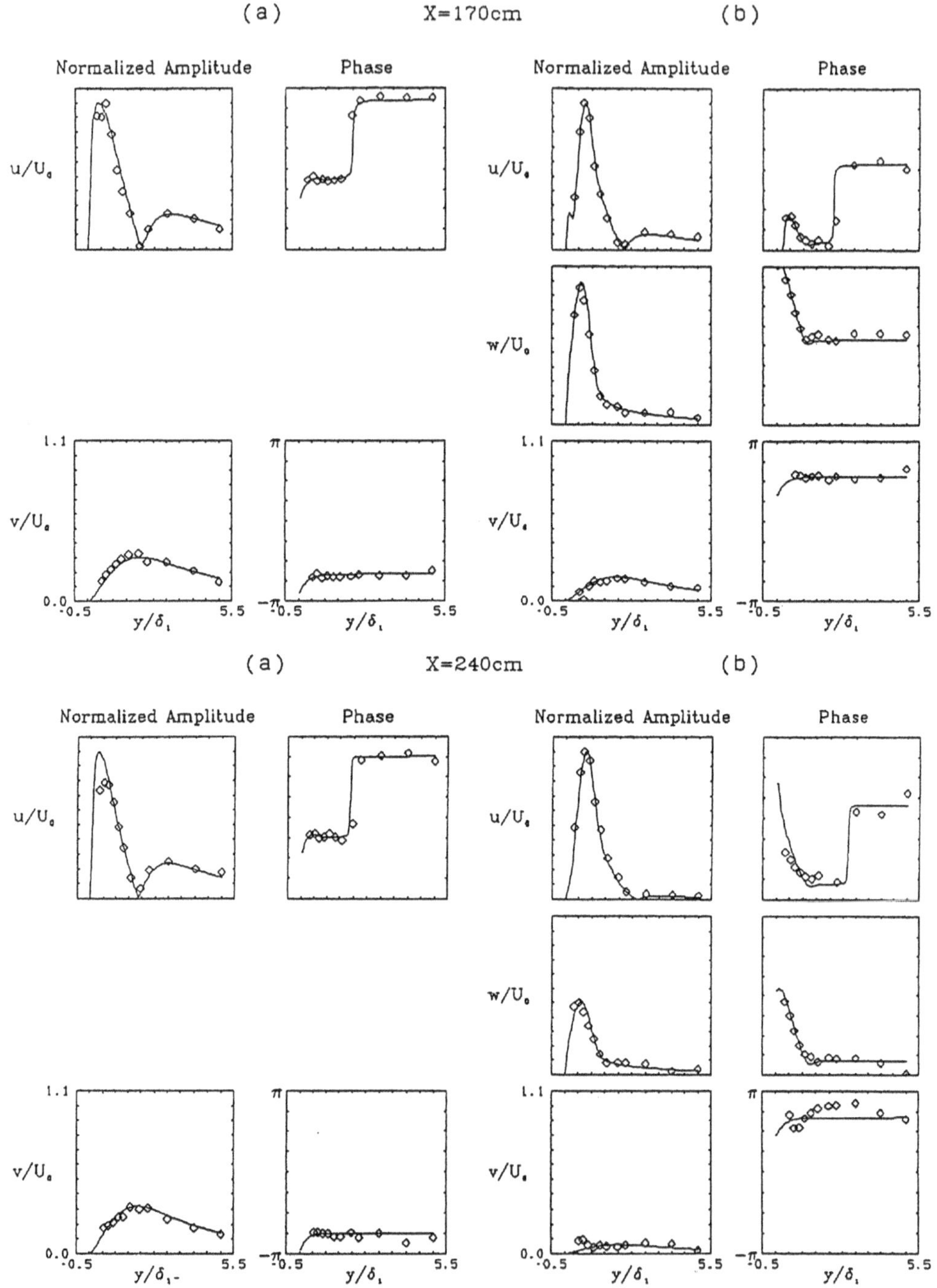

Figure 4: Experimental and theoretical distributions of the most energetic (a) two-dimensional and (b) three-dimensional modes in both the linear and subharmonic stages. Symbols indicate the experimental results while the solid lines show the theoretical mode structure.

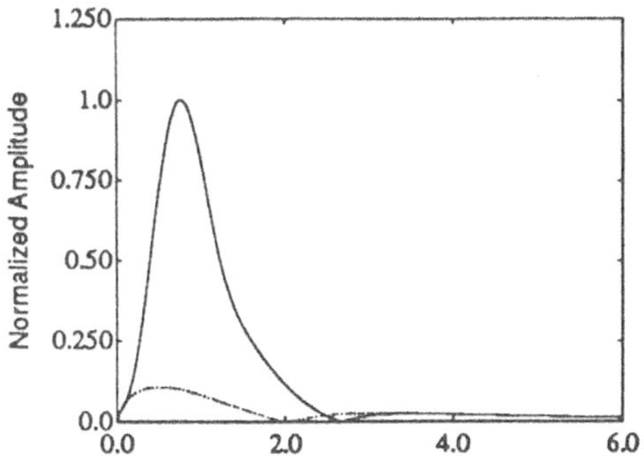

Figure 5: Computed eigenfunction for the most energetic three-dimensional mode in the subharmonic stage. The solid line indicates the amplitude of the entire eigenfunction while the dotted line indicates the contribution to the amplitude from the Orr-Sommerfeld mode only (i.e. without the contribution from η).

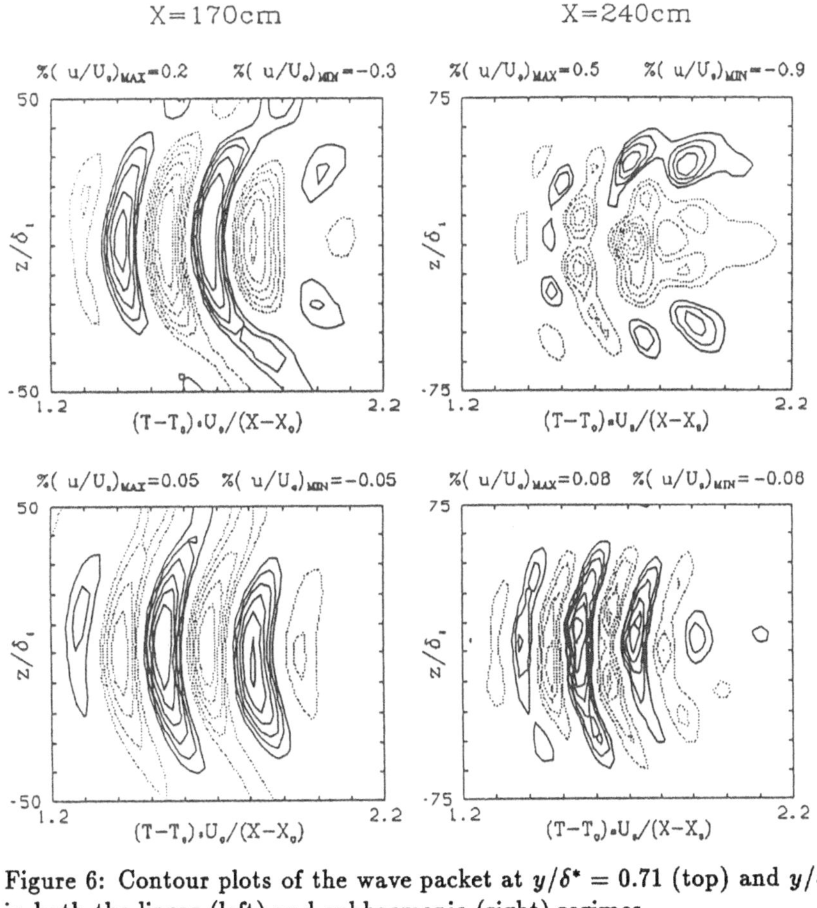

Figure 6: Contour plots of the wave packet at $y/\delta^* = 0.71$ (top) and $y/\delta^* = 4$ (bottom) in both the linear (left) and subharmonic (right) regimes.

wave crests are essentially intact, although there does appear to be some "peak splitting". The difference between the disturbance structure inside the boundary layer and above it can be attributed to the role of the three-dimensional modes. Due to the subharmonic resonance, the three-dimensional modes have become the most energetic and dominant modes in the disturbance. However, while they are very large inside the boundary layer, they decay very rapidly outside the boundary layer since the vertical vorticity, η, quickly goes to zero outside the region of mean shear, resulting in both the difference between both the structure and amplitude of the wave packet inside and outside the boundary layer.

This work was supported by the Bantrell Fellowship and AFOSR grant F49620-83-C0019 (JC), the ONR Graduate Fellowship and the Center for Fluid Mechanics at Brown University, supported by DARPA-URI grant N00014-86-K0754 (KSB). JHH enjoyed the support of ONR through grant N00014-90-J1302 during the preparation of this paper.

References

Cohen, J., Breuer, K. S. & Haritonidis, J. H., 1990. On the evolution of a wave packet in a laminar boundary layer. (In press).

Corke, T. C. & Mangano, R. A. 1989. Resonant growth of thre-dimensional modes in transitional Blasius boundary layers. *J. Fluid Mech.* **209**, 93 – 150.

Craik, A. D. D. 1971. Nonlinear resonant instability in boundary layers. *J. Fluid Mech.* **50**, 393–413.

Gaster, M. 1975. A theoretical model for the development of a wave packet in a laminar boundary layer. *Proc. Roy. Soc. Lond. Ser. A.* **347**, 271–289.

Gaster, M. & Grant, I. 1975. An experimental investigation of the formation and development of a wave packet in a laminar boundary layer. *Proc. Roy. Soc. Lond. Ser. A.* **347**, 253–269.

Herbert, T. 1988. Secondary instability of boundary layers. *Ann. Rev. Fluid Mech.* **20**, 487–526.

Kachanov, Y. S. & Levchenko, V. Y. 1984. The resonant interaction of disturbances at laminar turbulent transition in a boundary layer. *J. Fluid Mech.* **138**, 209–247.

Saric, W. S. & Thomas, A. S. W. 1984. Experiments on the subharmonic route to turbulence in boundary layers. In Tatsumi, I., editor, *Turbulence and chaotic phenomena in fluids*, p. 117. North-Holland.

Large-Scale Structures in Helical Turbulence

O.A. Druzhinin and G.A. Khomenko

Space Research Institute Academy of Sciences 117810, Moscow, USSR

The generation of large-scale vortices in the helical turbulence is usually called the hydrodynamical alpha-effect ($H\alpha$-effect) [1]. The equation for the mean field has the same form as appropriate α-effect equation in the mean-field MHD:

$$\partial_t \mathbf{\Omega} + \alpha curl \mathbf{\Omega} = \nu \Delta \mathbf{\Omega} \tag{1}$$

where $\mathbf{\Omega} = curl\mathbf{V}$ is a mean vorticity and the uniform coefficients α and ν are related to the random velocity field parameters. The effect of generation of the large-scale vortices is associated with the term $\alpha curl\mathbf{\Omega}$ in the equation (1) which leads to exponential growth of vorticity. The factor α in this term is expressed in the helicity of turbulent motion. The mean flow is also helical, i.e. $\mathbf{V} \cdot \mathbf{\Omega} \neq 0$. Helicity characterizes the topology of the velocity field and is related with the integral of motion in inviscid approximation.

Helical turbulence is not invariant under space reflection, its characteristics differ in left- and right-hand systems of reference. Usually the turbulence subjected to some pseudovector external field (e.g. , Coriolis forces or magnetic field) possesses this property. The possibility of large-scale instability of helical turbulence has been extensively discussed recently (see e.g. [2] and references therein). In 1983 [1] the example of such effect has been found for compressible fluid. Later [3-5] analogous effects were revealed in the incompressible case. Thus, the generation of large-scale motions is generic feature of helical turbulence.

Here we derive and analyze the averaged equations which govern the stages of development and saturation of instability in $H\alpha$-effect using the multiscale method of perturbation theory [3]. Two cases are of interest. First, when mean velocity W is weak in comparison with characteristic amplitude of the small- scale random motion and the large-scale instability develops [6]. The other case corresponds to the situation, when the amplitude of the mean flow is large and the instability saturates [7].

In dimensionless variables equations of motion of the compressible medium with pressure and density related by a polytropic law $P = \frac{1}{2}c^2\rho^2/\rho_0$ (c is the sound speed) can be written as

$$\partial_t V_i + Re V_j \partial_j V_i = -Re^{-1}\mu^{-2}\partial_i \rho + \partial^2 V_i + f_i \tag{2}$$

$$\partial_t \rho + Re\partial_j(V_j \rho) = 0,$$

where $\mu = M/Re = \nu/c\lambda_T \leq 1$ with the Mach number $M = V^{(T)}/c$, ν the viscosity and Re the Reynolds number of small-scale motion $Re = \lambda_T V^{(T)}/\nu$ and λ_T is characteristic spatial scale of the random motion.

Advances in Turbulence 3
Editors: A.V. Johansson · P.H. Alfredsson

A zero-mean ($<\mathbf{f}>=0$) random external force is assumed to be homogeneous, isotropic, stationary and helical. We also imply that $\partial_i f_i = 0$. The general form of the correlation tensor for such a process is given by

$$< f_m(\mathbf{k},\omega) f_n(\mathbf{k}',\omega') >= \delta(\mathbf{k}+\mathbf{k}')\delta(\omega+\omega')Q_{mn}(k,\omega)(k^4+\omega^2) \tag{3}$$

where

$$Q_{mn}(k,\omega) = D(k,\omega)(k^2\delta_{mn} - k_{mn}) - iH(k,\omega)\epsilon_{mns}k_s$$

. Here and below the angular brackets denote the ensemble average. Helicity is characterized by the quantity $H \sim <\mathbf{f}\cdot curl\mathbf{f}>$.

We use the multiscale formalism (see e.g. [3]), with the small-scale Reynolds number considered as a small parameter.

Let us define two different spatial and temporal scales as follows:

$$\mathbf{x} = Re^2\mathbf{X}, \tau = Re^4 T, \tag{4}$$

where $\mathbf{x},\tau$ and $\mathbf{X},T$ are small-scale and large-scale variables, respectively.

First, we consider the case when large-scale component of the velocity W is assumed to be weak as compared to small-scale turbulent background.

The velocity and the density fields can be expanded in small-scale Reynolds number Re (which is treated as a small parameter) as

$$V_i = V_i^{(0)} + V_i^{(1)}Re^1 + V_i^{(2)}Re^2 + V_i^{(3)}Re^3 + V_i^{(4)}Re^4 + V_i^{(5)}Re^5 ... \tag{5a}$$

$$\rho = \rho^{(0)} + \rho^{(1)}Re^1 + \rho^{(2)}Re^2 + \rho^{(3)}Re^3 + \rho^{(4)}Re^4 + \rho^{(5)}Re^5 \tag{5b}$$

Substituting (5) into (2) and taking into account the scale separation, one comes to the set of linear differential equations for $\rho^{(k)} and \mathbf{V}^{(k)}$ at various orders in Re.

The equation describing the evolution of the mean flow W, which corresponds to the first order in Re (i.e. $<\mathbf{V}^{(1)}>=\mathbf{W}$), is obtained as a solvability condition at fifth order in Re (see [6] for detailes) in the form

$$\partial_T\mathbf{W} + (\mathbf{W}\cdot\nabla)\mathbf{W} + \alpha_0 \cdot curl\mathbf{W} = -\mu^{-2}\nabla < \rho^{(4)} > + \nabla^2\mathbf{W} \tag{6}$$

with the condition

$$\nabla\cdot\mathbf{W} = 0, \tag{7}$$

which follows from lower orders. Note, that in this approximation the mean field is solenoidal. The uniform coefficient α_0 is given by

$$\alpha_0 = \frac{8\pi}{3}\int_{-\infty}^{+\infty} d\omega \int_{-\infty}^{+\infty} k^4 dk \frac{H(k,\omega)\omega^2}{(k^2\mu^{-2}-\omega^2)^2 + \omega^2 k^4} \tag{8}$$

where $H(k,\omega)$ is the helical part of the correlation function of the background turbulence. The term $\alpha_0 curl\mathbf{W}$ is responsible for the exponential growth of the solution at rate $\gamma = \alpha_0 k - k^2$.

When the amplitude of the mean flow is large, the velocity and the density fields can be expanded in small-scale Reynolds number Re as follows (cf. (5)):

$$\mathbf{V} = \mathbf{V}^{(-1)}Re^{-1}1 + \mathbf{V}^{(0)} + \mathbf{V}^{(1)}Re^1 + \mathbf{V}^{(2)}Re^2 + ... \tag{9}$$

$$\rho = \rho^{(-1)}Re^{-1} + \rho^{(0)} + \rho^{(1)}Re^{1} + \rho^{(2)}Re^{2}...$$

where $\mathbf{V}^{(-1)}$ corresponds to large-scale component of the velocity field, $\mathbf{V}^{(1)} = \mathbf{W}$.

The averaged equation for mean velocity $\mathbf{W}$ appears as a solvability condition at third order in Re (see [7] for details):

$$\partial_T \mathbf{W} + \alpha(|\mathbf{W}|) curl\mathbf{W} = -\mu^2 \nabla < \rho > + \nabla^2 \mathbf{W} \tag{10}$$

with the conditions

$$\nabla \cdot \mathbf{W} = 0, \qquad (\mathbf{W} \cdot \nabla)\mathbf{W} = 0 \tag{11}$$

The mean field is solenoidal and the advective term vanishes. Hence, (5) contains the single nonlinear term $\alpha(|\mathbf{W}|)$ namely

$$\alpha(|\mathbf{W}|) = \frac{8\pi}{3}\int_{-\infty}^{+\infty} d\omega \int_0^{\infty} dk \int_{-1}^{1} d\varphi \frac{H(k,\omega)k^4(\omega - kW\varphi)^2}{[\mu^{-2}k^2 - (\omega^2 - kW\varphi)^2]^2 + (\omega^2 - kW\varphi)^2 k^4} \tag{12}$$

When W is small $\alpha \approx \alpha_0 = const$ and $\alpha \to 0$ for large W. Hence, when W grows, $\alpha(W)$ decreases and the instability saturates.

When $\mathbf{W}$ depends only on one spatial coordinate, namely, $\mathbf{W} = \{W_x(z), W_y(z), 0\}$ the nonlinear equation (10) has an exact stationary solution. This solution represents the Beltrami flow (the special case of the ABC-flow) and has the form

$$\mathbf{W} = \{W_{st}\sin(k_{st}z + \varphi), W_{st}\cos(k_{st}z + \varphi), 0\}$$

$$\mathbf{W}_{st} \cdot curl\mathbf{W}_{st} = k_{st}W_{st}^2$$

Here k_{st} and W_{st} are stationary wave number and the amplitude of the solution, respectively, which fits the condition $\alpha(W_{st}) = k_{st}$ and φ is a constant, determined by initial conditions.

We analyzed equation (10) numerically using an explicit difference scheme. The numerical analysis shows that solution of the boundary problem with free boundaries evolves in the following way. First the amplitude of the mode with the wave number $k_0 = \alpha_0/2$, which corresponds to the maximum growth rate, increases at the rate $\gamma = \alpha_0^2/4$ until W reaches the value W_0 determined by identity $\alpha(W_0) = k_0$. Then the modes with smaller k grow up slowly to much larger values of the amplitude (the stable stationary solution corresponds to $k_{st} = 0$). The temporal evolution is similar to the energy transfer to the region of small wave numbers. Results are represented in figs. 1, 2.

Thus, the initial stage of large-scale structure formation is described by eq. (6). At this stage a weak vortex is amplified and acquires some topological structure. The saturation stage is described by eq. (10). Note, that the advective term vanishes and α becomes dependent on amplitude of the mean flow W, which corresponds to the Doppler shift (see (12)). The stationary scale of the structure is defined by boundary conditions. For free boundaries this scale tends to infinity.

Therefore, the random small-scale helical stirring force brings about instability of the large-scale helical ($\mathbf{W} \cdot curl\mathbf{W} \neq 0$) motion. These results can be interpreted as an inverse energy flux in helical turbulence which leads to formation of a stationary large-scale helical structure.

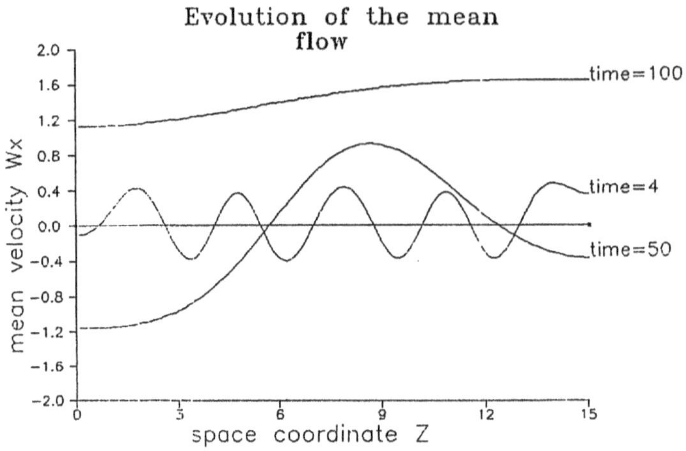

Fig.1

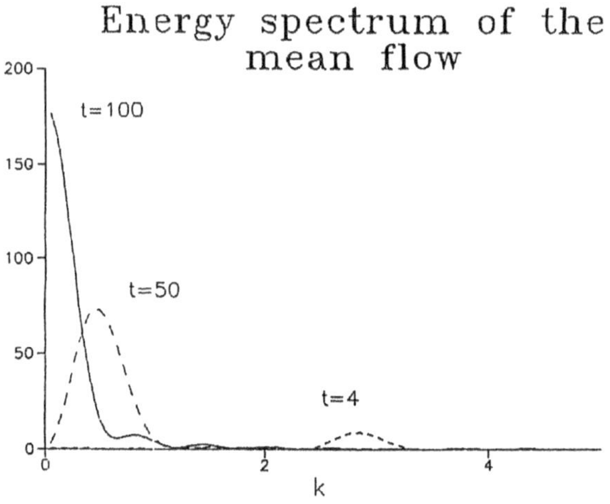

Fig.2

References

1. S.S. Moiseev, R.Z. Sagdeev, A.V. Tur, G.A. Khomenko and V:V. Yanovsky 1983 *Sov. Phys. JETP*, **58**, 1149.

2. Khomenko G.A., Moiseev S.S., Tur A.V. 1989 "Hydrodynamical alpha-effect in compressible fluid", submitted to *J. Fluid Mech.*

3. U. Frisch, Z.S. She, P.L. Sulem 1987 *Physica* **28D**, 382.

4. S.S. Moiseev, P.B. Rutkevich, A.V. Tur, V.V. Yanovskii 1988 *Z. Exp.Teor.Fiz.*, **94**, 144 (Engl. transl. Sov. Phys. JETP)

5. Gvaramadze, V.V., Khomenko, G.A. and Tur, A.V. 1989 *Geophys. Astrophys. Fluid Dyn.* **46**, 53-69.

6. O.A. Druzhinin, G.A. Khomenko and S.S. Moiseev 1990 "Turbulent forcing and large-scale instability of large-scale flows." (Submitted to *Physica D*)

7. O.A. Druzhinin and G.A. Khomenko 1990 "Nonlinear hydrodynamic α-effect in compressible medium" (Submitted to *Phys. Lett. A*)

Experiments on the Effect of Mean Flow Unsteadiness on the Subharmonic Growth in a Plane Mixing Layer

M.R. Hajj, R.W. Miksad, and E.J. Powers

College of Engineering, The University of Texas at Austin, Austin, Tx 78712, USA

1. Introduction

Shen (1961) considered the effect of mean flow unsteadiness on the growth of the instability fluctuations for the case where the mean flow is represented as a steady profile modified by a time function. The analysis resulted in a perturbation of the stream function that is similar to that under steady conditions but over a stretched time variable. Solis (1986) showed that in the case where the variation of the mean flow is periodic, the instability fluctuations are modulated in amplitude and phase at the frequency of the mean flow unsteadiness. These modulations were experimentally verified by Solis (1986) in a series of experiments on unsteady wakes. The present work focuses on the role of small-amplitude mean flow unsteadiness on the subharmonic generation and growth in plane mixing layers.

2. Experimental set-up

The mixing layer was formed by the merging of two streams that have different free stream velocities. The high speed stream was 7.17 m/sec and the low speed stream was 1.51 m/sec. This resulted in a velocity differential $\Delta U = U_1 - U_2 = 5.66$ m/sec and a velocity ratio, defined as $R = (U_1 - U_2)/(U_1 + U_2)$, equal to 0.65. The measured frequency of the fundamental instability under steady conditions was $f_o = 215$ Hz. The periodic mean flow unsteadiness was generated by varying the cross sectional area of the sonic throat of the tunnel. The magnitude of the sonic throat area variations was 1.1%. It is important to note here that the amplitude variations of the free stream velocities, besides being dependent on the variations in the area of the sonic throat, are also dependent on the acoustic characteristics of the wind tunnel. Desired peak to peak variations of the order of 1.5% in both streams were obtained when the unsteadiness frequency was set at 35 Hz. The velocity fluctuations at two downstream locations were measured using a special probe with two sensing elements (Jones et al. (1988)) and a DISA 56N/C hot wire anemometry system. The experiments were conducted under natural excitation conditions.

3. Effects of Mean Flow Unsteadiness

3.1 Spectral Development

The overall effect of mean flow unsteadiness on the development of the different components of the flow can be obtained by comparing the power spectra, under unsteady conditions, at two downstream locations, to those under steady conditions, as shown in figure 1. The immediate effect of the mean flow unsteadiness appears in the new

Advances in Turbulence 3
Editors: A.V. Johansson · P.H. Alfredsson

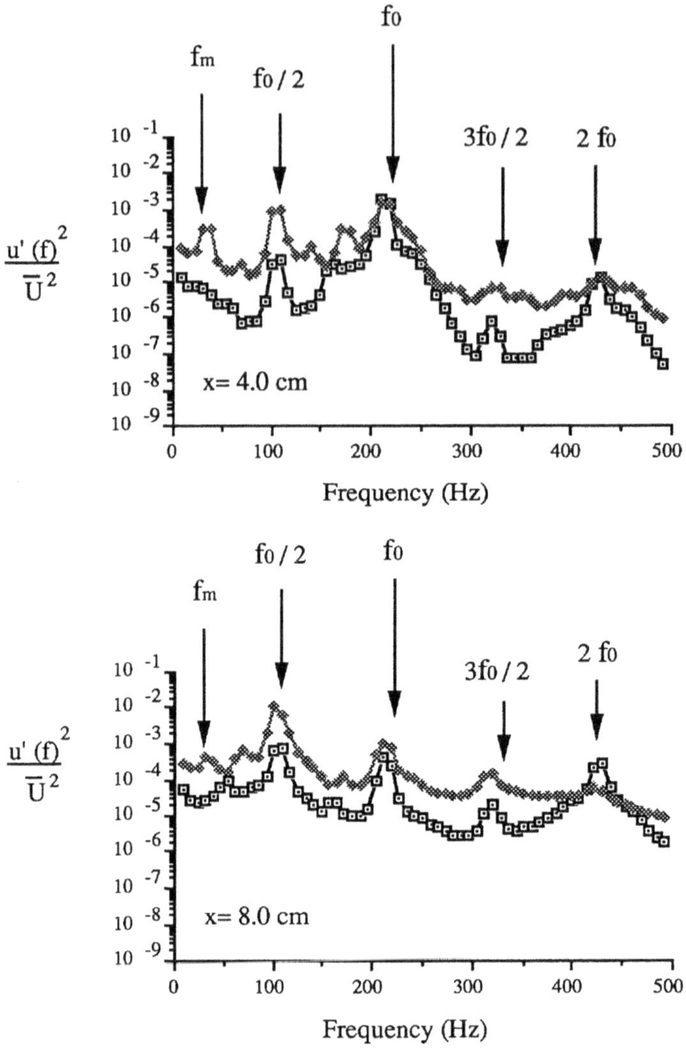

Figure 1. Velocity Power Spectra at x= 4.0 and 8.0 cm under steady —▫— and unsteady —◇— mean flow conditions.

peak at $f_m = 35$ Hz. Note that the energy of the fluctuations at the fundamental component is the same under steady and unsteady conditions. In contrast, the energy level of the fluctuations at the subharmonic frequency is much larger under the unsteady conditions. Note also that under unsteady conditions, the energy level in the valleys between the spectrum peaks is larger than under steady conditions. Therefore it appears that the mean flow unsteadiness enhances the nonlinear transfer of energy to the subharmonic and to valleys between the spectrum peaks. This energy transfer can best be detected and quantified by conducting polyspectral and transfer function measurements that estimate the level and strength of nonlinear three-wave interactions that are taking place.

3.2 Nonlinear Interactions

When two modes of frequencies f_i and f_j interact to form a third mode of frequency f_k, such that $f_k = f_i \pm f_j$, the phase of the resulting mode is equal to the sum (or difference) of the two interacting modes, i.e. $\theta_k = \theta_i \pm \theta_j$. This implies that the three modes become phase coherent over many realizations. Nonlinear couplings between the three modes can then be detected by measuring their phase coherence. The cross-bicoherence provides a measure of the phase coherence and subsequently of the three-wave coupling. It is defined as:

$$b^2(f_i, f_j) = \frac{| B(f_i, f_j) |^2}{<| X_{fi} X_{fj} |^2><| Y_{fk} |^2>}, \qquad f_i \pm f_j = f_k$$

where $B(f_i, f_j)$ is the cross-bispectrum and is equal to $< Y_{fk} X^*_{fi} X^*_{fj} >$, and $< ...... >$ denotes expected value over many realizations. The cross-bicoherence, as defined here, varies between 0 and 1. A more detailed explanation and calculation technique of the bicoherence is given in Kim and Powers (1979).

Figure 2 shows a comparison of the nonlinearly coupled modes under steady and unsteady mean flow conditions. These plots show that mean flow unsteadiness enhances the couplings between many modes that do not couple under steady conditions. For example, at $x = 4.0$ cm, under steady mean flow conditions, the level of nonlinear couplings is the largest between the fundamental and its subharmonic, $(f_0, \pm f0/2)$, and the fundamental and its harmonic, $(2f_0, -f_0)$ and (f_0, f_0). Under unsteady mean flow conditions, the level of the couplings between the fundamental and its harmonics is reduced. However, high levels of coherence are detected at triads that include the fundamental, f_0, the subharmonic, $f_0/2$, and the modulation frequency, f_m. These coherencies indicate that the fundamental and subharmonic modes are modulated by the low frequency component at f_m. This indicates the involvement of the modulation frequency in nonlinear couplings to transfer energy to the different components of the flow. By comparing the bicoherency plots under steady and unsteady mean flow conditions, at $x = 8$ cm, we notice that the mean flow unsteadiness increases the level of coupling between the fundamental and its subharmonic, $(f_0, \pm f_0/2)$ and the fundamental and the 3/2 harmonic, $(3f_0/2, \pm f_0)$. Also, notice the increased levels of coupling between many other triads. This indicates that the mean flow unsteadiness enhanced the nonlinear couplings and subsequently the energy transfer between the different components of the flow.

While the cross-bicoherence results provide an experimental evidence of the coupling between the mean flow unsteadiness and the fundamental and subharmonic modes, they cannot provide information on the efficiency of these couplings and their role in exchanging energy between these modes. These characteristics can best be investigated by modeling the governing equation by a transfer function system that is characterized by a parallel combination of linear, L_{fk}, and quadratic, $Q^{fk}_{fi;fj}$, transfer functions. The governing equations can then be expressed as a transfer function equation. Such an equation treats the fluctuations at two locations as an input-output system with the fluctuations at the upstream locations, X_{fk} , being the input and those at the downstream locations,Y_{fk}, being the output.

$$Y_{fk} = L_{fk} X_{fk} + \Sigma Q^{fk}_{fi;fj} X_{fi} X_{fj} \qquad f_i \pm f_j = f_k$$

In this equation, $L_{fk} X_{fk}$ is the amplitudecontribution of the linear growth to the output

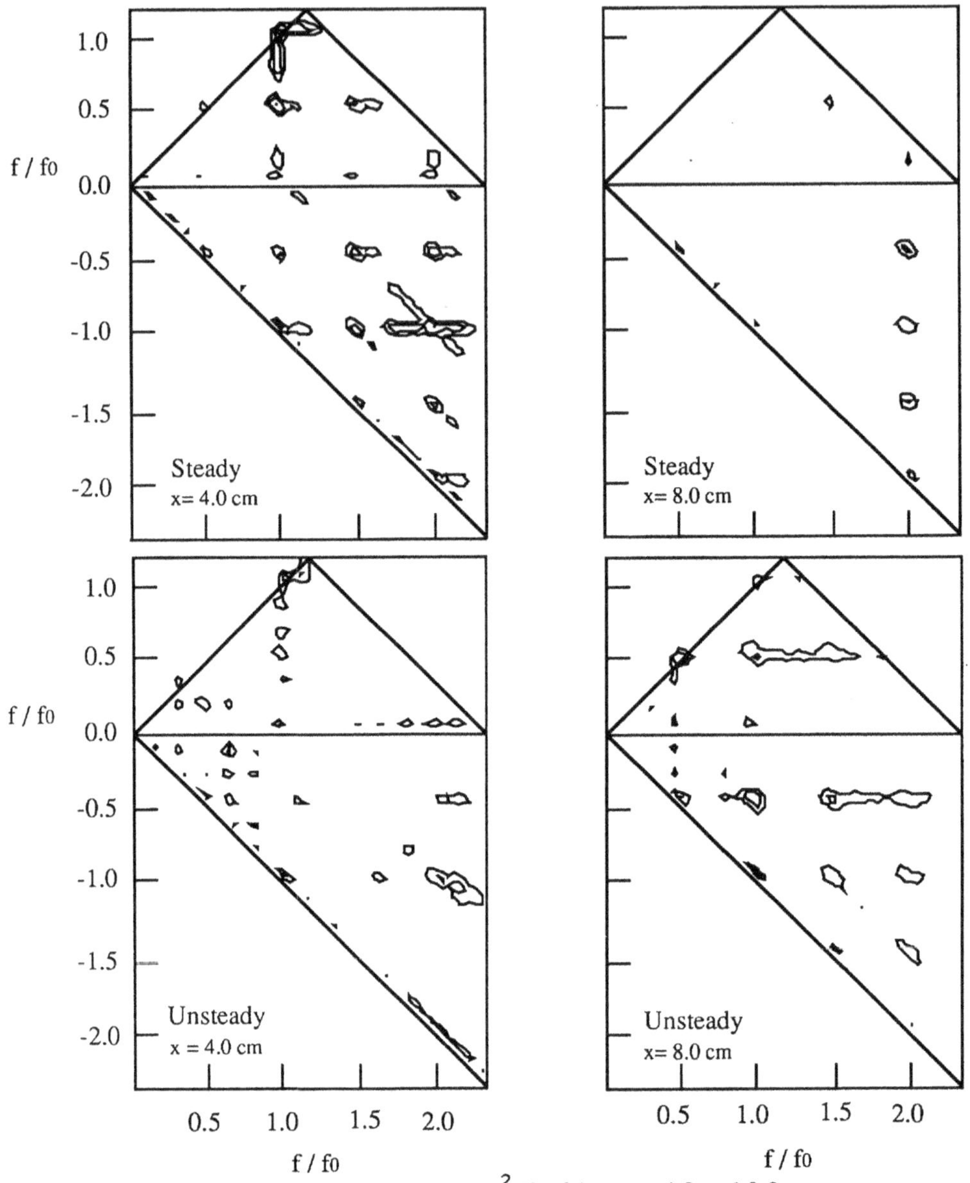

Figure 2. Contours of the cross-bicoherence, b^2 (fi , fj), at x= 4.0 and 8.0 cm under steady and unsteady mean flow conditions. Contour levels are set at 0.3, 0.6, and 0.9.

Y_{fk} and $\Sigma Q^{fk}_{fi;fj} X_{fi} X_{fj}$ is the contribution ofthe quadratic interactions of all input modes at f_i and f_j that could subtract or add to the output mode at f_k. A heuristic model that establishes the connection between the Navier-Stokes equations and our use of the transfer function measurements is given in Hajj (1989). Methods to estimate these functions are given by Kim and Powers (1988) and Ritz and Powers (1986). It is important to note here that while the linear transfer function, L_{fk}, is a dimensionless

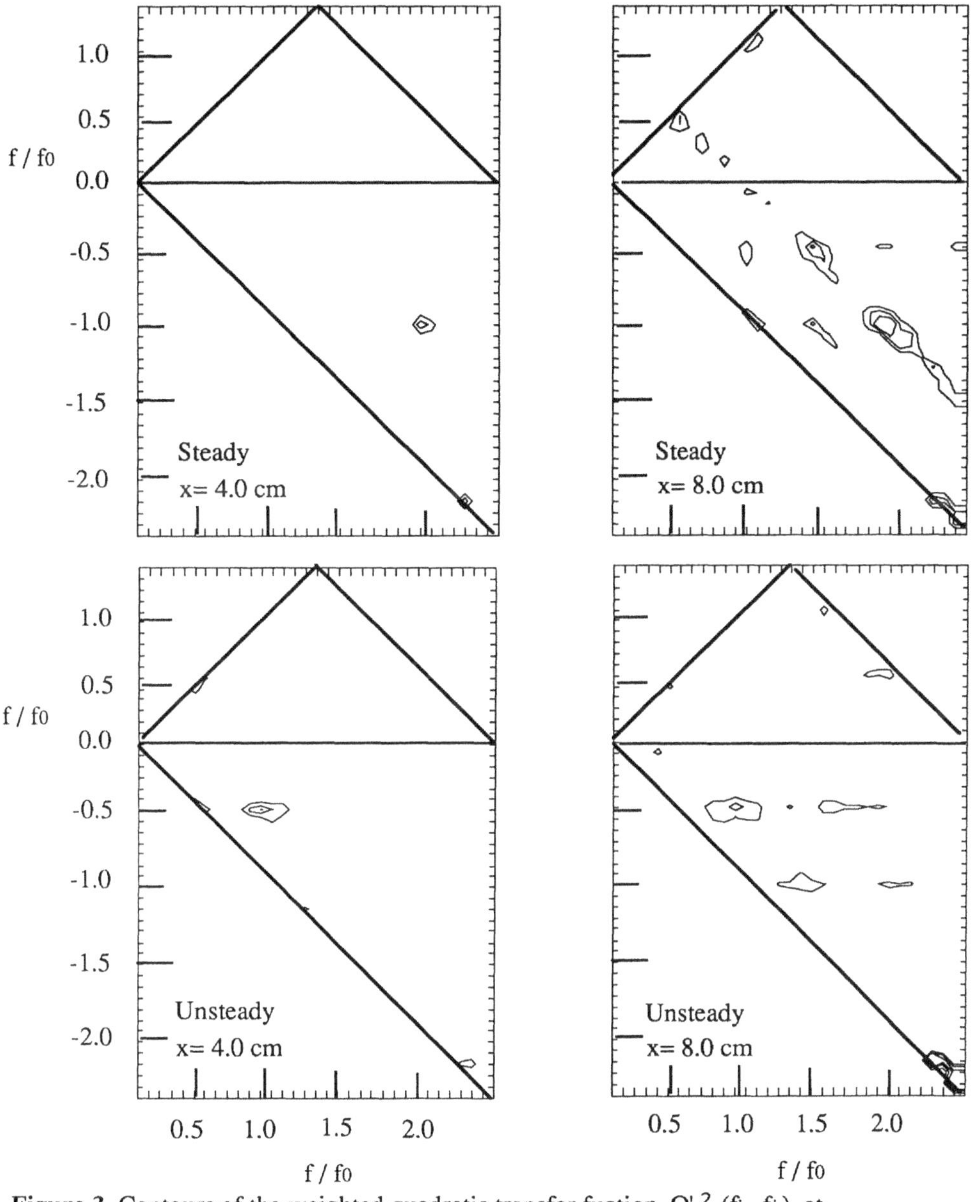

Figure 3. Contours of the weighted quadratic transfer fuction, Q' 2 (fi , fj), at x= 4.0 and 8.0 cm under steady and unsteady mean flow conditions.

function, the quadratic transfer function, Qfk,fifj, is dimensional. A good measure of the strength of the nonlinear interactions taking place between three modes can be determined by calculating a nondimensional, weighted quadratic transfer function defined as:

$$Q'^2(f_i, f_j) = \mid Q^{fk}_{fi;fj} \mid^2 \sqrt{\mid X_{fi} X_{fj} \mid^2} \qquad f_i \pm f_j = f_k$$

Notice that the weighted quadratic transfer function, $Q'^2(f_i, f_j)$, depends on the product of two quantities. The first is the amplitude of the quadratic transfer function. This quantity models the "underlying mechanism" of the nonlinear interaction. The second quantity is the "interaction potential". It represents the energy content of the interacting modes. For an interaction to be efficient, both quantities should be large.

The contour plots of the non-dimensional, weighted quadratic transfer function under steady and unsteady mean flow conditions is shown in figure 3. These plots show different efficient nonlinear interactions under unsteady mean flow conditions than under steady mean flow conditions. Under steady mean flow conditions, the most efficient energy transfer occurs between the fundamental and its harmonic. In contrast, under unsteady mean flow conditions, the most efficient energy transfer is that between the fundamental and its subharmonic. The effect of mean flow unsteadiness is also more pronounced further downstream. At $x = 8$ cm, under steady mean flow conditions, we see a pattern of interactions through which the fundamental is redistributing its energy to the different components of the flow. The fundamental-subharmonic energy exchange is also apparent. Under unsteady mean flow conditions, at the same location, we notice a more active role of the subharmonic in redistributing energy to the different components than under steady mean flow conditions.

4. Conclusions

In this paper, an experimental measure of the effect of mean flow unsteadiness on the transition to turbulence in a mixing layer is given. A comparison of the nonlinear couplings under steady and unsteady mean flow conditions show that mean flow unsteadiness plays a key role in enhancing the transfer of energy to the different components of the flow via nonlinear interactions. The results show that fluctuations introduced by the mean flow unsteadiness interact with the fundamental and subharmonic modes to produce multiple sideband fluctuations that can accelerate the redistribution of the energy the valleys via nonlinear interactions. Quadratic transfer function measurements indicate that the efficiency of energy transfer between the fundamental and its subharmonic increases under unsteady conditions. This results in a faster growth of the subharmonic and subsequently a more rapid transition to turbulence with downstream distance.

Acknowledgements

This paper is based in part upon work supported by the Texas Advanced Technology Program under Grant No. ATP 3280, and in part by the National Science Foundation under Grant No. MSM-8211205. The digital signal processing techniques were developed under the auspices of the Department of Defense Joint Services Electronics Program under the Air Force Office of Scientific Research (AFOSR) Contract No. F49620-89-C-0044.

References

Hajj, M. R. 1989 Experimental investigation of the nonlinear interactions associated with the subharmonic growth in the transition to turbulence in plane mixing layers, *Ph. D. Dissertation*, The University of Texas at Austin, Austin, Tx, 78712.

Jones, F. L., Ritz, C. P., Miksad, R. W., Powers, E. J. and Solis, S. R. 1988 Measurement of the local wavenumber and frequency spectrum in a plane wake, *Exp. in Fluids* **6**, 365- 372.

Kim, K. I. and Powers, E. J. 1988 A digital method of modeling quadratically nonlinear systems with a general random input, *IEEE Trans. on Acoustics, Speech, and Signal Processing* **36**, no.11, 1758-1769.

Kim, Y. C. and Powers, E. J. 1979 Digital bispectral analysis and its applications to nonlinear wave interactions, *IEEE Trans. Plasma Science* **PS-7**, 120-131.

Ritz, Ch. P. and Powers, E. J. 1986 Estimation of nonlinear transfer functions for fully developed turbulence, *Physica D, Nonlinear Phenomena* **20D**, no.2/3, 320-334.

Shen, S. F. 1961 Some considerations on the laminar stability of time-dependent basic flows *J. Aerosp. Sci.*, **28**, 397-404.

Solis, R. S. 1986 Experiments on the effects of mean flow unsteadiness on the transition to turbulence of a laminar wake, *Ph. D. Dissertation*, The University of Texas at Austin, Austin, Tx, 78712.

An Eigenfunction Expansion of Localized Disturbances

Dan S. Henningson
Department of Mathematics, MIT, Cambridge MA 02139, USA

1. Introduction

Turbulence inherently involves complicated three-dimensional motions, whereas laminar flow in many instances is two-dimensional. In investigations of the transition process from laminar to turbulent flow, the onset of three-dimensionality has therefore played a major role. The pioneering work of Klebanoff, Tidstrom & Sargent (1962) on the three-dimensional nature of transition in boundary layers started a process which led to the theory of secondary instability. Secondary instability, recently reviewed by Herbert (1988), is able to predict the onset of three-dimensionality as an instability of two-dimensional finite amplitude waves to infinitesimal oblique disturbances. The primary wave is usually taken as the least stable Orr-Sommerfeld (O-S) mode, which is two-dimensional in the Reynolds number range around the onset of growth, i.e. around the critical Reynolds number.

Once a disturbance is three-dimensional, large horizontal velocity components quickly develop. An illustrative description of that process can be made in terms of a *lift-up* effect. In a three-dimensional disturbance a major part of the horizontal momentum is retained when a fluid particle is lifted up or displaced vertically. This provides a contribution to the horizontal disturbance velocity if there is a mean shear present. Henningson (1988) pointed out that this lift-up part is directly proportional to the normal vorticity component and, using a numerical example, that its main effect on the flow is to create shear layers. Landahl (1980) showed the importance of the effect when low streamwise wavenumbers of the normal velocity are excited. He showed that the energy for such streaky disturbances will, in the inviscid case, grow linearly with time. We will refer to this growth mechanism as an algebraic instability.

Disturbances need not be initially two-dimensional. For subcritical Reynolds numbers in the channel flow problem, for example, the least stable mode is in many cases an oblique, i.e. three-dimensional, wave. Disturbances which are localized in space and time are also initially three-dimensional. Numerical simulations of the development of localized disturbances in channel flow by Henningson, Johansson & Lundbladh (1990) have shown that the algebraic instability can result in an increase of the energy up to one hundred times before dispersive and viscous effects dampen the development. An analysis of the energy content of the algebraically growing disturbances showed that essentially all of the energy was associated with the normal vorticity component.

A theoretical investigation of localized disturbances can be done by solving the initial value problem. For small amplitudes Schensted (1961) expressed the solution for two-dimensional channel flow using an eigenfunction expansion in O-S modes. In the case of three-dimensional disturbances Gustavsson & Hultgren (1980) used a Fourier-Laplace transform technique to solve the initial value problem. They pointed out that both O-S modes as well as Squire modes, i.e. eigensolutions of the normal vorticity equation,

Advances in Turbulence 3
Editors: A.V. Johansson · P.H. Alfredsson

were necessary for a complete description of the flow field. They also investigated the possibility of resonances between the two types of modes for Couette flow.

In this paper we will consider the complete eigenfunction expansion for three-dimensional channel flow. We obtain a compact expression which easily yields analytical as well as numerical results. The relation of the O-S and Squire modes to the lift-up effect and streak formation is illuminated and the expansion is also applied to the time development of localized disturbances.

2. The eigenfunction expansion

We consider the channel flow problem and start with the linearized, Fourier transformed, three-dimensional Navier-Stokes equations. If the pressure is eliminated, the problem is reduced to two equations governing the time development of the normal velocity (v) and the normal vorticity (η). We have

$$(\frac{\partial}{\partial t}+i\alpha U)(D^2-k^2)\hat{v}-i\alpha U''\hat{v}-\frac{1}{R}(D^2-k^2)^2\hat{v} = 0, \tag{1}$$

$$(\frac{\partial}{\partial t}+i\alpha U)\hat{\eta}-\frac{1}{R}(D^2-k^2)\hat{\eta} = -i\beta U'\hat{v}, \tag{2}$$

together with the boundary conditions $\hat{v}(\pm 1)=D\hat{v}(\pm 1)=\hat{\eta}(\pm 1)=0$. All quantities are non-dimensionalized using the channel half-height (h) and the centerline velocity (U_{CL}). U is the parabolic mean flow, $R=U_{CL}h/\nu$ the Reynolds number, $\mathrm{k}=(\alpha,\beta)$ the wavenumber vector and $D={}'=d/dy$ is the normal derivative. x and z are the streamwise and spanwise directions and α and β are the wavenumbers in the respective directions.

The eigenvalue problems obtained by assuming exponential time dependence for the solution of the operators above have a complete set of eigenfunctions (DiPrima & Habetler, 1969). Thus we may expand the solution in normal velocity and normal vorticity eigenfunctions. For the normal velocity we have

$$\hat{v}=\sum_{n=1}^{N} K_n\tilde{v}_n e^{-i\alpha c_n t}, \tag{3}$$

where $\tilde{v}_n$ is a solution to the O-S equation corresponding to the temporal eigenvalue c_n. A derivation of K_n gives

$$K_n=\int_{-1}^{1}\hat{v}_0(D^2-k^2)\tilde{\xi}_n^* dy, \tag{4}$$

where the $\tilde{\xi}_n$'s are solutions to the adjoint Orr-Sommerfeld equation and $\hat{v}_0$ is the initial normal velocity.

To solve the normal vorticity equation (2) we divide the solution into a homogeneous and a particular part. If the homogeneous part is expanded in Squire modes and the exponential time dependence of the driving $\tilde{v}_n$ modes are extracted out of the particular solution, we have

$$\hat{\eta}=\sum_{m=1}^{M} B_m\tilde{\eta}_m e^{-i\alpha\lambda_m t}+\sum_{n=1}^{N}\tilde{\eta}_n^{(p)}e^{-i\alpha c_n t}, \tag{5}$$

where the $\tilde{\eta}_m$ is the Squire mode associated with the eigenvalue λ_m and $\tilde{\eta}_n^{(p)}$ is the *particular* normal vorticity associated with the n:th O-S mode. Next we expand the particular normal vorticity in Squire modes

$$\tilde{\eta}_n^{(p)} = \sum_{m=1}^{M} D_{nm}\tilde{\eta}_m. \tag{6}$$

A derivation of B_m and D_{nm} give the total normal vorticity as

$$\hat{\eta} = \sum_{m=1}^{M} C_m \tilde{\eta}_m e^{-i\alpha\lambda_m t} + \sum_{m=1}^{M}\left(\sum_{n=1}^{N} D_{nm}(e^{-i\alpha c_n t} - e^{-i\alpha\lambda_m t})\right)\tilde{\eta}_m, \tag{7}$$

where

$$C_m = \int_{-1}^{1} \hat{\eta}_0 \tilde{\eta}_m dy \qquad D_{nm} = \frac{-\beta K_n}{\alpha\lambda_m - \alpha c_n}\int_{-1}^{1} U'\tilde{v}_n\tilde{\eta}_m dy.$$

3. The algebraic instability

If the initial normal vorticity ($\hat{\eta}_0$) is zero, it is apparent from Eq. (7) that each O-S mode excites a number of Squire modes. The excitation is determined such that at $t = 0$ the Squire modes cancel out the particular normal vorticity. When time increases, the Squire modes and the particular normal vorticity propagate with different speeds, as well as have different growth/decay rates. Thus they no longer cancel out, and the total normal vorticity grows. Since the particular normal vorticity in many instances is quite large, the initial growth of the total normal vorticity component can appear spectacular.

As examples of the growth possible we choose initial conditions consisting of the least stable O-S mode $\tilde{v}_1$ for $\alpha = 0,\ 0.1,\ 1,\ \beta = 1$ and $Re = 3000$. We assume that the normal vorticity is zero initially and calculate the constants D_{1m} in the expansion (7). The result is given in table 1, together with the eigenvalues of the corresponding Squire modes.

For $\alpha = 1$ the Squire modes with the most energy have quite high damping rates and will thus dissipate quickly, leaving the particular normal vorticity. Note that the coeffi-

Table 1: Excitation of Squire modes by the least stable O-S mode for $\alpha = 1,\ 0.1,\ 0,\ \beta = 1$ and $R = 3000$. Only odd Squire modes are included.

$\alpha = 1$ $\tilde{\eta}^{(p)}_{max} = 10.95$ $\alpha c_1 = 0.3559 - i0.0205$				$\alpha = 0.1$ $\tilde{\eta}^{(p)}_{max} = 43.43$ $\alpha c_1 = 0.0409 - i0.0094$			
m	$Re\{\alpha\lambda\}$	$Im\{\alpha\lambda\}$	$\lvert D_{1m}\rvert$	m	$Re\{\alpha\lambda\}$	$Im\{\alpha\lambda\}$	$\lvert D_{1m}\rvert$
1	0.9613	-0.0394	0.89	1	0.0877	-0.0126	17.43
2	0.9096	-0.0910	2.01	2	0.0460	-0.0241	46.22
3	0.2185	-0.1214	3.90	3	0.0724	-0.0283	30.42
4	0.8580	-0.1427	4.91	4	0.0673	-0.0494	7.96
5	0.8064	-0.1943	12.14	5	0.0670	-0.0801	0.61
6	0.3763	-0.2005	15.46				
7	0.7547	-0.2460	29.99	$\alpha = 0$ $\tilde{\eta}^{(p)}_{max} = 945.5$			
8	0.5022	-0.2573	31.10	$\alpha c_1 = 0.0000 - i0.0031$			
9	0.7036	-0.2964	68.81	m	$Re\{\alpha\lambda\}$	$Im\{\alpha\lambda\}$	$\lvert D_{1m}\rvert$
10	0.6111	-0.3009	78.79	1	0.0000	-0.0036	944.56
11	0.6759	-0.3525	39.55	2	0.0000	-0.0135	9.70
12	0.6736	-0.4323	2.76	3	0.0000	-0.0299	1.04

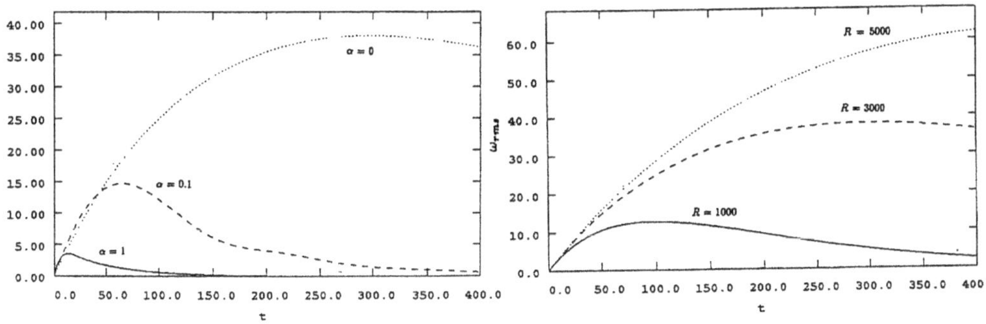

Figure 1: Time development of the normal vorticity using the least stable O-S mode as initial condition. a) $R = 3000$, $\beta = 1$, b) $\alpha = 0$, $\beta = 1$. The maximum amplitude of the driving O-S modes is one.

cients D_{1m} in the expansion for the particular normal vorticity (6) have large numerical values.

For $\alpha = 0.1$ a smaller number of Squire modes are excited and for $\alpha = 0$ only one is highly excited. The resulting time development of the normal vorticity in those cases is seen in the dashed and dotted curves in figure 1a. For the small streamwise wavenumbers the growth of the normal vorticity is larger than for $\alpha = 1$ because the particular normal vorticity cancelled out by the Squire modes is larger (see table 1).

For $\alpha = 0$ the large growth of the normal vorticity can also be viewed as a near resonance between an O-S mode and a Squire mode. Note that the eigenvalues of the most excited Squire mode and the driving O-S mode are almost equal. This near resonance is not present only for this special parameter combination but is rather the generic behaviour for waves with zero streamwise wavenumbers. In fact, if $\alpha = 0$ the following properties hold

$$\alpha c_n = -i\mu_n/R, \quad \alpha\lambda_m = -i\nu_m/R, \tag{8}$$

where μ_n and ν_m are both positive order one quantities, defined in terms of integrals over the eigenfunctions. If the properties (8) are substituted into the eigenfunction expansion of the normal vorticity (7), and the result is Taylor-expanded for large R, we find

$$\hat{\eta} = -i\beta \sum_{m=1}^{M} \left[\sum_{n=1}^{N} K_n \int_{-1}^{1} U' \tilde{v}_n \tilde{\eta}_m dy \left(t - (\nu_m + \mu_n) \frac{t^2}{R} + O(\frac{t^3}{R^2}) \right) \right] \tilde{\eta}_m. \tag{9}$$

Here we have assumed that the initial normal vorticity is zero. Eq. (9) shows that we have a large algebraic growth for disturbances which are elongated in the streamwise direction, conversely, disturbances with non-zero energy in small streamwise wavenumbers will develop a streaky structure. This is in agreement with Landahl's (1980) inviscid calculations and shows the relation between the algebraic instability in the viscous and the inviscid cases. Note that an exact resonance is also covered by the present approach, since the singularity in the expression for D_{nm} is canceled out by the first term in the Taylor expansion of the exponents. Figure 1b shows examples of the behaviour predicted by Eq. (9).

4. Application to localized disturbances

The development of small amplitude localized disturbances can be analyzed by the eigenfunction expansion developed here. We will calculate the development of a disturbance at the subcritical Reynolds number of 3000. We use a spectral Navier-Stokes code with an initial disturbance consisting of two pairs of counter rotating vortices. The initial condition has the following analytical form,

$$\psi = a(1-y^2)^2 xze^{-x^2-z^2},$$

where $v = \psi_z$ and $\eta = \psi_{yx}$. The amplitude a of the initial disturbance was chosen to 10^{-4} in order to insure that non-linear effects remained small.

Figure 2 shows the normal velocity and the normal vorticity obtained from the numerical simulations at $t = 30$, as well as their respective spectra. The normal velocity has developed into a wavepacket and decreased compared to its initial value, while the normal vorticity, which was comparable to v initially, has increased drastically. The increase of the normal vorticity is a result of the liftup effect discussed previously. The spectra also shows that the energy has been concentrated along the β-axis where the damping is low.

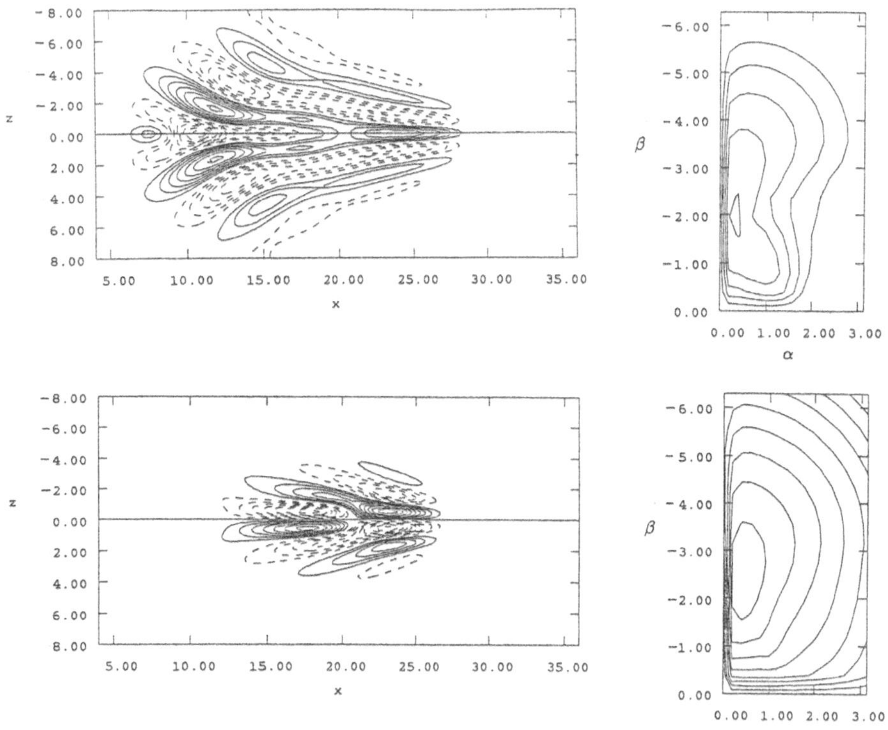

Figure 2: Localized disturbance from the numerical simulation with $R = 3000$, $t = 30$. a) Normal velocity at $y = -0.56$ with contour spacing 0.001a, b) spectrum of $|\hat{v}|^2$, c) normal vorticity at $y = -0.56$ with contour spacing 0.1a, d) spectrum of $|\hat{\eta}|^2$. Contour spacing for the spectra is 1 decade with the lowest contour representing the 10^{-19} level. Dashed lines in this and the following figures indicate negative contour levels.

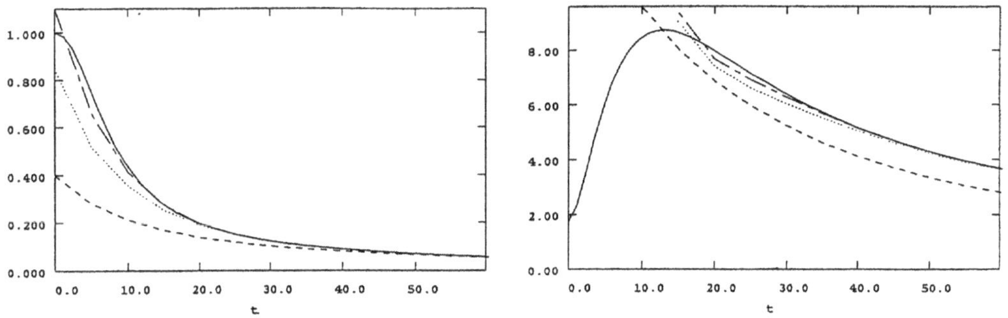

Figure 3: Time evolution of the integrated rms values for the a) normal velocity and b) normal vorticity. Solid curve represents the numerical simulations, while the dashed, dotted and chain-dashed curves represent the eigenfunction expansions using $N = M = 2$, 4, 6, respectively.

The results from the simulations can be reproduced using the eigenfunction expansion. We will include only a few of the least stable modes, since modes with a high damping rate will rapidly die out and only influence the behaviour for short times. For the normal velocity we find the expansion coefficients from (4) and for the normal vorticity we use (7), where we neglect the initial normal vorticity.

Figure 3 shows the time evolution of the rms values of the normal velocity and the normal vorticity for the localized disturbance. Two, four and six of the least stable modes are used in the eigenfunction approximations. Note that two modes do not capture the behaviour of the disturbance, while four and six modes are adequate for times larger than 15. Using the four least stable modes, there is also excellent agreement between the spectra obtained from the eigenfunction expansions (figure 4a,b) and the numerical simulations (figures 2b,d).

Another interesting question is the type of behaviour associated with the different O-S and Squire modes. As an example we will investigate the most important O-S modes

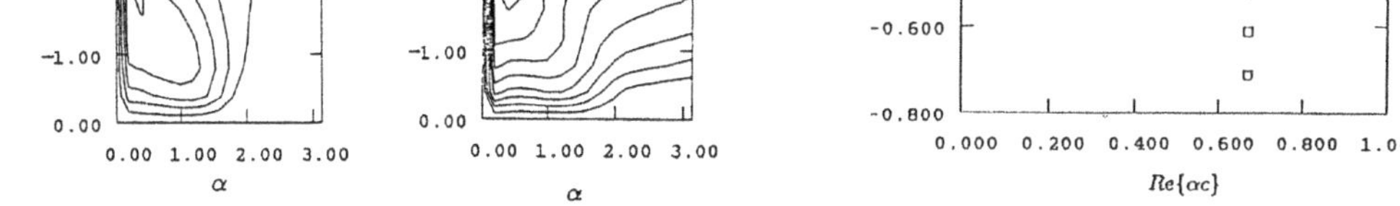

Figure 4: Spectra of the localized disturbance at $t = 30$ obtained from the eigenfunction expansions with $N = M = 4$ for a) $|\hat{v}|^2$, b) $|\hat{\eta}|^2$. c) Symmetric O-S eigenvalues for $\alpha = \beta = 1$ and $R = 3000$.

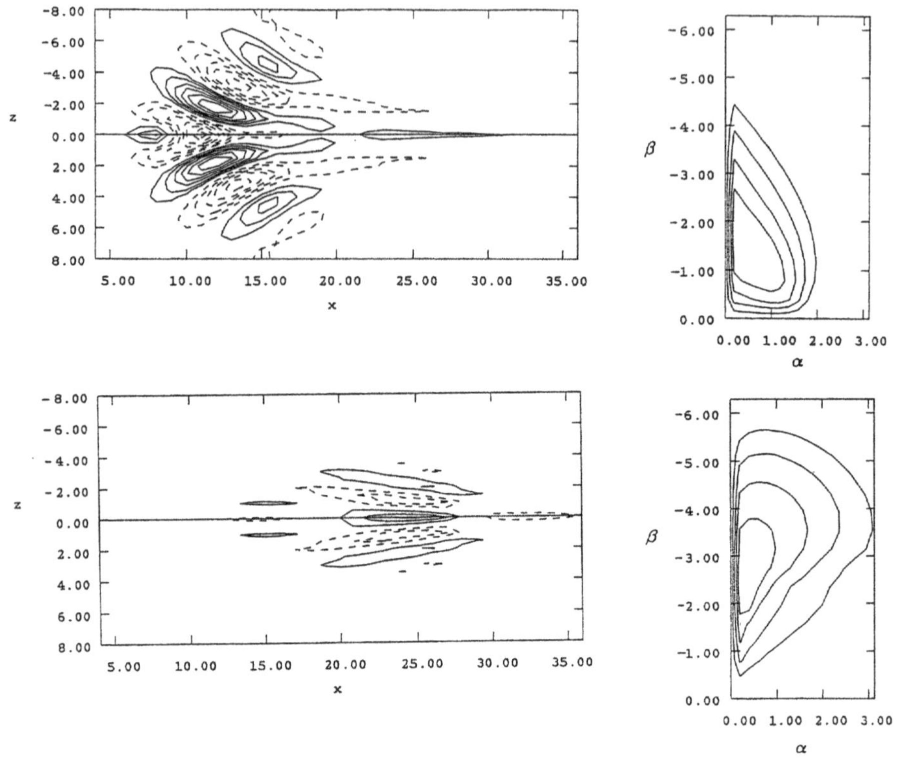

Figure 5: Normal velocity at $t = 30$ obtained from the eigenfunction expansion. a) Velocity contours using the $A1$-mode, b) spectrum of $|\hat{v}|^2$ using the $A1$-mode, c) velocity contours using the $P1$, $P2$ and $P3$-modes, d) spectrum of $|\hat{v}|^2$ using the $P1$, $P2$ and $P3$ modes. Contour spacing of the velocity is 0.001a.

excited at $t = 30$. The four least stable modes for a typical α, β combination are labeled $A1$, $P1$, $P2$ and $P3$ in figure figure 4c. Figure 5 shows the spectra and velocity associated with these modes. The $A1$-modes have a wavepacket character and move with a velocity around 0.4, while the $P1$–$P3$-modes show more of a streaky structure and move with a velocity around 0.75. Note also that the A and the P-modes are excited in different parts of wavenumber space. If the two velocity fields in figure 5 are added, they show good quantitative agreement with the numerically simulated velocity in figure 2a.

5. Summary and conclusions

An eigenfunction expansion for an arbitrary three-dimensional disturbance has been obtained for channel flow. The lift-up effect, by which large horizontal disturbance velocities are created in three-dimensional disturbances, has been related to the initial excitation of Squire modes. Their excitation is determined such that they cancel out the often numerically large particular normal vorticity. These modes subsequently propagate apart and cause growth of the normal vorticity component. Due to the near resonance and low damping of both O-S and Squire modes along the β-axis, the lift up effect was shown to be largest for disturbances with energy in that region in wavenumber space. This shows that the inviscid algebraic instability of Landahl (1980) can be extended to the viscous case and explained in term of the propagation of eigenmodes.

When the eigenfunction expansion was applied to the localized disturbance it revealed that more than the traditional Tollmien-Schlichting wave, i.e. the $A1$-mode, has to be taken into account in order to describe the time evolution. Modes on the P-branch of the spectrum, which represent a faster motion concentrated to the center of the channel, are also important. It is interesting to note that these modes correspond to the continuous spectrum in the boundary layer (Mack, 1976). Thus one may speculate that the often neglected continuous spectrum does play an important role in the evolution of disturbances in the boundary layer.

References

DiPrima, R.C. and Habetler, G.J. 1969 A completeness theorem for non-selfadjoint eigenvalue problems in hydrodynamic stability. *Arch. Rat. Mech. Anal.* 34, 218-227.

Gustavsson, L.H. and Hultgren, L.S. 1980 A resonance mechanism in plane Couette flow. *J. Fluid Mech.* 98, 149-159.

Henningson, D.S. 1988 The inviscid initial value problem for a piecewise linear mean flow. *Stud. Appl. Math.* 78, 31-56.

Henningson, D.S., Johansson, A.V. and Lundbladh, A. 1990 On the evolution of localized disturbances in laminar shear flows. In *Laminar - Turbulent Transition 3*, (Eds. Michel, R. and Arnal, D.), Springer.

Herbert, T. 1988 Secondary instability of boundary layers. *Ann. Rev. Fluid Mech.* 20, 487-526.

Klebanoff, P.S., Tidstrom, K.D. and Sargent, L.M. 1962 The three dimensional nature of boundary layer transition. *J. Fluid Mech.* 12, 1-34.

Landahl, M.T. 1980 A note on an algebraic instability of inviscid parallel shear flows. *J. Fluid Mech.* 98, 243-251.

Mack, L.M. 1976 A numerical study of the temporal eigenvalue spectrum of the Blasius boundary layer. *J. Fluid Mech.* 73, 497-520.

Schensted, I.V. 1961 Contributions to the theory of hydrodynamic stability. *Ph.D. Thesis*, Univ. Michigan, Ann Arbor.

The Role of Computation in Transition Research

Javier Jiménez
School of Aeronautics, Pl. Cardenal Cisneros 3, 28040 Madrid, Spain

The impact of numerical simulation on the study of turbulent transition is reviewed briefly, and its probable future evolution is discussed. It is argued that, in the next two decades, it should be possible to study many transition problems more economically and efficiently by means of direct numerical simulation than by actual experiments. This is already true today for some classes of flows. Some of the problems that will likely be studied in this way in the immediate future are then discussed. It is argued that much of the advantage of computational methods is due not only to their efficiency in simulating experiments, but on their ability to execute "thought" experiments that cannot be performed otherwise.

1. Introduction

Experiments on the transition to turbulence of shear flows have always been notoriously difficult. On the one hand, the results are very sensitive to the choice of initial conditions, as well as to the intensity and character of the external perturbations. On the other, they are expensive, specially when they deal with wall bounded flows, as they involve very long transition times or development lengths. This difficulty, together with the relatively low Reynolds numbers involved, makes transition a good subject for computational treatment.

Unfortunately, computations are not easy either and many of the problems found in experiments have their numerical counterpart. The long development lengths of experiments have their correspondence in very long integration times, which also makes computations expensive. The sensitivity to small perturbations, together with integration times involving hundreds of thousands of steps, makes the use of very high quality numerical methods mandatory. In most cases, this implies the use of spectral codes (e.g. Kleiser & Schumann 1984), although high quality finite difference schemes have recently become available which may become competitive with them. This is specially true in compressible problems, in flows involving sharp interfaces or flames, and in complex geometries (e.g. Lele, 1989)

Another difficulty is that many present transitional codes can only handle the temporal evolution of spatially periodic flows. This is again a consequence of the long development lengths of the spatial problems, and is only true for wall bounded flows. In contrast, the instability mechanisms of free shear flows are usually inertial and the development times are comparable to the eddy turnover time. In these flows, as a consequence, a few wavelengths are enough to observe the growth of the perturbations to their nonlinearly saturated stage, and spatial simulations become practical. In fact, the first simulations of a spatially growing free jet appeared relatively early (Acton, 1980), while equivalent computations of wall bounded flows are still rare.

The question of temporal versus spatial evolution is related to the distinction between convectively and absolutely unstable systems (Drazin & Reid, 1981). Roughly

Advances in Turbulence 3
Editors: A.V. Johansson · P.H. Alfredsson

speaking, temporal models are usually acceptable for closed systems (e.g. Taylor-Couette, Rayleigh-Benard), or for absolutely unstable flows, but can be misleading when applied to situations which are only convectively unstable. Still, even in those cases, temporal models are not necessarily useless. There is probably not a big difference between the evolution of an absolutely unstable flow and that of a convectively unstable one which is subject to constant forcing at the inlet. This is an open and interesting question, and one which can be explored most easily by computational means. In fact, it has been shown recently that locally convective instabilities can be made globally absolutely unstable if they are studied in a finite domain (Buell & Huerre, 1988). This interesting phenomenon was first observed computationally, and is doubtlessly related to the experimental problems associated to the influence of downstream outflow conditions on the behaviour of "unperturbed" free shear flows.

2. A short history

The "prehistory" of the use of computation for the study of transition starts with the linear eigenvalue calculations that appeared in the 1960's. This work clarified much of the stability of parallel incompressible flows, at least to small perturbations, and eigenvalue methods are today routine. Still there is much that can be learned from them, and new results keep appearing, especially in the area of compressible transition (see Mack, 1984 for a survey of applications to boundary layers, and the nice use of stability results by Papamoschou & Roshko, 1988).

Immediately afterwards, the study of transition was extended to the nonlinear range in two different computational directions that have continued to interact. On the one side are the weakly nonlinear representations of transition, pioneered by Stewartson & Stuart (1971), and the closely related theories of secondary instability (Herbert, 1983). On the other, the direct simulations of the initial value problem for the Navier Stokes equations (Orszag & Kells, 1980, Orszag & Patera, 1983). Both approaches are still active. Recent examples of the first one are the weakly nonlinear approximations of the free shear layer by Huerre (1987) and Balsa (1989), and the continuing investigations on boundary layers and channels by Herbert and collaborators.

Direct numerical simulation of the N-S equations is, of course, a whole industry by itself, and has been applied extensively to transitional flows, as shown by the work of Wray & Hussaini (1984) and Spallart (1988) in plane and three dimensional boundary layers, by the simulation of turbulent spots in plane channels by Henningson, Spalart & Kim (1987) and, specially, by the long series of papers of the Göttingen group (Kleiser & Schumann, 1984, Kleiser & Laurien, 1985, Meyer & Kleiser, 1988) on boundary layers and channels.

It is impractical to review the immense amount of work that has been done in this area during the last quarter of a century, and I will not attempt it here. Instead, I will try to give a more personal view of what is possible to do computationally at this moment, of what has been achieved that would have been difficult to do by other means, and of which are some of the outstanding interesting problems now under consideration.

3. The question of what is possible

Computations are expensive, and the question of whether to treat a problem computationally has to be weighted against the cost of studying it by more traditional "physical" experimental means.

The cost of the computations depends on many parameters, of which the two main ones are the spatial (and temporal) resolution, and the total (physical) time for which the simulation has to be run. We will restrict ourselves here to direct simulations of the temporal initial value problem. Spatial problems essentially trade integration time for grid size and are, at the moment, severely limited by the storage capacity of computers. A further related problem, that of the direct computation of nonlinear equilibrium solutions (or of other more complicated attractors) without recourse to the initial value problem, is even harder from the computational point of view. We will comment briefly on it in the next section.

We will take the view that a complete understanding of a transition problem implies following it to the initial stages of true turbulence and that, therefore, the numerical resolution requirements are equivalent to those of the corresponding turbulent flows. Define the basic "external" flow length and velocity scales as L and U, and the corresponding Reynolds number as Re. The resolution depends on the type of flow. The smallest scale that can be expected in a *two dimensional free* shear flow is $O(L\,Re^{-1/2})$, and this is the order of the smallest grid spacing needed, Δx. Because of numerical stability considerations it is usually necessary to keep the time stepping increment of the same order of magnitude, and we will assume that $\Delta t \approx \Delta x$. The equivalent estimate for *three dimensional free* shear flows is that $O(L\,Re^{-3/4})$.

Wall shear flows present special problems because thin, $O(L\,Re^{-1})$, Stokes vortex sheets appear near the walls (Jiménez, 1990). However, because of numerical considerations associated to the behaviour of spectral discretisation schemes near the domain boundaries, the near wall region is usually overresolved by the effect of the nonuniform grid, and the overall resolution is still limited by the smallest structures found in the outer flow. The estimates of the previous paragraph are still valid. This is also true when spectral methods are substituted by high quality (Padé) finite difference schemes, which have similar near wall problems.

The total time for which a simulation has to be run is dictated by the size of the instability eigenvalues. Free shear flow transition is governed by inertial instabilities of the Kelvin Helmholtz type, with characteristic times of $O(L/U)$. On the other hand, the primary instability of wall bounded flows is exemplified by the two dimensional Tollmien Schlichting waves, which have much longer development times, $O(L\,Re/U)$. When only this stage of transition is modelled, it is possible to use essentially two dimensional representations of the flow, but it is well known that true transition is mediated by the instability of a secondary three dimensional process, Λ-vortex formation, which needs to be adequately resolved. The eigenvalues associated with this secondary instability are again $O(U/L)$, and the computational requirements for this phase of wall flow transition are similar to those for three dimensional free shear flows.

It is now possible to characterise the cost of a given computation by introducing the concept of *site*, which is the cost of updating a single grid node for one time step. For a two dimensional grid whose evolution has to be followed for a time T, the number of required sites would be,

$$S \sim N^2 N_T \approx (L/\Delta x)^2 (T/\Delta t) \approx L^2 T/\Delta x^3. \tag{1}$$

Applying these arguments to the three types of flows studied in the previous paragraphs, we get the estimates,

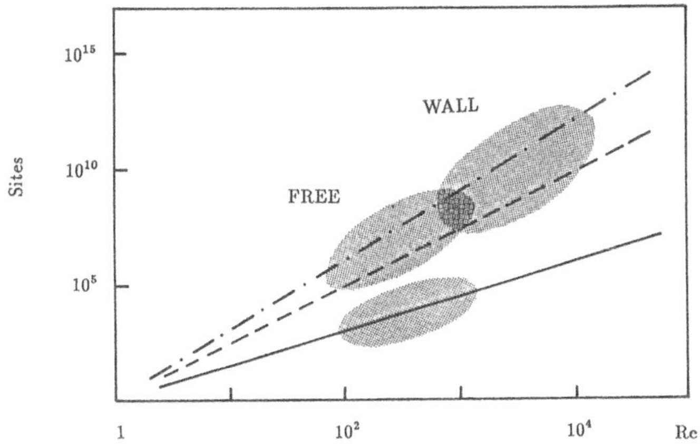

Figure 1.- Number of single grid point updates needed to simulate the initial value transition problem in each of the three types of flows described in the text. Shaded areas are typical transition Reynolds number ranges for those flows. —: 2-D free shear flows. - - : 2-D wall bounded flows. - · - : 3-D flows.

Free 2-D shear flows:

$$\Delta t \approx \Delta x \approx Re^{-1/2}; \qquad T \approx 1; \qquad S = N^2 N_T \approx Re^{3/2};$$

Wall bounded 2-D shear flows:

$$\Delta t \approx \Delta x \approx Re^{-1/2}; \qquad T \approx Re; \qquad S = N^2 N_T \approx Re^{5/2};$$

Free and wall bounded 3-D shear flows:

$$\Delta t \approx \Delta x \approx Re^{-3/4}; \qquad T \approx 1; \qquad S = N^3 N_T \approx Re^3.$$

These three estimates are represented as straight lines in the log-log plot in figure 1, which also incorporates the information that transition Reynolds numbers for free shear flows are $O(100 - 1000)$, while those for wall bounded flows are $O(1000 - 10000)$.

It is interesting to relate these estimates to actual total CPU time per job. For spectral codes using present day technology on a Cray Y-MP computer, a site is roughly equivalent to $10\mu s$. Using that number, and assuming that computer power increases roughly by a factor of 10 every decade, figure 2 represents the Reynolds number that could be simulated in a single ten hours job on the fastest available computer at each given year. Note that these are only order of magnitude estimates, and that actual computing times might easily be ten or even a hundred times longer.

Even so, some interesting remarks can be made about that figure. The first one is that the computation of transition in two dimensional free shear flows has been possible for as long as digital computers have been available, and probably before (see Rosenhead, 1932, for an early computation of a plane mixing layer). In fact, this problem is now in the *CPU minutes* category, and it is one case in which computations are probably cheaper than experiments. This is also almost true for two dimensional wall bounded flows, which are difficult to establish in the laboratory, and even for three dimensional free shear flows, which are beginning to enter in the "easily computable" class. Examples of the former type of computation are (Orszag & Kells, 1980, Jiménez, 1990) while, for the latter, one can think in the early papers by Lin & Corcos (1984), or in the most recent ones by Meiburg & Lasheras (1988) or Moser & Rogers (1990). Finally, the computation of fully three dimensional wall bounded flows is still largely in the future,

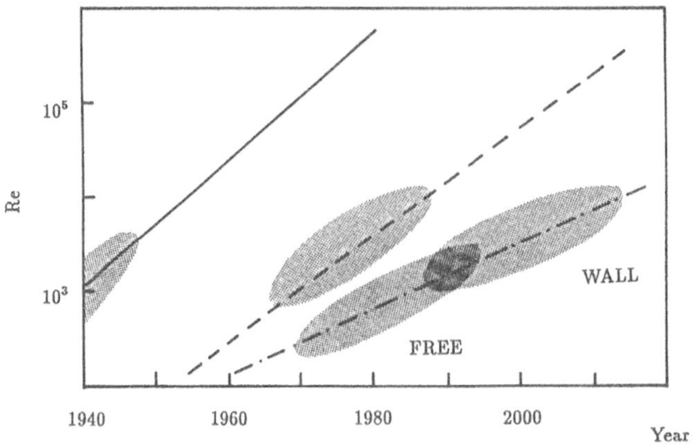

Figure 2.- Reynolds number achievable for a 10 hour simulation using the fastest computer predicted for a given year. Symbols as in figure 1.

although some initial successes have appeared in the the literature, both in the early transition stages (Kleiser & Laurien, 1985) or in the late transitional or fully turbulent flows (Kim, Moin & Moser, 1987).

A last consequence of the estimates in figure 2 is that in two decades it will be possible to compute any transitional flow within a reasonable expenditure of computer resources. This conclusion, even if it is restricted to the present assumptions of spectral codes (implying simple geometries), of temporal evolution, and of simple flow equations (newtonian fluids without chemical of compressibility effects), will have a profound influence in how transition experiments are planned and conducted in the future. In particular, it suggests that some physical experiments might no longer have to be conducted at all, and that some transition problems will be completely dominated by computations in the near future.

4. The question of what to do next

Any list of interesting problems is necessarily personal, and I will not pretend that the following one in anything else than a summary of the questions that interest me at the moment. Other workers would probably have different priorities, but this section is offered with the intention of pointing to some open problems which are either fashionable at this moment, or which I think should become so in the near future.

Bifurcation trees:

The present conceptual model for transition dates to Reynolds, who postulated that the origin of turbulence was due to an instability of the laminar flow. The later work of Ruelle & Takens (1971) introduced the concept of a short bifurcation tree, and of a generalised "route", leading from the primary laminar instability, through a cascade of successive bifurcations, to chaos and turbulence. It has become clear since then that no single route is enough to explain transition, and that different flows behave differently, but there is still hope that a limited number of scenarios may be enough to account for most transition problems. Whether or not this is true is one of the two most important

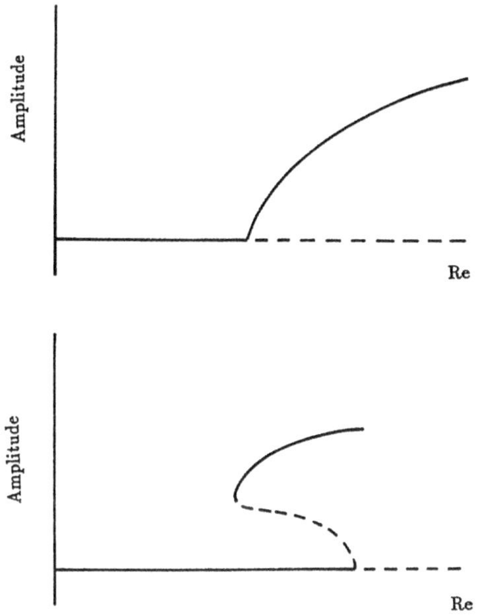

Figure 3.- Supercritical (top) and subcritical bifurcation diagrams. Stable branches are solid, unstable ones are dashed.

unsolved problems of transition (and of turbulence) at this moment (the other one being a clear understanding of the mechanics of the turbulent cascade).

Computing bifurcation trees is not easy, especially when some of the bifurcations are subcritical (see figure 3). In those cases, the bifurcated solution is initially unstable and cannot be observed through the solution of the initial value problem. In fact, for the same reason, it cannot be observed experimentally.

These unstable solutions, however, are as important to the continuity of the bifurcation tree as the stable ones, and a satisfactory way of computing them is one of the main numerical problems facing computational transition at the moment. The standard method is to continue the solutions from the bifurcation point using a Newton iteration scheme. The iteration normally involves the computation of the Jacobian matrix, and becomes impractical beyond a relatively small number of degrees of freedom. Moreover, after the first few bifurcations, the object that is being searched is no longer an equilibrium solution, but a more complicated attractor (a limit cycle, torus, etc.). Iteration into a limit point is now a well understood subject, but convergence to more complicated objects is still a very new field. (For example, how does one search for a set of hyperbolic, only partially attracting, points?)

It is sometimes possible to reduce significantly the number of modes involved in the calculation by working in the "centre" manifold, which is essentially the space spanned by the "important" modes in the neighbourhood of the bifurcation (see Guckenheimer & Holmes, 1983). This is a local concept, displaying the low dimensional behaviour of the system just after bifurcating, and is not easy to apply to the fully nonlinear case. In particular, it is usually not possible to use this technique to track subcritical bifurcated solutions far enough to reach, either a new bifurcation, or a change of stability that makes them stable and observable.

Still, some successes have been achieved. Many of the supercritical bifurcations in the transition of Taylor-Couette flow have been mapped (Andereck, Liu & Swinney, 1986, Golubitsky & Stewart, 1986), and the subcritical unstable solutions coming from the laminar bifurcation of *two dimensional* Poiseuille flow were computed very early (Zahn, *et al.*, 1974, Herbert, 1976). The same was done for other wall bounded flows in (Millinazzo & Saffman, 1985). Several secondary bifurcations of Poiseuille flow, still in the framework of the two dimensional problem, were also found, using the initial value problem approach, in (Jiménez, 1990). Other flows, like the circular pipe or plane Couette flow, lack an easy handle for their bifurcation trees, because no linear bifurcation from laminar flow is known. Many attempts have been done during the years to find a first nonlinear equilibrium solution in which to anchor the bifurcation map, but the results has been limited.

Another interesting development is the recent direct attempt to use the reduction of a complicated system to a few degrees of freedom in (Aubrey *et al.*, 1988). They study the dynamics of a highly truncated expansion of the near wall region of a plane three dimensional channel, and try to relate its behaviour to that of the complete system.

Most of these investigations are either preliminary or deal with very simplified model flows, but this direction promises to be one of the most fruitful in the field of transition in the next few years. Note that many of these works are as much related to fully turbulent flows as to transition, since the bifurcation tree is intended to lead eventually to fully developed turbulence. Some of them suggest that transitional processes are still important in the fully developed regime.

Very large scale aggregations:

A very interesting topic that has come to the fore in the last few years is the origin of the very large structures that dominate transition in some wall bounded flow, like turbulent "puffs" and "slugs", boundary layer spots, etc. These are structures whose scale is at least an order of magnitude larger that any of the turbulent scales of the flow (like pipe diameter, etc.).

Consider turbulent "puffs", which are transitional structures appearing as isolated permanent turbulent regions in an otherwise laminar flow. They were observed originally in certain regimes of the Taylor-Couette flow between counter-rotating cylinders (Coles, 1965, Van Atta, 1966, Andereck *et al.*, 1986), and later in circular pipe flow (Wygnanski, Sokolov & Friedman, 1975, Bandyopadhyay, 1986). Similar confined states were also found in the computations of two dimensional channels by (Jiménez, 1990), and in certain ranges of thermal convection in binary fluids (Kolodner, Bensimon & Surko, 1988). These are all subcritical situations. This ubiquity suggests that their origin may be independent of the detailed mechanisms that control turbulence, and may correspond to a more general property of transitional systems in general.

It has been suggested that these puffs may be related to the solitary wave solutions of some amplitude equation describing the modulation of the underlying "turbulent" structures (Pumir, Manneville & Pomeau 1983). In fact Thual & Fauve (1988) have shown that a modified subcritical Ginsburg Landau equation can sustain stable solitary structures that could be related to puffs. These pulses by themselves are not good models for the structures observed in a pipe flow, in that their interactions and secondary bifurcations seem to be different. In addition, there seems to be no clear way of deriving the relevant amplitude equation from the equations of pipe flow. Still, it can be shown that other relatively minor modifications to the original G-L equation result in solutions that are much more similar to those of the puffs, and also that many of the properties of the amplitude equation itself are not directly linked to its detailed form but only to its symmetry groups, which can be made to resemble those of the Navier Stokes equation (the appearance of stable solitary solution has still not been shown to be in this class of generic properties). Whether any of these results can really be related to the behaviour of shear flows remains to be seen, but this area is at present one of the most active and exciting areas of transition research.

"Dirty" flows:

It can be argued that any flow transition involves the breaking of some symmetry, such as the streamwise and spanwise spatial homogeneity in a laminar channel that goes turbulent. In most cases, more than one symmetry is eventually broken, and the question arises of what would have been the behaviour of the flow if some of the broken symmetries had been originally absent from the laminar flow. A case in point is the turbulent transition on flat plates with and without yaw. In the classical unyawed case, the longitudinal homogeneity is broken first, producing two dimensional Tollmien Schlichting waves, and only then can the spanwise homogeneity be broken, giving rise to the appearance of the much stronger processes of three dimensional transition. It is interesting to speculate what would happen if the original flat plate had not been uniform in the spanwise direction, and whether this would allow the 3D processes to appear earlier, bypassing the two dimensional stage. This seems to be indeed the case in yawed plates (and in aeroplane wings). The presence of streamline curvature produces streamwise rollers (a fast inertial instability) which immediately reinforces the three dimensional instabilities (Fischer & Dallmann, 1988), and precipitates early transition.

The question of the influence of extra symmetries is important, since many of the classical experimental flows are "too" symmetric, compared to the practical cases, and it is interesting to understand what the absence of this extra "cleanness" does to the onset of transition in industrial situations.

Other problems:

It is impossible to list here all the other interesting and open problems in transition. Besides the three "personal" choices discussed in the preceding paragraphs, the questions of compressible and chemically reacting flows comes immediately to mind. Their linear stability has been investigated quite thoroughly, but little is known of their nonlinear behaviour. Also, the computational techniques applicable to these problem are much less developed that those for incompressible flows. The same is true of flows with variable density (e.g. non-isothermal) or two phase flows.

From the numerical point of view, and besides the problems mentioned above, perhaps the most urgent need is for ways to treat spatially growing problems efficiently. Particularly important is the development of ways for dealing correctly with inflow and outflow boundary conditions.

A last interesting question is the influence of initial conditions. It has usually been assumed that, if transition simulations are initialised with random perturbations, the flow will eventually select by itself its most unstable mode, which will be representative of the results of "natural" experiments. While this is true in linear problems, it is not necessarily so in nonlinear situations, and there is an increasing awareness, from computational experience and from experimental evidence, that different initial conditions can give rise to very different results, lasting for very long times. None of these different outcomes is more "natural" than others, and a complete understanding of the transition process calls for a full mapping of the results of the different inputs.

5. Unrealistic simulations

Finally, we should look into the question of how close should numerical simulations be to experiments. Most of the computational works in transition try to simulate "realistic" flows, in most cases by studying their instabilities. They are therefore direct descendants of the classical theoretical and experimental tools previously used for those problems. It is clear that, by solving exactly the Navier Stokes equations, we end up repeating exactly the result of experiments. This is not the only way to study transition, specially computationally.

While experiments (of full numerical simulations) are undoubtedly "true", they have drawbacks, especially regarding the analysis of their results. Typically, the results of an experiment contain such an overwhelming amount of data, in the form of hot wire traces, flow visualisation pictures, etc., that understanding their meaning is an even harder task than obtaining them. Simulations, with their ability to measure "everything", generate even more data, and they may end up being no more useful than experiments in understanding particular basic questions about fundamentals, although partial, physical processes in a given phenomenon.

Numerical simulation, on the other hand, is ideally suited to the design of "partial" experiments to investigate these questions. It is relatively difficult to construct a physical experiment in which all but one symmetries are provided (think of the practical problems of building a yawed infinite cylinder, free from end effects). It is also practically impossible to force a flow to maintain a given symmetry while undergoing transition (for example, to bifurcate into two dimensional turbulence). Both things, on the other hand are easy to arrange numerically, and this unique ability of the simulations to work with "unphysical" models, and to study different effects in isolation is probably their biggest advantage over physical methods.

A good example is two dimensional flow. Except for some very limited experiments in which the effect of the third dimension has been somehow minimised, these are completely theoretical flows that do not exist in nature. Numerically, they have been widely studied, mostly perhaps because early numerical codes were not able to handle three dimensional situations. It was therefore a surprise to find in them, for example in two dimensional channels, many of the structures observed in three dimensional transition of real flows (Jiménez, 1987, 1990). The implication is that these structures do not depend on the details of channel flow but are generic to more general situations.

Another example are the simulations carried out purposely with inadequate resolution. May & Kleiser (1985) and Rozhdestvensky & Simakin (1984) both showed that it is possible to compute a turbulent channel using a surprisingly small number of spectral modes. Since these simulations essentially cut off the turbulent cascade by underresolving it, they suggest that the cascade might be a consequence, rather than an integral part of transition or turbulence in the channel. On a different approach, Jimenez & Moin (1990) used computational boxes for a channel that were purposely too small. Since the results in the near wall region were also approximately correct, that experiment suggests that the *large scales* are either unimportant or can be represented by a very simple (in fact, spatially periodic) model. Since these two sets of computations have essentially a single length scale in common, and since both of them "work", they convey a very specific message on which is the important scale for the maintenance of turbulence in a channel, which seems to be of the order of the width of the sublayer streaks.

None of these experiments could have been done without the help of computation. In fact they can not be repeated in the laboratory. This ability of computations for dealing with idealised flows tailored to answer specific questions is, more than their capacity to reproduce exactly the physical experiments, what makes them valuable in the study of fluid mechanics in general.

6. Conclusions

We have seen several examples of the use of computation in answering or, at least, framing questions in transition of shear flows into turbulence. Some of them have extended into the dynamics of fully turbulent flow, which was to be expected since the border between the two states is not sharp. We have also seen that, within the first part of the next century, it will be possible to compute exactly many transitional flows with a reasonable expenditure of computer time. In fact, this is true even today for whole classes of shear flows. We have argued, on the other hand, that this is not necessarily the best use of computers in studying transition. It is probably more useful to use their ability to do conceptual experiments directed at dissecting transition and turbulence into their elemental components. Transition is now in an exciting and fruitful period, and the availability of computers is already an integral part of its study.

Part of this work was carried out during a stay in the Centre for Turbulent Research at Stanford and NASA Ames. Their hospitality and support are gratefully appreciated.

References

Acton,E. 1980 A modelling of large eddies in an axisymmetric jet, *J. Fluid Mech.* **98,** 1-31.

Andereck,C.D., Liu,S.S. & Swinney,H.L. 1986 Flow regimes in a circular Couette system with independent rotating cylinders. *J. Fluid Mech.* **164,** 155-183.

Aubry,N., Holmes,P., Lumley,J.L. & Stone,E. 1988 The dynamics of coherent structures in the wall region of a turbulent boundary layer. *J. Fluid Mech.* **192,** 115-173.

Balsa,T.F. 1989 Amplitude equations for wave packets in slightly inhomogeneous unstable flows. *J. Fluid Mech.* **204,** 433-455.

Bandyopadhyay,P.R. 1986 Aspects of the equilibrium puff in transitional pipe flow. *J. Fluid Mech.* **163,** 439-458.

Buell,J.C. & Huerre,B. 1988 Inflow/outflow boundary conditions and the global dynamics of spatial mixing layers, in *Centre for Turbulence Research Report* **CTR-S88,** Stanford, Ca., pp. 19-27

Coles, D. 1965 Transition in circular Couette flow. *J. Fluid Mech.* **21,** 385-475.

Drazin,P.G. & Reid,W.H. 1981 *Hydrodynamic Stability,* Cambridge University Press, ch. 24

Fischer,T.M. & Dallman,U. 1988 Primary and secondary instability analysis applied to the DFVLR transition swept plate experiment, in *Fluid dynamics of 3D turbulent shear flow and transition,* **AGARD-CP-438,** 15.1-11.

Golubitsky,M. & Stewart,I. 1986 Symmetry and stability in Taylor Couette flow. *SIAM J. Math. Anal.* **17,** 249-288.

Guckenheimer,J. & Holmes,P. 1983 *Nonlinear oscillations, dynamical systems and bifurcations of vector fields*, Springer, p. 123.

Henningson,D.S., Spalart,P. & Kim,J. 1987 Numerical simultaions of turbulent spots in plane Poiseuille and boundary layer flows. *Phys. Fluids* **30**, 2914-2917.

Herbert,T. 1976 Periodic secondary motions in a plane channel, in *Proc. 5th. Internal Conf. on Numerical Methods in Fluid Dynamics* (A.I. Van de Vooren & P.J.Zahnbergen, eds.) Lecture Notes in Physics **59**, p.235, Springer.

Herbert,T. 1983 Secondary instability of plane channel flow to subharmonic three dimensional disturbances. *Phys. Fluids* **26,** 871-874.

Huerre,P. 1987 On the Landau constant in mixing layers, *Proc. R. Soc. Lond. A* **409,** 369-381.

Jiménez,J. 1987 Bifurcations and bursting in two-dimensional Poiseuille flow. *Phys. Fluids* **30,** 3644-3646.

Jiménez,J. 1990 Transition to turbulence in two dimensional Poiseuille flow. *J. Fluid Mech.*, **218**, 265-297

Jiménez,J.& Moin,P. (1990) The minimal flow unit in near wall turbulence, *CTR Manuscript* **105**, to appear *J. Fluid Mech.*

Kim,J. Moin,P. & Moser,R. 1987 Turbulence statistics in fully developed channel flow at low Reynolds numbers *J. Fluid Mech.* **177,** 133-166.

Kleiser,L. & Laurien,E. 1985 Three dimensional numerical simulation of laminar turbulent transition and its control by periodic disturbances, in *Laminar Turbulent transition* (V.V.Korlov, ed.), pp. 29-37, Springer.

Kleiser,L. & Schumann,U. 1984 Spectral simulations of the laminar turbulent transition in plane Poiseuille flow, in *Spectral methods for partial differential equations* (R.G.Voigt, D.Gottlieb and M.Y.Hussaini eds.) pp. 141-163, SIAM.

Kolodner,P., Bensimon,D. & Surko,C.M. 1988 Travelling wave convection in an annulus, *Phys. Rev. Lett.* **60**, 1723-1726.

Lele,S.J. 1989 Direct numerical simulations of compressible shear flows. *AIAA Paper* **89-0374**

Lin,S.J. & Corcos,G.M. 1984 The mixing layer: deterministic models of a turbulent flow: Part 3. The effect of plane strain on the dynamics of the streamwise vortices. *J. Fluid Mech.* **141,** 139-178.

Mack,L. 1984 Boundary layer linear stability theory, in *Stability and Transition of Laminar Flow,* **AGARD-R-709,** pp. 3.1-3.81.

May,C.L. & Kleiser,L. 1985 Numerical simulation of subharmonic transition in plane Poiseuille flow, *Bull. Amer. Phys. Soc.* **30**, 1748.

Meiburg,E. & Lasheras,J.C. 1988 Experimental and numerical investigation of three dimensional transition in a plane wake. *J. Fluid Mech.* **190,** 1-37.

Meyer,F. & Kleiser,L. 1988 Numerical investigation of transition in 3D boundary layers, in *Fluid dynamics of 3D turbulent shear flow and transition,* **AGARD-CP-438,** 16.1-17.

Millinazzo,F.A. & Saffman,P.G. 1985 Finite amplitude waves in plane viscous shear flows. *J. Fluid Mech.* **160,** 281-295.

Moser,R.D. & Rogers,M.M. 1990 Mixing transition and the cascade to small scales in a plane mixing layer. *Proc. IUTAM Symp. Stirring and Mixing*, La Jolla, 20-24 August, 1990. To appear Phys. Fluids A.

Orszag,S.A. & Kells,L.C. 1980 Transition to turbulence in plane Poiseuille and plane Couette flows. *J. Fluid Mech.* **96,** 159-205.

Orszag,S.A. & Patera,A,T. 1983 Secondary instability of wall bounded shear flows. *J. Fluid Mech.* **128,** 347-385.

Papamoschou,D. & Roshko,A. 1988 The compressible turbulent shear layer: an experimental study. *J. Fluid Mech.* **197,** 453-477.

Pumir,A., Manneville,P. & Pomeau,Y. 1983 On solitary waves running down an inclined plate. *J. Fluid Mech.* **135,** 27-50.

Rosenhead,L. 1932 The formation of vortices from a surface of discontinuity. *Proc. Roy. Soc.* A **134**, 170-192.

Rozhdestvensky,B.L & Simakin,I.N. 1984 Secondary flows in a plane channel: their relationship and comparison with turbulent flows. *J. Fluid Mech.* **147,** 261-289.

Ruelle,D. & Takens,F. 1971 On the nature of turbulence. *Comm. in Math. Physics* **20,** 167

Spalart,P.R. 1988 Direct numerical study of leading edge contamination, in *Fluid dynamics of 3D turbulent shear flow and transition,* **AGARD-CP-438,** 5.1-13.

Stewartson,K. & Stuart,J.T. 1971 A non-linear instability theory for a wave system in plane Poiseuille flow. *J. Fluid Mech.* **48,** 529-545.

Thual,O. & Fauve,S. 1988 Localised structures generated by subcritical instabilities. *J. Phys. France* **49,** 1829-1833.

Van Atta, C. 1966 Exploratory measurements in spiral turbulence. *J. Fluid Mech.* **25,** 495-512.

Wygnanski,I., Sokolov,M. & Friedman,D. 1975 On transition in a pipe, II: the equilibrium puff. *J. Fluid Mech.* **69,** 283-304.

Wray,A. & Hussaini,M.Y. 1984 Numerical experiments in boundary layer stability. *Proc. R. Soc. Lond. A* **392,** 373-389.

Zahn,J.P., Toomre,J., Spiegel,E.A. & Gough,D.O. 1974 Nonlinear cellular motion in Poiseuille channel flow. *J. Fluid Mech.* **64,** 319-345.

Experiments on the Evolution of a Point-like Disturbance in Plane Poiseuille Flow into a Turbulent Spot

Barbro G.B. Klingmann and P. Henrik Alfredsson

Department of Gasdynamics, Royal Institute of Technology
S-100 44 Stockholm, Sweden

The streamwise velocity field resulting from a point-like disturbance in plane channel flow in the transition regime was studied using hot-wire anemometry. Elongated structures associated with strong spanwise shear (i.e. normal vorticity) were observed preceding the appearance of 'spikes' and the subsequent formation of a turbulent spot. The streaks were found to have properties similar to those resulting from the solution of the 3-dimensional Orr-Sommerfeld problem for normal vorticity waves which are elongated in the streamwise direction. Although they eventually decay, the initial growth of these structures appears to have a destabilizing effect on the flow.

1. Introduction

Laminar-turbulent transition in plane channel flow is usually found at Reynolds numbers between 1000 and 3000 (Re based on the centreline velocity, U_{CL}, and the half channel height, h). Linear stability theory predicts a minimum critical Reynolds number (Re_{cr}) of 5772 for growth of Tollmien-Schlichting (TS) waves in plane Poiseuille flow, and even above this Re, growth rates are small. The discrepancy against naturally occurring transition in noisy environments is usually explained by non-linear effects. One such effect is the so-called secondary instability mechanism. It suggests that small amplitude three-dimensional (3-D) perturbations (secondary), superimposed on two-dimensional (2-D) Tollmien-Schlichting (TS) waves of finite amplitude (primary), may grow exponentially. This mechanism, elaborated by Orszag & Patera (1983) for plane channel flow, gives a correct prediction of the minimum transition Re, and the resulting flow has clearly turbulent features. However, at Re near transition, 2-D TS-waves which may appear naturally are damped, and therefore a primary flow consisting of such waves does not give a physically satisfactory explanation for the transition process in plane Poiseuille flow.

In the 3-D case, the Orr-Sommerfeld modes act as driving terms in the Squire equation for normal vorticity. In this coupled problem, a single (decaying) TS-wave will excite a set of associated Squire-waves with different phase velocities. These, although eventually decaying, exhibit an essential initial growth which may carry the disturbance to the non-linear stage. This has been studied by Gustavsson (1990), who found that the largest growth of normal vorticity is obtained for waves which have their wave crests aligned with the flow direction, i.e. elongated streaks, a structure which in turn delays non-linear effects (see also Landahl, 1980). The time it takes to reach the maximum amplitude is proportional to Re. This means that a state of relatively large flow perturbation, described by the linear equation, may evolve on a viscous time scale. Such

Advances in Turbulence 3
Editors: A.V. Johansson · P.H. Alfredsson

a state would provide a physically feasible condition for the explosive growth of small perturbations. This has hitherto not been experimentally verified.

There are, however, some interesting observations which may indicate that this kind of mechanism governs the evolution of a point-like disturbance into a turbulent spot. The incipient turbulent spot in a boundary layer was studied by Amini & Lespinard (1982), who noted that the propagation speed of the disturbance was near 0.5 for the rear and 0.95 for the front, in contrast to the least stable TS-wave, which propagates with about 0.4 times the free stream velocity. In a plan view, the streamwise velocity contours near the wall showed a pattern of elongated streaks, indicating the presence of normal vorticity. If the point disturbance had produced a packet of 2-D TS-waves, the wave fronts would be aligned in the spanwise direction and not appear as streamwise streaks. The observed pattern indicates the presence of elongated streaks of normal vorticity as expected from the mechanism proposed by Gustavsson. Near the boundary layer edge, turbulent spikes appeared preceding the formation of a turbulent spot.

In their study of turbulent spots in plane channel flow, Klingmann & Alfredsson (1989) found that there exists a critical initial disturbance amplitude above which a turbulent spot will develop. At the critical initial amplitude, 50 % of the disturbances develop into turbulent spots, and 50 % into decaying laminar disturbances. The rear of the laminar disturbances had a propagation speed near 0.7, i.e. they propagated faster than the spots. This indicates that the disturbance undergoes a phase of initial growth which is not exponential. The behaviour of the decaying disturbances may be explained by the linear growth - exponential decay which is characteristic for the Gustavsson mechanism, whereas those disturbances which become turbulent spots probably undergo transition near the maximum of the linear growth.

The present study is an attempt to clarify the process which may lead to these two distinctly different results, and to substantiate the hypothesis that the initial development of a point-like disturbance at low Re is governed by the linear growth of streamwise elongated normal vorticity structures.

2. Experimental set-up and procedure

Point-like disturbances were introduced in an air-channel flow in the form of an air jet from a 1 mm hole in the upper channel wall ($y = 1$). The jet was produced by a loudspeaker placed 200 h upstream of the channel outlet (see figure 1). The loudspeaker was fed with an input voltage (E_c) during a short time (2 ms), producing a jet strength proportional to E_c. The channel had an aspect ratio of about 100 and its height was 8 mm. A hot-wire was traversed in the streamwise, normal and spanwise directions. The triggering and data sampling were controlled by a PDP 11, and the data evaluation was carried out on a Macintosh SE/30. The sampling rate was 2 kHz.

3. Experimental results

In the present set-up, the critical input amplitude for spot formation at $Re = 1670$ is achieved by applying the input voltage $E_c = 1.4$, which corresponds to a jet velocity of about 1.5 U_{CL}. The evolution of the disturbance was followed by traversing a hot-wire in the downstream direction at $y = z = 0$. The velocity traces collected at the different x-stations are shown in figure 2 for two different initial disturbance amplitudes,

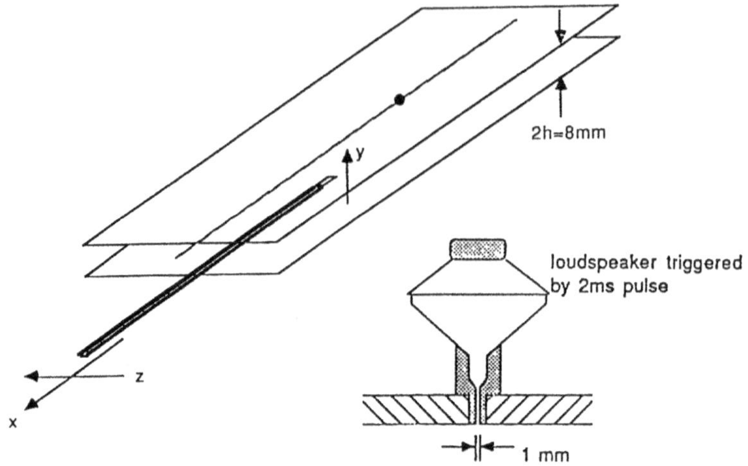

Figure 1. Experimental set-up.

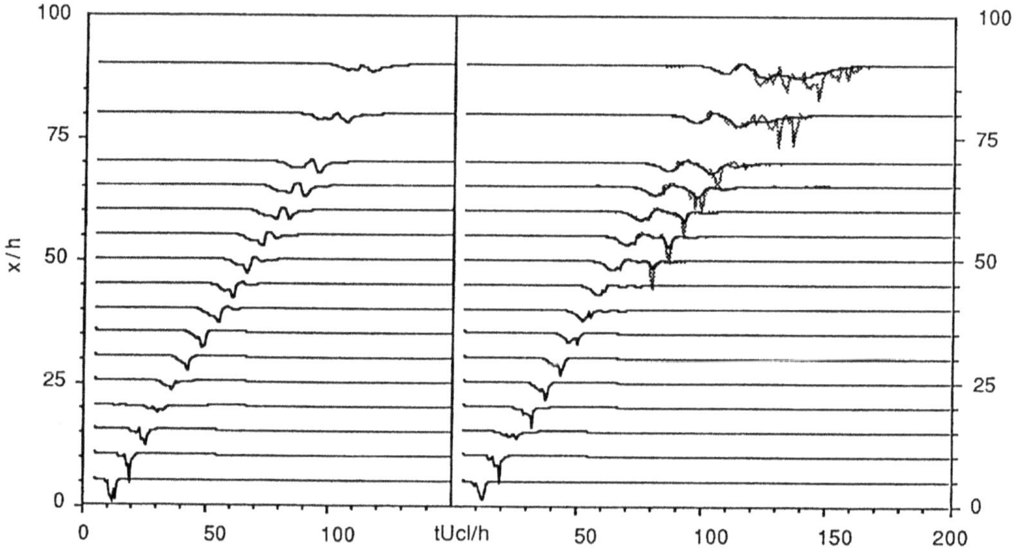

Figure 2. Downstream evolution of disturbances with an initial amplitude close to the critical value. $y = z = 0$, $Re = 1660$. (a) $E_c = 1.3$, (b) $E_c = 1.5$.

$E_c = 1.3$ and 1.5, the former below and the latter above the critical value of E_c. The full lines represent ensemble averages of 200 velocity signals, and the dotted lines are selected individual signals displaying enhanced non-deterministic features. Close to the source, a low velocity spike can be seen. The spike moves downstream with a velocity of $0.88U_{CL}$. Further downstream, a second spike appears, which propagates with the same velocity. Both spikes are completely deterministic and eventually decrease in amplitude. For $E_c = 1.5$, however, a spike with a different behaviour can be observed at the rear

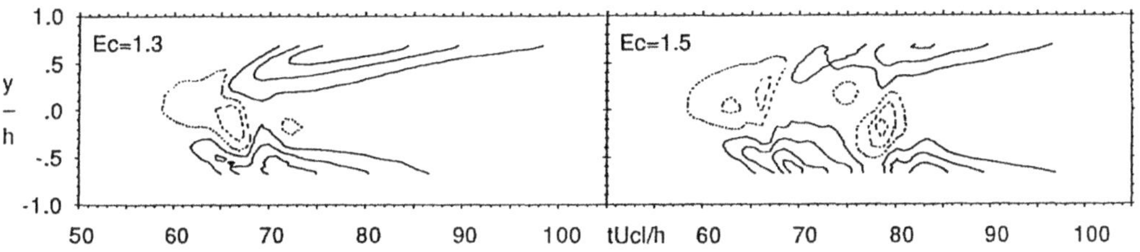

Figure 3. Cross channel view of streamwise velocity perturbations at $z = 0$, $x/h = 50$, $Re = 1670$. (a) $E_c = 1.3$, (b) $E_c = 1.5$. Levels are ± 0.025, ± 0.075, ± 0.125 and 0.175.

past $x/h = 50$. The time at which this spike first appears is deterministic, whereas its amplitude varies up to about 25 % of U_{CL}. Further downstream, this spike splits into two and three until it breaks down into small-scale turbulent motion. This cascade repeats itself, so that at $x/h = 90$ a fully turbulent rear part has formed. With increasing E_c, the process of spot formation takes place more rapidly. E.g. for $E_c = 4.0$, a turbulent rear begins to form downstream of $x/h = 20$. The further development of this turbulent spot has already been described in detail in previous works (see Klingmann & Alfredsson, 1990).

Figure 3a and b show contour lines of the streamwise disturbance velocity in a cross channel view at $x/h = 50$, $z = 0$ for $E_c = 1.3$ and 1.5. A decaying spike is seen as a low velocity lump near the CL of the channel at the front of the disturbance structure. For $E_c = 1.5$ (figure 3b) the rear spike is seen as a more intense low velocity region slightly below the CL near $tU_{CL}/h = 79$. This region is absent in the corresponding picture for $E_c = 1.3$ (figure 3a). Figure 3 is qualitatively similar to the corresponding result obtained by Amini & Lespinard (1982) in the boundary layer case. This may indicate that the process preceding transition is similar in the two cases.

Figure 4 shows a spanwise view of the streamwise velocity perturbations at $x/h =$ 30, 50 and 73, for $E_c = 1.3$. The normal position is approximately midway between the channel CL and the lower wall. Alternating high and low velocity streaks indicate the presence of normal vorticity ($\omega_y = \partial u/\partial z - \partial w/\partial x$). Again, there is a notable similarity with the results of Amini & Lespinard (1982). The length and intensity of the streaks increase up to about $x = 50$, and thereafter begin to decay. The "aspect ratio" ($\Delta t U_{CL}/\Delta z$) is about 20 at $x/h = 50$. Different coherent streaks are seen to travel downstream at different speeds, indicating that several vorticity waves are involved. They do not propagate in the spanwise direction, e.g. the maxima of the high velocity streaks placed $\Delta z/h = \pm 1$ from the symmetry line remain at the same spanwise position. At the opposite side of the channel (see figure 5 for $x/h = 73$), low velocity streaks are found at these spanwise positions, indicating that the normal vorticity is antisymmetric. The spanwise wave vector component (k) can be evaluated from the distribution of $\partial u/\partial z$ in figure 4. At all three x-positions, the wave numbers $k = 3$ and 5 are dominating.

Similar measurements for $E_c = 1.5$ show that the streaks begin to disrupt at $x/h = 50$, and that the the interior part of the disturbance is fully turbulent at $x/h = 73$.

Since the streaks travel downstream without essentially changing their spanwise position, it is possible to follow the excess velocity maximum point within one particular streak and to measure its amplitude (u) and propagation speed (c). This was done for the high velocity streak indicated by '+' in figure 4. Figure 6a shows that u increases linearly between $x/h = 15$ and 35. There are two maxima with an amplitude of about

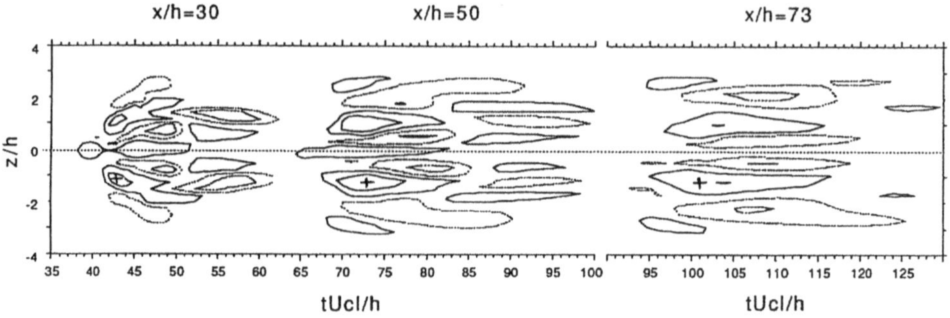

Figure 4. Spanwise view of disturbance evolution at Re near 1670, $E_c = 1.3$. Contour levels are ± 0.05 and ± 0.15. Negative disturbance levels are dashed.
(a) $x/h = 30$, $y/h = -0.5$
(b) $x/h = 50$, $y/h = -0.6$
(c) $x/h = 73$, $y/h = -0.45$

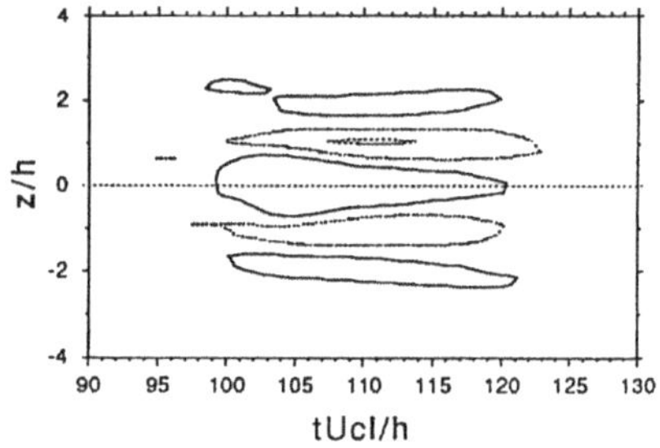

Figure 5. Spanwise view at $x/h = 73$, $Re = 1680$, $E_c = 1.3$, $y/h = +0.50$. Contour levels as in figure 4.

25 % of U_{CL} at $x/h = 42$ and 57 respectively. This is near the location where the pre-turbulent spikes begin to appear for the slightly higher input disturbance $E_c = 1.5$. The arrival times of the streak, plotted against the downstream location in figure 6b, give $c = 0.74 U_{CL}$. The values of $t\nu/h^2$ corresponding to the two maxima in u are 0.035 and 0.048.

These results are in fair agreement with the results for infinitely elongated normal vorticity structures described by Gustavsson (1990). These propagate with a speed near 2/3 of U_{CL}. For anti-symmetric vorticity modes with $k = 3$, he found the maximum at $t\nu/h^2$ near 0.065 and for $k = 5$ at $t\nu/h^2 = 0.35$. So encouraged, it should be possible to find similar combinations of disturbance amplitude maxima and spanwise spacing, which coincide with the linear theory. Following the above described path, investigations were undertaken for different Re and E_c. The result is listed in table I. E.g for $Re = 1710$, $E_c = 0.9$, the maximum was found at $t\nu/h^2 = 0.035$, and the wave number was estimated to 5. For $Re = 980$, which is close to the minimum transition Reynolds number, two maxima were found at $t\nu/h^2 = 0.028$ and 0.050. The dominating wave numbers were 3 and 4.5. At $Re = 2740$, a symmetric mode with $k \approx 5$ was found. It seems that the present configuration favours modes with wave numbers near 3 and 5, independently of Re and E_c. The parameter which might produce other wave-numbers is probably the geometrical shape and size of the triggering source. The fact that the maxima are found at $t\nu/h^2 = 0.05$ both for $Re = 1000$ and 1670 shows that the vorticity structures evolve on a viscous time scale. The propagation velocities range between 0.48 and 0.83.

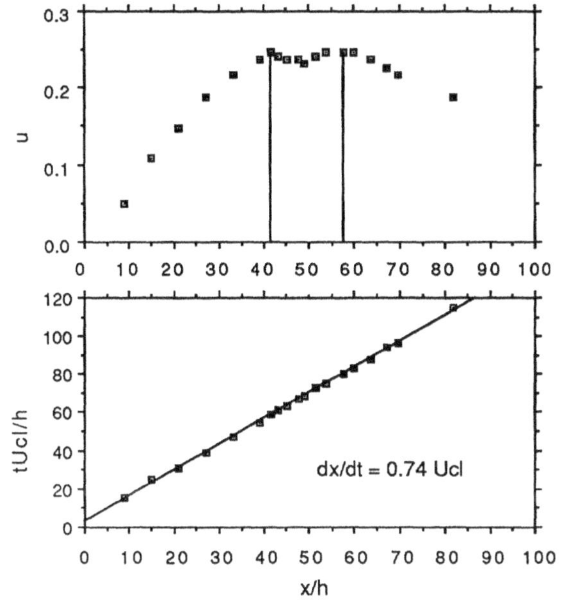

Figure 6.
(a) Evolution of the disturbance velocity amplitude u at a selected vorticity streak.
(b) Propagation speed c.
$E_c = 1.3, Re = 1680$,
$y/h = -0.55$

Table I. Wave properties of the normal vorticity for different Re

Re	E_c	mode shape	k	$(t\nu/h^2)_{max}$	$(y/h)_{max}$
980	0.9V/10ms	anti-symm	3	0.050	-0.50/+0.45
			4.5	0.028	
1670	1.3V/2ms	anti-symm	3	0.048	-0.70/+0.45
			5	0.035	
1710	0.9V/2ms	anti-symm	5	0.035	
2740	0.83V/2ms	symm	5		±0.55

4. Summary

The present study indicates that the linear growth mechanism suggested by Gustavsson (1990) may be generated by a point-like disturbance in plane Poiseuille flow. Streamwise elongated streaks of alternating high and low velocity indicate the presence of normal vorticity with properties similar to those described by Gustavsson for infinitely longs structures. The disturbance propagates at speeds between 0.48 and 0.83. Its local amplitude evolves on a viscous time scale $t \sim h^2/\nu$ and reaches a maximum level of about 25 % of U_{CL}. The streamwise location of maximum vorticity coincides with the location where non-deterministic motion is first observed for a slightly stronger initial disturbance. This indicates that the normal vorticity plays an important role in the evolution of a point-like disturbance into a turbulent spot. Although it does not in itself explain the appearance of turbulence, it may have the effect of making the flow more sensitive to non-linear motion on an inertial time scale. This is seen in the behaviour of the spikes, e.g. those which appear initially decay downstream, whereas the spike which triggers off the turbulent break-down appears when the amplitude of the normal vorticity is at its maximum. This spike is not directly related to the disturbance source and has a clearly inertial character. It appears at the rear, multiplies and finally breaks down to small-scale turbulent motion forming a turbulent spot.

The qualitative similarities between the present results and those of Amini & Lespinard (1982) allow us to hypothesize that a similar mechanism may also play a role in spot transition in a boundary layer.

Acknowledgements

This work is supported by the Swedish National Board for Technical Development (STU).

References

Amini, J. & Lespinard, G. 1982 Experimental study of an 'incipient spot' in a transitional boundary layer. *Phys. Fluids*, **25**, 1743.

Gustavsson, L.H. 1990 Energy growth of three-dimensional disturbances in plane Poiseuille flow. Accepted for publication in *J. Fluid Mech.*

Klingmann, B.G.B. & Alfredsson, P.H. 1989 On the development of turbulent spots in plane Poiseuille flow. In: *Proc. Third IUTAM Symp. on Laminar-Turbulent Transition*, Toulouse 1989.

Klingmann, B.G.B. & Alfredsson, P.H. 1990 Turbulent spots in plane Poiseuille flow - velocity measurements. To appear in *Phys. Fluids.* (December 1990)

Landahl, M.T. 1980 A note on an algebraic instability of inviscid parallel shear flows. *J. Fluid Mech.* **98**, 243.

Orszag, S. O. & Patera, A. T. 1983 Secondary instability in wall bounded shear flows. *J. Fluid Mech.* **128**, 347-385.

Direct Simulation of the Development of Turbulent Spots in Plane Couette Flow

Anders Lundbladh† *and Arne V. Johansson*

Department of Mechanics, Royal Institute of Technology,
S-100 44 Stockholm, Sweden

Turbulent spots in plane Couette flow were studied by means of direct numerical simulation in order to determine their basic characteristics. For instance, the lowest Reynolds number for which localized turbulent regions can persist, i.e. a *critical* Reynolds number, was found to be slightly lower than 375 (based on half the velocity difference between the two surfaces and half the gap width).

1. Introduction

Plane Couette flow represents one of the fundamental flow situations that has attracted much attention in terms of theoretical studies of stability (see e.g. Ellingsen, Gjevik & Palm, 1970) and transition mechanisms. The lack of curvature of the velocity profile simplifies the equations significantly, and it is now known to be linearly stable. Orszag & Kells (1980) carried out two- and three-dimensional calculations with a pseudospectral technique. They found two-dimensional disturbances to remain stable even at quite large amplitudes and high Reynolds numbers, in contrast to the stability calculations of Ellingsen et al. However, they observed transition to turbulence in three-dimensional calculations (with $16 \times 33 \times 16$ modes) at a Reynolds number, based on half channel width and half the difference in velocity between the two surfaces, of 1250. Due to spatial resolution effects they were not really able to carry the computations into the regime beyond initial breakdown.

Experimental investigations of plane Couette flow are associated with particular difficulties. The first study of turbulent spots in this situation has recently been initiated (Tillmark & Alfredsson, 1990). Experiments at transitional Reynolds numbers have been carried out by Reichardt (1956) in an oil channel and Leutheusser & Chu (1971) in an air flow between a stationary wall and a moving water surface. The *critical* Reynolds number, determined as that below which no turbulent fluctuations were observed, was found to be about 750 by Reichardt, whereas Leutheusser & Chu obtained a considerably lower value of about 280 (channel aspect ratios of 5 and 12, respectively).

2. Numerical method

The simulation code used for the present computations uses spectral methods to solve the Navier-Stokes equations, with Fourier representation in the streamwise (x) and spanwise (z) directions, and Chebyshev polynomials in the wall-normal (y) direction. For details see Lundbladh & Johansson (1990).

† Also at: FFA, P.O. Box 11021, S-161 11 Bromma, Sweden

Advances in Turbulence 3
Editors: A.V. Johansson · P.H. Alfredsson

For the bulk of the calculations the size of the computational domain was $128 \times 2 \times 64$ in the streamwise, wall-normal and spanwise directions, respectively, although symmetry with respect to the mid xy-plane was imposed. The surface velocities were taken as $\pm U_o$. If not explicitly stated otherwise, all quantities will be non-dimensionalized using the half channel height, h, and U_o. They were discretized using $256 \times 33 \times 256$ spectral modes for the computations at Reynolds numbers ($Re = U_o h/\nu$) up to 750. At $Re = 750$ the computations were also continued to large times by use of a larger box ($256 \times 2 \times 128$) for times between 90 and 160. The number of modes in the x and z directions was then doubled to retain the same spatial resolution as for earlier times.

For $Re = 1500$ the number of modes in the y direction was doubled and the box had a horizontal extent of 90×40 in order to keep the spatial resolution at nearly the same level as for $Re = 750$ in terms of wall-units, and still allow computation of the spot development up to a non-dimensional time, t, of 50.

A time step of 0.025 (0.01 for $Re = 1500$) was used for most of the computations, resulting in a CFL number at about half the empirical stability limit.

3. Results

The spots were triggered by an initial disturbance consisting of four pairs of counter-rotating vortices with axes of rotation in the x-direction. The maximum initial v was 0.5. The *effective* initial extents in the x and z directions are a few channel heights. The initial amplitude and initial energy were, hence, chosen quite large in order to obtain results valid in the limit of *very strong* disturbances. It is known from experimental investigations of spots in various types of flows that the spot characteristics become essentially independent of the specifics of the initial disturbance for strong enough disturbances.

The evolution of the disturbance into a growing region of turbulence at Reynolds numbers between 300 and 1500 is shown in figure 1. Here $\pm 2\%$ contours of v at the centreline ($y = 0$) are chosen to depict the spreading of the spot. The boundary between laminar and turbulent flow is quite distinct and relatively independent of actual choice of contour level. It should be noted that the centre remains at the same location since the two surfaces have velocities of equal magnitude but opposite sign. At $Re = 300$ the disturbed region grows initially, but the amplitudes and total energy decay, and the v-levels have fallen below 2 % at $t \approx 60$. The u-disturbance decays more slowly, but at large times the remaining u-disturbance consists merely of long longitudinal streaks of alternating low and high velocity, which are dominated by one specific spanwise wavelength (approximately 3.2).

At $Re = 375$ the turbulence is seen to be self-sustained as the region grows. The computations were here continued to $t = 160$ to verify the conclusion about self-sustained turbulence. Calculations were also carried out at $Re = 350$ in order to determine this *critical* Reynolds number (Re_{crit}) within a narrow interval. The behaviour at this Reynolds number was found to be essentially identical to that at $Re = 300$, although here the decay was somewhat slower and the v-amplitudes had fallen below 2 % at $t \approx 100$. Hence, Re_{crit} is within the interval between 350 and 375.

The box size and number of modes used in the computations were chosen such that the spatial resolution at $Re = 750$ corresponded to a horizontal cell size of approximately 27×14 wall-units, with the friction velocity, u_τ, taken from the central turbulent region

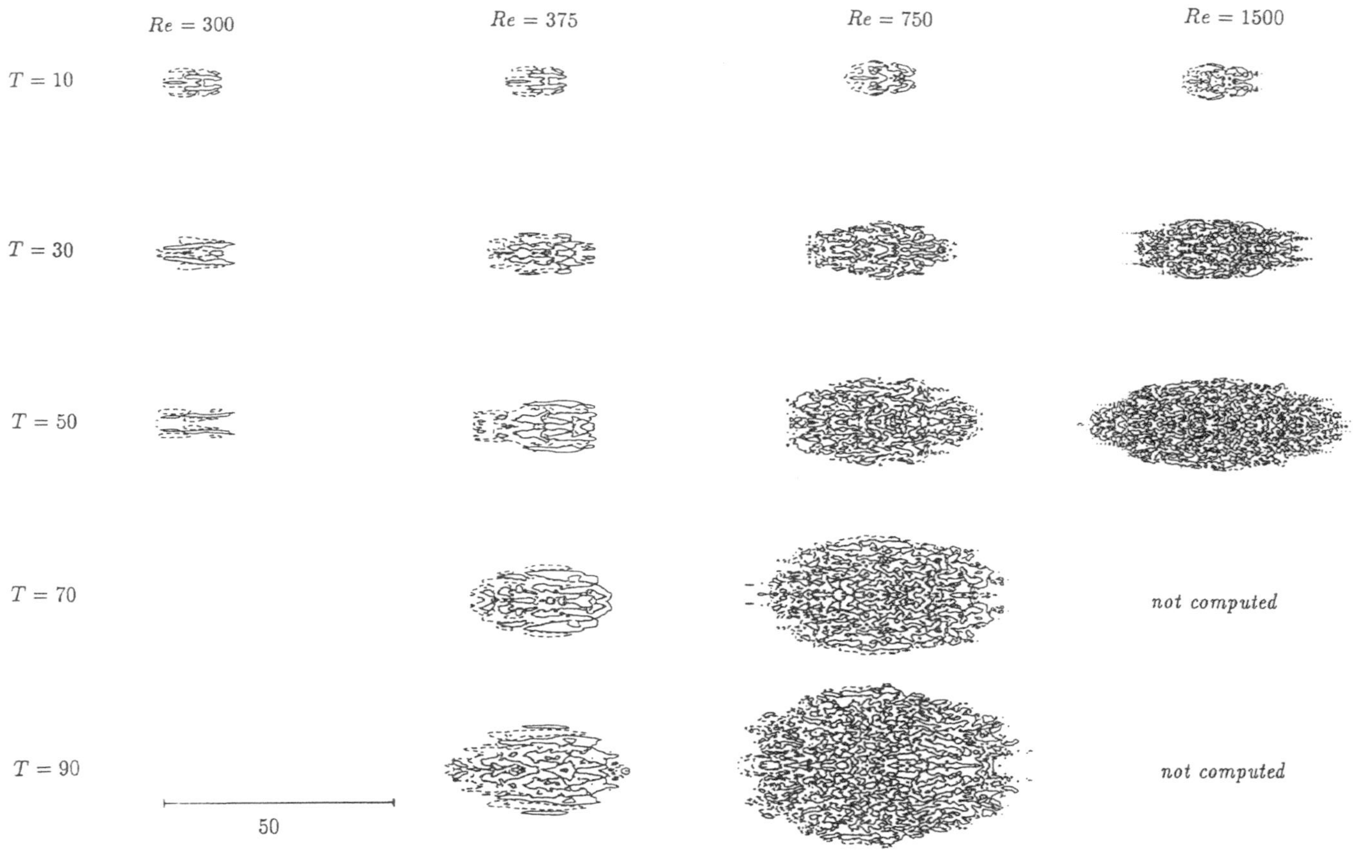

Figure 1 Spot development up to $t = 90$ for Reynolds numbers 300, 375 and 750, and up to $t = 50$ for $Re = 1500$. Contours of v at $\pm 2\%$ for the xz−plane at $y = 0$ are shown. Dashed contours represent negative values in this and following figures. A lengthscale is given in the figure for reference (same in both directions).

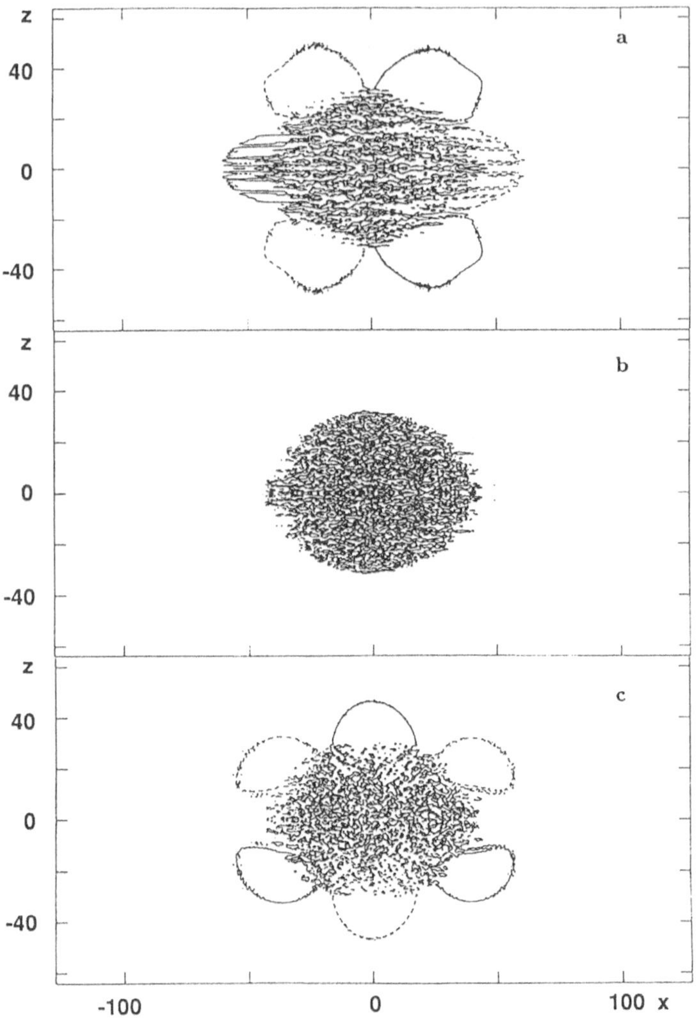

Figure 2 Contours of the velocity components at the centreline for $Re = 750$ at $t = 150$ **a** u at $\pm 5\%$, **b** v at $\pm 2\%$, and **c** w at $\pm 5\%$ are shown.

of the spot. At both the higher Reynolds numbers the maximum cell-size in the wall-normal direction was slightly above 5 in wall-units.

The $Re = 1500$ spot in figure 1, which was computed with nearly the same resolution in wall units as for $Re = 750$, was compared with a test run with the same box size and number of modes as for the lower Reynolds numbers. At this lower resolution the spot size at $t = 50$ was, within 5 %, the same as for that in figure 1, and turbulence statistics of the interior were found to be essentially identical.

The $Re = 750$ spot at $t = 150$ (figure 2) has here become very large with a total length of the order of a hundred half gap-widths. From comparison between figures 1 and 2 b it is seen that the length to width ratio of the spot has decreased substantially between $t = 90$ and $t = 150$. A closer study of the shape development yields that

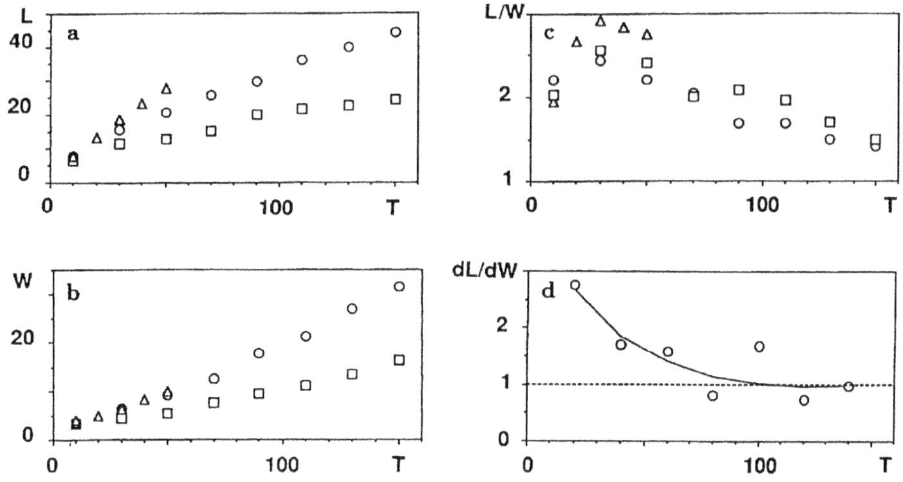

Figure 3 Length and width of the spot **a** spot half-length (L), **b** half-width (W), **c** aspect-ratio (L/W), and **d** growth rate ratio (dL/dW) as function of time for $Re = 375$ (□), $Re = 750$ (○) and $Re = 1500$ (△).

the growth-rate ratio is smaller than the length to width throughout the development and is equal to unity within the uncertainty of determination at the latest times. This implies that the half-length, L, to half-width, W, ratio, which is 2.2 at $t = 50$ and 1.4 at $t = 150$, decreases slowly but monotonously towards unity.

The streamwise spreading rate increases with increasing Reynolds number over the range investigated here (figure 3 a), whereas the results indicate that the spanwise growth rate levels off (figure 3 b), and is approximately the same at the two highest Reynolds numbers. At $Re = 750$ the rate of spread in the spanwise direction is 0.23 (of U_o). This corresponds to a spreading half-angle of approximately 13° for the case with one of the walls stationary. This spreading rate is roughly twice that at half the Reynolds number, but approximately the same as for $Re = 1500$. This bears resemblance to the increase in spreading angle found for spots in Poiseuille and free-surface water-table flows. For boundary layer spots the half-angle is 10° relatively independent of the Reynolds number. Figures 3 c-d show the development of the length to width ratio and the corresponding growth-rate ratio. The symbols in figure 3 c were obtained through direct differentiation and show considerable scatter. The solid curve, on the other hand was obtained through curve fits with third order polynomials to $L(t)$ and $W(t)$, whereafter these were differentiated before constructing the growth-rate ratio. The results indicate an asymptotic L/W close to unity.

An interesting feature of the spreading of the spot, which to some extent may be seen from figure 3 b and is evident from closer inspection of the data, is that the spanwise growth reaches a fairly constant level first after some time, and that this level is considerably higher than that at small times. This indicates that the mechanism for the spanwise spreading attains its full strength first when the spot has become large enough. A natural coupling would be to the modification and possible destabilization of the surrounding laminar velocity field.

The waves at the *wingtips* of Poiseuille spots do not have evident counterparts in the Couette spot. One may possibly say that a set of wave-like longitudinal streaks are

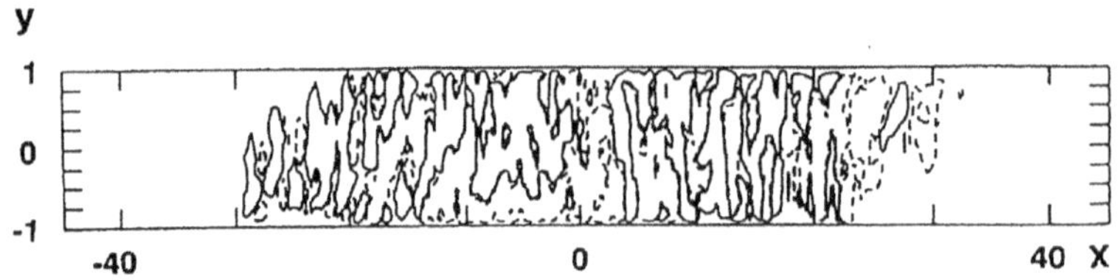

Figure 4 Contours ($\pm 2\%$) of v in the mid $xy-$plane at $t = 50$

present at the outermost (in the $z-$direction) part of the spot, at least for low Reynolds numbers. When present, the streaks move outward with a velocity slightly lower than the spreading rate of the turbulent region. Consequently they are not phase locked to the spot but new streaks are generated outside the spot and are overtaken by it.

The four lobes in the streamwise velocity (figure 2 a) and the six lobes in the spanwise velocity (figure 2 c) outside the turbulent region indicate the areas mostly affected in the external flow, and are present for all Reynolds numbers for which turbulence can be sustained. The extremum points of the four lobes for u were found to be just at the laminar-turbulent interface. This velocity is as large as 0.17 just outside the turbulent region, whereas the laminar Couette-value is zero. It is seen that the modified external flow is totally dominated by the horizontal components, with a quite negligible $v-$component, and that there is a motion out from the spot near the midpoint, $x = 0$, contrasted with motion towards the spot at the leading and trailing parts.

Close to the moving walls the spot extends further in the direction of the wall velocity than is seen in figure 1. The length of this *overhang* tends to a finite limit of 4 to 5 units for large times (figure 4). This is somewhat similar to the boundary layer spot, although there is a free stream above the overhang in that case.

In order to enable meaningful comparisons with other types of flows the total streamwise spreading rate should be divided by two since the overall velocity difference here is 2.0 and not 1.0 as in boundary layer and Poiseuille flows. Hence, an appropriate measure is the rate of increase of the half-length, dL/dt, which (from figure 2) for $Re = 750$ can be found to be close to 0.4 at early times, decreasing to about 0.24 (± 0.02) at $t = 160$. This may be compared with the difference between leading and trailing edge velocities of the boundary layer spot, which is about 0.4. The corresponding value for the Poiseuille spot is close to the Couette value (see Henningson, Spalart & Kim, 1987).

Local rms-values can be obtained through highpass filtering followed by squaring, lowpass filtering and taking the squareroot. In figure 5 are shown contours of the rms-values of u at a position ($y = -0.707$) that corresponds to a distance from the lower wall of 16 wall-units. Smaller details should not be regarded as significant, but maximum and typical levels are, as was verified from comparisons with the $y = +0.707$ position.

The intensities are significantly higher near the periphery of the spot (i.e. near the laminar-turbulent interface) as compared to the central part. This is also true for the $v-$fluctuations and at the centreline. The maximum local u_{rms} here corresponds to approximately $3.3u_\tau$, and typical values in the peripheral part are around $2.8u_\tau$. In the central part typical values are around $2.4u_\tau$. These values should be compared with the maximum u_{rms}/u_τ found in near-wall turbulence, which is 2.8-2.9.

The fluctuation intensities of u on the centreline are typically around $2.1u_\tau$, and hence only about 15% lower than at the near-wall position. This is in contrast to

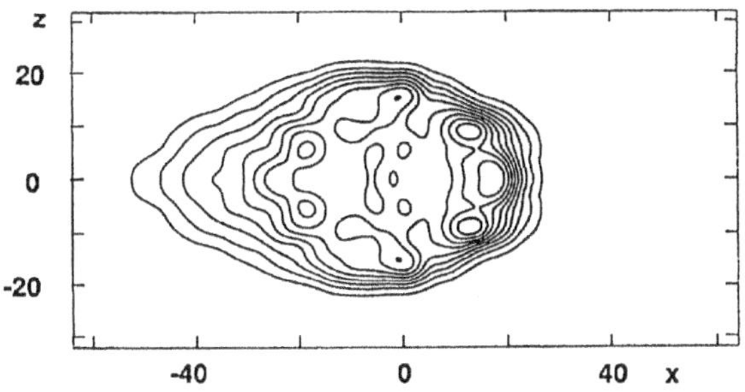

Figure 5 Fluctuation intensity at $t = 100$ ($Re = 750$) u_{rms} at $y = -0.707$ contour start at 0.025 with 0.025 increments)

e.g. high Reynolds number turbulent channel flow, where the centreline u−intensity is about $0.9u_\tau$. The low Reynolds number is one reason for the relatively small difference between the centre and the near-wall position. A further reason is the fact that for Couette flow, in contrast to e.g. channel flow, the mean velocity gradient is non-zero at the centreline. This also implies a non-zero turbulence production there. This effect will persist also at higher Reynolds numbers since the gradient at $y = 0$ decreases quite slowly. For the present spot this gradient varies between 0.2 and 0.4 (U_o/h) in the turbulent part at $y = 0$. It may here be worthwhile to note that measured values of u_{rms}/u_τ (Robertson & Johnson, 1970) at the centreline of fully developed turbulent Couette flow ($Re = 7 - 16 \times 10^3$) are about 1.5.

4. Conclusions

The lowest Reynolds number for which Couette spots can be sustained has here been determined within a narrow interval, and found to be slightly below 375. The horizontal extent of the spot has an elliptical character, where the streamwise elongation becomes less pronounced with increasing time.

The spanwise growth was found to increase with increasing Re for low Reynolds numbers, but levelled off to a constant rate at high Re. In the reference frame of one of the solid surfaces this spreading rate can be translated into a spreading half-angle of 13° for the Couette spot at high Reynolds numbers.

The overhang, as seen from vertical cuts of the spot, and the zero curvature of the velocity profile at the wall bear resemblance to the zero pressure-gradient boundary layer case. The Couette spot, hence, constitutes an interesting case that may be regarded as an intermediate one between the Poiseuille and boundary layer spots.

The modifications of the horizontal components of the laminar flow just outside the turbulent region were found to be quite large. Since no experiments existed when this study was carried out it represents a true numerical prediction of the basic features of turbulent spots in plane Couette flow.

Acknowledgement

This research has been partially sponsored by the Swedish State Board for Technical Development, the Aeronautical Institute of Sweden and The Göran Gustafsson Foundation. CRAY-time has been granted from the Swedish National Supercomputer Center. We also wish to thank Henrik Alfredsson for many fruitful discussions on the present topic, Dan Henningson for cooperation with the writing of the simulation code, and Lars-Göran Sundström for help in preparing graphics for the presentation.

5. References

Ellingsen, T., Gjevik, B. & Palm, E. 1970 On the non-linear stability of plane Couette flow. *J. Fluid Mech.* **40**, 97.

Henningson, D.S., Spalart, P. and Kim, J. 1987 Numerical simulations of turbulent spots in plane Poiseuille and boundary layer flows. *Phys. Fluids* **30**, 2914-2917.

Leutheusser, H.J. & Chu, V.H. 1971 Experiments on plane Couette flow. *J. Hydr. Div., ASCE*, **97**, 1269.

Lundbladh, A. & Johansson, A.V. 1990 Direct simulation of turbulent spots in plane Couette flow. Submitted to J. Fluid Mech.

Orszag, S.A. & Kells, L.C. 1980 Transition to turbulence in plane Poiseuille and plane Couette flow. *J. Fluid Mech.* **96**, 159.

Reichardt, H. 1956 Über die Geschwindigkeitsverteilung in einer geradlinigen turbulenten Couetteströmung *ZAMM* Sonderheft, 26.

Robertson, J.M. & Johnson, H.F. 1970 Turbulence structure in plane Couette flow. *J. Engineering Mech. Div., ASCE*, **96, EM6**, 1171.

Tillmark, N. & Alfredsson, P.H. 1990 An experimental study of transition in plane Couette flow. Paper to be presented at the Third European Turbulence Conference.

Transition from Order to Disorder in Elliptical Flow: A Direct Path to Shear Flow Turbulence

W.V.R. Malkus and F.A. Waleffe***

* Massachusetts Institute of Technology, Department of Mechanics
** Center for Turbulence Research, NASA-Ames MS202A-1, Moffet Field, CA 94035

ABSTRACT: Experiments on the instability of fluid in a rotating elliptically distorted elastic cylinder reveal that the origin of the observed broadband disordered flow is the collapse of the principal three-dimensional mode at a critical point in its growth. It is proposed here that the process responsible for this violent passage from completely ordered to disordered flow is a ubiquitous source of turbulence which by-passes lesser chaotic phases. Here we describe the sense in which the laboratory elliptical flow is a simplifying idealization of the observed transition to turbulence in channel shear flow. From third order amplitude equations describing the growth of the modes and evolution of the mean field, we estimate the positions at which critical conditions for collapse are first reached.

1. Introduction

This work describes the swift inertial growth of momentum transporting instabilities observed in elliptical vortex flow, and the sudden collapse of these organized large scale flows into small scale disorder. A finite amplitude Tollmien-Schlichting waves in a parallel shear flow is such an elliptical vortex, and is observed to exhibit both instability and collapse in boundary layers and channels. In a particularly clear example of this process, the initial "turbulent spot" in channel flow, like an advancing brush fire, burns its way through the field of shear by generating momentum transporting vortex waves (Gad-el-Hak et al. 1981, S. Widnall 1984). Behind the advancing front, the elliptic vortex waves exhibit 3D instability and collapse leaving a turbulent wake and only a fraction of the initial interior shear. Figure 1, from D. S. Henningson, et al. (1987) is a numerical simulation of centerline normal velocities in an advancing channel flow spot. The channel depth is comparable to the wavelength of the frontal waves. The secondary instability can be seen growing most rapidly in the oblique waves. Collapse occurs at the third wave, and is preceded by the most significant transfer of moment and alteration of the mean shear flow which occurs in the spot. This is the "burning front" and behind it appears, metaphorically speaking, only glowing cinders and ashes. The suggestion here is that the "turbulent" part of the "turbulent spot" plays only a small role in the spots complicated dynamics, and that a recognizable spot might be constructed numerically with the disordered aspect of the flow arbitrarily suppressed. However, in this study our principal search is for a mechanistic understanding of the origin of the sudden disorder, the "turbulence", due to the vortex collapse.

A first goal is to isolate, experimentally and theoretically, the elliptical vortex and its instabilities from the spatially inhomogeneous, time-dependent behavior of the spot.

Advances in Turbulence 3
Editors: A.V. Johansson · P.H. Alfredsson

Figure 1. A numerical simulation of centerline normal velocities in a channel flow spot. ($R = 1500$). One half of the field as seen from the top, the mean flow is from the right. Note the 3D instabilities in the outermost oblique vortex waves.

Figure 2. Experimental arrangement: A fluid filled rotating elastic cylinder distorted into elliptical shape in the laboratory frame by two freely rotating metal rollers. Length of cylinder is 15 cm, rotation rate is 80 rpm, and ellipticity is adjustable by changing the rollers' diameter. The spin-over mode is seen above, just two rotations before its violent collapse.

We have chosen to explore the temporal behavior of a single elliptical vortex, whose ellipticity can be altered at will. The boundaries of a fluid in a rotating elastic cylinder are distorted into elliptical shape by two fixed, but freely rotating rollers which are pressed against the cylinder. Figure 2 is a photograph of this experimental arrangement. Ideally, this confined steady elliptical flow is an exact solution to the full Navier

Stokes equations requiring no viscous boundary layer. Yet as anticipated, recent experiments (Malkus, 1989) have shown that flow in a close approximation to this ideal geometry exhibits inertial, three dimensional, instabilities. Here we select for study that most unstable mode, akin to the one observed in the Poiseuille flow spot, and report on progress to date in the interpretation of its growth and collapse into disorder. Theoretical study of the instability of elliptical flow in the infinite domain (Craik & Criminale 1986, Bayly, 1986, Waleffe 1990) provides the framework for this exploration of the similar instabilities observed in the bounded flow in the laboratory.
Application of this study of a single vortex, its 3-D instabilities and the subsequent disorder, to inhomogeneous turbulent flows, of course, remains speculative.

In the first section to follow, previous analysis of the linear stability problem in a bounded domain is outlined, attention being focused on the principal "spin-over" mode. Sub-harmonic, triad-like, bifurcation is characteristic of the many modes.

A second section addresses the post-critical, non-linear behavior of the growing modes. The first important question here is the energy source of the disturbance and its modification with increasing disturbance amplitude. Not only have frequency shifting and alteration of the mean field of rotation played roles in the approach to finite amplitude equilibration, but it is likely that "catalytic" behavior can exist in terms higher than cubic in the amplitude. Observations support this unusual finite amplitude behavior of small forcing leading to large response.

The third section contains a description of the collapse which follows the spin-over instability. At a critical point in its growth, this principal mode pours its energy into a cascade to smaller scale disordered motion. These smaller scales dissipate that energy and vanish. The fluid then spins-up to initiate the primary instability again. Several possible mechanisms for this collapse are explored. Our tentative conclusion is that the phase and amplitude of the growing mode have reached first extrema, permitting the exponential growth of a second triad-like mode. A possible consequence of this secondary growth is the transition from hyperbolic to elliptic behavior, leading one finite amplitude vortex to "see" the other as a field of caustics. Within one or two revolutions of the contained flow, the wave energy is dissipated. The locations in the fluid of first collapse depend on the spatial structure of primary and secondary mode, hence providing qualitative test of this interpretation. Details of the energy cascade in scale and time require further study of those two dramatic revolutions of the flow in which the collapse occurs.

Summary and Findings

In the following we report on first observations of the parametric dependence of the sub-harmonic "triad" instability process in bounded elliptical flow as well as the simpler spin-over modes.

A catalytic behavior of small ellipticity leading to a large response is observed for spin-over.

Typical of the non-linear dynamics is subcritical instability, with retuning destabilization of adjacent modes in the parameter space.

A shock-like transition of the principal mode to a disordered cascade is observed and related to local gradients of vorticity exceeding the condition for the maintenance of Poincaré waves. There is no evidence that centrifugal and inflexional instabilities play a role in this dissipation process.

The behavior of the single vortex isolated in this study appears to parallel the fate of flows with rotation in most non-axisymmetric geometries. Geophysical and astrophysical implications for tidal and precessional flow are discussed elsewhere.

2. Linear Theory of the Instability of Elliptical Flow In a Closed Container

Early theoretical study of precessional and tidal flow in oblate spheroids (Busse, F., 1968, Suess, S., 1969) and experimental observations of these elliptical flows emphasized instabilities due to the viscous boundary layers. An overlooked work (Hocking, L.M., 1965) clarified the mechanism for the inertial momentum transfer which occurs in non-axisymmetric rotating flows. The first study which clearly identified the spin over inertial stability of elliptical flow (seen, but not identified in the earlier experiments) was due to Gledzer, Y.E.B., et al. 1975. They first rotated, then suddenly stopped a rigid elliptical cylinder. Despite the boundary layer shear instabilities and centrifugal instabilities, they observed a strong spin-over in the body of the flow. Their motives apparently were unrelated to either tidal effects or shear flow vortices, but rather to the classical mechanical instability of rotating elliptical solids. A more recent study (V. A. Vladimirov, 1985) explores resonant excitation of non-axisymmetric waves. The renewed interest in the problem surely will lead researchers to provide a clearer view of the past contributions in less accessible parts of the literature. Here, the linear problem, based on an expansion in elliptical-polar coordinates and first written for a doctoral thesis (Waleffe, F., 1989) will be outlined to provide a footing for the post-critical non-linear study to follow.

For study of the flow in a bounded elliptical domain, the appropriate transformation proves to be

$$\begin{aligned} x &= \rho E \cos\theta \\ y &= \rho \sin\theta, \quad x^2/E^2 + y^2 = \rho^2 \end{aligned} \tag{1}$$

where ρ is constant along an ellipse and the aspect ratio determined by E is also constant. Note that these are not the usual elliptical coordinates, but instead are appropriate where streamlines of the basic flow are similar ellipses. Linearizing about the elliptical flow one finds:

$$\frac{\partial \mathbf{u}}{\partial t} + L\mathbf{u} + \nabla p = \frac{\beta}{2}(e^{i2\theta}N + e^{-i2\theta}N^*)\nabla p \tag{2}$$

where

$$N = \begin{pmatrix} 1 & i & 0 \\ i & -1 & 0 \\ 0 & 0 & 0 \end{pmatrix}, \beta = (E^2-1)/(E^2+1),$$

β is the ratio of the strain rate to the rotation rate of the basic flow. For $\beta = 0$, the left-hand operator in equation 2 has as solutions the stable inertial modes of solid rotation, $\mathbf{u} = U(\rho, z)e^{im\theta}e^{i\sigma t}$. For $\beta > 0$, only pairs of modes m and $m+2$ are coupled by elliptical effects. The condition for instability of a mode pair is written as a "triad resonance" with the basic elliptical state. For any length cylinder the inviscid theory predicts classes of unstable modes. However, viscous effects lead to the isolation of modes of low overall wave-number, and quantized cylinder lengths. The most unstable mode is spin-over, which in lowest radial order is resonated when cylinder length is approximately an integral multiple of cylinder diameter. The finite-amplitude behavior of this mode is discussed in the following section.

3. Post-Critical Non-linear Analysis

The boundedness and viscosity of the laboratory elliptical flow offer the possibility that a single eigenmode will have the minimum critical value of β, and can be followed to finite amplitude by weakly non-linear analysis. However, in contrast to convective and centrifugal instabilities whose initial modification of their own form and the mean field typically stabilizes both the growing mode and adjacent modes, elliptical instabilities are found to destabilize certain adjacent modes. To anticipate the observations one must track the growth of both the most unstable mode as well as potential instability of dangerous competitors. This is attempted here by a straightforward low order expansion in the disturbance amplitude of the triad-like eigenmodes with an assessment, at the amplitude cubed level, of the modification of stability properties of adjacent modes. The predictive power of this approach appears to be greater than warranted, in particular when observing the behavior of the principal spin-over mode. However, the details of the collapse of the larger modes into broadband disorder, to be discussed in the next section, do not appear accessible by any expansion procedure.

The amplitude expansion for a pair of modes, m and $m+2$, leads initially to five coupled equations, cubic in the mode amplitudes. However, for spin-over the second mode is the complex conjugate of the first. This simplifies the initial finite amplitude problem to the three equations.

$$\dot{A} = i(\delta + \sigma''|A|^2 + \chi A_0)A + i\beta C A^*, \dot{A}_0 = i\beta(A^2 - A^{*2}) \tag{3}$$

where the real part of δ is the "off resonance" frequency determined by cylinder length and the imaginary part of δ is due to the viscous boundary layer. C is the destabilizing

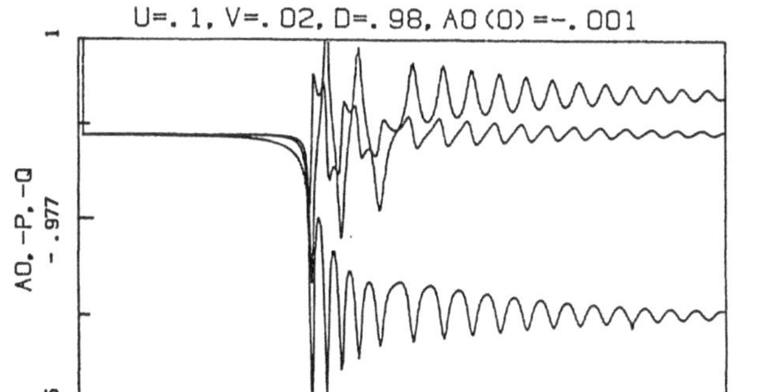

Figure 3. Growth, oscillation, and decay to fixed points of a disturbance to equation 3, for small βC and viscous damping, the upper curves are for p and q, where $A \equiv p + iq$. The lower curve is A_0.

term from the linear theory, σ'' and χ are stabilizing coefficients due to self-distortion of the mode and its modification of the mean flow A_0. An integration of these equations for parameter value just beyond the critical β for instability is seen in Figure 3. The appropriate viscous spin-up term has been added to the A_0 equation to permit the system to approach its fixed points. Observations determine that such finite amplitude equilibration is possible in only a small range of β beyond critical. Almost all of our observations for finite β led to the unrestricted growth and collapse described in the following section.

4. Collapse of the Wave Field

The finite-amplitude vortex wave growth and the low amplitude wave-wave interactions analyzed and discussed in the last section, in many cases observed exhibit a chaotic modification of amplitude and phase. However, nothing in the theory prepares one for the observed violent transition which generates small scale broad-band disorder. A quite similar transition occurs after the third frontal wave of the Poiseuille spot, leading to the "turbulence" part of the turbulent spot.

In the remainder of this section the process leading to the violent transition is dissected in order to isolate the principal mechanism for this dumping of energy from the large scale momentum transporting vortex wave to dissipative scales. However, details of the cascade after the transition remains unaddressed.

The similarity of Poincaré wave "collapse" (a term due to A. D. Mc Ewan 1970, also see R. Manasseh, 1990) and the breaking of water waves is considerable. In both cases the large scale wave losses its energy abruptly to a dissipative range of motions. However, the difference between the vortex wave field in a homogeneous fluid and a potential wave with a free surface is so large that one must resist overly simple comparison. One feature of the surface wave breaking which it is believed carries over to the vortex wave collapse, is that the transition represents a local failure of the restoring mechanisms of the wave. For potential waves, a shape criterion has long been known. Here we advance an analogous idea, that Poincaré waves break when the local field of rotation can no longer sustain them. This can occur, not directly due to the amplitude or shape of the wave, but due to gradients in the local field of rotation, ω. A representative, and plausibly principal, criterion is obtained for a local ω which varies uniformly but only at a right angles to its vector direction. This condition for non-existence may be rewritten in terms of a 'circulation' $\Gamma \equiv r^2|\omega|$ as

$$\frac{d\Gamma^2}{dr} < \frac{\Gamma^2}{r}, \tag{4}$$

and compared with the Rayleigh conditions for centrifugal instability, which requires a larger negative gradient of $\Gamma^2, d\Gamma^2/dr < 0$. However, only qualitative tests of this proposed mechanism are possible with the available data. The appearance of a second mode determines the few positions in the fluid where gradients of Γ^2 will be largest. Here is where collapse is first observed to occur. More quantitative tests are planned, for although the centrifugal criterion for stability is not nearby, further inertial triad interactions may be important. Hence one anticipates that collapse can start at locations where equation 4 is first realized.

For just post-critical β the collapse is followed by a period of recovery. Spin-up finally establishes enough of the field of rotation for the elliptical instability to reoccur, leading intermittently to another collapse. However, at sufficiently large β, one observes a continuous disordered form of the principal spin-over mode embedded in a disordered background of distinctly smaller scale motions. It would appear that the range of β leading to intermittent behavior offers the best region to explore the separate mechanisms of organized transport, dissipative cascade to smaller scales, and their short moment of interaction.

This ongoing work has been made possible by a grant from the National Science Foundation, ATM-8901473, for which we are most grateful. The experimental equipment is from the skillful hand of Edward Udas.

References

Bayly, B. J. 1986 Three-dimensional instability of elliptical flow, *Phys. Rev. Lett.*, **57**, 2160.

Busse, F. 1968 Steady fluid flow in a precessing spheroidal shell, *J. Fluid Mech.*, **33**, 739.

Craik, A.D.D. & Criminale, W.O. 1986 Evolution of wavelike disturbances in shear flows, *Proc. Roy. Soc. A*, **406**, 13.

Gad-el-Hak, M., Blackwelder, R. F. & Riley, J. J. 1981 On the growth of turbulent regions in laminar boundary layers, *J. Fluid Mech.*, **110**, 83-95.

Gledzer, Ye. B., Dolzhanskiy, F. V., Obukhov, A. M. & Pomonarev, V. M. 1975 An experimental and theoretical study of the stability of motion of a liquid in an elliptical cylinder", *Izv. Atmos. and Oceanic Phys.*, **11**, 981.

Henningson, D.S., Spalart, P. & J. Kim 1987 Numerical simulations of turbulent spots in Poiseuille and boundary layer flows, *Phys. Fluid*, **30**, 2914.

Hocking, L. M. 1965 On the unsteady motion of a rotating fluid in a cavity, *Mathematika*, **12**, 97-106.

Malkus, W. V. R. 1989 An experimental study of global instabilities due to tidal (elliptical) distortion of a rotating elastic cylinder, *Geophys. Astrophys. Fluid Dynamics*, **48**, 123-134.

Manasseh, R. 1990 Breaking Poincaré Waves. Thesis submitted to University of Cambridge.

McEwan, A. D. 1970 Inertial oscillations in a rotating fluid cylinder, *J. Fluid Mech.*, **40**, 603.

Suess, S. 1969 Effects of gravitational tides on a rotating fluid, Doctoral dissertation, Department of Planetary Science, UCLA.

Vladimirov, V.A. & V. Tarasov, V. 1984 in **Laminar-Turbulent Transition**, Proceedings of the IUTAM Symposium, Novosibirisk, edited by V. V. Kozlov (Springer, Berlin, 1985), 717-722.

Waleffe, F. 1990 On the three dimensional instability of strained vortex, *Phys. Fluids A*, **2** (1), 76-80.

Waleffe, F. 1989 Ph.D. Thesis, Massachusetts Institute of Technology.

Widnall, S.E. 1984 Growth of turbulent spots in plane Poiseuille flow, *Turbulence and Chaotic Phenomena in Fluids*, (ed. T. Tatsumi), 93-98, Elsevier.

Turbulent Spots in a Rotating Blasius Boundary Layer

M. Matsubara and S. Masuda
Keio University, 3-14-1 Hiyoshi, Kohoku-ku, Yokohama 223, Japan

The boundary layer stability is affected by a Coriolis force acting normal to a solid surface. The recent studies on the rotating Blasius boundary layer reveal that the transition of the positively rotating boundary layer, i.e. the Coriolis force acting toward the surface, is characterized by the streamwise vortex instability, while the transition of negatively rotating boundary layer is accompanied by the turbulent spot similar to the stationary one's. In the present investigation, the spot was artificially generated within the stationary and the negatively rotating boundary layer by introducing the pulsed jet from the small hole drilled on the test surface. The effects of the Coriolis force on the probability of the spot generation, on the shape and the growth of the spot as well as the velocity perturbations associated with the passing of the spot were investigated.

1. Introduction

A body force which acts normal to a solid surface affects the stability of a boundary layer. Examples are a boundary layer on a curved surface, a density-stratified boundary layer and a boundary layer with spanwise system rotation. The last example is the subject of the present investigation.

Masuda et al. (1984), (1989) studied the transition of the rotating Blasius boundary layer. The natural transition of the low pressure side boundary layer was found to be characterized by the turbulent spots similar to the stationary case. The areas of the turbulent spots decrease with increasing rotation number. This suggests that Coriolis force modulates the incipience and/or growth of the turbulent spots.

Conrad (1962) and Chawla (1969) performed the linear stability calculation and, reported that the effect of rotation changes the amplification rate of the spanwise wave disturbances and that the positive rotation destabilizes and the negative rotation stabilizes them.

The experimental study by Gad-el-Hak et al. (1981) concerns the turbulent spots in the stratified boundary layer, which is only one example to the knowledge of the present authors, of the investigation of the turbulent spot in the body force field. Their investigation in the stable boundary layer shows that the stratification inhibits the vertical growth of the turbulent spot.

In the present study, the turbulent spot was artificially generated in the rotating Blasius boundary layer, and the effects of Coriolis force stabilization on the generation and development of the spot have been investigated.

2. Experimental arrangement

The experimental facility is shown schematically in figure 1. The free stream turbulence intensity is 1.2%, and the laminar-turbulent transition dose not occur naturally with

Advances in Turbulence 3
Editors: A.V. Johansson · P.H. Alfredsson

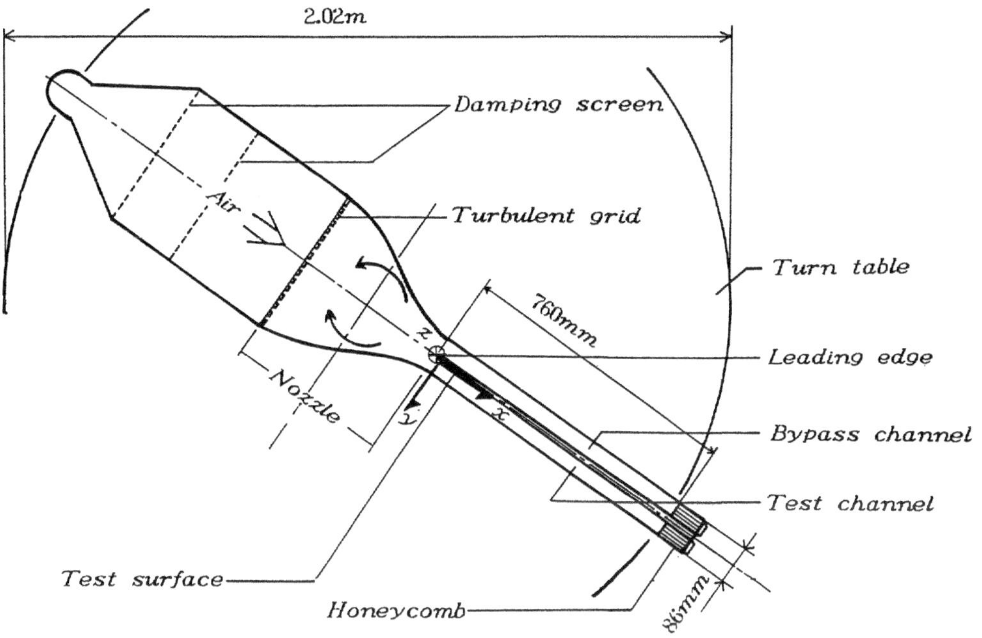

Figure 1. Schematic of the rotating wind tunnel facility

the free stream velocity below $10m/s$. The turbulent spot was initiated by a pulse jet type disturbance generated by a sound speaker and introduced from a 1 mm diameter hole located 279 mm downstream of the leading edge. The length Reynolds number at the hole, based on the free stream velocity of 8 m/s is 1.6×10^5, which is just above the critical Reynolds number.

For the visualization of the spots, the paraffin mist was introduced from the wall slit located 35 mm downstream of the leading edge and recorded on the monochromatic films. The amount of the mist was carefully controlled in order not to disturb the base flow by monitoring the video picture in the stationary frame. The video camera, 35 mm still camera as well as the lighting systems were mounted on the rotating table and remotely controlled from the stationary frame. The velocity field in the spot was measured with a single hot wire probe. The signal from the CTA system on the rotating table was transmitted to the stationary frame and then digitized and processed on a personal computer. The transmission of the electric power and the signals between the rotating and the stationary frames were performed via the carbon-copper slip rings.

3. Results and Discussions

Figure 2 shows the two typical velocity signals at 200 mm downstream from the initial disturbance point. The signal A containing the high frequency oscillation indicates the passage of the spot, while the signal B shows that the disturbance has not yet grown to the spot.

Figure 3 shows the probability of the generation of the spot, defined as the ratio of the number of the appearance of the signal A to the total number of realizations at the

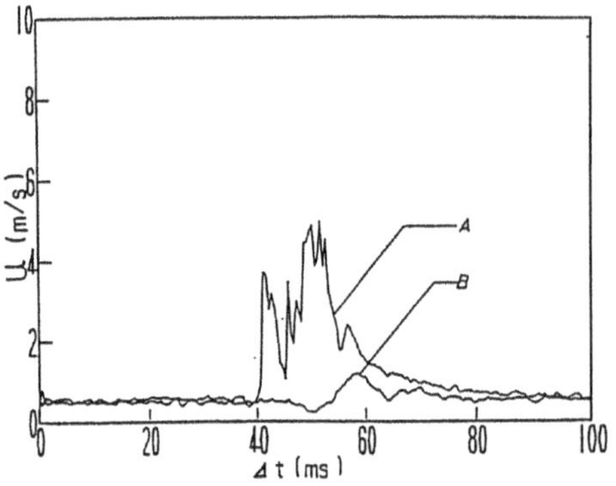

Figure 2. Velocity at $x = 500mm$, $y = 0.5mm$, jet velocity $v_j/U_w = 0.62$, A: disturbance has grown to spot, B: disturbance has not yet grown to spot.

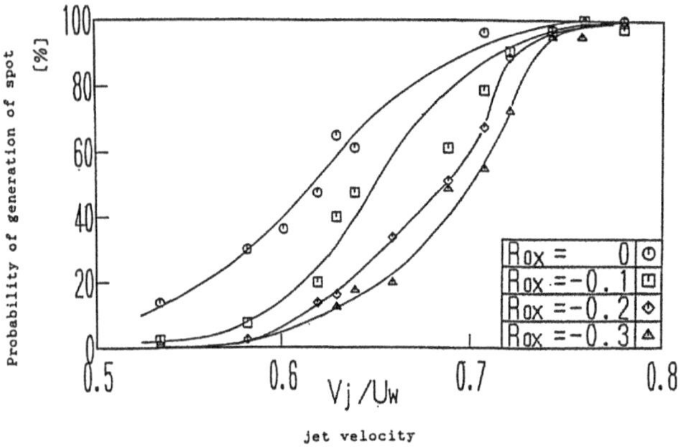

Figure 3. Probability of generation of spot vs. jet velocity.

fixed jet velocity v_j. For the jet velocity below $0.5U_w$, no spot was observed regardless of the system rotation, while v_j higher than $0.75U_w$, all the disturbances generated spots. In the intermediate range of v_j, the probability decreases with increasing negative rotational speed. This result indicates that the Coriolis force stabilization inhibits the onset of the spot, and explains the increase of the Reynolds number of the natural transition in the low pressure side boundary layer (see Masuda et al. 1989).

The photographs of the paraffin mist released from the wall slit are given in figure 4 for the stationary and negatively rotating boundary layers. Although the effect of Coriolis force is hardly recognized there, the typical shape of the spot can clearly be seen with the arrow-head and the trailing calmed region.

In order to examine the Coriolis force effects on the internal and the boundary structures of the spot, the measurements of the velocity during the transit of the spots were carried out. The single hot wire probe was located at 200 mm downstream from the initial disturbance point. The initial disturbance velocity of 8 m/s is generated by charging the speaker with the voltage of 5 V at $\Delta t = 0$. The disturbance of this intensity generates the spot in any rotation number. Figure 5 shows the ensemble-averaged velocity profiles in the stationary boundary layer. They are plotted in the

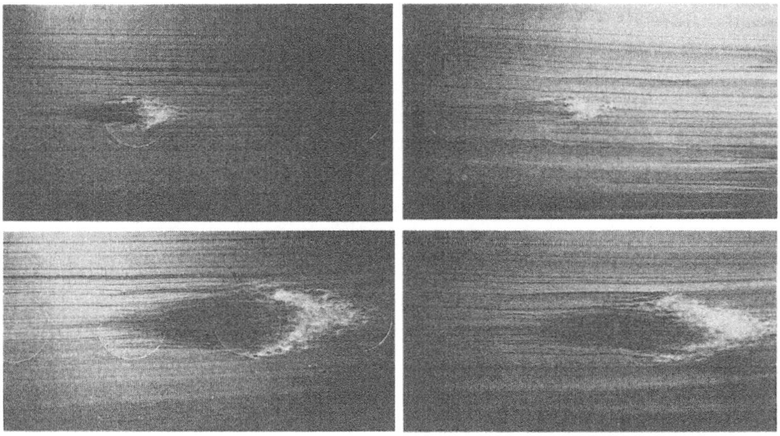

Figure 4. Visualization of turbulent spots.

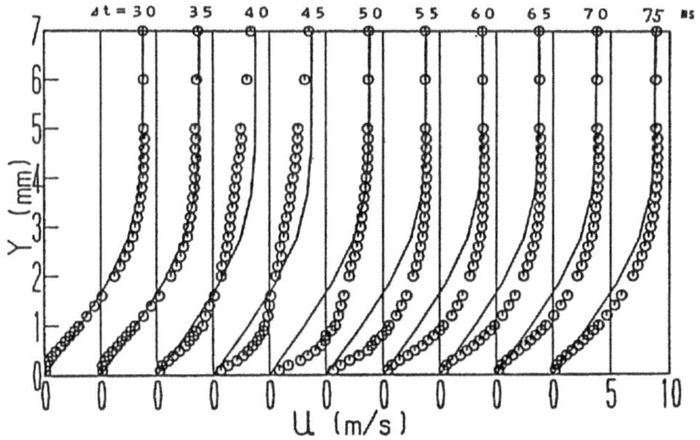

Figure 5. Ensemble averaged streamwise velocity profiles during transit of the spot at the symmetry plane (stationary boundary layer).

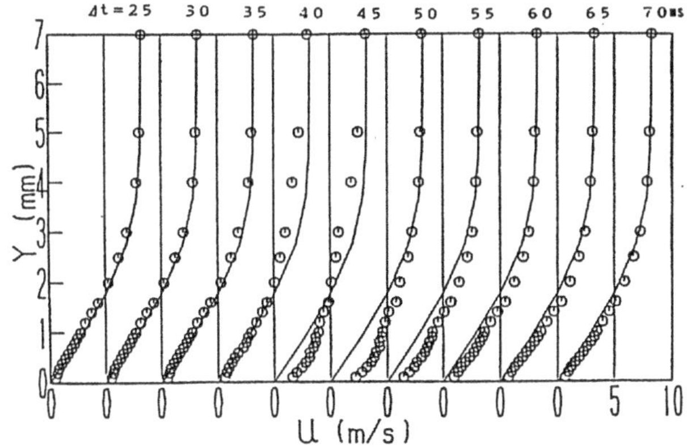

Figure 6. Same as figure 5 but with negative rotation ($N = -51$ rpm).

(y,t)-plane at the center line of the spot at time intervals of $10ms$. Before the spot hits the hot-wire ($\Delta t < 35$), the measured velocity profiles coincide with the Blasius profile in laminar boundary layer. When the head of the spot arrives at the measured station ($\Delta t = 40$ ms), the deceleration in the outer region first begins, followed by the acceleration in the inner region. As the spot is passing away ($\Delta t > 50$ m/s), the velocity profile slowly returns to the Blasius one. Coriolis force stabilization in the present range of the rotation parameter does not change this sequence as shown in figure 6.

For the purpose of visual presentation, the contours of the ensemble-averaged velocity perturbations are given in figure 7 (a) for the stationary boundary layer. The velocity perturbation is defined as the deviation of the ensemble-averaged velocity from the laminar one. The decelerated region (negative values) lies away from the wall and at the downstream of the accelerated region (positive values). This means that the momentum moves from the front part to the rear part in the turbulent spot. Contours of the turbulence intensity are shown in figure 7 (b). The turbulent region approximately coincides with the decelerated region, and corresponds to the front part of the spot.

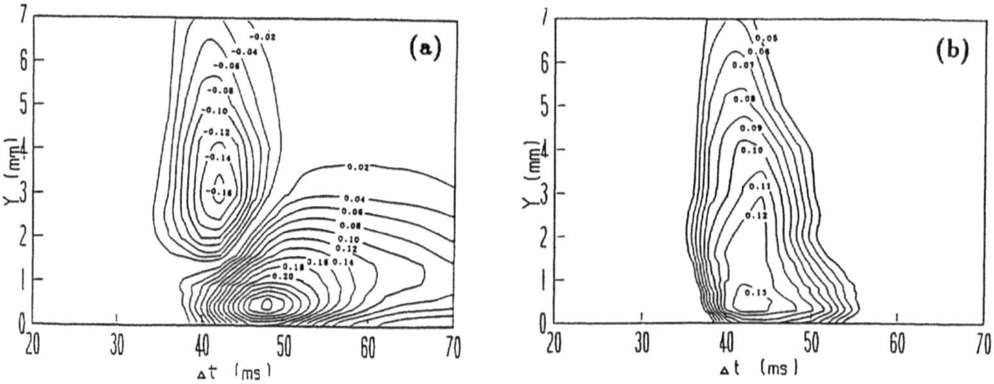

Figure 7. Contours at the center of the spot in stationary boundary layer, (a) velocity perturbation (b) rms velocity.

Contours of the velocity perturbation and the turbulence intensity in the boundary layer on the low pressure side are shown in figure 8. The overall shapes are weakly dependent on rotation except at the head of the spot. Both the velocity perturbation and the turbulence intensity vanish in the near wall region around $\Delta t = 40$ ms. This result indicates that the Coriolis force chips the head of the spot near the wall. This is in accordance with the collapse of the head in the stably stratified boundary layer found by Gad-el-Hak et al. (1981). Figure 9 (a) shows contours of the velocity perturbation and the turbulence intensity in (z,t)-plane at $y/\delta = 0.1$. Reversing the direction of the photographs in figure. 4 from left to right, one can find that the shape of the spots in the photograph well agrees with the contours seen in figure 9. The figure 9 (b) shows the same contours but with negative rotation. The width of the spot seems slightly less than that of the stationary case.

These results suggest that the Coriolis force stabilization may affect the instability near the wing-tip region and the near wall region underneath the spot. To gain more

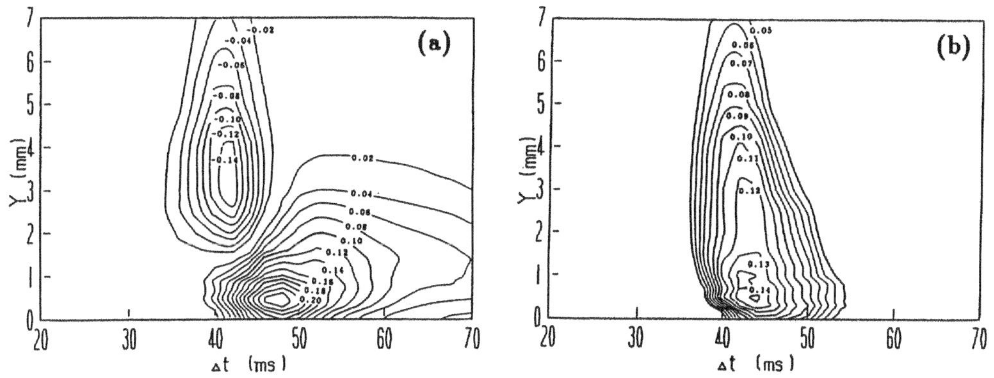

Figure 8. Same as figure 7 but with negative rotation ($N = -51$ rpm).

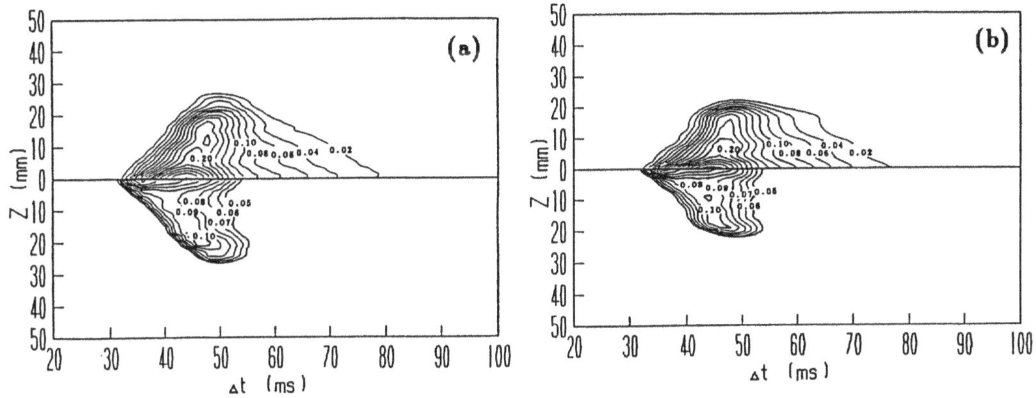

Figure 9. Contours of ensemble-averaged velocity perturbance (upper half), of rms velocity (lower half), at $x = 498$ mm at $y/\delta = 0.1$, (a) stationary, (b) $N = -51$ rpm.

insight into the mechanism of stabilization of the Coriolis force, it may be necessary to make measurements at different heights including positions closer to the wall and with wider range of rotation and Reynolds number. The development of the laminar disturbance preceding the spot is also to be investigated.

4. Conclusions

(1) Probability of the generation of the spot decreases with increasing negative rotational speed.
(2) The spanwise growth, especially in the wing-tip region of the spot is slightly decelerated by the Coriolis force stabilization.
(3) The Coriolis force stabilization also tends to inhibit the streamwise growth, especially in the bottom of the overhang region at the head of the spot.

References

Masuda, S., Okamae, K. & Ariga, I. 1984 Transition of boundary layer on rotating lat plate. In *Proc. IUTAM Symposium on Laminar-Turbulent Transition* (ed. V.V. Kozlov), Novosibirsk, USSR, Springer, 699.

Masuda, S. & Matsubara, M. 1989 Visual study of boundary layer transition on rotating flat plate. In *Proc. 3rd IUTAM Symposium on Laminar-Turbulent Transition*, Toulouse, France, to be published.

Conrad, P.W. 1962 The effects of rotation on the stability of laminar boundary layers on curved walls. Report FLD9, Mechanical Engineering Department, Princeton Univ.

Chawla, M. 1969 The stability of boundary layer flow subject to rotation. Ph.D. Diss., Michigan State Univ.

Gad-el-Hak, M., Blackwelder, R.F. & Riley, J.J. 1981 On the growth of turbulent regions in laminar boundary layers. *J. Fluid Mech.*, **110**, 73.

Structure and Statistics of Homogeneous Turbulence Obtained from a Direct Simulation

M. Meneguzzi and A. Vincent

CERFACS, 42 av. Gustave Coriolis, 31000 Toulouse, France

Direct numerical resolution of the three-dimensional Navier-Stokes equation on a 240^3 grid is used to obtain a three-dimensional homogeneous turbulent flow at Reynolds numbers $R_L \approx 1000$ and $R_\lambda \approx 150$, based respectively on integral scale and Taylor microscale. An inertial subrange appears in the energy spectrum over more than one decade in wavenumbers. The distributions of velocity derivatives are computed with high accuracy and found to be close to but not exponential. The distributions of velocity differences are also computed and used to evaluate the velocity structure functions. Visualization of the flow confirms that the vorticity is organized in very elongated thin tubes.

1. Introduction

Direct simulations of homogeneous turbulent flow at the highest presently reachable resolution are useful to gather some information on the statistical properties of such a flow and to explore its structure in space. Indeed, these "numerical experiments" allow measurement of many quantities unaccessible in the laboratory and visualization of the small scales structures of the flow. It becomes feasible, on computers of the last generation, to reach Reynolds numbers at which a genuine inertial subrange shows up in the energy spectrum (Kerr 1985, Yamamoto and Hosokawa 1988, She et al. 1988, 1990). The main goal of the present calculation was to obtain a more extended inertial subrange than in previous work, and to concentrate on the inertial domain properties.

2. The calculation

We solve the Navier-Stokes equation for incompressible fluids in rotational form

$$\frac{\partial \mathbf{v}}{\partial t} = \mathbf{v} \times \omega - \nabla(p + v^2/2) + \nu \nabla^2 \mathbf{v} + \mathbf{f} \tag{1}$$

with the continuity equation $\nabla \cdot \mathbf{v} = 0$. $\mathbf{v}$ is the velocity field, $\omega = \nabla \times \mathbf{v}$ the vorticity, p the pressure, ν the kinematic viscosity and $\mathbf{f}$ a force field. Since we are interested in (statistically) homogeneous turbulent flows, we take periodic boundary conditions in all directions. In Fourier space the two equations can be combined to give

$$\frac{\partial \mathbf{v_k}}{\partial t} = \mathbf{P}(\mathbf{k}) \cdot (\mathbf{v} \times \omega)_{\mathbf{k}} - \nu k^2 \mathbf{v_k} + \mathbf{f_k} \tag{2}$$

where the tensor $\mathbf{P}$ is the projector on the space of solenoidal fields, defined as $P_{ij}(\mathbf{k}) = \delta_{ij} - \frac{k_i k_j}{k^2}$. A pseudospectral method is used to compute the right-hand side of this

Advances in Turbulence 3
Editors: A.V. Johansson · P.H. Alfredsson

equation (see Gottlieb and Orszag 1977). The time marching is done using a second order finite-difference scheme. An Adams-Bashfort scheme is used for the non linear term while the dissipative term is integrated exactly. The resulting numerical scheme is

$$\frac{\mathbf{v_k}^{n+1} - \mathbf{v_k}^n e^{-\nu k^2 \delta t}}{\delta t} = \mathbf{P}(\mathbf{k}) \cdot \left[\frac{3}{2} (\mathbf{v} \times \omega)_\mathbf{k}{}^n e^{-\nu k^2 \delta t} - \frac{1}{2} (\mathbf{v} \times \omega)_\mathbf{k}{}^{n-1} e^{-2\nu k^2 \delta t} + \mathbf{f_k}^n e^{-\nu k^2 \delta t} \right]$$

To start (or restart) the calculation, we use a second order Runge-Kutta scheme.

We force the field at low wavenumbers in a deterministic way. All Fourier modes with $k < 1.5$ are forced with a constant amplitude $\mathbf{f}$ independent of $\mathbf{k}$ (the wavenumber $\mathbf{k}$ has integer components because the space period is 2π). This results in large fluctuations of the energy injection rate $\epsilon_i =< \mathbf{f} \cdot \mathbf{v} >$. One can only hope to reach a steady-state regime in the sense that ϵ_i fluctuates in time around a constant value.

The calculation presented here was done with 240^3 Fourier components, and a viscosity of 10^{-3}. One time-step takes 12 seconds on a Cray-2, using the 4 processors. Integration over one turnover time (defined below) takes of the order of 4 hours of Cray-2.

To estimate the degree of isotropy of the flow, we use the same method as Curry et al. (1984). For each wavenumber $\mathbf{k}$, we define two unit vectors $\mathbf{e_1}(\mathbf{k})$ and $\mathbf{e_2}(\mathbf{k})$ which form with $\mathbf{k}$ an orthogonal reference frame. Since $\mathbf{k} \cdot \mathbf{v} = 0$, each Fourier mode $\mathbf{v}(\mathbf{k})$ is defined by its two components in this frame $v_1(\mathbf{k})$ and $v_2(\mathbf{k})$. We define the isotropy I as $I = \left[\frac{<|v_1|^2>}{<|v_2|^2>}\right]^{1/2}$. In the calculations presented here, I fluctuates by a few percent around 0.95. Therefore, our flow is close to statistically isotropic.

Let us recall the definition of some characteristic quantities used in the following. Three characteristic lengths are used, the integral scale $L = \frac{\int_0^\infty k^{-1} E(k) dk}{\int_0^\infty E(k) dk}$, the Taylor microscale $\lambda = \left[\frac{\int_0^\infty E(k) dk}{\int_0^\infty k^2 E(k) dk}\right]^{\frac{1}{2}}$ and the Kolmogorov dissipation scale $l = \left(\frac{\nu^3}{\epsilon}\right)^{\frac{1}{4}}$, where ϵ is the mean energy dissipation rate per unit mass. The two characteristic time-scales of homogeneous turbulence are the eddy turnover time $\tau_0 = L/v_0$ where v_o is the root mean square velocity, and the dissipation time $\tau_\nu = L^2/\nu$. With these quantities one can define two Reynolds numbers, the integral scale Reynolds number $R_L = v_0 L/\nu$ and the Taylor microscale Reynolds number $R_\lambda = v_0 \lambda/\nu$.

3. The spectra of the flow

After integration over several tens of turnover times we obtain a statistically stationary regime in the above sense. The energy spectrum, shown in figure 1, displays a power law range for $k < 30$, with an exponent a little larger than -5/3. The fact that this is an inertial range is confirmed by inspection of the energy flux spectrum $\Phi(k) = \int_k^\infty T(k) dk$ where $T(k)$ is the energy transfer at wavenumber k. This function is found constant for k in the inertial range, as expected. After reaching this stationary regime, we integrate for 30 more turnover times in order to accumulate some statistics. The energy spectrum does not vary significantly during this period.

By fitting the spectrum with the Kolmogorov form $E(k) = C_K \epsilon^{2/3} k^{-5/3}$ we obtain a value of 2 for the Kolmogorov constant C_K, which does not vary by more than 3%

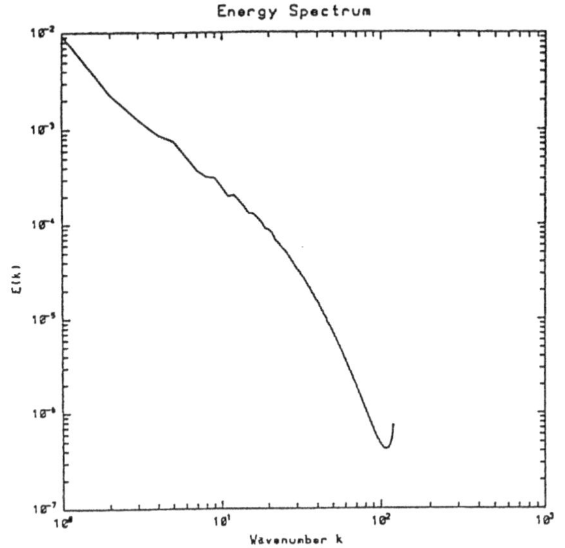

Figure 1: Energy spectrum averaged over 10 different times.

in the statistically steady-state period. This value of C_K is a little larger than the experimental value 1.5 (see Monin and Yaglom 1975 for references). Our Reynolds numbers are $R_L \approx 1000$ and $R_\lambda \approx 150$.

4. The statistical distribution of v and its derivatives

Rather than using time distributions of a velocity component v_i or its derivatives $\partial_i v_j$ at a given space location to obtain probability distributions, it is more efficient in our case to use space distributions, for which we have $\approx 1.4 \times 10^7$ grid point values at each time. Using approximately 70 different times, we reach a total of $\approx 10^9$ values. The resulting statistical distribution of v_x is close to gaussian, in agreement with what is known from experimental data (see Monin and Yaglom 1975 for a review).

Figure 2 shows the distribution of $\partial_x v_x$, known from experiments to be strongly non gaussian. Also shown (dashed line) is a gaussian, normalized so that $\sigma = 1$. It is seen that the distribution we obtain is closer to an exponential than to a gaussian, but our statistical sample allows one to see a departure from an exponential law. The wings are indeed not straight lines in this linear-logarithmic plot. The distribution is more intermittent than an exponential distribution. This is even more pronounced for the distribution of lateral derivatives $\partial_y v_x$, shown in figure 3.

We have also computed the distribution of velocity differences $\delta v_x(r) = v_x(\mathbf{x}+\mathbf{r}) - v_x(\mathbf{x})$, which depend only on r for homogeneous isotropic turbulence. Figure 4 shows the distribution of δv_x for $r = 0.39$, a value typical of the inertial sub-range scales. This distribution seems to display exponential wings. It is similar to the experimental distributions found by Anselmet et al.(1984) and Van Atta and Chen (1970). For values of r larger than L, the distribution of δv_x is gaussian. Therefore, when r goes from 0 to values $> L$, the δv_x distribution changes continuously from a function similar to the $\partial_x v_x$ distribution (Figure 2) to a gaussian. From the distributions of δv_x, one can compute the velocity structure functions, defined as $F_n(r) = < [\delta v_x(r)]^n >$, or the dimensionless structure functions, given by $f_n(r) = \frac{<[\delta v_x(r)]^n>}{<[\delta v_x(r)]^2>^{n/2}}$. To estimate the maximum value

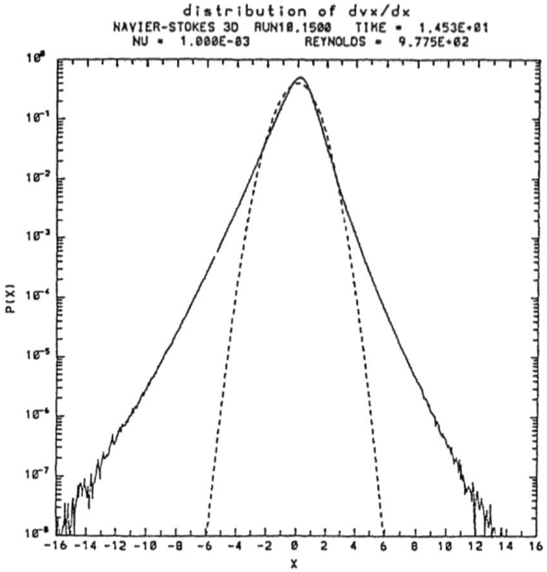

Figure 2: Probability distribution of $\frac{\partial v_x}{\partial x}$ normalized so that $\sigma = 1$, shown together with a gaussian distribution (dashed line).

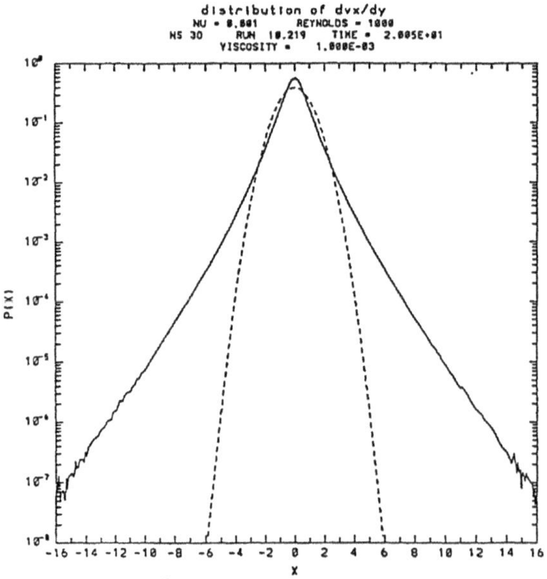

Figure 3: Probability distribution of $\frac{\partial v_x}{\partial y}$ normalized so that $\sigma = 1$, shown together with a gaussian distribution (dashed line).

of n for which this calculation can be done accurately, we use a method analogous to the one used by Anselmet et al.. The structure functions are computed in two different ways. Once with the distributions as they are, and a second time with a distribution obtained by extending the exponential wings to plus and minus infinity. This gives a rough estimate of the error due to our limited knowledge of these wings. The two

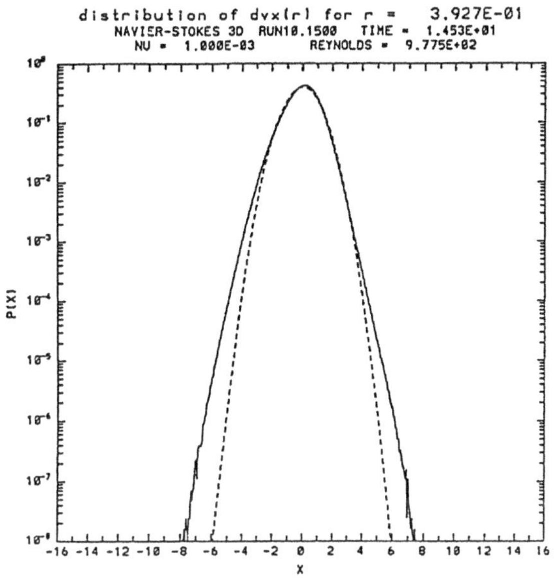

Figure 4: Probability distribution of $\delta v_x(r) = v_x(\mathrm{x}+\mathrm{r}) - v_x(\mathrm{x})$ for r = 0.39 (a value in the inertial sub-range), normalized so that $\sigma = 1$, shown together with a gaussian distribution (dashed line).

computed values agree to within a few percent for $n < 20$ and 20% for $n < 30$ when r is in the inertial range.

Intermittency models of turbulence predict power law forms of the structure functions for r in the inertial range $F_n(r) \sim r^{\zeta_n}$ with ζ_n given by $\zeta_n = \frac{1}{3}n - \frac{1}{18}\mu n(n-3)$ in the case of the Kolmogorov-Oboukhov (1962) Log-normal (LN) model , and $\zeta_n = \frac{1}{3}n - \frac{1}{3}\mu(n-3)$ in the case of the β-model (Frisch et al. 1978). Both predictions coincide for n = 6, for which $\zeta_6 = 2 - \mu$. The dimensionless structure function $f_n(r)$ also has a power law form r^{ξ_n} with $\xi_n = \zeta_n - \frac{n}{2}\zeta_2$. Figure 5 shows ζ_n as a function of n, and is compared with the predictions of the LN model, the β-model and the Kolmogorov 1941 theory ($\mu = 0, \zeta_n = n/3$). It is seen that none of the predictions fits our results. This figure is similar to figure 14 of Anselmet et al.(1984), who drew the same conclusion from their experimental data. Note that their Reynolds number $R_\lambda \approx 500$ is larger than ours.

The value we find for $\mu = 2 - \zeta_6$ is 0.25, a little larger than the value 0.2 found in the laboratory by Anselmet et al., also using ζ_6. In the above models, the parameter μ is identified with the power-law exponent of the energy dissipation correlation $< \epsilon(\mathbf{x}+\mathbf{r})\epsilon(\mathbf{x}) >\sim \left(\frac{L}{r}\right)^{\mu}$. We can obtain μ directly through this formula. The value we obtain is 0.15, smaller than the above value, but within the unceratinties.

An independent test of the LN model is a direct verification of Kolmogorov's third hypothesis, which states that the distribution of ϵ, the energy dissipation rate per unit mass, is log normal (Kolmogorov 1962). We have plotted in figure 6 the distribution of log ϵ, together with a properly normalized gaussian (dashed line). One can see that the distribution is definitely not normal, although it is close to normal for small fluctuations of log ϵ.

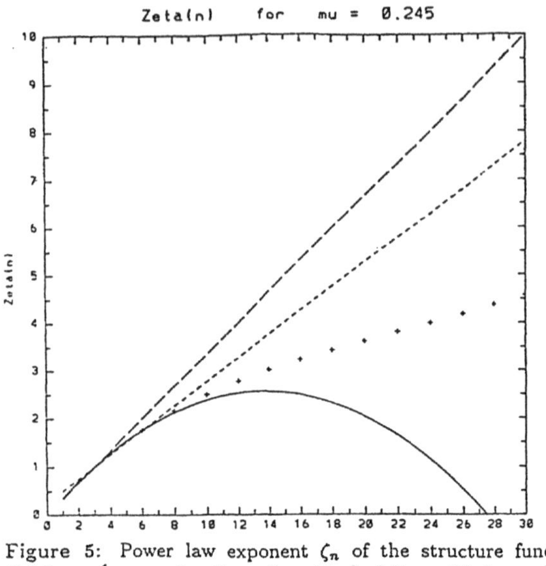

Figure 5: Power law exponent ζ_n of the structure functions $F_n(r) \sim r^{\zeta_n}$ as a function of n. Dashed line with large dashes : prediction of the Kolmogorov 1941 theory. Dashed line with small dashes : β model. Continuous line : Kolmogorv 1962 model. Crosses : values from our simulation.

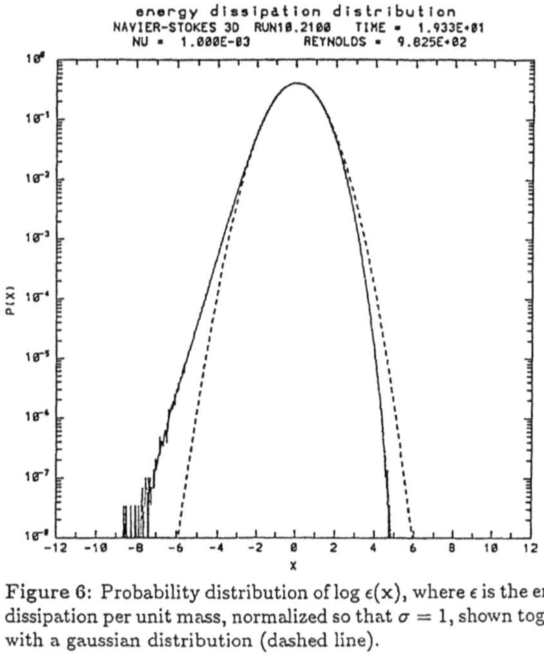

Figure 6: Probability distribution of log $\epsilon(x)$, where ϵ is the energy dissipation per unit mass, normalized so that $\sigma = 1$, shown together with a gaussian distribution (dashed line).

5. The spatial structure of the flow

Figure 7 shows a 3D picture of the vorticity field. The vorticity at each grid point is represented by a vector, here so small that individual vectors can hardly be seen. Vectors are only plotted if their modulus is larger than a given threshold. By varying this threshold and rotating the figure on a graphic workstation screen we can explore the structure of the field in detail. One can see that the vorticity is organized in thin elongated tubes, as previously reported by Siggia (1981), Kerr (1985), She et al. (1990). Figures 8 shows the effect of lowering the threshold, and therefore letting smaller amplitude vorticity vectors appear. The length of these tubes seems to be of order 1 , i.e. of the same order as the integral scale L (the cube size is 2π). Their thickness is of the order of a few dissipation scales, here a few grid points. This is confirmed by a more detailed analysis. The dissipation scale l, in our simulation, is of the order of the mesh size, while the Taylor microscale is approximately ten mesh sizes. Therefore, the characteristic tube thickness seems to be intermediate between these two lengths. Figure 9 shows a detailed view of a vorticity tube. It displays a sub-cube one sixth the size of the complete one, with 40 grid points on each side. Similar plots for the velocity field show mainly the forcing field if one uses a high threshold. But when the highest velocities are eliminated, the tubes are clearly visible.

From these visualizations, one is lead to the conclusion that these vorticity tubes, which seem to be the basic structure of three-dimensional homogeneous turbulence, involve all the scales of the flow. We have done the same kind of pictures of the vorticity field after removing all dissipation range scales. A smooth filter is applied in Fourier space in order to avoid spurious fluctuations. The large scales to which the forcing is applied were also removed. Figure 10 shows the same sub-cube as figure 9 when only inertial range scales are left. One can see the external regions of the tubes. Some helical structure can be seen, as noted by She et al. (1990).

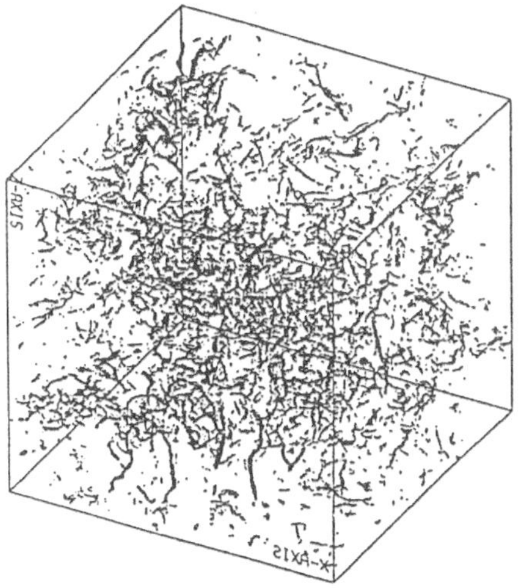

Figure 7: View of the vorticity field, represented by a vector of length proportional to the vorticity amplitude at each grid point. Only vectors larger than a given threshold value are shown.

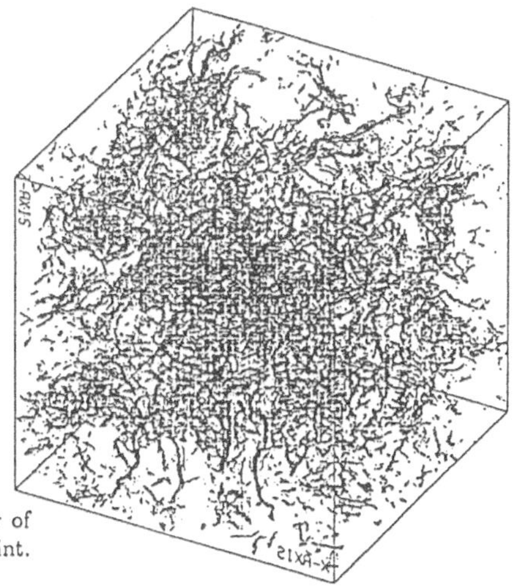

Figure 8: Same as figure 7, but with a lower threshold value, and therefore more vectors represented.

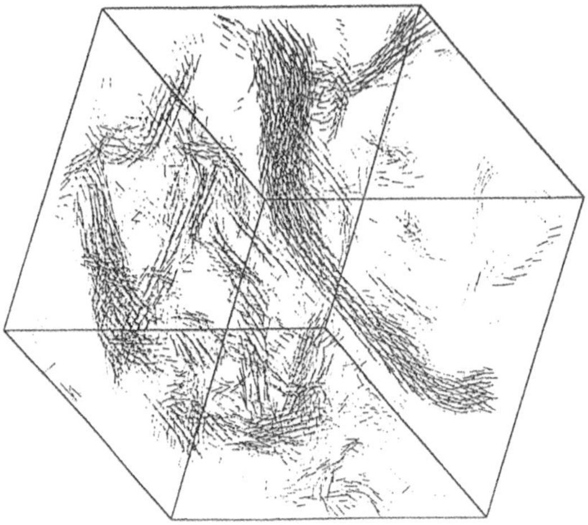

Figure 9: Detail of the vorticity field, showing a sub-cube of size one sixth of the complete cube, i.e. 40 grid points in each direction. The color are associated with vorticity amplitude. The red vectors are larger than the blue ones.

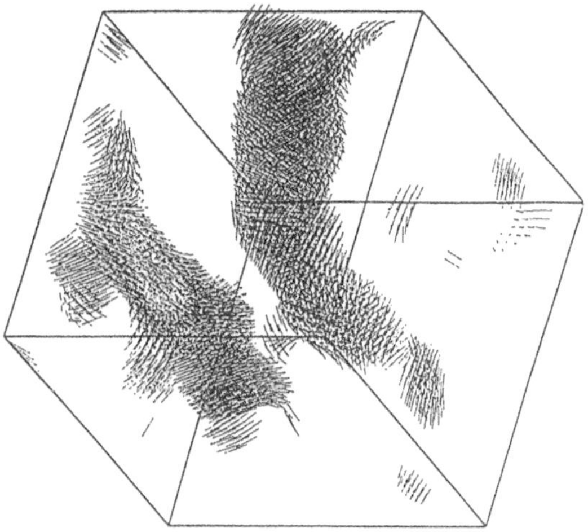

Figure 10: Same vorticity tube as in figure 9, but showing only inertial sub-range scales. Dissipation and energy injection scales have been removed by filtering in Fourier space.

We have examined the shape of many of these tubes in order to see wether they are in fact rolled-up vorticity sheets, as suggested by Lundgren (1982), but this does not seem to be the case. Figure 11 gives an example of the projection of the velocity field on a plane perpendicular to a vorticity tube, while figure 12 shows the curves of constant vorticity on such a plane. As one can see on these figures, there is little evidence of spiral structure, or any other two dimensional structure.

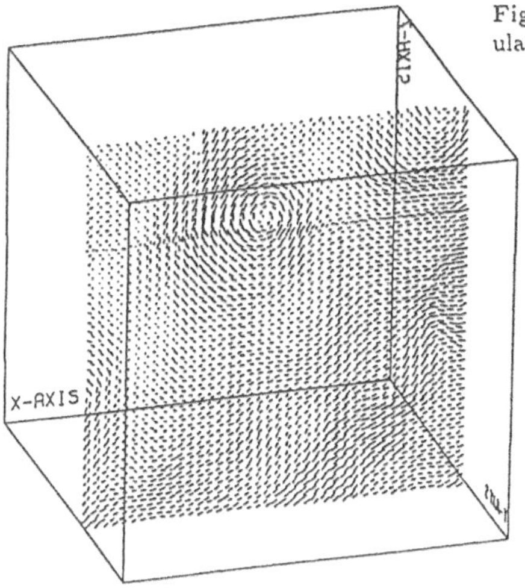

Figure 11: Projection of the velocity field on a plane perpendicular to a particular vorticity tube.

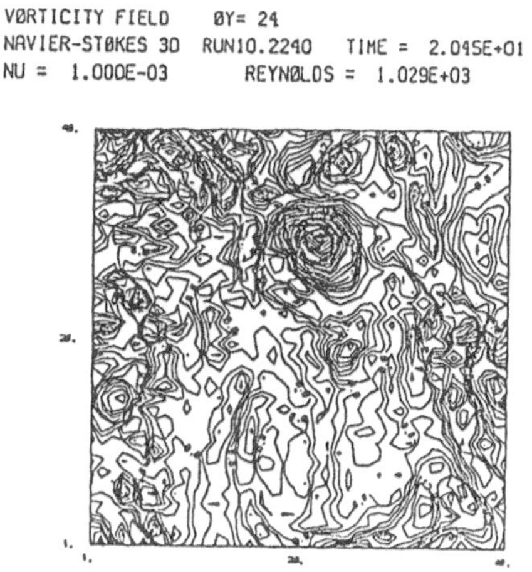

Figure 12: Curves of constant vorticity in a plane perpendicular to a selected vorticity tube.

6. Conclusion

We have obtained a turbulent homogeneous flow at $R_\lambda \approx 150$ and an inertial subrange more extended than in previous tridimensional simulations. The statistics of the velocity derivatives are strongly non gaussian both in the inertial and the viscous subranges. The distributions of velocity derivatives are more intermittent (i.e. it have larger wings) than an exponential distribution. The intermittency exponent μ is found to be 0.25. The Log-normal and β models of intermittency do not predict the nth order structure function scaling with n which we find. In addition, the Kolmogorov hypothesis of log-normality of the energy dissipation fluctuations is not confirmed by our calculations.

In physical space, we confirm the observation by Siggia (1981) that the vorticity is organized in thin tubes. We find that the thickness of these tubes is a few dissipation scales while their length is comparable to the integral scale of the flow. Therefore, the picture of intermittency that emerges in physical space from our simulations is that all the scales present in the flow are correlated through these elongated structures. The study of the time history of these vorticity tubes, which should allow us to understand the mechanism of their generation and their dynamical evolution, is under way.

References

Ashurst W.T., Kerstein A.R., Kerr R.M. & Gibson C.H. 1987 *Phys. Fluids* **30**, 2343.

Anselmet F., Gagne Y., Hopfinger E.J. & Antonia R.A. 1984 *J. Fluid Mech.* **140**, 63.

Benzi R., Patarnello S. & Santangelo 1987 ICE-0015 to appear in *J.Phys. A*.

Curry J.H., Herring J.R., Loncaric J. & Orszag S.A. 1984 *J. Fluid Mech.* **147**, 1.

Frisch U., Sulem P.L. & Nelkin M 1978 *J. Fluid Mech.* **87**, 719.

Gottlieb D. & Orszag S.A. 1977 CBMS-NSF Regional Conference Series in Aplied Math., **26**.*"Numerical Analysis of Spectral Methods : Theory and Applications."*

Kerr R. 1985 *J. Fluid Mech.* **153**, 31.

Kolmogorov A.N. 1962 *J. Fluid Mech.* **13**, 82.

Lundgren T.S. 1982 *Phys. Fluids* **25**, 2193.

Monin A.S. & Yaglom A.M. 1975 Statistical Fluid Mechanics, Vol 2, MIT Press.

Oboukhov A.M. 1962 *J. Fluid Mech.* **13**, 77.

She Z.S., Jackson E. & Orszag S.A. 1988 *J. Sci. Computing*, **3**, 407.

She Z.S., Jackson E. & Orszag S.A. 1990 *Nature* **344**, 226.

Siggia E.D. 1981 *J. Fluid Mech.* **107**, 375.

Yamamoto K. & Hosokawa I. 1988 *J. Phys. Soc. Japan* **57**, 1532.

Van Atta C.W. & Chen W.Y. 1970 *J. Fluid Mech.* **44**, 145.

On the Mechanism of 'Side-jet' Generation in Periodically Excited Axisymmetric Jets

S. Raghu[1], *B. Lehmann*[2] *and P.A. Monkewitz*[3]

[1]Department of Mechanical Engineering,
State University of New York, Stony Brook, NY 11794, U.S.A.
[2]DLR, Institute for Turbulence Research, Mueller-Breslau Strasse 8,
1000 Berlin 12, Federal Republic of Germany
[3]Department of Mechanical, Aerospace and Nuclear Engineering,
University of California, Los Angeles, CA 90024, U.S.A.

Transitional hot jets with self-excited oscillations and acoustically forced cold jets are investigated experimentally in order to understand the mechanism of 'side-jet' generation. The development of side-jets from the azimuthal instability of the primary vortex rings in the jet is visualized using a combination of laser sheet and phase-locked stroboscopic illumination techniques.

1. Introduction

Transitional jets with strong external forcing or with inherent global instabilities have been observed to have highly organized vortical structures in the near field of the jets, for egs., see Michalke (1964), Sreenivasan et al (1989) and Monkewitz et al (1989). Associated with the high strength of these primary structures are azimuthal instabilities which are vigorous enough to produce the interesting phenomenon of 'side-jets' not associated with the homogeneous unforced jets. 'Side-jets' are streams of jet fluid being ejected at large angles to the axis of the jet and they provide an efficient mixing mechanism that competes with the regular entrainment process in a jet.

Several hypotheses have been advanced to explain the mechanism of these side-jet productions. Monkewitz et al (1989) suggest that the primary mechanism is the azimuthal instability (Widnall instability) of the vortex rings. According to them, when the rings are sufficiently deformed, each lobe of the deformed vortex ring produced an outward jet of the central fluid due to the induced velocity in the lobe. A similar mechanism has also been suggested by Eickhoff et al (1989) when they noticed large spreading of hydrogen and methane flames at a certain distance downstream of the jet exit. Their visualizations of the flame cross-sections also show a multilobed structure similar to that of hot jets. It has also been suggested (Lehmann et al, 1988 and Martin and Meiburg, 1990) that stream-wise vorticity in the braid region of the jet plays an important role in the production of these side-jets. In order to get a clear picture of the scenario leading to the production of these side-jets we felt that a more detailed investigation was necessary and is described in this paper.

Advances in Turbulence 3
Editors: A.V. Johansson · P.H. Alfredsson

2. Experimental Arrangement

The experimental apparatus consists of a round jet with 15 mm diameter and provision to inject smoke into the entrainment region at the lip of the nozzle through means of a 1 mm annular slot. The jet fluid could also be heated upstream of the settling chamber to study the structure of hot jets with global instabilities. The details of the nozzle and the heating arrangement are described in Monkewitz et al (1989). A loudspeaker (35 cm diameter) mounted at 45 degrees to the jet axis was used to harmonically excite the cold jet at Strouhal numbers in the range of 0.3 to 0.6. By choosing the appropriate frequency, one could obtain either the helical or the axisymmetric modes of the jet instability. A stroboscope synchronized with the excitation frequency of the jet was used to view the structures 'stationary' in space. To visualize the deformation of the vortex ring, a thin laser sheet (obtained from a 5 watt Argon-ion laser) perpendicular to the jet axis was employed. Still pictures as well as video movies of the jet were taken in order to study the structures in detail.

3. Results and Discussion

Figure 1 shows an oblique view of the acoustically forced jet with strong 'side-jets'. The jet exit velocity was 8 m/s and the frequency of excitation was 210 Hz ($Re_D = 8000$ and St = 0.39). The sound level produced by the loudspeaker at the exit of the nozzle was around 103 dB. As one can see from the figure, the primary instability corresponds to the jet column mode and is axisymmetric in nature. The higher frequency shear layer mode can also be observed very near the jet exit but is soon overwhelmed by the columnar mode which persists for several diameters downstream. The axisymmetric rings develop azimuthal waviness within 1.2 to 1.5 diameters downstream of the exit.

In figure 1 one can also notice a weak stream of fluid ejected from the first fully formed vortex ring, and stronger streams or 'side-jets' ejected from the second and third vortex rings downstream of the nozzle at large angles to the main axis of the jet. Further downstream, the vortex rings are highly irregular due to the onset of turbulence and we do not see such distinct side jets beyond that region. When viewed continuously in time, it was observed that these jets move in the azimuthal plane in a slow and random fashion and hence when visualized in a two-dimensional plane parallel to the jet axis appear to be intermittent in time. Figure 2 shows the same jet at a different instant of time, where instabilties in the vortex rings are clearly visible. In order to investigate the spatial origin of these side-jets the phase of the illuminating stroboscope was slowly varied with reference to the loudspeaker phase. This enabled us to observe the various stages of the formation of the vortex rings as well as the azimuthal deformations. It clearly showed that the initial azimuthal instabilty occurs around $x/D = 1.2$ and the first side-jet occured at $x/D = 1.5$ to 2.10.

Figure 3 shows the cross-section of the jet at various downstream distances from the exit of the jet. In order to obtain these pictures, the whole jet fluid rather than the entrainment fluid was seeded with smoke and a laser sheet perpendicular to the main axis was traversed in the streamwise direction. The sequence of pictures clearly shows the various stages of the azimuthal instability and the formation of side-jets. The pictures have to be carefully interpreted keeping in mind figures 1 & 2. After the onset of the azimuthal instability and its growth to a finite amplitude, a portion of the central jet fluid is expelled through the azimuthal lobes through an induced velocity

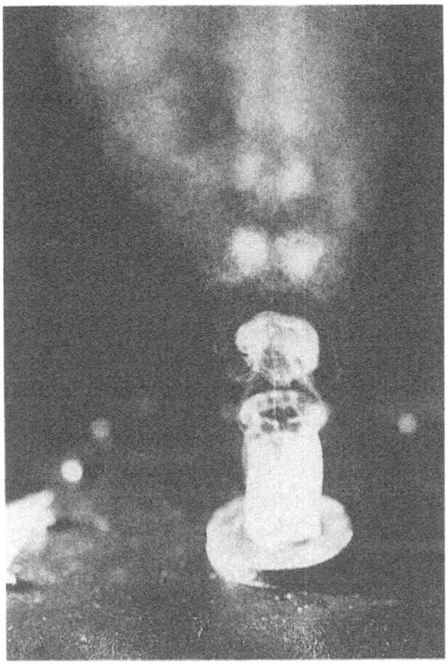

Figure 1.

Figure 2.

Figure 1. Side-jets in an acoustically forced cold jet with U_{jet} = 8000, and excitation frequency f = 120 Hz. Sound pressure level at exit of the jet = 103 dB.

Figure 2. Azimuthal instability of vortex rings. Same experimental conditions as fig. 1.

field produced by the orientation of these lobes. As the vortex rings move downstream, the size of these lobes increases and they get oriented in such a fashion so as to expel larger amounts of fluid outwards from the center of the jet. When the induced velocity is of sufficient magnitude relative to the streamwise velocity at the edge of the vortex rings, we see 'spurts' of fluid coming out at large angles to the main axis of the jet, which has been referred to as 'side-jets'. Further downstream, the sideward ejection occurs. At this point it has to be stressed again that the side-jets are not themselves the azimuthal deformations of the vortex ring as it may appear to be in the cross-section of the jet. Instead, they are the induced streams of fluid through the lobes. This can be particularly seen in figures 1 and 2 where the vortex rings are still intact while producing strong side-jets.

Similar side-jets have also been observed in hot jets which have a density ratio of less than about 0.7. In order to confirm the suggestion (Monkewitz et al, 1989) that the mechanism of side-jet production is the same in externally forced jets and in globally unstable hot jets, we attempted to visualize the jet structure in hot jets in a similar way as in cold jets. Interestingly, we found that the hot jet was very sensitive to the annular flow around the jet — even a small amount of annular flow completely suppressed the self-excited oscillations. Since the stroboscope was driven by the microphone signals from these oscillations, problems arose from this jet sensitivity. Figure 4 shows the photograph of a hot jet obtained by entraining still smoke into the hot

Figure 3. Cross-section of the jet at various downstream distances. Same experimental conditions as figure 1.

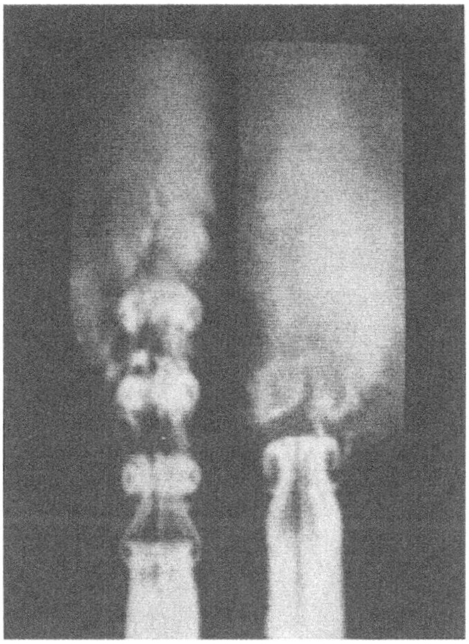

Figure 4. Front view of cold and hot jets. Left: Acoustically forced cold jet. Right: Self-excited hot jet.

jet. For comparison, a cold jet is also shown in the picture. The remarkable similarity in the organization of the vortices supports the hypothesis that the mechanism of side-jet generation is the same in cold and hot jets and that it depends only on the strength of the vortex rings formed in the near region of the jet.

From our visualization, we find that the mean velocity field of the jet determines the final shape of the vortex rings and hence the orientation of the azimuthal lobes. This can be particularly seen from figure 4 where the vortex rings assume a conical shape with the apex pointing in the downstream direction. This is in contrast to the computations of Meiburg and Martin (1990) where they obtained vortex ring deformations with the lobes inclined in the downstream direction. It is not yet clear how these computations relate to our observations, since strong side-jets seem only compatible with the lobes "hanging" back as we see them. From our visualizations, we also did not observe strong streamwise structures in the braid region between the vortex rings, possibly due to the thinning of the smoke in these regions. Further investigations are in progress to investigate this aspect of the flow.

Attempts were also made to determine some parameters corresponding to the azimuthal instability of the vortex rings in the jet. From our visualization, we could determine the diameter of the vortex ring (D), n, the number of azimuthal waves on the ring and a, the diameter of the ring cross-section. Using the experimental finding of Maxworthy (1978) that the ratio of the core diameter based on the maximum radial velocity to that of the moving fluid volume is approximately 0.3, we calculated the core diameter of the vortex rings in our experiments as 2.25 mm. Hence the value of the instability parameter (Ka) is found to be around 1.2, where K is the azimuthal wave number. Estimates from the hot jet visualization also gave a similar value for this parameter. This further confirms that the azimuthal instability is governed by the same mechanism as that of the single vortex ring suggested by Widnall and Sullivan (1973).

To summarize, we have investigated experimentally the mechanism of 'side-jet' generation in the transitional regime of externally forced cold jets as well as hot jets with self-excited oscillations and found evidence that the mechanism is the same in both cases: azimuthal lobes are produced through a 'Widnall instability' of the vortex rings in the near field region of the jet, their orientation being determined by the mean velocity field of the jet and the jet fluid is induced outward in each lobe of the vortex ring in the form of 'side-jets'. We do not see such a phenomenon in natural cold jets because the vortex rings are weak and less organized and turbulence sets in before the azimuthal instabilities grow to the amplitude which is needed for the production of side jets. This mechanism of mixing by secondary streams or 'side-jets' provides an alternate and efficient method of mixing and its application to practical situations where high levels of mixing are required needs to be explored in the future.

Acknowledgements

This work was done at the Hermann-Foettinger Institute of the Technical University of Berlin in cooperation with the Turbulence Research Department, DLR. We have benefitted from our discussions with Prof. H. Fiedler and Dr. D. Bechert and would like to acknowledge the same. The financial support of the Alexander von Humboldt Foundation for the stay of S. Raghu and P.A. Monkewitz is also gratefully acknowledged.

References

Eickhoff, H., Winandy, A. and Natarajan, R. 1989 Azimuthal instability in jet diffusion flames. *Experiments in Fluids*, **7**, 420.

Lehmann, B, and Barsikow, B. 1988 Sichtbarmachung von Instabilitatsformen eines runden Freistrahls durch eine Lichtschnittmethode. *DGLR Report* **88-04**.

Martin, J.E. and Meiburg, E. 1990 The three dimensional evolution of an axisymmetric jet: a numerical investigation. *Center for Fluid Mechanics Report*, **90-200**, Brown University, (Also submitted to *J. Fluid Mech.*).

Maxworthy, T. 1978 Some experimental studies of vortex rings. *J. Fluid Mech.* **81**, part 3, 465.

Michalke, A. 1964 Zur Instabilitat und nichtlinearen Entwicklung einer gestorten Scheersicht. *Ingenieur-Archiv*, **33**, 264.

Monkewitz, P.A., Lehmann, B., Barsikow, B. and Bechert, D.W. 1989 The spreading of self-excited hot jets by side-jets. *Phys. Fluids*, **A1**, 446.

Sreenivasan, K.R., Raghu, S. and Kyle, D. 1989 Absolute instability in variable density round jets. *Expts. in Fluids*, **7**, 309.

Widnall, S.E. and Sullivan, J.P. 1973 On the stability of vortex rings. *Proc. Roy. Soc.* **A332**, 335.

Jets of Air-Helium Mixtures in Air

R. Riva, G.Binder and M. Favre-Marinet

Institut de Mécanique de Grenoble, B.P. 53 X 38041, Grenoble-Cedex, FRANCE

1. Introduction

The motivation of the present research on flows with large density differences is the heating of the thermal shield of the European space shuttle HERMES during re-entry into the atmosphere. The question raised is whether the density differences, caused by the large temperature variations between the skin and the outer fluid which has been heated in passing through the shock, have specific effects on the turbulent rate of heat transfer to the surface. This situation will be simulated in the laboratory by injecting air or helium into boundary layers of helium or air respectively.

Our interest in jets less dense than the ambient fluid was raised by unexpected observations made on the facility designed to produce jets with known velocities and known air-helium mixtures for the calibration of hot-wires. The first intriguing observation was the perception of a pure tone sound which varied with velocity when the jet was run with pure helium. Later, when the flow visualization based on density differences such as shadowgraphy, schlieren and interferometry was tested, striking structures were observed in the initial region of the jet. It was then found that these were manifestations of absolute instabilities whose existence was recently investigated theoretically by Huerre, Monkewitz and coworkers (1-3) and, shortly after, confirmed in jets less dense than the ambient fluid by two experiments performed by Sreenivasan et al. (4) and by Monkewitz et al. (5-7). References (4) and (7) will later be referred to as SRK and MBBL.

This flow which depends upon the parameter $S = \rho_e/\rho_\infty$ in addition to those of the homogeneous jet, is evidently quite complex and the subject is far from being exhausted as revealed by some points of disagreement between the two cited experiments. The present experimental results, besides confirming the existence of absolute instabilities and of some of their consequences in such jets, also show some novel aspects of these flows.

2. Apparatus and flow conditions

The jet facility consists of the usual settling chamber with screens and honey-comb, a bell-shaped 25:1 contraction ending in a 15 mm nozzle. The 6 mm diameter jet was obtained by placing an orifice on the nozzle. The air helium mixtures were generated by a system of sonic throats which maintain rigorously constant rates of air and helium mass flows, hence constant jet velocity and composition. These characteristics were varied by selecting different throat combinations and by changing the generating pressures.

Velocities were measured with hot-wire anemometers and concentrations with an aspirating probe (8). Hot wires respond to velocity and concentration. Pure velocity

Advances in Turbulence 3
Editors: A.V. Johansson · P.H. Alfredsson

signals are, thus, only obtained when the wire is in the potential core of the jet. It was checked that the dominating frequencies in the spectra obtained from the raw hot-wire signals were the same irrespective of wire location near the exit section. In order to detect eventual probe interference effects, visualizations were made with and without probes in the flow. No observable differences were seen on the characteristics of the structures in the 15 mm jet.

The main flow conditions are summarized below ("e" for exit conditions):

$V_e = 0 - 50$ m/s $Re = V_e D_e/\nu_e = 0 - 3500$ in pure He ; $0 - 20000$ in air

$S = 0.14 - 1$ $Fr = V_e/(g' D_e)^{1/2} = 1 - 50,\ g' = g(1/S - 1)$

$\theta_e/D_e \approx 0.75 Re^{1/2}$ $(v_e)_{RMS}/V_e \approx 0.5\%$ for 15 mm jet; $\nu_{He}/\nu_{Air} = 7.93$

3. Results and discussion

The spectra of figure 1 (averages of 12 realizations) show the striking differences between the instability oscillations in helium and in air at four Reynolds numbers: while the spectral peaks are low level and broad band in air, they are high level and narrow band in helium. The difference in the peak levels is at least three orders of magnitude and reaches even five orders at $Re = 750$. At this Reynolds number the distribution is close to a pure line-spectrum. It is also noted that the frequencies of the major peaks are not the same in helium and in air. These preferred frequencies in helium are perfectly reproducible: the helium jets behaves also as a well tuned oscillator, a characteristic consequence of absolute instability.

It is recalled that a flow is termed absolutely or convectively unstable according to whether or not there is some region where a transient perturbation is amplified at the source. Thus, in a convectively unstable flow, a transient perturbation is everywhere convected away from the source and eventually dies out. In an absolutely unstable flow, on the contrary, such a perturbation will be amplified on the spot and will progressively contaminate the whole flow field. The most amplified frequency of the transient will eventually dominate all others and the flow acts as an oscillator as illustrated by Fig.1 and as was previously shown by SRK and MBBL.

The questions of the frequency actually selected by the jet and of the boundaries of absolute instability are still wide open. In order to contribute to these questions, experiments were performed by changing the density ratio S or the exit velocity V_e and by maintaining the other parameters constant.

The crossing of the boundary between absolute and convective instability with increasing S is illustrated on figure 2. It is seen that the signature of the power spectral density changes drastically when S increases from 0.14 to 0.22 but that the frequency of the peak varies progressively with S. The changes of the instability characteristics are less evident on the schlieren visualizations although the structures appear clearly weaker at $S = 0.22$. Yet, they are sharper again at $S = 0.315$. It would, thus, be difficult to detect the transition from absolute to convective instability from visualizations alone. Noteworthy on these visualizations are changes in the wave length of the structures between the flows $S = 0.14$ and 0.315 and, in the latter case, also the changes with downstream distance.

The evolution of the spectra with increasing velocities in the pure helium jet shown on figure 3 reveals no less than four transitions with bands of absolute instability approximately between $2.5 < V_e < 8$ m/s and $11 < V_e < 20$ m/s in the velocity range

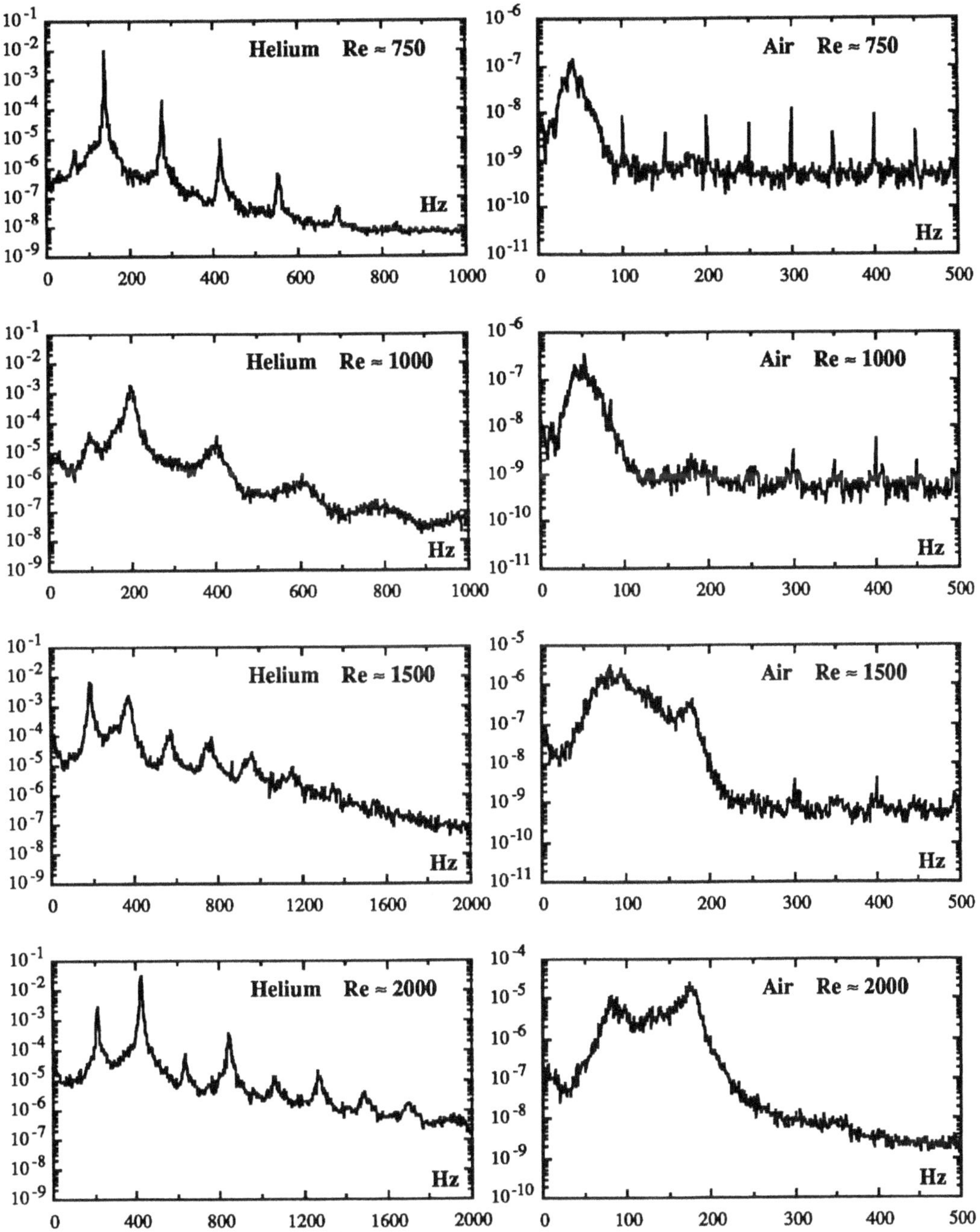

Figure 1 - Power spectra in Helium and Air at same Re . Hot wire at x ≈ 15 mm , r ≈ 5 mm from the nozzle exit . 15 mm dia. nozzle . Vertical arbitrary scale .

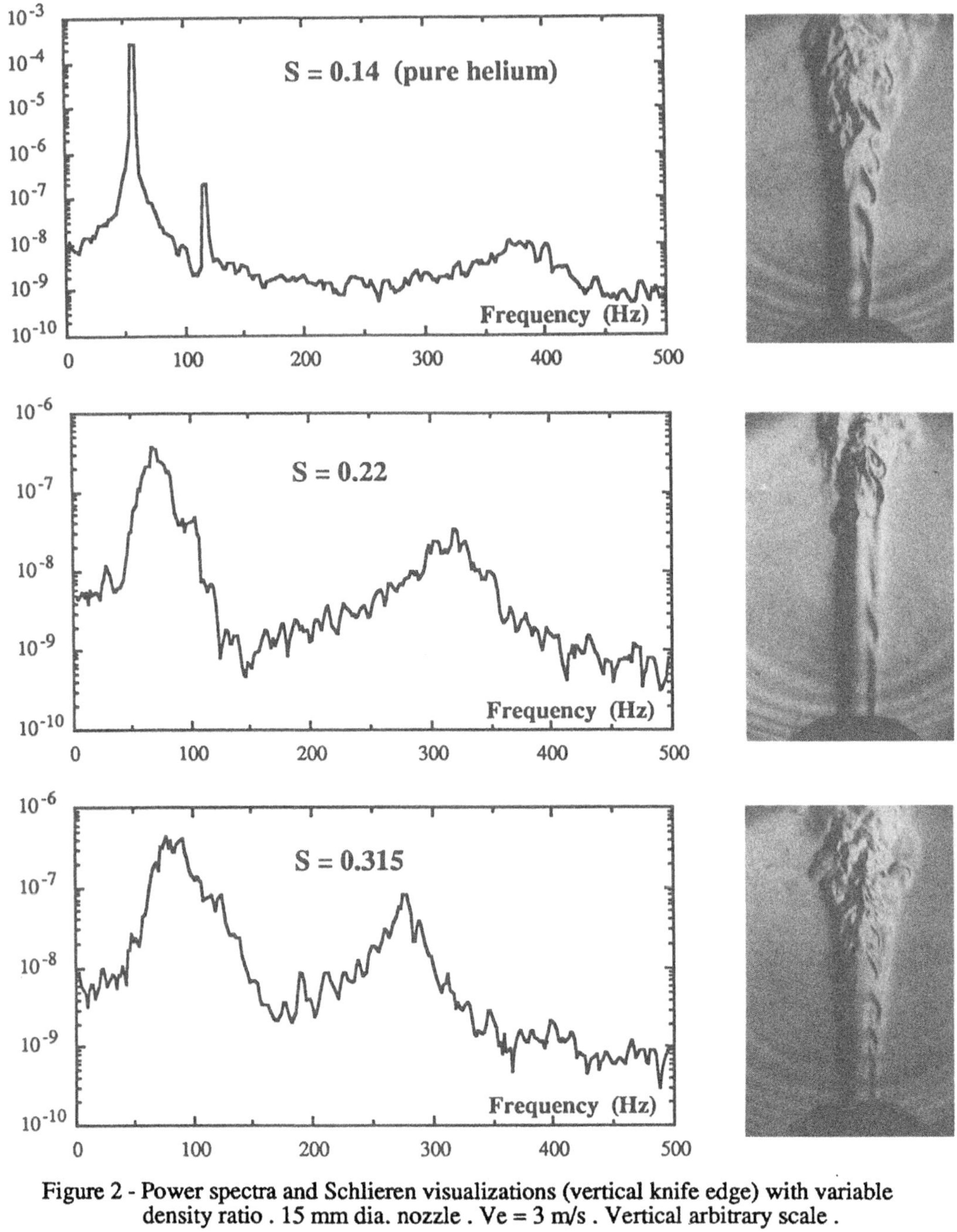

Figure 2 - Power spectra and Schlieren visualizations (vertical knife edge) with variable density ratio . 15 mm dia. nozzle . Ve = 3 m/s . Vertical arbitrary scale .

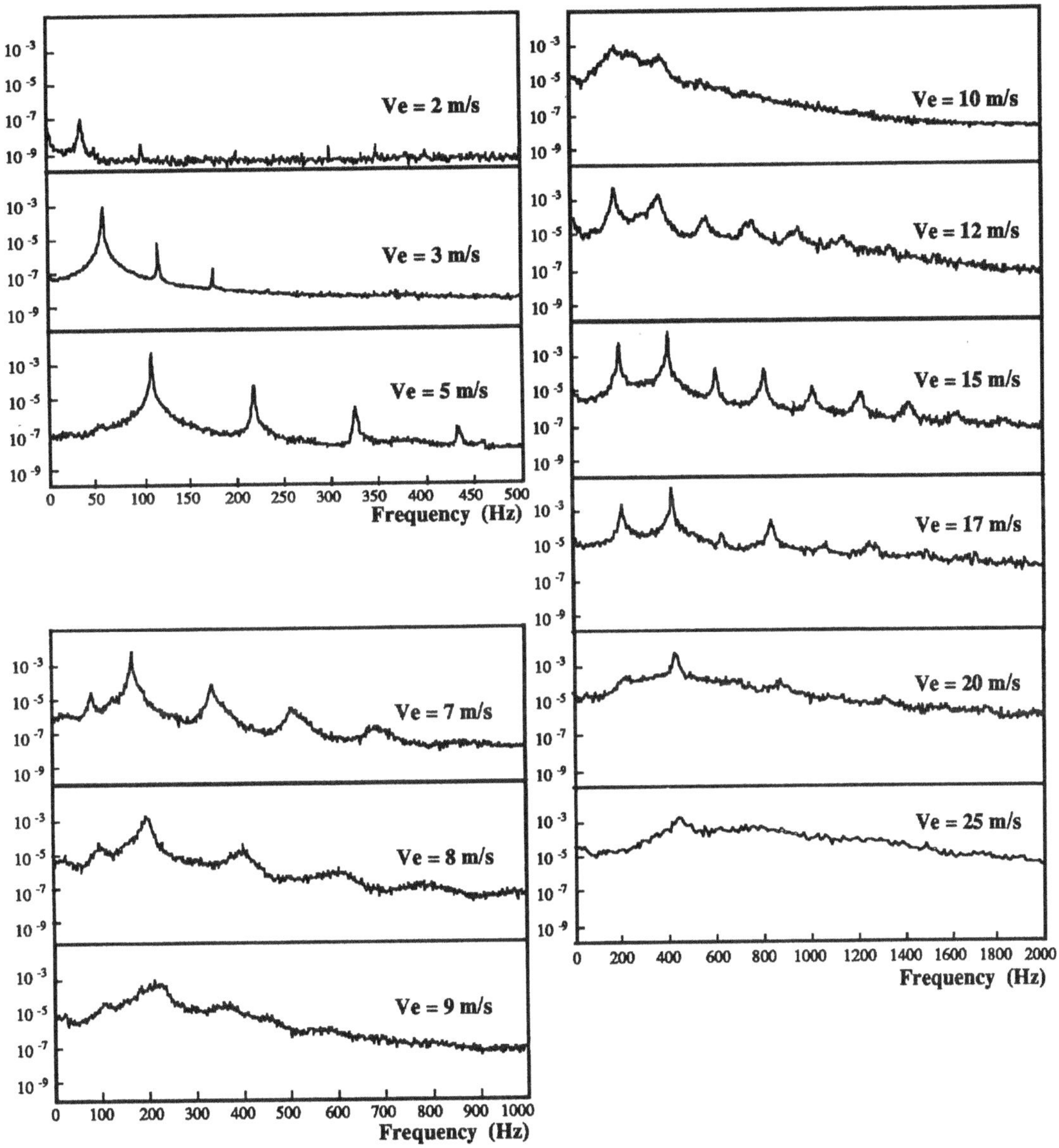

Figure 3 - Power spectra in pure Helium with increasing velocity. Hot wire at $x \approx 15$ mm, $r \approx 5$ mm from nozzle exit. 15 mm dia. nozzle. Vertical arbitrary scale.

(0 – 30 m/s) which can be covered with the apparatus. These findings contrast sharply with those of SRK who found absolute instability throughout the velocity range from 18 to 170 m/s in their 6 mm helium jet. In order to investigate this point further we also made some measurements in a 6 mm jet. The points are not sufficiently dense yet to define similar bands of absolute stability but they show a transition at $V_e \approx 30$ m/s which is very far from the upper limit of SRK.

It is clear from these results and the comparison with those of SRK that the $M - S$ plane (M being the Mach number) used by SRK to map the regions of absolute insta-

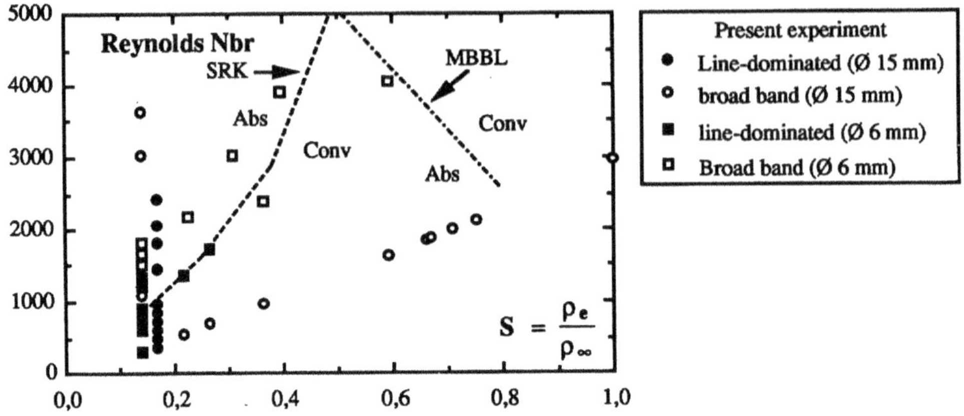

Figure 4 - Region of absolute instability in (Re,S) plane.

bility is not appropriate. Since the Reynolds number involves three major quantities (ρ, V and D) the $Re - S$ plane would seem a better choice. Indeed as shown by figure 4, the various boundaries for the lower values of S are at least in the same part of the graph while they would be considerably shifted in the vertical direction in the $M - S$ plane. For the sake of clarity the points of the 15 mm jet have been slightly shifted to the right. The correlation with the SRK data is, however, only limited since we have points of convective instability in their region of absolute instability. On the other hand, it is seen that the data of MBBL (15 mm jet) is not close at all and that their curve slopes are in the opposite direction. These discrepancies show that the transitions are not governed by a single parameter. Among the parameters which are likely to be be important, one may cite the momentum thickness, the turbulence level, asymmetries (as in the MBBL experiment), possibly probe interference effects from which the present hot wire measurements may not be completely exempt of.

Another noteworthy feature of Fig.3 is the progressive broadening or steepening of the the spectral peaks with changes in velocity across the boundaries between absolute and convective instability. This observation differs again with those of SRK who report sharp transitions between the two regimes.

Gradual changes are also observed on the Strouhal number across the transitions and over the whole range of Reynolds numbers covered as shown on Fig.5 where the transitions are indicated by arrows. It is also seen that there is fairly good agreement on the Strouhal numbers of the two jets which gives some weight to the Reynolds number as one of the control parameters. In the intermediate range of Reynolds number the peak of the first subharmonic is nearly as high as the one of the fundamental and the two points have been plotted on the graph. It looks as if the jet hesitates between two states as was observed by MBBL. This could be indicative of vortex pairing. However, no such pairing could be evidenced on any of the visualizations.

The existence of side jets, reported by both SRK and MBBL, is also observed on the present apparatus. Particularly striking examples from the 6 mm jet are shown on Fig. 6. It is found that these side jets occur above a Reynolds number of 700 in the 15 mm jet (the upper Re limit could not be reached with the apparatus) and for $300 < Re < 1500$ in the 6 mm jet. Since these side jets are also much more vigorous on the small jet it is clear that the nozzle geometry plays a role.

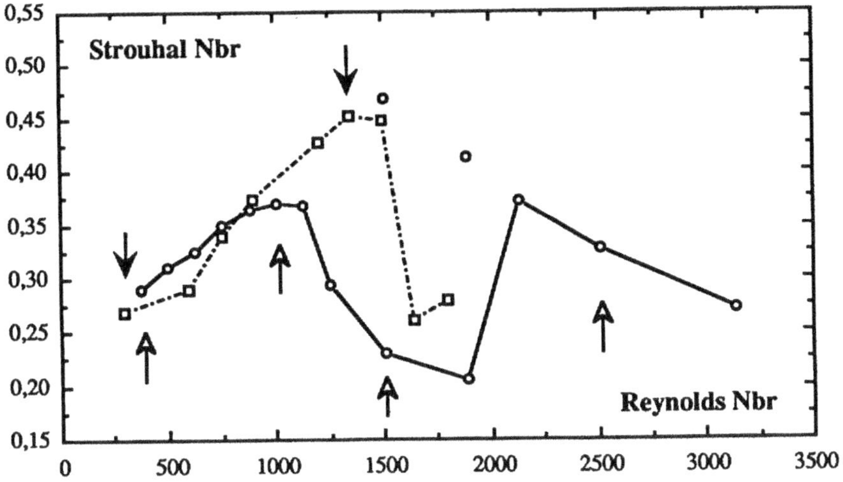

Figure 5 - Strouhal number of dominant frequencies versus Reynolds number . Pure helium (S=0.14).

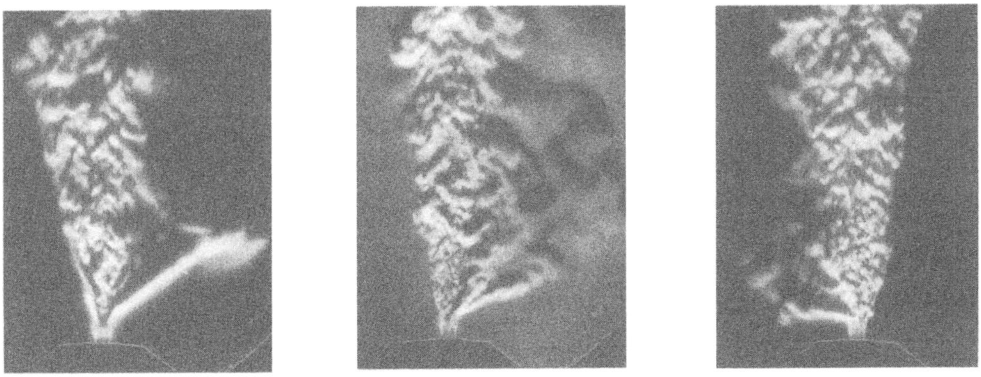

Ve = 8.7 m/s Re = 440 **Ve = 12.5 m/s Re = 620** **Ve = 14.7 m/s Re = 740**

Figure 6 - Schlieren visualizations (horizontal knife edge) . 6 mm dia. nozzle . Pure helium .

Some essential features of the present observations differ markedly from those of the Berlin Group: here, they occur closer to the nozzle exit plane and, more importantly, the lateral expulsion of fluid is much more violent, reaching distances up to 6 to 8 D_e from the jet axis as shown by Fig.6 as compared to at most 2 D_e. Furthermore, the previous results and the more recent ones as well as the analysis of the Berlin Group presented at this conference (9) demonstrate that their side jets result from the break-up of the vortex rings due to azimuthal instabilities and that about five or six side jets are thus simultaneously generated on the same vortex. Since the schlieren technique integrates across the flow, the present visualizations cannot give detailed views of the complete periphery of the jet. But considering that two side jets are hardly ever seen simultaneously on both sides at the same time on the 6 mm jet, it looks unlikely that there are more than two or three at the same time. Even more convincing is the fact that the present side jets persist at the same location during times which are incomparably longer than the periodicity of the vortex rings. Clearly, these side jets seem to be of a different kind than those observed by the Berlin Group on their heated jet. One should be reminded that in their flow $S > 0.5$, while our observations on side jets were made with pure helium, also with $S = 0.14$. This strongly suggests that the characteristics of side jets which occur naturally in jets less dense than the ambient fluid depend a great deal on the density difference.

4. Conclusion

The results of the experiments presented in this paper confirm the existence of absolute instabilities in jets lighter than the ambient fluid and reveal some novel features of these flows. They show, in particular, that there are several transitions from convective to absolute instability and vice versa when the velocity is increased and the density is kept constant at $S = 0.14$ and that these transitions are gradual. If the $Re - S$ plane is somewhat more appropriate than the Mach-S plane to correlate the boundaries of absolute instabilities of various experiments it is far from being completely satisfactory.The Strouhal number of the preferred mode also varies progressively with Reynolds number between about 0.25 and 0.50 with Reynolds number.

Very strong side jets have been observed in the 6 mm jet which differ markedly from those reported in other experiments.

Acknowledgments

This work was supported by the HERMES R&D Program through the Centre National dEtudes Spatiales and Dassault Aviation.

References

(1) Huerre P. and Monkewitz P.A. 1985 *J. Fluid Mech.* **159**, 151.

(2) Monkewitz P.A. and Sohn K.D. 1988 *A.I.A.A. J.* **26**, 911.

(3) Huerre P. and Monkewitz P.A. 1990 *Ann. Rev. Fluid Mech.*

(4) Sreenivasan K.R., Raghu S. and Kyle D. 1989 *Experiments in Fluids* **7**, 309.

(5) Monkewitz P.A., Bechert D.W.,Barsikow B. and Lehmann B. 1989 *Advances in Turbulence 2*, 455, Springer Verlag.

(6) Monkewitz P.A., Lehmann B., Barsikow B. and Bechert D.W. 1989 *Phys. Fluids A* **1**, no. 3, 446.

(7) Monkewitz P.A., Bechert D.W.,Barsikow B. and Lehmann B. 1990 *J. Fluid Mech.* **213**, 611.

(8) Brown G.L. and Rebollo M.R. 1972 *A.I.A.A. J.* **10**, 649.

(9) Raghu S.,Lehmann B. and Monkewitz P.A., 3 rd European Turbulence Conference, Stockholm 3-6 July 1990.

An Experimental Study of Transition in Plane Couette Flow

Nils Tillmark and P. Henrik Alfredsson

Department of Gasdynamics, Royal Institute of Technology
S-100 44 Stockholm, Sweden

The first flow visualization experiments of transition in plane Couette flow are reported. The transitional Reynolds number, i.e. the Reynolds number at which a turbulent spot survives, was determined to be 360 ± 10, based on half channel height and half the velocity difference between the walls.

1. Introduction

It is possible to predict from linear theory when a laminar flow first becomes unstable to small two-dimensional disturbances (so called Tollmien-Schlichting waves). In certain situations, as for instance plane Couette flow (see figure 1) and pipe flow, the linear theory shows that these flows are stable to all such disturbances although experimentally transition to turbulence is observed already at relatively low Reynolds numbers. However, transition to turbulence may occur through many other mechanisms beside growth of instability waves, all which are usually lumped together and referred to as a "by-pass"-mechanism. For instance, natural transition in plane Poiseuille flow at sub-critical Reynolds numbers occurs through the formation of turbulent spots (see e.g. Alavyoon, Henningson & Alfredsson, 1986)

In this paper we describe an experimental apparatus constructed to allow flow visualization of transition in plane Couette flow. We show that transition may occur through the formation of turbulent spots and we have determined the spreading of the turbulent region as a function of Reynolds number. The transitional Reynolds number has been determined to be 360 ± 10. In section 2 we review earlier experimental and theoretical work dealing with plane Couette flow. In section 3 we describe the flow apparatus and in section 4 the experimental results.

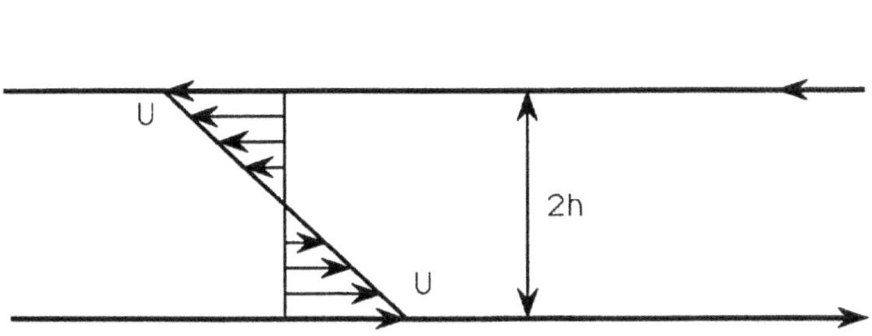

Figure 1. Geometry of plane Couette flow.

Advances in Turbulence 3
Editors: A.V. Johansson · P.H. Alfredsson

2. Experimental and numerical results on transition in plane Couette flow

The original work by Couette (1887) was done in order to determine the viscosity of water. He used an apparatus consisting of two vertical cylinders with a small gap and with the outer cylinder rotating. In his case the transitional Re * was around 480, but this value depends on the relation between the radius of the inner (R_i) and outer (R_o) cylinders or the ratio between the channel width and the radius of curvature $\gamma = (R_o - R_i)/R_o$. In the experiment by Couette γ was as low as 0.015. In the experiments by Taylor (1936) γ was varied and the transitional Re was shown to increase with increasing γ, i.e. the centrifugal force has a stabilizing effect when rotating the outer cylinder. Figure 2 shows the transitional Re from various investigations of flow between an inner stationary and an outer rotating cylinder. Also the data of Wendt (1933) and Coles (1965) fits into the picture. It should be remarked that the higher value from Coles study corresponds to the transitional Re when the Reynolds number is monotonously increased from zero, whereas the lower value is obtained when the Reynolds number is monotonously decreased from an already turbulent stage, i.e. there is a considerable hysteresis effect. From figure 2 one may extrapolate the transitional Re to be in the region 300-500 as $\gamma \to 0$. Wendt (1933) from a somewhat more limited data base approximated the transitional Re as 500 for $\gamma \to 0$.

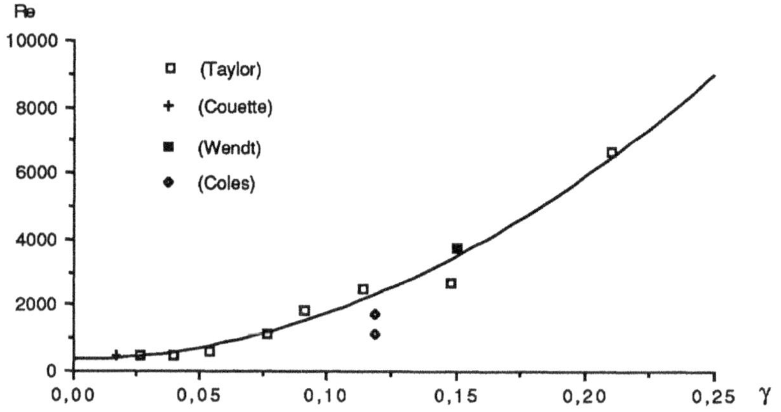

Figure 2. Transitional Re found in cylindrical Couette flow with outer cylinder rotating as function of the ratio between channel width and radius of curvature.

Only a few experimental studies of transition in plane Couette flow have been carried out, and the lack of experimental results is due to considerable experimental difficulties associated with the construction of, and measurement and flow visualization in a plane Couette flow. Reichardt (1956) used a running belt apparatus and observed turbulence at a Re of about 750. Leutheusser & Chu (1971) carried out transition

* By transitional Reynolds number we mean the lowest Reynolds number for which transition to turbulence occur, whereas the critical Reynolds number denotes the lowest Reynolds number for growing infinitesimal disturbances

experiments in a facility where a free surface water flow was used as the moving wall and the other wall was a stationary flat plate located above and parallel with the free water surface. In this way air was drawn into the channel formed by the moving water surface and the stationary plate. They determined the transitional Re to 280, by observing the signal of a hot wire located in the channel. However there are some problems also with such an apparatus, for instance surface waves gave rise to disturbances in the air flow. No flow visualization was performed of the transition process.

The numerical simulations of Orszag & Kells (1980) showed that turbulence was not sustained below a Re of approximately 1000, which is higher than in the experimental studies cited above. However, new numerical simulation of Lundbladh & Johansson (1990) give a transitional Re just below 375, for one specified finite, localized disturbance of high amplitude.

3. Experimental apparatus

The experiments were made in a running belt apparatus, where for the present experiments both walls are moving, and the velocity in the centre as well as the mean velocity in the channel is zero. The running belt is made of a transparent plastic material (type "overhead" film) so that optical access to the channel for flow visualization is possible. The working fluid is water and the flow is visualized by seeding the fluid with titanium-dioxide-coated-platelets (Merck Irodin 120, 5-20 μm in diameter) and illuminating the channel at an oblique angle with a 150 W slide projector. The particles are oriented by the flow and the reflected light clearly reveals the areas of turbulent and perturbed laminar flow.

The experimental apparatus is shown in figure 3, where the main parts are: the tank, the steel frame, the driving and steering cylinders and the plastic film formed into an infinite loop. The tank is mainly made of 10 mm acrylic plastic, except the part that forms the channel which is made from a 10 mm thick floated glass plate (#1). Its outer dimensions are 2.55 m $\times$ 0.45 m $\times$ 0.55 m and the length of the test section (and hence the glass plate) is 1.50 m. To form the Couette channel a second glass plate of the same size is placed in the tank parallel to the glass plate forming part of the outer wall of the tank. The distance between the facing surfaces is adjusted by gauge blocks at the bottom and top and could be chosen at will. In our experiments the gap was 10 mm. To produce the plane Couette flow between the glass plates a transparent polyester plastic film (#2) (0.1 mm thick and 360 mm wide) runs along the glass surfaces. The joint of the plastic film has been made by gluing the film together with silicone. No effect of the joint could be observed in the flow visualization. The film is driven and supported by two precision made brass cylinders (#3, #4) with a length of 450 mm and a diameter of 200 mm. The driving cylinder (#3) is connected to a servo coupled DC motor (#5). The second cylinder (#4) controls both the tension and the vertical position of the plastic film. An optical pulse generator (#6) connected to the shaft is used to measure the film speed. To ensure the right amount of lubricating fluid film between the glass surface and the plastic film, the position of the plastic film at the entrance of the Couette channel is controlled by a pair of freely rotating cylinders (#7). Each cylinder could be accurately positioned by fine thread screws. A microscope is used to check the exact position of the plastic film in the channel. LDV measurements

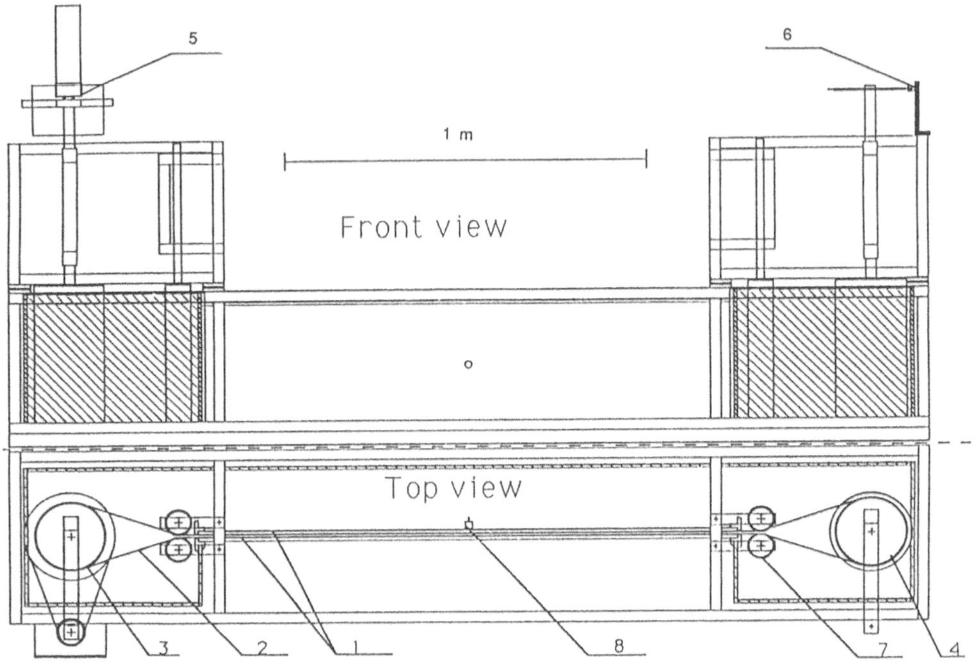

Figure 3. Experimental apparatus. 1: Floated glass plates, 2: polyester plastic belt formed into an infinte loop, 3: driving cylinder, 4: belt supporting and speed measuring cylinder, 5: servo coupled DC-motor with gear box, 6: belt speed indicator, 7: belt steering cylinders, 8: liquid jet injector for spot triggering.

were carried out at $Re = 345$ and showed an essentially ($\pm 2\%$) linear profile across the main part of the channel.

Forced disturbances can be made by a short duration jet of fluid which enters the channel from a 1 mm hole in the inner wall. A solenoid (#8) is activated and forces fluid through the hole when a 3 mm diameter opening in the plastic belt is just above the hole in the wall. The activation of the solenoid is effectuated via a photodetector and at the same time a clock is started. This disturbance will lead to a turbulent spot if the Reynolds number is high enough.

4. Results

One major objective of the present study was to accurately determine the transitional Reynolds number. This can be made essentially in two different ways. One is to introduce a sufficiently large disturbance so that any existing unstable non-linear mode can grow. The solenoid driven disturbance was not strong enough, but by introducing a large air bubble at the bottom of the channel which rose to the surface a sufficiently strong disturbance was introduced when the bubble reached the surface. The other method is to start out from a turbulent state, i.e. at a high Reynolds number and then gradually decrease Re until all turbulence has died away. Both methods were tried and gave identical results for the transitional Re, namely 360 ± 10. Note that the principle of counter running belts make it possible to view the spot during long times. This is necessary in order to accurrately determine the transitional Re as otherwise the decay may not happen within the time before the disturbance is convected out from a finite experimental apparatus. This should be seen in connection with the numerical simulation of plane Couette flow carried out by Lundbladh & Johansson (1990) who determined the transitional Reynolds number to be between 350 and 375.

In the following photographic sequencies we will show the spot developement at two different Re, namely 405 and 912. In both cases a spot was triggered by the jet disturbance.

At a $Re = 405$ figures 4a-d show a sequence of the spot from 1.4 - 20.5 sec (time in nondimensional units is 25, 95, 210, 390, respectively) after triggering. The square pattern seen in the photographs has side lengths of 50 mm. Initially a very weak disturbance is seen, however, in figure 4b it seems like breakdown has occurred in one region. In figure 4c the spot has grown and is now approximately as wide as it is long. The boundaries are quite irregular with fingering turbulent patches stretching in the streamwise directions. One also notes the turbulent scales seems to be fairly large. As time increases the spot becomes less symmetrical (figure 4d) and it now seems that its width is larger than its length.

At a $Re = 912$ figures 5a-d show a sequence of the spot from 0.6 - 4.6 sec (time in nondimensional units, scaled with belt speed and channel half-height, is 25, 60, 110, 195, respectively). The initial disturbance becomes quite elongated (figure 5a) even after breakdown (figure 5b). Note that it here that the centre of the disturbance is shifted slightly from the jet loctaion, probably due to the fact that the original disturbance is non-symmetrical. As the spot grows older its aspect ratio (defined as its length-to-width ratio) decreases and is about 2 in figure 5c, whereas in figure 5d it is about 1.5. This means that the spanwise growth is larger than the growth in the streamwise direction. The outer boundary of the spot is more regular and well-defined here than at $Re = 405$ and the turbulence scales within the spot is seen to be smaller. At this Re the turbulent spot will grow in the streamwise and spanwise directions until the channel is fully turbulent.

Experiments were carried out at several Reynolds numbers and the half-length (L) and half-width (W) of the spots were determined as function of time (figure 6a,b). We have here plotted data for all Re in one graph and shifted the time axis so that data from the "fully" developed spot region should collapse. The difference in the virtual time origin of the spots is probably due to the fact that the initial disturbance was the same for all cases, thereby becoming relatively stronger for the higher Re, leading to a faster initial growth. We see that data from all Re except the lowest, collapse nicely for long enough times after triggering. The spreading velocity is 0.17 in the streamwise and 0.20 in the spanwise direction. This means that the spot will finally have an aspect ratio smaller than one. The spreading rate is here constant at least for Re larger than 40 % above the transitional Re, which is different compared to spots in plane Poiseuille flow, but compares well with boundary layer spots. The corresponding half-spreading angle of the spot seen in a coordinate system at rest relative to one of the belts is 11°, close to what is observed for Blasius boundary layer spots.

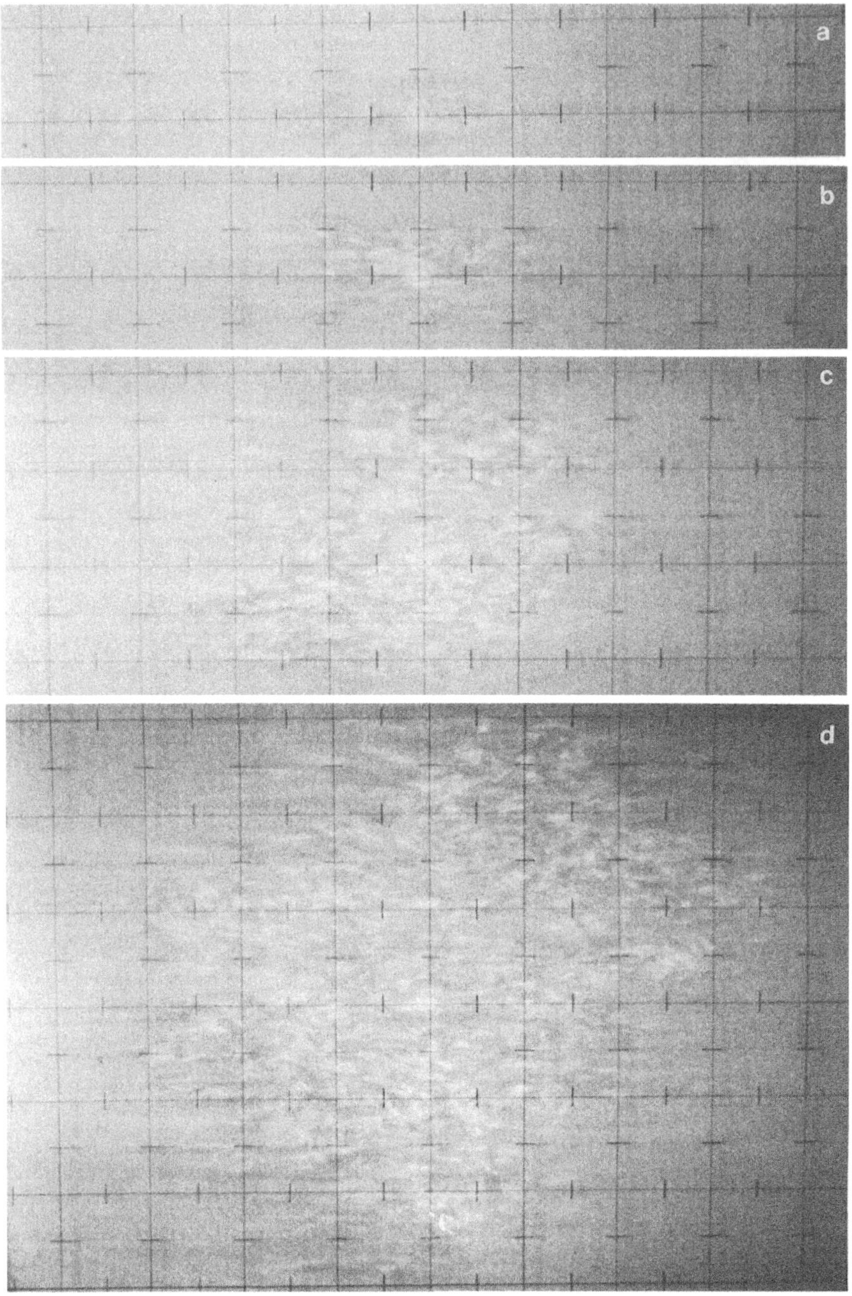

Figure 4. Turbulent spot development at Re=405. Belt velocity is 89 mm/s. Non-dimensional time after triggering a) 25, b) 95, c) 210, d) 390.

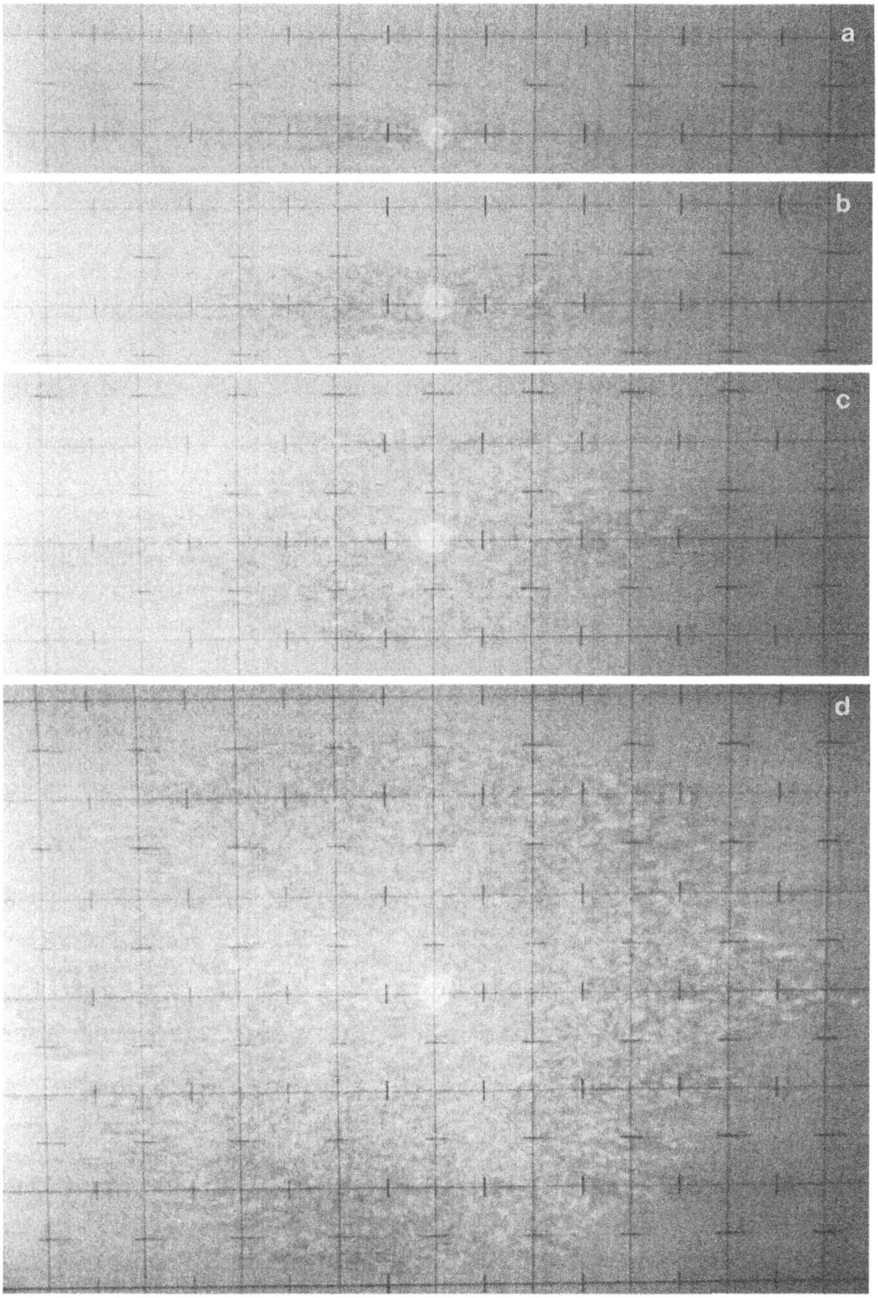

Figure 5. Turbulent spot development at Re=912. Belt velocity is 200 mm/s. Non-dimensional time after triggering a) 25, b) 60, c) 110, d) 195.

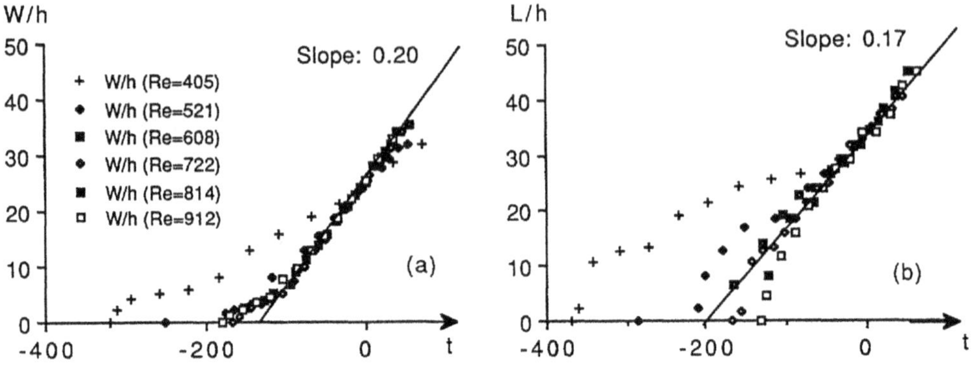

Figure 6. Width (a) and length (b) of a individual realization as function of time for six differnt *Re*. Note that time axis is shifted individually for each spot.

Acknowledgements

This study could not have been done without the skilful work of Erik Liljeholm, Department of Gasdynamics, during the planning, construction and manufacturing phases of the Couette flow apparatus. Fruitful discussions with Professors Arne Johansson and Håkan Gustavsson are also acknowledged. This work is funded by the National Swedish Board for Technical Development through its program for basic research (STUF).

References

Alavyoon, F., Henningson, D.S., & Alfredsson, P.H. 1986 Turbulent spots in plane Poiseuille flow-flow visualization. *Phys. Fluids* **29**, 1328.

Coles, D. 1965 Transition in circular Couette flow. *J. Fluid Mech.* **21**, 385.

Couette, M. 1890 Études sur le frottement des liquides. *Ann.Chim. Phys.* **21**, 433.

Leutheusser, H.J. & Chu, V.H. 1971 Experiments on plane Couette flow. *J. Hydr. Div., ASCE*, **97**, HY9, 1269.

Lundbladh, A. & Johansson, A.V. 1990 Direct simulation of turbulent spots in plane Couette flow. Submitted to *J. Fluid Mech.* See also the present Proceedings.

Orszag, S.A. & Kells, C. 1980 Transition to turbulence in plane Poiseuille and plane Couette flow. *J. Fluid Mech.* **96**, 159.

Reichardt, H. 1956 Über die Geschwindigkeitsverteilung in einer geradliningen turbulenten Couetteströmung. *Z. Angew. Math. Mech.* **36**, S26.

Taylor, G.I. 1936 Fluid friction between rotating cylinders, Part I. Torque measurements. *Proc. Roy. Soc.* **A**, **157**, 546.

Wendt, F. 1933 Turbulente Strömungen zwischen zwei rotierenden konaxialen Zylindern. *Ing. Arch.* **4**, 577.

Dynamical Theory and Mixing

G. M. Zaslavsky

Space Research Institute, Profsoyuznaya 84/32, Moscow 117810, USSR

Novel consideration of the transition to turbulence shows a multiplicity of ways and a variety of scenarios. All of them are realized in different forms of onset of a spatio-time chaos. Lagrangian chaos (chaos of passive particles) and chaos of streamlines begin to play more important role to understand physical and topological properties of flows in their evolution to turbulent regime. General properties of preturbulent states are discussed. Different patterns and Lagrangian chaos of them are considered. Special attention is applied to properties of passive particle transport and their mixing due to the chaos of streamlines. It is shown that the process of transport should be likely Levy random walk process. Abnormal diffusion properties are discussed and their connection with dynamical and topological properties of the flows is verified. The pattern of phase space of streamlines controls their diffusion parameters. We discuss the diffusion process dependence on the symmetry of patterns which affect the character of the passive random walk. Levy random walk on multifractals and its scaling properties are considered.

1. Introduction

The old attempts to apply dynamic systems theory to understanding of the turbulence have got new perspectives now promoted by the rise of the chaos theory. It is a theory describing how regular (or laminar) motion can transit to irregular (turbulent) one. The simple scenario of turbulence onset by Landau-Hopf is replaces now by more sofisticated scenarios by Ruell-Takens, Pomeau-Manneville, doubling bifurcations etc. Lorenz model gives an example of well-studied picture regular-chaotic motions transition (see discussions in [1]). Although experimental data do not confirm the existence of a unique scenario, attempts to find a universal transition behaviour are being continued. The form of universality of transition to turbulence is not clear and even there is no understanding why such a universality should exist.

In parallel to these efforts there exists an obvious scepticism in scientific opinion that theory of chaos should help to understand the turbulence. This scepticism is reasonable to some extent. But only to some extent. Let me say only a few words on the reason why I do think so. The most convenient way is to look at the Lorenz model. Its simplification is in introducing of the cut off procedure which gives a finite number of modes with different space scales. Onset of chaos in this model means the random behaviour of amplitudes of these modes. In other words, Lorenz model can describe the transition from a discret spectrum in time to a continuous one. But nothing happens to space scales and so space spectrum remains discrete and even finite. From the other point of view, the developed turbulence i characterised by in infinite space spectrum with complicated processes of space scale transformations. Up to now there is no model which can describe transition to continuous both space and time spectra. And that is just one of important problems in turbulence to be understood (see discussions in [2,3]).

Advances in Turbulence 3
Editors: A.V. Johansson · P.H. Alfredsson

It is known that the formation of ordered structures may precede the turbulence. There is a lot of examples in different kinds of convection, zonal flows in planetary atmospheres and others. Thus space disordering should follow as set of different patterns and their transformations for such cases.

Description of pattern transformations is a part of the scenario of transition to turbulence but there is an entirely different way of analysing the patterns. For a given vector field $v(x, y, z)$ one can study the problem of its streamline behaviour:

$$\frac{dx}{v_x} = \frac{dy}{v_y} = \frac{dz}{v_z} \tag{1}$$

A topology of streamlines is another clue to understand different properties of the velocity field and its level of ordering. It was mentioned that topology of streamlines can be nontrivial (that is chaotic) for Beltrami flows [4]. The celebrated example is ABC (Arnold-Beltrami-Childress) flow for which chaos of streamlines was confirmed in [5,6] by numerical simulations. It was shown based on general considerations of q-fold symmetric Beltrami flows that area of chaotic streamlines forms a connected web of the same symmetry [7,3]. This phenomena turns out to be relevant for the passive scalar transport. Chaos of passive particles is called Lagrangian turbulence. It plays an important role in various problems of fluid mechanics [8]. Right now we can show a correlation between Lagrangian chaos and velocity field.

It is a special problem to study Lagrangian chaos for arbitrary velocity field but some special accent should be done for Beltrami flows which meet the condition

$$\vec{v} \times curl\ \vec{v} = 0 \tag{2}$$

There are indications that the nonlinearity is suppressed in a strong isotropic turbulence [9-12]. And it means in some sense the Beltramization of a turbulent flow. So properties of particles involved in the Lagrangian turbulence process can reveal not only in the large scale dynamics of fluids but they can reflect some microscale featuring (small eddies advection for example). Studying of transformations of a topology of passive particle motion in a process of transition to turbulence was performed in [13,14].

Below the Lagrangian turbulence for stationary three-dimensional flows is treated as a nonlinear dynamical system problem. Special attention is applied to the problem of diffusion of streamlines or passive particles. Random walk in a fractal space-time is a new characteristic property of the problem leading to anomalous laws of diffusion.

2. Beltrami flows with Q-fold symmetry (Q-flows)

There is a well-known solution for Beltrami flows satisfying condition (2) and incompressibility condition [15]. From (2) we have an equivalent condition

$$curl\ \vec{v} = \kappa\vec{v} \tag{3}$$

where $\kappa = \kappa(\vec{r})$ is an arbitrary function. The solution of (3) is

$$\vec{v} = \kappa\nabla\ \vec{l}\Psi + \nabla \times \nabla\vec{l}\Psi \tag{4}$$

where $\vec{l}$ is an arbitrary unit vector and Ψ is an arbitrary solution of the Helmholtz equation

$$\Delta\Psi + \kappa^2\Psi = 0 \quad . \tag{5}$$

Actually, this solution was known very long ago as Trkalian field [16] and there is large review [17] on various kinds of solutions of (3) [18].

Below the case

$$\kappa = const = \pm 1 \tag{6}$$

will be considered only. The following simple corollaries take place from (4)-(6):

(i) For any Ψ in the form

$$\Psi(x,y,z) = \Psi(x,y) cos z \tag{7}$$

we find $v_z = 0$ if $\vec{l} = \vec{z}^0$

(ii) For any Ψ in the form,

$$\Psi(x,y,z) = \psi(x,y) + \varphi(x,y)\cos y \tag{8}$$

we find the same $v_z = 0$ if $\vec{l} = \vec{z}^0$

(iii) For $\vec{l} = \vec{z}^0$ and

$$\Psi(x,y,z) = \psi(x,y) \tag{9}$$

we find

$$\vec{v} = (\partial\psi/\partial y - \partial\psi/\partial x, \psi) \tag{10}$$

Let $\vec{l} = \vec{y}^0$ and

$$\Psi = cos z \tag{11}$$

Then we have

$$\vec{v} = (sin z, cos z, 0) \tag{12}$$

(iv) Linear combination of solutions with different $\vec{l}$ is also a solution. So sum of (10) and (12) gives the field:

$$\vec{v} = (\frac{\partial\psi}{\partial y} + \varepsilon \cdot sin z, -\frac{\partial\psi}{\partial x} + \varepsilon \cdot cos z, \psi) \tag{13}$$

where ε is an arbitrary parameter and ψ is a 2D function from (9).

Q-flow that was introduced in [7] is of the form (13) with

$$\psi_q = \sum_{j=1}^{q} A_j cos(\vec{R} \cdot \vec{e}_j) \tag{14}$$

where A_j are constants, $\vec{R} = (x,y)$ and $\vec{e}_j$ - the set of unit vectors forming a regular star. For $q = 4$ Eqs. (13) and (14) determine an ABC-flow. If $q = 3$ or 6 the flow is of hexagonal symmetry in the (x,y) plane. For all other flows their symmetries are quasicrystal (and that is why they are called Q-flows).

It looks like details of the formation of stationary flows are not important for studying of streamlines' behaviour and only symmetry of the flow is relevant. This fact separates Q-flows from all others.

3. Chaos of streamlines

The equations of streamlines (1) can be written in the form

$$\frac{dx}{dy} = \frac{v_x}{v_z}; \frac{dy}{dz} = \frac{v_y}{v_z} \tag{15}$$

Consider z as "time" and the right-hand side as a given function of x, y, z. Then (15) can be studied as a dynamical system. For $\vec{v}$ given in (13) the system can be reduced to a Hamiltonian form [7]

$$\dot{x} = \frac{\partial H}{\partial y}; \dot{y} = -\frac{\partial H}{\partial x} \tag{16}$$

where

$$H = \psi + \varepsilon(y sin z + x cos z)$$

$$\dot{x} \equiv \frac{dx}{d\tau}; \dot{y} \equiv \frac{dy}{d\tau} : \dot{z} \equiv \frac{dz}{d\tau} = \psi(x, y, z) \tag{17}$$

The last equation in (17) determines new time τ.

Existence of Hamiltonian representation is the consequence of incompressibility. Thus the Hamiltonian equations have the following form for general field $\vec{v}(x, y, z)$ if $div\ \vec{v} = 0$ [19]. Let z be the "time". Then the canonical representation of (15) can be done in the following form

$$\frac{dx}{dz} = \frac{\partial H}{\partial p}; \qquad \frac{\partial p}{\partial z} = -\frac{\partial H}{\partial x}$$

where

$$p = p(z, y) = \int^{y} v_z(y', z) dy'$$

$$H = \int_0^{y} v_x dy' - \int^{x} v_y(y = 0) dx'$$

where $x = x(z)$ in the expression for v_z.

All these preliminaries are very essential because the crucial property Hamiltonian systems can be directly applied now. That is the field (13) possesses an area of chaos of streamlines [7]. For the Q-flows the area of chaos has the appearance of a connected net called the stochastic web. Proof of the existence of webs for the flows (13), (14) and estimates of their width were made in [7]. For small $\varepsilon \ll 1$ the width is of order ε. Two examples of fluid stochastic webs are plotted in Fig. 1 based on numerical simulation [3]. This is a Poincare section of a streamline. This means that all points where the streamline crosses the plane $z = 0\ (mod\ 2\pi)$ are plotted on (x, y). The chaotic mixing inside the range of the web can be seen in Fig. 1. There are islands of stability inside meshs of the web. Streamlines wind around invariant cylindrical surfaces. Crossings of this surfaces with the plane $z = 0(mod 2\pi)$ form closed invariant curves in (x, y)-plane. There are different sets of islands. Some of them appeared due to crossings of meandering cylinders.

The system of islands is fractal. Its properties depend on the parameter ε.This affects also the kinetic properties of streamlines.

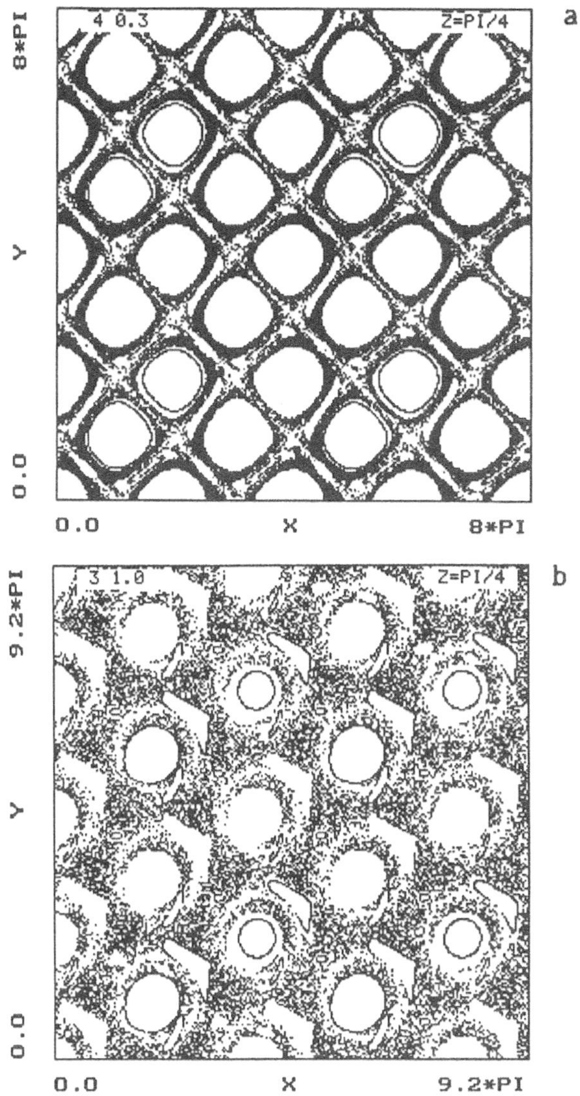

Fig 1 Poincare sections of a streamline by the plane $z = \pi/4$; (mod 2π: a) $q = 4$; $\varepsilon = 0.6$; b) $q = 3$; $\varepsilon = 1$.

4. Anomalous transport: trappings and flights

The law of liquid particle diffusion is determined by intrinsic properties of chaos. If chaotic motion is homogeneous inside the range of the web in some sense, then the particle kinetics should be considered as some kind of random walk along the net with q-symmetry. And that is just the case for some range of values of ε. Example of such a random walk diffusion is given in Fig. 2a. Left-side picture shows the Poincare section of a streamline on small part of (x, y)-plane which consists of two periods of the web.

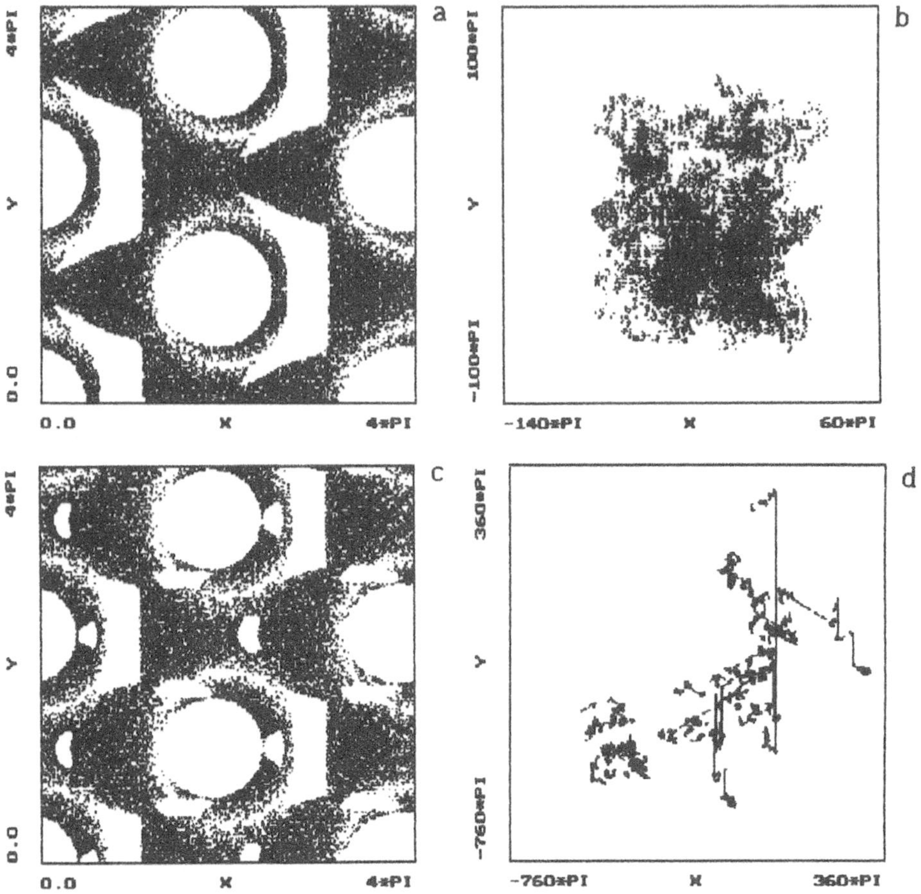

Fig. 2 Poincare sections of a streamline by the plane $z = 0 \pmod{2\pi}$ with $q = 3$ and: $\varepsilon = 0.7$ (a, b) and $\varepsilon = 1.1$ (c, d) for small (a, c) and larger (b, d) scales.

Right-side picture shows the same streamline section but on the full (x, y)-plane. It looks like a typical Brownian motion. If one writes an asymptotical law of diffusion in the form.

$$< R > \sim D_{\perp}^{1/2} \cdot t^{\alpha_{\perp}} \tag{18}$$

$$< L > \sim D_{\parallel}^{1/2} \cdot t^{\alpha_{\parallel}}$$

where R and L are correspondingly displacements of liquid particle in (x, y)-plane and z-direction, $D_{\perp}, D_{\parallel}, \alpha_{\perp}, \alpha_{\parallel}$, - some constants, then for the wandering in Fig. 2a numerical simulation gives $\alpha_{\perp} \approx 1/2; \alpha_{\parallel} \approx 1$. This means also almost free motion ("flight") along z-axis.

Another situation is displayed in Figs. 2b, c. There are flights in (x, y)-plane in these pictures and $x_{\perp}$ for these cases is between 1/2 and 1. The described phenomina of flights reminds us the phenomena of Le'vy random walk and Le'vy flights [20, 21]. That is the process with long-tails distribution function and the exponent of transport

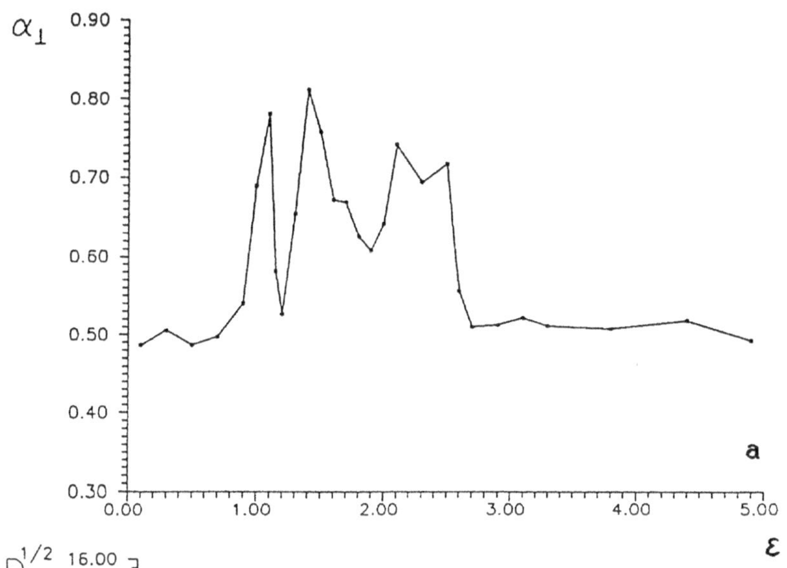

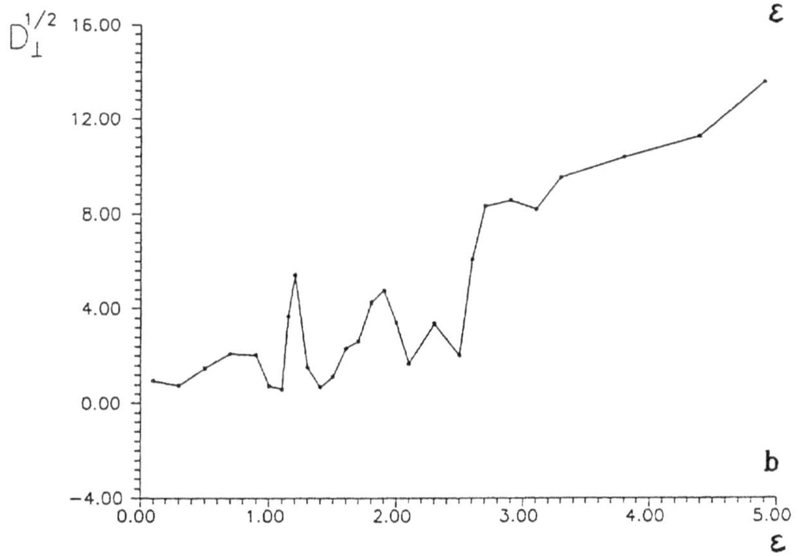

Fig. 3 The random walk parameters $\alpha_\perp$ (a) and $D_\perp^{1/2}$ (b) plotted as a function of ε for $q = 3$

in (18) should depend on the fractal dimension of the random walk space [22]. There are different confirmations that chaos of streamlines for Q-flows reveals in anomalous diffusion of Lvy-like random walk (yet maybe not exactly the same) [3, 23, 24].

Detailed numerical simulations gave possibility for two important statements. The first one is that the exponents $\alpha_\perp, \alpha_\parallel$ determining the asymptotic law of diffusion are sensitive to a topology of phase space. Fig. 3 gives a dependence of $\alpha_\perp$ and $D_\perp^{1/2}$ on ε for $q = 3$. There exists a range of anomalous diffusion at $2 \leq \varepsilon \leq 3$ and strong oscillations of $\alpha_\perp$. It seems that ABC-flow is anomalous for any not too large ε. The second statement concerns the origin of the flights. If liquid particle or streamline

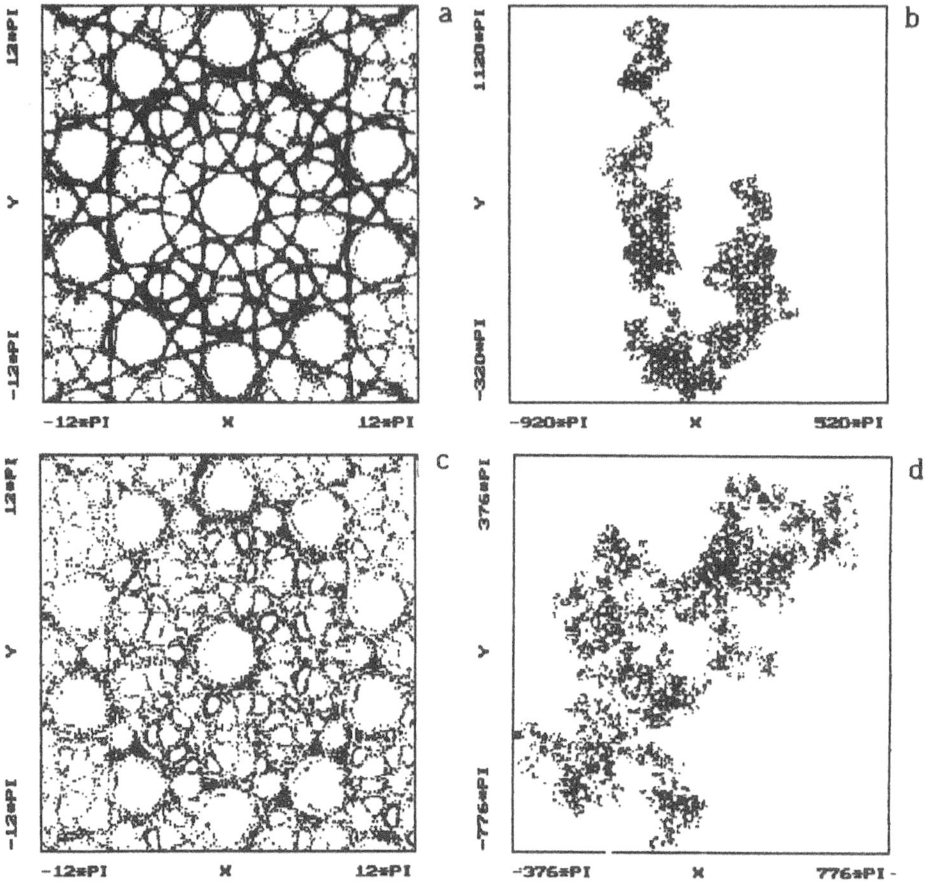

Fig. 4 Poincare sections of a streamline with $q = 5$ for $\varepsilon = 0.1$ (a, b) and $\varepsilon = 0.3$ (c, d) for small (a, c) and large (b, d) scales.

reaches the small vicinity of the boundary of an island it will be trapped in this area for a very long time. Two the longest flights in Fig. 3b are marked out and the points of the streamline belonging to these flights are plotted in mod 4π. The Fig. 4 makes evidence of the statement that the flight's onset is due to trapping in a boundary of some island. The trapping in some plane means the same time the flight in the perpendicular to the plane direction because the boundary of the island is a section of the corresponding tube surface along which the free motion takes place. So in real space (x, y, z) there are trappings and flights and the exponents $\alpha_\perp, \alpha_\parallel$ reveal as a result of their competition. Changing of ε produces transformation of topology of the phase space and for some critical ε bifurcations appear. The rise of a new set of islands reveals in changing of exponents $\alpha_\perp, \alpha_\parallel$ and that is the cause for the oscillations in Fig. 3.

5. Stochastic jets

Existence of flights and area of trappings reflects a property of strong inhomogeneity of space. It should be understood to which properties of observable physical values they are relevant. One of the feature properties is connected with a finite observation time in real experiments or in simulations. It means that for not too large range of initial conditions and for finite time one can receive only intermediate asymptotics of diffusional law for some domain of initial conditions and parameters. Of course this suggests also that contributions of flights or trappings are visible. In its turn this depends on the measure of anomalous orbits. Such kind of investigations was done in [25] for the Q-flow with 5-fold symmetry ($q = 5$).

It was shown that for the range $0.5 \leq \varepsilon \leq 1.5$ there is onset of stochastic jets [26]. The meaning of the last notion is the following. There are such orbits which have so huge flights (or trappings) that their duration time is almost of the order of simulation or observation ($10^4 - 10^5$ characteristic steps or periods). For the mentioned domain of ε such orbits form several percents of the full observed number of orbits (~ 500). But maybe the most essential property of the long flights is that they are very slightly unstable. So if a bundle of orbits is taken with initial conditions not too far from the long-flighting orbit then dispersing of the bundle will be small during the time of the flight. Such bundles survive for the same long time though there are some outlets and inlets of orbits. We have called these bundles stochastic jets as well as an only orbit with anomalously long flight (or trapping). Here are several demonstrations of jets [25].

Examples of stochastic web with pentagonal symmetry are given in Fig. 4. They are received for a number of orbits and reflect the projection on the (x, y)-plane. Figs, 4a, c show the Poincare section for small scale and Figs. 4b, d display the large scale view on the random walk. It is seen from Fig. 4 that the diffusional process is almost homogeneous like in usual Brownian motion. Averaging over 500 orbits gives:

$$\varepsilon = 0.1 : \alpha_\perp = 0.5; \alpha_\parallel = 0.97; D_\perp^{1/2} = 1.2; D_\parallel^{1/2} = 0.6$$

$$\varepsilon = 0.3 : \alpha_\perp = 0.49; \alpha_\parallel = 0.85; D_\perp^{1/2} = 1.6; D_\parallel^{1/2} = 0.9$$

That means the normal diffusion in the (x, y)-plane and almost free flights along z-direction.

All these results were obtained for small $\varepsilon < 0.5$. The situation changes strongly when jets appear for larger ε. A typical orbit is shown in Fig. 5. Flights are well resolved and responsible for anomalous diffusion with exponent $\alpha_\parallel \neq 1/2$. The visualization of jets is given in Figs. 6, 7. In Fig. 6 the Poincare section of a jet is presented. There is a small area in which chaos is "trapped" for a very long time of $4 \cdot 10^3$ (Figs. 6a, b). For this jet $maxR_\parallel = 8.5 \cdot 10^4$ though at the same time $maxR_\parallel = 3.6 \cdot 10^3$ for typical orbits. The location of the jet is sensitive to the dynamical singularity occurence nearby. Changing of parameter ε (Figs. 6c, d) and passing over a bifurcation explain the disappearance of the jet.

Another look at the jet is plotted in Fig. 7. It displays the stagnation of perpendicular motion of a particle in (x, y)-plane (Fig. 7a) due to the trapping in some area (see Fig. 6b) and a flight in z- direction (Fig. 7b) corresponding to the stagnation. The flight is performed during the time interval $(15000 - 45000)$. In the interval of time $(0 - 15000)$ there is normal diffusion which is plotted in the Fig. 7c. An example with several flights is given in Fig. 7d.

In general jets appear due to the inhomogeneity of chaos in phase space produced by cantori which look like effective barriers for a particle penetration to some area.

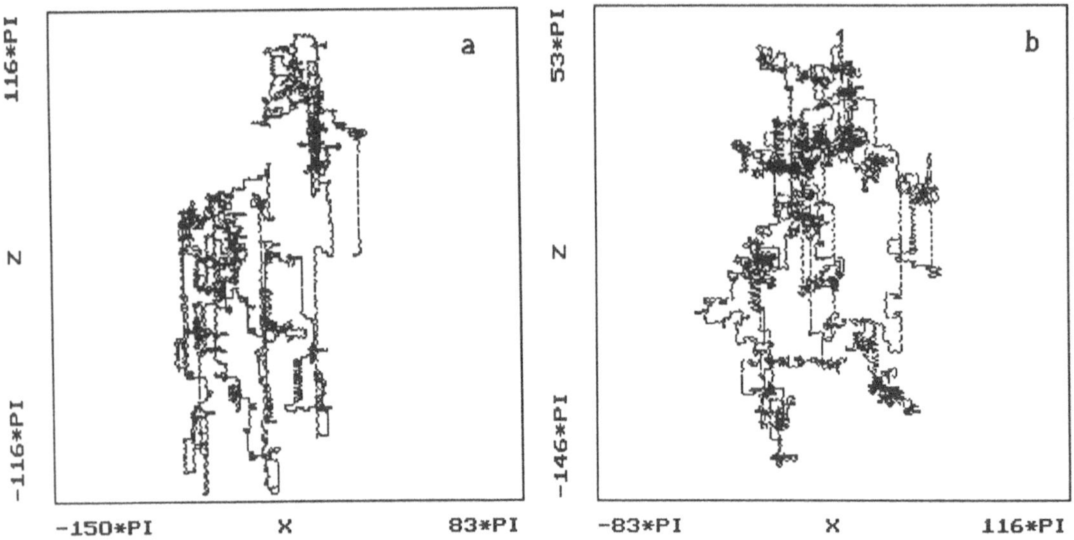

Fig. 5 Two examples of long flights in (a, z) plane for $q = 5$ and $\varepsilon = 0.9$ (a) and $\varepsilon = 1.5$ (b).

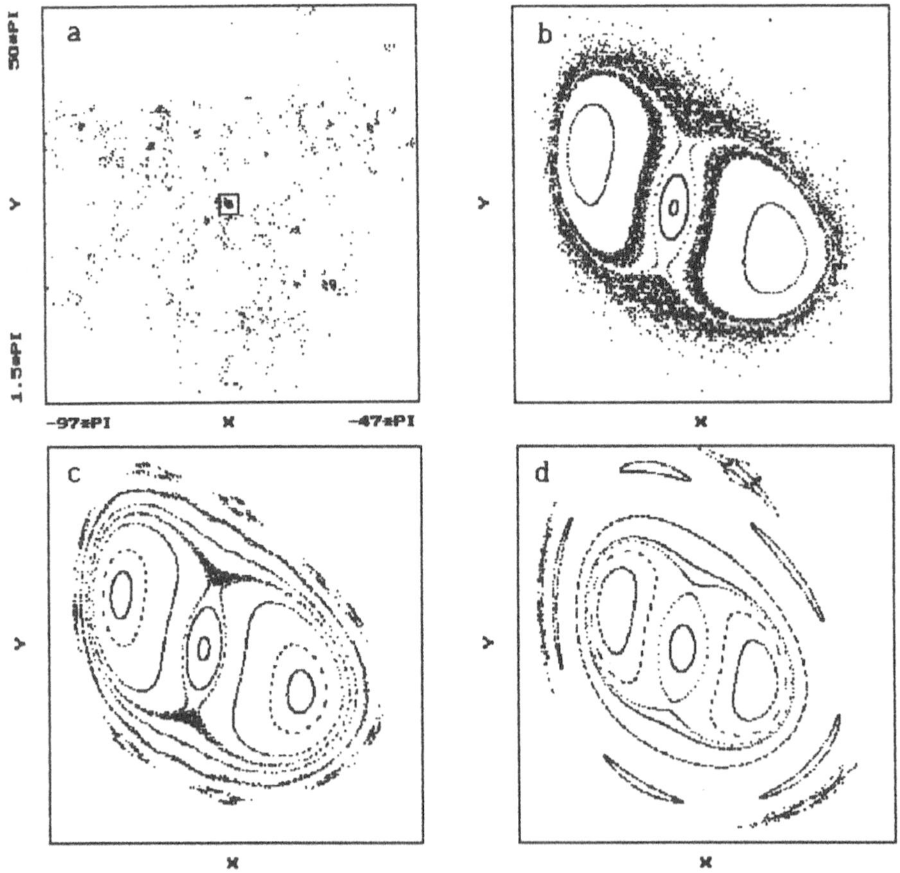

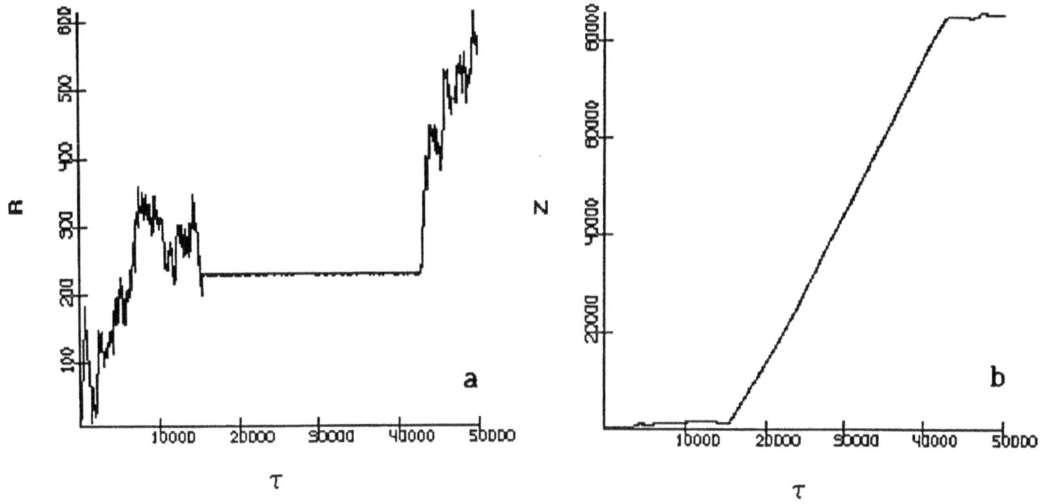

Fig. 7 Dependence of different displacements for jets: a) plot R versus τ for the jet in Fig. 3a,b; b) z versus τ for the same jet.

That is why the onset of jets is so sensitive to any reconstruction of the phase space. More correct is to say that jets exist in any case but their measure turns to be more observable in a special rang of parameters. Although the percentage of jets is small they could imply noticeably on the general exponents $\alpha_{\perp}, \alpha_{\|}$ if averaging is performed over all orbits.

6. Random walk in a fractal space-time

The main feature of new kind random walk is the existing of a pausing time at some positions of a liquid particle. Due to this fact the random walk should be characterised by two transition probabilities: $W(r)$ - the probability density for particle transition from position R to new position $(R+r)$; and $\psi(t)$ - the probability density for the interval between two consequent steps to equal t. Fourier and Laplace transforms for $W(r)$ and $\psi(t)$ correspondingly can be introduced.

$$W(q) = \frac{1}{2\pi} \int_{-\infty}^{\infty} e^{iqR} \cdot W(R) dR \tag{19}$$

where one-dimensional case for simplicity is considered, and

$$\psi(u) = \int_{0}^{\infty} e^{-ut} \cdot \psi(t) dt \tag{20}$$

Fig. 6 Poincare section in (x,y) plane of a jet for $\varepsilon = 0.9$; a) location of the jet; b) zoom of the square in a); c) the same as in b) but for $\varepsilon = 0.8$; d) the same as in b) but for $\varepsilon = 0.5$.

Let $P(R,t)$ be the probability density to find a particle in R at the moment t. Then the Montroll-Weiss result for $P(R,t)$ can be used [27, 22]:

$$P(R,t) = \frac{1}{2\pi i} \cdot \frac{1}{2\pi} \int_{c-i\infty}^{c+i\infty} du \frac{1-\psi(u)}{u} \cdot e^{ut} \int_{-\infty}^{\infty} dq \frac{e^{-iqR}}{1-\psi(u)W(q)} \tag{21}$$

Our main interest will be in the asymptotics at $t \to \infty, R \to \infty$. This is equivalent to the limit $u \to 0, q \to 0$ in (21). So the behaviour of the functions $\psi(u)$ for small u and $W(q)$ for small q is important. The simple expansion

$$\psi(u) = 1 - t_0 u \tag{22}$$

$$W(q) = 1 + iR_0 q - \frac{1}{2}\sigma q^2$$

where t_0, R_0, σ - are constants, leads to the regular Gaussian distribution

$$P(R,t) = (4\pi\sigma t/t_0)^{-1/2} exp[-(R - v_0 t)^2 t_0/4\sigma t]$$

where $v_0 = R_0/t_0$. It gives the simple law

$$< (R - v_0 t)^2 > \sim t \qquad (t \to \infty) \tag{23}$$

where the physical meaning of $R_0 =< R >, t_0 =< t >$ should be incorporated.

Nontrivial situation linked to the Le'vy random walk appears for nontrivial singular expansion of $\psi(u)$ and $W(q)$ instead of (22). Let us write the expansion in the form

$$W(q) = 1 + iR_0 q - A \mid q \mid^{\alpha} \tag{24}$$

$$\psi(u) = 1 + t_0 u - Bu^{\beta}$$

where A, B and α, β - some constants with restrictions

$$0 < \alpha < 2; \qquad 0 < \beta < 2 \tag{25}$$

provided for positive definition of $P(R,t)$. In general the fractal values of α and β are linked to fractal properties of space and time correspondingly [22]. It was shown in [26] that different asymptotics for particles transport can appear for different ranges of α and β. There are also various intermediate asymptotics if the transport observation is performed in a finite time.

Let us consider for example a simple situation when R_0 and t_0 are so small that can be neglected. Then $P(R,t)$ from (21) is reduced to

$$P(R,t) = \frac{1}{2\pi i} \cdot \frac{1}{2\pi} \int_{c-i\infty}^{c+i\infty} du \int_{\infty}^{\infty} dq \frac{Bu^{\beta-1}}{Bu^{\beta} + A \mid q \mid^{\alpha}} \cdot e^{ut-iqR} \tag{26}$$

After changing the variables

$$u' = u \cdot t, \; q' = q \cdot t^{\beta/\alpha}$$

it is possible to find from (26)

$$<| R |>= \int_{\infty}^{\infty} dR \mid R \mid P(R,t) = const \cdot t^{\beta/\alpha} \quad (27)$$

This result gives the scaling law for asymptotic displacement. If $\beta/\alpha \neq 1/2$ then a liquid particle exhibits the anomalous transport which is governed by fractal properties of the space-time wandering.

Yet more profound analysis of the result (27) shows that the scaling law (27) needs some nontrivial comments. The *const* in (27) can be infinite. So no finite moments $<| R |^n>$ exist for any $n \geq 1$. The crucial property regulating such a behaviour of moments is the asymptotic tails of the distribution function $R(R,t)$. If for example $B = 0$ in (24) then [22]

$$P(R,t) \sim t / \mid R \mid^{\alpha+1} \quad (28)$$

and $<| R |>$ is finite for $\alpha > 1$. Therefore the range of α is $1 < \alpha < 2$

Another comment concerns the possibility of exponent in (27) to be $\beta/\alpha > 1$. There is well-known Richardson's law in turbulence

$$<| R |> \sim t^{3/2}$$

with exponent larger than one. In the case of free motion $R \sim t$. The exponent larger than one means that there exists an acceleration process in microscale behaviour of a particle. The main difficulty is to describe in any form the connection between α, β and real properties of a particle motion and its phase space topology. This can be considered as a new serious problem of dynamics of liquid.

7. Conclusion

Real transition to turbulent flow can't be understood without obtaining a clear picture of onset of chaos in both space and time. Some properties of a so-called preturbulent state can be studied from the consideration of patterns' topology, streamlines' behaviour (including Lagrangian chaos) and liquid particle transport. The properties of passive scalars intermittency and transport may reflect not only the large scale behaviour of liquid but the microscale intermittent featuring of developed turbulence too.

References

1. D. Ruelle. In Proceedings of the meeting on turbulence. Eds. LSirovich and S. Orszag. Springer L(to be published).

2. R.Z. Sagdeev, D.A. Usikov, and G.M. Zaslavsky. Nonlinear Physics, Harwood Academic Publ. N.Y. 1989.

3. V.V. Beloshapkin, A.A. Chernikov, M.Ya. Natenzon, B.A. Petrovichev, R.Z. Sagdeev, and G.M.Zaslavsky. Nature 337. 133 (1989).

4. V.I. Arnold. Mathematical Methods in Classical Mechanics, Springer, N.Y., 1978.

5. M. Henon. C.R. Acad. Sci. Paris 262, 312 (1966).

6. T. Dombre, U. Frisch, J. Green, A. Mehr, and A.M. Soward. J. Fluid Mech 167, 353 (1986).

7. G.M. Zaslavsky, R.Z. Sagdeev, A.A. Chernikov. Sov. Phys. JETP 67, 270 (1988).

8. H. Aref. J. Fluid. Mech. 143, 1 (1984).

9. A. Tsinober, E. Levich. Phys. Lett. A 99, 321 (1983).

10. R.B. Pelz, V. Yakhot, S.A. Orszag, L. Shtilman, and E. Levich. Phys. Rev. Lett. 54, 2505 (1985).

11. H.K. Moffatt. J. Fluid Mech.. 159, 359 (1985).

12. R.H. Kraichnan and R. Panda, Phys. Fluids, 31, 2395 (1988).

13. K. Ishii and R. Iwatsu. In: Topological fluid mechanics. Eds. H.K. Moffatt and A. Tsinober. Cambridge Univ. Press, 1990, p. 54.

14. K. Ishii, R. Iwatsu, T. Kambe, and N. Matsumoto. In: Nonlinear World. Eds. V.G. Bar'yakhtar et.al. World Scientific, Singapore, 1990.

15. C. Chandrasekhar, P.C. Kendall. Astrophys. Journ. 126, 457 (1957).

16. V. Trkal. Poznamka k hydrodynamice vazkych tekutin. Casopis pro Pestovani Matematiky a Fisiky (Praha) vol. 48, p. 302 (1919).

17. O. Bjorgum. On Beltrami Vector Fields and Flows. Universitetet i Bergen, Arbok (1951), Naturvitenskapelig rekke, part 1, N1, p. 5; O. Bjorgum and T. Godal, part 2, ibid., N13, p. 4.

18. This information I received from A. Tsinober.

19. K.J. Whiteman. Repo. Progr. Phys. 40, 1033 (1977).

20. P. Le'vy. Theorie de l'Addition des variables Ele'atores. Gauthier-Villiers, Paris, 1937.

21. B.B. Mandelbrot. The fractal geometry of Nature. Freeman, New York, 1983.

22. E.W. Montroll, M.F. Shlesinger. In: Studies in Statistical mechanics., v. 11, p. 1. Eds. J. Leibovitz am? E.W. Montroll, North-Holland, Amsterdam, 1984.

23. A.A. Chernikov, B.A. Petrovichev, A.V. Rogal'sky. R.Z. Sagdeev, and G.M. Zaslavsky. Sherwood meeting on Plasma theory, 1990 (to be published).

24. A.A. Chernikov, R.Z. Sagdeev, and G.M. Zaslavsky. In: Topological Fluid Mechanics. Eds. H.K. Moffatt and A. Tsinober, Cambridge Univ. Press, 1990, p. 45.

25. B.A. Petrovichev, A.V. Rogal'sky, R.Z.Sagdeev. and G.M. Zaslavsky (to be published).

26. V.V. Afanas'ev, R.Z. Sagdeev, and G.M. Zaslavsky. Sherwood meeting on Plasma theory, 1990 (to be published).

27. E.W. Montroll and G.H. Weiss, J. Math. Phys. 6, 167 (1965).

Algebraization of 2-D Ideal Fluid Hydrodynamical Systems and Their Finite-Mode Approximations

V. Yu. Zeitlin

Institute of atmospheric physics, Pyzhevsky 3, 109017 Moscow, USSR

Starting from the description of ideal 2-D hydrodynamics in the framework of the Lie algebra of area-preserving diffeomorphisms sdiff two observations are made: 1st - this construction may be generalized by means of the central extension of the latter algebra, giving equations equivalent to variants of the vorticity equation on the β-plane; 2nd - a regular way to obtain finite-mode analogs of 2-D hydrodynamical equations preserving the essential algebraic features of these latter exists due to the relation between sdiff and $su(N), N-\succ\infty$, algebras recently pointed out in literature.

We shall treat the generalized Euler equations of the form:

$$\dot{\omega}_i = a^{lm} C^k_{im} \omega_1 \omega_k \tag{1}$$

where ω_i is a set of dynamical variables taking their values in a dual space g^* of some Lie algebra g, described in a suitable basis by its structure constants $C^k_{im} = -C^k_{im}$, and a^{lm} (metric) is a symmetric tensor defining the energy $E = 1/2a^{kl}\omega_k\omega_l$ (summation over repeated indices is assumed). The most familiar equations of this type are Euler ones for the rigid body rotations, where $g = so(3), \omega_i$ are identified with angular momenta, $a^{lm} = diag(I_1^{-1}, I_2^{-1}, I_3^{-1}), I_i$ are moments of inertia. Up to now an algebraic treatment of systems of the form of eq. (2) proceeded far - a generalized rigid body motion on various Lie algebras was studied and the entire classes of integrable dynamical systems were found this way (for a review see e.g. [1]). On the other hand, it is known from the pioneering work of Arnold [2] that the ideal fluid motion may be described in the framework of the infinite-dimensional generalization of eg. (1) (similar ideas were developed independently in [3]). We shall reformulate some results of the paper [2] in a somewhat different (and more appropriate for the following) form.

Consider the (complexified) Lie algebra of area-preserving diffeomorphisms of the torus $sdiffT^2$ ("torus" means nothing but periodic boundary conditions in a rectangular domain), that do not shift the torus surface as a whole. In the natural basis of Fourier-modes the generators of such diffeomorphisms take the form:

$$L_{n_1 n_2} = ie^{i(n_1 x_1 + n_2 x_2)}(n_2\partial_1 - n_1\partial_2) \equiv ie^{i\bar{n}\bar{x}}\bar{n} * \bar{\partial} \tag{2}$$

Here $n_{1,2}$ are integers $\neq 0$, $x_{1,2}$ are dimensionless coordinates that absorb characteristic lengths in both spatial directions on the torus. The Lie algebraic structure of $sdiffT^2$ is given by

$$[L_{\bar{n}}, L_{\bar{m}}] = (\bar{n} * \bar{m})L_{\bar{n}+\bar{m}} \tag{3}$$

Hence the structure constants of this algebra are

Advances in Turbulence 3
Editors: A.V. Johansson · P.H. Alfredsson

$$C_{\bar{n}\bar{m}}^{\bar{l}} = (\bar{n} * \bar{m})\partial_{\bar{n}+\bar{m}}^{\bar{l}} \tag{4}$$

Choosing

$$a^{\bar{k}\bar{l}} = \partial^{\bar{k},-\bar{l}}\bar{k}^{-2} \tag{5}$$

it is easy to see that eq. (1) becomes

$$\dot{\omega}_{\bar{m}} = \sum_{\bar{k}} \bar{k}^{-2}(\bar{m} * \bar{k})\omega_{\bar{m}+\bar{k}}\omega_{-\bar{k}} \tag{6}$$

which, evidently, is the Fourier-transform of the vorticity equation (to be precise, with zero mean vorticity that in the space of Fourier- modes means $\omega_O = O$, otherwise the formalL_O generator is to be added, commuting with all of $L_{\bar{n}}$ and corresponding to ω_O) in 2-D hydrodynamics

$$\dot{\omega} = \bar{\partial}(\Delta^{-1}\omega) * \bar{\partial}\omega \tag{7}$$

Comparison of eq. (7) with the canonical example of rigid body motion shows that Fourier-transforms of vorticity ω, stream-function $\Delta^{-1}\omega$ and Laplacian Δ play the roles of angular momenta, velocities and moments of inertia, respectively cf.[2]. Now, it is known that eq. (7) possesses an infinite number of conservation laws - any power of vorticity integrated over the torus is conserved. In terms of Fourier- modes these conserved quantities look as follows:

$$\Omega^{(N)} = \sum_{\bar{n}_i} \omega_{\bar{n}_1} \ldots \omega_{\bar{n}_N} \qquad \sum_{i}^{N} \bar{n}_i = O, N = 1 \ldots \infty \tag{8}$$

It is evident that their conservation is insensitive to the choice of $a^{\bar{k}\bar{l}}$(Δ^{-1} might be substituted by any integral operator) and is due to the specific form of the structure constants, eq. (4), i.e. the different $\Omega^(N)$ belong to the algebra of Casimir functions of $sdiffT^2$.

The inspection of the above algebraic construction for 2-D hydrodynamics leads to the conclusion that it may be immediately generalized by means of the central extension of the Lie algebra, eq. (3). The fact that this latter has a non-trivial (in a sense that it cannot be absorbed by a shift of generators) central extension has been intensively exploited recently in physical literature, but in a rather different context (cf. e.g. [4]). Indeed, it is easy to check that an identity generator may be introduced in the r.h.s. of eq. (3) without destroying the Jacobi identity, with a constant - 2-cocycle - fixed up to the arbitrary 2- vector $\bar{a}$:

$$[L_{\bar{n}}, L_{\bar{m}}] = (\bar{n} * \bar{m}) + L_{\bar{n}+\bar{m}} + \bar{a}\bar{m} + \partial_{\bar{n},-\bar{m}}E, \qquad [E, L_{\bar{n}} = O] \tag{9}$$

Adding a new generator means adding a new ω, say $\bar{\omega}$, in the context of eq. (1). However, it is easy to see from eq. (4) that $\dot{\bar{\omega}} = O$, and hence it remains constant leading to the appearance of the linear in the rest of ω terms in eq. (1). Repeating the same procedure as above for the case of the modified algebra, eg. (9), with the same $a^{\bar{k}\bar{l}}$ and using the freedom in the choice of constants, it is possible to get the Fourier-transform.

$$\dot{\omega}_{\bar{m}} = \sum_{\bar{k}} \bar{k}^{-2}[(\bar{m} * \bar{k})\omega_{\bar{m}+\bar{k}}\omega_{-\bar{k}} + i\beta k_1 \omega_{\bar{k}} \partial_{\bar{k},\bar{m}}] \tag{6'}$$

of the vorticity equation on the β - plane

$$\dot{\omega} = \bar{\partial}(\Delta^{-1}\omega) * \bar{\partial}\omega - \beta\partial_x(\Delta^{-1}\omega) \tag{7'}$$

that takes into account effects of the differential rotation of the system.

Eq (6') may be further modified by the choice of the screened inverse Laplacian: $a^{\bar{k}\bar{l}} = \partial^{\bar{k},-\bar{l}}(\bar{k}^2 + c^2)^{-1}1$ as metric instead of eq. (5), which leads to one more known equation, playing together with the previous one an important role in geo- and planetary physics (cf. e.g. [5]:

$$\dot{\omega} = \bar{\partial}[(\Delta - c^2)^{-1}\omega] * \bar{\partial}\omega - \beta\partial_x[(\Delta - c^2)^{-1}\omega]. \qquad (7'')$$

As the original eq. (7) both eqs (7', 7") have infinite number of conservation laws - the integrated powers of $[\omega(\bar{x}) - \beta x_2]$ in coordinate representation.

Thus, starting from the algebraization of 2-D ideal hydrodynamics it is possible to get an algebraic picture for the related equations on the β-plane, which enables one to use algebraic and geometric methods for their investigation (this is in progress now).

The next observation basing on the original construction concerns the long-standing problem (for a comprehensive discussion cf. [6]) of finding adequate finite-mode approximations to hydrodynamical equations - eqs (7, 7', 7") in our case. Some comments are in order in this connection.

Putting aside a general question, in what circumstances such approximations are justified, it seems reasonable to require, in any case, that they should inherit an algebraic structure of eq. (1) which, in particular, provides additional integrals of the type of eq. (8). In the absence of these latter the (non-dissipative) finite-mode system can hardly be identified as "hydrodynamical" - to deserve this name it is to remember of such typically hydrodynamical things as conservation of powers of vorticity in 2-D.

Thus, a problem arises how to find a finite-dimensional Lie algebra that can be connected in a regular way to the Lie algebra, eq. (3), the latter being in some sense a large N limit of it (such a limit in a rigorous sense has a number of subtleties that are beyond the scope of the present semi-heuristic discussion). Then, using this algebra with the same metric as before, the desired finite-mode system may be achieved. It would be, evidently, the particular case of the most general construction developed in [3],[6]. Fortunately, for $sdiffT^2$ a procedure of such finite-dimensional reduction exists. It is based on the recently introduced lattice algebras [7],[8]. Namely, let's make a following trick: replace (identically) the factor $\bar{n} * \bar{m}$ in the structure constant, eq. (4), by $\lim_{N\to\infty} N/2\pi \sin(2\pi\bar{n} * \bar{m}/N)$ delete lim, absorb $N/2\pi$ into normalization of $L_{\bar{n}}$ and consider the resulting commutation relations. Fortunately enough, they form another infinite-dimensional - lattice - algebra, that has a finite-dimensional ideal which is obtained when modulo N arithmetics for lattice vectors $\bar{n}, \bar{m}$ is used in eq. (3) instead of the ordinary one. This means that a fundamental NxN cell is to be chosen on the infinite 2-D lattice where vector indices take their values and the lattice is to be factorized over this cell (all the points separated by nN,n - integer, in both directions are to be identified). If the reality condition $\omega_{\bar{n}} = \omega^*_{-\bar{n}}$ is to be further imposed, then it is convenient to place the fundamental cell around the (dropped out) origin. The resulting $N^2 - 1$ algebra

$$[L_{\bar{n}}, L_{\bar{m}}] = \sin(2\pi\bar{n} * \bar{m}/N)L_{\bar{n}+\bar{m}} \qquad (10)$$

is especially simple in the case of odd N, when it coincides with the sl(N)- (or su(N)- when reality condition is imposed) algebra in a special basis, which in a matrix realization generalizes Pauli matrices to arbitrary N [9]. In the hydrodynamical context it is important that in this basis N Casimir operators of sl(N)- algebra are strings of $L'_{\bar{n}}s : \sum L_{n_1}...L_{n_i}, \sum_i^K n_i = 0 \bmod N, K = 1...N$, thus giving for the corresponding $N^2 - 1$ -mode dynamical system, eq. (1), N conserved quantities which in the limit

$N- \succ \infty$ reproduce the first N integrated powers of vorticity. Hence, to obtain a reasonable (in above formulated terms) finite-mode hydrodynamical-type system, it is necessary not only to cut down the Fourier decomposition of vorticity, but also to modify dynamical equations in a proper way (ones related to eq. (3) to ones related to eq. (10)).

The situation is slightly more complicated in the case of even N (say, $N = 2M$) when there are three generators besides L_O commuting with all of $L\bar{n}'s$. It is evident from eq. (10) which in this case takes the form

$$[L_{\bar{n}}, L_{\bar{m}}] = \sin(\pi \bar{n} * \bar{m}/M) L_{\bar{n}+\bar{m}} \tag{10'}$$

that those (trapped into the fundamental cell which now spreads from $-M$ to $M-1$ in both directions) are $L_{-M,O}, L_{O,-M}, L_{-M,-M}$.The remaining $4(M^2-1)$ generators in the fundamental cell form four copies of the Lie algebra sl(M). Since the algebra eq. (10') is not semi-simple, even without $L_{\bar{O}}$ - generator, contrary to the eq. (10) one, it admits non-trivial central extensions. However, it is not clear whether it is possible to relate them somehow with eq. (9) and to use the corresponding dynamical systems as a finite-mode approximations to the β - plane equations. Here we encounter one of the above-mentioned subtleties of the large-N limit. The algebra of eq. (10') itself may be used for ordinary hydrodynamics as well as eq (10), the latter being more economical (the minimal model constructed with the help of eq. (10') has 12 modes - 4 hydrodynamical triplets in terminology of ref. [6]. The systems based on eqs [10'] possess additional conservation laws as well.

Thus, the class of finite-mode systems is introduced that is intrinsically related to 2-D hydrodynamics and hence deserves a further study. The method used in course of their derivation might be a useful supplement to the traditional Galerkin approach. It is worth noting, however, that it has no straightforward generalization to the 3-D case due to the different structure of the volume- preserving diffeomorphisms algebra. But for the 2-D case the above scheme seems to be quite general although it was demonstrated here only for the technically simplest example of the torus. For another geometry (i.e. another boundary conditions) an appropriate basis for the sdiff algebra is to be chosen instead of eq. (2) and all the procedure is to be repeated (in this connection cf. refs [4,10] for realizations of sdiff on the sphere and its relation to su(N)). Note, that not for all geometries the central extension is possible.

References

1. V.V. Trofimov, A.T. Fomenko Russian Math. Surveys, 39 (1984) 3
2. V. Arnold Ann. Inst. Fourier, 16 (1966) 319
3. A.M. Obukhov Sov. Phys. Doklady, 184 (1969) 309
4. J. Hoppe Phys. Lett., 215B (1988) 706
5. J. Pedlosky "Geophysical Fluid Dynamics", Springer, 1978
6. E.B. Gledzer, F.V. Dolzhansky, A.M. Obukhov "The hydrodynamical-type systems and their applications", Nauka, 1981 (in Russian, in English the essential ideas of this book are presented e.g. in A.M. Obukhov, F.V. Dolzhansky Geoph. Fluid Dyn., 6 (1975) 195)
7. D.B. Fairlie, P. Fletcher, C.K. Zachos Phys. Lett., 218B (1988) 203
8. D.B. Fairlie, C.K. Zachos Phys. Lett., 224B (1989) 101
9. J. Patera, H. Zassenhaus Journ. Math. Phys., 29 (1989) 665
10. E.G. Floratos, J. Iliopoulos, Phys. Lett., 217B (1989) 285 G. Tictopoulos

Part III

Turbulent Combustion and Mixing

Reactive Mixing Layers and Turbulent Combustion

J.P. Chollet, R.J. Gathmann and M.R. Vallcorba
Institut de Mécanique de Grenoble, B.P. 53X, 38041 Grenoble cédex France

Flow composition of a fluid made of reactive species may be determined by a competition between chemical reactions and turbulent mixing. At first, the constant density approximation is used to examine the way turbulent coherent structures of a three dimensional mixing layer change the global evolution of a chemical reaction. Afterwards, a direct numerical simulation is developed to take account of all the interactions between flow dynamics, thermodynamics and chemistry. The detonation has been selected among combustion evolutions we have studied with this simulation.

Introduction

Dynamics, thermodynamics and chemistry interact in turbulent combustion problems. Direct numerical simulations provide modellers with powerful tools to study time and space evolutions for various variables. A complete simulation of turbulent combustion is much beyond the capabilities of existing numerical methods and computers. Additional fields must be associated to chemical species and large Reynolds numbers need high spatial resolution. The accurate description of fluid dynamics we want to develop means a drastic modelling of thermodynamics and chemistry kinetics. Specific characters of reacting flows are : *(i)* additional terms with typical time τ_c coming from chemistry and acting as sources of fluctuations *(ii)* strong interactions between dynamics, thermodynamics and chemistry

Section 1 starts with constant density mixing layers where coherent structures develop. In these well documented incompressible turbulent flows eddy structures are observed to have drastic effects on the evolution of fluid composition. Section 2 is devoted to the development of a numerical code to fully simulate reactive flows with heat release, density and compressibility effects. This code can be used for a variety of evolutions. Combustion with detonation illustrates the ability of the method to compute continuous evolutions as accurately as steep gradients.

1. Reactive incompressible mixing layer

When density ρ is constant, reacting flows remain close to classical fluid dynamics. The mixing layer is a flow prototype of particular interest because of turbulence which fully develops from initial instability mechanisms. Coherent eddy structures control most of large scale species mixing as well as smaller scale phenomena through scales interactions. Each stream of the mixing layer carries either species A or species B. Reacting species A and B undergo a chemical reaction which generates a product P. The reaction is assumed to be one-step, binary and irreversible : $\nu_A A + \nu_B B \rightarrow \nu_P P$

Advances in Turbulence 3
Editors: A.V. Johansson · P.H. Alfredsson

Chemical reactions in a mixing layer simulation need additional scalar fields. Mass fractions $Y_i(\vec{x},t)$ or concentrations $C_i(\vec{x},t)$ are associated to species i. Velocity and scalars evolve according to :

$$\frac{\partial \mathbf{U}(\vec{x},t)}{\partial t} + \vec{u}\ \nabla\ \mathbf{U}(\vec{x},t) = \mathcal{D}\nabla^2\ \mathbf{U}(\vec{x},t) + \mathbf{S}(\vec{x},t) \tag{1}$$

with

$$\mathbf{U} = \begin{Bmatrix} \vec{\omega} \\ C_A/\nu_A \\ C_B/\nu_B \\ C_P/\nu_P \end{Bmatrix} \quad \mathbf{S} = \begin{Bmatrix} \vec{\omega}\nabla\vec{u} \\ -\dot{w} \\ -\dot{w} \\ \dot{w} \end{Bmatrix} \tag{2}$$

Just like reacting species, vorticity $\vec{\omega}(\vec{x},t) = \nabla \otimes \vec{u}(\vec{x},t)$ is submitted to a source S_ω which is zero in two dimensions. (1) is written for $\mathcal{D} = \nu = D_A = D_B = D_P$, with molecular viscosity ν and molecular species diffusivities D_i.

The reaction rate $\dot{w}(\vec{x},t)$ which is used in this section is assumed to be isothermal although some of our computations (Chollet, Vallcorba & Gathmann, 1989) were carried out with a temperature dependent rate (as in section 2): $\dot{w} = C_A^{\nu_A}\ C_B^{\nu_B}/\tau_c$, where time τ_c characterizes the chemical reaction. The total amount of moles is conserved through reaction ($\nu_A + \nu_B = \nu_P$) in order to have a realistic model for constant density flows. Alternate formulations of (1) are used with passive conserved inert scalar $\Phi = C_A/\nu_A - C_B/\nu_B$ and the product concentration C_P as a tracer of reaction progress.

Spatial calculations with free outflow at downstream boundary (Chollet & al., 1989) are close to real flows and yield reliable statistics such as layer thickening $\delta(x)$. Nevertheless, spatial simulations need so many computation nodes in streamwise direction that they must often be restricted to two dimension approximations which do not correctly model real small scale behaviour. Three dimensional simulations must be used to look at details of interactions between eddies and species mixing. Accordingly, streamwise development of the domain must be be reduced. Temporal approximation may be useful: the layer is observed while following the streamwise direction at constant speed. Boundary conditions are periodic in streamwise direction ($x = 0$, L_x), periodic in spanwise direction and free slip in transverse direction ($y = \pm L_y/2$). A pseudo spectral code, (Picart, Chollet & Borghi, 1989), (Comte, Fouillet, Gonze, Lesieur, Métais & Normand, 1990), is used to get accurate spatial derivatives. Results of figure 1 and 2 were obtained with $64 \times 32 \times 32$ computational nodes and an initial Reynolds number $R_i = \frac{\delta_i U}{\nu} = 150$. Time unit is $\tau_{2D_i} = \delta_i/U$, where δ_i is the initial thickness of the layer and U the velocity difference between the two streams. The Damkhöler number $Da = \tau_{2D_i}/\tau_c$ is used to measure the initial ratio of turbulent time to chemical time.

Three dimensional graphics are helpful for the analysis of evolutions of reaction rate $\dot{w}(\vec{x},t)$ in space and time. $\dot{w}$ cannot be measured directly from experiments although it seems to be especially relevant for observation. Figure 1 shows an isosurface of the reaction rate when streamwise coherent structures begin to have a significant effect on space distribution of reacting activity. Similar pictures are also used to measure the distribution of the reaction rate $\dot{w}$ on the surface $\Phi = 0$ for various Damkhöler numbers. This surface is the infinitely thin reactive sheet in the approximation of the infinitely fast reaction (Williams, 1985). Various statistics have been calculated : mean values,

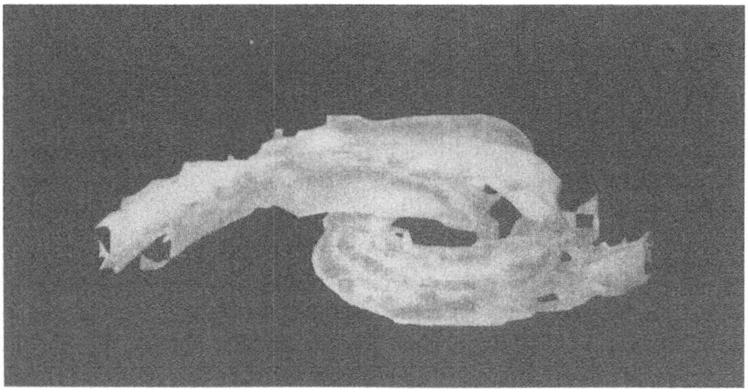

Figure 1. Three dimensional simulation of a constant density temporal mixing layer : an isosurface of the reaction rate $\dot{w}$

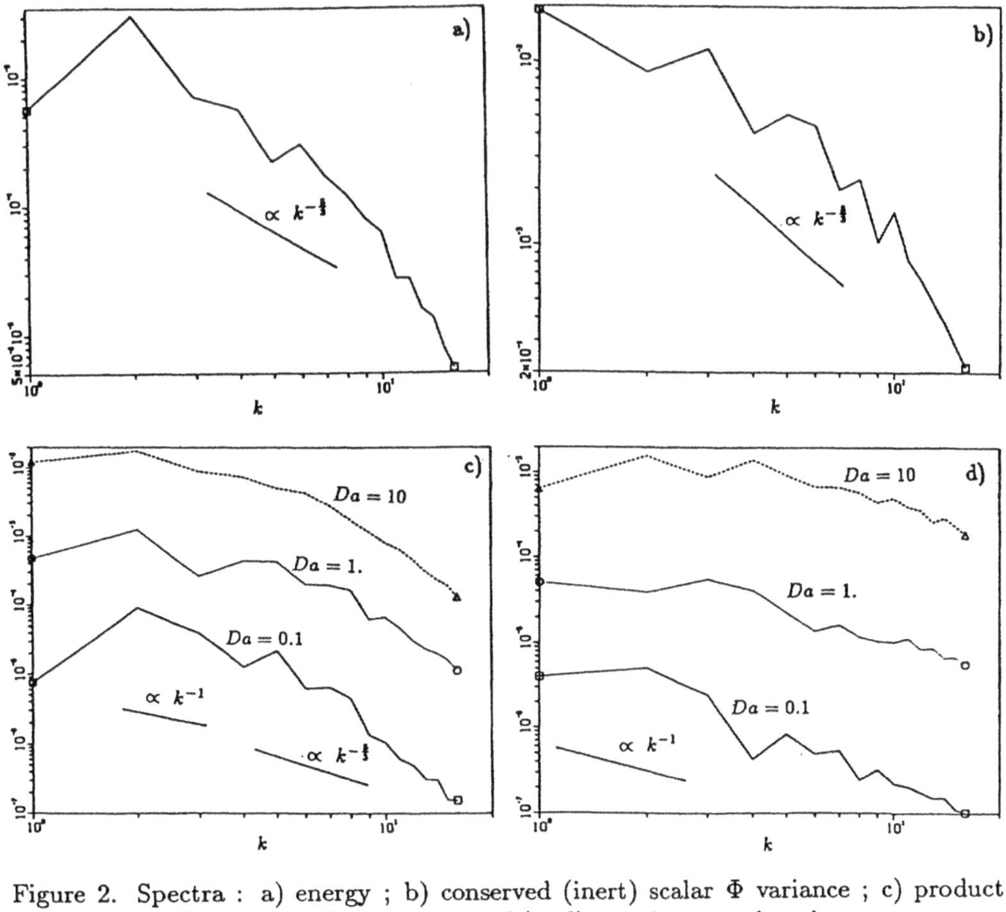

Figure 2. Spectra : a) energy ; b) conserved (inert) scalar Φ variance ; c) product concentration C_P variance (progress variable) ; d) reaction rate $\dot{w}$ variance

p.d.f., spectra. Typical spectra, (i.e. at time t_i of the evolution) are plotted in figure 2, for Damkhöler numbers ranging from 0.1 to 10. Energy and conserved scalar Φ exhibit the usual behaviour with almost $k^{-5/3}$ distributions despite the rather low Reynolds number. The fluctuations of product concentration C_P and, even more, the fluctuation of the reaction rate $\dot{w}$ exhibit a noticeable dependence with Da. The reaction appears to be a non local process in spectral $\vec{k}$ space, which is a counterpart of its localness in physical $\vec{x}$ space since reaction rates $\dot{w}$ do not depend on gradients.

Although they oversimplify the chemistry, constant density flows lead to sophisticated analysis of the competition between mixing and scalar production (or consumption). The velocity field does not depend on chemical reactions since all the scalars are treated as passive. The time step $\Delta t_{\dot{w}}$ used to compute species concentrations C_i can be different (smaller when attempting to increase Da values) from Δt_u which is used for the computation of the velocity field.

2. Full simulation, detonation

Heat release from chemical reactions must be considered in a full simulation. Low compressibility approximations are suitable for problems with heat release provided pressure wave effects are negligible (McMurtry, Riley & Metcalfe, 1989). The simulation presented below takes a complete account of fluid compressibility effects. With D dimensions in space and N species, the vector $\mathbf{U}(\vec{x},t)$ of $(D+N+2)$ unknown scalar quantities

$$\mathbf{U}(\vec{x},t) = (\rho,\ \rho u_x,\ \rho u_y,\ \rho u_z,\ \rho E,\ \rho Y_i,\ \rho Y_{i+1}, ...) \tag{3}$$

with : $E = \frac{1}{2}(u_x^2 + u_y^2 + u_z^2) + \frac{\bar{c}_v}{M}$,
evolves according to :

$$\frac{\partial \mathbf{U}}{\partial t} + \frac{\partial \mathbf{F}(\mathbf{U})}{\partial x} + \frac{\partial \mathbf{G}(\mathbf{U})}{\partial y} + \frac{\partial \mathbf{H}(\mathbf{U})}{\partial z} = \mathbf{D} + \mathbf{S} \tag{4}$$

with $\mathbf{S}$: source induced by chemical reactions and $\mathbf{D}$: diffusive (and viscous) terms computed explicitly.
The operator $\mathbf{F}$, and similarly $\mathbf{G}$ and $\mathbf{H}$, is :

$$\begin{aligned}\mathbf{F}(\mathbf{U}) = (&\rho u_x,\ \rho u_x^2 + p,\ \rho u_x u_y,\ \rho u_x u_z,\\ &(\rho E + p)u_x,\ \rho Y_i u_x,\ \rho Y_{i+1} u_x,\ ...)\end{aligned} \tag{5}$$

Chemistry is introduced under Arrhenius formulation :

$$S \propto \dot{w} = B Y_A^{\nu_A} Y_B^{\nu_B} T^{\alpha} \exp(-T_A/T) \tag{6}$$

Numerical methods have to be more specific than in incompressible flows because of fast variations: shocks and sound waves, steep temperature and mass fraction gradients, density discontinuities. Also the code we developed focuses on : *(1)* hyperbolic operators for accurate wave propagation and transport : u, p and ρ evolutions are associated to characteristics $u \pm c$ and ρ, when T and the various Y_i are associated to the characteristics u (c is the sound speed); *(2)* conservation of all the variables through the numerical scheme, especially Y_i or C_i which trace the composition of the fluid ; *(3)* a treatment of steep gradients and discontinuities due to strong heat release from chemical reactions; *(4)* modularity of the code in order to handle either binary or several steps reactions; *(5)* an explicit account of viscosity and diffusivities whose effect can be significantly enhanced by chemical reaction (e.g. in deflagration reaction, Williams, 1985)

Therefore, the numerical code is based on a splitting between an hyperbolic operator L_i and a chemical-diffusive operator Ψ; when going from time step $n\Delta t$ to time step $(n+2)\Delta t$:

$$\mathbf{U}^{(n+2)} = \Psi^{\frac{1}{2}} L_x L_y L_z \Psi L_z L_y L_x \Psi^{\frac{1}{2}} \mathbf{U}^{(n)} \tag{7}$$

L_i is a Riemann solver which is combined with detection and treatment of density discontinuities, through a PPM (piecewise parabolic method) approximation and the use of dependence domains associated to the three characteristic directions (Collela & Woodward, 1984). The PPM method by requiring monotonous parabolic appoximations avoids dispersivity and keeps mass fractions Y_i inside the range $[0,1]$. Ψ computes the chemical sources (for heat Q, species mass fraction Y_i) and the viscous-diffusive terms. These diffusive terms can peak at high values in the neighbourhood of steep temperature gradients. An optimal value of the time step $\Delta t(t)$ is computed at each $n\Delta t$ from the smaller of the times which characterize wave propagation, advection, chemical sources and diffusion

Detonation is selected as a test case with huge pressure gradients and shear zones as a secondary phenomenon. This case (Schoffel & Ebert, 1989) of practical interest uses free slip rigid boundaries which reflect pressure (Greenough, Riley, Soesstrino & Eberhardt, 1989). The two step reaction involves a premixed fuel A. Step 1 transforms species A into species B through an induction reaction $A \rightarrow B$ which lets the reaction start depending on the high activation temperature T_A. Step 2 is a reaction $B \longrightarrow C + Q$, which yields burnt gas C and strong heat release Q. Parameters have to be added to the usual ones already considered in section 1. Simulation of figure 3 uses: $T_A^{(1)} = 30T_0$, $B^{(1)} = 1400$ for step 1, and $T_A^{(2)} = 5T_0$, $B^{(2)} = 1000$, $Q = 15RT_0$ for step 2, with 50×100 computational nodes. A strong pressure discontinuity is used as the initial condition. Flow is uniform at the upstream boundary, with premixed species (A). The downstream boundary models free outflow. The characteristics formulations which is used in the algorithm makes the calculation of free outflow quite accurate. Boundary condition (along $y = \pm L_y/2$) reflects pressure, although being free slip. Pressure fields are observed at two time-steps on figure 3 with peak values at triple points. Strongly non linear transversal cellular instabilities are observed when time evolves. In the high pressure region, density dicontinuities (slipstream lines) develop from triple points (Williams, 1985) acting as vorticity sources.

Conclusions, large eddy simulations and subgrid models

Specific algorithms were developed in section 2 in order to deal with a full simulation of compressible reactive flows. Non compressible turbulence of section 1 remains useful when studying interactions between chemistry and large scale mixing. Reactive flow

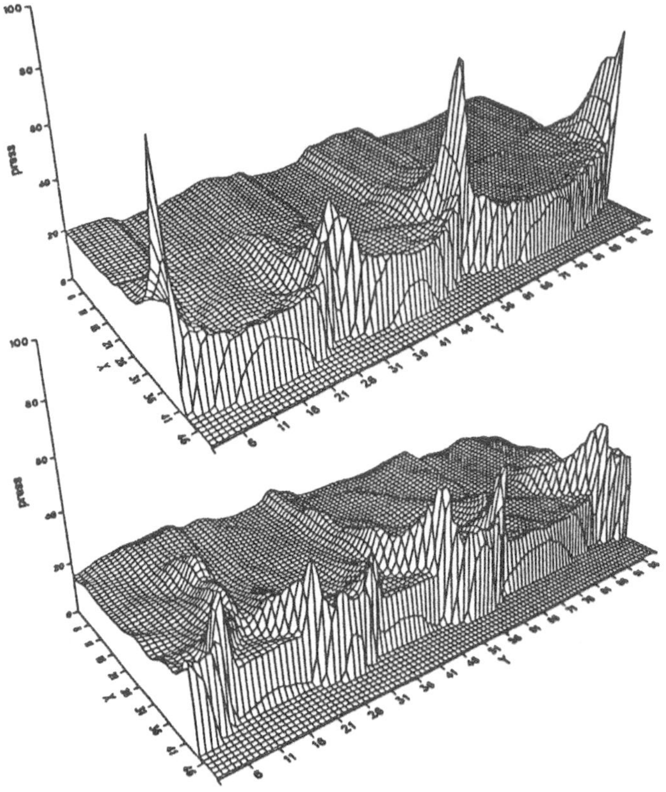

Figure 3. Evolution of cellular detonation from a two dimensional direct simulation of fully compressible premixed flows: pressure field at two times of the evolution

simulations without heat release can describe many laboratory experiments (e.g. Masutani & Bowman, 1986) and some problems of practical interest, for instance in the environmental field.

Numerical simulations are restricted to a relatively narrow range of time and space scales. They can be viewed as direct as far as turbulent scales are concerned. Large eddy simulations allow to increase the Reynolds number but require to solve two separate problems which involve scales smaller than the computational space grid Δx : *(i)* steep gradients modelling in narrow regions, *(ii)* small scales modelling for $\vec{u}$, T, Y_i fields everywhere else taking account of non linear reaction rate $\dot{w}$ (6).

Steep gradients may develop in reactive flows with shock waves (involving p, ρ) or flame sheets (involving Y_i, T). Flame sheets obtained from our direct simulations are rather thick. In order to deal with thin flames, space mesh should be refined drastically in locations where steep gradients develop. The grid $\vec{\Delta}(\vec{x}, t)$ has then to be redefined at each time step and in each space location. These adaptative grids become feasible for a two dimensional computation. Nevertheless two dimensional approximations may be far from real three dimensional flows, the vorticity source in (1) remains zero and no vorticity stretching is produced even with high resolutions. Furthermore two dimensional diffusive terms whose importance can be enhanced by a chemical reaction are not built on fluid molecular properties.

Subgrid modelling seems to be a more classical problem when dealing with properties more uniformly distributed in space when gradients are calculated on several computation nodes. Chemical reaction rates $\dot{w}(\vec{x},t)$ are more intermittent in space and time than any other quantity (C_i, Φ). Spectral eddy viscosity and diffusivity (Chollet, 1985), (Lesieur, 1987) help to understand the meaning of subgrid models, although they were originally designed for spectral codes. The underlying hypotheses are : *(i)* inertial energy and scalar variance spectra in unresolved subgrid scales ($< \Delta(\vec{x})$); molecular viscosity (and diffusivities) tends to 0, but remain always positive, *(2)* inertial spectra exist on both sides of the cut off scale in order to be able to compute transfers from large to subgrid scales. Simulations can be combined with statistical models (Chollet, 1985). The explicit computation of large scales uses a subgrid transfer term which is computed at each time step depending on the evolution of the subgrid scales energy spectra computed from an EDQNM equation (Lesieur, 1987). A combination of explicit large eddy simulations with statistical modelling of non stationary small scales need to solve additional problems when dealing with reactive shear flows. Non homogeneous and nonisotropic flows should depend on transfers from small to large scales which cannot be really represented in eddy viscosity formulations. In any case, nonhomogeneous flows have to be treated through non uniform space grids $\Delta\vec{x}(\vec{x},t)$ and subgrid models need to be derived in this frame. Several scalar fields must be computed when simulating reactive flows. In the constant density case (section 1) subgrid field of Φ can be modeled as a passive conserved scalar. C_P needs a specific model because of the contribution of subgrid scales to the reaction rate. If velocity subgrid models could be better expressed in terms of vorticity because of the vorticity stretching in three dimensions, scalar subgrid models are probably more dependent on strain tensor.

For low values of the Mach number, characteristic times for wave propagation are much larger than numerical time step, in such a way that compressible effects associated to wave propagation need not to be considered at a subgrid level. Nevertheless, because of heat sources scattered by chemical reactions, density varies at subgrid level, even without significant sound waves effects. Evolutions of temperature T and density are interrelated ($\rho \propto 1/T$ in perfect gas) and therefore a specific subgrid model should be developed from low Mach number equations.

References

Chollet, J.P. 1985 Spectral closures to derive a subgrid scale modelling for large eddy simulations, *lecture notes in Physics*, **230**, Springer-Verlag, 161.

Chollet, J.P., Vallcorba, M.R. & Gathmann, R.J., 1989 Direct numerical simulations of turbulent combustion in mixing layers, *Turbulence and Coherent Structures* (ed. M. Lesieur & O. Métais), Kluwer Acad.pub.

Collela, P. & Woodward, P.R. 1984 The piecewise method (PPM) for gas dynamical simulations, *J. Comput. Phys.*, **54**, 174.

Comte, P., Fouillet, Y., Gonze, M.A., Lesieur, M., Metais, O. & Normand, X. 1989 Generation of coherent structures in free shear layers, *Turbulence and Coherent Structures* (ed. M. Lesieur & O. Métais), Kluwer Acad.pub.

Greenough, J.H., Riley, J.J., Soestrino, M. & Eberhardt, D.S. 1989 The effect of walls on a compressible mixing layer, *AIAA-89*-**0372**.

Lesieur, M., 1987, *Turbulence in Fluids*, Kluwer.

Masutani, S.M., Bowman, C.T. 1986 The structure of a chemically reacting plane mixing layer, *J. Fluid Mech.*, **172**, 93.

McMurtry, P.A., Riley, J.J. & Metcalfe, R.W. 1989 Effects of heat release on the large-scale structure in turbulent mixing layers, *J.Fluid. Mech.*, **199**, 297.

Picart, A., Chollet, J.P.& Borghi, R. 1989 Turbulent diffusion from thermal or reactive sources: numerical experiment, *Advances in Turbulence 2*, Springer-Verlag, 192.

Schoffel,S.U. & Ebert, F. 1989 Numerical analysis concerning the spatial dynamics of an initial gaseous ZDN detonation, *in Dynamic of explosions, ed. by Kuhl, Bowen, Leyer and Borisov*, **114**, *Progress in Astronautics and Aeronautics AIAA*.

Williams, F.A. 1985 Combustion Theory, Benjamin Cummings Pub.

Concentration Statistics in Shear-free Grid-Generated Turbulence with a Second-Order Rapid Reaction

Satoru Komori, Takao Kanzaki and Yasuhiro Murakâmi

Department of Chemical Engineering, Kyushu University Hakozaki, Fukuoka 812, Japan

Effects of the turbulent mixing on a rapid second-order chemical reaction between the acetic acid and ammonium hydroxide (acid/base reaction) are investigated in shear-free grid- generated turbulence. Simultaneous measurements of the instantaneous concentrations of two reacting species are carried out using a combined laser-induced fluorescence and electrode-conductivity technique with the spatial and time resolutions smaller than or comparable to the Kolmogorov micro-scales. The concentration statistics; mean concentrations, concentration intensities, concentration correlation and segregation parameter, are estimated from the simultaneous measurements. The results show that the reaction progresses in the initial mixing region and the chemical products prevent the further progress of the reaction in the region far from the grids. The concentration statistics also show that the similarity for the concentration field collapses in a reacting flow and the correlation of the concentration fluctuations of two reacting species is weaken. In addition to the measurements, the concentration statistics are calculated using the two- particles Lagrangian stochastic model and the predictions are in qualitative agreement with the measurements.

1. Introduction

The diffusive-reactive mechanism between two nonpremixed reactants released into turbulent flows is of great practical interest in the areas of both engineering and environmental science. Such reacting turbulence can be seen in many industrial reactors and combustors, and it is deeply related to environmental problems such as the chemical transformations of exhaust gases from automobiles and power plants and the formation of acid rain.

One of the most important subjects for such a nonpremixed reacting flow is to investigate the interaction between the turbulent and molecular mixing and the chemical reaction, and a number of experimental studies (Komori & Ueda 1984, Mudford & Bilger 1984, Bennani et al. 1985, Saetran et al. 1989 and others) have showed the significant effects of the mixing on the chemical reaction rate in nonpremixed reacting flows with moderately or slow second-order reactions. Recently, Komori et al. (1991a) have reviewed the measurements and explained the effects of the turbulent mixing on the mean reaction rate by developing a Lagrangian stochastic model. However, these studies have not been enough to discuss a reacting flow with a very fast reaction, such as a combustion flame. The rapid reaction is a limiting case with an infinite reaction rate constant, and there the diffusion perfectly limits the reaction. Such a reacting flow is one of the main subjects of theoretical studies and a number of models and numerical computations have been developed. However, these theoretical studies have not been justified by experiments, since full measurements of concentration statistics have not

Advances in Turbulence 3
Editors: A.V. Johansson · P.H. Alfredsson

been given. Much depend on the fact that simultaneously measuring the instantaneous concentrations of two reacting species is a very difficult task.

The purpose of this paper is, therefore, to present the concentration statistics measured in a reacting turbulent flow with a rapid reaction. The experiments were carried out in shear-free grid-generated turbulence with an acid/base reaction and instantaneous concentrations of two reactants were measured by means of a combined laser-induced fluorescence and electrode-conductivity technique developed by Komori et al. (1991b). In addition, the measurements of the concentration statistics were compared with the predictions by the Lagrangian stochastic model of Komori et al. (1991a).

2. Experiments

The rapid, irreversible, second-order chemical reaction between the acetic acid (species A) and ammonium hydroxide (species B) with the same initial concentration of 10 mol/m^3 was used here;

$$CH_3COOH + NH_4OH \rightarrow CH_3COONH_4 + H_2O \quad \text{(acid/base reaction)} \qquad (1)$$

Figure 1 shows the experimental apparatus and measuring system. The test apparatus used was a water tunnel made of polymethylemethacrylate and it had the length of 1m and the cross section of 0.1m × 0.1m. The species A and B were homogeneously premixed with the fluorescence dye (uranin), respectively, and they were provided into two layers separated by a splitter plate. Turbulence-generation grids were installed at

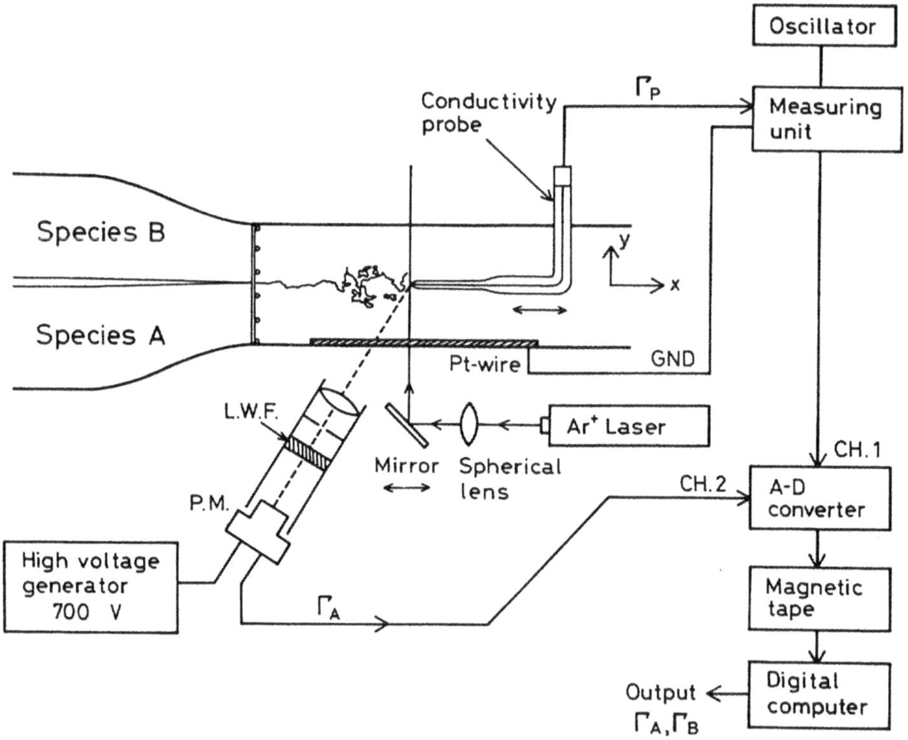

Figure 1. Measuring system and test apparatus.

the entrance of the test section, and they were of round-rod, square-mesh, single-biplane construction. The mesh size M and rod diameter d were 0.02 m and 0.003 m, respectively. Mean velocities of two streams of the species A and B were here set to the same value of 0.25 m/sec, so that the Reynolds number Re_M based on the mesh size was 5000. The velocity field is not described here, but it was investigated in detail by using a two- components laser-Doppler velocimeter.

To simultaneously measure the concentrations of two reactants in (1), we used the following technique. The instantaneous concentration of NH_4OH cannot be directly measured, and therefore the mass conservation equation was used here;

$$\Gamma_A + \Gamma_B + \Gamma_P = 1.0$$

(2) where $\Gamma_i(= C_i/C_{i0})$ is the i -species concentration normalized by the initial concentration of each species. Here it should be noted that (2) is valid only for the same initial concentration (i.e. $C_{A0} = C_{B0}$). If the two concentrations of CH_3COOH (species A) and chemical products of $CH_3COONH_4 + H_2O$ (species P), Γ_A and Γ_P , can be simultaneously measured, the instantaneous concentration of NH_4OH, Γ_B can be determined from (2). Therefore, a technique for the measurements of Γ_A and Γ_P was developed by Komori et al. (1991b), and it combined both laser-induced fluorescence (LIF) and electrode-conductivity (EC) techniques. The details of the technique are described in Komori et al. (1991b) and only the fundamentals are given here.

When the laser dye (uranin) premixed homogeneously in a solution is excited by a high-power argon-ion laser, the fluorescence is generated. The fluorescence intensity depends significantly on the local pH of the solution and it decays with decreasing pH in the range of $pH < 7$. Thus the concentration of CH_3COOH was measured by the laser-induced fluorescence (LIF) technique. The concentration of the chemical products of $CH_3COONH_4 + H_2O$ was measured by the electrode-conductivity technique with a single- electrode conductivity probe (Gibson & Schwartz 1963), and the technique used the difference of the electrode conductivity between CH_3COONH_4 and other species. The electrode conductivity is, however, affected by the reactants. The effect was compensated here by using the instantaneous signals of the above fluorescence (Komori et al. 1991b). The spatial resolutions of these concentration-measuring techniques were estimated to be 19 mm and 43 mm for the two concentration-measurements of CH_3COOH and $CH_3COONH_4 + H_2O$ from the power spectra of concentration fluctuations, and they were smaller than the Kolmogorov scale in the present turbulent flow field.

In addition to the concentration measurements in a reacting flow, the concentration statistics were also measured in a non-reacting flow with the same flow field as in a reacting flow, by means of a combined laser-induced fluorescence and Mie-scattering technique of Komori et al. (1989).

3. A Lagrangian stochastic model

A number of theoretical studies on reacting turbulence have been made, and both Eulerian and Lagrangian models have been developed. Recently Komori et al. (1991a) modified the marked-particle pair trajectory model of Sawford & Hunt (1986) and applied it to reacting flows with moderately fast or slow second-order reactions. The predictions have showed the good agreement with the measurements published in the previous papers. Although it remains to be discussed about the applicability of the model to a rapid reaction, the concentration statistics computed by the model were here compared with the present measurements. The details of the model are described in Komori et al. (1991a), and here only the computational results will be shown in the next section.

4. Results and discussion

Streamwise profiles of the mean concentrations normalized with the initial concentrations are shown in figure 2(a). The thin dashed-lines in the figures show the best-fit curves of the mean concentrations measured in a non-reacting flow with the same flow field as in a reacting flow. The concentrations of the reacting species A and B decrease with increasing x/M, whereas the concentration of the chemical products P increases. This means that the reaction distinctly progresses in the initial mixing region. However, the reaction rate rather decreases in the downstream region of $x/M > 10$, and the decay of the mean concentration of the reactant is suppressed. This suggests the importance of the molecular diffusion (see Koochesfahani & Dimotakis 1986). That is, the layer of the chemical products is rapidly produced in the interfacial region, and the mixing cannot follow the progress of the reaction in the downstream region of a decaying turbulence. In the same figure, the predictions of the mean concentrations by the Lagrangian stochastic model are shown by the thick lines. The predictions of the mean concentrations of the reacting species are a little larger than the measurements, but they are in the qualitative agreement with the measurements. The reason why the predictions show the larger values of the mean concentration than the measurements is discussed in Komori et al. (1991a).

Figure 2(b) shows the transverse profiles of the mean concentrations. The concentrations of the reacting species are smaller than those in a non-reacting flow and therefore the progress of the reaction is observed in the central region of $| y/M | < 0.5$. The predictions also show the similar behavior.

Both streamwise and transverse profiles of the mean squared values of the concentration fluctuations of the reacting species and chemical products are shown in figures

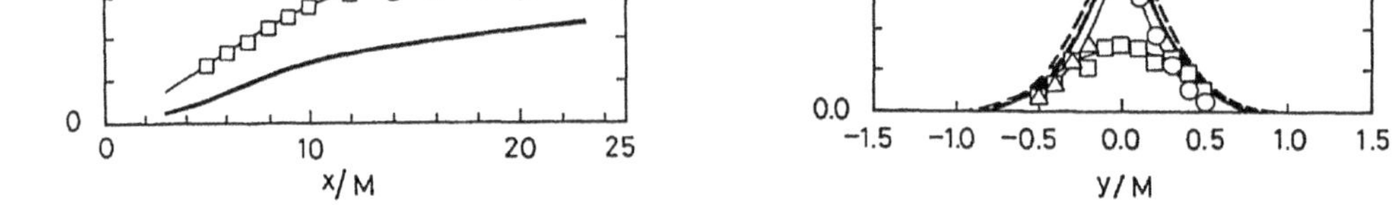

Figure 2. Streamwise and transverse profiles of the mean concentrations. (a) Streamwise profiles on the central plane of $y/M = 0$. (b) Transverse profiles at $x/M = 6$. ○, △, □: measurements of species A, B and P concentrations. - - - & — —: best-fit curves of the measurements and predictions in a non-reacting flow; —— : best-fit curves of the predictions in a reacting flow. A best-fit curve of streamwise profile of the concentration measured in a non-reacting flow is the same as that of the prediction.

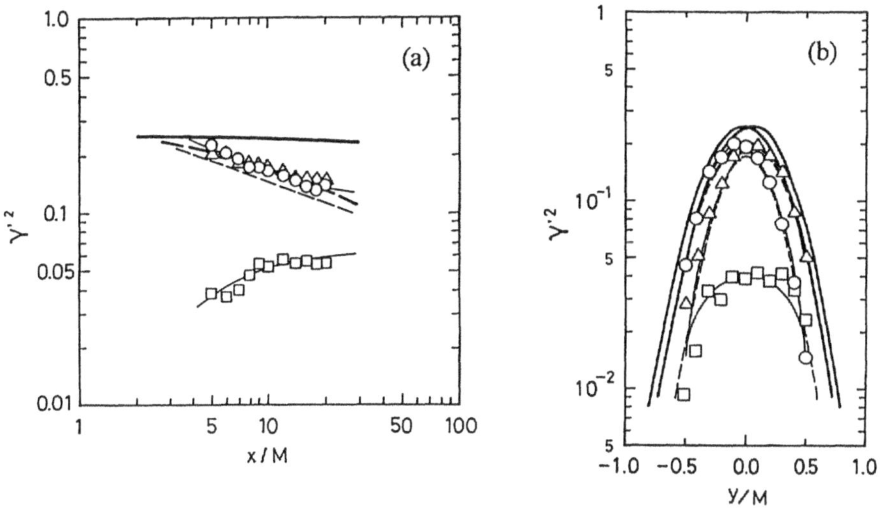

Figure 3. Streamwise and transverse profiles of the mean squared values of the concentration fluctuations. (a) Streamwise profiles on the central plane of $y/M = 0$. (b) Transverse profiles at $x/M = 6$. Symbols as in figure 2.

3(a) and 3(b). The intensities of the concentration fluctuations of the reacting species A and B decay with increasing x/M , and they are larger than the intensities in a non-reacting flow. The intensities linearly decay following the similarity in the initial mixing region, but they slightly deviate from the similarity in the downstream region, because of the occurrence of the chemical products. The predicted curve for the reacting case is rather larger than measurements, but it obviously shows the difference between the reacting and non-reacting cases. The transverse profiles show the asymmetry that a maximum point of the concentrations intensity shifts from the centre ($y/M = 0$) to each side. This behavior is also shown by the predictions.

Figures 4(a) and 4(b) show the streamwise and transverse profiles of the correlation coefficient between the concentration fluctuations of two reacting species, $R_{AB}(= \overline{\gamma_A\gamma_B}/\gamma'_A\gamma'_B)$. The correlation coefficient is always equal to -1 in a non-reacting flow, but the absolute value of the coefficient decreases with x/M and $\mid y/M \mid$ in a reacting flow. This means that the chemical products are generated in the interfacial region between A and B fluid-lumps, and they smears the high correlation between two concentration fluctuations. The predictions are in good agreement with the measurements.

Streamwise and transverse profiles of the segregation parameter $\alpha(= \overline{\gamma_A\gamma_B}/\bar{\Gamma}_A\bar{\Gamma}_B$ are shown in figures 5(a) and 5(b). The segregation parameter increases progressively with x/M in a non-reacting flow, whereas it is scattered around -1 in a reacting flow. The value of $\alpha = -1$ suggests that the mean reaction rate $R = k(\bar{\Gamma}_A\bar{\Gamma}_B + \overline{\gamma_A\gamma_B})$ should be zero. Here k is the reaction rate constant. However, this does not mean that the reaction is completely stopped. Because the segregation parameter a is a much sensitive quantity to both mean concentration and concentration correlation and it is very difficult to estimate α within a very small error. If α slightly deviates from -1, the mean reaction rate does not become zero, because of a very high reaction rate constant for a rapid reaction. Unfortunately, the present measurements cannot ensure

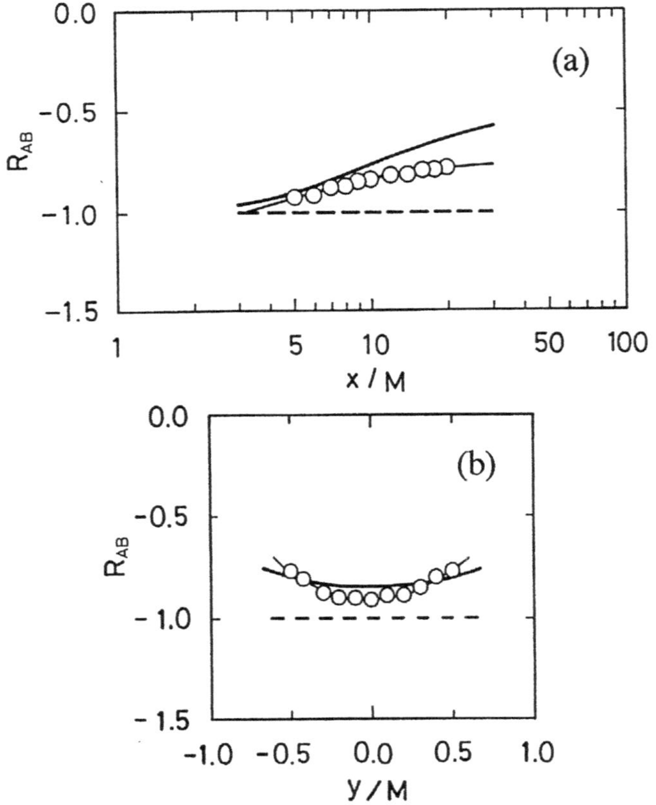

Figure 4. Streamwise and transverse profiles of R_{AB}. (a) Streamwise profiles on the central plane of $y/M = 0$. (b) Transverse profiles at $x/M = 6$. ○, measurements in a reacting flow; – – –, best-fit curve of the measurements and predictions in a non-reacting flow; ——, best-fit curve of the predictions in a reacting flow.

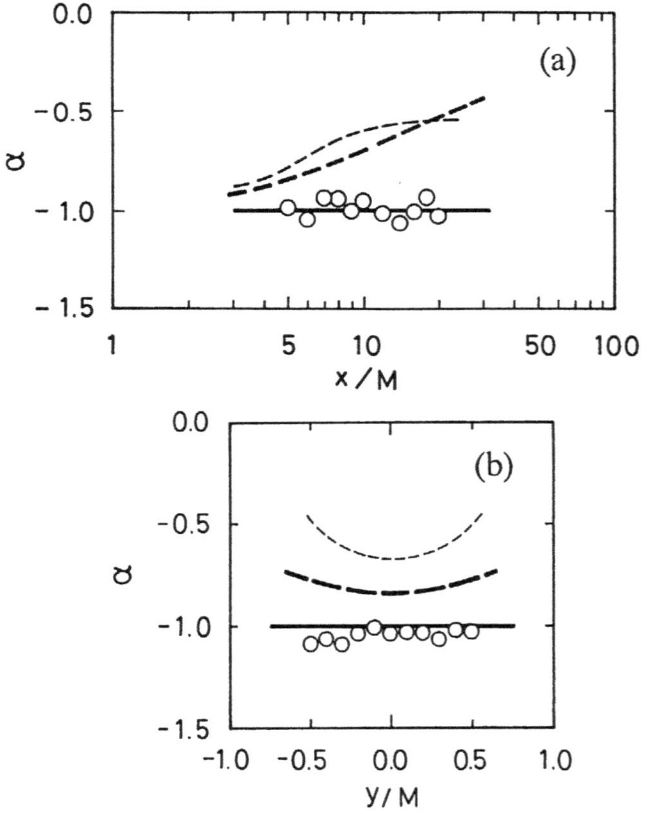

Figure 5. Streamwise and transverse profiles of α. (a) Streamwise profiles on the central plane of $y/M = 0$. (b) Transverse profiles at $x/M = 6$. - - - & — —: best-fit curves of the measurements and predictions in a non-reacting flow; —— : best-fit curves of the predictions in a reacting flow.

the accuracy of the order of 10^{-2} for α, and therefore we cannot strictly discuss it. As far as the mean concentration profiles in figure 2(a) show, the mean reaction rate is certainly decelerated in the downstream region.

5. Conclusion

To investigate the effects of the fluid mixing on a rapid second-order reaction, the instantaneous concentrations of two reacting species were measured using a combined laser-induced fluorescence and electrode-conductivity technique in shear-free grid-generated turbulence, and the concentration statistics were estimated.

The results show that the chemical reaction progresses in the initial mixing region and it is restrained by the gradual increase of the chemical products layer in the downstream developing region. The profiles of the concentration statistics follow the similarity in a non-reacting flow, but they deviate from it in a reacting flow. The absolute value of the correlation coefficient of the concentration fluctuations R_{AB} decreases in a flow with a rapid reaction and the segregation parameter α is almost equal to -1.

Acknowledgement

The authors gratefully acknowledge the financial support for this study by the NISSAN Science Foundation and Grant-in-Aid for Scientific Research on Priority Areas "Exploration of Combustion Mechanism". SK would like to thank the financial support for his travel expenses to Stockholm by the NISSAN Science Foundation.

References

Bennani, A., Gence, J.N. & Mathieu, J. 1985 The influence of a grid-generated turbulence on the development of chemical reactions. *AIChE J.* **31**, 1157.

Gibson, C.H. & Schwarz, W.H. 1963 Detection of conductivity fluctuations in a turbulent flow field. *J. Fluid Mech.* **16**, 357.

Komori, S. & Ueda, H. 1984 Turbulent effects on the chemical reaction for a jet in a nonturbulent stream and for a plume in a grid-generated turbulence. *Phys. Fluids* **27**, 77.

Komori, S., Kanzaki, T., Murakami, Y. & Ueda, H. 1989 Simultaneous measurements of instantaneous concentrations of two species being mixed in a turbulent flow by using a combined laser-induced fluorescence and laser-scattering technique, *Phys. Fluids A* **1**, 349.

Komori, S., Hunt, J.C.R., Kanzaki, T. & Murakami, Y. 1991a The effects of turbulent mixing on the correlation between two species and on concentration fluctuations in non-premixed reacting flows. *J. Fluid Mech.* in press.

Komori, S., Kanzaki, T. & Murakami, Y. 1991b Simultaneous measurements of instantaneous concentrations of two reacting species in a turbulent flow with a rapid reaction. *Phys. Fluids A* **3**, in press.

Koochesfahani, M.M. & Dimotakis, P.E. 1986 Mixing and chemical reactions in a turbulent liquid mixing layer. *J. Fluid Mech.* **170**, 83.

Mudford, N.R. & Bilger, R.W. 1984 Examination of closure models for mean chemical reaction rate using experimental results for an isothermal turbulent reacting flow. In *20th Symposium on Combustion*, The Combustion Institute.

Rilley, J.J., Metcalfe, R.W. & Orszag, S.A. 1986 Direct numerical simulations of chemically reacting turbulent mixing layers. *Phys. Fluids* **29**, 406.

Saetran, L.R., Honnery, D.R., Starner, S.H. & Bilger, R.W. 1989 Scalar mixing layer in grid turbulence with transport of passive and reactive species. In *Turbulent Shear Flow* **6** (ed. J.C. Andre et al.), Springer.

Sawford, B.L. & Hunt, J.C.R. 1986 Effects of turbulence structure, molecular diffusion and source size on scalar fluctuations in homogeneous turbulence. *J. Fluid Mech.* **165**, 373.

Turbulent Combustion Modelling — A Review *

C.H. Priddin
Rolls-Royce plc, PO Box 31, Derby, DE2 8BJ, UK

This review covers a brief description of methods available to model turbulent combustion problems. It is shown that the classification of flames by Damkohler number is limited and difficult to apply. For the case of gas turbine combustion, while high-power operation can be simulated effectively by simple combustion models, low-power simulations require the use of more detailed kinetic schemes and the coupling of these to the turbulence. Transport methods for the probability density function seem to offer the best prospect of achieving this but require more research to be used in practical situations.

1. Introduction

Many practical devices of engineering interest require an understanding of turbulent combusting flow behaviour, or some equivalent turbulent reacting flow. The requirements are many and various, depending upon the application. It may be that only the mean density is of interest, as this is the only parameter affecting the flow pattern directly. More likely the mean reaction rate (or product formation rate) is of interest, which may require a more detailed description of the processes. Further detail still is needed for estimating pollutant emissions and other trace species.

There are three ways to handle the difficulties imposed by turbulent reacting flows. One is to concentrate on a description of the turbulence and treat the reaction in a very simplified way. This is appropriate if the rates of reaction are orders of magnitude faster than the turbulent mixing times in the flow. Examples of this approach are the eddy-breakup model of Spalding (1971), the Magnussen model (1985) and the instantaneous equilibrium model of Jones and Priddin (1978).

The second approach is to represent the chemical processes in great detail so that the processes and trace species can be followed throughout the flow. If this is done however the turbulence must effectively be ignored as otherwise too much complication results to make the problem tractable. Effectively, the time mean properties are used to describe the chemical processes. This method is useful only when all reactions of interest are orders of magnitude slower than the turbulent eddy lifetime.

The third and most general approach is needed when chemical reaction times and turbulence time scales are of the same order. Here a method is needed which can encompass the complex interactions between the reacting species and their intermixing due to turbulence. All three of these possible schemes will be described in some detail, with the addition of a few hybrid schemes. Methods based on the solution of the joint probability distribution function (pdf) transport equations (Pope & Correa 1988, Jones & Kollman 1987, Chen, Kollman & Dibble 1989 and Pope 1985) fall into this category.

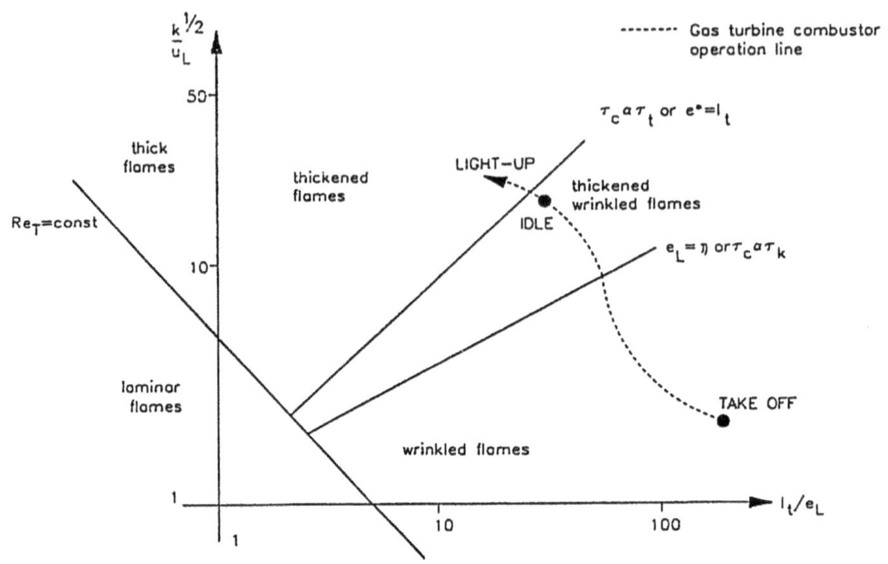

Figure 1. The different regimes of turbulent premixed flames

The parameter which distinguishes between these various reaction regimes is the Damkohler number (Da) which, together with a few other properties, notably the local strain rate, characterises the structure of any flame system. The state of a flame is conveniently plotted on a 'Borghi diagram' (Fig 1) taken from Borghi (1985) which shows the similarities and differences between flames in different circumstances. (Borghi (1988) gives a very useful survey of this approach which should be consulted to follow this subject further).

One very useful simplification which works quite well for flames in the first category is the laminar flamelet model (Bray, Libby & Moss (1985), Peters (1984), Veynante, Lucas & Candel (1989)). Here the full detail of the chemical reaction and diffusion processes can be simulated in a laminar flame, where it is found that all the properties of the flame can be reduced to a one-parameter description, regardless of physical location. In other words, knowledge of one parameter alone is enough to describe the state (composition, temperature, density) of the flame. In practice, heat loss and/or buoyancy effects introduce other parameters but there is often a significantly large portion of the flame for which this is true. The actual parameter chosen depends on whether the flame is a premixed one (in which case a reaction progress variable is used) or a diffusion flame (in which case a mixture fraction or similar is used). In this type of method the turbulent flame is thought of as an assembly of many wrinkled laminar flames whose properties are still locally determined by the same single parameter: the effect of turbulence is to introduce fluctuations in that parameter, the magnitude of which must be calculated. To complete this description (and other one-parameter descriptions) an assumption of the shape of the pdf for the parameter has to be made. In virtually every practical method a pdf with two degrees of freedom is chosen, whose first two moments must then satisfy the values derived for these quantities from transport equations. Typical approaches are Priddin & Coupland (1988) and Shyy, Braaten & Burrus (1989). However Pope(1979) maintains that at least three moments are required to define a pdf shape correctly.

Unfortunately, for consistency this would require closure of all the turbulent equations at the third level which introduces a large increase in the number of correlations to be solved (from 3 turbulence equations to 11 for a full second-moment closure, to 31 for a third moment closure). In addition, all of these 20 new equations introduce new terms for which no modelling work has yet been done and whose properties are not yet known. We are still a long way from verifying the assumptions made at the second-order closure level and the prospect of third-order modelling seems very remote as yet.

2. Global Turbulent Flame Classification

As described in the introduction many workers are concerned with the classification of flame types according to some global parameters such as Damkohler number (Damkohler 1940) which can be expressed as the ratio of typical turbulence lifetime to a characteristic chemical time, and the turbulence Reynolds number

$$Re_T = \frac{k^{\frac{1}{2}} l_T}{\nu} \tag{1}$$

Although the flame regimes may indeed be plotted as a function of Da and Re_T, it has become more popular recently to separate turbulent velocity and length scale effects by plotting Fig 1 in terms of the ratios $k^{\frac{1}{2}}/U_L$, the turbulent velocity to laminar flame speed ratio, and l_T/e_L, the turbulent length scale to laminar flame thickness ratio. The length scale of turbulence acts principally as a means of extending the surface area of the flame sheet by a wrinkling and distortion process; whereas the turbulent velocity is principally responsible for straining flames — if too high in relation to the flame speed, local extinction can occur.

Although useful for general classification, there are serious difficulties in applying such concepts to general industrial combustion devices. All four of the parameters k, U_L, l_T, e_L vary enormously throughout the device, and may not in general be known or even possible to estimate. The reaction parameters U_L, e_L can vary by many orders of magnitude (Pope 1990) depending on which of the many combustion reactions are considered. There will be simultaneously some processes occuring at very short timescales (radical/radical reactions) together with relatively slow ones (such as CO oxidation) and very slow ones (such as NO_x formation) which make it virtually impossible to perform such an analysis. Even the distinction often made between premixed and diffusion flames may not be clear cut. It is quite possible for a liquid fuel spray to evaporate while it mixes with air and to burn partially premixed, but principally as a diffusion flame. This preamble is intended to show that while such approaches and correlations based on them (turbulent flame speed, for example) do give a physical description of turbulent combustion process in single flames they are not suitable for the prediction of local properties as part of a CFD model. The rest of this paper will therefore focus on the modelling aspects of turbulent combustion.

3. Theoretical Description of Turbulent Flames

3.1 What is a turbulent flame?

Some definitions are appropriate at this stage. A turbulent flame is a reacting flow which, in classical turbulence terms is fluctuating quasi-randomly in time. Generally, although the combustion may modify the turbulence, the turbulence is pre-established by the flow field before reaction begins. We are considering a reacting flow which in the long time average or ensemble average over many realisations is steady. Rather than use conventional time averages to describe the properties of such flows, it is now well established that it is more convenient to follow Favre (1965) and work with density-weighted mean quantities for velocity u_i, species mass fractions Y_i, temperature T and enthalpy h. Conventional means are used for pressure p and density ρ.

3.2 Basic thermodynamic equations

The instantaneous equations which are satisfied exactly represent conservation of mass and momentum (the Navier-Stokes equations), energy and species concentration. In Cartesian tensor notation, these are

$$\frac{\partial \rho}{\partial t} + \frac{\partial}{\partial x_i}(\rho U_i) = 0 \tag{2}$$

$$\frac{\partial}{\partial t}(\rho U_i) + \frac{\partial}{\partial x_j}(\rho U_i U_j) = \frac{\partial}{\partial x_j}(T_{ij}) - \frac{\partial p}{\partial x_i} \tag{3}$$

Gravitation and other body forces are neglected here - for the high Reynolds number and low Grashof number of most industrial flows this is a good approximation. T_{ij} is the (laminar) viscous stress: for a Newtonian fluid this may be expressed in terms of the viscosity μ

$$T_{ij} = \mu \left(\frac{\partial U_i}{\partial x_j} + \frac{\partial U_j}{\partial x_i} \right) \tag{4}$$

It is not always necessary to solve an equation for the mixture enthalpy. Whether or not this is needed depends on the problem boundary conditions and the internal energy exchanges occuring in the problem of interest. Straightforward chemical reaction for instance, is best treated by simply allowing for the heat release in the definition of the base enthalpy of each species. With these definitions, the enthalpy of reactants and products is the same and an enthalpy equation is unnecessary.The enthalpy of each species i may be expressed as

$$h_i(T) = H_i^0 + \int_0^T C_i dT \tag{5}$$

The precision with which the enthalpy is represented will also depend on the application needs. However, for the case where enthalpy solution is not required, it is quite feasible to use accurate polynomials as these calculations are performed once only, to establish the relationships between species concentration, temperature and mixture fraction. If however enthalpy calculation is required to be coupled, it may be sufficient to assume a constant or linearly varying value for C_i; this will greatly ease the task of evaluating T from the local composition and h_i values. Further approximations are needed to fully decouple the enthalpy equation from the other equations to be solved. First, the Lewis numbers of all the species must be unity, so that heat and concentration diffuse at the same rate, otherwise an initially uniform enthalpy field can be made non-uniform by diffusion. Fortunately for most species this is not a bad approximation; exceptions are hydrogen and radical species which diffuse more rapidly, but are not generally present in large concentrations.

$$Le_i = \frac{\lambda / C_i}{\rho D_i} \tag{6}$$

Enthalpy equation is

$$\frac{\partial}{\partial t}(\rho e) + \frac{\partial}{\partial x_i}(\rho U_i h) = \frac{\partial}{\partial x_i}(\lambda \frac{\partial T}{\partial x_i} + H_j^0 J_i^j) \tag{7}$$

where for Fickian diffusion the flux J_i^j for species j in direction i can be represented by

$$J_i^j = \rho D_j \frac{\partial Y_j}{\partial x_i} \tag{8}$$

Equation (7) reduces to

$$\frac{\partial}{\partial t}(\rho e) + \frac{\partial}{\partial x_i}(\rho U_i h) = \frac{\partial}{\partial x_i}\left\{\frac{\lambda}{C_p}\left(\frac{\partial h}{\partial x_i} + (\frac{1}{Le_j} - 1)H_j^0 \frac{\partial Y_j}{\partial x_i}\right)\right\} \tag{9}$$

which if all $Le_j \to 1$ becomes

$$\frac{\partial}{\partial t}(\rho e) + \frac{\partial}{\partial x_i}(\rho U_i h) = \frac{\partial}{\partial x_i}(\frac{\lambda}{C_p}\frac{\partial h}{\partial x_i}) \tag{10}$$

so that an initially uniform h field with constant boundary conditions will remain constant. An additional constraint is via the boundary conditions. If any significant heat loss or addition at the walls occurs, (or throughout the flow by radiation exchange) the enthalpy equation will again be required.

The reason for trying to eliminate this equation is principally because for a diffusion flame the chemical state then reduces to functions of a single independent scalar. If the enthalpy has to be retained, there are two independent scalars and the turbulent description becomes an order of magnitude more complex. For the aero gas turbine combustor chamber at high power operation, all air inlets are at the same pressure and temperature, heat loss to the walls is small and is 'recycled', and hence the enthalpy can be assumed constant.

The fuel fraction f is a strictly conserved scalar, (provided that the Lewis numbers of all species are equal), so its transport equation has no source:

$$\frac{\partial f}{\partial t} + \frac{\partial}{\partial x_i}(\rho U_i f) = \frac{\partial}{\partial x_i}(\Gamma \frac{\partial f}{\partial x_i}) \tag{11}$$

For a premixed flame this is not useful, since f is constant everywhere in such a flame. In a diffusion flame however it plays a crucial role in determining the rate of spread of the fuel.

The mass fraction of species j, Y_j is given by

$$\frac{\partial}{\partial t}(\rho Y_j) + \frac{\partial}{\partial x_i}(\rho U_i Y_j) = \frac{\partial}{\partial x_i}(\Gamma \frac{\partial Y_j}{\partial x_i}) + \rho W_j \tag{12}$$

where W_j is the net formation rate by chemical reaction of species j.

3.3 Turbulent flow relationships

The above formulae are all valid relationships between instantaneous quantities, and in principle could be integrated exactly to determine the thermochemical state of a

mixture at any time after some prescribed initial conditions. For non-reacting flows where $W_j = 0$ this approach has been followed by some researchers (Schumann 1973) and is termed Direct Navier Stokes solution (DNS). However, because of the large range of length scales which have to be represented, a fine numerical grid must be used - this in turn requires small time steps, limiting the calculations to date to relatively low Reynolds numbers (3–4000) and small physical regions of flow. Rogers, Moin & Reynolds (1986) calculated a flow corresponding to a small region of a jet shear layer which consumed 5–10 hrs of CRAY XMP time. These calculations are nevertheless very useful in guiding the modelling of some terms which are difficult or impossible to measure experimentally. In the case of a turbulent flame however the problem is several orders of magnitude more difficult, because of the wide range of timescales covered by the various reactions which have to be tracked. As an example, the timescales of radical shuffle reactions such as $H+O \rightleftharpoons OH$ are of the order 10^{-5} sec, while the important CO burn out reaction $CO+OH \Rightarrow CO_2+H$ is of the order 10^{-3} sec, and NO formation via the mechanism of Zel'dovich (1946) (equations 26) has a characteristic time of 10^{-2} sec (gas turbine take off conditions). This results in the species equations being very stiff, and time steps of order 10^{-6} sec are required to resolve the temporal gradients generated: thus at least 10^5 time steps will be required to calculate the complete problem.

Since only time-mean properties are generally of interest and affect the combustion hardware directly virtually all the turbulent combustion work to date has proceeded by time averaging the equations 7 to 12.

It turns out that the resulting equations are much simpler if density weighted averages are used, defined for any variable ϕ by

$$\tilde{\phi} = \overline{\rho\phi}/\overline{\rho} \tag{13}$$

Only those quantities which appear in terms multiplied by the density are treated in this way: otherwise straight time averages are used (principally for p and ρ).

As well as conferring benefits of simplicity on the equations, these density weighted averages are generally what is measured by most instruments. Even LDV will tend to give a mean velocity which is density-weighted because more seed particles are present when the density is high. Thus results from density weighted modelling can be reliably compared with experimental data.

The corresponding turbulent combusting flow equations are

$$\frac{\partial}{\partial x_i}(\overline{\rho}\tilde{U}_i) = 0 \tag{14}$$

$$\frac{\partial}{\partial x_i}(\overline{\rho}\tilde{U}_i\tilde{U}_j) = -\frac{\partial p}{\partial x_i} + \frac{\partial}{\partial x_i}\left(\mu\left(\frac{\partial \tilde{U}_i}{\partial x_j} + \frac{\partial \tilde{U}_j}{\partial x_i}\right) - \overline{\rho}\widetilde{u_i u_j}\right)$$

$$\frac{\partial}{\partial x_i}(\overline{\rho}\tilde{U}_i\tilde{f}) = \frac{\partial}{\partial x_i}\left(\Gamma\frac{\partial \tilde{f}}{\partial x_i} - \overline{\rho}\widetilde{u_i f}\right) \tag{15}$$

$$\frac{\partial}{\partial x_i}(\overline{\rho}\tilde{U}_i\tilde{Y}_j) = \frac{\partial}{\partial x_i}\left(\Gamma\frac{\partial \tilde{Y}_j}{\partial x_i} - \overline{\rho}\widetilde{u_i Y_j}\right) + \overline{\rho}\tilde{W}_j$$

The time averaging has introduced the additional correlations $\widetilde{u_i u_j}, \widetilde{u_i f}$ and $\widetilde{u_i Y_j}$, and also requires the evaluation of W_j. The correlations are generally treated by analogy

with non-reacting flows, so for example

$$-\overline{\rho}\widetilde{u_i u_j} = \mu_T \left(\left(\frac{\partial \tilde{U}_i}{\partial x_j} + \frac{\partial \tilde{U}_j}{\partial x_i} \right) - \frac{2}{3}\delta_{ij}\frac{\partial \tilde{U}_l}{\partial x_l} \right) - \frac{2}{3}\delta_{ij}\overline{\rho}k \tag{16}$$

with $\mu_T = C_\mu \overline{\rho} k^2/\varepsilon$ is the two-equation model consistent with the isothermal $k - \varepsilon$ model. Here the mean density is no longer constant, and the dilation term

$$-\frac{2}{3}\delta_{ij}\frac{\partial \tilde{U}_l}{\partial x_l} \tag{17}$$

has been added (which is zero in isodensity flows) to preserve the contraction property.

For closure at the second moment level, using algebraic stress (Boysan, Erdogan, Ewan & Swithenbank (1981)) or Reynolds stress (La under, Reece & Rodi (1975)) models to describe these correlations, some additional work is needed. Some authors have applied isodensity Reynolds-stress models unchanged to variable density flows, but these do not then capture the (often important) effects of density gradients in a buoyancy-affected flow (including strongly swirling flows). The effect can be to stabilise (stratify) or to cause greatly enhanced mixing depending on whether the density gradient is aligned or opposed to the acceleration vector, and is of great importance in atmospheric flows. Similarly the scalar flux terms such as $\overline{\rho}\widetilde{u_i f}$ are modelled (at the first order level) by

$$-\overline{\rho}\widetilde{u_i f} = \frac{\mu_T}{\sigma_f}\frac{\partial \tilde{f}}{\partial x_i} \tag{18}$$

where σ_f is a turbulent Schmidt number for f, typically taken to be 0.7–1.0.

Although further work still needs to be done to improve the variable density Reynolds stress modelling these aspects are relatively straightforward. Much greater difficulty surrounds the evaluation of W_j, because of the extreme non-linearity of many of the reaction rate expressions required. W_j is likely (depending on what chemical species j is) to contain several terms of the form $K_j[A][B]$, which is the rate of production of species j from species A and B, and $[A]$ refers to the concentration of species A. The expression for K_j is typically:-

$$K_j = CT^n \exp(-T_A/T) \tag{19}$$

where C, n, and T_A are constants for a particular reaction. The expression for W_j becomes

$$W_j = CT^n \exp(-T_A/T)[A][B] \tag{20}$$

The obvious method to form W_j is to expand T,$[A]$ and $[B]$ using their mean and fluctuating parts, and to expand the exponential term as a power series in (T_A/T). This does eventually lead to an expression for W_j in terms of means $\tilde{T}, \widetilde{[A]}$ and $\widetilde{[B]}$ but also includes the second order correlations $\widetilde{[A]'[B]'}, \widetilde{T'[A]'}, \widetilde{T'[B]'}, \widetilde{T'^2}, \widetilde{[A]'^2}, \widetilde{[B]'^2}$. In fact because in general $T_A \gg 1$, the expansion does not converge until a large number of high order terms have been introduced. Borghi (1974) has attempted this approach but because of the above complications it is not generally reckoned to be helpful, except for special reactions which do not have high T_A.

4. Modelling Approaches for W_j

4.1 High Damkohler number flames

A high Damkohler number implies very short reaction times compared with turbulent mixing times (bearing in mind earlier comments about the evaluation of Da). This in turn implies that the mean rate of reaction is principally controlled not by any details of the chemical process but by the rate at which the fuel and air (for diffusion flames) can mix together. It is useful therefore in these situations to completely characterise the state of the gas by a single scalar which for diffusion flames is taken to be f, the mixture fraction, and for premixed flames a reaction progress variable, c defined by

$$c = \frac{T - T_{IN}}{T_{PRODUCTS} - T_{IN}} \tag{21}$$

If a reaction rate W_j is required the simplest expression available in the high Da limit is the Eddy Break Up model of Spalding (1971) for the fuel concentration:

$$\widetilde{W_{fuel}} = C_{EBU}(\varepsilon/k)\widetilde{Y'^2_{fuel}} \tag{22}$$

which uses a typical turbulent timescale (k/ε) and assumes the fuel will burn completely over the lifetime of one typical turbulent eddy. Because of the 'fast chemistry' assumption, however, it is not generally necessary to evaluate W_j. All properties (species concentration, temperature, density) are functions of the relevant scalar variable (f or c). Because of the non-linearity of these functions, fluctuations must be taken into account when mean values are calculated. The key to this is a knowledge (or assumption) of the pdf (probability density function) describing the variation of f with time. All means can then be evaluated since

$$\tilde{\phi} = \int_0^1 p(f)\phi(f)df \tag{23}$$

for any quantity ϕ.

The problem then reduces to the determination of $p(f)$. There have been many proposals for the form of $p(f)$ — from two delta functions to a clipped Gaussian (Naguib & Lockwood (1975): the clipping is necessary to account for the tails of the distribution which extend outside the allowable range of 0–1). A better, flexible function shape which is automatically constrained to these limits and also (Rhodes 1975) matches observed pdf shapes of scalars is the Beta function used by Jones et al. (1978) who define $p(f)$ as

$$p(f) = Af^a(1-f)^b \tag{24}$$

There are three constants to be prescribed to fully specify $p(f)$; A, a and b in the above expression. They are determined from the following three relationships.

$$\begin{aligned} \int_0^1 p(f)df &= 1 \\ \int_0^1 fp(f)df &= \tilde{f} \\ \int_0^1 f^2p(f)df &= \widetilde{f^2} \end{aligned} \tag{25}$$

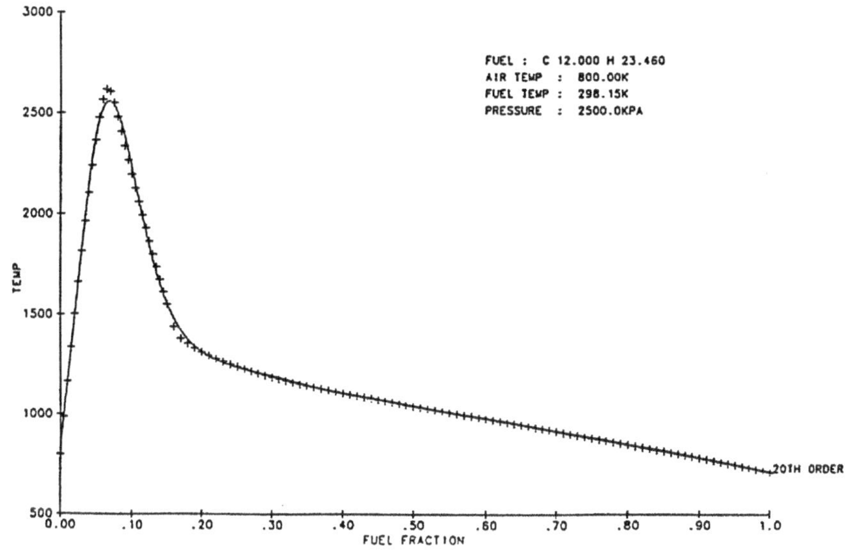

Figure 2. Equilibrium temperature as a function of fuel fraction

where $\tilde{f}$ and $\widetilde{f^2}$ are determined from the transport equations of the form of equation (15).

Pope (1979) has examined some real distribution functions and determined that to reproduce all their features three moments should be specified. Unfortunately this would require closure of the turbulence at the third level, involving third-order correlations of velocity and scalar concentration. Such modelling is currently not possible or practical and has hardly even been considered. We must therefore be satisfied with second order modelling. Fortunately the discrepancies are not serious in evaluation of the integrals for the mean quantities required, and the second order level is adequate for most purposes. It is therefore possible, by solving transport equations for $\tilde{f}$ and $\widetilde{f'^2}$ to completely define the chemical state in these mixing-controlled flames and hence determine the mean density $\overline{\rho}$ which is the only influence of the chemistry on the flow.

So far we have described how the effects of turbulence may be properly included in these flames. But how should one determine $\phi(f)$? There are two popular prescriptions. One, which has been used extensively (Priddin et al. 1988, Coupland et al. 1987) is to perform a full equilibrium calculation for a range of values of f between 0 and 1, and then for every ϕ of interest either tabulate or curve fit the results to provide $\phi(f)$. The equilibrium calculations use the method of Gordon and McBride (1971). There are significant advantages in making $\phi(f)$ a single polynomial over the whole 0–1 interval, even if this has to be a high order, (typically 18–20th order is used (Fig 2) with double precision) because the integration of equation 23 may then be performed analytically.

The same integration technique can be used with the second method for determining $\phi(f)$, the laminar flamelet model.

Laminar flamelet models (Bray et al. 1985) make the assertion that the effect of turbulence is principally to wrinkle and stretch the flame front, but its basic properties remain unchanged and hence $\phi(f)$ may be found by measurement or calculation of a

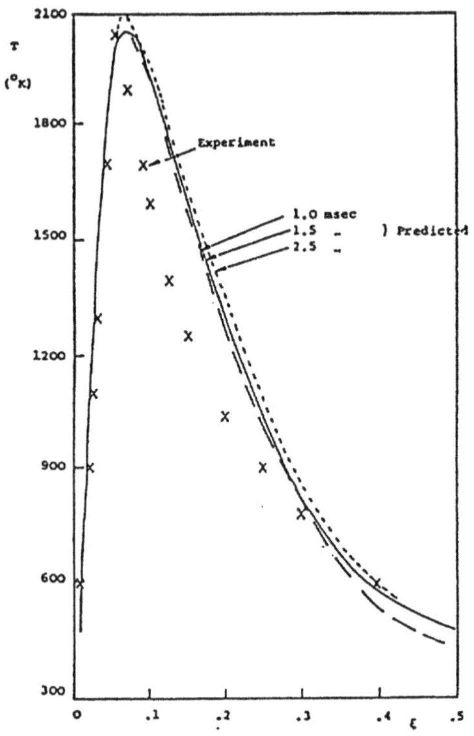

Figure 3. Flamelet calculations at various times from start

simple laminar flame at the same conditions. The theory relies on the plot of any property ϕ vs f collapsing to a single curve, regardless of the position of the point in the flame. Fig 3 from Bray et al. (1985) shows that this is indeed the case, and that most properties do collapse in this fashion and can therefore be treated in the same way as the equilibrium calculations reported earlier. Exceptions are the species involved in relatively slow reactions, such as soot formation/consumption, and NO formation. However, even for these the rate of formation may be characterised as a single function of f (Coupland et al. (1987), Moss, Stewart & Syed (1987)) since these rates are determined by the concentrations of other species which are themselves only functions of f.

There is clearly an effect of strain rate on a laminar flame which should be taken into account, since at high enough strain the flame will be extinguished. However, recent research (Abdel-Gayed, Bradley, Hamid & Lowes (1985)) has shown that the effects of strain are not great except for strain rates close to extinction, and since this is a local model with no history effects, when the strain is removed combustion instantly recurs. It is therefore of little benefit to introduce strain rate as an additional parameter.

$\phi(f)$ may be generated for a laminar flame by measurement. This is the best method if a suitable rig and instrumentation at the correct conditions of inlet temperature, pressure and fuel type exists. The approach is being followed by Moss et al. (1987) in the study of soot formation in kerosene flames.

For simple fuels, the reaction chemistry is now fairly well established. The burning of hydrogen in air for instance is well described by a few basic reaction steps (Warnatz (1981)), and that for methane and hydrocarbons up to propane, while requiring more

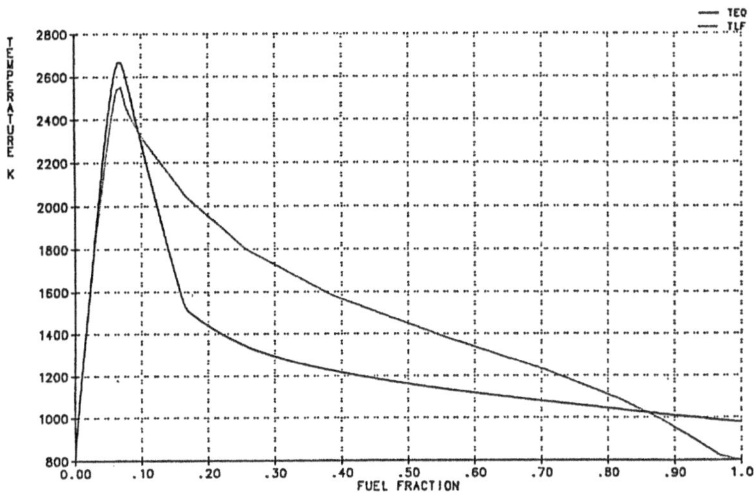

Figure 4. Laminar flamelet *vs* equilibrium temperature

reactions (Westbrook & Dryer (1979)) has been shown to describe observed behaviour quite well. It is now therefore feasible to use a one-dimensional computer program such as GENMIX (Spalding (1977)) to *calculate* the development of a laminar flame. Although straightforward this is not a trivial task, as there will be a large number of dependent variables to solve for all the individual species present. The laminar interdiffusion of all these species must also be computed, and this must be done using accurate binary diffusion coefficients and not Fickian diffusion. In view of these difficulties it is preferable to use a code which has been specialised for this task such as CHEMKIN (Kee, Miller & Jefferson (1980)). If such a code is available together with the relevant physical data, the functions $\phi(f)$ may be calculated this way instead of performing the laminar flame experiments. A substantial databank now exists for simple hydrocarbon flames at various strain rates as a result of the JOULE programme funded by the EEC which is still continuing. Consult Greenhalgh & Stopford (1989) for further details.

Fig 4 shows a comparison of laminar flamelet profile for temperature with the equilibrium profile at the same conditions (gas turbine take off) and it can be seen that on the lean side of stoichiometric they are virtually identical, and even on the rich side differences are limited to fuel fractions greater than 0.1. There are only small regions of the gas turbine combustor where this region is important as can be seen on Fig 5, which shows the temperature and fuel fraction fields in a typical gas turbine combustor.

The principal advantage of either of these methods over the simpler EBU models is that they allow a complete and consistent chemical description of the gas state, which can then be extended logically to cover other processes. As an example, consider the formation rate of NO, an important environmental pollutant. There are three major sources of NO_x in flames. The first is via fuel-bound nitrogen, in which any nitrogen contained in the fuel itself is directly converted to NO. There are some additive treatments which can reduce the conversion rate, but this is generally only a problem when burning pulverised coal, heavy oils and some natural gases. For aero engine use, kerosene is treated to remove all the nitrogen-containing species so this is not a problem for the engine designer. The second route for formation is the so called 'prompt'

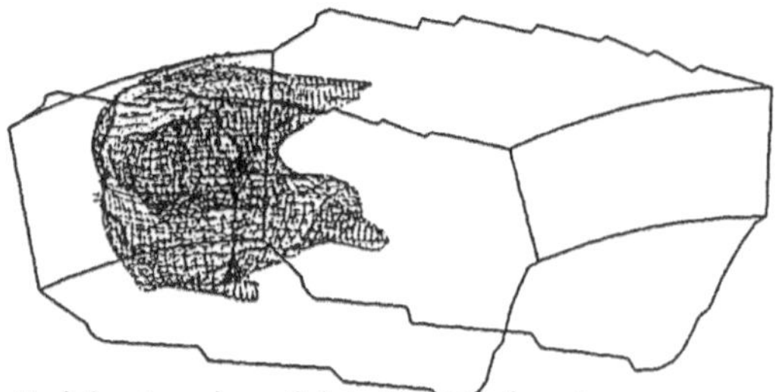

Fuel fraction above 0.1 inside shaded region

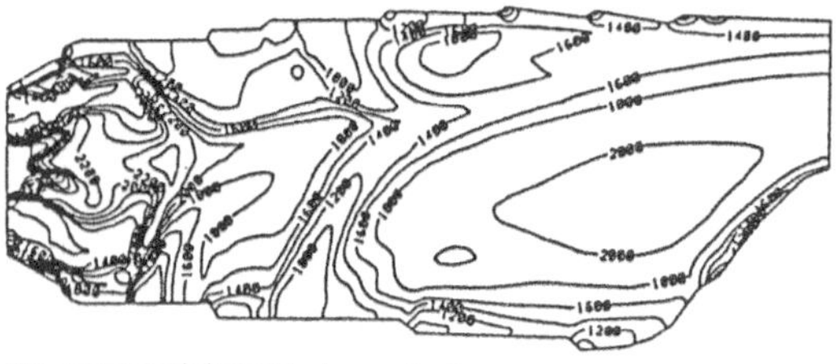

Temperature (K) on injector centreline

Figure 5. RB211 combustor fuel fraction and temperature distributions

NO_x phenomenon which appears to generate NO in the flame front itself. There have been several explanations for this process in the literature (eg Fenimore (1971)). Some of these rely on "super-equilibrium" oxygen atom concentrations found in flame-front flame regions, (which can be several orders of magnitude larger than equilibrium) to form NO, while others (Haynes (1977)) claim the dominant role is through the intermediate formation of HCN compounds. Whatever the mechanism, it seems that the quantity of NO formed in this way is roughly constant or falls as pressure is increased, such that at high pressure gas turbine conditions only the third mechanism produces significant NO. This third principal route is through the extended Zeldovich (1946) mechanism for flames (often termed 'thermal' NO_x).

$$N_2 + O \rightleftharpoons NO + N \tag{26.1}$$

$$O_2 + N \rightleftharpoons NO + O \tag{26.2}$$

$$N + OH \rightleftharpoons NO + H \tag{26.3}$$

To simplify this further, the N atom concentration is assumed to be in steady-state with the NO produced.

The implication of this is that for each NO molecule produced by reaction 26.1 an additional N atom is produced which then reacts via 26.2 or 26.3 to generate a second NO molecule. Thus the net NO formation rate from the process is just twice the rate of reaction 26.1.

$$\frac{d}{dt}[NO] = 2k_1[O][N_2] \tag{27}$$

with $k_1 = AT^n \exp(T_A/T)$. All of the quantities appearing on the right of equation (27) are known functions of the fuel fraction, f and hence the formation rate may be expressed as a single function ϕ_{NO} of f. This may then be integrated over the pdf to derive the time-mean formation rate:

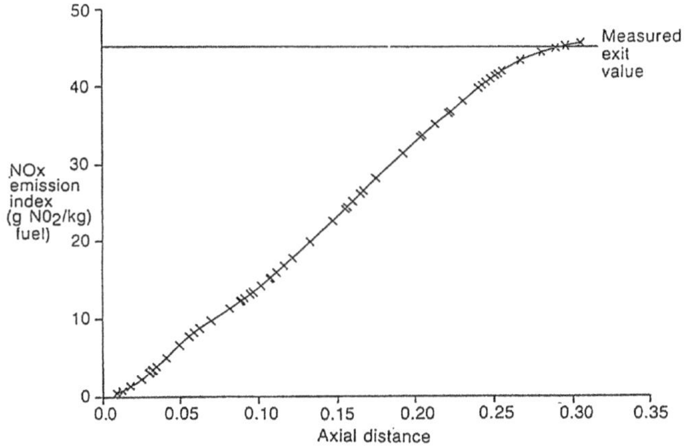

Figure 6. NO_x emission index *vs* distance along the combustor

$$\widetilde{R_{NO}} = \int_0^1 \phi_{NO}(f).p(f).df \tag{28}$$

Note that this is only possible since the rate of NO formation is *independent* of the NO level itself, i.e. is well away from equilibrium. If this assumption cannot be made then joint pdfs of NO and f again have to be determined and this simpler approach cannot be made. The above equation effectively specifies W_j in the conservation equation for NO mass fraction ($Y_j \equiv [NO]$) and hence the local NO distribution may be computed.

Fig 6 shows the resulting NO_x levels in the gas turbine combustor and the distribution of the source terms. This approach may be similarly extended to other species for which similar assumptions can be made.

4.2 Small Damkohler number

This case is really the situation of the NO formation reaction discussed above, which has a relatively long characteristic chemical time. The source term W_j can then be computed independently without considering the depletion of reactants, which occur over a time scale much longer than that of the turbulence. A single mean equation for Y_j is sufficient to define the chemical composition, using say a constrained equilibrium approach where the equilibrium composition is determined subject to the constraint that concentration of species j is kept to Y_j.

4.3 Intermediate Damkohler number

When neither extreme can be justifiably considered, a method must be chosen which allows both turbulent mixing and reaction rate processes to determine gas composition. An extension of the high Da limit method is to choose a two dimensional pdf form (Jones (1979)) with sufficient free parameters to fit the following six constraints. (f = fuel fraction, x = some species concentration)

$$\int_0^1 \int_0^1 p(f,x) df\, dx = 1$$
$$\int_0^1 \int_0^1 p(f,x).f df\, dx = \tilde{f}$$
$$\int_0^1 \int_0^1 p(f,x).x df\, dx = \tilde{x} \quad (29)$$
$$\int_0^1 \int_0^1 p(f,x).f^2 df\, dx = \widetilde{f^2}$$
$$\int_0^1 \int_0^1 p(f,x).x^2 df\, dx = \widetilde{x^2}$$
$$\int_0^1 \int_0^1 p(f,x).fx df\, dx = \widetilde{fx}$$

It is now virtually impossible to satisfy all the physical limits of the variables automatically by choice of $p(f,x)$, as was done for $p(f)$ by choosing a Beta function. The one constraint $p(f,x) > 0$ may be realised by a choice such as $p(f,x) = A\exp(af + bf^2 + cx + dx^2 + efx)$. Possible choices for the second variable x depend on the reaction scheme chosen — with two independent scalars only a two step scheme can be used, such as

$$C_nH_m + \frac{n}{2}O_2 \Rightarrow nCO + \frac{m}{2}H_2 \quad (30)$$
$$CO + OH \Rightarrow CO_2 + H$$

For this scheme we may take $x \equiv CO_2$, so that $\frac{dx}{dt}$ represents the CO consumption rate, and the model describes CO emissions. All other species may be calculated from constrained equilibrium.

Since $p(f,x)$ is not constrained to the physical bounds of f and x, these bounds must be imposed through the integration limits. f itself obviously always is to be integrated from 0 to 1, but what of the integration over x? Considerations of element conservation show that the allowable region for x is the shaded triangle on Fig 7.

It is simpler to transform the variables so that x' also is integrated from 0 to 1 ($x' \equiv x/x_{max}$), allowing the values of each function $\phi(f,x)$ which may be of interest (density,

Figure 7. Allowable region for 2D pdf $p(CO_2, f)$

temperature, composition, NO_x formation) to be tabulated in a two-way table. The problem than reduces to the determination of the six constants in the above expression for $p(f,x)$ and the evaluation of

$$\int_0^1 \int_0^1 p(f,x).\phi(f,x')df dx' \tag{31}$$

by numerical quadrature. Although straightforward this is not a trivial task and can take a significantly large proportion of the total cpu time to solve a flow problem as it must be evaluated for every point in the flow. The approach has been quite successful in predicting flame CO levels.

An alternative prescription for x is to use it to follow the radial recombination reactions such as

$$O + O + M \rightleftharpoons O_2 + M \tag{32}$$

These three-body reactions are relatively slow, which can be responsible for high radical levels seen in post-flame gases and hence potentially for prompt NO_x (in one description) and other post-flame processes. To do this x is taken to represent the molar concentration (moles/unit mass) which for the H_2/air flame of Correa, Gulati & Pope (1988) only has a source when a three-body reaction occurs: it is then a measure of the progress of such reactions, and the radical pool may be equilibriated with the local values of x.

4.4 Pdf transport methods

All the above schemes have their application in special circumstances, but as we have seen the most complete reaction scheme which could be considered is a two-step scheme. Most chemical kineticists maintain that an absolute minimum of 4–5 global steps are required to describe the main features of hydrcarbon combustion (and often use upwards of 100 in their calculations). How then can these two extremes be reconciled?

A method which has been pursued and refined over many years by Pope (1990) and others which may be used on more complex reaction schemes and does not require any assumptions about pdf shapes is the pdf transport method (or methods — there are several variants of this). In its simplest form, the joint pdf of n composition variables $Y_j (j = 1, n)$ is derived at every point in the flow from the Composition Joint pdf equation (taken from Pope (1990)).

$$\frac{\partial}{\partial t}[p(Y_j)] + \frac{\partial}{\partial x_i}[\tilde{U}_i p(Y_j)] + \frac{\partial}{\partial Y_k}[\rho W_k p(Yj)]$$

$$= \frac{\partial}{\partial Y_k}[< \frac{1}{\rho}\frac{\partial J_i^k}{\partial x_i}|Y_j > p(Y_j)] - \frac{\partial}{\partial x_i}[< u_i''|Y_j > p(Y_j)] \tag{33}$$

The left hand side of this equation contains all the terms which appear in exact, closed form (if the velocity field is known). This includes the mean convection, and (especially useful) all effects of chemical reaction. Since this latter term is exact, the chemistry may be treated in as much detail as desired with many steps. The only limitations are that increasing the number of species increases the dimensionality of the problem, with consequences for the speed of solution, and that increasing number of species also requires small timesteps, as the equations become stiffer.

The terms on the RHS represent transport in composition space due to molecular mixing, and transport in physical space due to turbulent mixing. Both of these require modelling. The molecular mixing term is usually represented by a stochastic mixing model (Chen & Kollman (1990)), the simplest of which is to choose two particles and replace their individual concentrations with the mean of them both. The probability of any particle pair exchanging concentrations in a time δt is taken to be $\delta t/\tau.(C_\phi N)$ where C_ϕ is a constant controlling the rate of decay of the variance, N is the number of particles and τ is the turbulence timescale $\tau = k/\varepsilon$. The turbulent mixing term can be modelled by a gradient diffusion hypothesis

$$\rho < u_i''|Y_j > p(Y_j) = \Gamma_T \frac{\partial p(Y_j)}{\partial x_i} \tag{34}$$

The resulting modelled equations could be solved by a conventional finite difference technique, where in addition to the space dimensions x_i the species compositions Y_j would be discretised over all possible values. However, the advantages of the pdf method in treating complex reactions are then quickly lost in that the total number of grid points required in m dimensions for n points in each dimension is n^m; thus a very coarse grid of 10 points in 3D with 7 species gives a total of 10^{10} points! This is some 5 orders of magnitude greater than todays most intensive computing task.

A different method is therefore required. Fortunately, Monte Carlo solution techniques require cpu time increasing only proportional to $m*n$; in the above situation this is of order 100 — eight orders of magnitude less. Such methods proceed by representing the pdf by a finite number of discrete 'conditional particles' each of which has a set of compositions and a position associated with it. The particles are transported to new locations by the mean velocity field and reaction proceeds in each particle, changing its composition. A second step performs the mixing operations. At the end of these steps mean quantities are evaluated as required from the ensemble of the particle properties at each point.

For the composition joint pdf solution to be successful, a separate (conventional) finite difference solution for the velocity field using a $k-\varepsilon$ turbulence model is required in order to determine U_i, τ and Γ_T in equations 33 & 34; accuracy of the overall scheme may be limited by this step.

A further development of this approach is to solve for the joint velocity-composition pdf, which contains all the above information plus the turbulent velocity field, so that turbulent diffusion may be treated exactly as well as reaction and convection. The only modelled term now is the molecular mixing term, which still requires the turbulent timescale. This has been finally removed by Pope (1990) in proposing the solution of a joint velocity-dissipation-composition joint pdf, which contains all the information required to close the equation, but introduces difficult modelling problems for modelling the rate of change of the dissipation following a particle.

In spite of the difficulties I believe this approach is the way forward to a realistic, general model of turbulent combustion. Pope (1990) gives as an example of what can be predicted for a $CO/H_2/N_2$ - air diffusion flame. The mean quantities reproduced in Fig 8 are well predicted, even though the rates of reaction are relatively low in this flame.

Through the pdf treatment however, much more detail can be extracted if required - even quantities such as the rms OH concentration are predicted reasonably well.

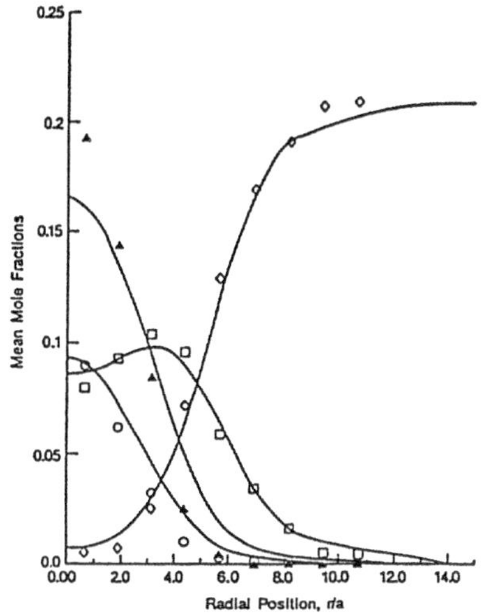

Figure 8. Example of a joint pdf calculation. Lines are predictions, points are data from Correa et al. (1988). Symbols: ○ H_2, △ CO, □ CO_2, ◇ O_2

An additional benefit of the Monte Carlo based methods is that they should be very easy to adapt to run on a highly parallel computer such as a transputer array. These machines offer of the order of 100 processors in parallel, each rated at some 20 Mflops. While conventional implicit solution procedures require a large number of inter-processor communications, much of the computation in these pdf transport schemes may be done independently with each processor working on a group of conditional particles. Only the diffusion step requires exchange of information, and then only between a small fraction of the particles. The other inhibiting factor to the widespread use of highly parallel computers has been the need to program in a new language: fortunately this is no longer necessary and Fortran and C compilers are now available. The time now seems right for increasing the research effort in this area.

What still needs to be done? First, a wider range of flames need to be studied, with more species and reactions to verify that statistically valid solutions can be obtained with more complex reaction schemes without requiring an excessive number of particles to be computed. Secondly, the full scheme of Pope (1990) must be investigated in more detail so that solutions do not require finite difference solutions to provide missing information. This is particularly important when using a highly-parallel processor to solve the particle equations, since good efficiency will only be maintained by every processor working on the same number of particles using the same method, but with different initial conditions. Thirdly, these assertions of the suitability of highly parallel architectures to this type of problem need to be checked by implementing a Monte Carlo solver for the pdf equations on a parallel processing computer. Fourthly, the models for the laminar mixing process need to be improved. There is a growing body of evidence from CARS data that the important species and temperature do not diffuse at the same rate (Fig 9) which makes the basic assumptions in most other models (that the fuel fraction f has no source, for example) highly suspect.

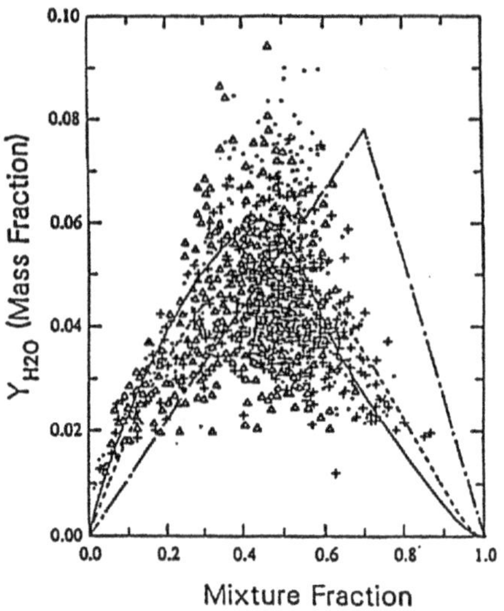

0

Figure 9. Instantaneous H_2O *vs* mixture fraction for a $CO/H_2/N_2/$air flame. Data should fall inside chain dotted line if all species diffusivities are equal. Solid and dashed lines show unstrained and strained laminar flames (Chen & Kollman (1990))

5. Summary

We have seen that the classification of flames on the basis of Da and Re_T, while offering some description of the processes occurring in flames, is in practice very difficult and arbitrary. The focus may change depending upon the parameters and conditions of interest. For some flame properties the assumed-shape pdf approach gives good results when used in conjunction with a laminar flamelet model or an equilibrium calculation to derive state relationships. Expansion of the reaction rate expression is not generally useful except for some very restricted circumstances. The only model which appears general enough for all conditions and does not impose assumptions is the modelled pdf equation. Although promising, there are still several areas of research to be pursued, particularly in the modelling of molecular diffusion. In view of its importance as the only method capable of generalising the calculation of turbulent flames to all Damkohler numbers, further research in this area would be of most benefit to the combustion community.

References

Abdel-Gayed, R.G., Bradley, D., Hamid, M.N. & Lowes M.(1985) p505, *20th Symposium (International) on Combustion.* The Combustion Institute Pittsburgh.

Aguirre-Saldivar, R.G.(1987) Two-scalar Probability density function models for turbulent flames. PhD thesis University of London.

Borghi, R.(1974) Chemical calculations in turbulent flows - application to CO containing turbojet flame. *Adv. Geoph* **18B** 349.

Borghi, R.(1985) On the Structure and Morphology of Turbulent Premixed Flames, in *Recent Advanced on Aerospace Sciences* 117 (Ed C.Casci) Plenum.

Borghi, R.(1988) Turbulent Combustion Modelling Progress, *Energy & Combustion Science* **14** 245.

Boysan, F. Erdogan, E. Ewan, B.C.R. & Swithenbank, J.(1981) Numerical Predictions of Confined Turbulent Flows with an Algebraic Reynolds Stress Closure. *Dept of Chem Eng & Fuel Tech report HIC 365*, Sheffield University.

Bray, K.N.C, Libby, P.A. & Moss J.B.(1985) Unified modelling approach for premixed turbulent combustion - part I: General formulation. *Combustion & Flame* **61** (1) 87.

Chen, J-Y & Kollman, W.(1990) Segregation Parameters and Pair-Exchange Mixing Models for Turbulent Non premixed flames. Paper 23-366, *23rd Symposium (International) on Combustion, Orleans.* To be published by The Combustion Institute.

Chen, J.Y, Kollman, W. & Dibble, R.W.(1989) *Combustion Science & Technology* **65**, 315.

Correa, S.M.(1985) A model for non-premixed turbulent combustion of CO/H2 jets. *Archivum Combustionis* **5** (3-4) 223.

Correa, S.M., Gulati, A. & Pope, S.B.(1988) *Combustion and Flame* **72**, 159.

Coupland, J. & Priddin, C.H.(1987) Modelling the Flow and Combustion in a production gas turbine combustor. *Turbulent Shear Flows 5*, Springer-Verlag.

Damkohler, G. Z.(1940) *Elektrochem* **45**, (11).

Favre, A.(1965) Equations des gaz turbulents compressibles. *J. Mechanique* **4** 361.

Fenimore, C.P.(1971) Formation of nitric oxide in premixed flames. *13th Symposium (International) on Combustion,* 373.

Gordon, S. & McBride, B.J.(1971) Computer program for calculation of complex equilibrium composition, rocket performance, incident and reflected shocks and Chapman-Jouguet detonations *NASA* SP-273.

Greenhalgh, D. & Stopford P. (Eds).(1989) Final report turbulent combustion and diagnostics project. *Energy Conservation Programme of EEC.*

Haynes, B.S.(1977) Kinetics of Nitric Oxide formation in Combustion, in *Alternative Hydrocarbon Fuels: Combustion and Chemical Kinetics* Ed. Bowman & Birkeland. Also see *Progress in Astro & Aeronautics* **62** 359.

Jones, W.P, Clifford, W.C, Priddin, C.H. & de Chair, R.(1978) A comparison between predicted and measured species concentrations and velocities in a research combustor. Paper 40, *AGARD* CP-229.

Jones, W.P. & Kollman, W.(1987) *Turbulent Shear Flows 5* (ed. F. Durst et al.) Springer-Verlag 296.

Jones, W.P. & Priddin, C.H.(1978) Predictions of the flowfield and local gas composition in gas turbine combustors. *Proc. 17th Symposium on Combustion,* The Combustion Institute, 399-409.

Jones, W.P.(1979) Models for Turbulent flows with variable density and Combustion. In *Prediction methods for turbulent flows* Ed. W.Kollman (VKI Lecture Series 1979-2).

Kee, R.J., Miller, J.A. & Jefferson, T.H.(1980) CHEMKIN: A general purpose, problem-independent transportable Fortran chemical kinetics code package. *Sandia rept no 80-8003* 204. Sandia National labs, California.

Launder, B.E. Reece, G.J. & Rodi W.(1975) Progress in the development of a Reynolds Stress turbulent closure. *J. Fluid Mech* **68** 537.

Magnussen, B.J.(1985) Heat transfer in gas turbine combustion. *AGARD 65th Symposium,* Bergen, Norway May 1985.

Moss, J.B., Stewart, C.D. & Syed K.J.(1987) Flamelet Chemistry Modelling of Soot formation for radiation prediction in combustor flow fields. Paper 15, *AGARD PEP 70th Symposium Chania, Crete* (CP442).

Naguib, A.S. & Lockwood, F.C.(1975) Prediction of the fluctuations in the properties of free round jet turbulent diffusion flames. *Combustion and Flame,* **24** 109.

Peters, N.(1984) Laminar diffusion flamelet models in non-premixed turbulent combustion. *Progress in Energy and Combustion Science* **10** (3) 319.

Pope, S.B. & Correa, S.M.(1988) *21st Symposium (International) on Combustion,* The Combustion Institute, 1341.

Pope, S.B.(1979) Probability Distributions of Scalars in turbulent shear flow. *2nd Symp on Turbulent Shear Flows.*

Pope, S.B.(1985) *Progress in Energy & Combustion Science* **11**, 119.

Pope, S.B.(1990) Computations of Turbulent Combustion: progress and challenges. *23rd Symposium (International) on Combustion, Orleans, France.*

Priddin, C.H. & Coupland, J.(1988) Impact of Numerical Methods on Gas Turbine Combustor Design and Development. *Combustion Science & Technology* **5 8** 119.

Rhodes, R.P.(1975) A probability distribution function for turbulent flows. In *Turbulent Mixing in Non-reactive and Reactive Flows.* ed. S.N.B.Murthy, Plenum.

Rogers, M.M., Moin P, & Reynolds W.C.(1986) The structure and modeling of the hydrodynamic and passive scalar fields in homogenous turbulent shear flow. *Stanford University report TF-25* (Thermosciences, Mech Eng Dept).

Schumann, U.(1973) Procedure for the direct numerical simulation of turbulent flows in plane channels and annuli and its application in the development of turbulence models. *Univ. of Karlsruhe report KFK-1854.*

Shyy, W, Braaten, M.E. & Burrus, D.L.(1989) Study of three dimensional gas turbine combustor flows. *Int J of Heat Mass Transfer,* **32** (6) 1155.

Spalding, D.B.(1971) Concentration fluctuations in a round turbulent free jet. *Chem Eng Science* **26**, 95.

Spalding, D.B.(1971) Mixing and chemical reaction in steady confined flames. *13th Symposium on combustion,* The Combustion Institute, Pittsburg, 649-657.

Spalding, D.B.(1977) *GENMIX: A general Computer Program for two dimensional Parabolic Phenomena.* Pergamon Press.

Veynante, D, Lucas, F. & Candel, S.M.(1989) A new flamelet combustion model combining premixed and non-premixed turbulent flames. *AIAA 27th Aerospace Sciences meeting,* paper 89-0487.

Warnatz, J.(1981) *Combustion Science & Technology* **26** 203.

Westbrook, C.K. & Dryer, F.L.(1979) *Combustion Science & Technology* **20** 125.

Zel'dovich, Ya.B.(1946) *Acta Physicochim.* (USSR) **21** 577.

Effects of Turbulence on the Mixing and Chemical Reaction for Cross Flow and Coflowing jets

H. Stapountzis[1], *P. Tzavellas*[1], *T. Moros*[2]

1. University of Thessaloniki, Dept. of Mechanical Engineering, Thessaloniki 540 06, Greece
2. BP Research, Sunbury Research Center, Middlessex TW16 7LN U.K.

1. Introduction and scope of the work

The subject of reactive mixing in turbulent flows is one of vast technological importance and arises in many practical situations such as combustion devices, batch, tubular and continuous stirred tank reactors, chemical lasers and photochemical reactions in the atmospheric boundary layer. Turbulence is known to influence strongly the diffusion process and as a result the chemical reaction in a flowing fluid mixture. If the time scale characterizing the reaction kinetics is much greater than any appropriate time scale of turbulent scalar transport in the flow, then the chemical reaction is said to be slow. In this case turbulence succeeds in mixing the constituents before any chemical reactions become effective. In the opposite case the reaction proceeds at a rate determined by the fluid transport process through turbulent and molecular diffusion.

The present work is motivated by the need to improve atmospheric pollution models which involve relatively fast chemical reactions. In a second order isothermal fast irreversible chemical reaction involving two species A and B, the time averaged reaction rate, reactant and product concentrations depend also on the concentration correlation, $\varrho = < c_A c_B >$, the concentration fluctuations c_A and c_B existing because of turbulence. In fact ϱ is negative in fast reactions, Komori and Ueda (1984), ϱ growing because fast chemistry prevents A and B from coexisting and ϱ diminishing because of destruction by molecular diffusion. A corresponding quantity $\alpha = < c_A c_B > /(c'_A c'_B)$ (where ' denotes RMS) for scalar mixing of A and B was introduced by Warhaft (1984) and termed cross correlation coefficient. For the case of the mid-region between two turbulent plumes of A and B produced by line sources α is initially negative since the plumes never overlap instantaneously and far from the sources is positive, $\alpha = 1$ indicating complete mixing, $\alpha \sim 0$ roughly indicating commencing of mixing.

In this paper we are concerned with the influence of background turbulence on the mixing and chemical reaction of two species present or injected into the flow. All three cases examined are shown in Fig. 1. By background turbulence we mean here the turbulence which exists in the oncoming flow in addition to that present in the flowfield of the injected scalars or reactants. The length scales of the oncoming turbulent flow were chosen to be greater or comparable to the length scales of the released species.

2. Experimental arrangement and conditions

A schematic representation of the test cases is shown in Fig. 1. Scalar (salinity) dispersion measurements were made in a pump driven water channel 1.5 m wide, 0.8 m deep and 5 m long at Cambridge University Engineering Department. Chemical reaction experiments were performed in a gravity driven water tunnel of 0.1 × 0.1 cross section

Advances in Turbulence 3
Editors: A.V. Johansson · P.H. Alfredsson

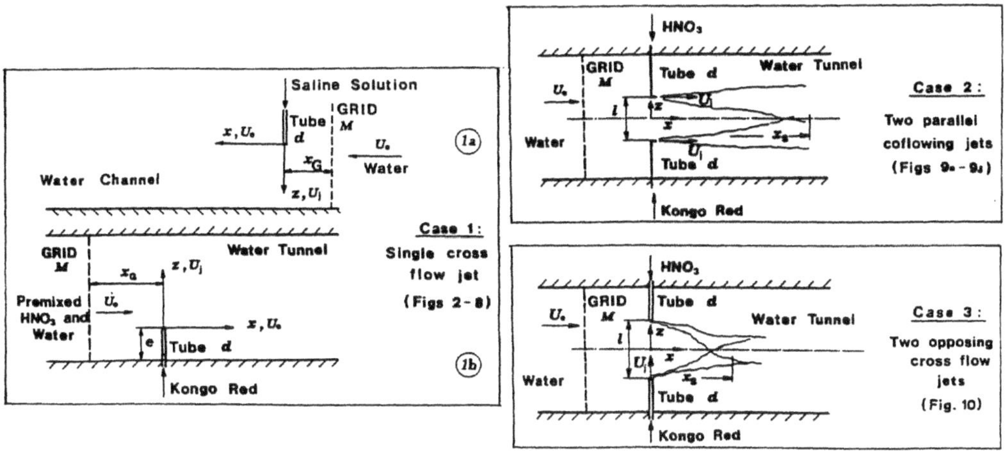

Fig. 1 Schematic representation of the test cases

and 1.9 m long at the Mechanical Engineering Department of Thessaloniki University. Cases 1a and 1b refer to a single cross flow jet, case 2 to two parallel coflowing jets and case 3 to two opposing cross flow jets. The scalar tracer and reactants were introduced into the flow by means of tubes of inner diameter d placed at distance x_G from biplanar square mesh grids of mesh size M (bar width to mesh size ratio = 5.3). The ratio $R = U_j/U_0$ of the injection velocity to the free stream velocity was varied in the experiments as well as the distance e of the tube top from the tunnel floor and the distance l between the tube ends.

The inert scalar marker (salty water) had a density of $1008 kg/m^3$ at 17 C and for the values of U_0 and R considered here ($R = 5$) it was essentially neutrally buoyant. For such low electrolyte concentrations, it was possible to use an aspirating conductivity probe for concentration measurements. Mean and fluctuating velocity signals were measured by a hot film. The reactants for the flow visualization were aqueous solutions of Nitric Acid and a Congo red indicator. The product of their reaction has a deep blue to black colour. The densities and kinematic viscosities of the acid and indicator solutions were at 17 C $1060 kg/m^3$, $997 kg/m^3$ and $1.063 \times 10^{-6} m^2/s$, $1.048 \times 10^{-6} m^2/s$ respectively. For the case 1b where the oncoming flow was premixed HNO_3 and water, the mixing took place upstream of the settling chamber.

The useful advantages of the visible reaction product are mentioned in Breidenthal (1981): For very fast chemical reactions, completed mixing implies completed reaction, and the amount of product formed is just equal to the amount of molecular scale mixing between the two species at the stoichiometric ratio. The axial position of commencing of reaction, x_s, was defined by the appearance of intense color of the product, above a preset threshold value ; it was estimated by image processing of the video recordings and from prints at long time exposure. A summary of the experimental flow conditions is shown in Table 1.

Table 1. Experimental flow conditions.

Case	U_0 (m/s)	R	M (mm)	x_G/M	d (mm)
1a	0.06	5.0	152	10	8.0
1b	0.30	5.0	25	10,18,34	2.5
1b	0.30	2.5	25	10	2.5
1b	0.30	2.0	25	10	2.5
1b	0.30	1.0	No Grid	--	4.0
1b	0.30	1.0	25	10,18,34	4.0
2	0.035	54.0	20	10	0.5
2	0.055	34.0	20	10	0.5
2	0.080	24.0	20	10	0.5
2	0.130	15.0	20	10	0.5
3	0.13	2	25	10,18,34	2.5

3. Experimental results and discussion

3.1 Non-reactive cross flow jet (Case 1a)

All axial velocity and scalar concentration measurements presented here refer to the symmetry plane $x - z$ and an injection ratio $R = U_j/U_0 = 5.0$. Fig. 2 shows the transverse distribution of mean concentration, C, at various streamwise positions x/d. The mean concentrations are nondimensionalized by the mean background concentration C_0 and the injected fluid mean concentration C_j. The mean profiles are approximately symmetric with diminishing maxima as x/d increases. In Andreopoulos (1983) non-symmetric profiles were noticed for about $R < 2$ this being a result of asymmetric pressure distribution around the exit imposed by the cross stream. This causes the flow at the upwind side of the pipe to decelerate and at the downwind side to accelerate.

Fig. 3 shows the jet centerline obtained from the locations of the maxima of mean velocity and concentration profiles. Pratte and Baines (1967), found that the centerline

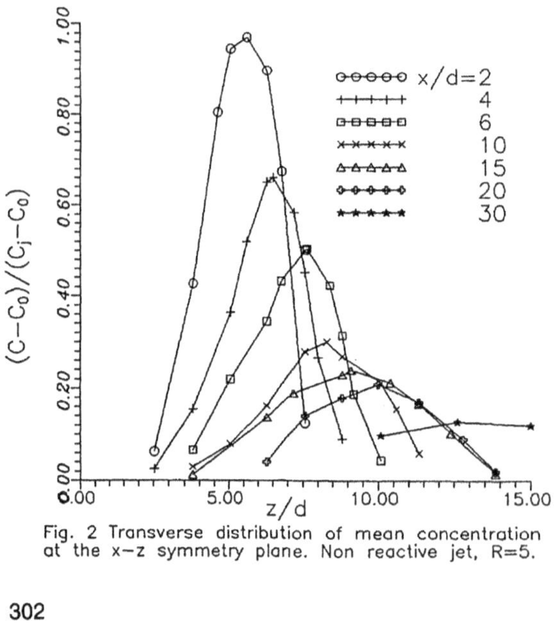

Fig. 2 Transverse distribution of mean concentration at the x−z symmetry plane. Non reactive jet, R=5.

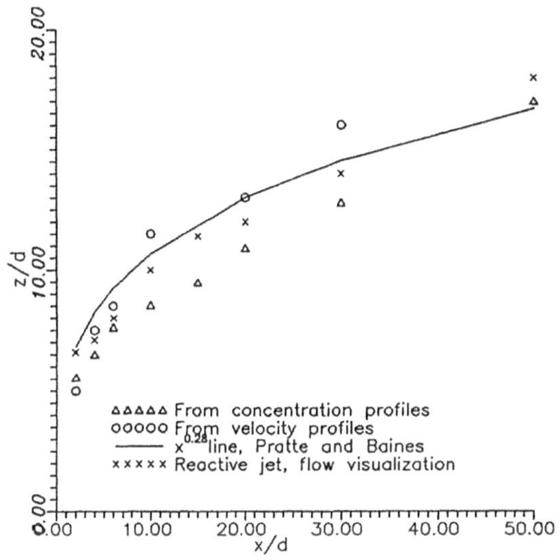

Fig. 3 Location of jet centerline at the symmetry plane, R=5.0

for velocity could be described by $z/d = 1.76R^{0.72}(x/d)^{0.28}$ and this equation is plotted as a solid line in Fig. 3. Also shown is a centerline obtained from an averaged picture of the flow by taking a long time exposure in the chemical reaction flow visualization tests. The velocity centerline is close to that given by Pratte and Baines both lying above the mean concentration centerline. This has been observed by other workers e.g. Haniu and Ramaprian (1989). One reason for this discrepancy could be the method of determining the centreline, since velocity profiles have two maxima while concentration profiles have one maximum (Fig. 2).

Non-marked fluid entrainment is caused by the actions of counter rotating bound vorticity, the shed vorticity (along the z axis, generated by interfacial shear on the sides of the jet) and the background turbulence. The rates of transport of momentum and mass appear to be different for this flow as demonstrated in Figs 4 and 5 ; they represent

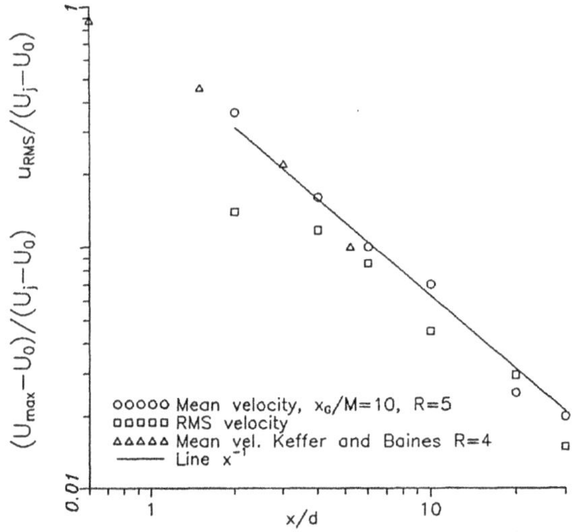

Fig. 4 Distribution of mean and RMS velocity fluctuations along the jet centerline

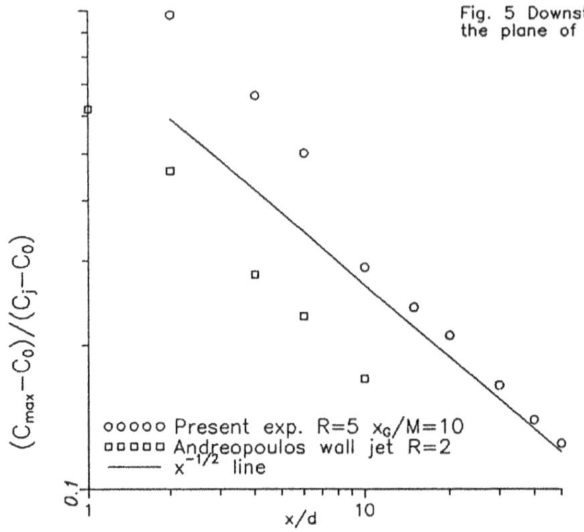

Fig. 5 Downstream decay of mean maximum concentration at the plane of symmetry along centerline. Non reactive jet.

the decay of mean maximum velocity U_{max}, the RMS velocity fluctuations along the centerline and the mean maximum concentration C_{max} with downstream distance x. While the mean and the RMS velocity tend to decay according to $1/x$ (as in Keffer and Baines (1963) for the mean velocity) the mean maximum concentration attains a self preserving form for about $x/d > 10$ and decays according to $x^{-1/2}$ as also found by Andreopoulos (1983) for a slightly heated cross flow jet. Axisymmetric coflowing jets are characterized by an $1/x$ rate of decay of non-reactive species concentration, while the U_{max} decay depends on both $1/x$ and R.

Although the RMS concentration fluctuations, c', along the centerline are found to decrease when they are non- dimensionalized by the maximum concentration $C_{max} - C_0$, they are seen to initially increase and for large x/d to flatten out at a level of about 0.8, see Fig. 6, resembling the trend of a turbulent plume from a line source, Stapountzis

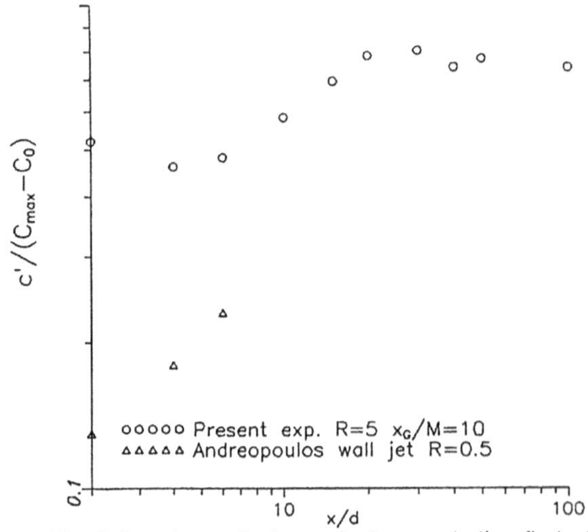

Fig. 6 Downstream development of concentration fluctuations at the plane of symmetry along centerline. Non reactive jet.

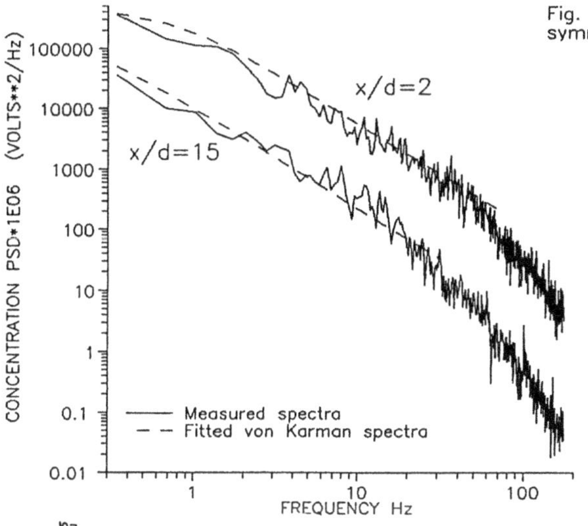

Fig. 7 Concentration spectra measured at the symmetry plane where c_{RMS} is max. R=5.

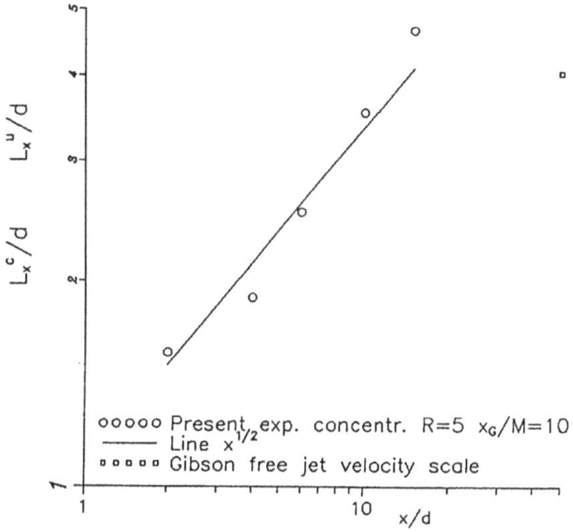

Fig. 8 Downstream development of concentration and velocity length scales at locations of max. RMS at the plane of symmetry.

et al. (1986). The data of Andreopoulos (1983) for $R = 0.5$ do not extend sufficiently downstream to reveal such behaviour.

The lowering of the overall level of the concentration fluctuations is also reflected in the concentration spectra for $x/d = 2$ and 15. Due to the relatively low Reynolds numbers of the present flow there is not a clear cut region of an inertial subrange. However, fitted von Karman spectra (strictly valid for isotropic conditions) up to 50 Hz and shown in Fig. 7 helped to get an estimate of the integral length scales L_x^c, plotted in Fig. 8. It is observed that the general tendency of the concentration length scale is to increase at nearly the same rate as the velocity scale in grid turbulence without the jet, Stapountzis et al. (1986).

3.2 Reactive cross flow jet (Case 1b)

The progress of the reaction, as affected by changes in the level of background turbulence, was studied by flow visualization. The jet was exposed to nominally smooth flow and to turbulence produced by the grid. Examples, for $R = 1.0$, are shown in plates 1 and 2 for smooth and turbulent ($x_G/M = 34$) background flow respectively. These are long time exposure prints and the region of initial strong product generation is marked by a white dotted line. The wider spread of the jet in the latter case is evident, in accordance with our earlier concentration measurements (Stapountzis 1989, Stapountzis and Moros 1989). The jet also "bends" less in turbulent flow.

In the potential core of the jet, where mixing is weak, no chemical reaction takes place in both smooth and turbulent background flow. x_s is also little affected, for $R = 1$ and the other values of R tested, its value being smaller for turbulent flow.

Next the reactive jet (with $R = 5.0$) was positioned at three downstream locations $x_G/M = 10, 18$ and 34. Plates 3 and 4 show the jet flow for the first and last position. Turbulence intensities there are about 4.5% and 2.6% respectively, while the vertical velocity fluctuation integral length scales, L_x^w, about 0.28M and 0.41M. The flow visualization employed here does not show drastic changes in the progress of the reaction for different x_G/M. When the injection takes place closer to the grid, where turbulence intensity and entrainment are higher, a stronger activity on the lee side of the jet and close to the tube wall is observed. Product formation is enhanced in this region. Similar increased activity is evident when the tube exit is immersed into the turbulent boundary layer of the tunnel wall, $e = 1$ mm. It is now manifested as inclined vortex shedding, which is more pronounced closer to the grid, see Plates 5 and 6 for $x_G/M = 10$ and 34 respectively.

The effect of changing the stoichiometry of the reaction, by decreasing the concentration of the Nitric Acid was to decrease the rate of formation of the product and increase x_s. This has been observed by other workers like Komori and Ueda (1984).

To summarize, it appears that for the range of turbulence intensities and length scales tested, the mixing and chemical reaction proceeds according to the turbulence generated by the jet itself, rather than the ambient turbulence.

3.3 Coflowing parallel reactive jets (Case 2)

Only one downstream location was tested, $x_G/M = 10$, at jet axis transverse distances $l/d = 2, 8, 16, 36$ and ratios $R = 54, 34, 24$ and 15. The background turbulence characteristics are therefore kept the same, but the scale of mixing varies relative to the turbulence length scale. Plate 7 shows a typical case. The variation of x_s with l/d and R is shown in Figs. 9a to 9d. As l/d increases, intense chemical reaction is postponed till further downstream. There is always an initial region where very little reaction occurs even when the two tubes are in contact because of the potential core of the jets. In Fig. 9, x_b is the distance from the jet exits where the half width b of the jet velocity profile is just equal to $l/2$, assuming that the velocity fields of the two jets are not significantly interfering with each other. The half width is defined as the distance from a single jet axis where the streamwise velocity is equal to $(U_{max} + U_0)/2$. Blevins (1984) gives an equation for $b = (d/2)(x/x_1)^{1-r}$ where $r = 1/R$ and x_1 is the length of an initial region, $x_1 = (5 + 12r)d$. We note that x_b (a measure of the distance where the two jets have a sizeable area of contact) correlates well with xs except for low l/d. As U_0 increases (or

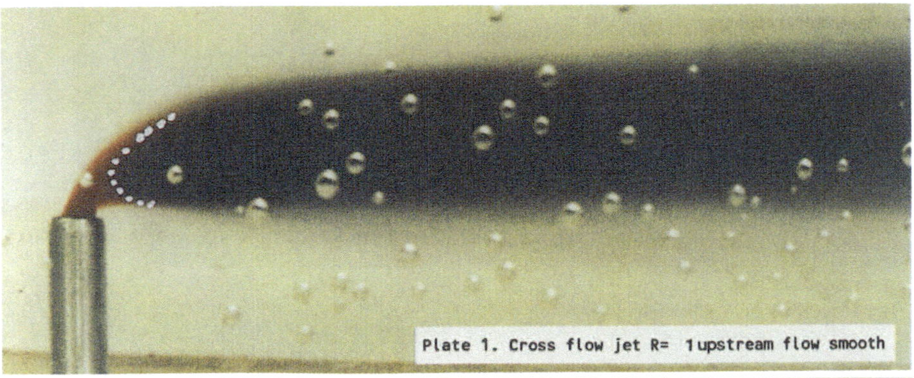

Plate 1. Cross flow jet R= 1 upstream flow smooth

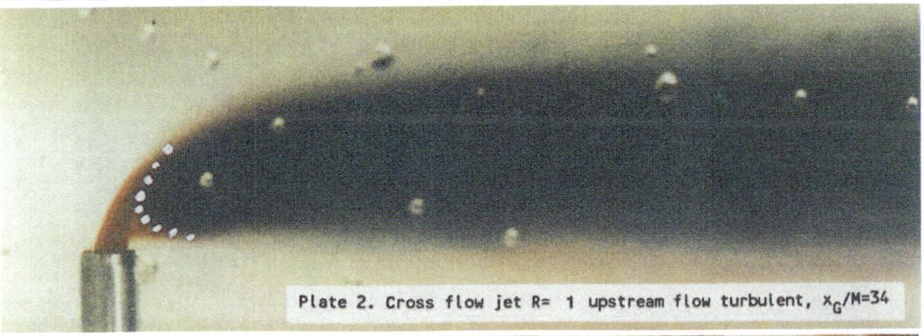

Plate 2. Cross flow jet R= 1 upstream flow turbulent, $x_G/M=34$

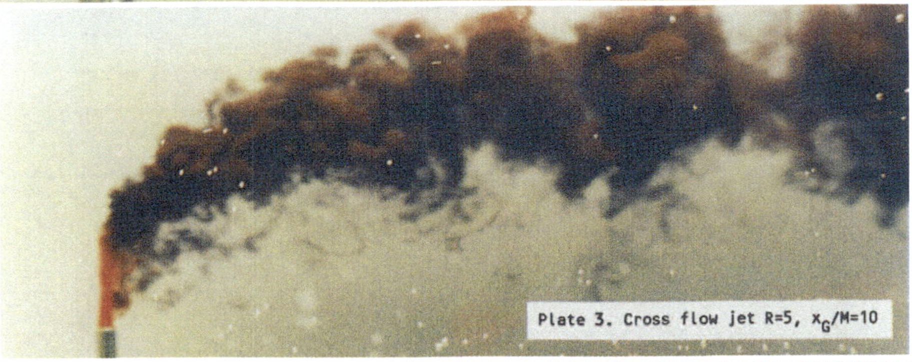

Plate 3. Cross flow jet R=5, $x_G/M=10$

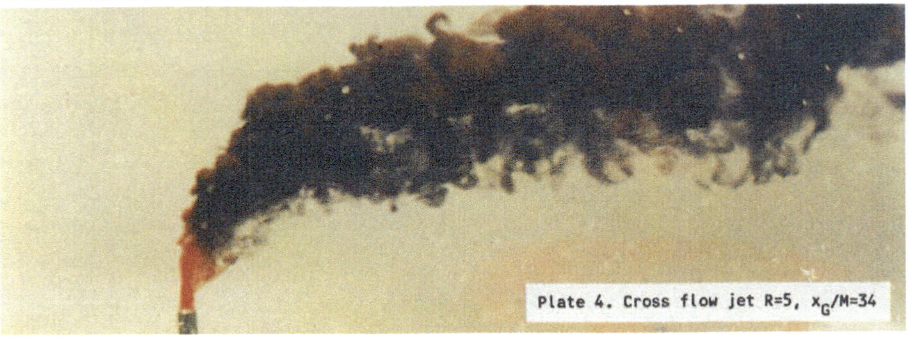

Plate 4. Cross flow jet R=5, $x_G/M=34$

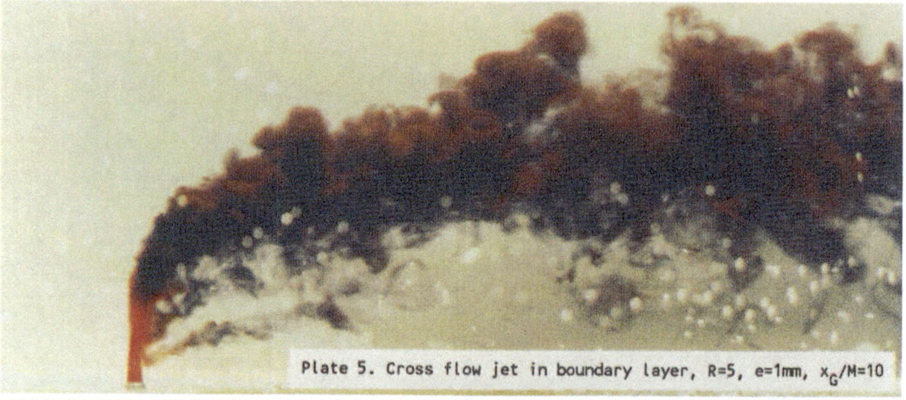
Plate 5. Cross flow jet in boundary layer, $R=5$, $e=1mm$, $x_G/M=10$

Plate 6. Cross flow jet in boundary layer, $R=5$, $e=1mm$, $x_G/M=34$

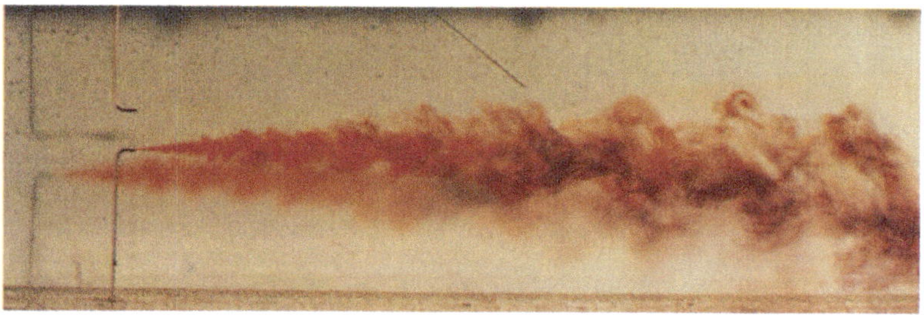
Plate 7. Two coflowing jets, $R=34$, $l/d=16$, $x_G/M=10$

Plate 8. Two cross flow opposing jets, $R=2$, $l/d=2$, $x_G/M=34$

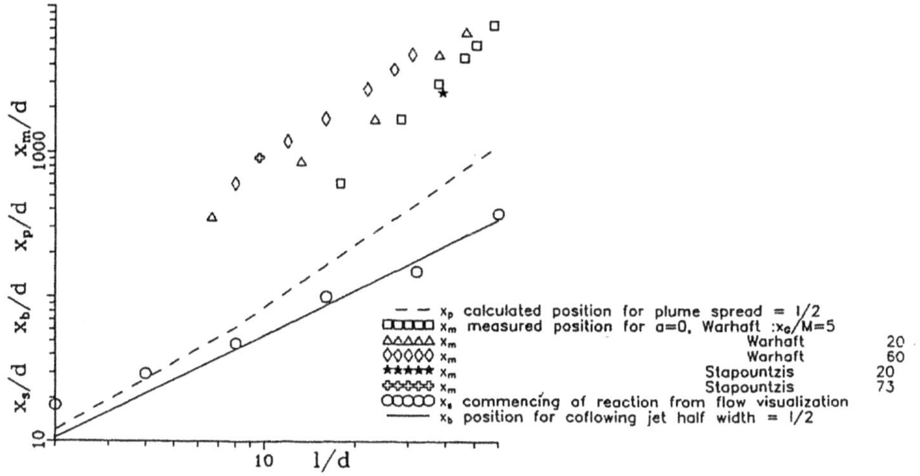

Fig. 9a Streamwise position of commencing of reaction for two coflowing jets and comparison with single jet and plume characteristics. $U_j/U_0=54$, $Re_M=700$, d=0.5mm

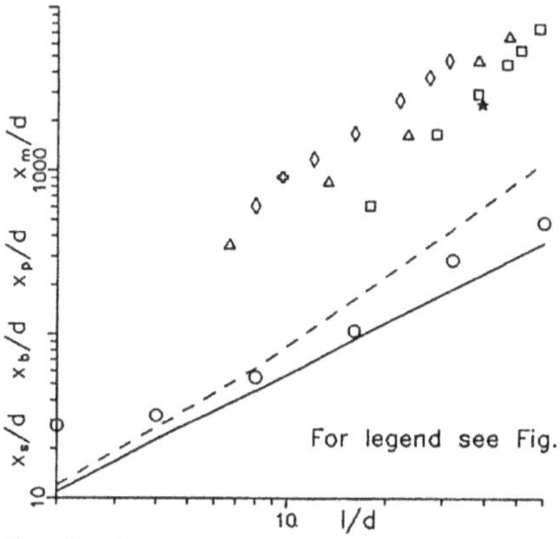

Fig. 9b Streamwise position of commencing of reaction for two coflowing jets and comparison with single jet and plume characteristics. $U_j/U_0=34$, $Re_M=1100$, d=0.5mm

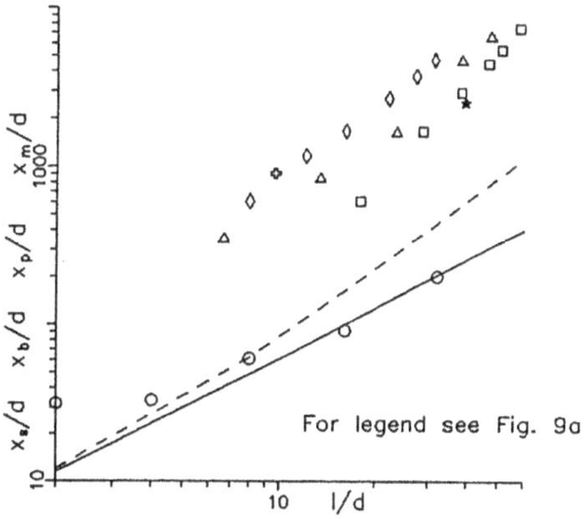

Fig. 9c Streamwise position of commencing of reaction for two coflowing jets and comparison with single jet and plume characteristics. $U_j/U_0=24$, $Re_M=1600$, d=0.5mm

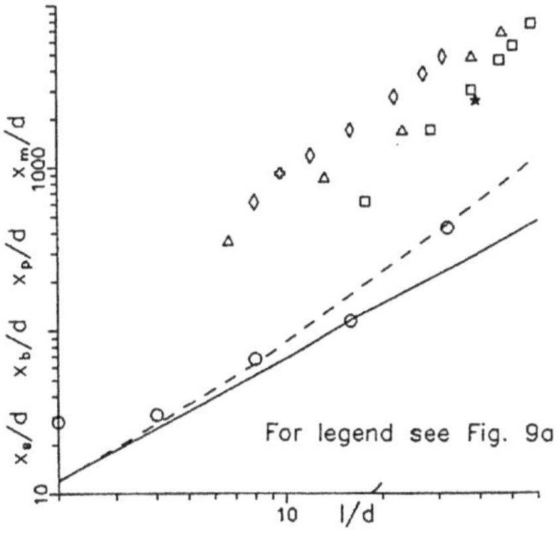

Fig. 9d Streamwise position of commencing of reaction for two coflowing jets and comparison with single jet and plume characteristics. $U_j/U_0=15$, $Re_M=2600$, d=0.5mm

R decreases) x_s becomes systematically greater than x_b, i.e. chemical reaction occurs later, this being probably a result of the greater dilution of the reactants. It is also noticed that for increasing U_0, x_b approaches x_p, the distance from the jet exits where the spread of a turbulent plume of initial size d becomes equal to $l/2$. This spread was computed from Taylor's theory of dispersion, see for example Durbin (1980), assuming isotropic turbulent flow conditions. As R tends to l, the coflowing jet is closer to a plume. Komori and Ueda (1984) found that turbulence in the ambient flow promotes the chemical reaction of a plume to a much greater extent than it does for a jet and this is in line with our findings, i.e. x_s correlates well with x_b (derived for non turbulent background flow).

The conjecture that two plumes released from point sources at distance l would probably mix not as efficiently as two jets at the same distance l can be seen in Fig. 9 where x_m is also plotted against l/d. x_m is defined as the streamwise position, downstream and midway between two line sources, where α (defined by Warhaft, see Introduction) becomes zero, a sign that the two plumes begin to interact. The experimental values of x_m from Warhaft (1984) and Stapountzis (1987) which are plotted in Fig. 9 are well above x_s. The similarity in the behaviour of x_b (related to smooth ambient flow) and x_s, suggests, as in the cross flow jet, that the present levels of background turbulence do not significantly influence the progress of mixing and reaction as much as the turbulence inherently present in the coflowing jets.

3.4 Two opposing crossflow reacting jets (Case 3)

An example of this flow type is shown in plate 8. Fig. 10 demonstrates the variation of x_s with l/d for $R = 2$ and $x_G/M = 10$. It is noticed that the chemical reaction occurs much earlier than for the two coflowing jets. At large distances x/d the values of xs for opposite jets approach those for two coflowing jets since for large l/d the penetration length of the cross jets are also small compared to l (for small fixed R) and the bent over cross jet resembles a coflowing jet from a fictious (virtual) origin. The value of x_s for a single cross flow jet (Fig. 10). For small l/d the formation of the reaction product is critically dependent upon the exact alignment of the two tubes. For $l/d = 2$ the product is formed almost immediately and is carried downstream. Again there is no appreciable effect of the background turbulence, apart from a wider spread of the reactants and the product as one gets closer to the grid. Due to the low injection ratio

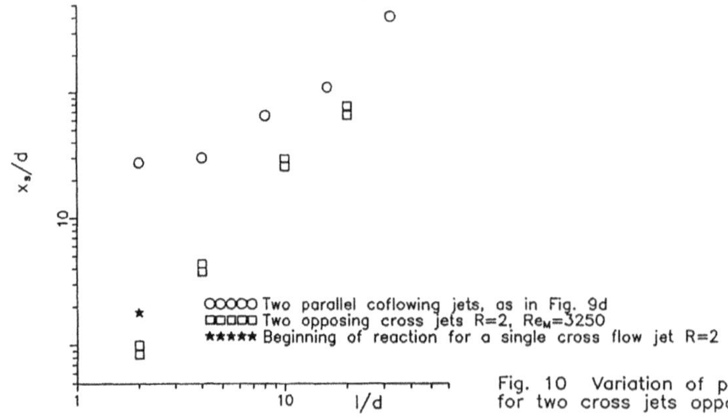

Fig. 10 Variation of position of beginning of reaction for two cross jets opposite to each other.

and the jet configuration the resulting flow has lower momentum in the x direction and it tends to behave more like a plume, which is more amenable to turbulence effects than the jet. For small l/d it is also observed that the injected fluids tend to overlap and occupy each other's regions.

4. Conclusions

The region of intense mixing and chemical reaction of a reactive jet in a turbulent cross flow corresponds to the region of high concentration fluctuations of a similar non reactive jet. For the range of intensities and length scales examined in the present work, the presence of turbulence in the cross stream was not found to exert a significant influence on the progress of the chemical reaction and this is also the case for two parallel coflowing reactive jets. It seems that the turbulence generated in the jets themselves has a more important role to play in the mixing and chemical reaction. Turbulence is likely to influence more the two cross flow opposing jets with low injection ratio of which the far field would resemble that of a plume.

References

Andreopoulos J. 1983 *Phys. Fluids* **26** no 11 p 3201.

Blevins R.D. 1984 " Applied Fluid Dynamics Handbook " Van Nostrand Reinhold
Breidenthal R. 1981 *J. Fluid Mech.* **109** pp 1-24.

Durbin P.A. 1980 *J. Fluid Mech.* **100** p161.

Gibson M.M. 1963 *J. Fluid Mech.* **15** pp 161-173.

Haniu H. Ramaprian B.R. 1989 *Trans. ASME J. Fluids Eng.* **111** pp 78-86.

Keffer J.F. Baines W.D. 1962, *J. Fluid Mech.* **15** p 481.

Komori S. Ueda H. 1984 *Phys. Fluids* **27** no 1 p 77.

Pratte B.D. Baines W.D. 1967 *ASCE J. Hydr. Div.* **92** pp 53-64, also **93** pp 815-816.

Stapountzis H. 1989 *5th Int. Symp. on Flow Visualization.*

Stapountzis H. 1988 *Int. Symp. on Transport Phenomena in Turb. Flows*, Tokyo pp 419-431.

Stapountzis H. Moros T. 1989, *Euromech 253* Brunel Univ. UK.

Stapountzis H. Sawford B.L. Hunt J.C.R. Britter R.E. 1986 *J. Fluid Mech.* **165** pp 401-424.

Warhaft Z. 1984 *J. Fluid Mech.* **144** pp 363-387.

Joint Statistics of Scalars and Their Gradients in Nearly Homogeneous Turbulence

Luis Valiño and César Dopazo

School of Mechanical Engineering, University of Zaragoza, M. de Luna, 3, 50015 Zaragoza, Spain

The mixing of dynamically passive inert and reactive scalars convected by a solenoidal field of nearly homogeneous turbulence is investigated. The transport equation for the joint probability density function (pdf) of scalars and their gradients is used. Closure approximations are required for those processes describing the turbulent straining of scalar gradients, the cross-dissipation of scalar/scalar-gradient and the scalar-gradient dissipation. Stochastic models in terms of random vectors are proposed and a Monte-Carlo simulation is then conducted. Predictions for a single inert scalar agree rather well with available experimental data. The simulation accurately predicts existing experiments for two moderately fast reacting scalars; large values of the flatness factors of species and species-gradients are strongly suggestive of spatial segregation and small-scale intermittency respectively. A far from gaussian nature of reactive scalars is well documented.

1. Introduction

The turbulent mixing of inert or reactive scalars exhibits features of great scientific and practical interest. The macromixing due to large structures leads to a phase in which small scale mixing follows. The turbulent straining of material lines and surfaces or iso-scalar surfaces significantly enhances molecular mixing. Should the scalars be reactive, the latter promotes the action of the chemical phenomena. Reactive processes with Damköhler numbers of order unity or larger display certain degrees of spatial segregation (O'Brien, 1971). The concentration flatness factors are then significantly larger than one. Likewise, large values of the flatness factors of concentration-gradients are traditionally associated with the small scale intermittency (Kerr, 1985) or spottiness of dissipation for both inert and reactive scalars.

Statistically averaged moment (Jones and Musonge, 1988, and references therein), or probability density function (pdf) (Hill, 1976) transport equations for scalar fields are conventionally used to describe turbulent mixing problems, the latter being naturally suited for reactive scalars. The modelled transport equation for the scalar dissipation rate is acknowledgedly the weakest point in second order closures (Jones and Musonge, 1988). This difficulty is related to the external specification of a time or length scale in pdf methods (Pope, 1985). The original pdf transport equation for one scalar (Dopazo and O'Brien, 1974, 1976) and for one scalar and velocity (Dopazo, 1976; Pope, 1981a) have subsequently been extended to multipoint statistics (O'Brien, 1980), two-time pdf's (Pope, 1983; Kollmann and Wu, 1987) and joint pdf's of one scalar and its gradient (Meyers and O'Brien, 1981). The probability mixing associated to molecular diffusion

Advances in Turbulence 3
Editors: A.V. Johansson · P.H. Alfredsson

terms is the main source of difficulties in closing the pdf transport equations (Dopazo, 1979). Models in the past partially solved the relaxation from some arbitrary initial conditions to an asymptotically gaussian scalar field (Pope, 1985). Recent ideas (Valiño, 1989; Valiño and Dopazo, 1990; Chen *et al.*, 1989) significantly improve upon existing closure hypothesis leading to a qualitatively reasonable and a correct asymptotically gaussian behavior when compared with full turbulence simulations (Eswaran and Pope, 1988). Another peculiarity of the pdf methodology has to do with the high dimensionality of the dependent variable. Given the large number of independent variables in most cases of interest, the pdf transport equations have to be solved by MonteCarlo techniques (Pope, 1981b, 1985). Stochastic models suited for MonteCarlo numerical simulations have also been proposed (Pope, 1985; Haworth and Pope, 1986; Valiño and Dopazo, 1990; Girimaji and Pope, 1990).

A pdf methodology to describe the turbulent mixing of inert or reactive scalars is presented in this paper. In Section 2 the joint pdf transport equation of scalar fields and their gradients is introduced. The closure of the turbulent straining of scalar gradients is considered in Section 3. Section 4 deals with the MonteCarlo simulation of the turbulent straining and the molecular transport terms. The simulation of pdf transport by mean velocity gradients and chemical reaction is exact within discretization and sampling errors. Numerical results for both inert and reactive scalars are presented in Section 5 and compared with available experimental data for homogeneous turbulent flows. Section 6 contains some concluding remarks.

2. Problem formulation

The equation governing the transport of a scalar field α is,

$$LC_\alpha = S_\alpha(\mathbf{C}) \tag{1}$$

where

$$L = \frac{\partial}{\partial t} + v_j \frac{\partial}{\partial x_j} - D\nabla^2 \tag{2}$$

is the convection/diffusion mixing operator, $v_j(\mathbf{x}, t)$ is the local, at point $\mathbf{x}$, and instantaneous, at time t, velocity field with mean $U_j(\mathbf{x}, t)$ and fluctuation $u_j(\mathbf{x}, t)$. D is the molecular diffusivity assumed identical for all the species α. $\mathbf{C} = \{C_\alpha\}$, for $\alpha = 1, 2, ..., M$, is the ensemble of M concentration fields, driven by the turbulence and simultaneously undergoing molecular mixing and chemical reaction. The local and instantaneous $C_\alpha(\mathbf{x}, t)$ has mean $< C_\alpha(\mathbf{x}, t) >$ and fluctuation $c_\alpha(\mathbf{x}, t)$. The chemical source term, S_α, depends, in general, upon all the scalar variables $\mathbf{C}$.

Let the gradient of C_α be defined as $C_{\alpha,i} = \partial C_\alpha / \partial x_i$. It obeys the conservation equation

$$LC_{\alpha,i} = -\frac{\partial v_j}{\partial x_i} C_{\alpha,j} + \frac{\partial S_\alpha}{\partial C_\beta} C_{\beta,i} \tag{3}$$

Define the joint pdf of $\mathbf{C}$ and $\nabla\mathbf{C}$ as

$$P(\underset{\sim}{\Gamma}, \underset{\approx}{\chi}; \mathbf{x}, t) = \langle \delta[\underset{\sim}{\Gamma} - \mathbf{C}(\mathbf{x}, t)] \delta[\underset{\approx}{\chi} - \nabla\mathbf{C}(\mathbf{x}, t)] \rangle \tag{4}$$

where $\underset{\sim}{\Gamma}$ and $\underset{\approx}{\chi}$ stand for $\mathbf{C}(\mathbf{x},t)$ and $\nabla\mathbf{C}(\mathbf{x},t)$, respectively, in probability space and the angular brackets denote ensemble average. $\underset{\approx}{\chi}$ is the tensor with $3M$ components $\chi_{\alpha i}$. The first subscript, α, identifies the specific scalar field and the second, i, the gradient vector component. Restricting the investigation to nearly homogeneous turbulence and scalar fields with, possibly, a mean velocity gradient, it is straightforward to obtain the transport equation for $P(\underset{\sim}{\Gamma},\underset{\approx}{\chi};\mathbf{x},t)$,

$$\frac{\partial P}{\partial t}-\frac{\partial U_j}{\partial x_i}\frac{\partial}{\partial \chi_{\alpha i}}(\chi_{\alpha j}P)+\frac{\partial}{\partial \Gamma_\alpha}[S_\alpha(\underset{\sim}{\Gamma})P]+\frac{\partial}{\partial \chi_{\alpha i}}[\frac{\partial S_\alpha(\underset{\sim}{\Gamma})}{\partial \Gamma_\beta}\chi_{\beta i}P]$$
$$+D\frac{\partial^2}{\partial \Gamma_\alpha \partial \Gamma_\beta}(\chi_{\alpha i}\chi_{\beta i}P)=\frac{\partial}{\partial \chi_{\alpha i}}(\langle\frac{\partial u_j}{\partial x_i}|\underset{\sim}{\Gamma},\underset{\approx}{\chi}\rangle\chi_{\alpha j}P)$$
$$-2D\frac{\partial^2}{\partial \Gamma_\alpha \partial \chi_{\beta j}}(\langle\frac{\partial C_{\beta,j}}{\partial x_i}|\underset{\sim}{\Gamma},\underset{\approx}{\chi}\rangle\chi_{\alpha i}P)$$
$$-D\frac{\partial^2}{\partial \chi_{\alpha j}\partial \chi_{\beta k}}(\langle\frac{\partial C_{\alpha,j}}{\partial x_i}\frac{\partial C_{\beta,k}}{\partial x_i}|\underset{\sim}{\Gamma},\underset{\approx}{\chi}\rangle P) \quad (5)$$

where $< V|\underset{\sim}{\Gamma},\underset{\approx}{\chi} >$ is the expectation of the variable V conditional upon $\mathbf{C}(\mathbf{x},t)=\underset{\sim}{\Gamma}$ and $\nabla\mathbf{C}(\mathbf{x},t)=\underset{\approx}{\chi}$. All the terms on the left hand side (LHS) of Eq. (5) are closed. The second term on the LHS physically represent the straining of scalar gradients by the mean velocity gradient. The following two terms produce transport of pdf in concentration and concentration gradient spaces, respectively due to chemical reaction. The dissipation of the scalar fields is expressed by the last term on the LHS. The terms on the right hand side (RHS) of Eq. (5) are not closed. The straining of scalar gradients by the fluctuating velocity gradient is ascribed to the first term on the RHS. The last two terms represent the cross-dissipation of scalar/gradient and the dissipation of gradient/gradient correlations, respectively, due to molecular diffusitity. In order to use Eq. (6) to predict the joint statistics of scalars and their gradients the three terms on the RHS must be approximated.

3. Turbulent straining of scalar gradients

The first term on the RHS of Eq. (5) represents the action of the turbulent strain rate and rotation upon scalar gradients. A sequential MonteCarlo simulation of Eq. (5) will be explained in the following section. According to this approximation, the turbulent straining can be treated in isolation from the molecular dissipation and chemical processes. All of them sequentially and separatelly act upon the pdf or its discrete representation by an ensemble of stochastic particles. Using a cumulant expansion technique (Kubo, 1963) and an assumption on the gaussianity of $\partial u_j/\partial x_i$ (Meyers and O'Brien, 1981), it is easy to derive the model

$$\frac{\partial P}{\partial t}=\omega_s(4\chi_{\alpha k}\chi_{\beta k}\delta_{ij}-\chi_{\alpha i}\chi_{\beta j}-\chi_{\alpha j}\chi_{\beta i})\frac{\partial^2 P}{\partial \chi_{\alpha i}\partial \chi_{\beta j}} \quad (6)$$

where ω_s is termed the straining characteristic frequency. For a single scalar, $\alpha=\beta$, Eq. (6) becomes

$$\frac{\partial P}{\partial t} = 2\omega_s(2\chi_k\chi_k\delta_{ij} - \chi_i\chi_j)\frac{\partial^2 P}{\partial\chi_i\partial\chi_j} \tag{7}$$

Eq. (7) displays a tendency to an isotropic growth of $< c_{,i}\, c_{,j} >$. The off-diagonal terms decrease, while the diagonal ones increase, the smallest growing the fastest. The proper identification of the strong perturbation (Brissaud and Frisch, 1974) and the small time limits allows a rigorous derivation of Eqs. (6) and (7), avoiding the gaussianity assumption and yielding $\omega_s = t\langle\epsilon\rangle/30\nu$, where $\langle\epsilon\rangle$ is the average turbulent kinetic energy dissipation rate and ν is the kinematic viscosity.

4. MonteCarlo simulation

A sequential MonteCarlo simulation (Pope, 1981.b) of Eq. (5) is next considered. Every physical process represented by the various terms in Eq. (5) act, separately and sequentially, upon N stochastic particles that approximate $P(\underset{\sim}{\Gamma}, \underset{\approx}{\chi}; \mathbf{x}, t)$.

A. Turbulent straining

The objective is to devise a stochastic process that replicates Eqs. (6) or (7), the pdf rate equation under the action of turbulent straining. In order to achieve that, Eq. (6), for instance, is multiplied by an arbitrary function of $\underset{\sim}{\Gamma}$ and $\underset{\approx}{\chi}$, $f(\underset{\sim}{\Gamma}, \underset{\approx}{\chi})$ and, integrated over $\underset{\sim}{\Gamma}$ and $\underset{\approx}{\chi}$. Using then the discrete stochastic particle pdf representation, expanding in Taylor series up to second order in scalar and scalar-gradient increments and equating like powers of $\Delta C_\alpha^{(n)}$ and $\Delta C_{\alpha,i}^{(n)}$ suggest the following stochastic Langevin model

$$\Delta C_{\alpha,i}^{(n)}(t) = \omega_s^{1/2}(\delta t)^{1/2}(x\eta_i^{(n)}\xi_j^{(n)} + y\eta_j^{(n)}\xi_i^{(n)} + z\eta_k^{(n)}\xi_k^{(n)}\delta_{ij})C_{\alpha,j}^{(n)}(t) \tag{8}$$

where $\underset{\sim}{\eta}$ and $\underset{\sim}{\xi}$ are two statistically independent dichotomic or gaussian random vectors with zero mean and unity correlation matrix and $x \approx -0.3564$, $y \approx 2.8059$ and $z \approx -0.8165$.

A single random vector could alternatively be used to model the turbulent straining action upon the gradient for both single or multiple scalar fields. Eq. (8) could be replaced by

$$\Delta C_{,i}^{(n)}(t) = 2\omega_s^{1/2}(\delta t)^{1/2}|(\nabla C)^{(n)}|[\sqrt{2}\delta_{ij} - (\sqrt{2}-1)\frac{C_{,i}^{(n)}(t)}{|(\nabla C)^{(n)}|}\frac{C_{,j}^{(n)}(t)}{|(\nabla C)^{(n)}|}]\xi_j^{(n)} \tag{9}$$

B. Molecular mixing

The binomial sampling model (Valiño, 1989; Valiño and Dopazo, 1990) is used to evaluate the evolution of the ensemble of N MonteCarlo or stochastic particles replicating the joint pdf. Values of $C_\alpha^{(n)}$ and $(\nabla C_\alpha^{(n)})$ are ascribed to each particle. During a time step, δt, the ensemble undergoes an LMSE process changing the mean, $\langle C\rangle$, and the variance, $\langle c^2\rangle$ and $N\beta\omega_v$ randomly selected particles modify their values according to a binomial sampling. β is the relaxation coefficient, a positive dimensionless constant to be obtained by comparing the predictions and the experimental data, ω_v is the "mixing frequency" for the variable $V^{(n)}(t)$. The cross-dissipation of two scalars, α and β, is controlled by the number of MonteCarlo particles simultaneously participating in the two binomial samplings of α and β, respectively.

The gradient characteristic mixing frequency is in principle unkown. It is readily shown that

$$\frac{\omega_{\nabla \mathbf{c}}}{\omega_{\mathbf{c}}} = D\frac{\langle c^2\rangle\langle\epsilon_{\nabla \mathbf{c}}\rangle}{\langle\epsilon_{\mathbf{c}}\rangle^2} \tag{10}$$

where $\omega_{\mathbf{c}}$ and $\omega_{\nabla \mathbf{c}}$ are the characteristic mixing frequencies for scalar and gradient dissipation, respectively. $\epsilon_{\mathbf{c}} = Dc_{,i}\, c_{,i}$ and $\epsilon_{\nabla \mathbf{c}} = D\nabla c_{,i}\,\nabla c_{,i}$ are the local and instantaneous scalar and scalar-gradient dissipation rates, respectively. It is intuitively clear that the dissipation rate of the scalar gradient must have a characteristic time somehow smaller than that of the scalar, since smaller eddy structures are involved in the first process.

C. Straining by the mean velocity gradient

This process corresponds to the second term on the LHS of Eq. (5). In a sequential MonteCarlo simulation increment rate equation corresponding only to this term is

$$\Delta C^{(n)}_{\alpha,i}(t) = -\delta t(S_{ij} + R_{ij})C^{(n)}_{\alpha,j}(t) \tag{11}$$

S_{ij} and R_{ij} are the symmetric and the skew-symmetric parts of $\partial U_i/\partial x_j$, respectively.

D. Chemical reaction

The third and fourth terms on the LHS of Eq. (5) represent transport of probability in scalar and gradient spaces, respectively, due to chemical reactions. The increment equation for these isolated phenomena are

$$\Delta C^{(n)}_{\alpha}(t) = (\delta t)S_{\alpha}(C^{(n)}(t)) \tag{12}$$

$$\Delta C^{(n)}_{\alpha,i}(t) = (\delta t)\frac{\partial S_{\alpha}(C^{(n)}(t))}{\partial C^{(n)}_{\beta}}C^{(n)}_{\beta,i}(t) \tag{13}$$

For a two-species second-order bimolecular reaction, $S_{\alpha}(C) = -K_c C_1 C_2$ for $\alpha = 1,2$. K_c is the chemical rate constant.

5. Numerical results

No experimental data on the pdf of reactive scalars and their gradients for incompressible homogeneous turbulent flows exist. Bennani et al. (1985) produced a limited set of data on the homogeneous turbulent mixing of inert and reactive scalars; these data were used by Arrojo et al. (1988) to test the numerical predictions of velocity and concentration moments with a plausible prescription of the starting statistics. The interested reader is referred to Bennani et al. (1985) for a description of the experimental apparatus and flow parameters. Two different cases are considered.

A. Single inert scalar turbulent mixing

Bennani et al. (1985) injected dilute sodium chloride at the grid (mesh size M) generating turbulence within a main water flow with velocity $U = 0.7$ m/s. At $x = 10$ M statistical homogeneity was reached. The ratio of frequencies in Eq. (10) has been taken equal to 3.7 and the straining characteristic frequency $\omega_s = 0.0815\omega$, ω being the turbulent mixing frequency.

Figure 1 presents the evolution with time (equivalent to downstream variation) of the concentration variance. The agreement with experimental data is perfect. The scalar characteristic mixing frequency is plotted in Figure 2 together with the measured values showing moderately small deviations. The flatness factors of both concentration and concentration-derivative in the x-direction are represented in Figure 3. No experimentally measured values are available. The concentration flatness factor maintains a value of 3.0, indicating a preservation of the initial gaussianity, while that of the concentration-derivative stabilizes at a value around 4.0. The latter might be suggestive of the scalar-field internal intermittency of the small scales. The fluctuations in Figure 3 are connected with the sampling errors and could be reduced by increasing the number of MonteCarlo particles in the simulations.

B. Turbulent mixing of two reactive scalars

Bennani *et al.* (1985) injected a dilute solution of methyl formiate, species A, at the grid-generating turbulence while sodium hydroxide, species B, was convected in the main flow. The reaction rate constant, K_c, for the homogeneous chemical kinetics of the two species is 47 l/mole.s. The initial conditions are identical to those used by Arrojo *et al.* (1988). The value 1.5 for the relaxation coefficient, β, has been used in most simulations. A few runs are also presented for $\beta = 0.5$. The CPU time on a VAX 8300 is slightly under one hour for the smallest relaxation coefficient and almost two hours for the largest value.

Figure 4 compares the predicted evolution of the average concentrations of the two reactants for $\beta = 0.5$ and $\beta = 1.5$ with the experimental data. The agreement is excellent. The species variance evolutions for $\beta = 0.5$ and $\beta = 1.5$ are plotted in Figure 5. The sodium hydroxide variance unequivocally displays curves with a maximum for both values of the relaxation coefficient. This feature qualitatively agrees with the measurements; the chemical source term seems to dominate the variance generation during the first part of the evolution while the molecular dissipation controls the process decay after the maximum. There is however one order of magnitude difference between the measured values and those predicted by Arrojo *et al.* (1988). Moreover, using the actual values of Bennani *et al.* (1985) for first and second order concentration moments Schwartz inequality is not satisfied.

The experimentally determined segregation coefficient for A and B species stabilizes at a value around 0.7. The evolutions shown in Figure 6 for two values of the relaxation coefficient displays the tendency to level off between 0.6 and 0.7. Figure 7 represents the reactant characteristic mixing frequencies.The difference in the initial statistics for the two reactants due to the injection conditions in combination with the chemical kinetics lead to substantially different evolutions for the mixing frequencies of the two reactive scalars. This is a call of attention to reactive system modellers; in general, not enough attention is being paid to the modification of the characteristic mixing times due to the chemistry.

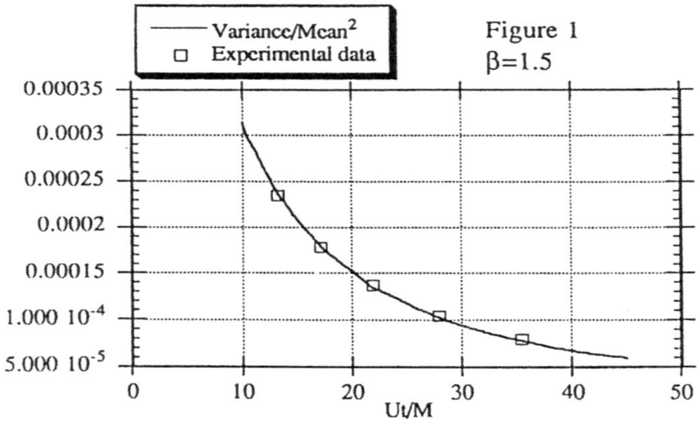

Time evolution of the variance of a passive scalar. The symbols represent the experimental data of Bennani (1983).

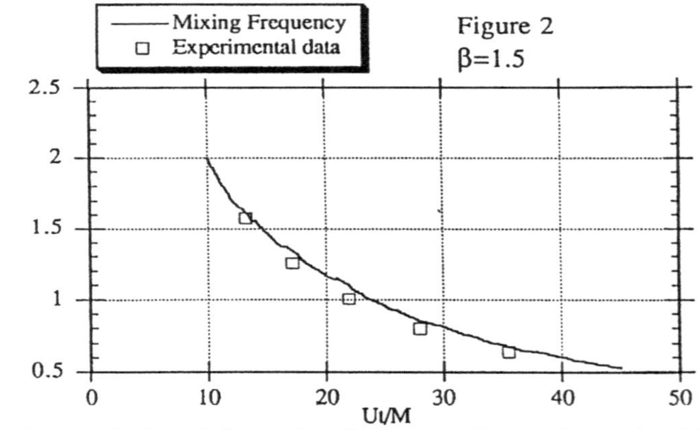

Time evolution of the mixing frequency of a passive scalar. The symbols represent the experimental data of Bennani (1983).

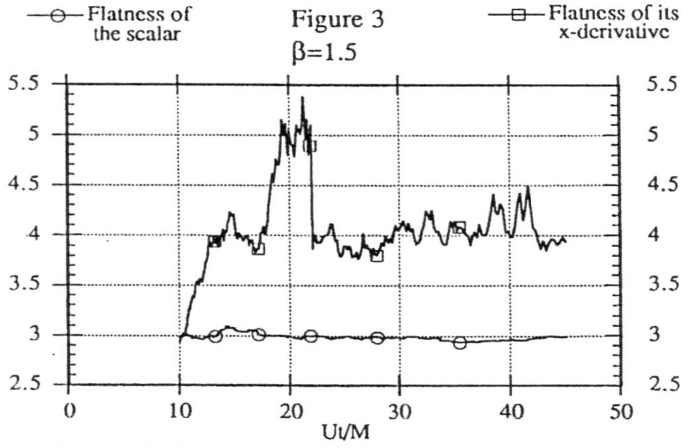

Time evolution of the flatness factor of a passive scalar and its x-derivative.

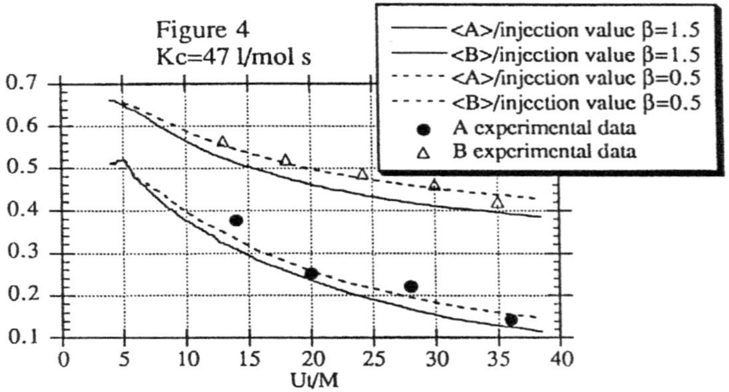

Time evolution of the means of two reactive scalars for two different values of the relaxation parameter, β.
The symbols represent the experimental data of Bennani (1983).

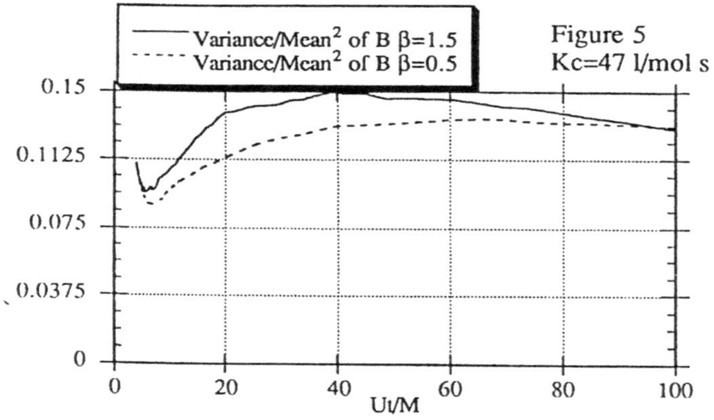

Time evolution of the variances of two reactive scalars for two different values of the relaxation parameter, β.

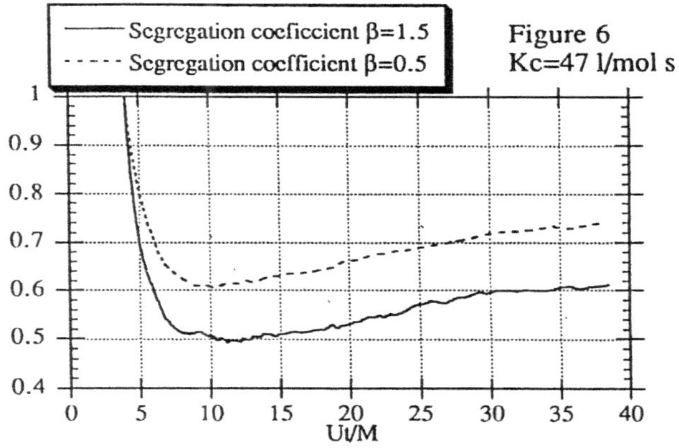

Time evolution of the segregation coefficient of two reactive scalars for two different values of the relaxation parameter, β.

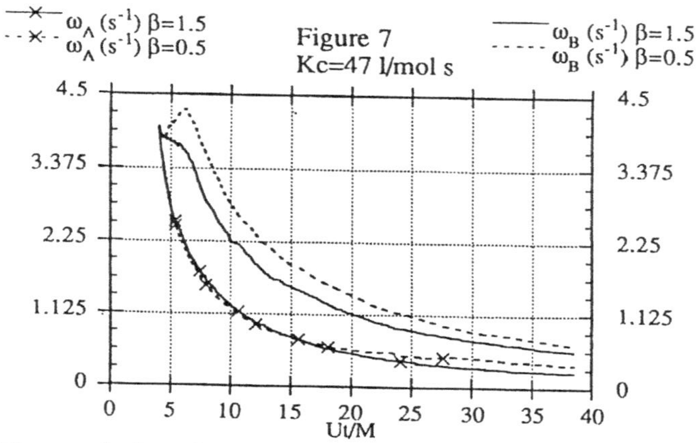

Time evolution of the mixing frequencies of two reactive scalars for two different values of the relaxation parameter, β.

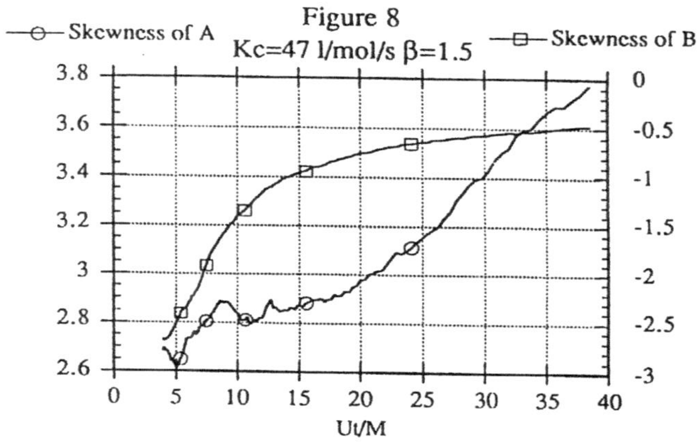

Time evolution of the skewness coefficients of two reactive scalars.

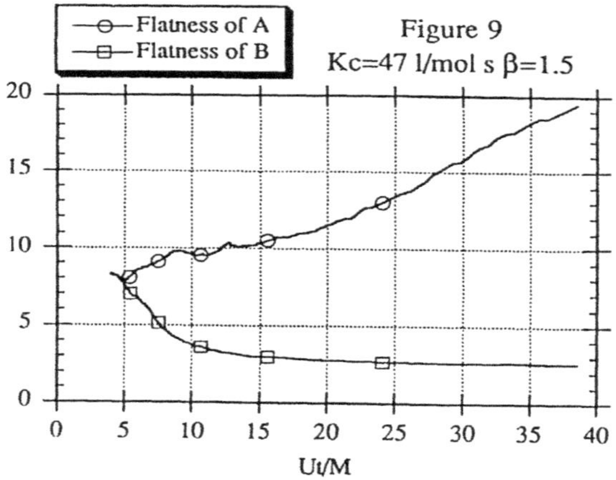

Time evolution of the flatness factors of two reactive scalars.

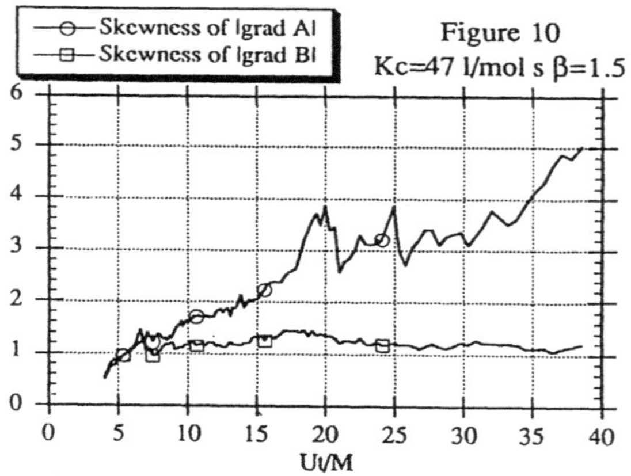

Time evolution of the skewness coefficients of the gradient magnitudes of two reactive scalars.

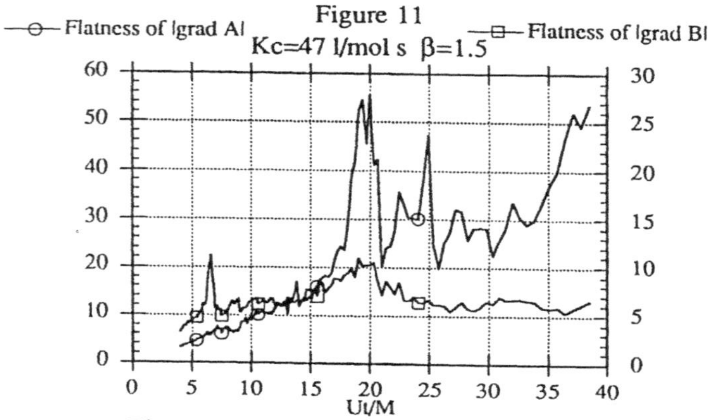

Time evolution of the flatness factors of the gradient magnitudes of two reactive scalars.

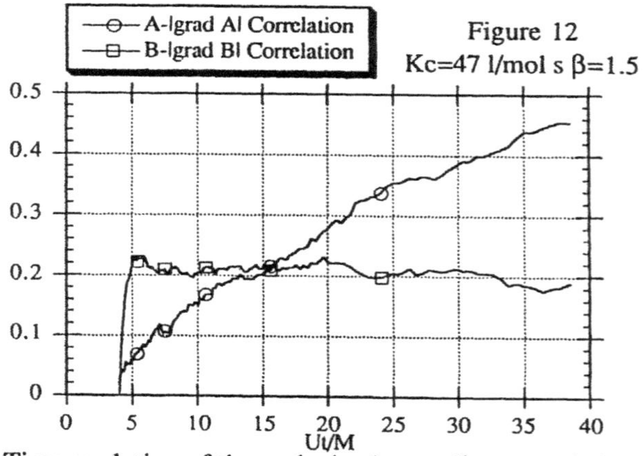

Time evolution of the scalar/scalar-gradient correlations in the turbulent mixing of two reactive scalars.

Figures 8 and 9 are plots of the reactant skewness coefficients and flatness factors respectively. The methyl formiate skewness monotonically increases from around 2.7 to 2.9. The negative values for the sodium hydroxide skewness are also significant. Similar considerations hold for the reactant flatness factors. Chemistry tends to dramatically increase the flatness of the injected species and substantially reduce that of the species diluted in the main flow. In any event, the spatial segregation of species is indicative in those predictions.

The chemical reaction has a dramatic effect upon the evolution of the skewness coefficient and flatness factor of the injected species (methyl formiate) gradient magnitude while that on the sodium hydroxide is a minor one. This can be observed in Figures 10 and 11, respectively. The large values of the flatness factor of A-gradients are suggestive of the internal intermittency of the small scales for that reactant.

Within statistical or sampling error the correlation induced, exclusively by the mixing process, over initially uncorrelated concentration and their corresponding gradient fields is zero (Valiño, 1989). Figure 12 indicates that the chemical reaction leads to significant positive correlation between the reactants and their gradient fields.

6. Concluding remarks

The mixing of dynamically passive inert and reactive scalars convected by a solenoidal field of nearly homogeneous turbulence has been investigated using a transport equation for the joint pdf of scalars and their gradients.

Both inert and reactive scalar predictions agree well with available experimental data for statistically homogeneous turbulent flows. Inert scalar-gradient flatness factors greater than 3.0 seem to be a mild indication of the small scale intermittency of the scalar field. Reactant flatness-factors significantly greater than 3.0 clearly suggest spatial segregation, particularly for the injected species; the chemistry has thus an important influence upon the pdf shape, specially on that of the stoichiometrically deffective reagent, A. The chemistry has also a drastic effect on the scalar-gradient skewness and flatness factors, in particular, those of the injected species, A; very large values of the methyl-formiate gradient flatness factor are a strong indication of small- scale intermittency; the nature of the injection process in combination with the chemical kinetics seems to lead a spotty distribution of scalar dissipation. The chemistry can induce significant cross-correlations between a reactant and its gradient when prescribed initially uncorrelated. The characteristic mixing times or frequencies are critically modified by the chemistry; this is a justification for the inclusion of scalar gradients in the present formulation; it is also a call of attention to some extended practices among turbulent reactive flow modellers who overlook this fact. The evolution of the pdf shape is decisive; the use of different relaxation coefficients leads to radically different results and confirms this important influence.

Acknowledgements

Most of the work reported here forms a part of the Ph.D. Dissertation of L. Valiño. L. V. would like to acknowledge a CONAI-DGA research fellowship during part of the duration of this work. C. D. gratefully thanks the CICYT for its contribution to his sabbatical stay at the Universities of California-Davis and Southern California. Ms. P. Ezquerra did as usual an excellent work in typing and editing the manuscript.

References

Arrojo, P., Dopazo, C., Valiño, L. and Jones, W.P. 1988 Numerical Simulation of Velocity and Concentration Fields in a Continuous Flow Reactor, *Chem. Engng. Sci.* **43**, 1935-1940.

Bennani, A., Gence, J.N. and Mathieu, J. 1985 The Influence of a Grid-Generated Turbulence on the Development of Chemical Reactions, *AIChE J.* **31**, 1157-1166 (1985).

Brissaud, A. and Frisch, U. 1974 Solving Linear Stochastic Differential Equations, *J. Math. Phys.* **15**, 524-534.

Chen, H., Chen, S. and Kraichnan, R.H. 1989 Probability Distribution of a Stochastically Advected Scalar Field, *Phys. Rev. Lett.* **3**, 2657-2660.

Dopazo, C. and O'Brien, E.E. 1974 An Approach to the Autoignition of a Turbulent Mixture, *Acta Astronautica* **1**, 1239-1266.

Dopazo, C. and O'Brien, E.E. 1976 Statistical Treatment of Non-Isothermal Chemical Reactions in Turbulence, *Combust. Sci. Technol.* **13**, 99-122.

Dopazo, C. 1976 A Probabilistic Approach to Turbulent Flame Theory, *Acta Astronautica* **3**, 853-878.

Dopazo, C. 1979 Relaxation of Initial Probability Density Functions in the Turbulent Convection of Scalar Fields, *Phys. Fluids* **22**, 20-30.

Eswaran, V. and Pope, S.B. 1988 Direct Numerical Simulations of the Turbulent Mixing of a Passive Scalar, *Phys. Fluids* **31**, 506-520.

Girimaji, S. S. and Pope, S. B. 1990 A Diffusion Model for Velocity Gradients in Turbulence, *Phys. Fluids* **A 2**, 242-256.

Haworth, D.C. and Pope, S.B. 1986 A Generalized Langevin Model for Turbulent Flows, *Phys. Fluids* **29**, 387-405.

Hill, J.C. 1976 Homogeneous Turbulent Mixing with Chemical Reaction, *Ann. Rev. Fluid Mech.* **8**, 135-161.

Jones, W.P. and Musonge, P. 1988 Closure of the Reynolds Stress and Scalar Flux Equations, *Phys. Fluids* **31**, 3589-3604.

Kerr, R.M. 1985 Higher-Order Derivative Correlations and the Alignment of Small-Scale Structures in Isotropic Numerical Turbulence, *J. Fluid Mech.* **153**, 31-58.

Kollmann, W. and Wu, A. 1989 Equation for the Probability Density Function of Velocity and Scalar for Turbulent Shear Flows, *AIAA J.* **27**, 1052-1058.

Kubo, R. 1963 Stochastic Liouville Equations, *J. Math. Phys.* **4**, 174-183.

Meyers, R.E. and O'Brien, E.E. 1981 The Joint PDF of a Scalar and its Gradient at a Point in a Turbulent Fluid, *Combust. Sci. Technol.* **26**, 123-134.

O'Brien, E.E. 1971 Turbulent Mixing of Two Rapidly Reacting Chemical Species, *Phys. Fluids* **14**, 1326-1330.

O'Brien, E.E. 1980 The Probability Density Function (pdf) Approach to Reacting Turbulent Flows, pp. 185-218. In: *Turbulent Reacting Flows*, Eds. P.A. Libby and F.A. Williams, Springer-Verlag.

Pope, S.B. 1981a Transport Equation for the Joint Probability Density Function of Velocity and Scalars in Turbulent Flow, *Phys. Fluids* **24**, 588-596.

Pope, S.B. 1981b A MonteCarlo Method for the PDF Equations of Turbulent Reactive Flow, *Combust. Sci. Technol.* **25**, 159-174.

Pope, S. B. 1983 A Lagrangian Two-Time Probability Density Function Equation for Inhomogeneous Turbulent Flows, *Phys. Fluids* **26**, 3448-3450.

Pope, S.B. 1985 PDF Methods for Turbulent Reactive Flows, *Prog. Energy Combust. Sci.* **11**, 119-192.

Valiño, L. 1989 Computation of Homogeneous Turbulent Reactive Flows. Monte-Carlo Simulation of Velocity, Concentrations and Concentration-Gradients, (in Spanish). Ph.D. Dissertation, Univ. Zaragoza.

Valiño, L. and Dopazo, C. 1990 A Binomial Sampling Model for Scalar Turbulent Mixing, *Phys. Fluids* **A 2**, 1204-1212.

Part IV

Turbulence Affected by Body Forces

Transport of Passive Scalars in Neutrally and Stably Stratified Homogeneous Turbulent Shear Flows

H.-J. Kaltenbach, T. Gerz and U. Schumann

DLR, Institut für Physik der Atmosphäre, W-8031 Oberpfaffenhofen Germany

The dispersion of a passive scalar in a turbulent flow can be described by the turbulent diffusivity tensor D_{ij} which relates the turbulent flux of a scalar fluctuation c to the gradient of its mean value C according to $\overline{u_i c} = -D_{ij}\partial C/\partial x_j$. By means of Direct Numerical Simulation we show how components of the unsymmetric tensor behave under the influence of shear and thermally stable stratification. From an analysis of three-dimensional flux spectra it turns out that weakly stratified flows are dominated by large scale motions. The total fluxes can then sufficiently be explained by linear theory like RDT. However, in moderatly stratified flows we find significant small-scale contributions with signs opposing to the large scales. These are due to non-linearity and should be considered in scalar dispersion models.

1. Introduction and method

We extend the study of Rogers, Mansour & Reynolds (1989), who calculated the development in time of the components of D_{ij} for an unbounded shear flow, to stably stratified flows. They derived an algebraic flux model which predicts D_{ij} from the Reynolds stresses and the mean strain-rate.

Our Direct Numerical Simulation captures all relevant scales of a turbulent flow at low Reynolds number. The time-dependent, three-dimensional Navier-Stokes equations in the Boussinesq approximation for the velocity fluctuations u, v and w and the balances of the temperature fluctuation T and of three passive scalar fluctuations c_1 , c_2 and c_3 are integrated in space and time. Figure 1 shows the computational domain and profiles

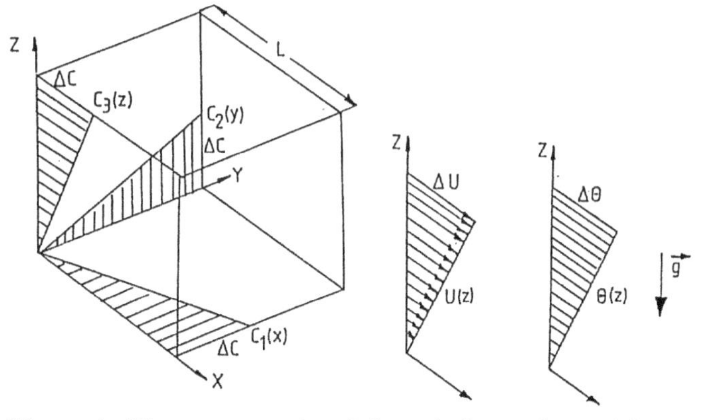

Figure 1. The computational domain is a cubus with axes x, y and z. Note the uniform gradients of the mean quantities U, Θ, C_1, C_2 and C_3

Advances in Turbulence 3
Editors: A.V. Johansson · P.H. Alfredsson

of the mean quantities. Velocity U and temperature Θ have uniform vertical gradients S and s. whereas the mean scalar fields C_1, C_2, C_3 possess gradients s_1, s_2, s_3 in the streamwise (x-), spanwise (y-) and vertical (z-) direction, respectively. Temperature fluctuation T and scalar fluctuation c_3 behave identically because we consider only cases with equal molecular diffusivities ($Pr = Sc$).

The horizontal boundaries are periodical, the vertical is shear-periodic. Our code uses central differences, an explicit integration scheme in time (Adams-Bashforth) and a pseudospectral approximation of the advection by the mean flow. The grid resolution is 128 cells in each direction. For details see Gerz, Schumann & Elgobashi (1989).

Firstly, we performed a simulation under neutral stratification (temperature with vertical gradient as passive scalar) starting from an isotropic velocity field with a given energy spectrum. When this flow has reached self-similar behaviour with characteristic properties of a developed shear flow, we then take these data as initial fields for runs labeled as A to D in Table 1.

Table 1. Nondimensional parameters of developed shear flow at the onset of buoyancy forces at shear time $St = 6$. Results from wind tunnel experiments carried out by Tavoularis & Karnik (1989) with Re_λ between 120 and 400 are given in brackets.

Box length, integral length	$L/L, \ell/L$		1, 0.09	
Taylor micro-l., Kolmogorov l.	$\lambda_{11,1}/L$, l_K/L		0.084, 0.004	
rms of velocity	$v' = (\overline{u_i u_i}/3)^{1/2} = (2E_{kin})^{1/2}$		0.025	
rms of temperature	$T' = (\overline{TT})^{1/2} (= c_3')$		0.049	
rms of concentration	$c_1' = (\overline{c_1 c_1})^{1/2}, c_2'$		0.099, 0.067	
Shear number	$Sh = (\Delta U/L)(\ell/v')$		3.2	
Reynolds number	$Re_\lambda = v'\lambda_{11,1}/\nu$		64	
Prandtl, Schmidt number	Pr, Sc		1, 1	
Reynolds stresses	$\overline{u^2}/(2E_{kin}), \overline{v^2}/(2E_{kin})$		0.56 (0.51),0.29(0.27)	
	$\overline{w^2}/(2E_{kin}), \overline{uw}/(2E_{kin})$		0.15(0.22),-0.16(-0.16)	
Richardson number for case	A	B	C	D
$Ri = \alpha g(\Delta\Theta/L)/(\Delta U/L)^2$	0	0.16	0.33	0.66

2. Results

2.1 Development of kinetic energy and turbulence variances

The shear number $Sh = (\Delta U/L)(\ell/v')$ approximately measures the ratio of energy production by shear and destruction by dissipation ϵ. For $Sh = 3.2$ the kinetic energy grows in the neutral case. The balances of the kinetic energy $\partial E_{kin}/\partial t = -S\overline{uw} + Ri\overline{wT} - \epsilon$ and the potential energy $\partial E_{pot}/\partial t = -Ri\overline{wT} - 0.5Ri\epsilon_{TT}$ reveal the role of the vertical heat flux $\overline{w}T$. It transfers energy from its kinetic form $E_{kin} = 0.5(v')^2$ into the potential form $E_{pot} = 0.5Ri(T')^2$. Depending on the strength of stability, the kinetic energy grows or decays according to an exponential law (Figure 2a). Due to the low Reynolds number of the flow, the critical Richardson number, at which the flow reaches steady state, has a value of 0.16, which is smaller than the typical value of 0.25 for inviscid linear cases as discussed by Gerz & Schumann (1990).

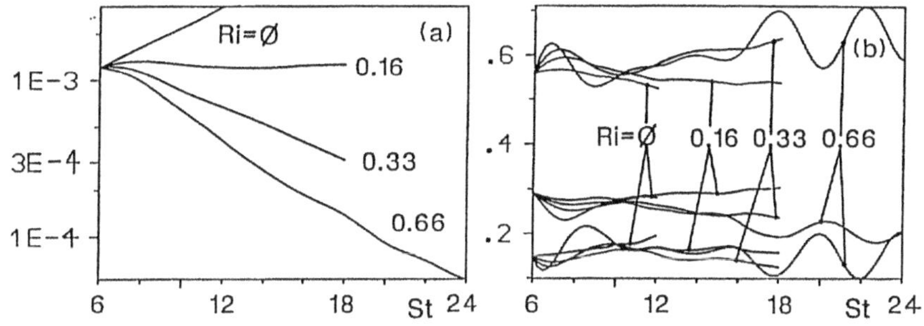

Figure 2. (a) Logarithmic plot of the kinetic energy versus shear time St. (b) Components of the Reynolds stress tensor versus St. Upper, middle and lower curves correspond to $\overline{uu}/(2E_{kin})$, $\overline{vv}/(2E_{kin})$ and $\overline{ww}/(2E_{kin})$, respectively.

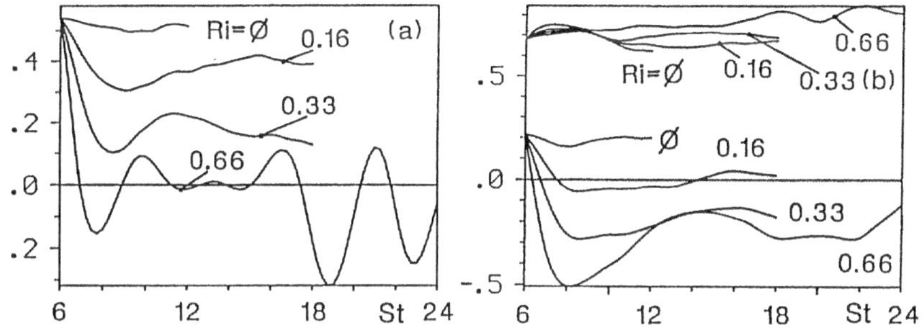

Figure 3. Correlation coefficients of the vertical momentum flux $-\overline{uw}/u'w'$. (a) and the scalar fluxes $\overline{uT}/u'T'$ (upper values) and $\overline{wc_1}/w'c_1$ (lower values) (b) versus time St.

The strong anisotropy of the flow is not much changed by the stratification (Figure 2b). Except for the strongly stable case D, all cases show steady normalized variances; i.e. the flow is self-similar. The anisotropy extends over all scales of the velocities because of the considerable viscosity of the flow. With increasing stability the vertical momentum flux vanishes (Figure 3a) because vertical motions which cause temperature fluctuations are damped by buoyancy forces. Oscillations with twice the Brunt-Väisälä frequency $N = S\sqrt{Ri}$ result from the linear forcing terms by shear and buoyancy.

2.2 Scalar fluxes and diffusivity tensor

The three fluxes $\overline{uc_j}$, $\overline{vc_j}$, $\overline{wc_j}$ of each of the three scalars c_1, c_2, c_3 divided by the mean gradients $-dC_j/dx_j = -s_j$ form the nine components of the turbulent diffusivity tensor $D_{ij} = -\overline{u_ic_j}/s_j$. Due to normalization, $s_j = 1$ and therefore $D_{ij} = -\overline{u_ic_j}$. In our case D_{12}, D_{21}, D_{23} and D_{32} are zero because of flow symmetry. The diagonal components D_{11}, D_{22} and D_{33} describe down-gradient transport along the axes whereas $D_{13} = -\overline{uc_3} = -\overline{uT}$ and $D_{31} = -\overline{wc_1}$ express fluxes orthogonal to the mean gradients which cause them. The velocity variances times the mean scalar gradients s_i form the production terms in the scalar flux balances (1a, 1c and 1e; Φ denotes pressure terms, ϵ denotes molecular smearing). Their anisotropy gives rise to the strong anisotropy of the diffusivity tensor. D_{31} and D_{13} both contain the momentum flux or shear stress

$\overline{uw}$ in the production term (1b and 1d). The additional term $\overline{wT}S$ in the $\overline{uT}$-balance causes the asymmetry of the diffusivity tensor in the neutral case. $Ri\overline{Tc_1}$ enhances the asymmetry for cases with $Ri > 0$.

$$D_{11}: \quad \frac{\partial \overline{wc_1}}{\partial t} = -\overline{uu}s_1 - \overline{wc_1}S \qquad + \phi_{1c_1} - \epsilon_{1c_1} \tag{1a}$$

$$D_{31}: \quad \frac{\partial \overline{wc_1}}{\partial t} = -\overline{uw}s_1 \qquad + Ri\overline{Tc_1} + \phi_{3c_1} - \epsilon_{3c_1} \tag{1b}$$

$$D_{22}: \quad \frac{\partial \overline{vc_2}}{\partial t} = -\overline{vv}s_2 \qquad + \phi_{2c_2} - \epsilon_{2c_2} \tag{1c}$$

$$D_{13}: \quad \frac{\partial \overline{uT}}{\partial t} = -\overline{uw}s - \overline{wT}S \qquad + \phi_{1T} - \epsilon_{1T} \tag{1d}$$

$$D_{33}: \quad \frac{\partial \overline{wT}}{\partial t} = -\overline{ww}s \qquad + Ri\overline{TT} + \phi_{3T} - \epsilon_{3T} \tag{1e}$$

Because of the tensor asymmetry there is no analogy between turbulent and molecular diffusivity (e.g. heat conduction in an anisotropic medium) which is always symmetric according to the Onsager-Casimir theorem.

The time series of the scalar flux correlation coefficients become more or less stationary as to be expected in a self-similar flow (Figure 3b). If the turbulent diffusivity is an universal property of the flow, the ratios of the scalar fluxes should become steady. Other than Rogers et al. (1989) we use the spanwise flux (or D_{22}) in order to normalize the other fluxes because the vertical scalar (heat) flux reaches very small values in strongly stable cases (Figure 4a). Although these time series do not show exact steady state, they give us some approximate values of the normalized tensor components as a function of Ri (Figure 4b). For D_{13} and D_{31} measured values are available in the neutral case. As Rogers et al. (1989) pointed out, it is practically impossible to realize a stationary mean scalar gradient in the streamwise direction in a laboratory experiment.

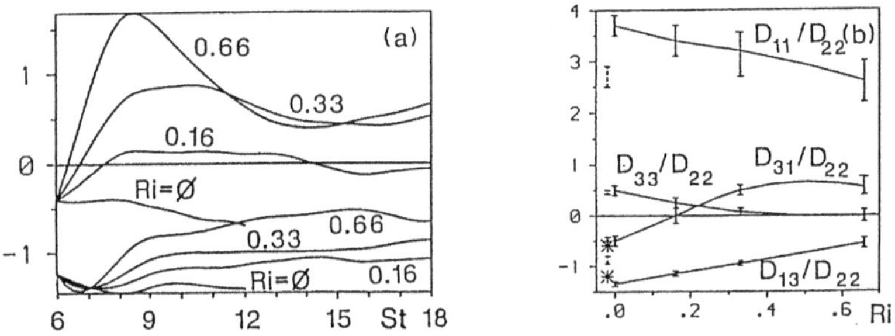

Figure 4. (a) Components of the diffusivity tensor, normalized by the spanwise component D_{22} versus time St; upper values correspond to D_{31}/D_{22}, lower values to D_{13}/D_{22} (b) Normalized diffusivity tensor versus Richardson number at $St > 12$. The dashed error bars mark values from Direct Numerical Simulations of Rogers et al. (1989). Stars mark measured tensor components from Tavoularis & Corrsin (1981 and 1985).

The stability has significant influence on the tensor components D_{33} and D_{31}, both of them expressing vertical fluxes. The sign change of D_{31} results from the fact that the production term $-\overline{uw}s_1$ is counterbalanced by the sink $Ri\overline{Tc_1}$ (1b). The large correlation of the scalar fluctuations T' and c_1 extends over all scales and is not much affected by the stability, except for case D. Explanation of scalar flux behaviour is limited to simple qualitative statements if only the balances (1a-1e) are used. However, Hunt, Stretch & Britter (1988) showed, that important aspects of the flow behaviour are derivable using linear RDT. From a detailed discussion of the distribution of flux contributions on several scales - which is done in the next section - it turns out that some of the scalar fluxes show opposing contributions on large and small scales which are due to non-linearity.

2.3 Spectral analysis of flux behaviour for neutral stratification

In order to understand how the flow is affected by the buoyancy forces it is useful to discuss the neutral flow first. We extend Rogers et al.'s work by analysing the three-dimensional flux spectra of momentum and scalars of our neutral case A, which is comparable to their flow cases. Results from Brasseur & Lin (1990) who analysed the data of Rogers' simulations, are very useful to interprete the spectra. By means of a sampling method they looked out for spots with strong local momentum flux uw which they called "events". They found that the most intense uw-events are completely dominated by negative uw-fluctuations - which can be expected from mixing-length theory - whereas weak events cause as often negative as positive uw-fluxes. In our simulation we find positive uw at 33 % of all grid points and a value of 2.7 for the ratio of rms-values of negative and positive uw-samples, which is consistent with the results of Brasseur.

The shear-stress spectrum Φ_{uw} has strong negative contributions at large scales (see Figure 5a). We find a sharp decrease at medium wave numbers and almost no contributions at small scales. Analysis of the isocontour plot of uw illuminates that the low wave-number contributions mark rather the distance between strong negative events than their spatial extension. The absence of small-scale contributions in Φ_{uw} does not say that there is no small-scale momentum flux. It expresses rather the fact

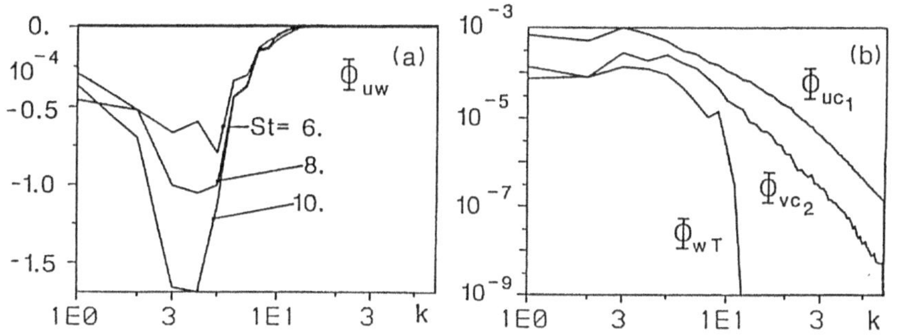

Figure 5. Cospectra versus wave number k of the momentum flux Φ_{uw} (a) at several times and of $-\Phi_{uc_1}$, $-\Phi_{vc_2}$ and $-\Phi_{wT}$ (b) at $St = 9$. The countergradient values (originally positive) of Φ_{wT} are omitted due to the logarithmic axis. Wave-number $k = 1$ corresponds to a wave-length L.

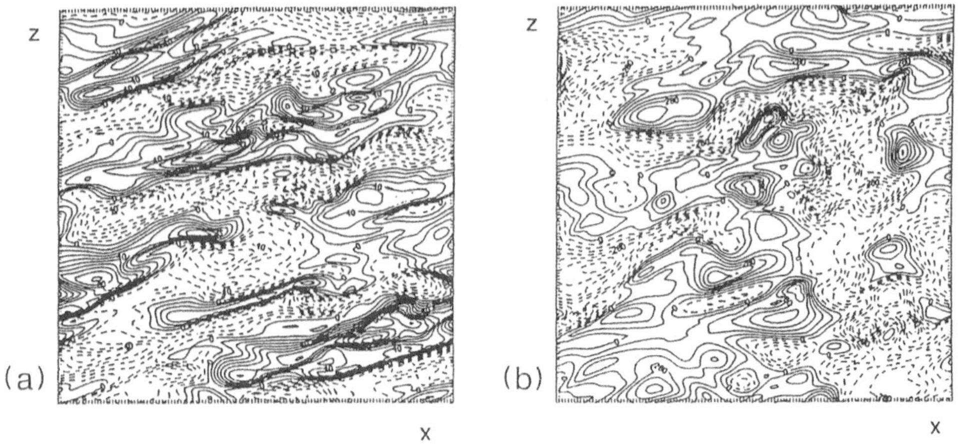

Figure 6. Isocontours of velocitiy fields of u (a) and w (b) from case A with $Ri = 0$ in a x-z plane at $St = 6$.

that at small scales the positive and therefore counter-gradient shear-stress is as big as the negative one. Vanishing shear-stress on small scales, however, does not indicate local isotropy in our case, since we observe an anisotropic flow behaviour at **all** scales. Further, in case of local isotropy, there would be no cause for the strikingly different shapes of the cospectra of the down-gradient scalar fluxes shown in Figure 5b. Since the three scalars c_1, c_2 and $c_3 = T$ differ only with regard to the orientation of their mean gradients, one would expect, that the cospectra look similar at high wave numbers in an locally isotropic flow.

The vertical velocity fluctuation w behaves differently than horizontal fluctuations u and v. Comparison of vertical cross-sections of the fields of u and w (Figure 6) gives rise to the conjecture that the w-fluctuations play the key-role in this flow, since the w-field looks more random and, hence, changes sign more frequently, whereas the u-fluctuations show some inclined streaky coherent structures (see also Gerz, 1990). This is consistent with the fact that the skewness-coefficients of the velocity derivatives show strong skewness of vertical and horizontal derivatives of u but only weak values of the corresponding coefficients of w. The physical reason for this behaviour is the orientation of the gradient of the mean velocity $U(z)$.

Thus, if we accept the key-role of w-fluctuations, we are able to present a concept of the evolution of a probable flow event. Strong events which caused negative uw and other scalar fluxes according to mixing-length theory break down into small-scale fluid parcels which carry their properties for some time, e.g. horizontal velocity fluctuation u and scalar properties like T and c_1. If w changes sign before the fluid parcels have lost their properties from the past event, vertical fluxes uw and wT will become counter-gradient. Also wc_1 will change sign. On the other hand, small-scale fluxes or correlations which do not involve vertical motion, e.g. uT, uc_1 and c_1T will retain the same values as in the event. In Figure 7 it is tried to sketch the evolution of a "typical" event.

The figure illustrates how properties of a fluid parcel, which starts with mean values of its surroundings according to its initial position, change during a vertical upward displacement $\Delta z(t)$ which is followed by a break-up into several small-scale fluid parcels under the action of turbulence. Some of the small-scale "relics" - still

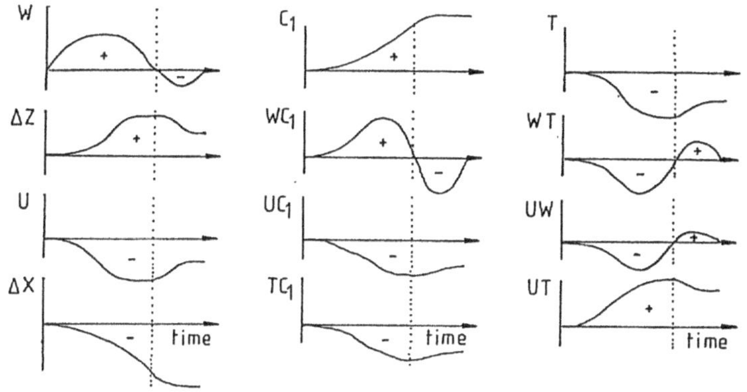

Figure 7. Sketch of properties of a fluid parcel versus time during a "typical" flow event. The dotted lines mark the time when the fluid parcel breaks up into small-scale parcels.

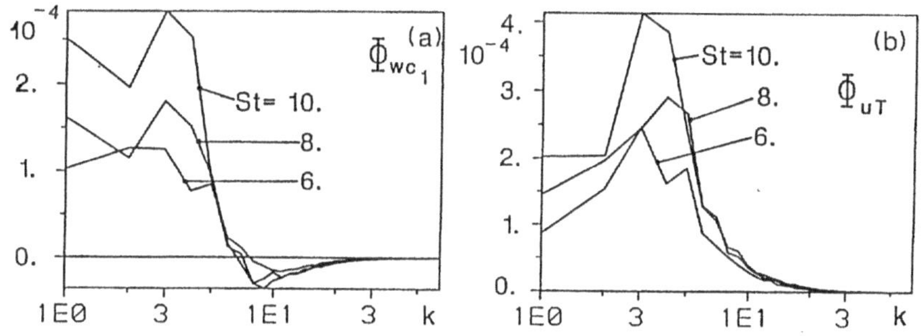

Figure 8. Cospectra from case A with $Ri = 0$ at several times of Φ_{wc_1} (a) and of the streamwise heat flux Φ_{uT} (b).

carrying some of their initial properties - undergo a sort of "compensating" downward motion. It is possible to derive all other fluctuations from the development of w, if one uses the relations $w = d(\Delta z)/dt$, $u = d(\Delta x)/dt \approx -\Delta z dU/dz$, $T \approx -\Delta z d\Theta/dz$, $c_1 \approx -\Delta x dC_1/dx$, where Δx only considers the horizontal displacement caused by the fluctuation u. This presentation is consistent with the moving reference frame due to the mean advection $U(z)$.

So we can subdivide the correlations into two groups. The fluxes uw, wT and wc_1 change their sign during the break-down of the event whereas uT, uc_1 and c_1T retain their signs. It turns out that the cospectra of these correlations can be separated into the same two classes. Cospectra Φ_{uT}, Φ_{uc_1} and Φ_{c_1T} have similar shapes and do not change sign (Figure 8b) whereas Φ_{uw}, Φ_{wT} and Φ_{wc_1} show sharp decreases at medium wave numbers followed by opposing or vanishing contributions at small scales (Figure 8a).

Whereas Φ_{wT} has small-scale counter-gradient contributions (not shown here) Φ_{uw} does not (Figure 5). This can be explained by the longer life time of scalar fluctuations T than of velocity fluctuations u for $Pr \approx 1$. The negative small-scale parts in Φ_{wc_1} have two causes. Firstly, due to $c_1'/T' \approx 2$, the flux wc_1 is greater than wT on average; $(wc_1)' \approx 2(wT)'$. Secondly, the different orientation of the mean gradients of u and T on one side and of c_1' on the other side leads to the distinct small-scale behaviour. From the sketch of an event (Figure 7) it is evident that a vertical backward motion reduces the values of T and u whereas it does not affect c_1. Therefore wc_1 stays relatively large on small scales. The shape of the cospectrum Φ_{wc_1} emphasizes that non-linear effects can have considerable contributions. The subdivision into two groups also holds if one regards the correlation coefficients. The first group with $\overline{uw}/u'w' = -0.5$, $\overline{wT}/w'T' = -0.4$ and $\overline{wc_1}/w'c_1' = 0.2$ has smaller values than the second group with $\overline{uc_1}/u'c_1' = -0.75$, $\overline{uT}/u'T' = 0.65$ and $\overline{c_1T}/c_1'T' = -0.75$.

2.4 Buoyancy influence on the spectra

The moderately stratified cases B and C show similar shape of spectra as the neutral case. However the large-scale contributions are smaller, the high wave-number parts of Φ_{uw}, Φ_{wT} and Φ_{wc_1} are distinctly counter-gradient (not shown here). We conclude that buoyancy acts on **all** scales of a shear flow at the given (small) Reynolds number. The large-scale upward motion in our flow example (Figure 7) is diminished by the stratification, whereas the small-scale "compensating" downward displacement is enhanced due to the body force which draws the fluid-parcels with negative T-fluctuation downwards. T and c_1 stay strongly correlated in stratified flows except for case D which indicates that the mechanism of flow events given above is adequate even for the description of moderately stratified flows. From the distinct opposing behaviour at different scales we conclude that accurate modeling of scalar dispersion under stable stratification will be difficult.

References

Brasseur, J.G. & Lin, W.-Q. 1990 Structural and statistical characteristics of intermittency in homogeneous turbulent shear flow. In: *Proc. Third European turbulence conference*, Stockholm 1990.

Gerz, T. 1990 Coherent structures in stratified turbulent shear flows deduced from direct simulations. In: *Turbulence and Coherent Structures.* (M. Lesieur & O. Metais, eds.), Kluwer Academic Publishers, in press.

Gerz, T. & Schumann, U. 1990 Direct simulation of homogeneous turbulence and gravity waves in sheared an unsheared stratified flows. In: *Turbulent Shear Flow 7* (F. Durst et al., ed.), Springer-Verlag, in press.

Gerz, T., Schumann, U. & Elghobashi, S.E. 1989 Direct numerical simulation of stratified homogeneous turbulent shear flows; *J. Fluid Mech.* **200**, 563-594.

Hunt, J.C.R., Stretch, D.D. & Britter, R.E. 1988 Length scales in stably stratified turbulent flows and their use in turbulence models. In: *Stably Stratified Flows and Dense Gas Dispersion* (J.S. Puttock, ed.), Clarendon Press, Oxford, pp. 285-321.

Rogers, M.M., Mansour, N.N. & Reynolds, W.C. 1989 An algebraic model for the turbulent flux of a passive scalar. *J. Fluid Mech.* **203**, 77-101.

Tavoularis, S. & Corrsin, S. 1981 Experiments in nearly homogeneous turbulent shear flow with a uniform mean temperature gradient, Part 1. *J. Fluid Mech.* **104**, 311-347.

Tavoularis, S. & Corrsin, S. 1985 Effects of shear on the turbulent diffusivity tensor. *Int. J. Heat Mass Transfer* **28**, 265-276.

Tavoularis, S. & Karnik, U. 1989 Further experiments on the evolution of turbulent stresses and scales in uniformly sheared turbulence. *J. Fluid Mech.* **204**, 457-478.

A Model for the Turbulent Breakdown of Internal Gravity Waves

S.D. Mobbs and M.J. Rabbitt

Department of Applied Mathematical Studies, University of Leeds, Leeds LS2 9JT, U.K.

A model is proposed for the breakdown of internal gravity waves in the middle and upper atmosphere. It follows earlier models in using an eddy diffusivity which is just sufficient to limit the overturning of isentropic surfaces. However, it extends previous models in allowing indpendent calculation of eddy viscosity and conductivity coefficients. Results are presented for the interaction of a single wave mode with a critical level.

1. Introduction

Internal gravity waves play a major role in the momentum balance of the mesosphere (50-80km altitude) and they also contribute significantly in the lower stratosphere (10-20km altitude). The waves only cause permanent changes to the larger scale flow if they are dissipating; thus decay mechanisms are of great interest.

Internal waves break due to shear instability or convective instability when:

(i) the waves approach a critical level (where phase speed and wind speed are equal),
(ii) due to the decrease of density with height, the amplitude becomes large.

Both processes involve the transfer of energy to small scales and turbulence generation. Major contributions were made by Lindzen (1967, 1981) and Hodges (1967) who proposed that dissipation should be just sufficient to prevent the waves from overturning. Refinements have been made by Fritts and Dunkerton (1985) who included the effect of localised turbulence and by Lindzen (1988) who accounted more precisely for convective instability. The model assumes an eddy Prandtl number of unity whereas a simple argument due to Chao and Schoeberl (1984) suggests that the momentum flux is much greater than the heat flux, i.e. the Prandtl number should be large. In this paper the restriction on the Prandtl number is removed. An application involving gravity wave-critical layer interaction is presented.

2. Formulation of the Eddy Diffusion Model

The Boussinesq equations for stratified flow, incorporating eddy diffusion effects, are

$$\frac{\partial \mathbf{u}}{\partial t} + \mathbf{u}.\nabla \mathbf{u} = -\frac{1}{\rho}\nabla(p-\overline{p}) + \frac{g(\theta-\overline{\theta})}{\theta_0}\mathbf{k} + \nabla.(\nu\nabla\mathbf{u}), \tag{1}$$

$$\frac{\partial \theta}{\partial t} + \mathbf{u}.\nabla\theta = \nabla.(\kappa\nabla\theta), \tag{2}$$

$$\nabla.\mathbf{u} = 0. \tag{3}$$

Advances in Turbulence 3
Editors: A.V. Johansson · P.H. Alfredsson

$\mathbf{u}$ is the velocity with components (u, v, w), ρ is the density, p is the pressure, θ is the potential temperature (with basic state $\theta_0(z)$) and ν and κ are the eddy viscosity and conductivity. $\mathbf{k}$ is the upward unit vector. We consider a single wave mode which propagates in the x direction. The y equation of motion may then be ignored.

Let the horizontal wavenumber of the disturbance to the basic state be k. Mean and perturbations may be partitioned using an average over a horizontal wavelength (denoted by an overbar). The perturbation parts of (1) and (2) are

$$u'_t + (uu_x)' + (wu_z)' = -P'_x/\rho + (\overline{\nu}u'_x)_x + (\overline{\nu}u'_z)_z + (\nu'\overline{u}_z)_z, \tag{4}$$

$$w'_t + (uw_x)' + (ww_z)' = -P'_z/\rho + (\overline{\nu}w'_x)_x + (\overline{\nu}w'_z)_z, \tag{5}$$

$$\theta'_t + (u\theta_x)' + (w\theta_z)' = (\overline{\kappa}\theta'_x)_x + (\overline{\kappa}\theta'_z)_z + (\kappa'\overline{\theta}_z)_z, \tag{6}$$

where $P = p - p_0$, primes denote perturbations and subscripts denote differentiation.

Consider the potential temperature variance equation formed by multiplying (6) by θ', averaging and neglecting products of perturbation quantities (the quasi-linear limit):

$$\overline{\theta'\theta'_t} + \overline{u}(\overline{\theta'\theta'_x}) + (\overline{w'\theta'})\overline{\theta}_z = \overline{\kappa}(\overline{\theta'\theta'_{xx}}) + \overline{\kappa}_z(\overline{\theta'\theta'_z}) + \overline{\kappa}(\overline{\theta'\theta'_{zz}}) + (\overline{\kappa'\theta'})\overline{\theta}_{zz} + (\overline{\kappa'_z\theta'})\overline{\theta}_z. \tag{7}$$

Since θ' and θ'_x are $\pi/2$ out of phase the second term in (7) is zero. For linear, conservative waves and constant $\overline{u}$, θ' and θ'_z are also $\pi/2$ out of phase and so the second term on the right is also negligible. The Lindzen type wave breaking hypothesis assumes that once wave breaking has begun, there is no further growth in wave amplitude, i.e.

$$\overline{(\theta'^2)}_t = 2\overline{\theta'\theta'_t} = 0. \tag{8}$$

The first, third and fifth terms on the right of (7) are integrated by parts giving

$$(\overline{w'\theta'})\overline{\theta}_z = -\overline{\kappa}(\overline{\theta'^2_x}) - \overline{\kappa}(\overline{\theta'^2_z}) + (\overline{\kappa'\theta'})\overline{\theta}_{zz} - (\overline{\kappa'\theta'_z})\overline{\theta}_z. \tag{9}$$

The horizontal velocity variance equation formed by multiplying (4) by u' is

$$\overline{u'u'_t} + \overline{u}(\overline{u'u'_x}) + (\overline{w'u'})\overline{u}_z$$

$$= -(\overline{u'P'_x})/\rho + \overline{\nu}(\overline{u'u'_{xx}}) + \overline{\nu}_z(\overline{u'u'_z}) + \overline{\nu}(\overline{u'u'_{zz}}) + (\overline{\nu'u'})\overline{u}_{zz} + (\overline{\nu'_z u'})\overline{u}_z. \tag{10}$$

This may be simplified in exactly the same way as (7) giving

$$(\overline{w'u'})\overline{u}_z = -(\overline{u'P'_x})/\rho - \overline{\nu}(\overline{u'^2_x}) - \overline{\nu}(\overline{u'^2_z}) + (\overline{\nu'u'})\overline{u}_{zz} + (\overline{\nu'_z u'})\overline{u}_z. \tag{11}$$

(9) and (11) give the eddy diffusivities under two different assumptions:

(a) *Constant Diffusivities* In this case, (9) and (11) give

$$\overline{\kappa} = -\frac{(\overline{w'\theta'})\overline{\theta}_z}{(\overline{\theta'^2_x}) + (\overline{\theta'^2_z})} \quad \text{and} \quad \overline{\nu} = \frac{(\overline{u'w'})\overline{u}_z + (\overline{u'P'_x})}{(\overline{u'^2_x}) + (\overline{u'^2_z})} \tag{12, 13}$$

(b) *Localised Diffusion* The small-scale turbulent diffusion will be greatest where θ surfaces are overturning (i.e. the shaded regions in figure 1). Following Fritts and Dunkerton (1985) we suppose that

$$\kappa = \kappa_0 \left(1 - \frac{\theta'_z}{\theta'_{zmax}}\right)^n \tag{14}$$

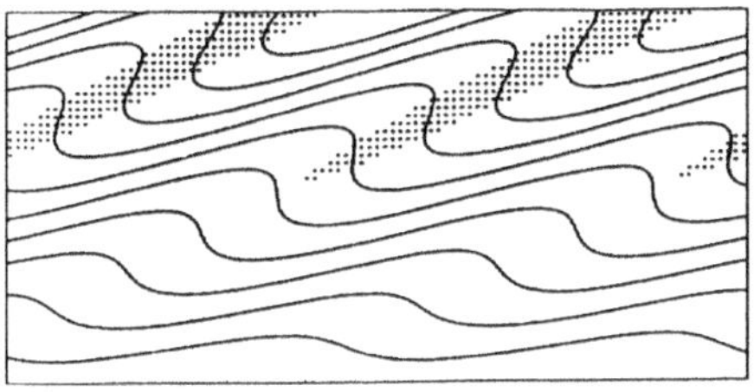

Figure 1. Schematic diagram showing a wave increasing in amplitude with height. The shaded regions indicate where isentropes are overturning.

where $\theta'_{zmax}(z)$ is the maximum (for all x) of θ'_z and n can take any value. κ' may be obtained from (14) and used to evaluate the final term in (9) (see Rabbitt (1989) for details). The result is $\overline{\kappa'\theta'_z} = -\overline{\kappa}\beta\theta'_{zmax}$ where $\beta = n/(n+1)$ represents the "degree of localisation" of turbulent diffusion. Inserting this into (9) gives

$$\overline{\kappa} = -\frac{(\overline{w'\theta'})\overline{\theta}_z}{(\overline{\theta_x'^2}) + (\overline{\theta_z'^2}) - \beta\theta'_{zmax}\overline{\theta}_z} \; . \qquad (15)$$

For the effect of localised eddy viscosity, note that from (1)

$$\frac{\partial \overline{u}}{\partial t} + \frac{\partial}{\partial z}(\overline{u'w'}) = \frac{\partial}{\partial z}(\overline{\nu}\,\overline{u}_z + (\overline{\nu' u'_z})). \qquad (16)$$

If ν' and κ' are in phase then by (14), ν' and θ' are $\pi/2$ out of phase. For linear waves, u' and θ' are $\pi/2$ out of phase. Hence ν' and u' are in phase or π out of phase and ν' and u'_z are $\pi/2$ out of phase. Thus, in (16), $(\overline{\nu' u'_z})$ is negligible and ν' is not needed.

3. The Numerical Model

The numerical model is described fully by Mobbs (1987). It uses a spectral representation in the horizontal and a finite difference representation in the vertical and in time. Two regimes (non-breaking and breaking) can be identified:

Non-Breaking Regime

There are two alternative conditions for this regime to exist:

$\frac{g}{\theta_0}\left(\frac{\partial \theta}{\partial z}\right)_{min} > \alpha$ (ii) $\frac{g}{\theta_0}\left(\frac{\partial \theta}{\partial z}\right)_{min} < \alpha$ and $(\overline{\theta'\theta'_t}) < 0$.

$(\partial\theta/\partial z)_{min}$ is the minimum value over a horizontal wavelength and α is a constant specifing the amount of overturning allowed. Condition (ii) above means that waves which are decaying in amplitude are assumed not be be generating significant turbulence. At non-breaking heights (1) and (2) are integrated in time with $\nu = \kappa = 0$.

Breaking Regime

This regimes is defined by $\frac{g}{\theta_0}\left(\frac{\partial \theta}{\partial z}\right)_{min} < \alpha$ and $(\overline{\theta'\theta'_t}) > 0$.

For example, if $\alpha = 0$, there would be wave breaking at levels where there is shading in figure 1. If the wave breaking conditions are satisfied at a particular height, the wave amplitude is held constant and the perturbation components of (1) and (2) are not used. The diffusivities are calculated using (13) and (15) and used in the horizontally averaged parts of (1) and (2), allowing continued evolution of the mean flow.

4. Model Atmosphere and Wave Generation

We consider the problem of internal gravity waves approaching a critical level (a similar problem with constant diffusivities was studied by Fritts (1978)). The model has:

(i) A domain height of 30km.

(ii) A critical level at 18km, arranged by having an initial velocity profile u_0 given by

$$u_0(z) = 40 \tanh\left(\frac{z - z_c}{D}\right) \text{ ms}^{-1} \tag{17}$$

where $z_c = 18\,\text{km}$ and $D = 8\,\text{km}$.

(iii) Constant initial Brunt-Väisälä N such that $N^2 = 3 \times 10^{-4}\ \text{s}^{-2}$.

(iv) Waves with a horizontal wavelength of 50km were forced at $z = 0$.

(v) Waves were forced at $z = 0$ via w' as follows (with $A = 1.5$ and $t_r = 12000\text{s}$):

$$w' = Af(t)\sin(kx) \text{ ms}^{-1} \qquad \text{and} \qquad f(t) = \begin{cases} \sin^2\left(\frac{\pi t}{t_r}\right), & \text{if } t \le t_r; \\ 0, & \text{if } t > t_r. \end{cases} \tag{18}$$

(vi) A vertical grid spacing $\Delta z = 75\text{m}$ and a time step $\Delta t = 12.5\text{s}$ were used.

5. Results

Results are first given for the case without any turbulence generation (i.e. $\nu = \kappa = 0$ everywhere for all time). Figure 2 shows contours of total potential temperature θ after 9000s. Considerable overturning of θ surfaces has occurred. Figure 3 shows the mean horizontal velocity $\overline{u}$ at the same time. There has been considerable acceleration of the mean flow towards the phase speed (zero) in the critical layer below 18km.

Figure 4 shows contours of θ after 9000s with parametrised wave breaking ($\alpha = \beta = 0$). Comparing with figure 2, it can be seen that overturning is considerably reduced. Some overturning still occurs due mean flow changes which continue after wave breaking begins. Figure 5 shows the corresponding mean flow. Limiting the wave overturning clearly reduces the effect on the mean flow. However, the mean flow changes in this case are due both wave transience and dissipation and are thus permanent; those shown in figure 3 are due to wave transience alone and cannot be permanent.

Figure 6(a) shows θ contours after 9000s for the case where $\alpha = -7.5 \times 10^{-5}\text{s}^{-2}$ and $\beta = 0$. This allows some overturning of isentropes to occur before turbulence generation begins. Figure 6(b) shows the same quantity but with $\alpha = 7.5 \times 10^{-5}\text{s}^{-2}$. Here, steepening of the isentropes is quite severely reduced. The effect on the mean flow in these two cases in shown in figure 7. It is evident that the thickness of the "critical layer" where significant modification of the mean flow occurs is reduced as α is reduced. This strongly affects the eddy Prandtl number $Pr = \overline{\nu}/\overline{\kappa}$ (see below).

Figure 8 shows the eddy viscosity $\overline{\nu}$ and conductivity $\overline{\kappa}$ at 8000s, 9000s and 10000s with $\alpha = \beta = 0$. In the early stages (figure 8(a)) wave breaking begins close to the critical level, where the intrinsic frequency $\hat{\omega} = -\overline{u}k$ is small. Suppose that $\theta' = \Theta(z)\exp[i(kx - \omega t)]$, where $\omega = \omega_r + i\omega_i$ ($\omega_r = 0$). The linearised equation for θ' is

$$-i\hat{\omega}\theta' + \omega_i\theta' + \overline{\theta}_z w' = 0. \tag{19}$$

Since $|\hat{\omega}| \ll \omega_i$ where wave breaking occurs, θ' and w' are in phase or π out of phase. Hence $\overline{w'\theta'}$ will be significant. Continuity requires that u' and w' should be either in

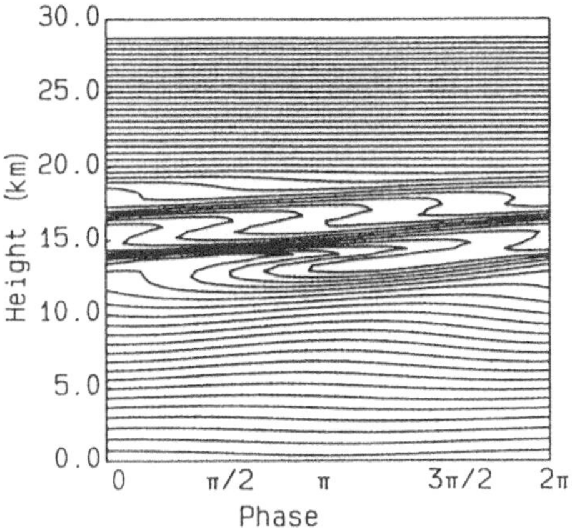

Figure 2. Isentropes at $t = 9000$s with no parametrisation applied.

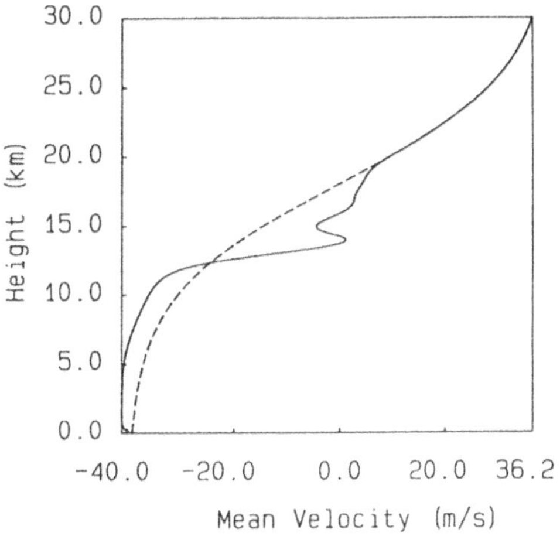

Figure 3. The mean flow at $t = 0$ (dashed) and $t = 9000$s with no parametrisation.

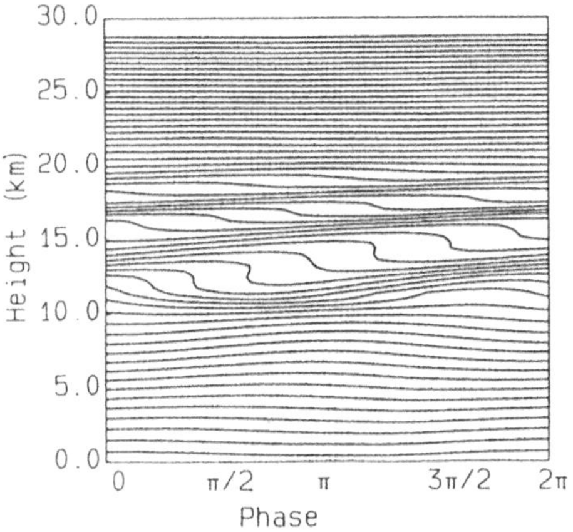

Figure 4. Isentropes at $t = 9000$s with $\alpha = 0$ and $\beta = 0$.

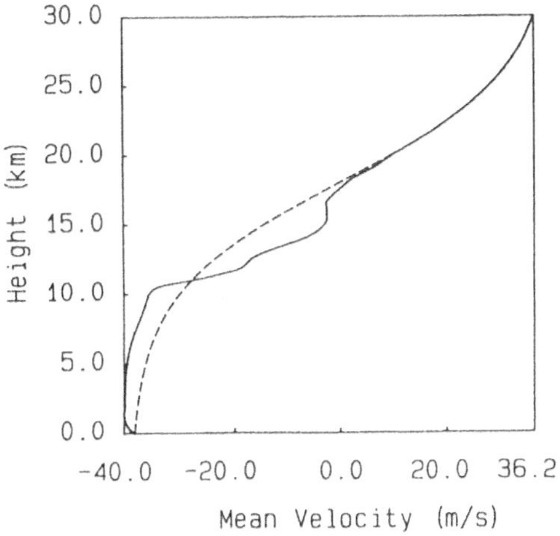

Figure 5. The mean flow at $t = 0$ (dashed) and $t = 9000$s with $\alpha = 0$ and $\beta = 0$.

phase or π out of phase. Hence $\overline{u'w'}$ will not be small. From (13) and (15) we then see that both $\overline{\nu}$ and $\overline{\kappa}$ will be significant, giving a Prandtl number of order unity (figure 9(a)). Later (9000s), larger amplitude waves reach the vicinity of the critical layer and break some distance below the critical level. Here $|\hat{\omega}| \gg \omega_i$ and so θ' and w' are now $\pi/2$ out of phase, giving small values of $\overline{w'\theta'}$ and hence small $\overline{\kappa}$. As there is no similar dependence of $\overline{\nu}$ on $\hat{\omega}$, $\overline{\nu}$ is not small, giving rise to a large Prandtl number (figure 9(b)). Later still (10000s), a deep critical layer has developed where $|\hat{\omega}| \ll \omega_i$, due to modification of the mean flow. Thus Pr is once again of order unity (figure 9(c)).

The effect of localising diffusion near certain phases of the wave is shown in figure 10. The value of $\beta = 0.3$ used here was obtained by Rabbitt (1989) by means of a linear stability analysis. Figures 4 and 10 show that localising diffusion more efficiently prevents wave overturning. Figure 11 shows the average temperature flux between 6000s and 12000s for the cases $\alpha = \pm 7.5 \times 10^{-5}\text{s}^{-2}$ and $\beta = 0$ or 0.3. The most significant

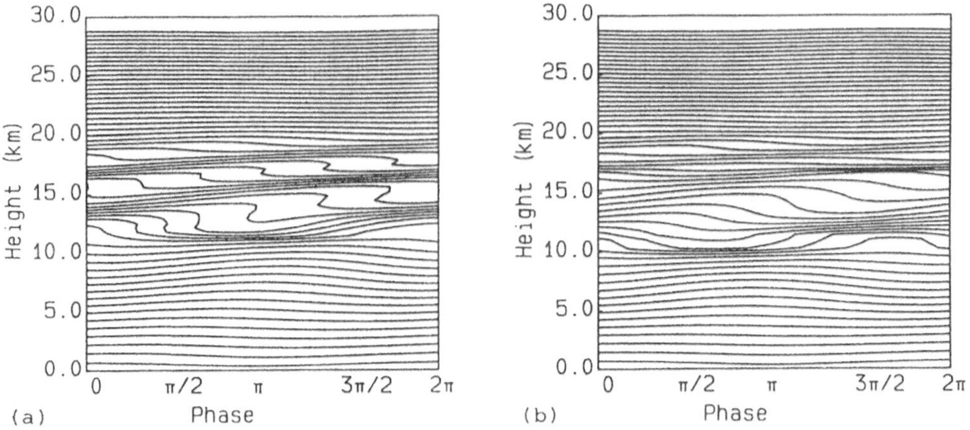

Figure 6. Isentropes at t = 9000s with $\beta = 0$ and (a) $\alpha = -7.5 \times 10^{-5}\mathrm{s}^{-2}$ (b) $\alpha = 7.5 \times 10^{-5}\mathrm{s}^{-2}$

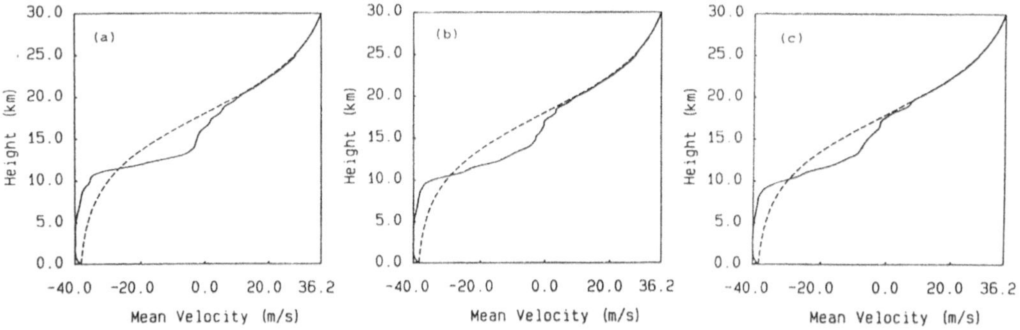

Figure 7. $\overline{u}$ at $t = 12000$s for (a) $\alpha = -7.5 \times 10^{-5}\mathrm{s}^{-2}$ (b) $\alpha = 0$ (c) $\alpha = 7.5 \times 10^{-5}\mathrm{s}^{-2}$.

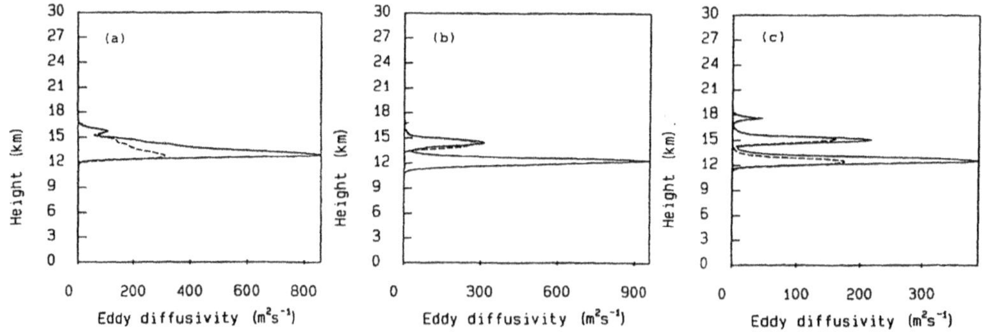

Figure 8. $\overline{\nu}$ (full line) and $\overline{\kappa}$ (dashed line) at (a) $t = 8000$s, (b) $t = 9000$s, (c) $t = 10000$s with $\alpha = \beta = 0$.

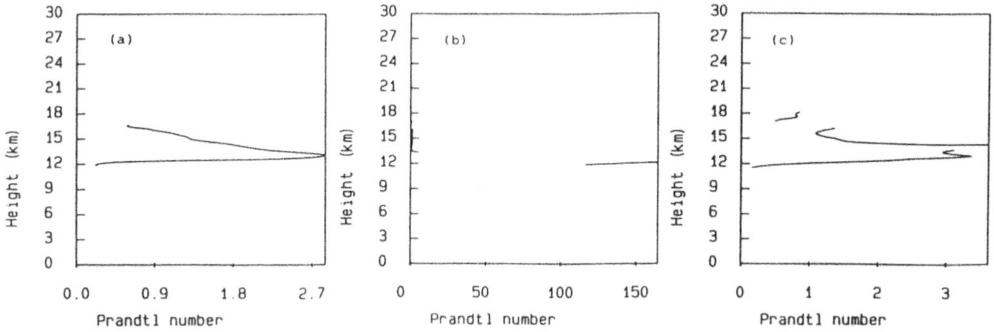

Figure 9. The Prandtl number $\overline{\nu}/\overline{\kappa}$ at (a) $t = 8000$s, (b) $t = 9000$s, (c) $t = 10000$s with $\alpha = \beta = 0$. Only values where $\overline{\kappa} > 5\mathrm{m}^2\mathrm{s}^{-1}$ are plotted.

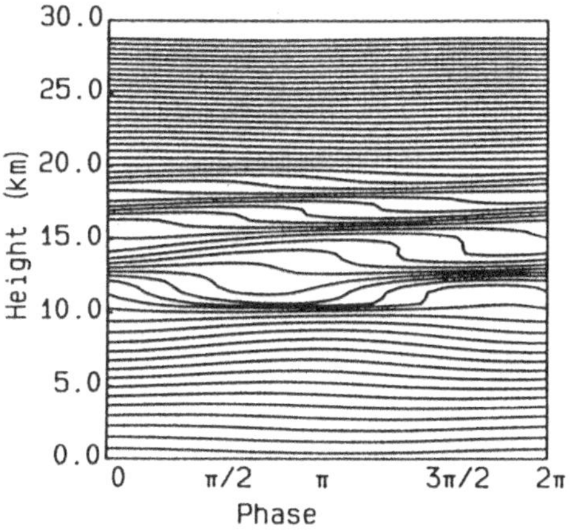

Figure 10. Isentropes at $t = 9000$s with $\alpha = 0$ and $\beta = 0.3$.

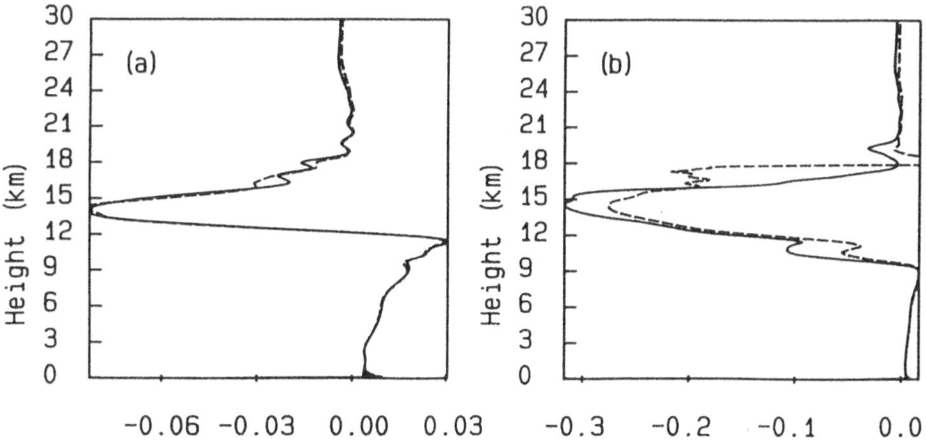

Figure 11. Average temperature flux per second ($\mathrm{ms}^{-2}\mathrm{K}$) between 6000s and 12000s with $\beta = 0$ (full) and $\beta = 0.3$ (dashed). (a) $\alpha = -7.5 \times 10^{-5}\mathrm{s}^{-2}$ (b) $\alpha = 7.5 \times 10^{-5}\mathrm{s}^{-2}$.

effects of localising diffusion occur when $\alpha > 0$, i.e. isentropes never become vertical. Recall that positive values of α lead to the thickest critical layers where $\overline{u}$ is significantly modified. This in turn leads to the largest heat fluxes, as explained in the previous paragraph. Hence the effect of localised turbulence is most noticeable for $\alpha > 0$.

6. Conclusions

A model for the breaking of internal gravity waves has been demonstrated. It assumes that there is just sufficient turbulence generation to prevent wave overturning. The model extends previous ones in allowing a variable turbulent Prandtl number.

The interaction of an internal gravity wave with a critical layer has been studied. The wave breaking model efficiently limits the wave amplitude. The interaction process undergoes three stages. At early times the wave breakdown occurs very close to the critical level and the Prandtl number is small. Later, the waves overturn at greater distances from the critical level and the Prandtl number becomes large. Finally, the wave-mean flow acceleration gives rise to a thick critical layer and the Prandtl number again becomes of order unity. Localising the turbulence around certain phases of the wave has greatest effect when turbulence generation begins before isentropes overturn.

References

Chao, W.C. & Schoeberl, M.R. 1984 One the linear approximation of gravity wave saturation in the mesosphere. *J. Atmos. Sci.* **41**, 1893.

Fritts, D.C. 1978 The nonlinear gravity wave-critical level interaction. *J. Atmos. Sci.* **35**, 397.

Fritts, D.C. & Dunkerton, T.J. 1985 Fluxes of heat and constituents due to convectively unstable gravity waves. *J. Atmos. Sci.* **42**, 549.

Hodges, R.R. 1967 Generation of turbulence in the upper atmosphere by internal gravity waves. *J. Geophys. Res.* **74**, 4087.

Lindzen, R.S. 1967 Thermally driven tide in the atmosphere. *Quart. J. R. Met. Soc.* **93**, 18.

Lindzen, R.S. 1981 Turbulence and stress due to gravity wave and tidal breakdown *J. Geophys. Res.* **86**, 9707.

Lindzen, R.S. 1988 Supersaturation of vertically propagating internal gravity waves. *J. Atmos. Sci.* **45**, 705.

Turbulence and Gravitation

A. Pouquet[1], *J. Léorat*[2] *and T. Passot*[3]

[1] Observatoire de la Côte d'Azur, BP 139, Nice cedex 06003, France
[2] Observatoire de Meudon, 1 Place Jansen, 92 Meudon, France
[3] Applied Mathematics, University of Arizona, Tucson 87527, USA

A marginal equilibrium at all scales is found with phenomenology and numerical simulations in self-gravitating turbulent molecular clouds. The influence of magnetic fields is also investigated

1. Introduction

Molecular clouds which are the site of star formation are in a state of supersonic turbulence, with *rms* Mach numbers around 4, typical velocity of 1 kms^{-1}, length of a few parsecs (where 1 $pc \sim 3.2$ light-years), and temperature of $20K$. Such clouds will hold of the order of 10^3 solar masses, and may undergo gravitational collapse. The collapse is not as efficient as the linear analysis of the Jeans' instability indicates for a static cloud. Turbulence and magnetic fields, which are observed but whose scale and intensity are not directly known, have been invoked as possible agents of the slowing down or stopping of the collapse, but the precise mechanisms are still unknown (Lada & Shu, 1990). Recent observations *eg* in CO lines, also indicate that down to the smallest resolved scales of the order of $.02pc$, the clouds are highly filamentary and clumpy. Intermittency may be present (Scalo, 1990), as well as fractality (Henriksen, 1990), and departures from a gaussian profile in the wings of molecular lines have been reported (Falgarone & Phillips, 1990).

The observed distribution of cloud velocity versus scale corresponds to a -2 energy spectrum spectral index, over a wide distribution of clouds and three orders of magnitude in scales for the latest observations (Perrault, 1987), although a Kolmogorov-like $-5/3$ index has been reported (Larson, 1981) and may also appear at the smallest scales observed. One may then ask whether the evolution of such clouds and the slowing-down of their collapse cannot be understood in a fluid context. We present here a phenomenological analysis (Section 2) which indicates that supersonic turbulence may indeed stop the collapse, backed by numerical evidence for 2D flows using spectral methods and periodic boundary conditions on a 128^2 grid. Section 3 gives preliminary results for the coupling of turbulence, magnetic fields and self-gravitation and Section 4 is the Conclusion.

2. Phenomenology

There are several models that have been derived to describe the way by which turbulent eddies can stabilize an otherwise collapsing flow. They basically take into account turbulent pressure, but neglect any other non-linear effect. The phenomenological analysis presented succintly below leads to a simple model of an unmagnetised cloud (Léorat et

Advances in Turbulence 3
Editors: A.V. Johansson · P.H. Alfredsson

al., 1990). The essence of the argument is to encompass in a single parameter the various effects at play by examination of their characteristic times : transfer of energy to small scales due to non–linear coupling, exchange between turbulent eddies and acoustic waves, and finally collapse. The model shows that the energy transfer to small scales can be balanced by the collapse, the two phenomena having the same lack of dependence of their temporal scales with spatial scale. In brief, let us first define the several time scales entering the problem, in a flow of characteristic length and velocity L_* and u_*, the latter with a compressive component u_c*. The free–fall time is $\tau_{ff} = 1/\sqrt{G\rho_0} = L_J/c_0$ where G is the Gravitational constant and ρ_0 the mean density, c_0 the speed of sound and L_J the Jeans' length obtained by comparing τ_{ff} and the acoustic time $\tau_{ac} = \ell/c_0$. The non–linear eddy turn–over time is $\tau_{NL} = \ell/u_\ell$ where u_ℓ is the typical turbulent velocity at scale ℓ. The transfer time of energy by turbulence to the small scales τ_{tr} is the time whose determination requires some care. In the simplest case of incompressible turbulence, it is the eddy turn–over time computed on an inertial range index $\alpha = 5/3$ with $E(k) \sim k^{-\alpha}$ the kinetic energy spectrum. When turbulence and acoustic waves are coupled, α is modified (Moisseev et al., 1983). However, in molecular clouds at high Mach number, sound waves steepen into shocks and the compressible part of the velocity u_c has a spectrum $E^c(k) \sim k^{-2}$. This leads to a characteristic transfer time L_*/u_{c*} independent of the local scale ℓ (Léorat et al., 1990) which corresponds to the fact that a shock is phase–locked and dissipates equally at all scales. In subsonic turbulence, τ_{tr} depends on ℓ and the Jeans' length is simply augmented (Chandrasekhar, 1951; Sasao, 1973), whereas when $\alpha > 3$, the criterion is inverted (Bonazzola et al., 1987).

In the high compressibility limit, the dissipation of energy due to shocks may dominate the transfer to small scales through the classical cascading in the Kolmogorov spirit of local mode–coupling. Indeed, if we consider an isolated one–dimensional shock wave, transfer in the usual sense is absent and we are faced with a self–similar decay of the spectrum, the characteristic time of energy dissipation being the same for every harmonic of the shock. In the following, we will determine this time considering the one–dimensional Burgers' equation. It can be considered a good approximation, since for highly supersonic flows dissipation is plausibly that of shocks, and the pressure term with a M_a^{-2} scaling (where $M_a = u_*/c_0$ is the Mach number) is weak. Let us start with a train of sawtooth waves with period L and amplitude $u_c(0)$. Since $u_c(dt) = \frac{u_c(0)}{1+\frac{dt}{L_*}u_c(0))}$, the total kinetic energy at $t = 0$ verifies $(1/E)\partial_t E = -(1/T_d)$ where $T_d = L_*/4u_c(0)$, and $E(t) = \int \frac{1}{2}u_c^2(x,t)dx$, with the same notations as before.

This time of self–similar decay is directly found scale–independent and is nothing else than the characteristic time of decay of a strongly supersonic turbulence. Since the free–fall time as well is scale–independent when only the mean density is taken into account, this leads to a **marginal equilibrium** at all scales between turbulence and collapse. This marginal equilibrium, which depends on the presence of shocks in the flow, can be locally broken by the over–densities in the vicinity of shocks which are important when radiative transfer is taken into account (Zeldovich & Raizer, 1966). Such a density enhancement leads to a significantly shorter free–fall time allowing for local collapse. This in turn leads to clumps along the filamentary shocks, as suggested by observations. In the simplest case, the control parameter can be written as

$$\tilde{r} = \tau_{tr}/\tau_{ff} = (L_*/L_J)c_0/u_c \ . \qquad (1)$$

We conjecture that there is a critical value $\tilde{r}_c$ of order unity below which the flow becomes stable. For a typical cold supersonic cloud, this model indicates that the collapse is stopped at all scales, contrary to both the linear analysis of Jeans and to a two-point closure of turbulence (Chandrasekhar, 1951). The latter model predicted a simple rescaling, due to turbulent pressure, of the length beyond which collapse prevails.

Numerical experimentation on two-dimensional flows supports the prediction of the model. In the presence of a developed velocity field, the collapse can be withheld up to ten free-fall times, as observations suggest, and more. In the marginal case, when the parameter $\tilde{r}$ is close to its critical value estimated numerically at $\tilde{r}_c \sim 0.3$, local small-scale gravitational instabilities can indeed be triggered because of the local enhancement of density contrasts in the vicinity of shocks.

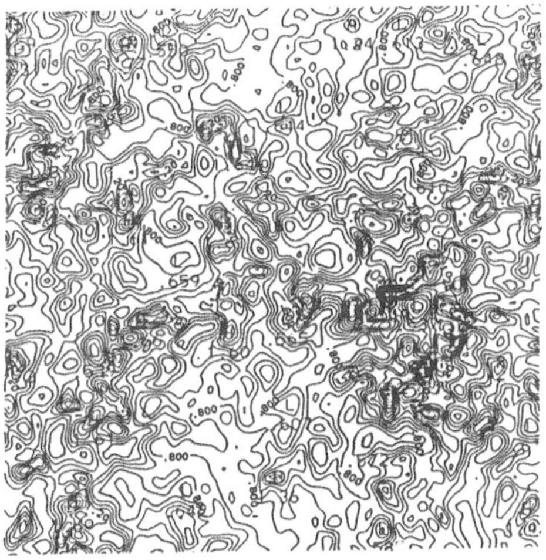

Figure 1: Density contours : stable flow

Figure 2: Density contours : marginally unstable flow

In Figures 1 (left) and 2 (right) are displayed contour lines of the density field at 9.4 and 3.0 free-fall times respectively, with a density contrast $\Delta\rho = \rho_{max}/\rho_{min}$ (in units of its initial value) of 1.02 in the first stable case ($\tilde{r} = 0.26$), and 6.0 in the second marginal case ($\tilde{r} = 0.58$). The prevalent scale for the stable flow is that of the turbulent velocity, whereas clumpiness is noticeable in a range of intermediate scales for the marginal flow, as exemplified by a flat density spectrum in that range of scales. By comparison, in a similar collapsing cloud at lower Mach number, $\Delta\rho \sim 970$ at 0.9 free-fall time.

In three dimensions, the condensations may be planar, following planar (possibly warped) shocks, or filamentary. In the absence of self-gravitation, preliminary results using both a Navier-Stokes solver (Passot and Pouquet, 1990) and a Piecewise-Parabolic Method (Porter et al., 1990) indicate a clear filamentation of the density, may be linked to shock intersections, as well as for the vorticity and its production term.

3. Interaction with a magnetic field

Magnetic fields are observed in the interstellar medium. Their magnitudes are scarcely known directly, the Zeeman splitting being small. Synchrotron radiation yields the product of the density along the line of sight by their longitudinal component. A consensus appears as to the dynamical importance of the magnetic field. On the other hand, it is not clear what is its characteristic scale. It may be argued, using concepts for incompressible flows of inverse cascading to large scales of the magnetic helicity $H^M = \int \mathbf{a} \cdot \mathbf{b} \, d^3\mathbf{x}$ where $\mathbf{b} = \nabla \times \mathbf{a}$, that such fields are large–scale quasi–uniform. Some observations point out to helical structures (Heiles, 1990; Scalo, 1990) as well as bubbles (Bally, 1990).

What rôle can the magnetic field play in the collapse ? One criterion (Mouschovias, 1987) stipulates that, when the field is strong enough, it will be able to prevent collapse, because of flux conservation in the limit of small magnetic diffusivity. Reconnection processes (invoked *eg* in the heating of the solar corona), shocks and instabilities such as the tearing–mode are not taken into account in this criterion.

In the absence of a gravitational term, preliminary numerical simulations in two dimensions (Pouquet et al., 1990) indicate that the magnetic field has a non–trivial effect on the dynamics. In particular, for weak magnetic fields, shocks are slightly hindered, whereas when the turbulent magnetic field is strong, current sheets are formed which heat the flow locally, producing expanding bubbles at the edge of which new thin shocks appear. Furthermore, because there is now more than one mechanism of heating (*viz* viscous and Joule), the temperature and density fields may be markedly different, inducing a strong production of vorticity. Unless radiative transfer occurs on a much shorter time scale than any of these mechanisms (which may not be the case for strong magnetic fields), the isothermal approximation will be of less validity in MHD than for a neutral fluid.

When coupling gravity and MHD, one may conjecture that pressure is modified by both the turbulent pressure, and the magnetic one (see also Pudritz, 1990). For a strong quasi–uniform magnetic field B_0 which stiffens the medium and favors the propagation of Alfvén waves, a somewhat *ad hoc* modification of the phenomenological argument presented above leads to a $(L_*/L_J)M_{sr}^{-1}$ dependence of $\tilde{r}$, where $M_{sr} = u_{sr}/v_a$ with $u_{sr} = \sqrt{c_0^2 + v_a^2}$, *ie* a Mach number based on the fast magnetosonic waves; $v_a = B_0/\sqrt{\rho\mu_0}$ where μ_0 is the magnetic permeability in free space, is the Alfvén velocity. Note that the characteristic times of both an acoustic wave $\tau_{ac} \sim \ell/c_0$ and of an Alfvén wave $\tau_{Al} \sim \ell/v_a$ have a similar scale dependence. Thus a supersonic but sufficiently sub–alfvenic flow may be more prone to collapse than its neutral counterpart at the same Mach number M_a. However, geometrical factors and flux–conservation constraints are not included in such a simple phenomenology. Moreover, it is not clear what spectral index for the kinetic energy spectrum will emerge in the compressible MHD regime. If the magnetic field is strong enough and effectively prevents the formation of shocks so that the Burgers' analysis is only weakly relevant, then one plausible modification of the critical parameter for collapse may be the one just quoted whereby the Jeans' length will be augmented through pressure effects.

A study of the interaction of a compressible turbulent flow with a magnetic field and self–gravitation, as a function of both its intensity relative to the velocity field, and its geometry and characteristic scale relative to the typical size of the cloud leads to a large parameter space, larger when radiative transfer is included (Monaghan and Lattanzio, 1990); early work in 3D with rotation was done by Dorfi (1982).

We now give the results of 2D numerical simulations performed as an extension of the non–magnetic runs described in the preceding Section. We show the density for two identical runs without (Figure 3, left) magnetic field and with it (Figure 4, right); in the latter case, the Alfvenic Mach number $M_b = u_*/v_a$ is $M_b = 1.8$. For this weak–field regime, there are few differences. The density maximum is slightly larger in the presence of the magnetic field, but at a different location in the flow, namely where the magnetic potential is the highest. The strongest condensation in the case $\mathbf{B} = 0$ is prevented by the presence of a complex structure in the magnetic current, more bubble–like than sheet–like, that locally heats the flow, acts as a pressure, and makes it expand. The contrast of density is 1.6 times higher when $\mathbf{B} = 0$, and the density field appears less structured at that time.

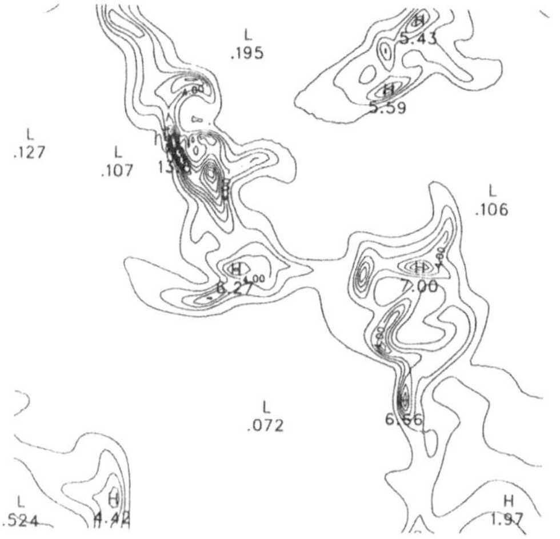

Figure 3: Density contours : unstable flow

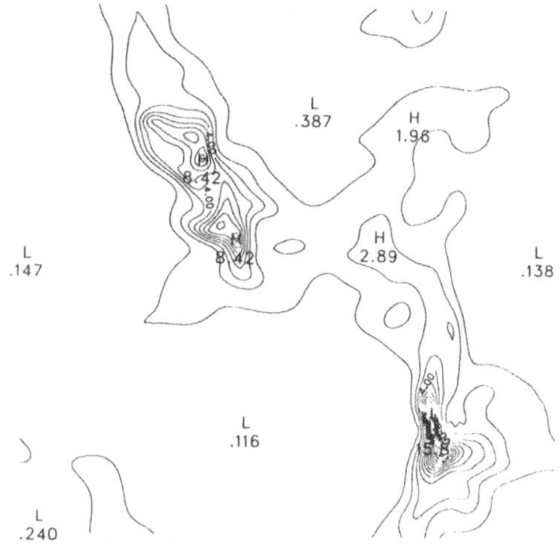

Figure 4: Density contours : magnetised unstable flow

To reach higher values of the magnetic field, we introduce an independent forcing term in the induction equation, leading to supersonic and sub–Alfvenic flows. Its amplitude relative to that of the kinetic forcing is 0, 2, 7 and 35 (twice) respectively for the Runs listed in the Table below. For all flows, the length scale of the turbulence k_*^{-1} is large; the mean density ρ_0 is equal to unity; $\Delta\rho = \rho_{max}/\rho_{min}$ is the density contrast, $\chi = E^c/E^v$ gives the percentage of the compressive component of the kinetic energy, and $\tilde{r}$ is the parameter defined in eq. (1), all at the onset of gravity switching, once the flow has achieved a statistical steady–state (a dozen turn–over times). $\Delta\rho$ at the final time t_f is given in units of its initial value. N is the run number.

N	v_a/u_* t=0	M_a 0	$\Delta\rho$ 0	k_* 0	χ 0	$\tilde{r}$ 0	v_a/u_* $t = 2.6$	$\Delta\rho$ t_f	t_f
1	0 .01	2.5	3.4	0.7	.6	16	.004	388	2.6
2	2.1	4.1	10.	1.0	6	2.2	1.0	34	2.6
3	4.3	4.9	24.	2.1	36	0.4	2.5	15	3.7
4	4.8	1.3	11.	2.7	45	0.9	2.3	5.4	3.1
5	4.7	3.6	81.	3.2	39	0.3	3.9	1.7	0.5

In all cases, the magnetic field at the onset is quasi–uniform, with $b_{max}/b_{min} \sim 2$, whereas the velocity has large excursions ($\sim 10^2$); the Jeans' wavenumber is $k_J = 2.2$

(the number of Jeans' masses in the flow is equal to five), and the free–fall time is $\tau_{ff} = 2.2$ except for Run 4 for which it is $\tau_{ff} = 0.7$; the forcing terms are purely solenoidal. The final time t_f is in units of the free–fall time. In Run 5, the computation stops early probably due to strong initial density fluctuations.

For a typical cloud of length $2pc$ at $20K$, with 10^3 solar masses, and for a speed of sound of $0.3kms^{-1}$, the values of the magnetic field, in $\mu Gauss$, are for Runs 2 through 5 respectively $B_2 = 45$ [40], $B_3 = 115$ [30], $B_4 = 28$ [30] and $B_5 = 86$ [84], the second value in brackets being at the final time, whereas the former value is that at the onset of the runs. For such a cloud, the classical value above which collapse is prevented (Mouschovias, 1987) is $30\mu Gauss$. All the magnetic fields are comparable or stronger, and yet collapse does occur albeit slowly. This criterion assumes no dissipation through Ohmic losses or ambipolar diffusion (flux conservation), whereas in our calculations magnetic diffusivity is non–negligible although balanced by forcing to maintain developed spectra with excitation at all scales. Moreover, the values of the magnetic field, which depend on $\sqrt{\rho}$, are evaluated on the mean density, not on the local peak values, although at the end of the evolution the magnetic field, in its small–scale current structures, is concentrated in the condensations of density.

We find that the stronger the magnetic field, the weaker the collapse, and that the parameter $\tilde{r}$ diminishes as B_0 increases. This may be attributed mainly to the growth of the compressive component of the velocity. Indeed, a compressible MHD flow has a much higher E^c/E^v ratio than its neutral counterpart (Pouquet et al., 1990), because of the magnetic pressure term. Also, the ratio of the Jeans' length to the characteristic length of the turbulent flow is here $\mathcal{O}(1)$, whereas in Léorat et al. (1990) for the neutral case, there is a scale separation. In three dimensions without magnetic field, collapse is favored when $L_* \sim L_J$ and $\tilde{r} > 1$ (Chantry et al., 1990).

The normalised correlation between the velocity and magnetic field is low (of the order of 10%) for all runs except at the final time of Run 3, where it is equal to -0.63. This should slow down the evolution, and may be indicative of an ordering, through gravity, of the fields towards a few central condensations. During all computations, the *rms* Mach numbers have increased by roughly a factor two, both through an acceleration due to the onset of gravitational collapse, and through an input due to the magnetic energy, which systematically decreases during the same lapse of time, to reach values closer to (but above) equipartition. Finally, we note that in runs with a strong magnetic field, oscillations occur in the density contrast $\Delta\rho$ and the Alfvenic Mach number M_b. For example, in Run 4, their time scale is of the order of the free–fall time, with a peak of $\Delta\rho \sim 58$ at $t = 2.7\tau_{ff}$.

A range of parameters for which the collapse of the cloud will be stopped may require stronger magnetic fields, and probably smaller values of the critical parameter $\tilde{r}$. Clearly, two effects are competing. On the one hand, a magnetic field may suppress efficiently the formation of shocks; non–linear transfer to small scales may now be better evaluated through the usual Kolmogorov–like formula, and the slowing–down of the collapse may occur in the way envisioned by Chandrasekhar, using an MHD–turbulent pressure, and leading to rescaling of the Jeans' length. On the other hand, the flux freezing condition at high magnetic conductivity will tend to suppress the collapse for strong magnetic fields. Flux freezing may be partly destroyed by turbulent diffusivity, through reconnection and tearing modes for example, and shocks. The interplay of all these phenomena is intricate.

4. Conclusion

We have presented a simple model that takes into account the interaction of eddies, waves, gravitational collapse and non–linear tranfer to small scales, by examining the characteristic times for such phenomena. This model leads to the prediction that collapse is stopped at all scales when the flow is supersonic, except in strong over–densities in the vicinity of shocks. The model also tends to support small–scale injection of energy in the clouds, such as through stellar winds emanating from a first generation of stars.

Two–dimensional numerical simulations confirm the analysis of the model, although the power of present–day computers and the sophistication of the numerical methods (or lack thereof) do not clearly allow for a rejection of prior models, such as the modification of the Jeans' length by turbulent pressure, as predicted by Chandrasekhar. The presence of a magnetic field slows the collapse down, but may not be sufficient *per se* to stop it completely, unless it is substantially stronger to prevent shock formation. Amplitudes higher than the classical criterion involving the concept of flux freezing seems required, but Joule dissipation is important in our calculations.

Three–dimensionality (Chantry et al., 1990), and a sub–grid scale modelling (Passot & Pouquet, in preparation) are presently being included in the computations. In the latter, one modifies the dissipative terms in order to extend the region of wavenumbers for which the non–linear terms are prevalent, without treating explicitely the small–scale interactions. The turbulent viscosity thus introduced can be based on a model of turbulence (Yoshizawa, 1987; Zeman, 1990), or on a model of dissipation (Passot & Pouquet, 1988) leading to transport coefficients with a non–linear dependence on the fields. These non–linear transport coefficients, reminiscent of the classical Smagorinsky (1963) modelling of convective flows, may introduce a dissipative range with an algebraic tail rather than an exponential one. Spectral methods then lose their accuracy, and the Gibbs–like ubiquitous noise introduced may prove a hindrance.

Support from the DRET contract 500–276 is acknowledged. Numerical simulations were performed at NCAR (Boulder, CO) and at CCVR (Saclay).

References

J. Bally; 1990, in Proceedings IAU Symposium **147**, Eds. F. Boulanger, G. Duvert and E. Falgarone, Kluwer Dordrecht.

S. Bonazzola, E. Falgarone, J. Heyvaerts, M. Perrault and J.L. Puget; 1987, Astron. Astrophys. **172**, 293.

S. Chandrasekhar; 1951, Proc. Roy. Soc. **A 210**, 18; **A 220**, 26.

G. Chantry, R. Grappin & J. Léorat; 1990, in Proceedings IAU Symposium **147**, *op cit.*

E. Dorfi; 1982, Astron. Astrophys. **114**, 151.

E. Falgarone and T.G. Phillips; 1990, to appear, Astrophys. J.

C. Heiles; 1990, in Proceedings IAU Symposium **147**, *op cit.*

L. Henriksen; 1990, in Proceedings IAU Symposium **147**, *op cit.*

C.J. Lada and F.H. Shu; 1990, Science **248**, 564.

R.B. Larson; 1981, Monthly Not. Roy. Astron. Soc. **194**, 809.

J. Léorat, T. Passot and A. Pouquet; 1990, Monthly Not. Roy. Astron. Soc. **243**, 293.

S. Moisseev, R. Sagdeev, A. Tur, G. Khomenko & V. Yanovski; 1983, JETP **58**, 1149.

J.J. Monaghan and J.C. Lattanzio; 1990, in Proceedings IAU Symposium **147**, *op cit.*

T. Mouschovias; 1987, in *Physical Processes in Interstellar clouds*, eds. G.E. Morfill & M. Scholer, Dordretch, Reidel, 453.
T. Passot and A. Pouquet; 1988, J. Comp. Phys. **75**, 300.
T. Passot and A. Pouquet; 1990, submitted to J. of Mechanicś **B**/ Fluids.
T. Passot, A. Pouquet and P. Woodward; 1988, Astron. Astrophys. **197**, 228.
D. Porter, A. Pouquet and P.R. Woodward; 1990, in preparation.
A. Pouquet; 1989, in Les Houches, J.P. Zahn Ed., Gordon and Breach.
A. Pouquet, T. Passot and J. Léorat; 1990, in Proceedings IAU Symposium **147**, *op cit.*
R. Pudritz; 1990, in Proceedings IAU Symposium **147**, *op cit.*
T. Sasao; 1973, Pub. Astr. Soc. Japan **25**, 1.
J. Scalo; 1990, in *Physical Processes in Fragmentation and Star Formation*, R. Capuzzo–Dolcetta, C. Chiosi and A. Di Fazio Eds., Kluwer Dordrecht.
J. Smagorinsky; 1963, Monthly Weather Rev. **91**, 99.
B.U. Sonnerup; 1988, Computer Physics Communications **49**, 143.
P. Woodward and P. J. Collela; 1984, J. Comp. Phys. **54**, 115.
Yoshizawa; 1987, Phys. Fluids **30**, 1089.
Ya. B. Zeldovich and Yu. P. Raizer; 1966 *Physics of Shock Waves and High Temperature Hydrodynamic Phenomena*, Academic Press.
O. Zeman; 1990, in *Proceedings of the Third European Turbulence*, Stockholm.

Spectral Properties of Turbulence and the Structure of the Atmospheric Boundary Layer

J.M. Rees

Department of Applied and Computational Mathematics,
University of Sheffield, Sheffield S10 2TN, U.K.

A discussion of the spectral and cross-spectral properties of eddies in the nocturnal boundary layer is presented. The dependence of the spectra on the vertical structure of the boundary layer is studied. Meteorological data collected by the British Antarctic Survey during the austral winter of 1986 at Halley station, Antarctica are used to obtain profiles of wind, temperature and the local Richardson number (Ri) up to 40m. Cross-spectral analysis is used to determine whether the nature of the observed eddies is primarily turbulent of whether their structure is characteristic of coherent internal gravity waves. It is found that the cross-spectral phases indicate the presence of turbulent eddies only in regions where Ri is less than the critical value of $\frac{1}{4}$.

1. Introduction

Several methods for distinguishing between coherent eddies (or waves) and turbulent eddies are described in the literature. According to linear theory, waves are not responsible for the vertical transport of heat, i.e. the vertical velocity component (w) and the temperature (T) are $\frac{\pi}{2}$ radians out of phase. However, heat is transported vertically by turbulent eddies and nonlinear waves (Axford, 1971). The cross-spectral phase (ph) gives a measure of the phase difference between two time series. Hence it can be deduced that for waves, ph(wT)= $\pm\frac{\pi}{2}$ and for turbulent eddies, ph(wT)$\approx$0 or π.

During the southern winter of 1986, the British Antarctic Survey conducted a series of boundary layer experiments at Halley station, Antarctica (75.6°S, 26.7°W). During the months of darkness, wavelike disturbances are observed during periods of stable stratification (King, Mobbs, Darby & Rees, 1987). Several mechanisms for the generation of these waves have been proposed:

1. Topographic forcing: Rees & Mobbs (1988) showed that when the wind direction is from over the land, the low frequency disturbances often have the characteristics of trapped, neutral modes, which could have been forced either by the undular ridges in the ice shelf or by the irregular terrain in the vicinity of the Hinge Line (this is the region where the ice shelf joins the Antarctic plateau). Mobbs & Rees (1989) showed further that these modes were only present when the surface layer is stably stratified and the wind speed is less than 7ms^{-1}.

2. Under conditions of high wind speed and near-neutral stratification, locally generated instabilities, i.e. turbulent eddies are observed in the surface layer (Mobbs & Rees, 1989).

Advances in Turbulence 3
Editors: A.V. Johansson · P.H. Alfredsson

3. On several occasions, short bursts of large amplitude waves have been detected in the surface layer. These disturbances have been identified as Kelvin-Helmholtz type instabilities which have been generated above the surface layer in regions of low Ri and then radiated to the ground (King et al., 1987).

In this paper, the earlier work is extended to determine the circumstances under which the disturbances in the surface layer have the properties of turbulent eddies or waves. In particular, the significance of the value of local Richardson in the surface layer will be discussed.

2. Instrumentation and Data

The extensive data set supplied by the British Antarctic Survey enabled us to make a detailed study of the surface layer. Instrumentation included a 32m mast supporting several anemometers, wind vanes and thermometers. Full details of the site and instrumentation are given in King & Anderson (1988). Data were logged at several frequencies - the data used in this study comprises 10s averages from the following instruments:

1. Sonic anemometers at 5m, 16 and 32m - these gave measurements of the three velocity components (u,v,w) and temperature fluctuations.

2. Cup anemometers at 2m, 8m, and 24m.

3. Wind vane at 8m.

4. Absolute temperature at 5m from sonic anemometer.

5. Measurements of the temperature difference between 2-4m, 4-8m, 8-16m and 16-32m.

The data was processed to correct for experimental errors in accordance with the details given in King & Anderson (1988).

3. Methods of Data Analysis
(a) Calculation of Cross-spectra

Cross-spectral analysis can be used to determine the phase difference between two times series. Descriptions of the statistical properties of cross-spectra can be found in Chatfield (1984). Mobbs & Rees (1988) showed that certain different types of wave motion could be distinguished by calculating the cross-spectral phases of the velocity components (u,v,w) and temperature (T). This gives rise to six cross-spectra: uv, uw, vw, uT, vT, wT. In particular, they showed that in order to distinguish between horizontally propagating, trapped waves and locally generated instabilities, it is sufficient to calculate just ph(wT). According to linear theory, for trapped waves

$$ph(wT) = \begin{cases} \frac{\pi}{2} & \text{if } c > \overline{u} \\ -\frac{\pi}{2} & \text{if } c < \overline{u} \end{cases}$$

and for locally generated instabilities (i.e. turbulence)

$$ph(wT) = \begin{cases} \pi & \text{if } \sigma > 0 \\ 0 & \text{if } \sigma < 0 \end{cases}$$

where c denotes phase speed, $\overline{u}$ is the mean wind speed and σ is the growth rate. Thus an additional advantage of cross-spectral analysis is that it can be used to distinguish between growing and decaying turbulence.

Cross-spectra were calculated using data from the sonic anemometers obtained between 24th May and 17th June 1986. Results were computed for sections of data of 10000 seconds duration - details of the method used can be found in Rees (1988). To estimate the statistical significance of the results, the squared cross-coherence (henceforth coherence) was used as a measure of accuracy. For the chosen method of analysis, if the coherence is greater than 0.06, there is a 95% chance of the phase being significantly different from 0.

(b) Mean Profiles and Richardson Number

In order to obtain profiles of the mean velocity and temperature fields, use was made of data from all the instruments listed in S2. This gave us measurements of wind speed at 6 heights between 2 m and 32m, wind direction at 4 heights and temperature at 5 levels (to obtain absolute measurements for temperature we made the assumption that the temperature reading at 5m was approximately equal to that at 4m).

Following Nieuwstadt (1984), log-linear profiles were fitted to the wind speed and temperature using a least squares procedure. Since the error in the direction measurements may be as large as 5°, a linear profile was fitted to the wind direction. Under most conditions, Monin-Obukhov similarity theory is a good approximation for the mean boundary layer flow, however, under certain conditions at Halley it is not always appropriate (see King, 1990). For all of the examples presented herein, every observation lay within one standard deviation from the fitted profile. From the derived profiles of speed, direction and temperature, we were able to estimate the local Richardson number (Ri) up to about 40m.

4. Results

Figure 1 shows a section of the 5m wind speed record comprising 10 second averages of wind speed for 14th June 1986. Note the presence of considerable wave activity. The corresponding wT cross-spectrum is shown in figure 2. For this period, ph(wT) is consistently $\frac{\pi}{2}$, indicating the presence of coherent waves. Most of the waves have periods in the range 0.01 Hz to 0.02 Hz, that is, towards the low frequency end of the spectrum. For this frequency range, the coherence is high which is again a feature associated with coherent

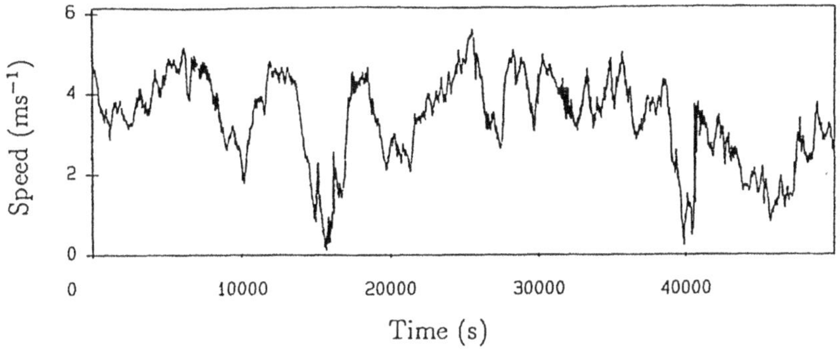

Figure 1. Section of wind speed record from sonic anemometer at 5m for 14th June 1986.

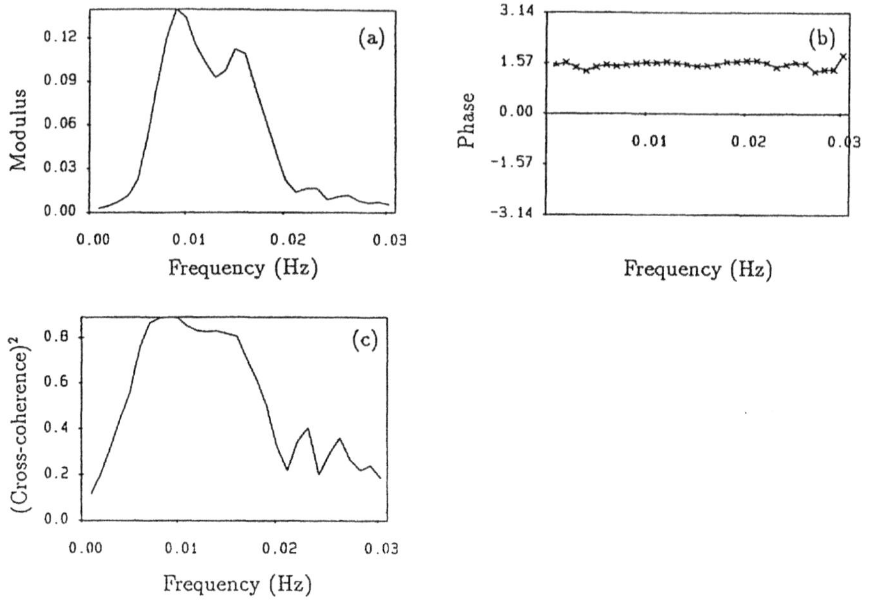

Figure 2. wT cross-spectrum at 32m for 14th June 1986; (a) modulus, (b) phase, (c) coherence.

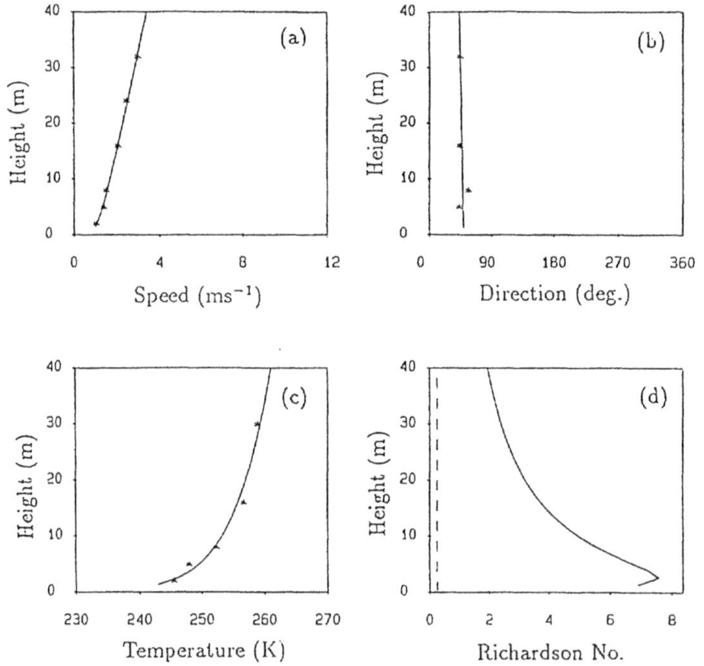

Figure 3. Profiles of mean (a) wind speed, (b) wind direction, (c) temperature, (d) Richardson number for 14th June 1986. Observations are denoted by *.

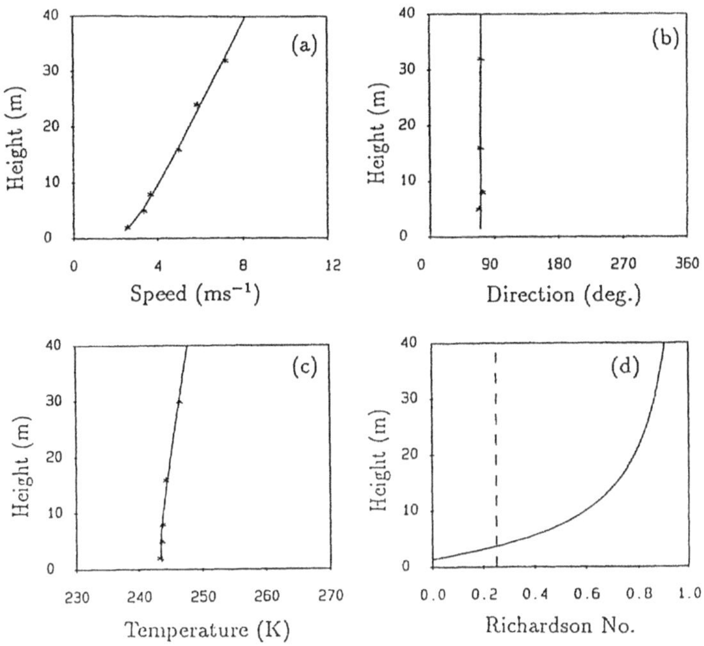

Figure 4. Profiles of mean (a) wind speed, (b) wind direction, (c) temperature, (d) Richardson number for 24th May 1986.

wave motions. Profiles of wind speed, direction and temperature at 45000s are shown in figure 3, together with the Richardson number. (The critical value of Ri for instability, i.e. Ri = $\frac{1}{4}$ is also shown.) Throughout the lowest 40m, the wind speed is low (less than $4ms^{-1}$) and there is strong stable stratification. Ri is greater than $\frac{1}{4}$, which supports the conclusion drawn from the cross-spectra that the observed eddies are coherent waves. The wind is from a north-easterly direction, hence it is likely that these eddies are of topographic origin (see Rees & Mobbs, 1988 for further discussion on this point).

On occasions when the surface layer is not as strongly stably stratified as in the above case, it is possible for Ri to become subcritical near the ground with the result that ph(wT)=0 or π at 5m and $\pm\frac{\pi}{2}$ at 16m and 32m. An example of this occurred on 24th May 1986. Profiles of the mean flow are shown in figure 4. Again, the wind is from over the land, suggesting that topographic effects may influence the surface layer flow. Over the lowest 10m, the surface layer is near-neutral and the overlying stratification is only weakly stable. Hence the Ri is significantly lower than for 14th June example. Up to about 5m, it is below the critical value of $\frac{1}{4}$. The wT cross-spectra calculated from the 5m and 32m sonic readings are shown in figures 5 & 6 respectively. At 5m, ph(wT)=, π and the coherence is low, indicating the presence of turbulent eddies. However, at 32m, ph(wT) is predominantly $\frac{\pi}{2}$ and the coherence is highly significant (~0.6), hence in this region the eddies again possess the characteristics of coherent waves. Note the distinct differences in the modulus of wT in each case. At 5m, the higher frequencies dominate whereas at 32m, there is a spectral peak between 0.01Hz and 0.02Hz.

The final case study presented is for 7th June 1986. On this occasion strong on-shore winds prevailed and the surface layer was weakly stably stratified (see figure 7). Below

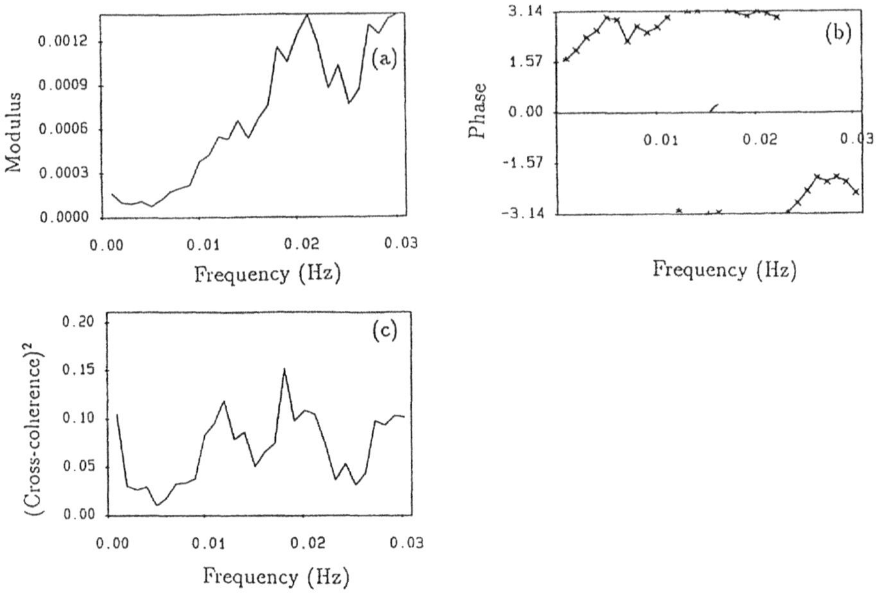

Figure 5. wT cross-spectrum at 5m for 24th May 1986; (a) modulus, (b) phase, (c) coherence.

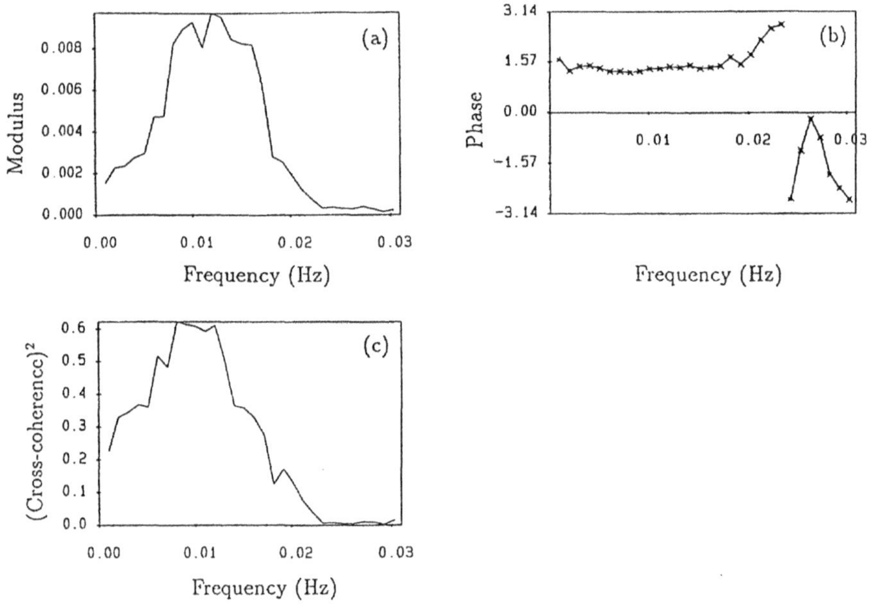

Figure 6. wT cross-spectrum at 32m for 24th May 1986; (a) modulus, (b) phase, (c) coherence.

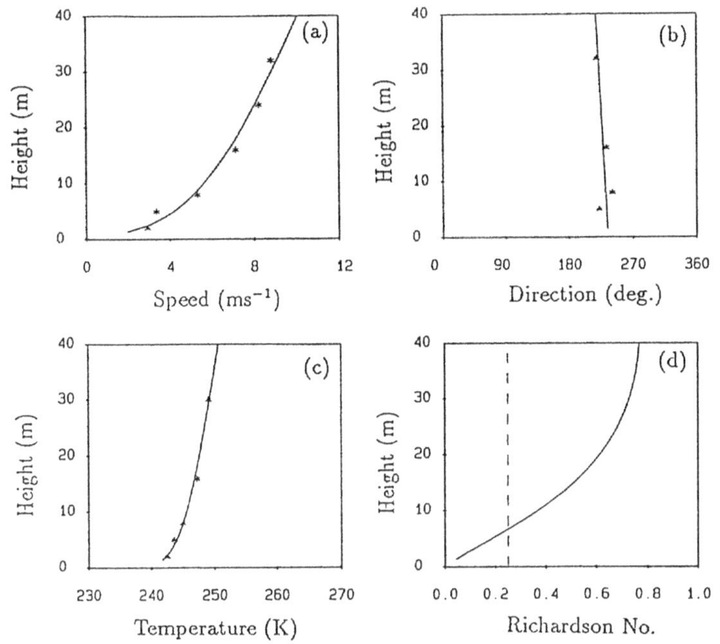

Figure 7. Profiles of mean (a) wind speed, (b) wind direction, (c) temperature, (d) Richardson number for 7th June 1986.

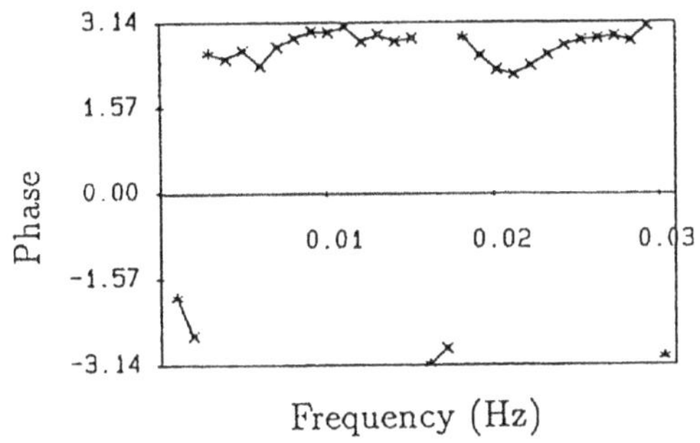

Figure 8.
Ph(wT) at 5m for 7th June 1986.

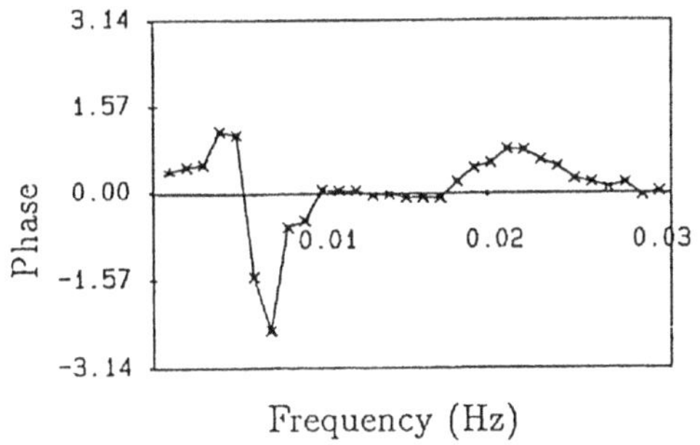

Figure 9.
Ph(wT) at 16m for 7th June 1986.

about 8m, Ri fell below $\frac{1}{4}$. wT cross-spectra calculated from 5 and 16m sonics are shown in figures 8 & 9. At 5m, ph(wT)=0,π which, according to linear theory indicates the presence of growing instabilities. Just above the region of subcritical Ri, however, ph(wT)=0 which is indicative of decaying turbulence. At 32m, no wave activity was discernible. This is to be expected due to the existence of an off-shore wind and hence no topographically forced modes.

5. Conclusions

The nature of the observed eddies in the surface layer over the Antarctic ice-shelf at Halley is closely associated with the mean flow conditions. When the local Richardson number (Ri) is above that required for the onset of instability, and an off-shore wind prevails, cross-spectral phases indicate that the eddies have the characteristics of coherent wave motions. These waves are probably generated by topographic effects. When strong winds prevail the surface layer becomes more mixed and Ri can become less than $\frac{1}{4}$ at lower levels. During such conditions ph(wT)=0 or π which is an indication of the turbulent nature of the eddies.

Acknowledgements

The author would like to thank the British Antarctic Survey for providing the data used in this study and in particular, Dr. J.C. King and Mr. P.S. Anderson. I am also grateful to Dr. S. Mobbs and Mr. I. McConnell for the suggestions they have made.

References

Axford, D.N. 1971 Spectral analysis of an aircraft observation of gravity waves. *Quart. J. R. Met. Soc.*, **97**, 313-321.

Chatfield, C. 1984 The analysis of time series, 3rd edition. *Chapman and Hall.*

King, J.C. 1990 Measurements of turbulence in the stably-stratified surface layer over an Antarctic ice shelf. *Quart. J. R. Met. Soc.*, **116**, 379-400.

King, J.C. & Anderson, P.S. 1988 Installation and performance of the STABLE instrumentation at Halley. *Br. Antarc. Survey Bull.*, **79**, 65-77.

King, J.C., Mobbs, S.D., Darby, M.S. & Rees, J.M. 1987 Observations of an internal gravity wave in the lower troposphere at Halley, Antarctica. *Boundary-Layer Met.*, **39**, 1-13.

Mobbs, S.D. & Rees, J.M. 1989 Cross-spectral properties of large eddies in the stably stratified atmospheric boundary layer. *Advances in Turbulence 2*, edited H.-H. Fernholtz and H.E. Fiedler, Heidelberg, 180-185.

Nieuwstadt, F.T.M. 1984 The turbulent structure of the stable, nocturnal boundary layer. *J. Atmos. Sci.*, **41**, 2202-2216.

Rees, J.M. 1988 Ph.D. Thesis, University of Leeds.

Rees, J.M. & Mobbs, S.D. 1988 Studies of internal gravity waves at Halley Base, Antarctica using wind observations. *Quart. J. R. Met. Soc.*, **114**, 939-966.

Self-similar Structures in 2-D Turbulence: Experimental and Theoretical Study of Vortex Multipoles

S.I. Voropayev

Institute of Oceanology, Academy of Science USSR, Krasikova 23, Moscow, 117218, USSR

Experimental evidence for dipole-like vortex structures in turbulence affected by body forces in a stratified fluid is given. The results from an experimental investigation into the dynamics of the interactions between vortex dipoles and also between vortex dipoles and a vertical barrier are presented. Theoretical consideration reveals that the result of interaction is determined by two main governing parameters: dipolar (J) and quadrupolar (Q) vorticity distributions of the flow. The connection between J, Q and forces applied to the viscous fluid is obtained and a self-similar solution for a vortex quadrupole is found.

1. Introduction

A large family of various irregular two-dimensional flows is associated with the term "2D turbulence". Two-dimensional or quasi-planar flows are realized in liquid systems where some physical mechanism suppresses the motion in one direction. Examples include the flows in a rotating homogeneous fluid [1], in a density stratified resting fluid [2, 3], in a soap film [4], and in a layer of mercury in a magnetic field [5].

To illustrate specific features of the 2D chaotic flow consider a simple experiment. In a layer of a dyed fresh water laying on the layer of salt water we generate the irregular motions moving a vertical row of wires back and forth several times in horizontal directions. In fig. 1 it is clearly visible that fundamental elements of the flow are not single vortices but dipole-like vortex structures. Similar dipolar structures (mushroom-like currents) were discovered recently [6, 7] on the satellite images of the upper layer of the ocean which is a typical example of nearly two-dimensional system where various natural factors act as irregular forces. It seems that vortex dipoles and also their combinations are the universal product of any irregular forcing in two-dimensional systems.

Different aspects of vortex dipoles dynamics in a stratified fluid were studied experimentally and theoretically in [2, 3, 8-12]. Our present aim is to do the next step in this direction: to study the fundamental types of vortex dipoles interactions and explain these interactions in terms of forces acting in a viscous fluid.

2. Theory

2.1 Multipole expansion

Following [11] consider the planar flow induced by a system of forces $\vec{F}(\vec{x}', t)$ (force per unit volume), which acts on the fluid inside the area $S'(\vec{x}')$. The forces and associated vorticity distribution $\vec{\omega}(\vec{x}', t) = (0, 0, \omega) = \vec{k}\omega$ are assumed to occupy a finite area $S'(\vec{x}')$ and the fluid extends to infinity ($\vec{k}$ being the unit vector along z-axis perpendicular to

Advances in Turbulence 3
Editors: A.V. Johansson · P.H. Alfredsson

Fig. 1. Numerous dipole-like vortex structures in initially turbulent upper layer of density stratified water. The length scale of motion (typical diameter of dipoles) is ten times greater than the depth of the upper layer and increases with time.

the plane of motion). The vector potential $\vec{B}(\vec{x},t)=(0,0,\Psi)=\vec{k}\Psi$, where $\Psi(\vec{x},t)$ is the stream function at point $\vec{x}$, can be found from the vector Poisson equation $\Delta\vec{B}=-\vec{\omega}$ and for $|\vec{x}|\gg|\vec{x}'|$ the vector potential may be approximated by a few terms of a multipole expansion.

$$\vec{B}=\vec{k}[\frac{\ln|\vec{x}|}{2\pi}\Gamma+\frac{n_\alpha}{2\pi|\vec{x}|}D_\alpha+\frac{n_\alpha n_\beta}{4\pi|\vec{x}|^2}Q_{\alpha\beta}+O(\frac{1}{|\vec{x}|^3})] \tag{1}$$

where

$$\Gamma=-\int_{S'}\omega(\vec{x}',t)dS' \tag{2}$$

$$D_\alpha=-\int_{S'}x'_\alpha\omega(\vec{x}',t)dS' \tag{3}$$

$$Q_{\alpha\beta}=\int_{S'}(2x'_\alpha x'_\beta-|\vec{x}'|^2\delta_{\alpha\beta})\omega(\vec{x}',t)dS' \tag{4}$$

and

$$n_\alpha=\frac{x_\alpha}{|\vec{x}|},\quad \vec{x}=(x_1,x_2).$$

The fact that forces apply no net moment to the fluid implies that $\Gamma=0$. Making use of the fact that $n_\alpha D_\alpha\vec{k}=\vec{J}\times\vec{n}$ where

$$\vec{J}(t)=\int_{S'}\vec{x}'\times\vec{\omega}(\vec{x}',t)dS' \tag{5}$$

and is equal to the total momentum $\vec{K}(t)=\int_0^t\vec{I}/\rho dt$ applied to the fluid by the total force $\vec{I}(t)=\int_{S'}\vec{F}dS$ since the onset of the motion leads (1) to

$$\vec{B}=\frac{\vec{J}\times\vec{n}}{2\pi|\vec{x}|}+\vec{K}\frac{n_\alpha n_\beta}{4\pi|\vec{x}|^2}Q_{\alpha\beta}+O(\frac{1}{|\vec{x}|^3}) \tag{6}$$

For the first (dipolar) term in (6) see [11], the second (quadrupolar) term is a new one.

2.2. Flow induced by a point force

Consider the planar flow of a viscous incompressible fluid induced by a point momentum source which starts at time $t = 0$ and thereafter exerts on the fluid total force I in the direction $\varphi = 0$ in a polar coordinates system (r, φ). The source is at the point $r = 0$ and the fluid is initially at rest.

Equation (5) gives an additional condition connecting the characteristics of the flow with the source intensity

$$\int_0^{2\pi} \int_0^{\infty} \omega(r, \varphi, t) \sin \varphi r^2 dr d\varphi = J(t) \tag{7}$$

where $J(t) = \int_0^t I/\rho \, \mathrm{dt}$. The self-similar Stokes solution of the problem with the condition (7) is [2, 11]

$$\Psi = \frac{Ir \sin \varphi}{2\pi \rho \nu} [\frac{\nu t}{r^2}(1 - \exp(\frac{r^2}{4\nu t})) + \frac{1}{4} E_1(\frac{r^2}{4\nu t})] \tag{8}$$

where E_1 is an integral exponent and ν is the viscosity.

2.3. Flow induced by a point force dipole

Consider now two point momentum sources applying to the fluid equal forces I. The sources are at the points $(\frac{\varepsilon}{2}, \pi), (\frac{\varepsilon}{2}, 0)$ and act on the fluid in the directions $\varphi = 0, \varphi = \pi$ respectively. Using the first term in (6) we can present the stream function in the form

$$\Psi(r, \varphi) = \Psi_1 + \Psi_2 = \frac{J \sin \varphi_1}{2\pi r_1} + \frac{J \sin \varphi_2}{2\pi r_2} \tag{9}$$

and for $\frac{\varepsilon}{r} \ll 1$ with the accuracy of order $(\frac{\varepsilon}{r})$ we have

$$\Psi(r, \varphi) = -\frac{J\varepsilon}{4\pi r^2} \sin 2\varphi \tag{10}$$

Decreasing ε and increasing I so that the product $\varepsilon I = M$ remains constant we obtain the force dipole at the point $r = 0$. The stream function for this source at large r is

$$\Psi(r, \varphi) = -\frac{Q}{4\pi r^2} \sin 2\varphi \tag{11}$$

where $Q(t) = \int_0^t M/\rho dt$. On the other hand the point force dipole applies no net momentum to the fluid and (6) gives for large r

$$\Psi(r, \varphi) = \frac{n_\alpha n_\beta}{4\pi r^2} Q_{\alpha\beta} \tag{12}$$

From (11), (12) and (4) we obtain the connection between the force dipole intensity and the appropriate vorticity distribution

$$Q(t) = -\int_0^{2\pi} \int_0^{\infty} \omega(r, \varphi, t) \sin 2\varphi r^3 dr d\varphi \tag{13}$$

For simplicity consider the case when the source acts impulsively and its intensity can be presented by the Dirac delta-function: $M(t) \sim \delta(t)$ so that $Q(t) = \int_0^t M/\rho dt =$

const. The self-similar Stokes solution of the problem with the integral "boundary" condition (13) is

$$\Psi(r,\varphi) = \frac{Q\sin 2\varphi}{16\pi\nu t}[1 + \frac{r^2}{4\nu t} + \frac{4\nu t}{r^2} + (1 - \exp\frac{r^2}{4\nu t})]\exp\frac{-r^2}{4\nu t} \quad (14)$$

3. Experimental results

3.1. Description of the experiment

The experiments were performed in a tank filled by a linearly stratified salt water. Typical value of the buoyancy frequency $N = 1 - 2\ s^{-1}$. The flows were generated by horizontal jets from one or two round nozzles (nozzle diameter D = 0.1 cm). Jets were injected horizontally into a stratified fluid on the level of its own density (ρ). The jet exerts on the fluid a localized force $f = 4\rho q^2/\pi D^2$ and imparts to the fluid a localized impulse given by $P = \int_0^t f/\rho dt$ (q being the volume flux from the nozzle). When the force acts continuously the flow of increasing intensity is generated. When the force acts for a short period of time the impulsive flow is generated. Taking into consideration that in a stratified fluid the flow rapidly becomes quasi-planar with a typical thickness H at the intermediate asymptotic stage (for details se [2, 12]) in our following analysis we regard the flow as planar and discuss experimental results in terms of forces $I = f/H$ and impulses $J = P/H$ per unit depth of the flow. In some experiments we used narrow ($10^{-2} cm$) slotted nozzle and planar flow was induced in a layer of water - glycerin solution placed on a low viscosity tetra-chlorous-carbon. The depth of the upper layer was equal to the height of the nozzle ($H = 0.6 cm$).

3.2 Vortex dipole

Consider first the planar flow induced by the weak momentum source acting continuously (fig. 2a). Integration of the equation of motion for marked particles with the stream function [8] gives the distribution shown in fig 2b. In fig 2 one can see that the model correctly reproduces the general features of the real flow.

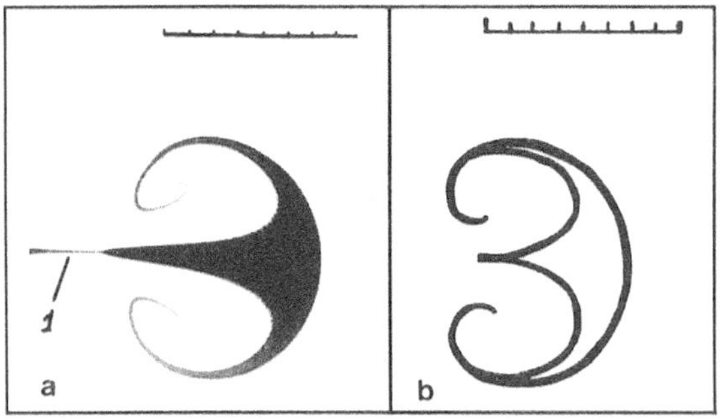

Fig. 2. Planar vortex dipole of increasing intensity: (a) experiment, b) calculations with the stream function (8). For both cases: $\nu = 2.7\ 10^{-2} cm^2 s^{-1}$, $I/\rho = 2.8 cm^3 s^{-2}$, scales in centimeters, $t = 24s$. 1-nozzle.

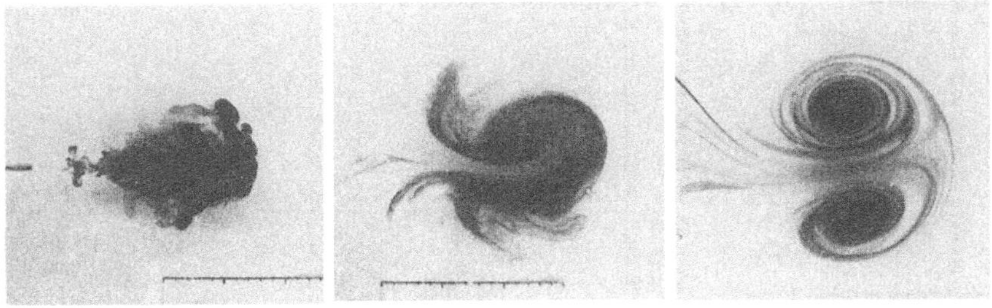

Fig. 3. Impulsive vortex dipole in a stratified fluid. Top view.

The photographic sequence in fig. 3 shows the evolution of the flow induced in a stratified fluid by the impulsive strong jet. Initially the jet forms a chaotic turbulent cloud. Owing to the stratification the vertical perturbation rapidly decays. The flow becomes quasi-planar and impulsive vortex dipole appears. Note that the final result of the force action does not depend on the force intensity. Thus the localized momentum source generates in a viscous fluid a vortex dipole governed by only one parameter J (for details see [12]).

Below we present a qualitative interpretation of the experimental results on dipole interactions. In our analysis we use two parameters governing the flow dynamics: dipolar (J) and guadrupolar (Q) vorticity distributions of the flow and make no difference between impulse applied to the fluid and momentum of the flow because they are proportional to one another [11]. For the length scale of the flow we use typical length scale (R) of the dyed water region and take into account that R increases with time.

3.3 Collision of two dipoles of equal momentum

Consider two sources of momentum which start simultaneously and thereafter apply to the fluid equal forces in the direction opposite to each other ($J_1 = J_2$, fig. 4). For this case $J = J_1 - J_2 = 0$ and $Q = \varepsilon J_1$ where ε is the distance between two sources. In the initial period of the motion when $R \leq \frac{\varepsilon}{2}$ (small distance asymptotic) two "small distance" dipoles governed by their intensities (J_1, J_2) appear and develop independently (fig. 4a). With time R increases and when $R > \frac{\varepsilon}{2}$ (large distance asymptotic) the flow turns into the "large distance" quadrupole (fig. 4b, c) because for large distance only one parameter governs the flow dynamics - the vortex quadrupole intensity Q.

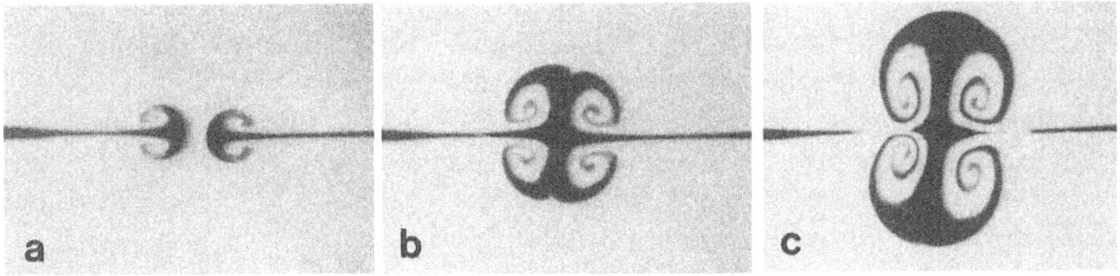

Fig. 4. Photographic sequence showing the collision of two vortex dipoles of equal momentum and the formation of vortex quadrupole in a linearly stratified fluid. Top view.

Fig. 5

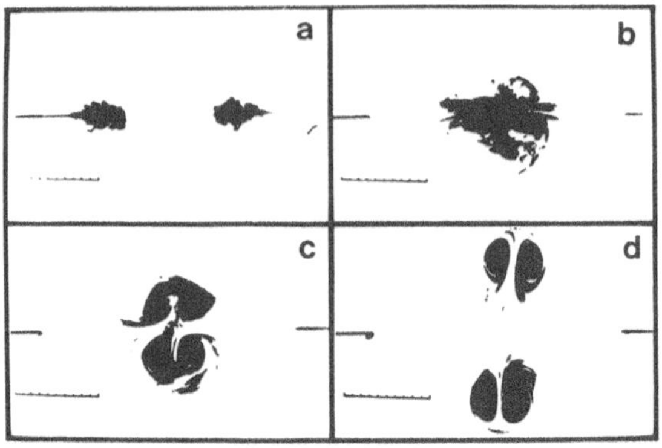

Fig. 6

Fig. 5. Planar vortex quadrupole: calculations with the stream function (14).
$Q = -2cm^4 s^{-1}, \nu = 10^{-2} cm^2 s^{-1}, t = 5s\ (a), 10s\ (b)$.

Fig. 6. Photographic sequence showing the collision or two impulsive strong jets of equal momentum and the formation of a vortex quadrupole.

For comparison the distribution of marked particles given by the stream function (14) is shown in fig. 5. We did not perform special experiment to verify the solution (14) quantitatively but a simple comparison of fig. 4 and fig. 5 demonstrates that the model correctly reproduces the dynamics of the real quasi-planar flow in a stratified fluid. Note that for strong sources of momentum (fig. 6) the initial motion is three-dimensional and "small distance" dipoles are not developed clearly (fig. 6a) but the final result of collision ("large distance" quadrupole) does not depend on the intensity of the sources (fig. 6d).

3.4. Other types of dipole interactions

Consider two dipoles moving one after another ($J_2 > J_1$, fig. 7). The second more intensive dipole can be seen to penetrate the first one. The vortices of the first dipole orbit around the intruding second dipole vortices. This continues until the new vortex dipole forms as a result of the vortex merging. The interpretation is very simple. For this case $J = J_1 + J_2$ and the final result of interaction is the "large distance" dipole with the intensity J.

Consider now the collision of two dipoles of different momenta ($J_1 > J_2$, fig. 8). For this case $J = J_1 - J_2, Q = J_2$ and an additional length scale equal to $L = \frac{Q}{J} = \frac{J_2}{J}\varepsilon$ appears (for the experiment shown in fig. Neglecting the "small distance" dipoles which are not developed here one may predict that for intermediate distance $\frac{\varepsilon}{2} \leq R \leq L$ guadrupolar flow must prevail (fig. 8b, c). When $R > L$ the "intermediate distance" the quadrupole transforms into the "large distance" dipole (fig. 8d) because for large distance velocities induced by the dipole and quadrupole decay as r^{-2} and r^{-3} respectively and the flow induced by the dipole prevails (fig. 8d).

Consider at last the oblique interaction of two dipoles of equal momentum ($J_1 = J_2$, fig. 9). For this case $J = 2J_1 \cos\alpha$, $Q = J_1 \varepsilon \sin\alpha$ and $L = \frac{\varepsilon}{2} tg\alpha$ where 2α is the angle between dipole directions. The final result of interaction (fig. 9c) in this case "large distance" dipole) is the same as for collision of two dipoles of different momenta (fig.

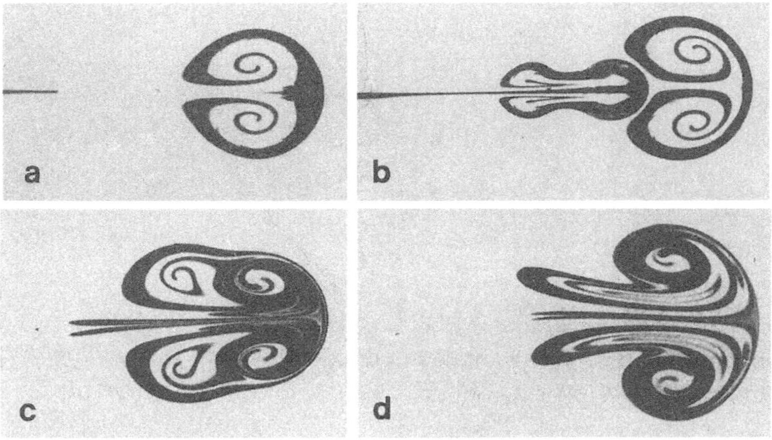

Fig. 7. Photographic sequence showing the interaction of two impulsive vortex dipoles moving one after another.

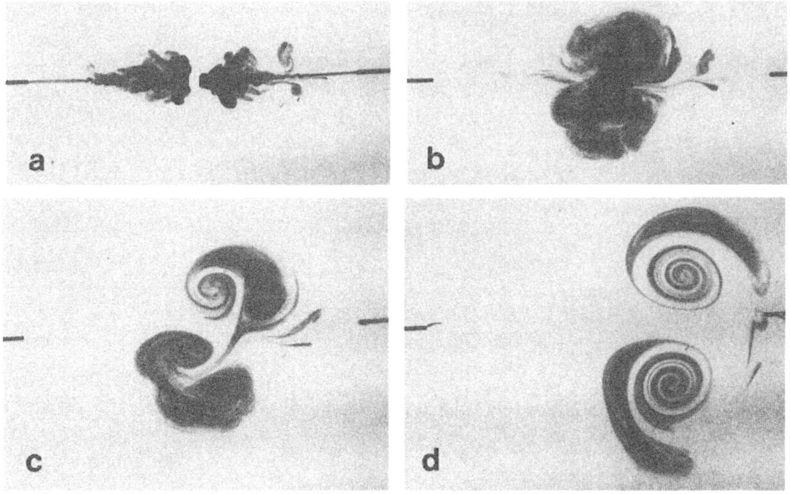

Fig. 8. Collision of two impulsive strong jets of different momenta.

8d). But the intermediate distance asymptotics in these two cases are different. For oblique interaction the intensity Q of the "intermediate distance" quadrupole is positive (contrary to the previous case where it is negative). Hence this quadrupole produces two dipoles directed to the left and to the right. These two halfs of the "intermediate distance" quadrupole develop in the field of the "large distance" dipole (directed to the right). The velocities are summed up on the right side and are subtracted on the left side. Thus the right dipole becomes more intensive and the left dipole becomes weaker (fig. 9c).

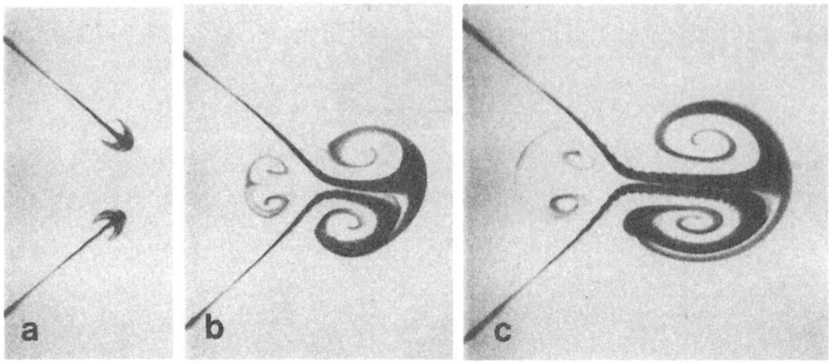

Fig. 9. Photographic sequence showing the oblique interaction of two vortex dipoles of increasing with time equal momentum.

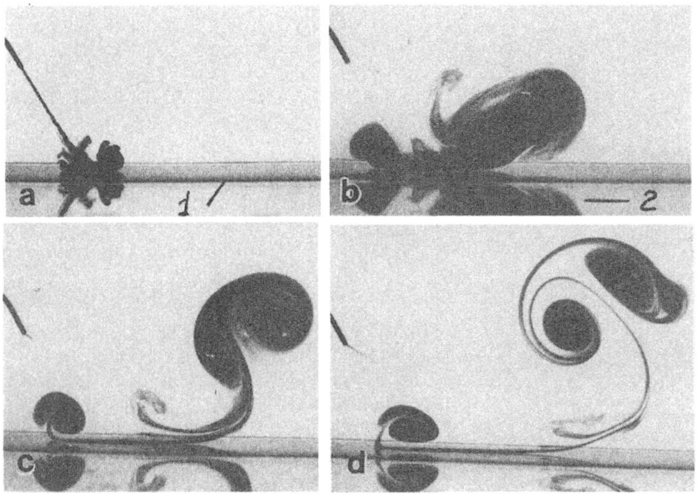

Fig. 10. Oblique collision of the vortex dipole with vertical glass plate (1), 2-image (compare with fig. 9).

Note that in our qualitative analysis we make no difference between impulsive dipoles ($J = const$) and dipoles of increasing intensity ($J \sim t$). All arguments used above are valid in both cases.

The same consideration can be done for the vortex dipole interaction with a vertical barrier (fig. 10). Neglecting the friction on the barrier surface and using the fact that $\Psi = 0$ on the barrier one may introduce the mirror image of the flow and do similar consideration.

4. Concluding remarks

We have considered the fundamental types of interactions between vortex dipoles and also the interactions of vortex dipoles with a barrier. Note that there are several obvious extensions to the work reported here. Examples include the whole family of triple or multiple interactions of vortex dipoles, as well as the interactions between single vortices (vortex monopoles) and also between monopoles and dipoles.

References

1. Flierl, G.R., Stern, M.E. and Whitehead, J.A.: Dyn. Atm. and Oceans 7(4), 2333-263 (1983)

2. Voropayev, S.I.: Proc. 20th Intern. Liege Colloquium on Ocean Hydrodynamics. Elsevier Oceanography Ser. 50, 671-690 (1989).

3. Van Heijst, G.J.F. and Flor, J.B.: Proc. 20th Intern. Liege Colloquium on Ocean Hydrodynamics. Elsevier Oceanography Ser. 50, 591- 608 (1989).

4. Couder, Y. and Basdevant, C.: J. Fluid Mech. 173, 225-251 (1986). 5. Nguen Duc, J.-M. and Sommeria, J.: J. Fluid Mech. 192, 175-192 (1988).

6. Fedorov, K.K. and Ginzburg, A.I.: Surface layer of the ocean. Leningrad. Gidrometeoizdat (1986).

7. Nihoul, J.C.J. and Lamart, N.B. (edrs): Mesoscale/synoptic coherent structures in geophysical turbulence. Elsevier Oceanography Ser. 50. Amsterdam (1989).

8. Afanasyev, Ya. D. and Voropayev, S.I.: Reports USSR Acad. of Sci. 308(1), 179-183 (1989).

9. Barenblatt, G.I., Voropayev, I.I. and Filippov, I.A.: Reports USSR Acad. Sci. 307(3), 720-724 (1989).

10. Voropayev, S.I. and Filippov, I.A.: Izvestia USSR Acad. of Sci. Atm. and Ocean Phys. 21(9), 964-972 (1985).

11. Cantwell, B.J.: Fluid Mech. 173, 159-189 (1986).

12. Afanasyev, Ya. D. and Voropayev, S.I.: Izvestia USSR Acad. of Sci. Atm. and Ocean Phys. 25(8), 843-851 (1989).

Large Eddy Simulation of MHD Flows in Metallurgy

Said Zahrai and Fritz H. Bark

Dept. of Hydromechanics, Royal Institute of Technology
Stockholm, Sweden

The turbulent motion of an electromagnetically stirred liquid metal is computed for low magnetic Reynolds numbers by using large eddy simulation. Two cases are considered. An electromagnetic stirrer and an electromagnetic furnace. The results compare favourably with experimental data published in the literature.

1 Introduction

The solidification process of a pure liquid metal is basically controlled by heat transfer. The solidification fronts are locally plane and the growth direction is in the direction of the temperature gradient. In the case of an alloy, the solidification process is more complicated. The growth rate of the solidified zone depends on both the local temperature and the local concentration. The initially plane solidification fronts are usually unstable. Small disturbances can grow and eventually form dendrites. During this process, small regions of high concentration may form which destroy the homogenity of the cast and result in defects in the product.

The quality of the cast can be improved by stirring of the liquid metal. Stirring can be brought about in different ways such as mechanical stirring by rotating blades, injection of gas bubbles or by using an electromagnetic stirrer (EMS). Electromagnetic stirring is usually used in continous casting. The main advantage of this method is that there is no physical contact between the hot melt and foreign matter except, of course, the walls of the mould.

An EMS has a number of coils that generate a moving magnetic field. The laminar flow of a conducting metal, driven by a rotating magnetic field, in a long cylinder was studied analytically by Moffat (1965) and Dahlberg (1972). In applications, however, the dynamical Reynolds number is usually very high due to the low viscosity of the fluid. Experiments with mercury, see e.g. Robinson & Larsson (1973), indicate that the fluctuating velocity is typically about 10% of the mean velocity.

The turbulent motion of an electromagnetically stirred liquid metal has been computed by, among others, Spitzer, Dubke & Schwerdtfeger (1986) and Davidson & Hunt (1987), using $k-\epsilon$ model. The latter authors also developed an approximate analytic model, results of which were shown to be in quite reasonable agreement with results obtained from the $k-\epsilon$ model and experiment.

MHD phenomena also occur in electromagnetic furnaces (EMF). In this application, the main purpose is heating by ohmic dissipation, but the liquid metal is stirred as well. Moore (1982) and Taberlet et al. (1985) have investigated such cases experimentally and reported measurments of the mean and fluctuating velocity fields. The former experiment were compared with numerical simulations by El-Kaddah & Szekely (1981), who also used the $k-\epsilon$ model to compute the velocity field.

Advances in Turbulence 3
Editors: A.V. Johansson · P.H. Alfredsson

In the present work, large-eddy simulation (LES) is used to compute approximately the turbulent motion of a liquid metal in a cylindrical vessel. Two cases are considered, the motion by an electromagnetic stirrer and that in an electromagnetic furnace. The use of LES-methodology was chosen because more detailed and more accurate results can be obtained compared with those obtained by using the $k-\epsilon$ model. The numerical predictions are compared with the measurements by Robinson & Larsson (1973) in the EMS case and by Moore (1982) for the electromagnetic furnace.

2 Problem Description

This section gives brief descriptions of the two problems considered. More specified problem statements will be given elsewhere.

The EMS considered is sketched in figure 1.a. There are four coils, which are located symmetrically around the container, that generate a rotating magnetic field. The magnetic flux density is periodic in time, approximately axisymmetric and almost independent of the vertical coordinate. The Lorentz force is mainly in the circumferential direction as the largest velocity component. However, as is well known from the theory of rotating fluids, the presence of the walls will set up a secondary meridional circulation.

In figure 1.b, a sketch of the EMF considered is given. In this case there is an axisymmetric coil that generates a force field, and hence a mean flow, in the meridional planes.

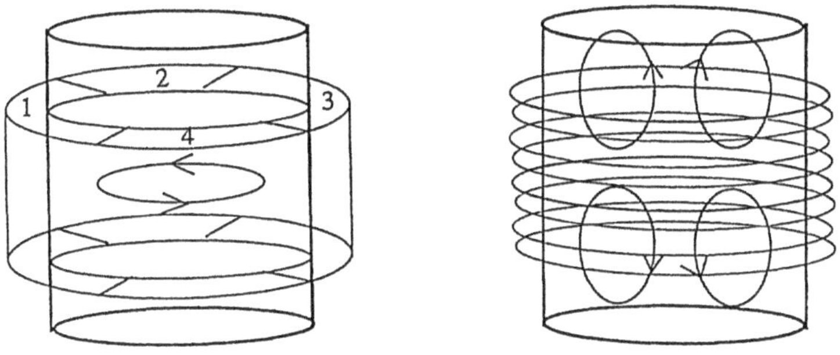

Figure 1. a, left) Electromagnetic stirring of liquid metals. b, right) Electromagnetic furnace.

In both experiments the height of the container is about twice the radius. In the simulation it is set exactly to that value.

The dimensional parameters which can affect the motion are, see Hunt & Maxey (1978), the skin depth of the Lorentz force, $\delta \sim \frac{1}{\sqrt{\omega_0 \mu \sigma}}$, the current density, $\mathcal{J} \sim \mathcal{B}_0 \mu \sqrt{\omega_0 \mu \sigma}$ and the Lorentz force, $\mathcal{F} \sim \mathcal{B}_0^2 \mu \sqrt{\omega_0 \mu \sigma}$. We have chosen $\mathcal{U} \sim \sqrt{\frac{\mathcal{F}\mathcal{R}}{\rho}} = \sqrt{\frac{\mathcal{B}_0^2 \mathcal{R}}{\mu \rho \delta}}$, $\mathcal{R}$ and $\frac{\mathcal{R}}{\mathcal{U}}$ as typical scales for velocity, length and time to write the Navier-Stokes equations in their non-dimensional form. Here, $\mathcal{B}_0 \sim 10^{-3} - 10^{-2}\ T$ and $\omega_0 \sim 2\pi \times 10^2\ s^{-1}$ are the magnitude and the cycle frequency of the external magnetic flux density, $R \sim 10^{-1}\ m$ is the radius of the container and $\rho \sim 10^4\ kg\ m^{-3}$, $\mu \sim \mu_0 = 4\pi \times 10^{-7}\ H\ m^{-1}$, $\sigma \sim 1.04 \times 10^6\ mho\ m^{-1}$, $\eta = \frac{1}{\mu\sigma} \sim 1\ m^2\ s^{-1}$ and $\nu \sim 10^{-7} - 10^{-5}\ m^2\ s^{-1}$

are the density of the fluid, the magnetic permeability of the fluid, the conductivity of the fluid, the magnetic diffusivity of the fluid and the kinematic viscosity of the fluid. The use of the typical scales as given above results in appearance of three different Reynolds numbers in the equations. The magnetic Reynolds number based on the fluid velocity, $Re_M = \frac{U\mathcal{L}}{\eta} \sim 10^{-3} - 10^{-2}$, The magnetic Reynolds number based on the cycle frequency of the external field, $Re^*_M = \frac{\omega_0 \mathcal{L}^2}{\eta} \sim 10^0 - 10^1$ and the dynamic Reynolds number $Re = \frac{UR}{\nu} \sim 10^4 - 10^5$.

3 Numerical Method, Subgrid Model and The Force Field

The filtering procedure of the Navier-Stokes equations used in the present work is the one proposed by Deardorf (1970). The only difference is the refinement of the mesh close to the vertical walls. As shown by Schumann (1975), this procedure leads to a set of equations that are discretized on a staggered mesh. The effects of subgrid scale motions on the motion of the resolved scales is modelled according to Smagorinsky, see Deardorff (1970). There are strong reasons to believe that the small scale motions near the vertical wall are important. Therefore the mesh used in the computations is refined close to the periphery and it was decided to focus on the small scale motion near the vertical wall which could be resolved reasonably well. For the motion near the horizontal walls, which is not accurately resolved, a simple viscous no-slip condition is used on the coarse grid. In the present LES model, this condition is in fact incorrect but was chosen because the usual wall models could not give satisfactory results for the mean flow close to the horizontal walls. Furthermore, because of the computer limitations the Reynolds number in the computations is lower than in the experiments. For the Reynolds number used in the computations (10^3) a typical grid distance near the wall is 2 wall length units. Note that a lower Reynolds number is equivalent to a weaker stirring.

A cylindrical coordinate system is used and the discretized equations are integrated in time by using a second order explicit Adam-Bashforth scheme for all terms except the pressure term, which is treated implicitly. Figure 2 shows the computational domain and the coordinate system.

It is assumed that the frequency is high enough, and the magnetic Reynolds number is low enough that the Lorentz force can be modelled by a given steady body force, see Hunt & Maxey (1978). The following expressions are reasonable approximations of measurements

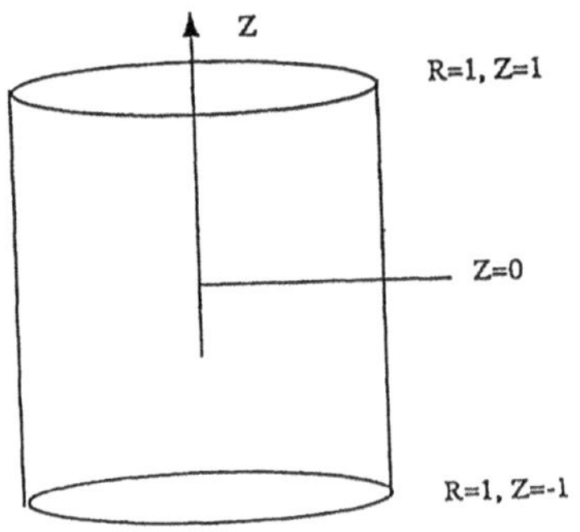

Figure 2. The computational domain.

$$\vec{F}_{EMS} = (F_r, F_\phi, F_z) = (0, r, 0)$$

$$\vec{F}_{EMF} = (F_r, F_\phi, F_z) = (-e^{10(r-1)} \times \sin \pi \frac{z+1}{2}, 0, e^{10(r-1)} \times \sin(\pi z))$$

Note that in the EMS case the vertical variation of the body force is neglected.

4 Results

4.1 EMS

The computation is carried out for $Re = 10^3$ and the results are compared with Robinson's and Larsson's data, which were measured at three different Reynolds numbers, $Re = 10^4,\ 7 \times 10^4,\ 3 \times 10^5$. Robinson & Larsson measured the mean and the fluctuating part of the circumferential velocity component at $r = 0.2$, 0.4, 0.8 and $z = 0$, 0.5, 0.8. The differences between different heights are small both in the experiments and in the numerical simulation. The mean profile and the RMS levels of the circumferential velocity are shown in figures 3.a and 3.b.

According to these graphs, the mean velocity profile varies significantly with Re, whereas the RMS levels are approximately independent of Re. Figure 4 shows the RMS levels of the axial and radial velocities.

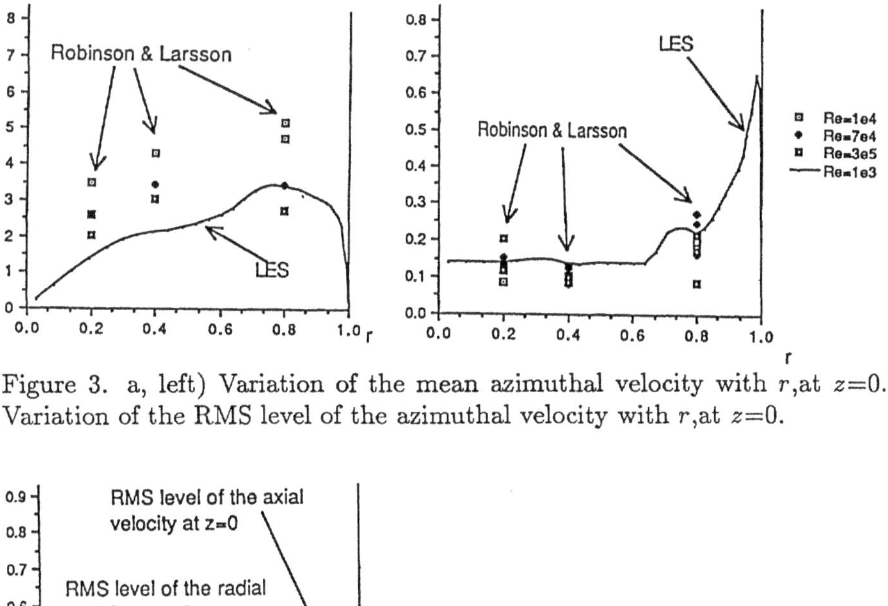

Figure 3. a, left) Variation of the mean azimuthal velocity with r,at z=0. b, right) Variation of the RMS level of the azimuthal velocity with r,at z=0.

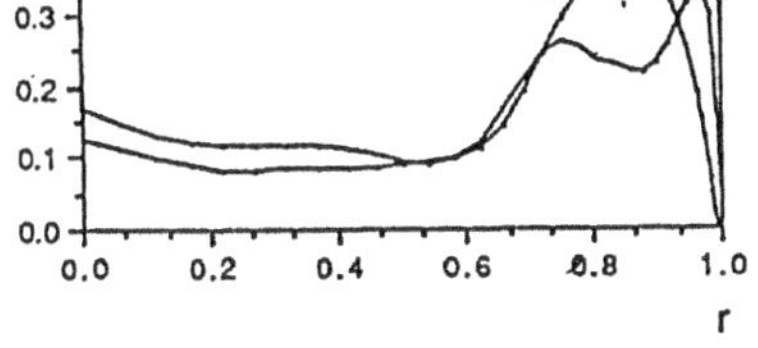

Figure 4.The RMS levels of the radial and axial velocity components.

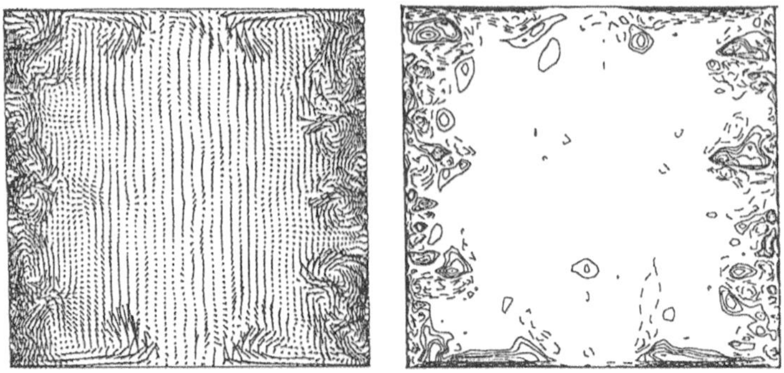

Figure 5. a, left) The instantaneous meridional velocity field. b, right) Level curves of the instantaneous circumferential vorticity field.

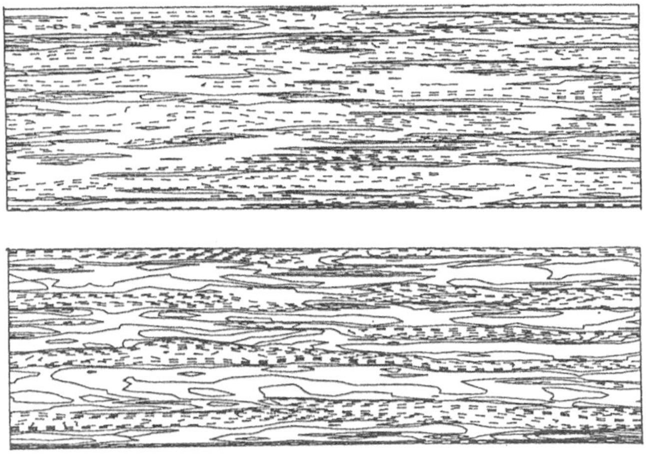

Figure 6. a, above) Iso-lines for the radial velocity at r=0.975. b, below) Iso-lines for the radial velocity at r=0.88.

An important point is that the mean profile has a boundary layer character. Due to the curvature of the vertical wall, streamwise vortices appear in the boundary layer, see figure 5.a. These vortices can be identified more clearly in figure 5.b, which shows instantaneous isolines of the circumferential vorticity.

Figures 6.a and 6.b, which show instantaneous level curves of the radial velocity field close to the vertical wall, demonstrate that the vortices are strongly coherent in the circumferential direction. In the statistically steady state, the vertical location of the vortices is not fixed but a slow, oscillatory vertical motion takes place. The structure of the fluctuating velocity field provides an effective transport mechanism close to the vertical wall.

4.2 EMF

The instantaneous and the mean velocity fields in a meridional plane are presented in figures 7.a and 7.b. The experiments by Moore were carried out at $Re = 2 \times 10^5$. As in the EMS case, the mean flow appears depend significantly on Re, see figures 8.a

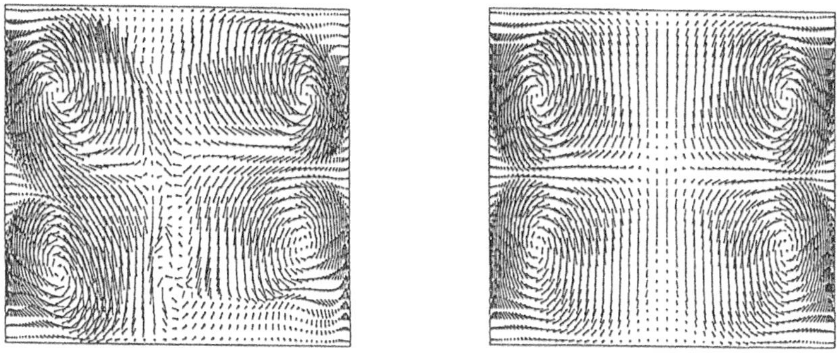

Figure 7. a, left) The instantaneous meridional velocity field. b, right) The mean meridional velocity field.

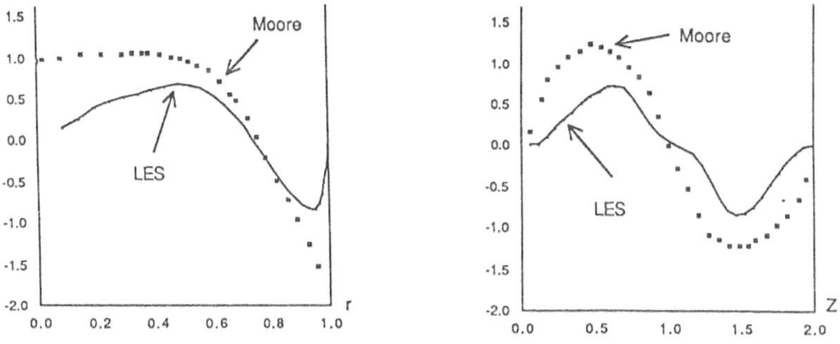

Figure 8. a, left) The radial profile of the axial velocity through the eye of the upper vortex, $z = 1.52$. b, right) The variation of the mean axial velocity with z at $r = 0.95$.

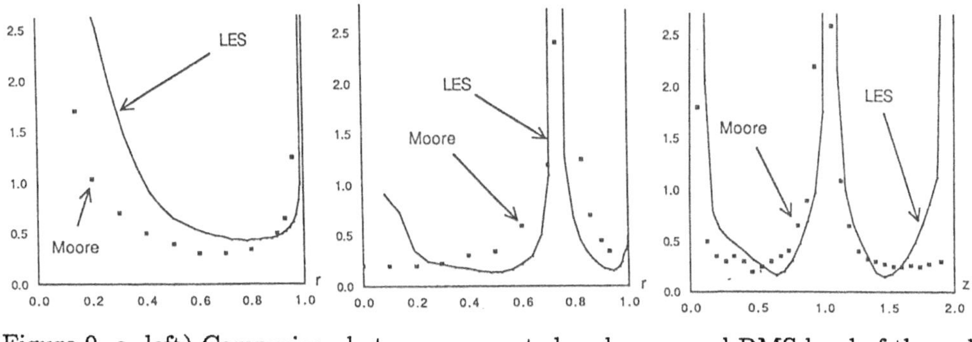

Figure 9. a, left) Comparison between computed and measured RMS level of the radial velocity, normalized with the local mean radial velocity with r at z=1.2. b, middle)

and 8.b, but the RMS levels seem to depend only weakly on *Re* as can be infered from figures 9.a-9.c.

Comparison between computed and measured RMS level of the axial velocity, normalized with the local mean axial velocity with r at z=1.5. c, right) Comparison between computed and measured RMS level of the axial velocity, normalized with the local mean axial velocity with z at r=0.95.

5 Discussion and Conclusions

The mean and fluctuating velocity fields have been computed for a flow driven by an electromagnetic stirrer and in an electromagnetic furnace. The magnetic Reynolds numbers considered are low implying that the force field is not affected by the fluid motion. The force fields were modelled by approximate analytic expressions that simulate experimental data.

The numerical results are compared with the experimental data by Robinson & Larsson (1973) and Moore (1982). For computational reasons, the Reynolds number in the computations was lower than those in the experiments.

The predicted mean velocities are consistently lower than those observed in experiments although the qualitative behaviour agrees with the experimental data. This trend is confirmed by the data of Robinson & Larsson. The fluctuating velocity field , however, is predicted quite well by the numerical computations.

It can be concluded that LES is a promising prediction methodology for metallurgical application of MHD-flows and may probably be used to approximately compute technologically important processes such as transport of heat and mass in liquid metals during turbulent stirring.

References

Dahlberg E, AB Atomenergi (1972), Sweden, Rep. AE-447.

Davidsson P. A. & Hunt J. C. R., *J. Fluid Mech.* (1987), **185**, 67.

Deardorff J. W., *J. Fluid Mech.* (1970), **41**, part 4, 453.

El-Kaddah, N. & Szekely J., *J. Fluid Mech.* (1983), **133**, 37.

Hunt J. C. R. & Maxey M. R., *Proc. Beer Sheva Conf.* (1982), 249.

Moffat H. K., *J. Fluid Mech.* (1965), **22**, part 3, 521.

Moore D. J., PhD thesis (1982), Trinity College, University of Cambrige.

Robinsson T. & Larsson K., *J. Fluid Mech.* (1973), **60**, part 4, 641.

Spitzer K. H., Dubke M., and Schwerdtfeger K., *Metal. Trans. B* (1986), **17B**, 119.

Schumann U., *J. Comp. Physics* (1975), **18**, 376.

Taberlet E. & and Fautrelle Y., *J. Fluid Mech.* (1985), **159**, 409.

Part V

Turbulence Modelling

The Modelling of Perturbed Turbulent Boundary Layers

S.E. Belcher, W.S. Weng, D.J. Carruthers and J.C.R. Hunt

Department of Applied Mathematics and Theoretical Physics,
University of Cambridge, Cambridge CB3 9EW, United Kingdom.

Abstract We examine, using theoretical and computational techniques, how a turbulent boundary layer responds to an external pressure gradient. Simple scaling arguments are developed which show how different aspects of the turbulence dynamics determine the changes to the flow in different regions above the rigid boundary. These ideas lead to analytical solutions for the effect of the turbulence on the perturbations to the mean flow, which performs well when the changes to the boundary layer are small. When the applied pressure gradient is stronger, mean flow separation may occur and the turbulence length scale in the separating region is reduced by the elevated layer of mean shear. We examine a new shear–blocking mixing length, 1–equation model which accounts for this effect and apply it in a numerical model. This is shown to improve on the poor results used with the usual mixing length, 1–equation model.

1. Introduction

What are the dominant physical mechanisms that govern how a turbulent boundary layer responds to an external pressure gradient and how should these processes be modelled? Using theory and computations we investigate accelerating and decelerating turbulent boundary layer. As well as modelling these flows, one of our aims is to provide a physical explanation for why turbulent boundary layers are so sensitive to external pressure gradients, a phenomena which has been noted by experimentalists, eg. Dengel & Fernholtz (1990).

We partcularly focus on very high Reynolds number flows when the horizontal length scale, L, of the variation of the pressure gradient is small compared with h_*, the natural length scale of adjustment of a turbulent boundary layer, which is given by $h_* = hU(L)/u_*$, where h is the boundary layer depth, $U(L)$ is the approach flow mean velocity at the height L and u_* is the upstream friction velocity; furthermore, since the Reynolds number is high, $u_*/U(L)$ is small (in the atmosphere it is typically 0.03). The significant changes to the flow then occur well within the boundary layer, in a region of depth $\ell \sim Lu_*/U(L)$ (i.e. $\ell \ll h \sim L \ll h_*$). Hence we focus on the turbulent processes which occur close to a rigid wall.

In practical problems the varying pressure gradient may arise from a variety of different sources, but the best specific example is the flow of the atmospheric boundary layer over a hill (fig. 1): perhaps surprisingly to laboratory experimentalists, the fine detail of these flows is at least as well defined as laboratory studies especially in the inner region! The hill has the effect of accelerating the flow on the upwind slope (a favourable pressure gradient) and decelerating the flow on the downwind slope (an adverse pressure gradient).

Advances in Turbulence 3
Editors: A.V. Johansson · P.H. Alfredsson

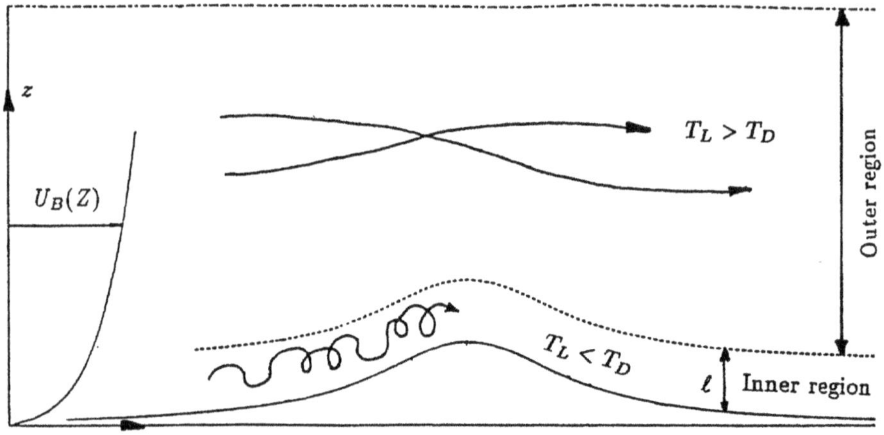

FIGURE 1. Schematic diagram of the paths of fluid elements in flow over a hill.

2. Processes controlling the perturbations to a turbulent boundary layer

The pressure gradient changes the structure of the turbulence through two processes. Firstly, the mean flow is altered by the pressure gradient and hence strains the turbulent eddies. The time scale, T_D, of these straining distortions is the time taken for an eddy to be advected by the mean flow over the distance L, i.e.

$$T_D = L/U(z) = L/(u_*/\kappa)\ln(z/z_0), \tag{2.1}$$

assuming the approach flow profile is logarithmic. Secondly, the turbulence is changed by the interactions between the eddies, so that the second fundamental time scale is the time, T_L, taken for an eddy to decorrelate or turnover. Assuming that at a height z above the surface the large, energy containing eddies have a vertical size of order z (an idea which is refined in §3), T_L may be estimated as $T_L = \kappa z/u_*$, where κ is the von Karman constant whose value is taken as 0.4.

These two time scales are of the same order of magnitude at a height ℓ defined implicitly by $\ell \ln(\ell/z_0) \sim L$, which separates the perturbed flow into two regions (fig.1), whose properties can be defined with aid of the kinetic energy equation and thence the appropriate models can be chosen.

a. Inner region

In the inner region, where $z < \ell$, $T_L/T_D < 1$ and the eddy–eddy interactions are more important than the straining by the distorted mean flow. Furthermore, as $z \to z_0$, $T_L/T_D \to 0$: the turbulence tends to a local equilibrium where production of turbulent kinetic energy is balanced by its dissipation and the turbulence adjusts to the local mean velocity gradient (Townsend, 1961). The changes to the turbulence statistics (eg. the streamwise velocity variance perturbation, $\Delta\sigma_u^2$) may then be estimated using mixing length theory,

$$\Delta\sigma_u^2 \propto u_* \ell_m \partial \Delta U/\partial z. \tag{2.2}$$

The detailed analysis (Hunt et al. 1988a) shows that, near the surface, the streamwise velocity perturbation, ΔU, varies logarithmically, so that $\partial \Delta U/\partial z \sim \Delta U/\ell \ln(\ell/z_0)$ and

$$\Delta \sigma_u^2 \sim u_* \ell \Delta U/\ell \ln(\ell/z_0) \sim u_*(\ell/L)\Delta U \sim (H/L)u_*^2. \quad (2.3)$$

b. Outer region

In the outer region, $z > \ell$, $T_D/T_L < 1$, i.e. the 'time of flight' of an eddy is less than the eddy–eddy interaction time. The changes to the turbulence are then 'rapid' (Hunt, 1973), i.e. dependent on the history of the mean flow strain and *not* on the local velocity gradient. The leading order changes to the mean flow in the outer region are inviscid. Rapid distortion theory shows that

$$\Delta \sigma_u^2/\sigma_u^2 \sim \Delta U/U(L) \sim H/L. \quad (2.4)$$

The effect on the mean flow of these changes to the turbulence may be estimated by considering the x-momentum equation:

$$U\partial \Delta u/\partial x \sim \partial \Delta \sigma_u^2/\partial x \quad \Longrightarrow \quad U(L)\Delta U/L \sim \Delta \sigma_u^2/L. \quad (2.5)$$

Hence the perturbed Reynolds stress gradients lead to mean velocity perturbations of order

$$\Delta U \sim (H/L)(u_*^2/U^2(L))U(L). \quad (2.6)$$

Many authors have assumed that, because a mixing length model can satisfactorily be used to model a fully developed boundary layer, it can also be used to model the *perturbations* to a turbulent boundary layer. This analysis shows that mixing length can be used for perturbations *only* in the inner region, which does not mean that the profile is logarithmic throughout this region.

If the mixing length model is *erroneously* used in the outer region the predicted change in the Reynolds stresses have the wrong sign *and* magnitude! This may be explained with the aid of figure 2. Figure 2*b* shows the mean streamwise velocity calculated by Newley (1986), who used the second order closure model of Launder et al. (1975). In the outer region at the top of a hill the vertical gradient of ΔU is negative so that using mixing length,

$$-\Delta \overline{uw} \propto u_* \ell_m \partial \Delta U/\partial z, \quad (2.7)$$

implies a *negative* change in the shear stress, whereas figure 2*a* shows that the shear stress perturbation in this region is positive. By considering the change in turbulent kinetic energy this idea may be extended to show that *any* eddy viscosity model that relates the turbulence changes to the *local* gradients gives Reynolds stress perturbations of the wrong sign. If the mixing length model is erroneously used in the outer region, not only is the sign incorrect, but also the magnitude of the effect of the Reynolds stress gradients on the mean flow is too large by a factor $U(L)/u_*$.

c. Limitations of the mixing length model in the inner region

Even within the inner region, basing the mixing length on the height above the surface does not produce accurate predictions of the mean flow in the accelerating and decelerating regions (i.e. on the sloping parts of a hill). Firstly, the advection of turbulent kinetic energy is of the same magnitude as shear production and dissipation in the up-

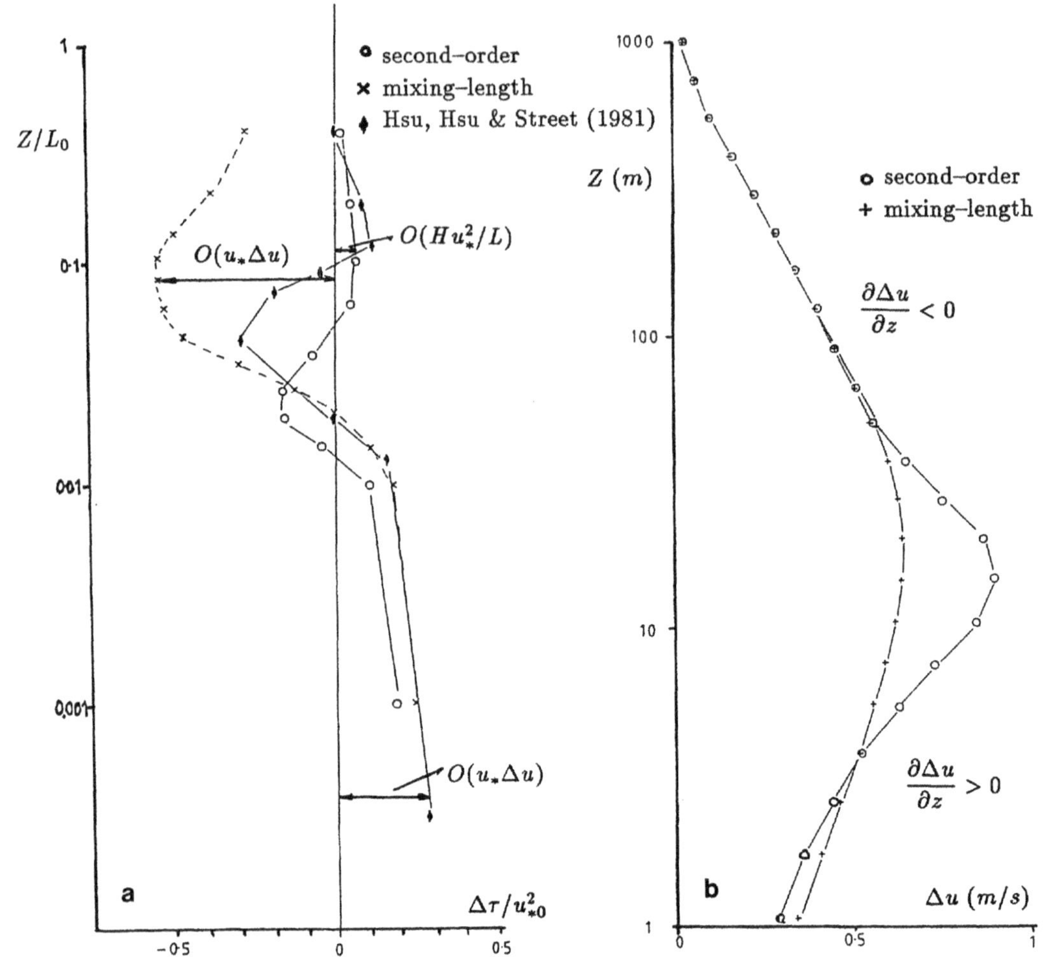

FIGURE 2. Numerical computation of airflow over undulating surfaces (Newley, 1986). (a) normalised shear stress, (b) mean streamwise velocity perturbation.

per two-thirds of the inner region. Secondly, when the strength of the adverse pressure gradient is increased beyond about 1/3, (i.e. a hill of slope greater than 1/3), mean flow separation occurs and there is a layer of strong shear elevated above the surface. This mean shear reduces the vertical extent of the turbulent eddies, thereby reducing the efficiency of the momentum transfer.

d. Comparison of linear theory surface mixing length with numerical and computational results

The arguments leading to (2.6) show that for a boundary layer passing over a hill, a mixing length model is appropriate in the inner region, whilst the analysis for the perturbations in the outer region is (to leading order) inviscid. This analysis, by Hunt et al. (1988a), has been extended by Belcher et al. (1990) to second order in $u_*/U(L)$ to calculate the drag force on the topography, which is of $O(\frac{H^2}{L^2}Lu_*^2)$ and depends on the very small out of phase pressure field of $O\left(\frac{H}{L}U^2(L).u_*^2/U^2(L)\right)$. In figure 3 this theoretical prediction of the drag force is compared with the results computed by Newley

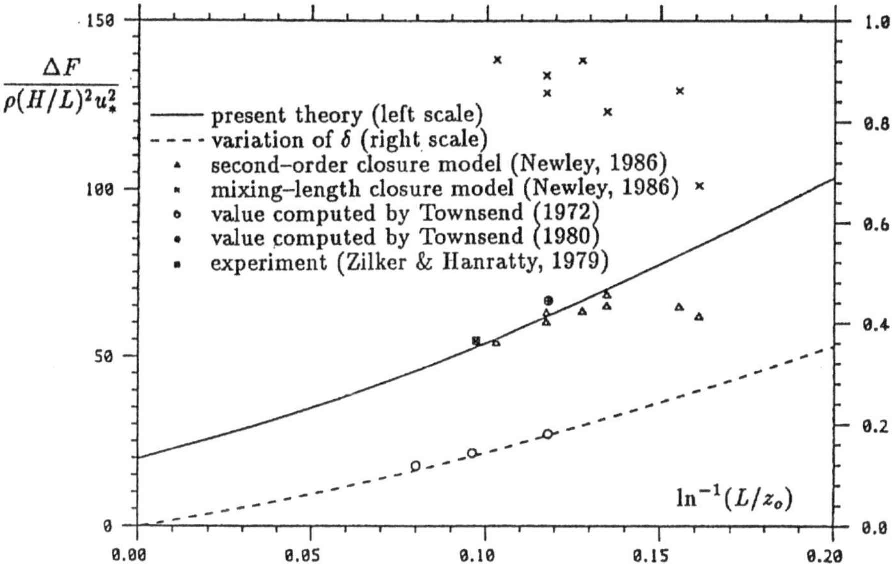

FIGURE 3. Perturbation drag force coefficient (sinusoidal topography).

(1986), on this plot there is also one point from the experiments of Zilker and Hanratty (1979). This shows that truncating the mixing length model at the top of the inner region leads to a drag force which is in good agreement with the results obtained using the second order closure model, whereas using the mixing length model throughout the flow leads to a 100% over-prediction of the drag (an incorrect approach used by Jacobs (1987) for estimating wave growth).

3. Effects of the perturbation on the turbulence structure

a. Turbulence models for perturbed shear layers

Following Kolmogorov, Prandtl, Spalding and others, turbulent shear stresses have been modelled using a mixing length formulation, viz.

$$\tau_{ij} = k^{1/2}\ell_\tau \partial U_i/\partial x_j. \tag{3.1}$$

The kinetic energy per unit mass, k, is calculated from a separate equation, in which there are two terms that cannot be explicitly described by in terms of the mean flow and the Reynolds stresses and have to be modelled. These are the energy dissipation term, ϵ, and the transfer of energy by turbulence itself, T, (often misleadingly called diffusion of energy). The equation for the transport of k can then be written

$$\underbrace{U_j\frac{\partial k}{\partial x_j}}_{A} = \underbrace{-\overline{u_i u_j}\frac{\partial U_i}{\partial x_j}}_{Pr} \underbrace{- \frac{\partial}{\partial x_j}\left[\frac{1}{\rho}\overline{u_j p} + \frac{1}{2}\overline{u_i u_i u_j} - 2\nu\overline{u_i s_{ij}}\right]}_{T} - \underbrace{2\nu\overline{s_{ij}s_{ij}}}_{\epsilon} \tag{3.2}$$

In 1-equation models (eg. Taylor and Gent,1974), ϵ is estimated in terms of an energy dissipation length scale, ℓ_ϵ, (see Hunt et al. 1987), as

$$\epsilon = k^{3/2}/\ell_\epsilon \qquad \text{or} \qquad \epsilon = (\overline{w^2})^{3/2}/\ell_\epsilon \tag{3.3}$$

Thence the problem reduces to defining the two functional length scales, ℓ_τ and ℓ_ϵ, which in some cases can be related to physical length scales defined by the correlation functions.

In 2-equation models, such as $k - \epsilon$, a supplementary equation is constructed for ϵ in terms of $\partial U_i/\partial x_j$ and k; it is assumed that the length scale constructed from ϵ and k is the appropriate length scale for ℓ_τ, i.e.

$$\ell_\tau \sim \ell_\epsilon \sim k^{3/2}/\epsilon. \tag{3.4}$$

For boundary layers the usual assumption with the 1-equation method has been that these two length scales are the same and determined by (i) the distance, z, from the rigid surface at the bottom of the boundary layer and (ii) by the boundary layer depth, h, in the upper part of the layer. Since ℓ_ϵ determines the scale for the straining of smaller eddies and the dissipation, it is determined by the *smallest* macroscopic length scale. Therefore the harmonic mean is chosen so that

$$1/\ell_\tau \sim 1/\ell_\epsilon = A_B/z + A_h/h. \tag{3.5}$$

Britter et al. (1981) have shown, however, that this model does not account for the marked *reduction* in length scale that occur above the surface in regions of high adverse pressure gradient. Nor does this formula account for the marked changes in ϵ that occur in stably stratified flows. Either special ***ad hoc*** formulae have to be introduced for 'extra strain rates', eg. Bradshaw (1971), or for buoyancy forces, eg. Frost and Wyngaard (1976).

Alternatively an implicit model for the length scale ℓ_ϵ can be used which accounts for the sensitivity of ℓ_ϵ to local shear, $\partial U/\partial z$. Such a formula was proposed by Hunt et al. (1988b) for high Reynolds number flows near a wall,

$$1/\ell_\epsilon = A_B/z + A_S(\partial U/\partial z)/\sqrt{\overline{w^2}}. \tag{3.6}$$

This formula was found to agree well with numerical simulation of turbulent boundary layers and channel flows (over smooth walls) by Hunt et al. (1987). More recently, it has been found to agree even near points where $\partial U/\partial z = 0$ so that, at a height z, $\partial U/\partial z$ is taken as the average value over a length $\ell_\epsilon/2$ above and below z.

The essential physics of the ϵ equation and the implicit for of ℓ_ϵ are quite similar for perturbed shear layers. The advantage of the latter is that it provides a method for obtaining explicit solutions to perturbed layers near the surface *and* a method for computing the flow that is numerically more straightforward then using the $k-\epsilon$ model.

b. The surface structure of a perturbed turbulent boundary layer

Even in perturbed turbulent boundary layers, where the shear stress τ varies with x, very near the surface, it is possible to neglect the transfer (T) terms in the kinetic energy equation. Since $\overline{w^2}$ or k are finite at the surface and only vary *linearly* with distance, it follows that, substituting (3.5) or (3.6) into (3.2) leads to

$$U = \frac{\tau}{\kappa u_*}\left\{\ln\frac{z}{z_0} + \alpha z\left(\ln\frac{z}{z_0} - 1\right)\frac{\partial\tau/\partial x}{\tau}\right\}. \tag{3.7}$$

The coefficient α has contributions from the advection of turbulent kinetic energy, and strain production (and a small contribution from mean streamline curvature – not con-

∘ ∘ ∘ second–order closure model (Newley, 1986)
—— log–law
.... mixing–length (log–linear)
- - - - shear–blocking mixing–length (log–linear)

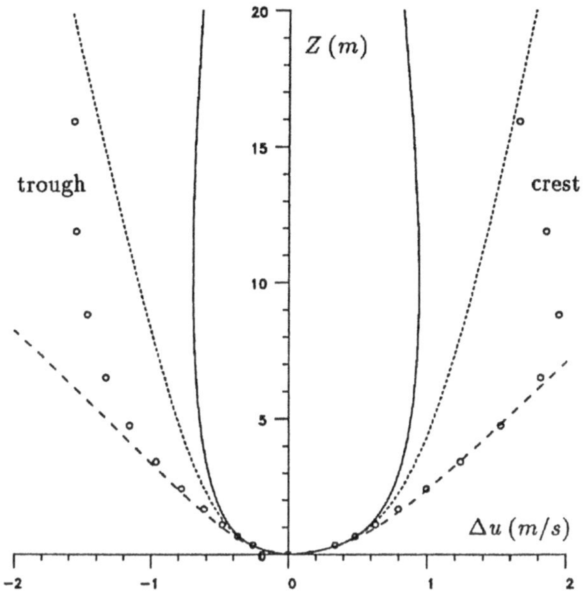

FIGURE 4. Comparison of the mean streamwise velocity perturbation at the crest and trough of a sinusoidal undulation. Surface shear stress fixed using the results of the second–order closure model (Newley, 1986).

sidered here). Using the OML (i.e. 3.5) for L_ϵ, $\alpha = 17$, whereas using the SBML (i.e. 3.6) $\alpha = 71$ – about four times larger. These large values of α mean that small pressure gradients, i.e. small $\partial\tau/\partial x$, lead to large mean velocity perturbations. This is entirely consistent with the qualitative concepts suggested by Bradshaw (1971).

Figure 4 shows a comparison of the mean streamwise velocity perturbation at the crest and trough of a sinusoidal undulation computed by Newley (1986) and using equation (3.5). For reference the mean velocity perturbation computed with $\alpha = 0$ is also shown; this shows that the profiles depart from the logarithmic law of the wall very close to the surface. In order to evaluate (3.5) for each value of α we used the values of τ and $\partial\tau/\partial x$ computed by Newley (1986). The profile calculated using the SBML is in excellent agreement with the full numerical results.

The SBML provides an estimate for the sensitivity of the boundary layer to pressure gradients and a simple physical explanation, viz. the dominant balance in the equation for the perturbed turbulent kinetic energy is between shear production and dissipation for a given value of surface turbulence, say u_*. The existence of a favourable pressure gradient ($\partial\tau/\partial x > 0$) reduces the production of turbulence and changes must be balanced by a change in the mean shear and an increases in the shear production. In the traditional mixing length formula this would be the whole story! But the SBML formula (3.6) shows that the dissipation is increased by the increase in shear. Therefore the production and $\partial U/\partial z$ have to be further increased (this would not happen if ℓ_ϵ was independent of the shear). This process also occurs with stratification and explains the sensitivity of turbulent boundary layers to small changes in surface heat flux (Hunt et al. 1988b).

c. A one equation closure model

Using the SBML formula (3.6) in (3.2) together with (3.1) for $\overline{uw}$ provides a model for the physical processes controlling k, ϵ and $\overline{uw}$ in a perturbed turbulent boundary layer. A closed set of equations is obtained by using $k^{1/2}$ in (3.6) instead of $\sqrt{\overline{w^2}}$. In the turbulent kinetic energy equation, the transfer of turbulence is modelled as a diffusive process in a traditional way using the specification of Gent and Taylor (1974). Details of the calculations are reported in Weng (1989), and preliminary results in Weng et al. (1989).

This 1–equation model for the turbulence has four main advantages over traditional $k-\epsilon$ and other models: (i) the ϵ transport equation is not well understood, especially near rigid boundaries (the inclusion of 'wall correction terms' can worsen the mean flow predictions, Schnieder 1990, private communication); (ii) it avoids the awkward boundary condition on ϵ at a rigid wall; (iii) the new model works well for separated flows, when some ϵ transport equations fail to converge, Newley (1986), and (iv) it is computationally cheaper since one less equation needs to be solved.

The non-linear momentum equations with the 1–equation closure model have been solved numerically. The results show that, because the turbulence length scale is *reduced*

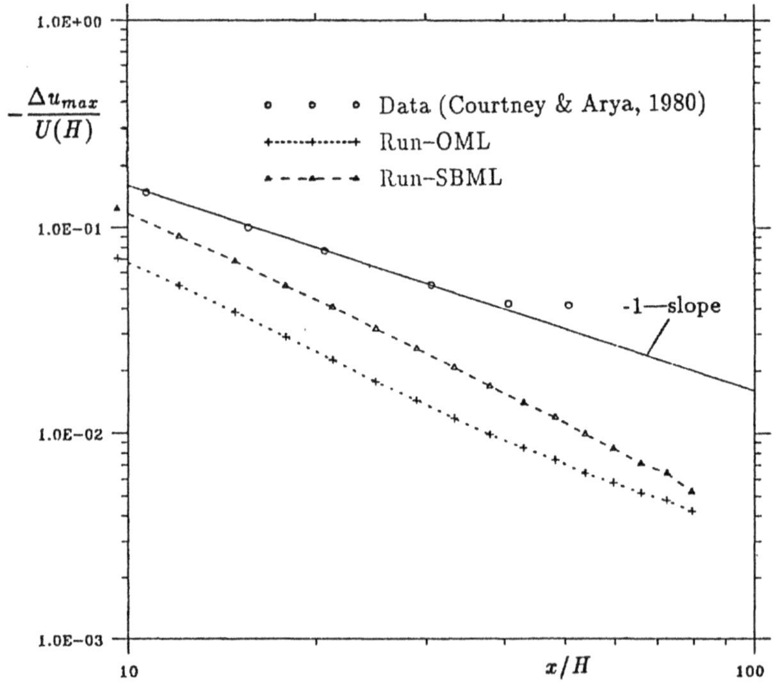

FIGURE 5. Maximum perturbation in mean velocity as a function of the downwind distance from the hilltop.

in regions of adverse pressure gradient there is a greater tendency for the boundary layer to separate and for recovery regions to extend some distance downstream of the region of adverse pressure gradient. The computed streamwise decay of the wake is improved using the SBML as compared with OML, as shown in fig.5. However in this situation, since the turbulence is advected from a region of high shear to a region of lower shear, the length scale is not just determined by the local shear by z. Hence the length scale should be estimated from a differential equation as in the $k - \epsilon$ model or a specific heuristic models specially developed for wakes (eg. Counihan, Hunt & Jackson, 1974).

4. Conclusions

We have shown how the changes to the turbulence in a perturbed boundary layer are controlled by fundamentally different processes in inner and outer regions of the flow. This mean that eddy viscosity models predict changes to the shear stress in the outer region of the wrong sign and, furthermore, using a mixing length based on height above the surface predicts changes of the wrong magnitude.

A 1–equation model for the effects of the turbulence on the mean flow has been developed which captures the essential physical processes. This model should be of particular value since near the rigid walls many traditional turbulence models are inadequate even when 'wall correction terms' are included.

References

Belcher, S.E., Newley, T.M.J. and Hunt, J.C.R. 1990 The structure of the linear perturbations to turbulent boundary layers. *Submitted to J. Fluid Mech.*.

Bradshaw, P. 1971 Variations on a theme of Prandtl. *I.C. Aero Report*, **71-13**.

Britter, R.E., Hunt, J.C.R. and Richards, K.J. 1981 Air flow over a 2-d hill: studies of velocity speed-up, roughness effects and turbulence. *Q. J. roy. Met. Soc.*, **107**, 91-110.

Counihan, J., Hunt, J.C.R. & Jackson, P.S. 1974 Wakes behind two–dimensional surface obstacles in turbulent boundary layers. *J. Fluid Mech.*, **64**, 529–563.

Dengel, P. and Fernholtz, H.H. 1990 An experimental investigation of an incompressible turbulent boundary layer in the vicinity of separation. *J. Fluid Mech.*, **212**, 615-636.

Hunt, J.C.R. 1973 A theory of turbulent flow round 2-d bluff bodies. *J. Fluid Mech.*, **61**, 625-706.

Hunt, J.C.R., Spalart, P., and Mansour, N. 1987 A general form for the dissipation length scale in turbulent shear flows.In *Proc. 1987 Summer Program, CTR.*

Hunt, J.C.R., Leibovich, S., and Richards, K.J. 1988a Turbulent shear flows over low hills. *Q. J. Roy. Met. Soc.* **114**, 1435-1471.

Hunt, J.C.R., Stretch, D.D., and Britter, R.E. 1988b Length scales in stably stratified turbulent flow and their use in turbulence models. *Proc. I.M.A. Conf. on Stably stratified flow and dense gas dispersion* (ed. J.S.Puttock), Clarendon.

Jacobs, S.J. 1987 An asymptotic theory for the turbulent flow over a progressive wave. *J. Fluid Mech.*, **174**, 69-80.

Launder, B.E., Reece, G.T. and Rodi, W. 1975 The development of a Reynolds stress turbulent closure. *J. Fluid Mech.*, **68**, 537-566.

Newley, T.M.J. 1986 Turbulent air flow over low hills. *Ph.D Thesis*, University of Cambridge.

Taylor, P.A. and Gent, P.R. 1974 a model of atmospheric boundary layer flow above an isolated hill: an example of flow over gentle topography. *Boundary Layer Met.*, **7**, 349-362.

Townsend, A.A. 1961 Equilibrium layers and wall turbulence. *J. Fluid Mech.*, **11**, 97-120.

Zilker, D.P. and Hanratty, T.J. 1979 Influence of the amplitude of a solid wavy boundary on a turbulent flow, part 2 separated flows. *J. Fluid Mech.*, **90**, 257-272.

Weng, W.S., Richards, K.J., and Carruthers, D.J. 1989 Some numerical studies of turbulent wake over hills. *Proc. Second European Turbulence Conf.*, (ed. Fernholtz and Fiedler), Springer–Verlag, Heidelberg.

Weng, W.S. 1989 Turbulent air flow and fluxes over low hills. *Ph.D Thesis*, University of Cambridge.

Towards an EDQNM-Closure for Inhomogeneous Turbulence

A.D. Burden

Department of Mechanics, Royal Institute of Technology,
S-100 44 Stockholm, Sweden

An EDQNM model is formulated for the wave-number spectra of moderately inhomogeneous turbulent flow. The derivation of the governing spectral equations supports recent trends in the singlepoint modelling of the 'rapid' part of the pressure-strain-rate redistribution of Reynolds stress. The model for the third-order wave-number spectrum generalizes both the EDQNM model for homogeneous turbulence and a singlepoint model for the turbulent flux of Reynolds stress.

1. Introduction

The Eddy-Damped, Quasi-Normal, Markovian closure procedure is an established two-point closure for homogeneous turbulence (Orszag, 1977, Lesieur, 1987) and has previously been applied to the spectral modelling of inhomogeneous turbulence by Hoyer & Sadourny (1982), Ménoret (1982), Dolez, Legait & Poyer (1984) and Bertoglio & Jeandel (1987). In this paper the EDQNM closure procedure is extended to moderately inhomogeneous turbulence in a general self-consistent manner paying particular attention to the singlepoint limits of the dynamical equations and to the constraints imposed by the continuity conditions in the product space of wave-number vector with position.

The assumptions made in this paper are:

1) The flow is only moderately inhomogeneous in the sense that the characteristic length-scale of the spatial inhomogeneity is larger than that of the turbulence thus allowing;
   a) Fourier analysis with respect to separations between points and
   b) Taylor-series expansions around positions.
2) Mean-velocity terms in the third-order dynamical equation can be neglected.
3) EDQN(M);
   a) The third-order spectrum is in a state of quasi-equilibrium.
   b) The terms in the third-order dynamical equation which contain the fourth-order cumulant are modelled by linear relaxation.

The long-term goal of this research is the development of axiomatic derivations of singlepoint computational models of turbulent flow, such as DRST-models and the k-ε model, and the detailed forms of these empirically derived yet relatively successful models lend some support to the first two assumptions. In this context it is also worth noting that the EDQNM procedure is conceptually similar to Hanjalić & Launder (1972)'s derivation of a model for the turbulent flux of the Reynolds stress.

Editors: A.V. Johansson · P.H. Alfredsson

2. Multipoint Equations

When the two points, $\mathbf{x}_1$ and $\mathbf{x}_2$, are rewritten in terms of their separation,

$$\mathbf{r} = \mathbf{x}_2 - \mathbf{x}_1, \tag{1}$$

and their 'position' chosen to be defined by,

$$\mathbf{x} = \tfrac{1}{2}(\mathbf{x}_1 + \mathbf{x}_2), \tag{2}$$

the two-point, single-time correlation between two components of the fluctuating velocity can be written in the form,

$$\left\langle u_i(\mathbf{x}-\tfrac{1}{2}\mathbf{r},t)\, u_j(\mathbf{x}+\tfrac{1}{2}\mathbf{r},t)\right\rangle = \Phi_{ij}(\mathbf{r};\mathbf{x},t). \tag{3}$$

A dynamical equation for this second-order correlation can be derived from the fluctuating part of the Navier-Stokes equation (Hinze, 1975, Monin & Yaglom, 1975). The resulting equation includes terms which contain the mean velocity at the original two points, $\mathbf{x}_1$ and $\mathbf{x}_2$, i.e. $\mathbf{U}(\mathbf{x} \pm \frac{1}{2}\mathbf{r},t)$. When these terms are expanded in a Taylor series in $\pm\frac{1}{2}\mathbf{r}$ around $\mathbf{x}$, the governing dynamical equation for the two-point, second-order velocity correlation can be written in the form,

$$\begin{aligned}\frac{\partial}{\partial t}\Phi_{ij} + U_l\frac{\partial}{\partial x_l}\Phi_{ij} = P_{ij} + \Lambda_{ij} + \Gamma_{ij} + \Pi_{ij} \; &- \frac{\partial}{\partial x_l}\left\{J_{ijl} - \nu\frac{\partial}{\partial x_l}\Phi_{ij}\right\} \\ &+ \nu\left\{2\frac{\partial^2}{\partial r_l \partial r_l} - \frac{1}{2}\frac{\partial^2}{\partial x_l \partial x_l}\right\}\Phi_{ij}.\end{aligned} \tag{4}$$

The various terms in the right-hand side of this equation have been defined after taking into consideration the rôles known or believed to be played by (1) their Fourier transforms in the wave-number formalism for homogeneous turbulence and (2) their singlepoint limits ($\mathbf{r} \to \mathbf{0}$) in the transport equation for the Reynolds stress tensor. The terms are: production,

$$\begin{aligned}P_{ij}(\mathbf{r};\mathbf{x},t) = &-\frac{\partial U_i}{\partial x_l}(\mathbf{x},t)\left\langle u_l(\mathbf{x}-\tfrac{1}{2}\mathbf{r},t)\, u_j(\mathbf{x}+\tfrac{1}{2}\mathbf{r},t)\right\rangle \\ &-\frac{\partial U_j}{\partial x_l}(\mathbf{x},t)\left\langle u_i(\mathbf{x}-\tfrac{1}{2}\mathbf{r},t)\, u_l(\mathbf{x}+\tfrac{1}{2}\mathbf{r},t)\right\rangle + \cdots,\end{aligned} \tag{5}$$

where '$+\cdots$' denotes the omission of higher-order terms in the Taylor-series expansion of the mean velocity; mean-velocity-gradient induced 'spectral' transfer (Hinze, 1975),

$$\Lambda_{ij}(\mathbf{r};\mathbf{x},t) = -r_m\frac{\partial U_l}{\partial x_m}(\mathbf{x},t)\frac{\partial}{\partial r_l}\left\langle u_i(\mathbf{x}-\tfrac{1}{2}\mathbf{r},t)\, u_j(\mathbf{x}+\tfrac{1}{2}\mathbf{r},t)\right\rangle + \cdots; \tag{6}$$

non-linear inertial 'spectral' transfer,

$$\begin{aligned}\Gamma_{ij}(\mathbf{r};\mathbf{x},t) = &\frac{\partial}{\partial r_l}\left\langle u_i(\mathbf{x}-\tfrac{1}{2}\mathbf{r},t)\, u_j(\mathbf{x}+\tfrac{1}{2}\mathbf{r},t)\, u_l(\mathbf{x}-\tfrac{1}{2}\mathbf{r},t)\right\rangle \\ &-\frac{\partial}{\partial r_l}\left\langle u_i(\mathbf{x}-\tfrac{1}{2}\mathbf{r},t)\, u_j(\mathbf{x}+\tfrac{1}{2}\mathbf{r},t)\, u_l(\mathbf{x}+\tfrac{1}{2}\mathbf{r},t)\right\rangle;\end{aligned} \tag{7}$$

and turbulent flux of $\mathbf{\Phi}$,

$$J_{ijl}(\mathbf{r};\mathbf{x},t) = \frac{1}{2}\Big\langle u_i(\mathbf{x}-\tfrac{1}{2}\mathbf{r},t)\,u_j(\mathbf{x}+\tfrac{1}{2}\mathbf{r},t)\,u_l(\mathbf{x}-\tfrac{1}{2}\mathbf{r},t)\Big\rangle + \frac{1}{2}\Big\langle u_i(\mathbf{x}-\tfrac{1}{2}\mathbf{r},t)\,u_j(\mathbf{x}+\tfrac{1}{2}\mathbf{r},t)\,u_l(\mathbf{x}+\tfrac{1}{2}\mathbf{r},t)\Big\rangle. \tag{8}$$

The transfer terms, $\mathbf{\Gamma}$ and $\mathbf{\Lambda}$, vanish in the singlepoint limit and do not contribute to redistribution amongst the components of the Reynolds stress tensor. The pressure terms in (4) are,

$$\Pi_{ij}(\mathbf{r};\mathbf{x},t) = \frac{\partial}{\partial r_i}\Big\langle \rho^{-1}p(\mathbf{x}-\tfrac{1}{2}\mathbf{r},t)\,u_j(\mathbf{x}+\tfrac{1}{2}\mathbf{r},t)\Big\rangle - \frac{\partial}{\partial r_j}\Big\langle \rho^{-1}p(\mathbf{x}+\tfrac{1}{2}\mathbf{r},t)\,u_j(\mathbf{x}-\tfrac{1}{2}\mathbf{r},t)\Big\rangle + \Pi_{ij}^{X}(\mathbf{r};\mathbf{x},t), \tag{9}$$

where,

$$\Pi_{ij}^{X}(\mathbf{r};\mathbf{x},t) = -\frac{1}{2}\frac{\partial}{\partial x_i}\Big\langle \rho^{-1}p(\mathbf{x}-\tfrac{1}{2}\mathbf{r},t)\,u_j(\mathbf{x}+\tfrac{1}{2}\mathbf{r},t)\Big\rangle - \frac{1}{2}\frac{\partial}{\partial x_j}\Big\langle \rho^{-1}p(\mathbf{x}+\tfrac{1}{2}\mathbf{r},t)\,u_j(\mathbf{x}-\tfrac{1}{2}\mathbf{r},t)\Big\rangle. \tag{10}$$

The viscous terms in (4) have been split into a gradient diffusion expression and a remainder, the last line of (4). The trace of the singlepoint limit of this remainder, when divided by two, is the so called isotropic rate of viscous dissipation of the turbulent kinetic energy. All the terms in (4), as defined in (5)–(10), are invariant under the simultaneous interchanges $[\mathbf{r} \rightarrow -\mathbf{r}]$ and $[i \leftrightarrow j]$ as is $\mathbf{\Phi}$ itself.

Corresponding to the two-point coordinate transformation, (1) and (2), the three-point transformation, $\{\mathbf{x_1},\mathbf{x_2},\mathbf{x_3}\} \rightarrow \{\mathbf{r_1},\mathbf{r_2},\mathbf{x}\}$, is defined by,

$$\mathbf{r_1} = \mathbf{x_2} - \mathbf{x_1} \qquad \text{and} \qquad \mathbf{r_2} = \mathbf{x_3} - \mathbf{x_1}, \tag{11}$$

together with (2) for $\mathbf{x}$, so that the three-point, single-time, third-order velocity correlation can be written in the form,

$$\Big\langle u_i(\mathbf{x}-\tfrac{1}{2}\mathbf{r_1},t)\,u_j(\mathbf{x}+\tfrac{1}{2}\mathbf{r_1},t)\,u_k(\mathbf{x}-\tfrac{1}{2}\mathbf{r_1}+\mathbf{r_2},t)\Big\rangle = \Psi_{ijk}(\mathbf{r_1},\mathbf{r_2};\mathbf{x},t). \tag{12}$$

A dynamical equation for $\mathbf{\Psi}$ can be derived from the fluctuating part of the Navier-Stokes equation (Monin & Yaglom, 1975). Detailed analysis of both singlepoint DRST-models of turbulent flow and of the EDQNM two-point closure for homogeneous turbulence suggests that mean-flow terms in the third-order equation can be neglected, at least in a first approximation. Under this assumption, the dynamical equation for $\mathbf{\Psi}$ can be written in the form,

$$\frac{\partial}{\partial t}\mathbf{\Psi} = \mathbf{\Omega} + \mathbf{\Pi}^{(1)} + \mathbf{Q} + \mathbf{D}. \tag{13}$$

The terms in this equation are given below in forms in which fourth-order velocity correlations or moments are written in terms of the four-point, fourth-order cumulant, $\Theta_{ijkl}(\mathbf{r_1},\mathbf{r_2},\mathbf{r_3};\mathbf{x},t)$, where $\mathbf{r_3}$ is given by the fourth point, $\mathbf{x_4} = \mathbf{x_1}+\mathbf{r_3} = \mathbf{x}-\frac{1}{2}\mathbf{r_1}+\mathbf{r_3}$. The first inertial term, $\mathbf{\Omega} = \mathbf{\Omega}^{\mathrm{C}} + \mathbf{\Omega}^{\mathrm{QN}}$, vanishes in the singlepoint limit and is given by,

$$\Omega^C_{ijk}(\mathbf{r_1},\mathbf{r_2};\mathbf{x},t) = -\frac{\partial}{\partial r_{0l}}\Theta_{ijkl}(\mathbf{r_1},\mathbf{r_2},\mathbf{0};\mathbf{x},t) - \frac{\partial}{\partial r_{1l}}\Theta_{ijkl}(\mathbf{r_1},\mathbf{r_2},\mathbf{r_1};\mathbf{x},t)$$
$$- \frac{\partial}{\partial r_{2l}}\Theta_{ijkl}(\mathbf{r_1},\mathbf{r_2},\mathbf{r_2};\mathbf{x},t), \tag{14}$$

with $\partial/\partial r_{0i} = -\partial/\partial r_{1i} - \partial/\partial r_{2i}$, and,

$$\Omega^{QN}_{ijk}(\mathbf{r_1},\mathbf{r_2};\mathbf{x},t) = -\Big\langle u_k(\mathbf{x_3},t)\,u_l(\mathbf{x_1},t)\Big\rangle\frac{\partial}{\partial r_{0l}}\Big\langle u_i(\mathbf{x_1},t)\,u_j(\mathbf{x_2},t)\Big\rangle$$
$$+ \text{ similar terms.} \tag{15}$$

The second inertial term, $\mathbf{Q}$, vanishes for homogeneous turbulence and is given by $\mathbf{Q} = \mathbf{Q^C} + \mathbf{Q^{QN}}$ where,

$$Q^C_{ijk}(\mathbf{r_1},\mathbf{r_2};\mathbf{x},t) = -\frac{1}{2}\frac{\partial}{\partial x_l}\Theta_{ijkl}(\mathbf{r_1},\mathbf{r_2},\mathbf{0};\mathbf{x},t) - \frac{1}{2}\frac{\partial}{\partial x_l}\Theta_{ijkl}(\mathbf{r_1},\mathbf{r_2},\mathbf{r_1};\mathbf{x},t), \tag{16}$$

and,

$$Q^{QN}_{ijk}(\mathbf{r_1},\mathbf{r_2};\mathbf{x},t) = -\frac{1}{2}\Big\langle u_k(\mathbf{x_3},t)\,u_l(\mathbf{x_1},t)\Big\rangle\frac{\partial}{\partial x_l}\Big\langle u_i(\mathbf{x_1},t)\,u_j(\mathbf{x_2},t)\Big\rangle$$
$$+ \text{ similar terms.} \tag{17}$$

The terms in (13) which contain pressure-double-velocity correlations are,

$$\Pi^{(1)}_{ijk}(\mathbf{r_1},\mathbf{r_2};\mathbf{x},t) = -\frac{\partial}{\partial r_{0i}}\Big\langle \rho^{-1}p^{(1)}(\mathbf{x_1},t)u_j(\mathbf{x_2},t)\,u_k(\mathbf{x_3},t)\Big\rangle$$
$$-\frac{1}{2}\frac{\partial}{\partial x_i}\Big\langle \rho^{-1}p^{(1)}(\mathbf{x_1},t)u_j(\mathbf{x_2},t)\,u_k(\mathbf{x_3},t)\Big\rangle \tag{18}$$
$$+ \text{ similar terms,}$$

where $p^{(1)}$ is the part of the fluctuating pressure field whose governing Poisson equation does not explicitly contain the mean velocity gradient. Finally, the viscous terms in (13) are,

$$D_{ijk}(\mathbf{r_1},\mathbf{r_2};\mathbf{x},t) = \nu\left\{\frac{\partial^2}{\partial r_{0l}\partial r_{0l}} + \frac{\partial^2}{\partial r_{1l}\partial r_{1l}} + \frac{\partial^2}{\partial r_{2l}\partial r_{2l}} + \cdots\right\}\Psi_{ijk}(\mathbf{r_1},\mathbf{r_2};\mathbf{x},t), \tag{19}$$

where '$+\cdots$' denotes the omission of terms which contain derivatives with respect to position.

3. Wave-Number Spectra

The multipoint velocity correlations, $\mathbf{\Phi}$ and $\mathbf{\Psi}$, are assumed to have the spectral representations,

$$\Phi_{ij}(\mathbf{r};\mathbf{x},t) = \int \widehat{\Phi}_{ij}(\mathbf{k};\mathbf{x},t)\,\exp(\mathrm{i}\mathbf{k}.\mathbf{r})\,d^3\mathbf{k}, \tag{20}$$

and,

$$\Psi_{ijk}(\mathbf{r_1},\mathbf{r_2};\mathbf{x},t) = \iint \widehat{\Psi}_{ijk}(\mathbf{k_1},\mathbf{k_2};\mathbf{x},t)\,\exp(\mathrm{i}\mathbf{k_1}.\mathbf{r_1}+\mathrm{i}\mathbf{k_2}.\mathbf{r_2})\,d^3\mathbf{k_1}\,d^3\mathbf{k_2}. \tag{21}$$

The second- and third-order wave-number spectra, $\widehat{\mathbf{\Phi}}$ and $\widehat{\mathbf{\Psi}}$, and the fourth-order spectral cumulant, $\widehat{\mathbf{\Theta}}$, will be given, at least approximately, by the Fourier transforms of $\mathbf{\Phi}$, $\mathbf{\Psi}$ and $\mathbf{\Theta}$ at positions which are many correlation lengths away from confining surfaces of the flow.

The continuity condition for the fluctuating velocity field, $\nabla.\mathbf{u} = 0$, yields continuity conditions for the wave-number spectra,

$$- k_i\widehat{\Phi}_{ij}(\mathbf{k};\mathbf{x},t) - \tfrac{1}{2}\mathrm{i}\frac{\partial}{\partial x_i}\widehat{\Phi}_{ij}(\mathbf{k};\mathbf{x},t) = 0, \tag{22}$$

$$k_j\widehat{\Phi}_{ij}(\mathbf{k};\mathbf{x},t) - \tfrac{1}{2}\mathrm{i}\frac{\partial}{\partial x_j}\widehat{\Phi}_{ij}(\mathbf{k};\mathbf{x},t) = 0, \tag{23}$$

and corresponding conditions for the higher-order spectra.

4. Fourier Analysis of Pressure Correlations

For homogeneous turbulence, the Poisson equation for the fluctuating pressure can be solved exactly using Fourier analysis to yield expressions for the Fourier transforms of terms containing correlations with the fluctuating pressure. This method is extended to weakly inhomogeneous turbulence by employing Taylor-series expansions with respect to position. The resulting expressions ensure that the dynamical equations for the wave-number spectra are compatible with the continuity conditions, e.g. (22) and (23). In the second-order equation (4), the Fourier transform of $\mathbf{\Pi}$ can be written in the form $\widehat{\mathbf{\Pi}} = \widehat{\mathbf{\Pi}}^{(1)} + \widehat{\mathbf{\Pi}}^{(2)}$ where,

$$\widehat{\mathbf{\Pi}}^{(1)} = \widehat{\mathbf{\Pi}}^{\mathbf{\Gamma}} + \widehat{\mathbf{\Pi}}^{\mathbf{J}} + \widetilde{\mathbf{\Pi}}^{1\mathbf{X}}, \tag{24}$$

and

$$\widehat{\mathbf{\Pi}}^{(2)} = \widehat{\mathbf{\Pi}}^{\mathbf{A}} + \widehat{\mathbf{\Pi}}^{\mathbf{P}} + \widehat{\mathbf{\Pi}}^{\mathbf{\Lambda}} + \widetilde{\mathbf{\Pi}}^{2\mathbf{X}} \tag{25}$$

with the result that, for each term $\mathbf{F} = \mathbf{P}$, $\mathbf{\Lambda}$ and $\mathbf{\Gamma}$, the expression $\widetilde{\mathbf{F}} = \widehat{\mathbf{F}} + \widehat{\mathbf{\Pi}}^{\mathbf{F}}$ satisfies (22) and (23). These continuity conditions are also satisfied by the expressions $(\mathbf{U}.\nabla)\widehat{\mathbf{\Phi}} - \widehat{\mathbf{\Pi}}^{\mathbf{A}}$ and $\nabla.\widetilde{\mathbf{J}} = \nabla.\mathbf{J} + \widehat{\mathbf{\Pi}}^{\mathbf{J}}$ as well as by the terms $\widetilde{\mathbf{\Pi}}^{1\mathbf{X}}$ and $\widetilde{\mathbf{\Pi}}^{2\mathbf{X}}$. In the third-order equation (13), the Fourier transform of $\mathbf{\Pi}^{(1)}$ can be written in the form,

$$\widehat{\mathbf{\Pi}}^{(1)} = \widehat{\mathbf{\Pi}}^{\mathbf{\Omega}} + \widehat{\mathbf{\Pi}}^{\mathbf{Q}} + \widetilde{\mathbf{\Pi}}^{\mathbf{C}} + \widetilde{\mathbf{\Pi}}^{\mathbf{QN}}, \tag{26}$$

with the result that $\widetilde{\mathbf{\Omega}} = \widehat{\mathbf{\Omega}} + \widehat{\mathbf{\Pi}}^{\mathbf{\Omega}}$ and $\widetilde{\mathbf{Q}} = \widehat{\mathbf{Q}} + \widehat{\mathbf{\Pi}}^{\mathbf{Q}}$ each satisfy the third-order continuity conditions. These continuity conditions are also satisfied by $\widetilde{\mathbf{\Pi}}^{\mathbf{C}}$ and $\widetilde{\mathbf{\Pi}}^{\mathbf{QN}}$.

The above results are a straightforward consequence of the way in which the continuity condition is used together with the Navier-Stokes equation to derive Poisson equations for various parts of the pressure field. While the detailed derivations of the expressions behind (24)–(26) are of limited validity, any alternative expressions for these pressure terms in the dynamical equations for the wave-number spectra must also ensure that these dynamical equations are compatible with the continuity conditions.

5. Dynamical Equations for the Spectra

The spectral form of the second-order dynamical equation (4) is now,

$$\frac{\partial}{\partial t}\widehat{\Phi}_{ij} + U_l\frac{\partial}{\partial x_l}\widehat{\Phi}_{ij} = \widehat{\Pi}^A_{ij} + \widetilde{P}_{ij} + \widetilde{\Lambda}_{ij} + \widetilde{\Gamma}_{ij} - \frac{\partial}{\partial x_l}\left\{\widetilde{J}_{ijl} - \nu\frac{\partial}{\partial x_l}\widehat{\Phi}_{ij}\right\} + \widetilde{\Pi}^X_{ij} - \left\{2\nu k^2 + \frac{1}{2}\nu\frac{\partial^2}{\partial x_l \partial x_l}\right\}\widehat{\Phi}_{ij}, \tag{27}$$

where:

$$\widehat{\Pi}^A_{ij}(\mathbf{k};\mathbf{x},t) = \tfrac{1}{2}\mathrm{i}\frac{k_i}{k^2}\frac{\partial U_k}{\partial x_l}(\mathbf{x},t)\frac{\partial}{\partial x_k}\widehat{\Phi}_{lj}(\mathbf{k};\mathbf{x},t) - \tfrac{1}{2}\mathrm{i}\frac{k_j}{k^2}\frac{\partial U_k}{\partial x_l}(\mathbf{x},t)\frac{\partial}{\partial x_k}\widehat{\Phi}_{il}(\mathbf{k};\mathbf{x},t) + \cdots \tag{28}$$

denotes pressure terms associated with advection;

$$\widetilde{P}_{ij}(\mathbf{k};\mathbf{x},t) = -\,\Delta_{ik}(\mathbf{k})\frac{\partial U_k}{\partial x_l}(\mathbf{x},t)\,\widehat{\Phi}_{lj}(\mathbf{k};\mathbf{x},t) - \Delta_{jk}(\mathbf{k})\frac{\partial U_k}{\partial x_l}(\mathbf{x},t)\,\widehat{\Phi}_{il}(\mathbf{k};\mathbf{x},t) + \cdots, \tag{29}$$

with $\Delta_{ij}(\mathbf{k}) = \delta_{ij} - k_i k_j/k^2$, denotes production together with associated pressure terms;

$$\widetilde{\Lambda}_{ij}(\mathbf{k};\mathbf{x},t) = -\mathrm{i}k_k\frac{\partial U_k}{\partial x_l}(\mathbf{x},t)\left(\mathrm{i}\frac{\partial}{\partial k_l}\right)\widehat{\Phi}_{ij}(\mathbf{k};\mathbf{x},t) + \frac{k_i k_k}{k^2}\frac{\partial U_k}{\partial x_l}(\mathbf{x},t)\,\widehat{\Phi}_{lj}(\mathbf{k};\mathbf{x},t) + \frac{k_j k_k}{k^2}\frac{\partial U_k}{\partial x_l}(\mathbf{x},t)\,\widehat{\Phi}_{il}(\mathbf{k};\mathbf{x},t) + \cdots \tag{30}$$

denotes mean-velocity-gradient induced spectral transfer and associated pressure terms;

$$\widetilde{\Gamma}_{ij}(\mathbf{k};\mathbf{x},t) = \mathrm{i}k_l\Delta_{ik}(\mathbf{k})\int\widehat{\Psi}_{kjl}(\mathbf{k},\mathbf{k}';\mathbf{x},t)\,d^3\mathbf{k}' - \mathrm{i}k_l\Delta_{jk}(\mathbf{k})\int\widehat{\Psi}_{ikl}(\mathbf{k}-\mathbf{k}',\mathbf{k}';\mathbf{x},t)\,d^3\mathbf{k}' + \cdots \tag{31}$$

denotes non-linear inertial spectral transfer;

$$\widetilde{J}_{ijl}(\mathbf{k};\mathbf{x},t) = \tfrac{1}{2}\Delta_{ik}(\mathbf{k})\int\widehat{\Psi}_{kjl}(\mathbf{k},\mathbf{k}';\mathbf{x},t)\,d^3\mathbf{k}' + \tfrac{1}{2}\Delta_{jk}(\mathbf{k})\int\widehat{\Psi}_{ikl}(\mathbf{k}-\mathbf{k}',\mathbf{k}';\mathbf{x},t)\,d^3\mathbf{k}' + \cdots \tag{32}$$

denotes the turbulent flux of $\widehat{\mathbf{\Phi}}$ together with associated pressure terms and,

$$\widetilde{\Pi}^X_{ij}(\mathbf{k};\mathbf{x},t) = \tfrac{1}{2}\Delta_{im}(\mathbf{k})\frac{k_k k_l}{k^2}\frac{\partial}{\partial x_m}\int\widehat{\Psi}_{kjl}(\mathbf{k},\mathbf{k}';\mathbf{x},t)\,d^3\mathbf{k}' + \mathrm{i}\Delta_{im}(\mathbf{k})\frac{k_k}{k^2}\frac{\partial}{\partial x_m}\left\{\frac{\partial U_k}{\partial x_l}(\mathbf{x},t)\,\widehat{\Phi}_{lj}(\mathbf{k};\mathbf{x},t)\right\} + \text{similar and higher-order terms}, \tag{33}$$

are 'extra' inhomogeneous pressure terms. All the terms in (27) are invariant under the simultaneous interchanges $[\mathbf{k} \to -\mathbf{k}]$ and $[i \leftrightarrow j]$ as is $\widehat{\boldsymbol{\Phi}}$ itself.

The dynamical equation for the third-order spectrum can be derived from (13) in an identical manner when use is made of (26).

6. Closure

When the third-order spectrum is in a state of quasi-equilibrium, its dynamical equation can be written in the form,

$$\mathbf{0} = \widetilde{\boldsymbol{\Omega}}^{\mathbf{QN}} + \widetilde{\boldsymbol{\Omega}}^{\mathbf{C}} + \widetilde{\mathbf{Q}}^{\mathbf{QN}} + \widetilde{\mathbf{Q}}^{\mathbf{C}} + \widetilde{\boldsymbol{\Pi}}^{\mathbf{QN}} + \widetilde{\boldsymbol{\Pi}}^{\mathbf{C}} + \widehat{\mathbf{D}}, \tag{34}$$

where $\widetilde{\boldsymbol{\Omega}}^{\mathbf{C}}$, $\widetilde{\mathbf{Q}}^{\mathbf{C}}$ and $\widetilde{\boldsymbol{\Pi}}^{\mathbf{C}}$ all contain the fourth-order cumulant. A closed set of equations for $\widehat{\boldsymbol{\Phi}}$ and $\widehat{\boldsymbol{\Psi}}$ is obtained by modelling these fourth-order terms as functions of the lower-order spectra, i.e. $\widetilde{\boldsymbol{\Omega}}^{\mathbf{C}} + \widetilde{\mathbf{Q}}^{\mathbf{C}} + \widetilde{\boldsymbol{\Pi}}^{\mathbf{C}} = \mathcal{C}[\widehat{\boldsymbol{\Phi}}, \widehat{\boldsymbol{\Psi}}]$. The dynamical equation for the fourth-order cumulant indicates that $\mathcal{C}[\widehat{\boldsymbol{\Phi}}, \widehat{\boldsymbol{\Psi}}]$ is linear in $\widehat{\boldsymbol{\Psi}}$ and can be modelled as a local operator with respect to the wave-number vectors $\mathbf{k_0}$, $\mathbf{k_1}$ and $\mathbf{k_2}$, where $\mathbf{k_0}$ is given by $\mathbf{k_0}+\mathbf{k_1}+\mathbf{k_2} = \mathbf{0}$. The third-order continuity conditions indicate that $\mathcal{C}[\widehat{\boldsymbol{\Phi}}, \widehat{\boldsymbol{\Psi}}]$ can be expected to be a differential operator with respect to $\mathbf{x}$. Further, in the established EDQNM closure for homogeneous turbulence $\mathcal{C}[\widehat{\boldsymbol{\Phi}}, \widehat{\boldsymbol{\Psi}}]$ is linear in $\widehat{\boldsymbol{\Psi}}(\mathbf{k_1}, \mathbf{k_2}; \mathbf{x}, t)$. Consequently the dynamical equation for the third-order spectrum is modelled by,

$$\mathcal{L}\widehat{\boldsymbol{\Psi}} - \widehat{\mathbf{D}} = \widetilde{\boldsymbol{\Omega}}^{\mathbf{QN}} + \widetilde{\mathbf{Q}}^{\mathbf{QN}} + \widetilde{\boldsymbol{\Pi}}^{\mathbf{QN}}, \tag{35}$$

where $\mathcal{L}$ is a function of $\widehat{\boldsymbol{\Phi}}$ and can be a differential operator with respect to $\mathbf{x}$. When the spatial inhomogeneity of the turbulence is relatively weak, (35) can be formally inverted to yield an expansion of $\widehat{\boldsymbol{\Psi}}$ in ascending powers of derivatives of the right-hand side with respect to position, $\mathbf{x}$. Bearing in mind the third-order continuity conditions and the detailed forms of the terms in (34), this closed expression for the third-order wave-number spectrum must take the general form,

$$\begin{aligned} \widehat{\boldsymbol{\Psi}} = {} & \mathbf{T}^{(0)}.\{\widetilde{\boldsymbol{\Omega}}^{\mathbf{QN}} + \widetilde{\mathbf{Q}}^{\mathbf{QN}} + \widetilde{\boldsymbol{\Pi}}^{\mathbf{QN}}\} \\ & + \left(\tfrac{1}{2}\mathrm{i}\frac{\mathbf{k_0}}{k_0^2}\nabla\cdot + \tfrac{1}{2}\mathrm{i}\frac{\mathbf{k_1}}{k_1^2}\nabla\cdot\right)\left\{\mathbf{T}^{(0)}.\{\widetilde{\boldsymbol{\Omega}}^{\mathbf{QN}} + \widetilde{\mathbf{Q}}^{\mathbf{QN}} + \widetilde{\boldsymbol{\Pi}}^{\mathbf{QN}}\}\right\} + \cdots \\ & + \mathbf{T}^{(1)}.\left\{\nabla\{\widetilde{\boldsymbol{\Omega}}^{\mathbf{QN}} + \widetilde{\mathbf{Q}}^{\mathbf{QN}} + \widetilde{\boldsymbol{\Pi}}^{\mathbf{QN}}\}\right\} + \cdots, \end{aligned} \tag{36}$$

where $\mathbf{T^0}$ can be taken to be given by the EDQNM closure for homogeneous turbulence, e.g. $T^{(0)}_{ijklmn} = \delta_{il}\,\delta_{jm}\,\delta_{kn}\left\{\mu(\mathbf{k_0}, \mathbf{k_1}, \mathbf{k_2}) + \nu(k_0^2+k_1^2+k_2^2)\right\}^{-1}$. Equation (36) generalizes both the EDQNM model for homogeneous turbulence (at large times) and Hanjalić & Launder (1972)'s model for the turbulent flux of the Reynolds stress.

7. Implications for Singlepoint Modelling

Equation (25) implies that the so called rapid part of the pressure-strain-rate redistribution of Reynolds stress is linked to both the explicit mean-flow terms, advection and production, in the Reynolds-stress transport equation and also contains a contribution from the mean-velocity-gradient induced spectral transfer. This result is relevant to more empirical approachés to singlepoint modelling such as *Isotropization of Production* and *Isotropization of Convection* — see, e.g., Fu, Launder & Leschziner (1987).

The model of Hanjalić & Launder (1972) for the turbulent flux of the Reynolds stress can be derived (approximately) from (12), (21) and (36) when 'pressure transport' can be neglected, the eddy-damping tensor, $\mathbf{T}^{(0)}$, is diagonal while contributions from $\mathbf{T}^{(n)}$ with $n > 0$ can be neglected, and either the eddy-damping rate, μ, is independent of the wave-number vectors or turbulent 'diffusion' is dominated by the energy-bearing eddies. See Weinstock (1989) for a recent discussion of the relationship between the EDQNM two-point closure and the singlepoint model of Hanjalić & Launder (1972).

Acknowledgements

The author wishes to thank Professor P. Hutchinson for encouragement and advice and for the original suggestion that established singlepoint models for turbulent flow could be developed into computational models for turbulent spectra. Financial support from the Swedish National Board for Technical Development and National Energy Administration is gratefully acknowledged.

References

Bertoglio, J.-P., & D. Jeandel, 1987, A simplified spectral closure for inhomogeneous turbulence: application to the boundary layer, in *Turbulent Shear Flows 5*, Eds: F. Durst *et al*, Springer.

Dolez, N., A. Legait & J.P. Poyer, 1984, Beyond mixing-length theory in convection, *Mem.Soc.Astron.It.* **55**, 293.

Fu, S., B.E. Launder & M.A. Leschziner, 1987, Modelling strongly swirling recirculating jet flow with Reynolds stress transport closures, in *Proc. of the 6th Symp. on Turbulent Shear Flows*, Toulouse, France.

Hanjalić, K., & B.E. Launder, 1972, A Reynolds stress model of turbulence and its application to thin shear flows, *J. Fluid Mech.* **52**, 609.

Hinze, J.O., 1975, *Turbulence*, McGraw-Hill.

Hoyer, J.-M., & R. Sadourny, 1982, Closure modeling of fully developed baroclinic instability, *J.Atmos.Sci.* **39**, 707.

Lesieur, M., 1987, *Turbulence in fluids*, Martinus-Nijhoff.

Ménoret, L., 1982, *Contribution à l'étude spectrale des écoulements turbulents faiblement inhomogènes et anisotropes*, Thèse Univ. Cl. Bernard, Lyon I, France.

Monin, A.S., & A.M. Yaglom, 1975, *Statistical fluid mechanics, vol. 2*, MIT Press.

Orszag, S.A., 1977, Lectures on the statistical theory of turbulence, in *Fluid dynamics*, Eds: R. Balian and J.-L. Peube, Gordon and Breach.

Weinstock, J., 1989, A theory of turbulent transport, *J. Fluid Mech.* **202**, 319.

Defining the Zonal Structure of Turbulence Using the Pressure and Invariants of the Deformation Tensor

Fung J.C.H., Hunt J.C.R. & Perkins R.J.
Department of Applied Mathematics and Theoretical Physics,
University of Cambridge, UK

Wray A.A. & Stretch D.
Center for Turbulence Research, CA 94035, USA

A set of objective criteria based on the local strain rate, vorticity and pressure have been found to describe regions in which the streamlines circulate, converge or diverge, and form streams of high velocity flow. The homogeneous and sheared turbulent flow fields are made up of characteristic flow zones – eddy, shear, convergence and streaming zones. These are studied in turbulent velocity fields produced by different methods of simulation, including the novel method of Kinematic Simulation of homogeneous isotropic turbulence are summarised. We derive and explain the zonal algorithm to classify structures and then use this classification to compare the results of two different numerical simulations (DNS, KS) both qualitatively (turbulence structure, physical processes) and quantitatively (turbulence statistics) for homogeneous isotropic turbulence and then we apply the zonal algorithm to a turbulent shear flow. New conclusions are reached about the significant regions in turbulent flows for dynamical and kinematical processes.

1. Introduction

Experiments and measurements show that in turbulent flows there are large scale motions with forms that recur in most realisations of the flow and persist over relatively long distances in the dominant flow direction (e.g. Hussain 1986). In addition, turbulent flows also contain small scale random fluctuations that may or may not be correlated with the large scales (see Hunt *et al.* 1990). The identification and description of such significant regions in a flow provides an important method for analysing the dynamics of the flow and might enable us to simplify the way in which we model some flows. Flow visualisation can be used to identify structures and infer their qualitative topology; but structures typically undergo cycles of formation, growth and decay. Detection and detailed investigation of these structures is complicated by the large variety of shapes, sizes, orientations, strengths and convection velocities. Consequently statistical methods are more reliable, such as those proposed by Lumley 1967 (see Aubry *et al.* 1988), Mumford (1983), Hussain (1986), Adrian & Moin (1988) for deriving in different ways the forms of eddy structures in various turbulent flows; but they do not completely agree with each other as to the form of these structures. None of these methods have been applied to complete flow fields over a period of time for both homogeneous and inhomogeneous turbulence, and therefore do not describe how different kinds of structure in a flow relate to each other spatially and temporally.

One of the reasons for studying these characteristic flow pattern and structures are that they are important locations for certain processes occurring in the flow. For

Advances in Turbulence 3
Editors: A.V. Johansson · P.H. Alfredsson

instance, there is evidence that in reacting flows fast chemical reactions are concentrated in regions where streamlines converge (Leonard & Hill 1988), which we shall refer to as convergence regions, and slow chemical reactions in recirculating eddying regions (Broadwell & Breidenthal 1982). Vortical regions (which we refer to as eddies) are also important in certain flame and combustion problems (Peters & Williams 1988), probably because of their effects on material surfaces (Vassilicos 1989). Recent experimental and computational research (Maxey 1987, Hunt *et al.* 1988, Fung & Perkins 1989) has shown that bubbles tend to concentrate in low-pressure regions, whilst denser particles, especially if buoyancy forces are important, tend to concentrate in the streaming regions between eddies. If particles are entrained into eddy regions, however, they can remain there for long periods.

This paper describes an algorithm recently proposed by Wray & Hunt (based on earlier work by Perry & Chong 1987) for defining the zonal structure of turbulence, using the kinematics of the flow. The algorithm is invariant under mean rotation and translation and is applicable to different *types* of flow – homogeneous and inhomogeneous; also it involves conditional sampling depending on the local strain rate, pressure and fluctuating velocity. But is this or other descriptions of the eddy structure only possible using simulations or multi-point measurements of the flow, or is it possible to use ensemble statistics, such as spectra or correlations? Whereas the approach of Lumley (1967) makes the latter assumption, the other methods require conditional sampling plus local velocity measurements (Munford 1983, Hussain 1986). To answer this question we compare two velocity fields one obtained from DNS and the other constructed solely from the spectra using random Fourier modes (KS). This demonstrates how the different flow zones can be used to compare the velocity fields produced by different simulation techniques. Also the zones can be useful in developing simpler idealised flow models for particular applications. For example, many simulations of particle and bubble motions in turbulent flows model the flow as a collection of eddies (Brown & Hutchinson 1979, Broadwell & Breidenthal 1982) and this type of analysis can suggest characteristic scales for these eddies.

2. Simulation of flow fields

2.1 ***Direct Numerical Simulation***: The events, structures, and processes characterizing the four flow zone types are defined at one instant of time but it is instructive to study how the flow zones evolve in time. Consequently a statistically stationary velocity field was used rather than a decaying one, which for homogeneous isotropic turbulence requires an energy-input mechanism. The 3-D Navier-Stokes equations were solved using the spectral code of Rogallo (1981) modified to include a steady body force $\mathbf{F}$. It was found that $\langle u_i u_j \rangle$ initially increased or decreased, but eventually oscillated around a stationary value. Then a time series $\widehat{\mathbf{u}}(\mathbf{k}, t)$ was collected on time interval of 10 time steps, or about 1 minimum Eulerian time scale $T^E_{\min} \sim \ell_{\mathrm{Kol}}/u_o$. In our computational units, $u_o \approx 1.2$ and $L = 0.63$; the Taylor microscale $\lambda_T \approx 0.416$. Details were reported by Hunt, Buell & Wray (1987).

2.2 ***Kinematic Simulation***: This is a method for simulating a turbulent velocity field in space and time using only the information of the space-time Eulerian spectra. It is described in detail in Fung *et al.* 1990 and is just summarised here. Any velocity field $u_i(\mathbf{x}, t)$ can be represented as the sum of N_k modes with random amplitudes a_{n_i} and mode functions $\phi_{n_i}(\mathbf{x}, t)$ whose covariance $E_{n_{ij}}$ and functional form, respectively, are both defined by the spatial-temporal spectra. The amplitude variables and the mode functions are not directly correlated with each other. The simulated turbulence is a stationary, isotropic and homogeneous velocity field and it is assumed that the pdf is Gaussian. Then the mode function ϕ_{n_i} consists of sine and cosine functions for each wavenumber k_n and frequency ω_n so that the velocity field is

$$\mathbf{u}(\mathbf{x},t) = \sum_{n=1}^{n=N_k} \left[(\mathbf{a}_n \wedge \widehat{\boldsymbol{\kappa}}_n)\cos\{\boldsymbol{\kappa}_n.\mathbf{x} + \omega_n t\} + (\mathbf{b}_n \wedge \widehat{\boldsymbol{\kappa}}_n)\sin\{\boldsymbol{\kappa}_n.\mathbf{x} + \omega_n t\}\right]. \tag{1}$$

The essential features of the time dependence of the velocity field have been simulated in two ways:

(i) by allowing each spatial Fourier component to vary with time at a frequency $\omega_n = \lambda \varepsilon^{\frac{1}{3}} k_n^{\frac{2}{3}}$;

(ii) by dividing the velocity into a large scale velocity field $\mathbf{u}_\ell$ $(k < k_c)$ and a small velocity field $\mathbf{u}_s$ $(k_c < k)$ at some critical wavenumber k_c and allowing the large scale field to advect the small scale field following Tennekes (1975). The 'large scale' velocity field can be obtained from dynamical simulation or from an idealised representation of large eddies moving randomly without much change of form.

The Eulerian and Lagrangian frequency spectra, space structure function and pressure spectra obtained from KS all compare well with DNS and experimental data are not sensitive to the choice of k_c and λ.

KS has the same advantage as spectral methods over finite difference or finite element methods in representing the velocity field. But in a *dynamical* simulation it is necessary to include modes for every wavenumber in the representation where the number of points $N_p \propto k_\eta^3 \sim Re^{\frac{9}{4}}$, whereas in KS a fewer number of modes are needed, to represent all the major *kinematical* features of the flow field. It is found empirically that $N_k \propto k_\eta \sim Re^{\frac{3}{4}}$.

Even though this is a random superposition of Fourier modes, it can generate localised regions of high vorticity or strain; two waves of different frequency travelling in the same direction will 'beat' together to give high, local energy, whilst two orthogonal waves can form a vortical structure. In a straining region, this vortical structure will give helical motion and spiral streamlines (Fig. 1).

3. Defining the zonal structure of homogeneous, isotropic turbulence

3.1 *Zone Algorithms*: We use three quantities to classify the zones; the second invariant Π, the pressure p and the kinetic energy $|\mathbf{u}|^2$.

(i) The second invariant can be decomposed into strain rate s_{ij} and vorticity ω_k:

$$\Pi = \frac{\partial u_i}{\partial x_j}\frac{\partial u_j}{\partial x_i} = \frac{1}{4}\left\{\overbrace{\left(\frac{\partial u_i}{\partial x_j} + \frac{\partial u_j}{\partial x_i}\right)^2}^{\text{strain rate}} - \overbrace{\left(\frac{\partial u_i}{\partial x_j} - \frac{\partial u_j}{\partial x_i}\right)^2}^{\text{vorticity}}\right\} = s_{ij}^2 - \frac{1}{2}\omega_k^2,$$

(ii) Pressure: This can be obtained from Π using Poisson equation

$$\nabla^2 p/\rho = -\Pi.$$

Therefore Π is equivalent to specifying higher derivatives of the pressure p, and of course, a maximum in p occurs where Π is positive and minimum in p occurs where

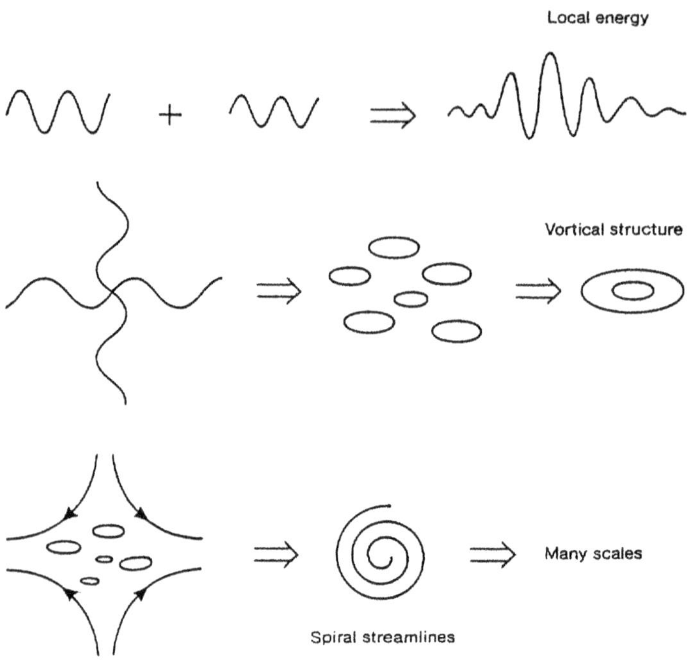

Figure 1. It shows a random superposition of Fourier modes can generate localised regions of high vorticity or strain.

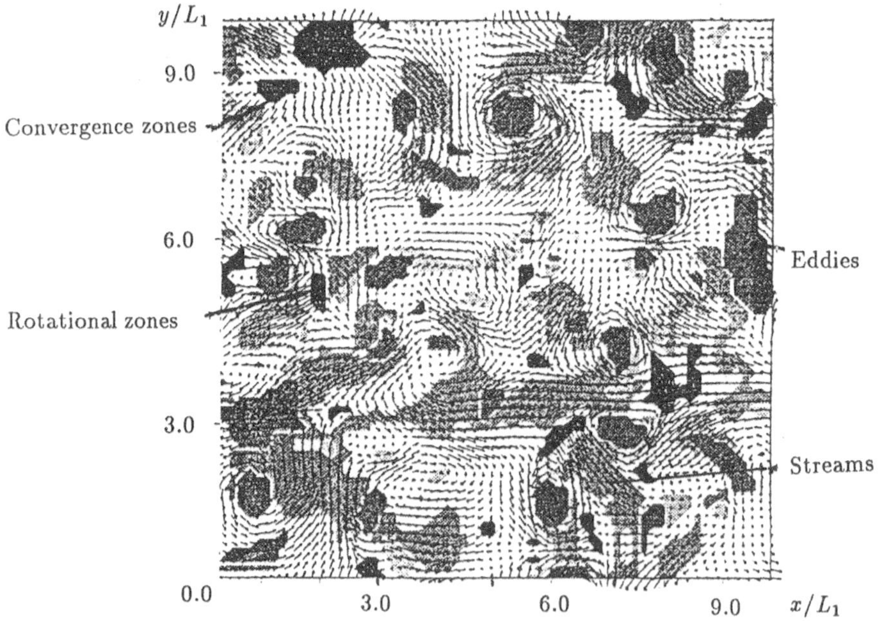

Figure 2. Zone classification in a typical plane obtained from DNS.

Π is negative. However, to ensure that the zones occupy a significant volume, a second criterion was used based on pressure which is determined by the average volume or Π over a certain regions around any point.

(iii) Kinetic energy – the unfiltered local velocity $\mathbf{u}$ is used to derive $|\mathbf{u}|$.

To examine the *nature* of the strain in different regions, we also evaluate the third invariant of the strain tensor, s_{ij}, defined as

$$\mathrm{III} = s_{ij}s_{jk}s_{ki}.$$

When $\mathrm{III} > 0$ a material sphere will elongate (becoming prolate or 'cigar-shaped') and when $\mathrm{III} < 0$ it will flatten (becoming oblate or 'pancake-shaped')

Following Wray & Hunt (1990), we identify four classes of flow structure, and distinguish them as follows:

Eddies (E)	**Shear** (SH)	**Streaming** (S)	**Convergence** (C)
$\Pi < -\frac{1}{2}\Pi_{\mathrm{rms}}$	$\Pi < -\frac{1}{2}\Pi_{\mathrm{rms}}$	$\lvert\Pi\rvert < \frac{1}{2}\Pi_{\mathrm{rms}}$	$\Pi > \Pi_{\mathrm{rms}}$
$p < -\frac{1}{2}p_{\mathrm{rms}}$	$-\frac{1}{2}p_{\mathrm{rms}} < p < p_{\mathrm{rms}}$	$\lvert u\rvert > u_{\mathrm{rms}}$	$p > p_{\mathrm{rms}}$

where the rms values are taken to be the *global* average in homogeneous, isotropic flows and the average over a plane of constant shear rate in shear flow. These criteria are mutually exclusive and in fact much of ($\sim 50\%$) the whole flow field is unclassified.

The kinematic properties of the zones were chosen so that in the eddy (E) zones, the flow is highly rotational, i.e. the vorticity is much larger than the strain rate, and so that these are regions of low enough pressure that the local streamlines are highly curved. In contrast the convergence (C) zones are regions in which the dominant straining motion is irrotational, and where, because the pressure is defined to be high there are likely to be converging and diverging streamlines around a local stagnation point (in frame moving with the local flow). Shear (SH) zones are highly vortical like E zones, but since the pressure are moderate, these zones cannot be at the centres of swirling regions. Streaming (S) zones are region where the flow is fast and the straining rate is small, which should define the high speed, rather straight, 'highways' or 'rivers' through any turbulent flow.

These criteria were defined so that the kinematic zones define large scale regions of flow in which kinematic properties are all of one kind or another; but the actual spatial extent of the zones and how they are related to each other spatially can only be determined by applying the criteria to a given velocity field.

A typical plane through a 3-D velocity field of isotropic turbulence obtained from our DNS is presented in figure 2. The zonal classification is indicated by grey shades as noted in the legend, and the arrows are velocity vectors in the plane. There is generally good agreement between the flow structure in the zones and the flow structure that was postulated in defining the zonal algorithm, in particular the swirling flow in E zones with SH zones outside them and stagnation flow in the C zones. It is necessary to examine the flow in 3-D or else the appropriate orientation of the plane is chosen in examining any one zone. The simulation shows that these kinematic zones, in which these are the same kinematic properties, are generally separated from each other, and that their typical dimension are about the same order as the length scale of the turbulence.

Using KS we have also simulated a flow field with approximately the same low Reynolds number spectrum as the DNS, where $E(k) \propto k^2 e^{-ak}$, and applied the same

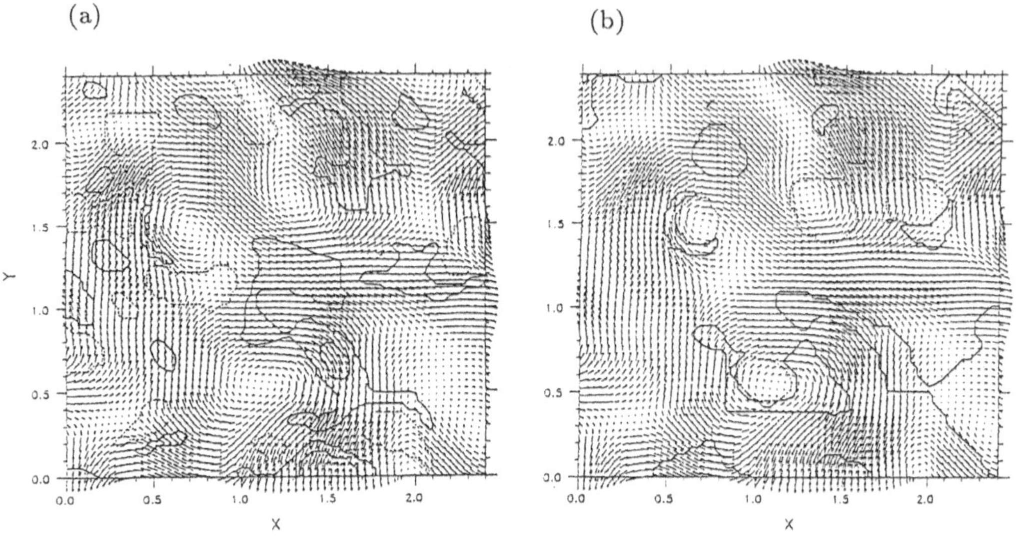

Figure 3. Zone classification in a typical plane obtained from KS with $E(k) \propto k^2 e^{-ak}$. (a) Dashed lines: convergence regions; solid lines: streaming regions. (b) Dashed lines: eddy regions; solid lines: shear regions.

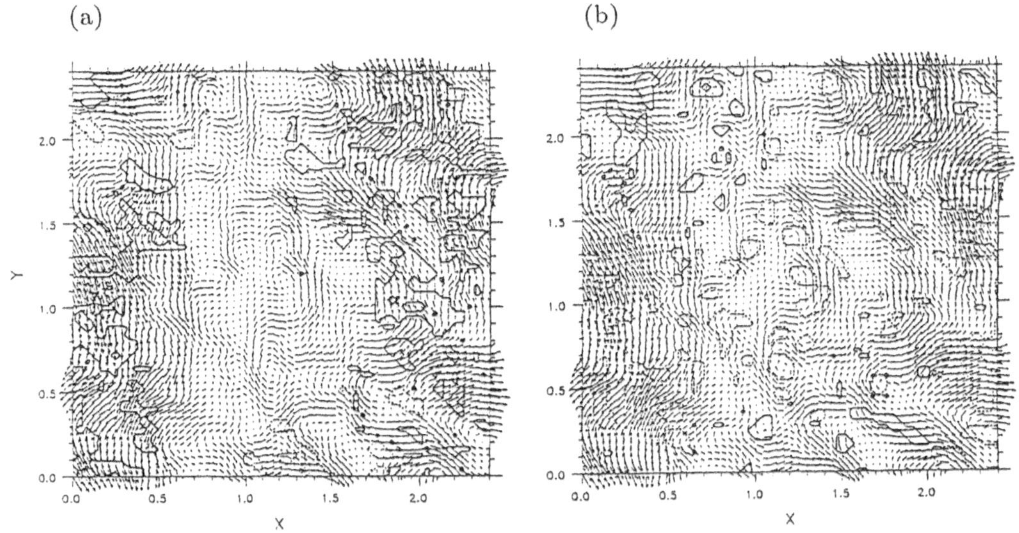

Figure 4. Zone classification in a typical plane obtained from KS with $E(k) \propto k^4(1 + L^2k^2)^{\frac{17}{6}}$. (a) Dashed lines: convergence regions; solid lines: streaming regions. (b) Dashed lines: eddy regions; solid lines: shear regions.

Table 1. Quantitative comparison between KS and (DNS).

Regions	Eddy		Shear		Streaming		Convergence		Total	
volume (%)	13	(13)	14	(6)	16	(25)	7	(4)	50	(48)
energy contributions (%)	13	(13)	12	(4)	26	(42)	7	(3)	58	(38)
enstrophy $\|\omega\|^2$(%)	28	(37)	24	(12)	13	(15)	4	(2)	69	(66)
strain rate squared (%)	9	(11)	10	(5)	13	(17)	15	(11)	47	(44)
energy contributions/volume	1.0	(1.0)	0.9	(0.7)	1.6	(1.6)	1.0	(0.8)	1.2	(0.8)
enstrophy/volume	2.1	(2.8)	1.7	(2.0)	0.8	(0.6)	0.5	(0.5)	1.4	(1.4)
strain rate squared/volume	0.7	(0.8)	0.7	(0.8)	0.8	(0.7)	2.0	(2.7)	0.9	(0.9)

zonal criteria (see Fig. 3a,b). It is found that the four zones have about the same length scales, flow patterns and spatial distributions as for the DNS. However, the shape of the E zones are not on average as elongated as the E zones in the DNS. Using KS, the velocity field with a typical high Reynolds number spectrum $E(k) \propto k^4/\left(1+L^2k^2\right)^{\frac{17}{6}}$ was also simulated and its flow zones evaluated. Figures 4a,b show that the flow zones are more 'spotty' compared with the low *Re* turbulence, i.e. they have a wider range of length scales and separation, and more varied relative spatial relationships between each of the zones.

3.2 *Dynamical properties*: By evaluating the time average statistics of various quantities of the flow field in different zones and comparing them with each other, new insight is provided into which kinds of eddy structure are most important for different dynamical and kinematical processes, e.g. production of kinematic energy, enstrophy, etc.

As shown in table 1, the streaming regions occupy the largest space (20% of the total for DNS and 16% for KS). E and SH zones together contribute most of the total enstrophy, and to the rate of enstrophy productions, while the C zone, though quite small in size dominate contributions to the total strain rate. In the E zones, III is generally positive (both in DNS and KS) which is consistent with local straining motions stretching the vorticity. In C zones, $\mathrm{III} < 0$ which indicates flattening of material surfaces and good local mixing.

It is interesting to note that although the contributions from the four zones to the total energy, enstrophy and strain rate squared are different in the two simulations, the corresponding 'densities' (energy/unit volume, enstrophy/unit volume, strain rate/unit volume) are approximately the same. This demonstrates that for many applications (chemical reactions, combustion processes, particle and bubble motions) Kinematic Simulation may be a 'cheap' way of simulating the important features of a turbulent velocity field. These results give a dynamical and kinematical explanation of the fundamental turbulence phenomenon of intermittency, or spatial inhomogenenity, of the rates of dissipation and enstrophy production.

4. Reliability of the algorithms for different flows – Shear flow

Are the same criteria applicable to different types of turbulent flows , such as those which are inhomogeneous and close to boundaries. So far the discussion has centred on homogeneous, isotropic turbulence but a general kinematical analysis can only be con-

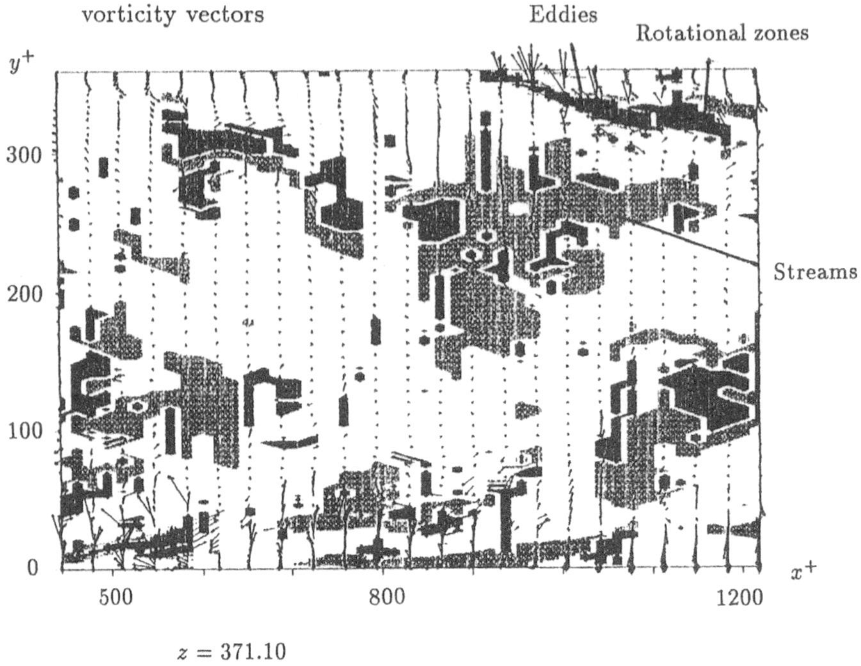

Figure 5. It shows the zones in an $x - y$ (streamwise-normal) plane of the turbulence channel flow.

sidered useful if it is consistent with the flow structure in a wide class of the flows. (Note that the algorithm was originally developed for homogeneous, isotropic turbulence.)

We have therefore applied the zonal algorithm to the highly inhomogeneous velocity field of a turbulent channel flow computed obtained using DNS by Kim, Moin & Moser (1987) at $Re_\theta = 287$. It is found that many of the well-known features of turbulence near a rigid wall and in the interior of the flow correspond to the zone types as defined here. For example, the near-wall 'streaks' are found to be streaming zones and near-wall streamwise vortices are classified as eddy zones; hairpin-like structures are seen to be a mixture of eddy and rotational zones.

Figure 5 shows the zones in an $x - y$ (streamwise-normal) plane of the turbulence channel flows. The arrows are projections of the vorticity onto this plane. There are two clear inclined vortices marked by E zones projecting out from the wall and leaning downstream, one on the lower wall ar x^+ $(= xu_*/\nu)$ between 500 and 600 and one on the upper wall at x^+ between 950 and 1200. These vortices slope upwards and are parallel but offset, in the z-direction, to the streaks. In many cases both these flow features (corresponding to E, S zones) emanate from the same complex region near the surface. In these complex regions there are different kinematic zones on a small scale and close to each other. This is quite consistent with the generation of small scale near the surface. (Note how the kinematic zonal analysis indicates spatial relationships between different zones, which is a particular advantage of the method, see also Stretch 1990)

The streaky structure of S zones above the wall is also found in computation of uniform shear flow either using DNS (Lee *et al.* 1987) or by applying Rapid Distortion Theory to initially isotropic velocity fields, whether defined by KS (Carruthers *et al.* 1990) or by DNS (Lee *et al.* 1987).

In the interior of the flow ($y^+ \approx 100$) the eddy zones indicate fluid volumes containing swirling flow moving away from the wall and interacting with the mean shear flow, such as analysed by Moin *et al.* 1986 and Hunt 1987. These may correspond to the mushroom shaped regions of smoke photographed by Head & Bandyopadhyay (1981).

5. Conclusions

(i) The kinematic zones analysed here are consistent with many but not all the eddy structures proposed by other investigators using different techniques.

(ii) The advantages of the method unlike previously proposed technique, is that it can be used straight forwardly to analyse different kinds of turbulent flow with the same assumptions in all cases. The flow zones are found to have a dynamical significance, and have given rise to new dynamical and kinematical insight.

(iii) A likely disadvantage of this general method is that it cannot extract the same detail about the eddy structures as methods specific to particular flows (e.g. VITA, pattern recognition).

(iv) The quantitative description of turbulence using flow zones shows that kinematic simulation can approximately (and 'cheaply') simulate the instantaneous flow structure in homogeneous and isotropic turbulence; but the zonal analysis also shows how certain key features (such as elongated vortex tubes) are not correctly simulated!

(v) In future, the time dependence of the flow zones needs to be investigated, which might provide a test for the non-linear models of these processes (such as those of Aubry *et al.* 1988).

References

Aubry, N., Holmes, P., Lumley, J.L. & Stone, E. 1988, *J. Fluid Mech.*, **192**, 115-173.

Adrian, R.J. & Moin, P. 1988, *J. Fluid Mech.*, **190**, 531-559.

Broadwell, J.R. & Breidenthal, R.E. 1982, *J. Fluid Mech.*, **125** 397-410.

Brown, D.G. & Hutchinson, P. 1979, *J. Fluid Eng.*, **101**, 265-269.

Carruthers D.J., Fung J.C.H., Hunt J.C.R. and R.J. Perkins (1990) Turbulence and coherent structures, Proc. Grenoble conference 1989. Eds.*Mesieur M. and Métais O.* Kluwer academic publishers.

Fung, J.C.H. & Perkins, R.J. 1989, Advances in Turbulence 2, pp. 322-328 (ed. H.-H. Fernholz & H.E. Fiedler).

Fung, J.C.H., Hunt, J.C.R., Malik, N.A. & Perkins, R.J. 1990, submitted to *J. Fluid Mech.*.

Head, M.R. & Bandyopadhyay, P. 1981, *J. Fluid Mech.*, **107**, 297-338.

Hunt, J.C.R., Buell, J.C. & Wray, A.A. 1987, Proceedings of the 1987 Summer Program, Stanford University.

Hunt J.C.R. 1987 *Trans. Can. Soc. Mech. Eng.*, **11**, 21-35.

Hunt, J.C.R., Aution, T.R., Sene, K., Thomas, N.H. & Kowe, R. 1988, *Proc. Conf. Transient Phenomena in Multiphase Flow*, Hemisphere.

Hunt, J.C.R., Fung, J.C.H., Malik, N.A., Perkins, R.J., Vassilicos, J.C., Bertoglio, J.P., Wray, A.A., Buell, J.C. & Rogallo, R. 1990 *Proc. 3rd European Turbl. Conference*, Springer-Verlag, Stockholm.

Hussain, A.K.M.F. 1986, *J. Fluid Mech.*, **173**, 303-356.

Kim, J., Moin, P. & Moser, R.D. 1987 *J. Fluid Mech.*, **177**, 133-166.

Lee, M.J., Kim, J. & Moin, P. 1987 In: *Sixth Symposium on Turbulence Shear Flows*, Toulouse, France (ed. F. Durst *et al.*).
Leonard, A.D. & Hill, J.C. 1988 *Proc. CTR Summer Program.*
Maxey, M.R. 1987, *J. Fluid Mech.*, **174**, 441-465.
Moin, P, Leonard, A.D. & Kim, J. 1986 *Phys. Fluid*, **29**, 955-963.
Mumford, J.C. 1983, *J. Fluid Mech.*, **137**, 447-456.
Perry, A.E. & Chong, M.S. 1987, *Ann. Rev. Fluid Mech.*, 125-155.
Peters, N. & Williams, F.A. 1988, *22nd Symposium on Combustion*, Seattle.
Rogallo, R.S. 1981, Numerical experiments in homogeneous turbulence, NASA TM81315,
Stretch, D. 1990 Annual Research Briefs -1989, Stanford University.
Tennekes, H. 1975, *J. Fluid Mech.*, **67**, 561-567.
Vassilicos, J.C. 1989, Advances in Turbulence 2, pp.404-411 (ed. H.-H. Fernholz & H.E. Fiedler).
Wray, A.A & Hunt, J.C.R. 1990, *ProcIUTAM, symposium on 'Topological Fluid Mechanics'*. Cambridge, CUP.

Modelling the Flow in Droplet-Driven Sprays

S. Ghosh, J.C.Phillips and R.J.Perkins
Department of Applied Mathematics and Theoretical Physics
University of Cambridge
England

This paper describes an investigation into some of the fundamental aspects of the motion of droplets, and the induced air velocities, in vertical droplet-driven spray jets using standard integral and differential methods of tubulence processes. We first present a robust 1-dimensional model for the average air velocity within a spray, the only empirical input being the droplet radius. We then extend the 1-d model to a detailed axi-symmetric model that predicts the radial variation of the air velocity profile. We also describe experiments to examine some of the important mechanisms and also to verify the theoretical results. We find that a droplet-driven spray jet is different from a regular turbulent jet - in such jets we only find small scales of turbulence in contrast to regular jets which are characterised by large turbulent structures.

1. Introduction

Spray-driven jets (when liquid is ejected from a nozzle at high speed) are used in many processes as a way of dispersing liquid droplets; typical examples include the spray deposition of molten metals and the agricultural spraying of fertilisers and pestcides. In order to exploit these processes fully, we need a better understanding of the physics of droplet-driven spray jets. In the spray deposition process, for example, we need to understand and model the mixing of droplets, so that new alloys can be created; in agricultural spraying we need to be able to predict and control the drift of spray droplets caused by the natural wind and by disturbed flows such as the wake of the spraying vehicle.

Initially in a spray jet a liquid sheet forms, but instabilities cause the break-up of this sheet just downstream of the nozzle, and spray droplets are formed. Although there has been much work on the formation and break-up of liquid sheets (Dombrowski and Johns (1963)), there has been much less research on the formation of the spray jet once the droplets have formed, and this is the subject of our study. It is believed that as the droplets travel downstream they entrain the ambient gas and set up an overall induced gas flow field. This results in the creation of a turbulent air jet within the spray. It is important to study the behaviour of this air jet because the induced air motion modifies the trajectories of the small droplets. The analysis here may also be applicable to water flows induced by bubble columns.

2. Interaction of droplets and induced air

In order to examine the origin of the induced air velocity, we first consider the motion of a single droplet through air. As a droplet moves forward, it carries with it some of

Advances in Turbulence 3
Editors: A.V. Johansson · P.H. Alfredsson

the surrounding air, and displaces a certain volume of air (Kowe et al. (1988); Darwin (1953)). Thus in an ideal flow without viscosity the spray droplets would induce air motion dependent on the number of droplets and their speed. Real droplets experience a drag force which slows them down and this causes a reaction on the air flow in the direction of the droplet movement and thence drives the air motion along the jet. When there is a population of drops, each of them induces it's own air flow with an associated profile. As a result, each of these individual velocity profiles interact with each other to give rise to a resultant air velocity profile with the greatest contribution at the centre because effectively the wakes add together. Both the individual and the total air velocity profile have points of inflexion and this promotes transition to turbulence on a local droplet scale as well as on the scale of the jet. The turbulence sets into motion the surrounding air, which can only happen if additional air is entrained in from the edges. As the droplets move forward and are acted upon by drag, they lose momentum and slow down.

3. Model formulation

3.1 1-Dimensional Analysis over different Zones of a Spray Jet

The aim here is to find how the air jet width l_a and the cross-section average velocity $\langle V_3 \rangle$ varies as a function of downstream distance. It is convenient to study these variations over 3 different zones (Figure 1), which we classify below :

Zone I

This is the region immediately below the nozzle, where the air jet is actually *driven* by the droplets. It can be assumed that in the initial development of the jet, in Zone I, the droplet velocity V_l is constant, and thereafter it decays exponentially until it attains a terminal velocity. Observations from sprays show that the droplets are ejected within a cone of radius $l = c(Z + Z_0)$ where c is the spray angle.

Let Q_l be the liquid discharge from the nozzle, then at a downstream cross-section with area $A \left(= \pi \ell^2\right)$ and radius ℓ,

$$Q_l = A\langle \alpha \rangle V_l = \pi \ell^2 \langle \alpha \rangle V_l \tag{1}$$

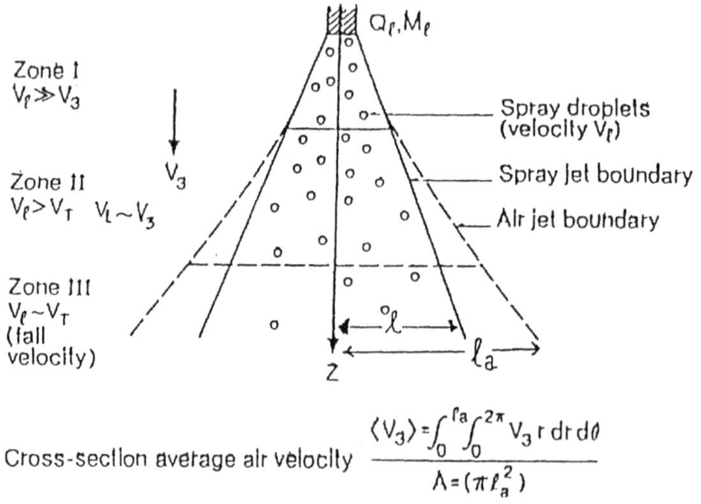

Figure 1. Definition sketch of 1-d model

where $\langle\alpha\rangle$ is the average volume fraction of the droplets and can be expressed as

$$\langle\alpha\rangle = L^2/(Z+Z_0)^2$$

where

$$L^2 \equiv Q_l/\pi c^2 V_l \qquad (2)$$

and L is a length scale proportional to the initial diameter of the spray. Similarly if we define a characteristic air jet velocity in the spray

$$V_s = \sqrt{Q_l{}^2/aL^3} \qquad (3)$$

where a is the spray droplet radius, then the average forcing/unit volume on the air by the droplets can be expressed as

$$\langle F_z\rangle/\rho_a = 3V_s^2 LC_D/\left((Z+Z_0)^2\, 8\pi^2 c^4\right) \qquad (4)$$

Having estimated the forcing term, we can now calculate the rate of change of average momentum flux of the air driven by the droplets :

$$d\left(\pi l_a^2 \langle V_3\rangle^2\right)/dz = \langle F_z/\rho_a\rangle \pi l^2 = 3C_D Q_L{}^2 \pi \ell^2/\left(8a\pi^2 c^4 L^2 (Z+Z_0)^2\right) \qquad (5)$$

The entrainment equation is given by :

$$d\left(\pi l_a^2 \langle V_3\rangle\right)/dz = 2\pi l_a \left(\beta\langle V_3\rangle\right) \qquad (6)$$

where β is the entrainment coefficient.

Boundary conditions

At the virtual origin of the jet where the width is zero and there is no net flux of air
$Z = -Z_0 \qquad l_a = 0\ ; \quad \langle V_3\rangle l_a{}^2 = 0$

Solutions

Solving (5) and (6) using the above boundary conditions we obtain

$$\langle V_3\rangle = V_s\sqrt{L/(Z+Z_0)} \qquad (7)$$

and

$$l_a = (4/3)\beta(Z+Z_0) \qquad (8)$$

(8) is true for a narrow spray where the air jet width is controlled by entrainment. However, for a wide spray the air jet width is controlled by the spray angle and we have $l_a = c(Z+Z_0)$

From (7) we find that within Zone I, the air jet velocity grows more slowly as $z^{-1/2}$ rather than z^{-1} as in an ordinary air jet.

We now define the droplet stopping distance z^* as

$$z^* = a\rho_l/\rho_a \qquad (9)$$

which is about 0.1 m for a typical droplet radius of 100μ . From experimental analysis including photographic observation of spray jets entraining air, it was observed that, up to a distance of about $3z^*$, the air velocity decayed as $z^{-1/2}$ rather than z^{-1} as in an ordinary jet, and therefore we can say that zone I extends up to a distance of $3z^*$.

Zone II
In this zone the droplet velocity is of the same order as the air velocity and thus there is no forcing on the air. The rate of change of average momentum flux is now given by :

$$d\left(\pi l_a^2 \langle V_3\rangle^2\right)/dz = 0 \qquad (10)$$

The entrainment equation is the same as in zone I (equation 6). The momentum flux of the droplets is also small compared to that of the air jet.
Boundary conditions
l_a and V_3 match at $Z = Z_I$ ($3z^*$). This also implies that dQ_A/dz is also continuous at $Z = Z_I$, where $Q_A = {l_a}^2 V_3$.
Solution
Using the above boundary conditions as before we obtain the following solution :
$\langle V_3\rangle l_a = C_1$
$\langle V_3\rangle {l_a}^2 = Q_A = 2\beta C_1 Z + C_2$
(Q_A is linear in Zone II unlike in Zone I where Q_A grows as $Z^{3/2}$.)
This gives,

$$V_3 = C_1/\left(2\beta Z + (C_2/C_1)\right) \qquad (11)$$

From (11) we find that in zone II the air jet behaves like an ordinary jet.
Zone III
Finally when the droplet velocity equals the terminal fall velocity, we enter the final zone where there is no further reduction of the droplet velocity; the rate of change of average momentum flux is now constant with downstream distance. Using the analysis as in the previous zones we find that within zone III, the average air velocity again varies as $z^{-1/2}$. Since the fall velocity of the spray droplets of 100μ radius is about $0.7ms^{-1}$, we estimate that the droplet velocity equals this value at a downstream distance of about 1.3m.

4. 2-Dimensional Axi-symmetric Model

In order to be able to predict the radial variation of the air velocity profile and also to confirm the predictions from the 1-d analysis, we developed a 2-d numerical model, where we obtain the velocity profile in the form of a similarity solution as we show below.
From the Reynolds' averaged equations of motion for steady axi-symmetric flow with external forcing F_z for the air flow we have :

$$V_1\frac{\partial V_3}{\partial r} + V_3\frac{\partial V_3}{\partial z} = \frac{-1}{\rho_a}\frac{\partial p}{\partial z} - \left(\frac{\partial \overline{v_r' v_z'}}{\partial r} + \frac{\partial \overline{v_z'^2}}{\partial z} + \frac{\overline{v_r' v_z'}}{r}\right) + F_z/\rho_a \qquad (12)$$

where terms have their usual meanings and v_r', v_z' are velocity fluctuations in the r and z directions respectively. From the usual thin shear layer appproximation, gradients in the radial direction are much larger than in the axial direction and the external pressure gradient is zero. Then with the Oseen approximation and Boussinesq's eddy viscosity hypothesis, we reduce our governing equation to the following simplified form :

$$V_3\frac{\partial V_3}{\partial z} = \nu\left(\frac{\partial^2 V_3}{\partial r^2} + \frac{1}{r}\frac{\partial V_3}{\partial r}\right) + F_z/\rho_a, \qquad (13)$$

ν being the eddy viscosity.

Boundary conditions
The appropriate boundary conditions on V_3 are that :

$$V_3 \rightarrow 0 \ as \ z \rightarrow \infty$$
$$V_3 \rightarrow 0 \ as \ r \rightarrow \infty \tag{14}$$

The governing equation was then transformed to obtain the solution in the similarity form using the following transformation:
$f(\eta) = V_3/V_0$,
where V_0 is the centre-line velocity and $\eta = r/l$ the similarity variable.

4.1 2-d Model Results

The governing second-order non-linear differential equation was solved numerically using a Runge-Kutta algorithm and the radial variation of the air-velocity profile within an axisymmetric spray was obtained. In figures 2 and 3 these profiles are shown for two cases corresponding to :
c = 0.03 for a narrow spray angle (i.e. less than the angle of a tubulent jet)
c = 0.2 for a wide spray angle (i.e. greater than the angle of a turbulent jet)
From these figures we find that with a given distribution function for the radial droplet spread (indicated in the figures), the shape of the profiles depend on the spray angle. For a narrow spray the profile spreads out wider than the spray whereas for a wide spray it follows the droplet distribution profile, because the spreading by turbulent eddies is no greater than those of the droplets.
The 2-d model was run to confirm the above results of the 1-d model, and by effectively calculating the entrainment coefficient a value of 0.11 was obtained which is in excellent agreement with the value of 0.1 for simple jets.

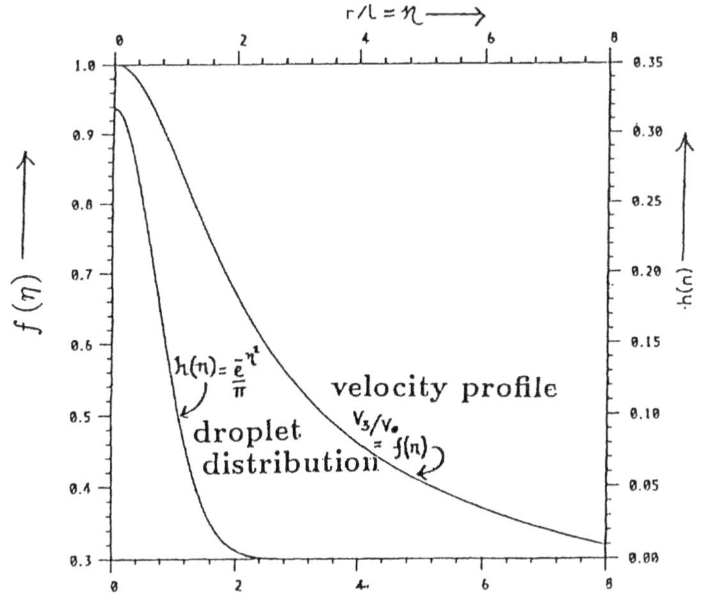

Figure 2. Air velocity profile and droplet distribution for narrow spray

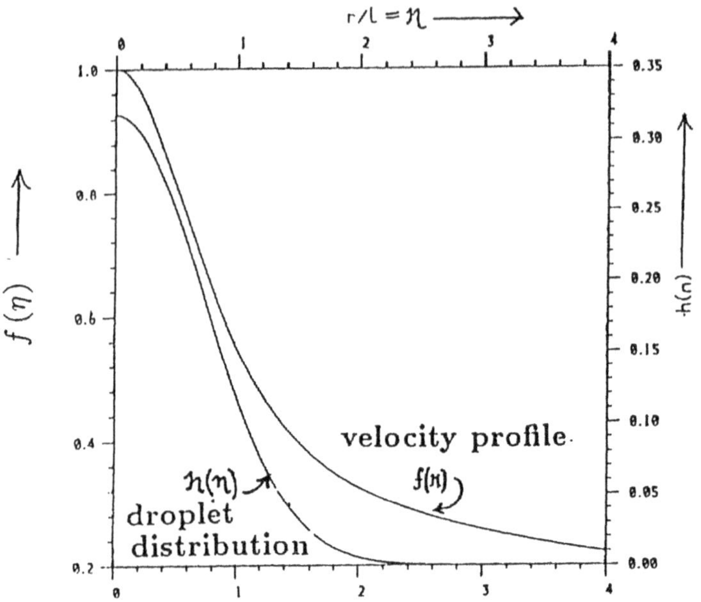

Figure 3. Air velocity and droplet distribution for wide spray

5. Experiments and Model Results

5.1 Experimental methods

The experimental measurements of the air velocity within a *flat-fan* spray presented as figure 5 have been obtained using phase-doppler anemometry at A.F.R.C. Engineering Silsoe in collaboration with Dr. Miller (Miller et al. (1990)). This technique shares the same basic principle as laser-doppler anemometry, but in this case we measure both the phase and frequency of the doppler shift in the scattered light. Thus, for a droplet traversing the measuring volume, both size and velocity may be determined.

Previous numerical models of a spray jet (Lee and Tankin (1984)) and experimental observations (Miller et al. (1990)) suggest that the smallest droplets within a spray jet have low enough inertia to move with entrained air velocities. Thus small droplets may be used as tracers for the air velocity profiles within a spray. Using phase-doppler anemometry in forward scatter, Eulerian measurements of the velocity of 50μ (collection range 45μ - 55μ) spray droplets have been made on the centreline of a flat-fan spray. Smaller droplet ranges are not suitable as seeding because of the low signal-to-noise ratios obtained. Each velocity measurement represents the mean of at least 10,000 data points.

In order to study the scales of turbulence in the spray jet, smoke was introduced on one side of the jet (Figure 4). The smoke did not rapidly spread across the jet, and on the edge of the jet only small eddy motion was seen, in contrast to a regular jet with large scale structures. This is because of the fact that the air jet within the spray is a *forced* jet. The relative motion of the droplets through the air can cause intense small scale turbulence on the scale of the droplets; also the presence of droplets can act as a resistance to large scale turbulent motions. However both effects are likely to *reduce* the

Figure 4. Photograph of spray jet entraining smoke (Courtesy AFRC Engineering)

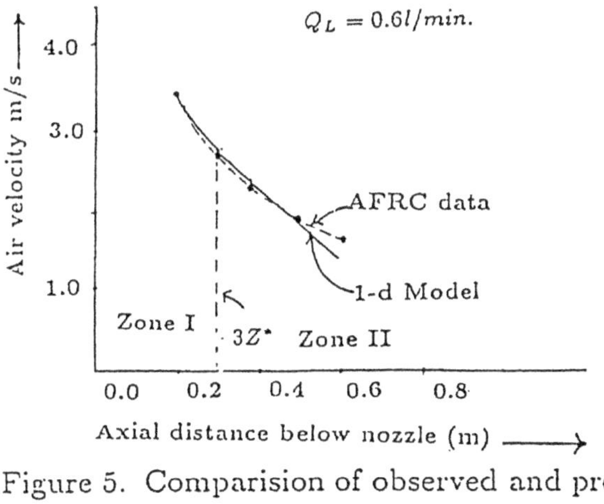

Figure 5. Comparision of observed and predicted centre-line air velocity

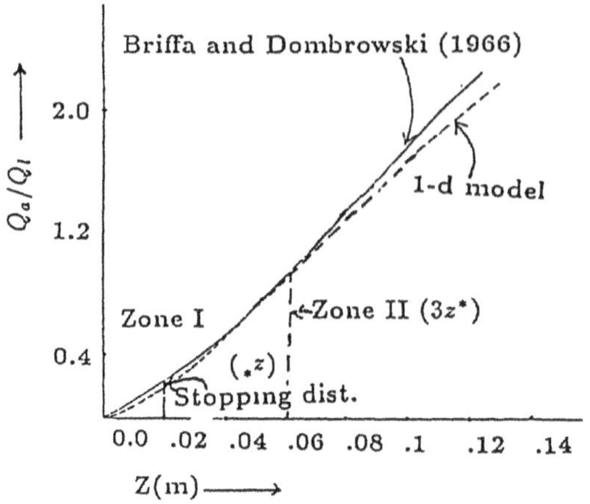

Figure 6. Comparision of observed and predicted entrained air masses

scale of turbulence (by analogy, putting a grid into a turbulent pipe flow also reduces the length scale and diffusion (Hinze (1975))).

5.2 Comparison of Model Results with Experiments

We used the 1-d model to compare the theoretical reslts with available experimental data and also with our own measurements. With the internal measurement of the mean droplet radius the model was used to evaluate the entrained air velocity along the centre-line of a flat-fan spray. The results compared very well with the experimental data (A.F.R.C. Engineering) and are shown in figure 5. We also compared our theoretical results with those of Briffa and Dombrowski (1966) for the variation of the mass of entrained air with downstream distance for a flat-fan spray (Figure 6) where again we find good agreement. Experimental work is in progress to obtain the radial variation of the air velocity profile within agricultural sprays.

6. Conclusions

This paper addresses fundamental aspects of the motion of droplets and air motion in droplet-driven sprays. A 1-d model was developed for the air jet that is created within the spray jet due to the movement of the droplets through air. The only internal data supplied was the mean droplet radius and the spray angle. The model was applied over 3 distinct zones of the jet :

(i) Zone I - the 'near' region where the air jet is *forced* and extends upto a distance of about thrice the droplet stopping distance

(ii) Zone II - the intermediate region where the air jet is no longer forced and is closest to that of a regular jet

(iii) Zone III - the 'far' region where the drops attain terminal velocity.

We find that though this air jet shows some charcteristic traits of a regular jet like entrainment and turbulence, in other ways it is different being a *forced* jet. For instance, the average air velocity in zones I and III decay as $z^{-1/2}$ rather than z^{-1} as in a regular jet. The most striking difference however is the presence of only small scale

structures. Nevertheless, the numerical model assumed the same turbulence shear stress parameterisation as a regular jet and produced consistent results.
Predictions from the 1-d model agreed very well with experimental results. Work is in progress to extend the present model to spray jets under the influence of a cross-flowing atmospheric wind.

Acknowledgement
We are grateful to Dr. J.C.R. Hunt for directing this research and to Dr. P.C.H Miller for providing us with experimental data.

References

Dombrowski, N. & Johns, W.R. 1963 The aerodynamic instability and disintegration of viscous sheets. *Chem. Engng. Sci.* **18**, 203

Kowe, R., Hunt, J.C.R., Hunt, A., Couet, B. & Bradbury, L.J.S. 1988 The effects of bubbles on the volume fluxes and pressure gradients in unsteady and non-uniform flow of liquids. *Int. J. Multiphase Flows* **14**, 5, 587.

Darwin, C. 1953 Note on hydrodynamics. *Proc. Camb. Phil. Soc.* **49**, 342.

Lee, S.Y. & Tankin, R.S. 1984 Study of liquid spray (water) in a non-condensable environment (air). *Int. J. Heat and Mass Transfer* **27**, 351.

Miller, P.C.H., Tuck, C.R. & Weston, N. 1990 In preparation.

Briffa, F.F.J. & Dombrowski, N. 1966 Entrainment of air into a liquid spray . *A. I. Ch. E. Jl* ,708.

Hinze, J.O., 1975 Turbulence. McGraw-Hill Book Company.

Anisotropic Dissipation Rate - Implications for Reynolds Stress Models

M. Hallbäck, J. Groth and A.V. Johansson

Department of Mechanics, Royal Institute of Technology, S-100 44 Stockholm, Sweden

Improvement of the modelling of the intercomponent energy transfer (pressure-strain rate) terms is of central importance for the development of Reynolds stress closures of turbulence. However, also the degree of anisotropy of the dissipation rate tensor has a significant influence on the evolution of the Reynolds stresses whenever dissipative effects are important. Here, homogeneous anisotropic turbulence was studied in order to elucidate the effects of anisotropic dissipation in strained and relaxing flows. Experiments were carried out where the streamwise dissipation rate component was measured in and downstream of a contraction, and the results indicate an anisotropic dissipation rate tensor. It was found through comparisons with direct numerical simulation data that the dissipation rate anisotropy could essentially be accounted for by an algebraic model that relates the anisotropy of the dissipation rate tensor to the anisotropy of the Reynolds stress tensor. The information provided by the model can in turn be used in a modified modelling of the total dissipation rate transport equation.

1. Introduction

Many flow situations exhibit a high degree of anisotropy in the Reynolds stresses. This is the case in, e.g., the near-wall region of turbulent boundary layers and in turbulence that is or has been subjected to rotation or to a mean strain field in a contraction or a shear flow. The anisotropy of the Reynolds stresses is dominated by the energy distribution on the large scales, and leads to an energy transfer between the velocity components. The small scale energy distribution has normally been assumed to be essentially unaffected by that on large scales, at least for high turbulent Reynolds numbers ($Re_\Lambda \equiv 4k^2/\nu\varepsilon$, k is the turbulent kinetic energy and ε is the total dissipation rate), and the dissipation rate tensor is consequently usually taken to be approximately isotropic. Several recent direct numerical simulations (see, e.g., Lee & Reynolds 1985 or Gerz, Schumann & Elgobashi 1989), indicate that the small scale anisotropy may be significant, particularly in flows with large mean strain rates. However, these simulations were conducted at rather low Re_Λ (≈ 50). Studies of the non-local triadic interactions, due to the nonlinear terms in the Fourier transformed Navier-Stokes equations, show that there is a strong coupling between the distribution of energy among the high wave numbers and the energy in the low wave number region (Brasseur & Corrsin 1987). This suggests that a significant degree of anisotropy in the dissipative scales can evolve due to large scale anisotropies also at high Reynolds numbers, despite the large separation of scales.

The Reynolds stress transport equations include terms describing the intercomponent energy transfer and the different dissipation rates, and hence models that are

Advances in Turbulence 3
Editors: A.V. Johansson · P.H. Alfredsson

based on these equations have the basic prerequisites necessary to address the problems associated with the modelling of anisotropic turbulence. From the transport equation for the Reynolds stress tensor a transport equation for the corresponding anisotropy tensor

$$a_{ij} \equiv \frac{\overline{u_i u_j}}{k} - \frac{2}{3}\delta_{ij} \tag{1}$$

where $\overline{u_i u_j}$ is the Reynolds stresses, can be derived. In homogeneous turbulence this equation reads

$$\frac{Da_{ij}}{Dt} = P^a_{ij} + \frac{1}{k}\{\varepsilon\,(a_{ij} - e_{ij}) + \Pi_{ij}\} \tag{2}$$

where P^a_{ij} is a mean velocity gradient dependent production term, e_{ij} is an anisotropy tensor defined in the same manner as a_{ij} but from the homogeneous dissipation rate tensor $\varepsilon_{ij} \equiv 2\nu\,\overline{u_{i,k}u_{j,k}}$ and Π_{ij} is the pressure-strain rate correlation. It is clear that if dissipative effects are important for the evolution of a_{ij}, so is also the degree of anisotropy, e_{ij}, of the dissipation rate tensor. If, as in most RST-models used today, the dissipation rate is assumed to be isotropic, the effects of an anisotropic dissipation rate can be regarded as included in the modelling of the pressure-strain rate terms (see, e.g., Lumley 1978). However, in order to improve the applicability of RST-models to a wide class of flow situations it is important to distinguish between effects of the physically different phenomena of anisotropic dissipation rate and intercomponent energy transfer (e.g., εe_{ij} and Π_{ij} behave quite differently in the vicinity of a wall). Explicit modelling of anisotropic dissipation rate has been tried in very simple forms, in which the anisotropy of ε_{ij} is assumed to be proportional to the corresponding anisotropy of the Reynolds stress tensor (Hanjalić & Launder 1976).

2. Experimental indications of dissipation rate anisotropy

There exists a number of experimental indications of anisotropy in the dissipative scales. Measurements by Uberoi (1956,1957) in axisymmetric nozzles, and in a boundary layer showed a significant departure from the isotropic relations for some of the spatial derivatives of the fluctuating velocities. Similar results were obtained by Antonia, Shah & Browne (1987) in the selfpreserving wake of a circular cylinder and by Hussein & George (1989) in an axisymmetric turbulent jet.

Measurements of the Reynolds stress tensor and the dissipation rate of the streamwise velocity component have been carried out by Hallbäck, Groth and Johansson (1989) in a small wind tunnel. In these experiments nearly isotropic grid turbulence ($Re_\Lambda \approx 1000$) was subjected to the mean strain field of an axisymmetric contraction and then allowed to relax towards isotropy in a straight duct with a constant mean velocity. Downstream of the contraction the turbulence was highly anisotropic with an initial value of a_{11} of about -0.45. The streamwise dissipation rate component was measured with a probe consisting of two parallel hot-wires (see figure 1). With this probe the necessary instantaneous cross-stream and streamwise (via Taylor hypothesis) derivatives could be determined. In figure 2 it is clearly seen that ε_{11} remains nearly constant in the straight duct, whereas it from measurements of the turbulent kinetic energy was deduced that the total dissipation rate decays as the downstream distance raised to a

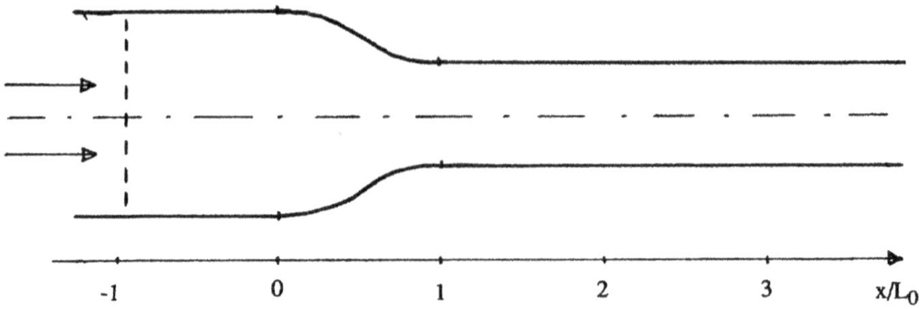

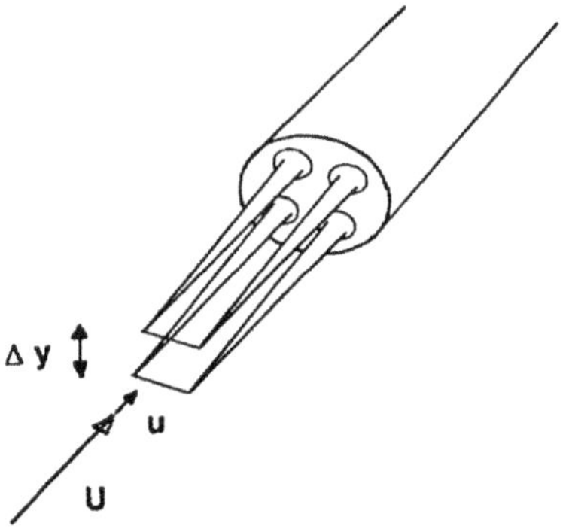

Figure 1. A two-sensor probe was used for measuring e_{11}. The wire lengths and the cross-stream separation were about 1.5 Kolmogorov scales. At the grid the mean velocity was about 4 m/s. The turbulence level was less than 5% in the contraction, making Taylor's hypothesis applicable. The contraction length (L_0) was 45 cm and the area ratio was 4.

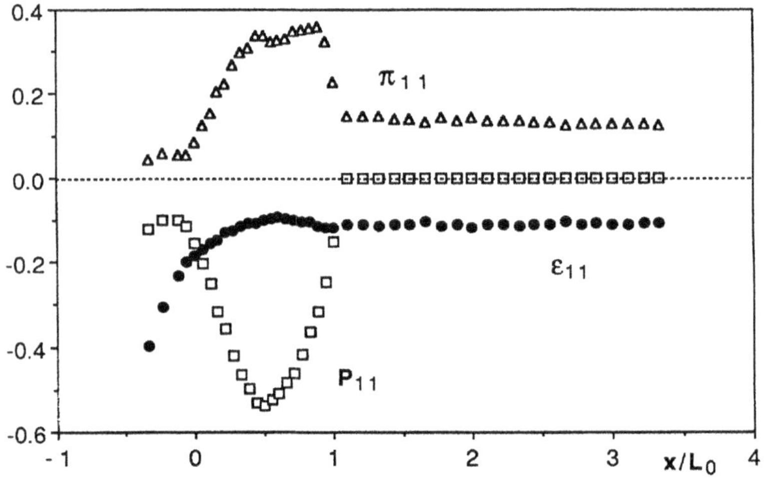

Figure 2. The terms of the streamwise Reynolds stress transport equation as determined from experiments. In this case the diffusion terms were negligible. The ε_{11} component is seen to be nearly constant in the straight duct ($x/L_0 > 1$) downstream the contraction ($0 < x/L_0 < 1$).

power of about -2. This means that ε decreases with a factor of about 3 in the straight duct, which in turn implies that during the relaxation phase e_{11} varies correspondingly, from an initially negative value towards zero. In figure 2 the streamwise pressure-strain rate term Π_{11} is also shown. Since ε_{11} and the downstream development of the streamwise Reynolds stress component were both measured the pressure-strain rate term could be calculated from the RST-equations under the approximation of negligible diffusion (this approximation was in part confirmed by measurements).

The experimental indications of strong anisotropy in the dissipation rate tensor clearly motivates the search for a model that may accurately take this anisotropy into account. The most straightforward way within the transport modelling concept would be to use the transport equations for the dissipation rate tensor. This however is quite a complicated approach since these equations contain many new terms that need modelling. Here another method is used in which an algebraic relation for e_{ij} is derived through considerations of the mathematical structure of (2) and the physical constraints that should be fulfilled by any dissipation rate model.

3. Dissipation rate anisotropy model

From (2) it is seen that e_{ij} (and Π_{ij}) can be written as functionals of a_{ij}, k, ε, the mean velocity derivatives $U_{i,j}$ and ν (included for generality). The definition of e_{ij} suggests that the mean velocity gradients enters the functional relationship for e_{ij} only as invariants of $U_{i,j}$. Assuming that the memory time of the turbulence is sufficiently short the functionals can be expanded in time (Lumley & Newman 1977) and, as a first approximation, reduced to functions of the present state. To proceed from this towards a tensorially correct expression for e_{ij} one might construct a general series expansion of e_{ij} in 'powers' of a_{kl}. This implies an implicit assumption that the Reynolds stress anisotropies have a primary influence on the e_{ij}:s whereas the other dependencies are assumed to be secondary. Such an expansion that satisfies the symmetry condition, $e_{ij} = e_{ji}$, and the zero trace condition, $e_{ii} = 0$, can be written (see also Hallbäck et al. 1989, 1990)

$$e_{ij} = \{c_1 + c_3\frac{1}{2}II_a + c_4\frac{1}{3}III_a + c_5\frac{1}{4}II_a^2 + ...\}\ a_{ij} + \tag{3}$$

$$+\ \{c_2 + c_4\frac{1}{2}II_a + c_5\frac{1}{3}III_a + ...\}\ (a_{ik}a_{kj} - \frac{1}{3}II_a\delta_{ij})$$

Here the Cayley-Hamilton theorem has been used to express higher order terms of the tensor products and their traces in terms of lower order quantities. In this expression both the dependence on the tensor a_{kl} and its invariants II_a ($= a_{ik}a_{ki}$) and III_a ($= a_{ik}a_{km}a_{mi}$) is explicitly accounted for. The model parameters c_k may depend on Re_Λ and suitably defined mean strain rate parameters.

To ensure that the model does not give rise to unphysical results (such as negative energies) we may impose a realizability condition on expression (3). A condition of this kind states that the right hand side of (2) must be greater than or equal to zero when $a_{\alpha\alpha}$ is equal to -2/3 (since $a_{\alpha\alpha} \geq -2/3$ always). (No summation over Greek indices.)

This condition implies, for the dissipation rate term, that

$$e_{\alpha\alpha} = a_{\alpha\alpha} \quad \text{if} \quad a_{\alpha\alpha} = -2/3 \tag{4}$$

In homogeneous turbulence this condition can be shown to follow strictly from the definition of the dissipation rate term. It also holds for turbulence at a solid wall. It simply states that if there is no energy in a velocity component there can be no dissipation either.

To reduce the number of unknown constants expression (3) has to be truncated. At least third order terms must be retained in order to satisfy (4) and still have a possibility to adjust the remaining model parameters. From this it follows that the simplest nonlinear relation between e_{ij} and a_{kl} that fulfills the symmetry, trace and realizability conditions can be written as

$$e_{ij} = \{1 + \alpha(\frac{1}{2}II_a - \frac{2}{3})\}\ a_{ij} - \alpha\ (a_{ik}a_{kj} - \frac{1}{3}II_a\delta_{ij}) \tag{5}$$

The model parameter α could in principle be determined from experiments. Another method would be to require that the model should correctly predict the initial behaviour of e_{ij} in a fundamental but simple flow situation such as isotropic turbulence subjected to a sudden rapid strain. In this case the relation between a_{ij} and e_{ij} can be determined using rapid distortion theory (see, e.g., Batchelor 1953). For small total strains the result, which holds for an arbitrary strain rate tensor, is that $e_{ij} = 1/2\ a_{ij}$. For the model to give this behaviour for small strains (and thereby small anisotropies) α must assume the value of 3/4. Thus the model parameter has been set to a specific value, which is not based on calibration against experimental or numerical results.

The model has been tested against data from direct numerical simulations of various homogeneous flow cases, such as turbulence subjected to irrotational strains or homogeneous shear, and relaxation from anisotropic states. In all cases the value of α was set to 3/4. The model was not complemented with models for the other terms of the RST-equations to form a complete closure, but was directly compared to data on

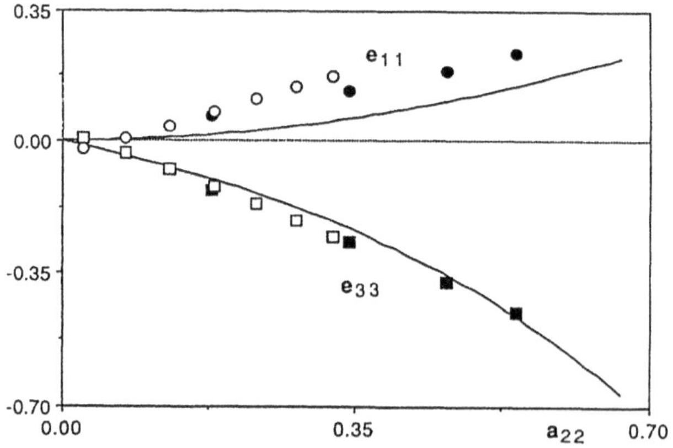

Figure 3. Model predictions (solid curves) of e_{11} and e_{33} vs a_{22} in a plane strain with $a_{11} \approx 0$, compared with direct simulation data of Lee & Reynolds (1985) (●, ■) and Larsson (Dept. of Mech., KTH, Stockholm) (○, □).

the dissipation rate anisotropy. Very good agreement was found for axisymmetric turbulence, in which there is only one independent anisotropy measure of each tensor, and reasonably good agreement was found for the other more complex cases. In figure 3 the anisotropies of the dissipation rate tensor of turbulence subjected to a plane irrotational strain are shown. Here the anisotropy measure for the streamwise velocity component (a_{11}) is nearly equal to zero whereas e_{11} differs significantly from this value, i.e. neither an isotropic nor a linear model could at all capture this behaviour. However, the model (5) still gives a fairly accurate prediction of e_{11}. Hence, the nonlinear part of the model, that describes the influence on one component of e_{ij} from the complete anisotropic state as given by a_{kl}, appears to model the physics to a reasonable approximation. The dissipation rate anisotropies in the case of a homogeneous shear, which can be regarded as a generic case for any shear flow, were equally well described by the model.

Even though the linear part ($e_{ij}^{linear} = 0.5\ a_{ij}$) of the proposed model, which was calibrated against rapid distortion analysis, in many cases is sufficient to capture the behaviour of the e_{ij}-tensor the nonlinear part enables a correct prediction of the dissipation rate of a stress component tending to zero. Furthermore, the model seems to capture the behaviour of the dissipation rate tensor in flow cases characterised by $III_a \approx 0$ and $III_e \neq 0$ (e.g., plane strain or homogeneous shear), something a linear model is incapable of.

4. The total dissipation rate equation

In order to improve RST-closures it is not only essential to find better models of the different terms in the stress transport equations but also to have a proper model of the total dissipation rate equation. For homogeneous situations the exact total dissipation rate transport equation reads

$$\frac{D\varepsilon}{Dt} = -(\varepsilon_{kl} + 2\nu\ \overline{u_{i,k}u_{i,l}})\ U_{k,l} - 2\nu^2\ \overline{u_{i,km}u_{i,km}} - 2\nu\ \overline{u_{k,m}u_{i,k}u_{i,m}} \tag{6}$$

The first term is a production or generation term, the second a viscous destruction term and the third can be interpreted as a generation of vorticity fluctuations by the turbulent field itself. The standard modelling of the production term is (see, e.g., Hanjalić & Launder 1972, 1976)

$$(\varepsilon_{kl} + 2\nu\ \overline{u_{i,k}u_{i,l}})\ U_{k,l} = C_{\varepsilon 1}\ \varepsilon a_{kl} U_{k,l} \tag{7}$$

If the RST-equations are supplemented with a model for the dissipation rate tensor anisotropy, such as that described above, it is possible to model the production term in (6) in a manner that appears closer to the original expression. One may note that the first part of the production term can be rewritten as $\varepsilon e_{kl} U_{k,l}$ in an incompressible fluid. In the limit of isotropic turbulence the two terms within the parentheses become identical. A natural modelling of the complete term would therefore be

$$(\varepsilon_{kl} + 2\nu\ \overline{u_{i,k}u_{i,l}})\ U_{k,l} = (1 + C_{\varepsilon p})\ \varepsilon e_{kl} U_{k,l} \tag{8}$$

The value of the model constant $C_{\varepsilon p}$ may be expected to be about 1 in view of the identity of the two parts in the isotropic limit. Preliminary results from comparing the model (8) with direct numeric simulation data of an axisymmetric strain case and a plane strain case indicate that $C_{\varepsilon p}$ should assume a value of about 1.3. As can be seen

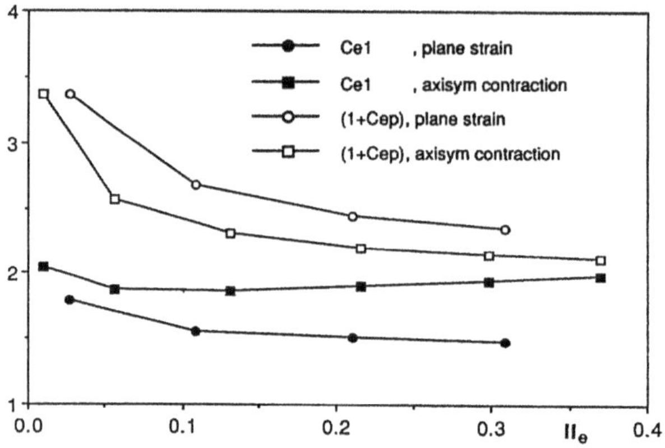

Figure 4. The model parameters $C_{\varepsilon 1}$ and $C_{\varepsilon p}$ as determined from direct simulation data of Lee & Reynolds (1985). Here plotted vs $II_e(= e_{ij}e_{ji})$ which quantifies the anisotropy of the e_{ij}-tensor.

in figure 4 the value of $C_{\varepsilon p}$ varies rather strongly with the degree of anisotropy of the dissipation rate tensor but on the other hand rather weakly between the two different cases. This feature is quite opposite to that of $C_{\varepsilon 1}$ of the standard modelling. Even though the two terms are identical in isotropic turbulence their degree of anisotropy, which is the essential property in the interaction with the mean field gradients, is initially significantly different when the turbulence is subjected to strain. In fact, RDT-analysis of the initial behaviour of the two tensors shows that a value of $C_{\varepsilon p}$ of about four should be expected (instead of one) for small II_e.

5. Concluding remarks

A significant degree of anisotropy of the dissipation rate tensor was observed in an experiment on axisymmetrically strained homogeneous turbulence.

A proposed algebraic model for e_{ij} was shown, by comparison with direct numerical simulation data, to accurately describe the anisotropy of the dissipation rate tensor in several different kinds of flows.

An alternative way of modelling the production rate term in the total dissipation rate equation was investigated in which use was made of the algebraic e_{ij}-model.

Support from the Swedish State Board for Technical Development is gratefully acknowledged. We also want to thank Mr. T. Larsson for help with the direct numerical simulations done at the department.

References

Antonia, R.A., Shah, D.A. & Browne, L.W.B. 1987 Spectra of velocity derivatives in a turbulent wake. *Phys. Fluids*, **30**, 3455.

Batchelor, G.K. 1953 *The theory of homogeneous turbulence*, Cambridge U. P.

Brasseur, J.G. & Corrsin, S. 1987 Spectral evolution of the Navier-Stokes equations for low order couplings of Fourier modes. In *Advances in Turbulence* (eds. G. Comte-Bellot & J. Mathieu), Springer.

Gerz, T., Schumann, U. & Elgobashi, S.E. 1989 Direct numerical simulation of stratified homogeneous turbulent shear flow. *J. Fluid Mech.*, **200**, 563.

Hallbäck, M., Groth, J. & Johansson, A.V. 1989 A Reynolds stress closure for the dissipation in anisotropic turbulent flows. In *Proc. of the Seventh Symposium on Turbulent Shear Flows*, Stanford University, USA.

Hallbäck, M., Groth, J. & Johansson, A.V. 1990 An algebraic model for non-isotropic turbulent dissipation in Reynolds stress closures. To appear in *Phys. Fluids.*

Hanjalić, K. & Launder, B.E. 1972 A Reynolds stress model of turbulence and its application to thin shear flows. *J. Fluid Mech.*, **52**, 609.

Hanjalić, K. & Launder, B.E. 1976 Contribution towards a Reynolds-stress closure for low Reynolds-number turbulence. *J. Fluid Mech.*, **74**, 593.

Hussein, H.J. & George, W.K. 1989 Measurement of small scale turbulence in an axisymmetric jet using moving hot-wires. In *Proc. of the Seventh Symposium on Turbulent Shear Flows*, Stanford University, USA.

Lee, M.J. & Reynolds, W.C. 1985 Numerical experiments on the structure of homogeneous turbulence. Rep. no. TF-24, Thermosciences Division, Dept. of Mech. Eng., Stanford University, USA.

Lumley, J.L. & Newman, G.R. 1977 The return to isotropy of homogeneous turbulence. *J. Fluid Mech.*, **82**, 161.

Lumley, J.L. 1978 Computational modelling of turbulent flows. *Adv. Appl. Mech.*, **18**, 123.

Uberoi, M.S. 1956 Effect of wind-tunnel contraction on free-stream turbulence. *J. Aero. Sci.*, **23**, 754.

Uberoi, M.S. 1957 Equipartition of energy and local isotropy in turbulent flows. *J. Appl. Phys.*, **28**, 1165.

Kinematics of Small Scale Motions in Homogeneous Isotropic Turbulence

J.C.R. Hunt, J.C.H. Fung, N.A. Malik, R.J. Perkins, J.C. Vassilicos, A.A. Wray[†], *J.C. Buell*[†], *and J.P. Bertoglio*[*]

DAMTP, University of Cambridge, Silver Street, CB3 9EW, U.K.
[†] NASA Ames Research Center, Moffett Field, CA94035, U.S.A.
[*] Ecole Centrale de Lyon, Ecully, France.

Using direct numerical simulation and 'kinematic simulation' (henceforth KS) of velocity fields, measurement, flow visualisation and novel kinematic analysis, the following aspects of small scale motions in turbulence were investigated: (i) Random advection and distortion of small scale motions by larger scale motions; (ii) the Lagrangian spectrum at high frequency of particles moving in small scale eddies and the effects of the time dependence of the eddies; (iii) the relative velocities Δu and the separation distance l between pairs of particles, to find the decorrelation time scales of Δu and the relation between the mean square separation $\overline{l^2}$ and the conditional displacements of single particles; (iv) how the forms of these motions can be inferred from the asymptotic forms of Fourier series and spectra; (v) the specific implications for Eulerian and Lagrangian spectra of the small scales being mostly associated with elongated regions of spiralling motions in which there are different orders of discontinuous derivatives of velocity normal to streamline surfaces. These studies suggest that small scale motion in isotropic turbulence has a characteristic spiralling structure, which is generally consistent with statistics, such as fractals, spectra and probability distributions.

1. Introduction

A major goal of current research in turbulence is to relate statistical description and analysis (e.g. in terms of moments, spectra, probability distributions etc) to deterministic descriptions of velocity fields within local regions of the random motions, and vice versa. Not only is it intellectually appealing to integrate these two approaches to turbulence, but it may provide methods for improving statistical models of turbulence, for 'designing' turbulent flows, and for developing approximate methods for simulating turbulent flow fields (e.g. as vortices, random modes, truncated modes etc). Working towards this goal is now practical not only because of the improved techniques of measurement and computer simulation, but also because of the recent findings that, in most kinds of turbulent flow, characteristic and recurring forms of the velocity field do exist in particular characteristic regions of the flows. Although this limits the task, nevertheless, there are many different forms of characteristic eddies dependent on the large scale nature of the flows and on their initial and boundary conditions.

In this paper we consider from both statistical and deterministic points of view the kinematical description and simulation of the small scales of motion in high Reynolds number turbulence.

The main aspects of the current 'picture' of turbulence (e.g. Monin & Yaglom, 1975) that may need some changing as a result of these and other recent investigations

Advances in Turbulence 3
Editors: A.V. Johansson · P.H. Alfredsson

are (a) large scales advect and distort smaller scales, but this distortion is largely caused by stretching the vorticity of the smaller scales (i.e. the $(\boldsymbol{\omega} \cdot \nabla)\mathbf{u}$ term in the vorticity equation); (b) the latter effect explains why the derivatives of turbulence are skewed (e.g. $\overline{(\partial u_1/\partial x_1)^3} < 0$); (c) small scale motions are unevenly distributed in turbulent flows, but are not particularly linked to the location of larger scales; (d) as a result of (c) the relative motions of particles separated by a distance l decorrelate rather fast on a time scale equal to the inverse of the strain rate over a distance l (viz. $\varepsilon^{-1/3}l^{2/3}$ where ε is the rate of energy dissipation per unit mass), which implies that models for two particle dispersion can be related to those of one particle (Monin & Yaglom, 1975); (e) the time scale on which the small scale eddies change (in a frame moving with the larger scales) does not significantly affect the small scale motions of fluid particles; (f) the criterion for the existence of high Reynolds number turbulence is that at high wave numbers the energy spectrum has the Kolmogorov inertial subrange form $E(k) \propto \varepsilon^{2/3}k^{-5/3}$.

Evidence and analysis presented here (building on studies of previous work e.g. Lundgren (1982)) indicates that the above concepts are not entirely consistent with the alternative description of small scale turbulence as local regions of unsteady spiralling motions in which many different scales of motions exist within the same eddy structure.

2. Time dependence of eddy motions

2.1 Big whirls carry little whirls

Consider velocity fluctuations $\mathbf{u}_s$ with length scales less than l_s that are much smaller than the integral scale L (i.e. $l_s \ll L$), but the fluctuations are strong enough that they are not affected by viscous stress (i.e. $|\mathbf{u}_s|l/\nu \gg 1$). The velocity fluctuations with scales greater than l_s are denoted by $\mathbf{u}_L$, since the spectra decrease rapidly with wavenumber when $k \gtrsim L^{-1}$, $\overline{|\mathbf{u}_s|^2} \ll \overline{|\mathbf{u}_L|^2} \simeq 3U_0^2$ where U_0 is the rms velocity in one direction. Note that the pressure fluctuations on this scale are of order $\rho|\mathbf{u}_s|^2$, where ρ is the density.

Since the rate of change with time of the small scales is given by

$$\frac{\partial \mathbf{u}_s}{\partial t} = -(\mathbf{u}_L \cdot \nabla)\mathbf{u}_s - (\mathbf{u}_s \cdot \nabla)\mathbf{u}_L - \{(\mathbf{u}_s \cdot \nabla)\mathbf{u}_s\} - \frac{1}{\rho}\nabla p_s + \nu\nabla^2\mathbf{u}_s \ , \qquad (2.1)$$

where $\{\ \ \}$ denotes the contribution of the term on the length scales l_s, the above definition of $\mathbf{u}_s$ implies that this equation can be approximated by (Tennekes 1975)

$$\partial \mathbf{u}_s/\partial t \simeq -(\mathbf{u}_L \cdot \nabla)\mathbf{u}_s \ . \qquad (2.2)$$

In other words the variation of $\mathbf{u}_s$ at a point is largely determined by the advection of small eddies by the larger scale more energetic motions (see figure 1).

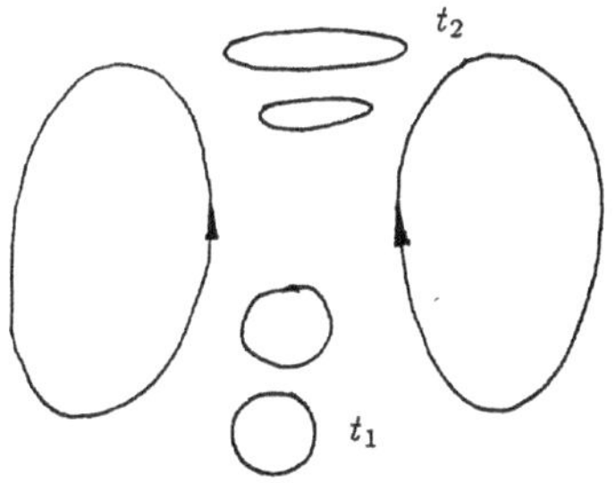

Figure 1. Little eddies being advected and distorted by larger eddies.

Since the large scales $\mathbf{u}_L$ have a Gaussian probability distribution and since they are, by definition, slowly varying over the scale of the small eddies the wavenumber-frequency energy spectrum $\mathcal{E}(k,\omega)$ can be calculated in term of the energy spectrum $E(k)$ where $k = |\boldsymbol{\kappa}|$ and $\boldsymbol{\kappa}$ is a three dimensional wavenumber. From Fourier analysis, (2.2) implies that

$$\mathbf{u}_s = \sum_n \mathbf{a}_n \exp i[\boldsymbol{\kappa}_n \cdot (\mathbf{x} - \mathbf{u}_L t)] \tag{2.3}$$

where $\overline{|\mathbf{a}_n|^2} \propto E(k)$. Thence by standard methods (e.g. Batchelor 1953), it follows that if $\omega \gg kU_0$,

$$\mathcal{E}(k,\omega) = E(k)\exp(-\omega^2/(2ak^2U_o^2))/(\sqrt{2\pi}akU_0) \ , \tag{2.4}$$

where the coefficient $a = 1$.

The result (2.4) was tested at the Center for Turbulence Research (CTR) using simulation of homogeneous stationary isotropic turbulence on a grid of 64^3, with the turbulence being created by instabilities in the flow driven by steady non-uniform body forces $\mathbf{F}(\mathbf{x})$. For more details, see Hunt et al (1987). It was found that (2.4) described the relation between $\mathcal{E}(k,\omega)$ and $E(k)$ for $kL \geq 1$ but that $a \simeq 0.5$, probably because of the low Reynolds number of the flow ($R_e = 25$). In fact (2.3) and (2.4) had been obtained previously by Chase (1970), who computed the time delayed covariances e.g. $\overline{u_1(\mathbf{x},t)u_2(\mathbf{x}+\mathbf{r},t+\tau)}$ and showed that they agreed well with the grid turbulence measurements of Favre *et al.* (1957) (in the case $a = 1$). Note that the result (2.4) agrees with experimental results even though $\mathbf{u}_L$ is *not* effectively uniform over the scale of $\mathbf{u}_s$, as assumed in the analysis.

2.2 Time scales of smallest eddies

As the small 'eddies' with length $l < l_s$ are advected by the larger scales they are not frozen, but their velocity $\mathbf{u}(l)$ changes on a timescale $\tau^*(l)$ that depends on l. Following Richardson (1926), and Obukhov (1941), it has generally been assumed that

$$\tau^*(l) \sim l^{2/3}\varepsilon^{-1/3} \ . \tag{2.5}$$

This is consistent with the unique experiments of Fermigier (1980) which showed that the mean square distance $\overline{p^2}$ between lines of marked fluid (using laser induced striping) increases as $d\overline{p^2}/dt \propto \overline{p^2}(t)^{2/3}$.

Two specific definitions for $\tau^*(l)$ have been proposed based on the velocity of pairs of fluid particles (figure 2) α, β located at time t at $\mathbf{x}$ and $\mathbf{x} + \mathbf{l}(t)$, which at time $t+\tau$ become α', β' at $\mathbf{x}', \mathbf{x}' + \mathbf{l}'(t+\tau)$. If

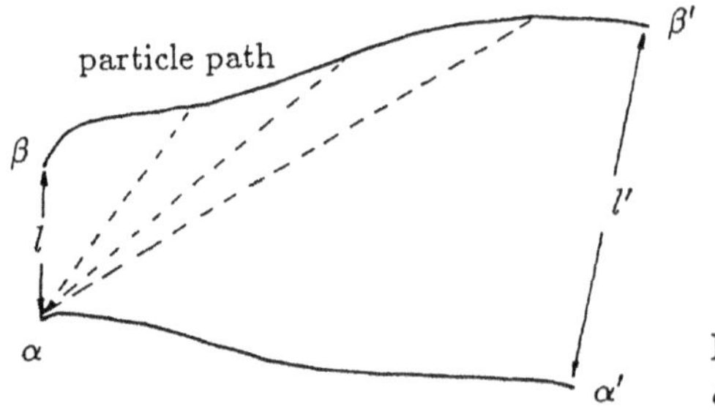

Figure 2. Two fluid particles starting from α and β and moving to α' and β'.

$$R^*_\triangle(\tau, l) = \overline{\Delta u(t+\tau)\Delta u(t)}, \qquad R_\triangle = R^*_\triangle / [\overline{\Delta u^2(t+\tau)}\ \overline{\Delta u^2(t)}]^{1/2}, \tag{2.6a}$$

where $\Delta u = |\mathbf{u}(\mathbf{x}' + \mathbf{l}') - \mathbf{u}(\mathbf{x})|$ is the modulus of the velocity difference between the particles, then the timescales of relative motion can be defined in terms of the slope of $R_\triangle$, i.e.

$$\tau^*_\triangle(l)^{-1} = (dR_\triangle(\tau)/d\tau) \qquad (as \quad \tau/T_L \to 0)\ , \tag{2.6b}$$

or as the integral scale of $R_\triangle$, i.e. $T_\triangle(l) = \int_0^\infty R_\triangle(\tau, l) d\tau$. (2.6c) These definitions are relevant for modelling relative diffusion (e.g. Thomson 1989). An important question is whether or not these relative motions correspond to a Markovian process where $R_\triangle = \exp(-\tau/\tau^*_\triangle)$, in which case the definitions (2.6b) and (2.6c) are equivalent. For estimating certain correlations, for wavenumber components k or eddies of scale $l \sim k^{-1}$, in the dynamical equations used in models such as EDQNM (e.g. Bertoglio 1986), the relevant time scale $\tau^*_{ij}(l)$ is defined in terms of the cross correlation between the velocities of β' and α' namely

$$R'_{ij}(\tau; l) = \overline{u_i(\mathbf{x}' + \mathbf{l}', t+\tau) u_j(\mathbf{x}, t)} / \overline{(u_i(\mathbf{x} + \mathbf{l}, t) u_j(\mathbf{x}, t))} \tag{2.7a}$$

where

$$\tau^*_{ij}(l)^{-1} = (dR'_{ij}/d\tau)(\tau/T_L \to 0) \propto \overline{u_i(\mathbf{x}) \frac{\partial p}{\partial x_j}(\mathbf{x} + \mathbf{l})/\rho}\ . \tag{2.7b}$$

In the EDQNM model it is assumed that $\tau^*_{ij}(l)$ is given by (2.5).

2.3 Simulation of velocity fields with given spatio-temporal statistics

The eddy structure in time and space and the motions of particles, cannot be derived directly from the 2-point spectra and correlation functions, but can only be investigated using the velocity field either computed directly (if the Reynolds number is small enough) *or* by using an approximate simulation of the velocity field. We have constructed an isotropic field, given the spectrum $E(k)$, the results (2.4) and (2.5), the probability distribution $\mathbf{P}(\mathbf{u})$ and an assumption about $\mathcal{E}(k, \omega)$ when $\omega T_L \lesssim 1, kL \leq 1$. The resulting velocity field can be tested against experiments and direct numerical simulations. In the 'kinematic simulation' proposed by Fung et al (1990), (see also Fung (1990)), the velocity field $\mathbf{u}(\mathbf{x}, t)$ is divided into large scale ($\mathbf{u}_L$) and small scale ($\mathbf{u}_s$) components, corresponding to $k < k_c$ and $k > k_c$, where $k_c \sim L^{-1}$; defined by

$$\mathbf{u}_L = \sum \{(\mathbf{a}_n \wedge \boldsymbol{\kappa}_n) \cos[\boldsymbol{\kappa}_n \cdot (\mathbf{x} - \boldsymbol{\chi})] + (\mathbf{b}_n \wedge \boldsymbol{\kappa}_n) \sin[\boldsymbol{\kappa}_n \cdot (\mathbf{x} - \boldsymbol{\chi})]\}, \tag{2.8a}$$

where $\boldsymbol{\chi} = \int \mathbf{W}(t) dt$ and $\mathbf{W}(t)$ is a Gaussian random vector, with rms value equal to $0.58 U_0^2$. Also

$$\mathbf{u}_s = \sum_{m,n} (\mathbf{a}_{mn} \wedge \boldsymbol{\kappa}_{mn}) \cos[\boldsymbol{\kappa}_{mn} \cdot (\mathbf{x} - \mathbf{x}_L) - \omega'_{mn} t] + (\mathbf{b}_{mn} \wedge \boldsymbol{\kappa}_{mn}) \sin[\boldsymbol{\kappa}_{mn} \cdot (\mathbf{x} - \mathbf{x}_L) - \omega'_{mn} t], \tag{2.8b}$$

where $\mathbf{x}_L$ is the random displacement by the large scale motions of a fluid element reaching $\mathbf{x}$ at time t, i.e., (from (2.2)) $\mathbf{x}_L = \int_0^t \mathbf{u}_L[\mathbf{x}'(t'; \mathbf{x}), t'] dt'$, (2.9). To satisfy (2.5), the decorrelation of $\mathbf{u}_s$ with time, as it is advected by u_L, is simulated by letting $\omega'_{mn} = \omega'_n \delta_{mn}$, where $\omega'_n = \lambda \varepsilon^{1/3} k_n^{2/3}$, (2.10). (Similar results are obtained by allowing ω'_{mn} to have a spectrum of values distributed around ω'_n (Malik 1991)). The coefficient λ is chosen to be 1.0. The amplitude of the Fourier coefficients $\mathbf{a}_n$

and $\mathbf{b}_n$ are chosen to have a Gaussian distribution with variance determined by $E(k)$, i.e. $\overline{|\mathbf{a}_n|^2}, \overline{|\mathbf{b}_n|^2} \propto E(k) = \alpha\varepsilon^{2/3}k^{-5/3}$ for $k_c < k < k_\eta$ (2.11). The significant statistical properties of this simulation have been evaluated and found to compare well with relevant experiments and simulation. The weak sensitivity to the assumptions has been checked.

2.4 New results from the 'kinematical simulation'

(i) From (2.8) to (2.11) the wavenumber-frequency spectrum $\mathcal{E}(k,\omega)$ can be deduced analytically (for $k \gg k_c$) viz.

$$\mathcal{E}(k,\omega) = E(k)\left[\exp\left(-1/2\,(\omega_+/kU_0)^2\right) + \exp\left(-1/2\,(\omega_-/kU_0)^2\right)\right]/2\sqrt{2\pi}kU_0 \quad (2.12)$$

where $\omega_+, \omega_- = \omega \pm \lambda\varepsilon^{1/3}k^{2/3}$. This is identical to (2.4), with $a = 1.0$, when $k \geq k_c$.

(ii) From the simulation or from (2.12) the Eulerian frequency spectrum is found to agree with Tennekes' result that $\phi_{11}^E(\omega) = C^E u_0^{2/3}\varepsilon^{2/3}\omega^{-5/3}$ (2.13), where the coefficient $C^E \simeq 0.8$. (See figure 3).

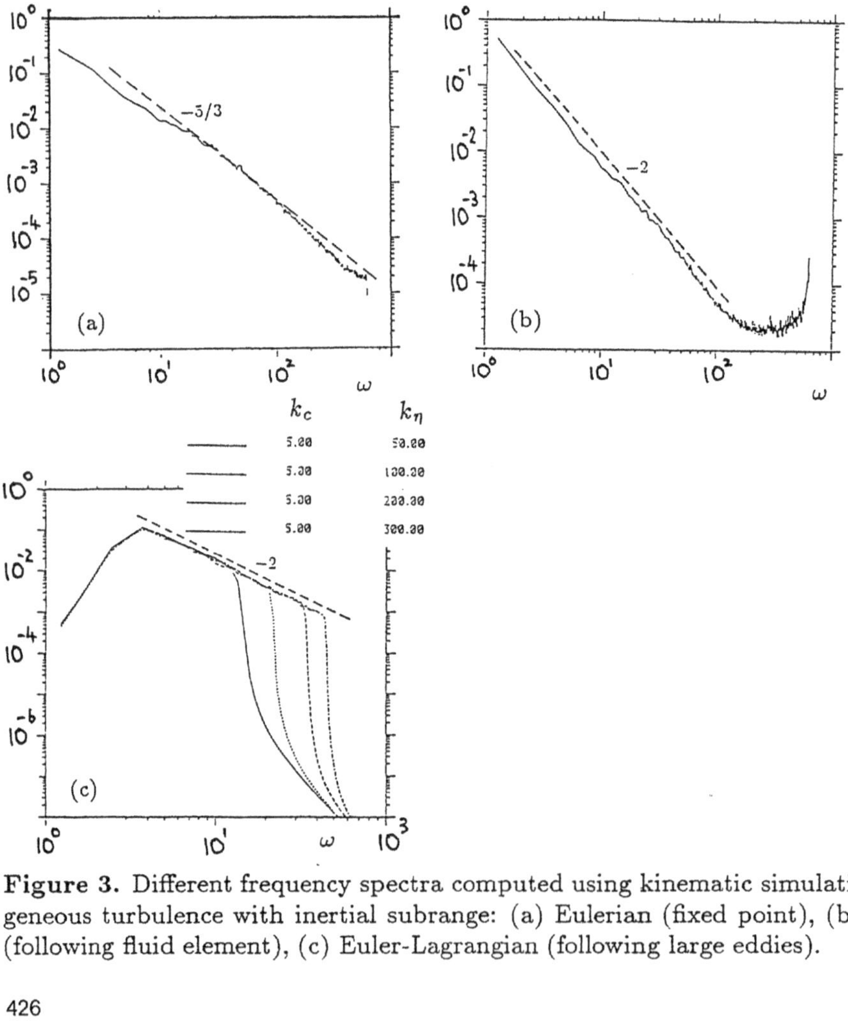

Figure 3. Different frequency spectra computed using kinematic simulations of homogeneous turbulence with inertial subrange: (a) Eulerian (fixed point), (b) Lagrangian (following fluid element), (c) Euler-Lagrangian (following large eddies).

(iii) We can define an 'Eulerian-Lagrangian' velocity seen by an observer at $\mathbf{x}_L(t)$ travelling with the large scales, so that the smallest scales move past the observer i.e. $\mathbf{u}^{EL}(t) = (\mathbf{u}_L + \mathbf{u}_s)(\mathbf{x}_L(t), t)$ (2.14). This velocity is different from the Lagrangian velocity of a fluid particle $\mathbf{v}^L$ which is determined by the changes in $\mathbf{u}_s$ and $\mathbf{u}_L$ seen by the particle as it moves i.e. $d\mathbf{X}/dt = \mathbf{v}^L(\mathbf{x}, t)$. Since the particle is not cutting through the eddies, the time scale for $\mathbf{v}^L$ to change is expected to be less than $\mathbf{v}^{EL}$. Thirdly, it is instructive to consider the velocity $\mathbf{v}^{LF}$ seen by a particle at $\mathbf{X}(t)$ travelling through a frozen *small scale* field at $t = t_0$. Therefore

$$\mathbf{v}^L(t) = (\mathbf{u}_L + \mathbf{u}_s)(\mathbf{x} = \mathbf{X}(t); t_0), \quad \text{where} \quad d\mathbf{X}(t)/dt = \mathbf{u}(\mathbf{X}(t), t_0) \quad (2.15a)$$

$$\text{and} \quad \mathbf{v}^{LF} = \mathbf{v}^L \quad \text{if, in (2.8),} \quad \omega'_{mn} = 0. \quad (2.15b)$$

From (2.8) to (2.11), if $E(k) = \alpha_k \varepsilon^{2/3} k^{-5/3}$, the form of the spectra of these velocities $\mathbf{v}^{EL}$, $\mathbf{v}^L$, $\mathbf{v}^{LF}$ which can be derived analytically for $\omega \gg u_0/L$, are

$$\phi_{11}^{EL}(\omega) \sim C^{EL}\varepsilon\omega^{-2}, \; \phi_{11}^{L}(\omega) \sim C^{L}\varepsilon\omega^{-2}, \; \phi_{11}^{LF}(\omega) \sim C^{LF}\varepsilon^{2/3}u_0^{2/3}\omega^{-5/3}. (2.16a, b, c)$$

The value of C^{EL} can be obtained analytically (Malik 1991) and is equal to 0.81 (in the limit of infinite inertial subrange). These are confirmed by direct computation of particle trajectories and velocities in the KS velocity field; the coefficients $C^{EL} \approx 0.75$, and $C^L \approx 0.54$, 0.81 for different inertial ranges (see figure 3). The coefficient C^L is close to the values found in the atmospheric boundary layer (0.6) by Hanna (1980) and by Yeung & Pope (1989). The difference between (2.16b) and (2.16c) is new and reveals the importance of the time dependence of the turbulence to obtain the correct spectra of particle motion and thence scalars. (See §4.3).

(iv) The decorrelation of the relative motion of fluid particles initially separated by l was investigated by computing $R^*_\triangle(\tau)$ and its Fourier Transform, $\Phi_\triangle(\omega, l)$, (noting that $\triangle u(t)$ is not a stationary random process, but $R^*_\triangle$ is such that $\Phi_\triangle(\omega, l)$ can be defined). The results in figures (4a,b) show that for $\tau \leq l^{2/3}\varepsilon^{-1/3}$, where $0.1 < l/L < 0.7$,

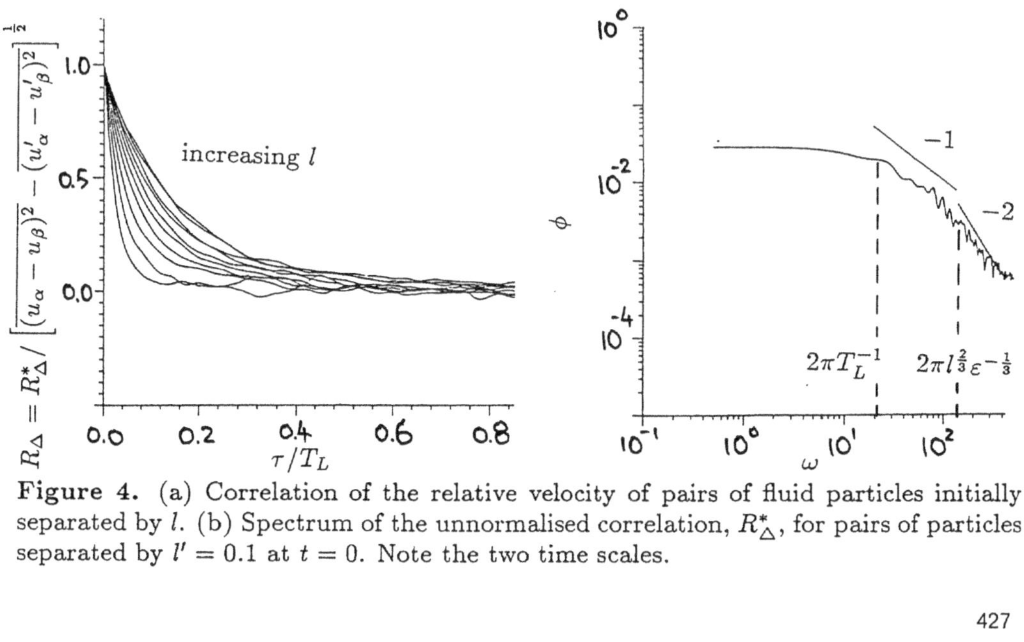

Figure 4. (a) Correlation of the relative velocity of pairs of fluid particles initially separated by l. (b) Spectrum of the unnormalised correlation, $R^*_\triangle$, for pairs of particles separated by $l' = 0.1$ at $t = 0$. Note the two time scales.

$$R^*_\triangle(\tau,l) \approx \overline{\triangle u^2(t)} - \pi C_\triangle \varepsilon\tau. \tag{2.17}$$

Significantly for larger times ($\tau \sim T_L$), the relative motion still remains weakly correlated.

Equivalently

$$\Phi_\triangle \sim C_\triangle \varepsilon\omega^{-2} \qquad \text{for} \qquad \omega \gtrsim l^{-2/3}\varepsilon^{-1/3} \tag{2.18}$$

and for $l \ll L$, $\Phi_\triangle \sim \delta \cdot \overline{\triangle u^2}(t)/\omega$ for $l^{-2/3}\varepsilon^{1/3} \gtrsim \omega \gtrsim T_L^{-1}$, and $\Phi_\triangle \sim \delta \cdot \overline{\triangle u^2}(t) T_L$ for $\omega \lesssim T_L^{-1}$ where the scaling factor $\delta \sim 1/\ln(T_L\varepsilon^{1/3}l^{-2/3})$, to ensure $2\int_0^\infty \Phi_\triangle(\omega)d\omega = \overline{\triangle u^2}(t)$. With reference to the discussion in section 2.2, these simulations imply that there are two time scales $\tau_\triangle \sim l^{2/3}\varepsilon^{-1/3}$ and $\tau_\triangle \sim \delta T_L$. (Meneguzzi inferred a similar long time scale in his DNS). For the purpose of dynamical modelling in EDQNM, the smaller time scale is consistent with the current assumptions. But the persistence of the correlation of relative motion of particles starting close together is not consistent with the picture of turbulence consisting of eddies of different scales with the acceleration on different scales being uncorrelated (e.g. Novikov 1963). This result may not be quite consistent with some stochastic models of two particle diffusion (e.g. Thomson 1986, 1989).

(v) Although the large scale and small scale motions (in their moving frame) are each Gaussian, the total velocity field (in a fixed frame) is not Gaussian because the small scales are advected by the non-uniform large scale motion. In fact the skewness of the derivatives are non zero.

$$S = 3\overline{(\partial u_1/\partial x_1)^3}/\overline{(\partial u_1/\partial x_1)^2}^{3/2} = 9\overline{\partial u_{L1}/\partial x_1(\partial u_{s1}/\partial x_1)^2}/\overline{(\partial u_1/\partial x_1)^2}^{3/2} < 0 \tag{2.19}$$

which means that regions of high wavenumber of $\mathbf{u}_s$ are associated with regions of low wavenumber of $\mathbf{u}_L$ and generally with stagnation regions where $\mathbf{u}_L$ is small. This is a fundamental process in all turbulent flows (including 2-dimensional turbulence) but is usually explained in terms of vortex stretching, rather than non-uniform advection. It is likely that this relative non-uniform advection is significant, even in high Reynolds number turbulence; but is particularly significant at the low Reynolds number of most numerical simulations and many experiments. With the single 'cut' between large and small scales used in this version of KS, $S \sim -0.15$ (see Fung *et al* 1990).

3. Localised regions of small scale dynamics and kinematics

3.1 Direct numerical simulations

Recent study of two dimensional turbulence for example on grids of 512×512 at $Re = 1000$, by McWilliams (1984) and Briscoli *et al.* (1989) have shown that the vorticity accumulates as streaks wound up into small isolated regions which are rounded but also have trailing 'tails'. For theory and model simulations of simpler models of these regions see Gilbert (1988) and Dritschel (1989).

In three-dimensional homogeneous isotropic turbulence the Reynolds numbers of DNS is lower (typically $Re \lesssim 100$), but in all cases the vortex lines are intermittently concentrated into thin tube like regions, around which there is locally significant axial streamwise motion, (e.g Meneguzzi (1990), Frisch & Orszag (1990), Fung *et al.* (1990), Wray & Hunt (1990)). But it appears that the main kinematic features are that, in a cross sectional plane across the zones, they are nodes or foci, and that the streamlines have a helical form both entering and leaving these regions. See figure 5.

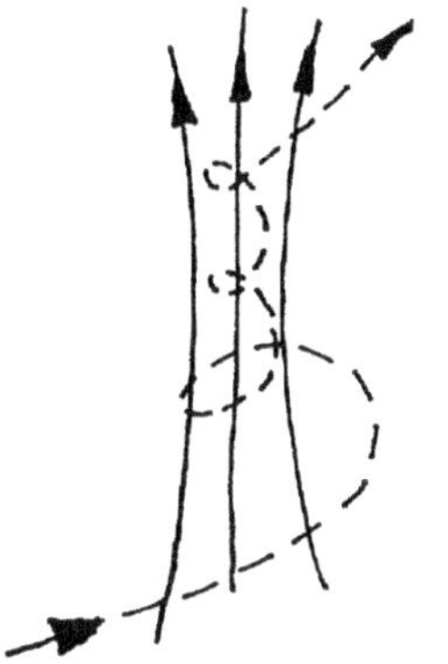

Figure 5. Typical vortex lines and stream lines in spiralling regions.

3.2 Interpreting spectra from the properties of Fourier Series

The forms of spectra $\Phi(k)$ at high wavenumber or frequency convey significant information about the form of the signal, say $u(x)$, and especially about a particular range of x where the derivatives (e.g. du/dx) are large (see figure 6). This information comes from the properties of Fourier series of $u(x) = \sum a_n \exp(ik_n x)$ where $k_n = 2\pi n/x$ and the fact that $\Phi(k) \propto |a_n|^2$ where $k = k_n$.

(i) If the function has *distinct* singularities in its Nth derivatives, then the magnitude of the Fourier coefficients vary with n, as $n \to \infty$, in proportion to $n^{-(N+1)}$ e.g. for any function which is discontinuous (such as vortex sheets) $|a_n| \propto n^{-1}$ and $\Phi(k) \propto k^{-2}$, or with discontinuities in its first derivatives (e.g. where the vorticity jumps), $|a_n| \propto n^{-2}$ and $\Phi(k) \propto k^{-4}$ (e.g. Courant & Hilbert vol. I, 1953 p.74).

(ii) Fourier coefficients also decrease like n^{-N} or $\phi(k) \propto k^{-2N}$ for functions that are not discontinuous, but are singular in other ways, such as having an infinite number of oscillations in a finite distance near an accumulation point. e.g. if $u(x) = x\cos(L/x)$, $|a_n| \propto n^{-1}$ as $n \to \infty$.

(iii) As Moffatt (1984) pointed out, it is possible to have a spectrum $\phi(k) \propto k^{-p}$ where $p < 2$, for a function $v(x)$ that is everywhere finite, if there are an infinite number of discontinuities in $v(x)$ (or its derivatives) within a finite distance of one or more accumulation points. If a function is constructed from a Fourier Series, as in KS, whose coefficients decrease in proportion to n^{-p}, and are random, then, the function does not usually contain discontinuities but rather is highly oscillatory near accumulation points. The effect of straining by larger scales or non-linear inertial processes is to change the

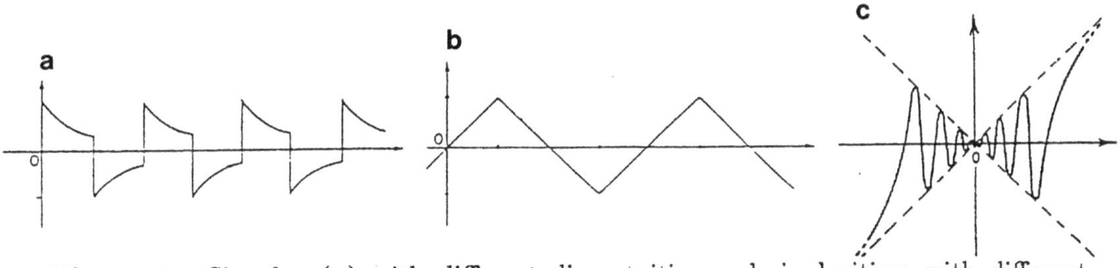

Figure 6. Signals $u(x)$ with different discontuities and sigularities; with different Fourier series: (a) discontinuities in $u(x)$, (b) discontinuities in $\partial u/\partial x$, (c) infinite number of oscillations.

value of p *and* the phase, so that, for example in sheared turbulence (Hunt & Carruthers 1990), where $p = 2$ (for low Reynolds number turbulence) sharp changes in derivatives occur in thin layers rather than as oscillations.

(In other words, a spectrum of the form $\phi \propto k^{-p}$ does not necessarily imply a homogeneous distribution of smooth 'eddies' with a range of length scales of $O(l)$ and velocity scales proportional to $l^{(p-1)/2}$ – the usual picture of turbulence).

3.3 Spiralling motions and spectra

In turbulent flows the elongated regions of high vorticity, or (E) zones, are the most singular regions (with the highest derivatives and the highest oscillation of derivatives), and therefore determine the spectra of the velocity fields. If the motions in these regions are helical they are determined by surfaces of the form $r = C\theta^{-\alpha}$ on which there are discontinuities in the velocity or scalar derivatives and which must be streamline surfaces if the vortices are evolving slowly (see figure 7). Consequently on a line $y = 0$ through the centre of the spiral, discontinuities of the mth derivative, $\partial^m v/\partial x^m$, and delta functions of $\partial^{m+1}v/\partial x^{m+1}(> 0)$, of a velocity component $v(x)$, occur where $\theta = 2M\pi(M = 1, 2, \ldots)$, and $x = x_M \propto M^{-\alpha}$. From the Fourier coefficients (a_n^m) of $(\partial^{m+1}v/\partial x^{m+1})$ the spectrum is obtained $\phi_v^{(m)}(k) \propto k^{+2/\alpha}$. Thence the spectrum of the velocity $\phi_v(k) \propto k^{-2(m+1)+2/\alpha}$ (3.1*a*). If the discontinuity of $\partial^m v/\partial x^m$ oscillates in sign, $\phi_v^m(k) \propto k^{1/1+\alpha}$ (3.1*b*). If the observed or simulated spectrum decreases as k^{-p}, equation (3.1) implies that, if the spiral consists of single sign *vortex sheets* (as suggested by Moffatt (1984) and Lundgren (1982)), i.e. $m = 0$, then $p = 2 - 2/\alpha$. So if $p = 5/3, \alpha = 6$. But if in the spiral the velocity is continuous, but the vorticity is discontinuous across helical streamlines, then $m = 1, p = 4 - 2/\alpha$; if $p = 5/3, \alpha = 7/6$. Note that the weaker the discontinuity in velocity then the spiral has to be more tightly wound for the spectrum to decrease with a given power law exponent.

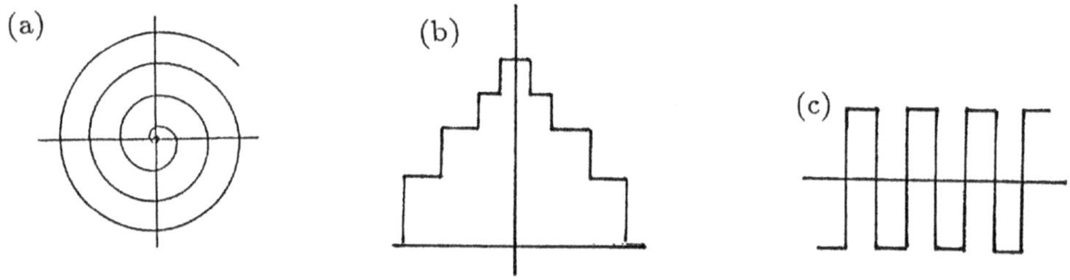

Figure 7. (a) Spiral form with vortex sheet, (b) section through spiral, (c) section through spiral of a scalar interface.

4. Lagrangian trajectories and scalar interfaces

4.1 Is $\phi^L(\omega)$ at high frequency consistent with a spiral structure?

From similarity theory (Inoue 1951), field experiment and the K.S. results of §2, at high Reynolds number the Lagrangian spectrum of any component, say u_1, of the velocity of a fluid particle is given by

$$\phi_{11}^L(\omega) = C^L \varepsilon\omega^{-2}. \tag{4.1}$$

This has sometimes been explained (with reference to §3.2) as particles suddenly changing their velocity in time ω^{-1} on a length scale much less than $\varepsilon^{1/2}\omega^{-3/2}$, but such

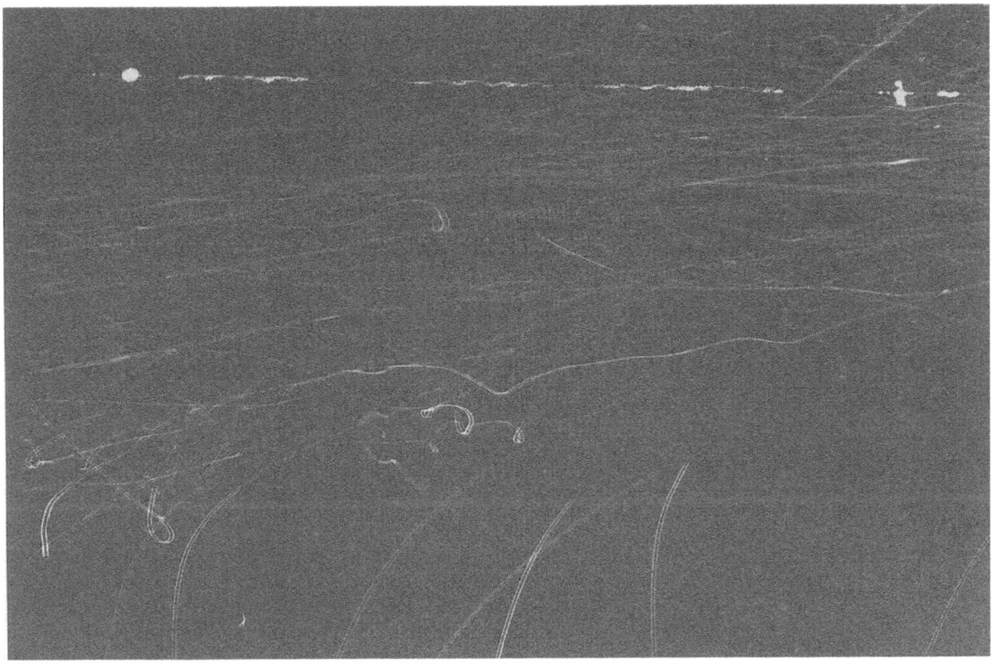

Figure 8. Helium bubbles in a jet, showing that it is spiralling regions where particles change direction rapidly.

changes of velocity in the direction of motion imply that particles pass through vortex sheets, which does not occur.

An alternative explanation (with reference to §3.2) is that the particle's velocity oscillates rapidly at certain periods during its trajectory, which correspond to its location in spiralling regions. (eg. photographs in Van Dop *et al* 1985, and figure 8)

4.2 Relative motion of particle pairs

If turbulence is a highly intermittent flow field where the smallest scales are concentrated in spiral eddy regions (which contains many scales), as opposed to a collection of eddies of different scales scattered through the flow, this has an effect on the relative motion of pairs of particles (figure 9a). The latter 'cascade' concept naturally leads to the idea that, if Y is the distance between a particle at time t after its release in a turbulent flow and the tangent line to its initial velocity (figure 9b), (i.e a conditional displacement) then if a second particle is released at time $t = 0$ its motion is quickly decorrelated from the motion of the first particle and therefore the mean square distance $\overline{l^2}$ between them at time t must satisfy $\overline{l^2} = 2\overline{Y^2}$ (4.2) (Novikov 1963). On the other hand if the particles separate mainly in the topologically singular, spiralling, regions in the flow field, it means that their motions are correlated over a longer time scale (as discussed in §2.). Outside these regions where the small scale motions are weak the particles do not separate and tend to move together (figure 10). The results of KS (Malik 1991, Fung et al (1990)) show how Y often increases without l being significantly changed. Over

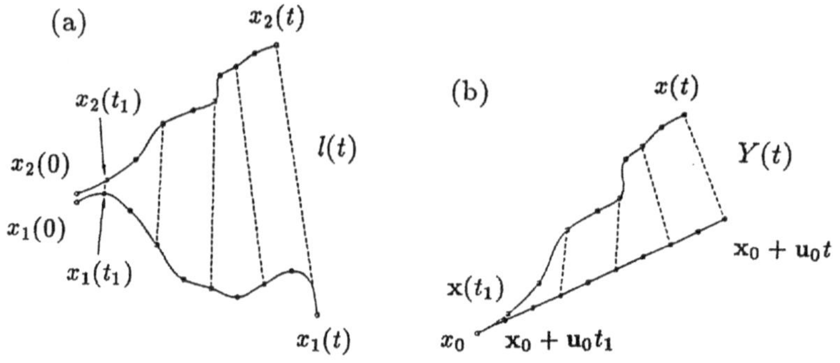

Figure 9. (a) Two particle relative dispersion, l. (b) One particle conditional dispersion, Y.

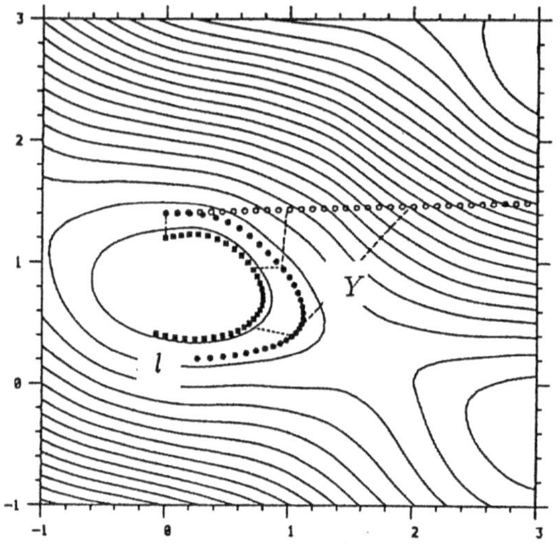

Figure 10. Two real particles (shaded) are released in a vortex. The conditional distance, Y, increases rapidly, but the separation of the particles remains small.

many realisations it is found that $\overline{l^2} \ll \overline{Y^2}$. In KS with a high Reynolds number inertial range spectrum ($k_\eta/k_c \sim 300$), it was found that $\overline{l^2} \simeq 0.1\varepsilon t^3$ (4.3). This coefficient lies at the lower end of the range estimated by Tatarski (1960) from atmospheric refractive index fluctuations (which gave ε) and the width of meteor tails (which gave $\overline{l^2}$). Note that this verification of the Obukov-Richardson law, (as defined precisely by Batchelor (1952)) requires a large inertial range where $k_\eta/k_c \simeq 400$, (perhaps because only then are there enough turns in the spiral eddies for a self similar pattern of relative motion to occur, (See §4.3)).

4.3 Scalar interfaces

Measurements of fluctuations of scalar concentrations and flow visualisation studies of turbulent flows indicate indirectly how the most significant differences in velocity on small scales often occur as part of spiralling motions, because the scalar (eg. concentration) fluctuations on the inertial scale are associated with sharp changes of concentration across sharp interfaces rolling up in the form of spirals (eg. Vassilicos 1989). The idealised model analysis of a scalar interface rolled up into spiral of the form $r = c\theta^{-\alpha}$, shows that its one-dimensional spectrum $\phi_\theta(k_1) \simeq k_1^{-p_\theta}$ (4.4), where $p_\theta = 2 - \frac{1}{1+\alpha}$. See (3.1b). The three-dimensional spectrum $E_\theta(k)$ has the same form.

Vassilicos & Hunt (1991) have shown that if such spirals form, they cannot be identified from the spectrum, unless they have many turns ($\sim 10^2$) which are not observed in practice. However, the structure of a spiral of a few turns (~ 2) is well defined by the value of the 1-dimensional fractal dimension D'_K (or Kolmogorov capacity) as a cut through the spirals. For a general interface $D'_K = 2 - p_\theta$ (4.5) and for a spiral $D'_K = \frac{1}{1+\alpha}$ (4.6) which leads to (4.4). Scalar surfaces in three-dimensional flows have been measured and values of $D'_K (= 0.33)$ were found for a wide range of turbulent flows and values of Reynolds number ($Re = u_0 L/\nu > 10^2$); which is consistent with $p_\theta = 5/3$ (from 4.1).

These scalar results are consistent with the proposed structure of the velocity field, in that the small scalar variations usually occur in larger scale eddies with complex internal structures, but scalars have the special property that their asymptotic form at high Reynolds number is indicated by the fractal dimensions D_K at lower Reynolds numbers (Vassilicos & Hunt 1991).

References

Batchelor, G.K. 1952 *Proc. Camb. Phil. Soc.*, **48**, 345–362.

Batchelor, G.K. 1953 *The theory of homogeneous turbulence.* CUP, New York.

Bertoglio, J.P. 1986 *These d'Etat, Universite C. Bernard, Lyon.*

Briscoli, M., Maggiore, P., Piccolo, F., Santongelo, P., Succi, S., Vitaletti, M., Zecca, V. 1989 *ERCOFTAC Bulletin* **III**, 3-7.

Chase, D.M. 1970 *Acustica* **22**, 303–320.

Courant & Hilbert 1953 *Methods Of Mathemayical Physics Vol.1* Interscience, New York.

Dritschel, D.G. 1989 *Comput. Phys. Rep.* **10(3)**, 77-146.

Favre, A.J., Gaviglio, J.J. & Dumas, R. 1957 *J. Fluid Mech.* **2**, 313–342.

Fermigier, M. 1980 Ph.D. dissertation. L'Universite Pierre Et Marie Curie, Paris.

Frisch, U.& Orszag, S. 1990 *Physics Today* **43(1)**, 24–32.

Fung, J.C.H. 1990 Ph.D. dissertation. University of Cambridge.

Fung, J.C.H., Hunt, J.C.R., Malik, N.A. & Perkins, R.J. 1990 Submitted to J. Fluid Mech.

Gilbert, A.D. 1988 *J. Fluid Mech.* **193**, 475-497.

Hanna, S.R. 1980 *J. Appl. Meteor.* **20**, 242–249.

Hunt, J.C.R. & Carruthers, D.J. 1990 *J. Fluid Mech.* **212**, 497-532.

Hunt, J.C.R., Buell, J.C. & Wray, A.A. 1987 NASA Report CTR-S87.

Inoue, E. 1951 *On turbulent diffusion in the atmosphere. J. Met. Soc. Japan* **29**, 246–252.

Lundgren, T.S. 1982 *Phys. Fluids* **25**, 2193-2903.

Malik, N.A. 1991 Ph.D. dissertation. University of Cambridge.

McWilliams, J.C. 1984 *J. Fluid Mech.* **146**, 21-43.

Meneguzzi, M. 1990 In *Advances In Turbulence* **3** Springer Verlag.

Moffatt, H.K. 1984 In *'Turbulence And Chaotic Phenomena In Fluids', pp. 223-230* North-Holland.

Monin, A.S.& Yaglom, A.M. 1975 *Statistical Fluid Mechanics.* Vol. 2, M.I.T. Press.

Novikov, E.A. 1963 *Soviet Phys. JETP* **17**, 1449–1453.

Obukhov, A. 1941 Bull. Acad. Sci. U.S.S.R., Geog. & Geophys., Moscow p453.

Richardson, L.F. 1926 *Proc. Roy. Soc., London, A.* **110**, 709–737.

Tatarski, V.I. 1960 *Izv. Vyssh. Uchebn. Zaved., 3 Radiofizika* **4**, 551-583.
Tennekes, H. 1975 *J. Fluid Mech.* **67**, 561-567.
Thomson, D.J. 1986 *Quart. J. Roy. Met. Soc.* **112**, 890–894.
Thomson, D.J. 1989 In *Advances In Turbulence.* **2** Springer Verlag.
Van Dop, H., Nieuwstadt, F.T.M. & Hunt, J.C.R. 1985 *Phys. Fluid* **28**, 1693-1653.
Vassilicos, J.C. 1989 In *Advances In Turbulence.* **2** Springer Verlag.
Vassilicos, J.C. & Hunt, J.C.R. 1991 *Submitted to Proc. Roy. Soc.* , .
Wray, A. & Hunt, J.C.R. 1990 In *'Topological Fluid Mechanics', pp. 95–104.* CUP.
Yeung, P.K. & Pope, S.B. 1989 *J. Fluid Mech.* **207**, 531–586.

Experimental and Theoretical Investigation of a Homogeneous Turbulent Shear Flow

O. Leuchter, J.P. Benoit, J.P. Bertoglio and J. Mathieu*

Office National d'Etudes et de Recherches Aerospatiales (ONERA) 92320 - Chatillon, France
*Ecole Centrale de Lyon (ECL) 69130 Ecully, France

A new configuration of homogeneous turbulent shear flow is investigated experimentally and numerically. The flow is created by superposition of plane strain and solid body rotation of equal rates. This provides uniform shear in planes perpendicular to the main flow direction. Very satisfactory homogeneity conditions are thus achieved. The evolution of the Reynolds stresses and associated integral length scales are determined by hot-wire measurements. The shear gives rise to a definite decrease of the turbulent energy decay and a pronounced anisotropisation of the Reynolds stresses and length scales, as compared to the evolution without shear. The experimental results are compared with numerical predictions based on a two-point closure approach of the EDQNM type. The numerical results are in good qualitative agreement with the measurements, especially for the evolution of the length scales.

1. Experimental method

The shear flow investigated in the present study is a direct superimposition of a solid body rotation and a plane strain, with the rotation rate Ω equal to the strain rate D [1]. With the rotation vector parallel to the flow direction, this results in a uniform shear at a rate $S = \Omega + D$ in planes perpendicular to the flow direction, in contrast to common practice where homogeneous shear lies in planes parallel to the flow (see ref.[1] for a short revue). The present flow belongs to a class of uniformly distorted plane flows characterized by a strain rate matrix of the form:

$$(\lambda_{ij}) = \begin{pmatrix} 0 & 0 & 0 \\ 0 & 0 & D-\Omega \\ 0 & D+\Omega & 0 \end{pmatrix} \quad \text{with} \quad \lambda_{ij} = \frac{\partial U_i}{\partial x_j} \tag{1}$$

Here the U_i are the mean velocity components. As the axial velocity U_1 is uniform, one needs only to consider the submatrix for the distortion plane x_2, x_3. Due to the particular coordinate system (chosen here for experimental convenience), the principal axes of the symmetrical part of the strain rate tensor are inclined $\pm 45°$ with respect to the axes x_2 and x_3. The stretching is in the $+45°$ direction (see below):

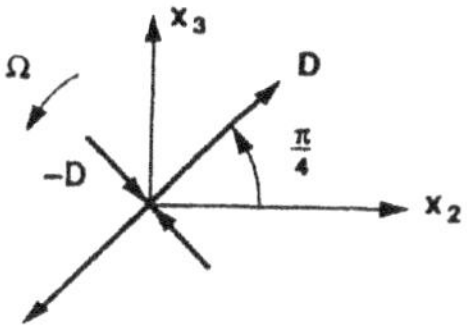

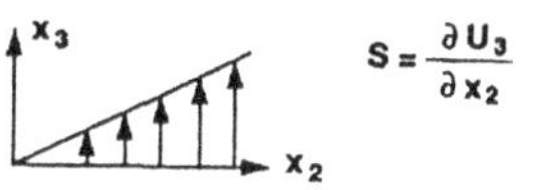

Advances in Turbulence 3
Editors: A.V. Johansson · P.H. Alfredsson

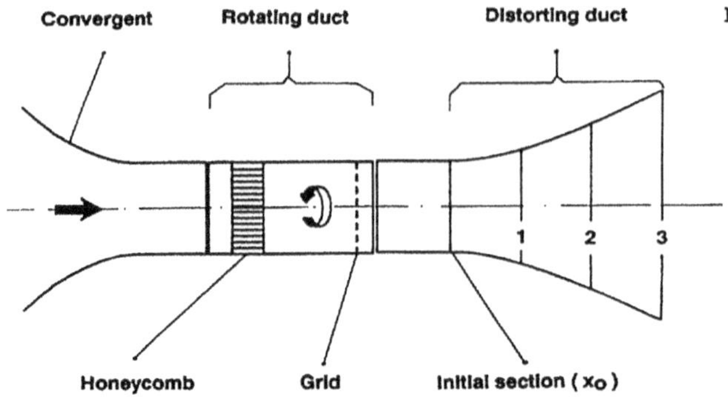

Figure 1. Schematic view of the experimental setup

As can be seen from eq. (1), pure shear is obtained for $D = \Omega$ and the shear rate $S = \partial U_3/\partial x_2$ is then equal to $D + \Omega$. The limiting case of shear separates the domain of rotation-dominated (or "elliptical") flows ($\Omega > D$) from that of strain-dominated (or "hyperbolic") flows ($\Omega < D$) with quite different behaviours. The streamlines, determined by the stream function

$$\Psi = -\frac{1}{2}[(D+\Omega)x_2^2 - (D-\Omega)x_3^2] \tag{2}$$

indeed have elliptical shapes in the former case and hyperbolic ones in the latter. For the limiting case of shear, they degenerate into straight lines parallel to the x_3 axis.

The experimental facility used to provide this type of rotation-strain coupling is described in ref.[1] and is illustrated schematically in figure 1. It is composed of a rotating pipe producing the solid body rotation, plus a distorting duct of appropriate shape which superimposes the strain at the required rate. The distorting duct has elliptical cross sections of constant area, but of varying aspect ratio and orientation. This geometry is determined from the requirement that the distorting duct takes a particular stream tube originating from the upstream cylindrical pipe and distorts it as defined by eq. (1). The characteristics of the ellipses are determined from the quadratic form (see [2], [3]):

$$C_{ij}x_ix_j = R^2 \quad (i,j = 2,3) \tag{3}$$

where R is the radius of the (circular) initial section of the duct. The tensor $\underline{C}$ is defined by $\underline{C} = (\underline{F}\underline{F}')^{-1}$, where $\underline{F}$ is the deformation gradient tensor associated with the strain rate tensor introduced above (the symbol $'$ stands for transposition). Considering the form of eq. (1), the restriction of $\underline{F}$ to the distortion plane ($i,j = 2,3$) is expressed:

$$F_{ij} = (\cosh\sigma t)\delta_{ij} + \frac{1}{\sigma}(\sinh\sigma t)\lambda_{ij} \quad \text{with} \quad \sigma^2 = D^2 - \Omega^2 \tag{4}$$

where t is taken equal to zero at the initial section of the duct.

From these general relations one readily deduces the aspect ratio a/b of the elliptical sections (i.e. the ratio of the major to minor axis) and the inclination angle ψ with respect to the x_2-axis, by:

$$\begin{aligned} \frac{a}{b} &= 1 + \frac{1}{2}G^2 + G\sqrt{1+\frac{1}{4}G^2} \text{ with } G = \frac{2D}{\sigma}\sinh\sigma t \\ \psi &= \frac{1}{2}[\pi - \tan^{-1}(\frac{\sigma}{\Omega}\coth\sigma t)] \end{aligned} \tag{5}$$

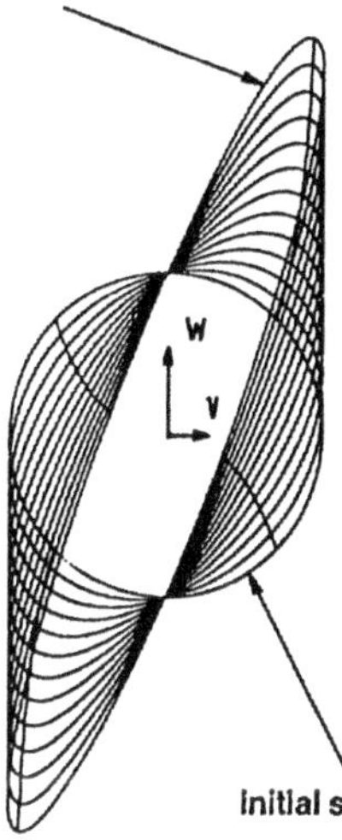

Figure 2. Evolution of the cross section of the distorting duct

For elliptical flows ($\Omega > D$) σ becomes imaginary and the hyperbolic functions become trigonometric ones. Then $(\sinh \sigma t)/\sigma$ is replaced in eq. (5) by $(\sin \sigma' t)/\sigma'$, with $\sigma' = -i\sigma$. In this case the gradient tensor $\underline{F}$ becomes periodic, the period being $2\pi/\sigma'$, and the corresponding period of the distorting duct is $\pi U_1/\sigma'$. In all other cases the aspect ratio is an increasing function of time (i.e. of the distance) and ψ lies within the interval $[\pi/4, \pi/2]$.

The limiting case of shear can be approached from either the hyperbolic or the elliptical side, by letting σ approach zero. Then $(\sinh \sigma t)/\sigma$ as well as $(\sin \sigma' t)/\sigma'$ become equal to t, and G equal to St. Finally we get:

$$\begin{aligned} \frac{a}{b} &= 1 + \frac{1}{2}(St)^2 + St\sqrt{1 + \frac{1}{4}(St)^2} \\ \psi &= \frac{1}{2}[\pi - \tan^{-1}(\frac{2}{St})] \end{aligned} \tag{6}$$

These relations show that the geometry of the shear duct depends only on the total shear St, that the aspect ratio approaches infinity as $(St)^2$ and that the orientation of the major axis varies from 45° for $t = 0$ to 90° for $t = \infty$.

The variation of the elliptical cross sections corresponding to (6) is shown in figure 2. These represent the distorting duct used at ONERA [1], with a maximum St of 2.25 at section 3, (where the aspect ratio is nearly 7). Due to the rapidly increasing eccentricity of the elliptical sections, the total distortion attainable with this type of shear generator is severely limited, compared with the more classical devices mentioned above.

One of the principal advantages of the present method lies in the spatial uniformity of the streamwise velocity U_1 (and of the gradients) which provides particularly favourable conditions for achieving a good transverse homogeneity in the turbulence. Moreover, the turbulent field is generated independently of the mean flow distortion, by grids (see figure 1) in the rotating duct, before strain is applied. Grids of different mesh sizes can be used, to examine different initial conditions of the turbulent field for a fixed shear rate. The shear rate itself can easily be varied by varying the mean velocity in the axial direction (and proportionally the rotation rate). More details concerning the experimental aspects can be found in [1].

Lastly we would like to emphasize the capability of this test facility to cover a wide range of homogeneously distorted flows of the type represented by eq.(1), either in the elliptical or in the hyperbolic flow domain. This can be achieved by simply substituting a new distorting duct of appropriate geometry as discussed above, for the one presently used for the shear. The case of elliptical flows appears to be a particularly interesting one, especially in connection with the problem of 3D-instabilities that develop in strained vortices (see e.g. the paper of Malkus and Waleffe at the present conference [4]). New elliptical and hyperbolic distorting ducts are presently being prepared at ONERA and the corresponding flows will be investigated in future stages of this study.

Mean flow and turbulence are measured by hot wire techniques with two- and four-wire probes, the latter yielding the three components of the velocity simultaneously. The evolution of the Reynolds stresses (including spectral distributions) and of various integral length scales are commonly measured. The qualification of the flow field revealed very satisfactory results with regard to homogeneity, with a domain of reasonable extent in which the shear flow is perfectly homogeneous [1]. The theoretical shear rate (34.1 s^{-1} for $U_1 = 10$ m/s) has been exactly duplicated in the measurements, and the streamlines were found to be straight lines parallel to the x_3-axis, in accordance with eq. (2). The most significant experimental findings concerning the turbulent field will be discussed in the third part of this paper, together with the numerical results.

2. Numerical method

The numerical method is based on a two-point closure approach of the EDQNM type developed at the ECL by Bertoglio [5],[6]. The model operates in spectral space and describes the evolution of the spectral tensor Φ_{ij} defined as the Fourier transform of the twopoint velocity correlation tensor. As can be seen from the equation for Φ_{ij}:

$$(\frac{\partial}{\partial t} + 2\nu k^2)\Phi_{ij}(\vec{k},t) - \lambda_{lm} k_l \frac{\partial \Phi_{ij}(\vec{k},t)}{\partial k_m} +$$

$$\lambda_{lm}[(P_{ij}(\vec{k}) - \frac{k_l k_l}{k^2})\Phi_{mj}(\vec{k},t) + (P_{ji}(\vec{k}) - \frac{k_l k_l}{k^2})\Phi_{lm}(\vec{k},t)] = T_{ij}(\vec{k},t) \tag{7}$$

the evolution of this quantity is first determined by the (linear) action of the mean velocity gradients, represented by the second and third terms on the left hand side, and also by the nonlinear transfer term T_{ij}. In the above equation $P_{ij}(k)$ is the solenoidal projector defined as:

$$P_{ij}(\vec{k}) = \delta_{ij} - \frac{k_i k_j}{k^2} \tag{8}$$

Only the transfer term $T_{ij}(k)$ needs to be closed. The closed form for the transfer can be written as follows, by splitting T_{ij} into two terms, an "input" term T_{ij}^{+} and an "output" term T_{ij}^{-}

$$T_{ij}(\vec{k},t) = T_{ij}^{+}(\vec{k},t) + T_{ij}^{-}(\vec{k},t) \tag{9}$$

where

$$\begin{aligned} T_{ij}^{+}(\vec{k},t) &= -2\int_{\vec{p}}\int_{\vec{q}} \theta_{kpq}(t) S_{ij}^{+}(\vec{k}|\vec{p},\vec{q},t)\delta(-\vec{k}+\vec{p}+\vec{q})\, dp^3 dq^3 \\ T_{ij}^{-}(\vec{k},t) &= -2\int_{\vec{p}}\int_{\vec{q}} \theta_{kpq}(t) S_{ij}^{-}(\vec{k}|\vec{p},\vec{q},t)\delta(-\vec{k}+\vec{p}+\vec{q})\, dp^3 dq^3 \end{aligned} \tag{10}$$

with :

$$
\begin{aligned}
S^{+}_{ij}(\vec{k}|\vec{p},\vec{q},t) &= P_{ikl}(\vec{k})P_{jmn}(-\vec{k})\Phi_{km}(\vec{p},t)\Phi_{ln}(\vec{q},t)+\\
&+ P_{jkl}(-\vec{k})P_{lnm}(\vec{k})\Phi_{kn}(-\vec{p},t)\Phi_{lm}(-\vec{q},t)\\
S^{-}_{ij}(\vec{k}|\vec{p},\vec{q},t) &= P_{ikl}(\vec{k})[P_{kmn}(\vec{p})\Phi_{lm}(\vec{q},t) + P_{lmn}(\vec{q})\Phi_{km}(\vec{p},t)]\Phi_{jn}(-\vec{k},t)+\\
&+ P_{jkl}(-\vec{k})[P_{kmn}(-\vec{p})\Phi_{ln}(-\vec{q},t) + P_{lmn}(-\vec{q})\Phi_{kn}(-\vec{p})]\Phi_{lm}(\vec{k},t)
\end{aligned} \tag{11}
$$

Here $P_{ijl}(\vec{k})$ is given by

$$
P_{ijl}(\vec{k}) = \frac{1}{2}[k_l P_{ij}(\vec{k}) + k_j P_{il}(\vec{k})] \tag{12}
$$

and θ_{kpq} is the characteristic time defined as

$$
\theta_{kpq}(t) = [\eta(k,t) + \eta(p,t) + \eta(q,t)]^{-1} \tag{13}
$$

$$
\text{with : } \eta(k,t) = \lambda[\int_0^k p^2 E(p,t)dp]^{1/2} + \nu k^2 \tag{14}
$$

as proposed by Pouquet et al. [8]. The value of λ, which is the unique constant of the model, is taken here to be equal to 0.36.

The closure model represented by the above relations constitutes a rather straightforward extension of the classical EDQNM model derived for isotropic turbulence (Orszag, [7]). The effects of the mean velocity gradients are taken into account in the equation for the double velocity correlations, but they are neglected in the equation for the triple correlations, as suggested by Cambon *et al.* [9]. It must be emphasized that the numerical solution of the above equations requires the calculation of the **double** volume integrals of equation (10) taking into account the triadic relations between the wave vectors k, p and q. This leads to much more extensive computing efforts than in the isotropic case, where the integration domain covers only a limited region of the $p-q$ plane. To alleviate this problem, a "modelled" form of the 3D-transfer can be used which consists in evaluating the complete 3D-transfer only at a limited number of time steps, between which the input term $T^{+}_{ij}(\vec{k},t)$ is assumed to vary as its isotropic form t^{+}_{r}, according to the relation:

$$
\{T^{+}_{ij}(\vec{k},t)\}_{\text{mod.}} = \frac{t^{+}_{r}(k,t)}{t^{+}_{r}(k,t_0)} T^{+}_{ij}(\vec{k},t_0). \tag{15}
$$

Here, t_0 is the value of t where the last full 3D calculation has been made. The quantity $t^{+}_{r}(k,t)$ is calculated from the energy spectrum $E(k,t)$ using the isotropic formulation of the EDQNM model. As for the output term $T^{-}_{ij}(k,t)$, it is written in a form where the spectral tensor elements for the wave vector k appear as factors, namely

$$
T^{-}_{ij}(\vec{k},t) = \eta_{ik}(\vec{k},t)\Phi_{kj}(\vec{k},t) + \eta_{jk}(-\vec{k},t)\Phi_{ik}(\vec{k},t), \tag{16}
$$

and the modelling assumptions are then made for the functions $\eta_{ij}(\vec{k},t)$, instead of for the output term $T_{ij}(\vec{k},t)$ itself:

$$
\{\eta_{ij}(\vec{k},t)\}_{\text{mod.}} = \frac{t^{-}_{r}(k,t)/t^{-}_{r}(k,t_0)}{E(k,t)/E(k,t_0)}\ \eta_{ij}(\vec{k},t_0) \tag{17}
$$

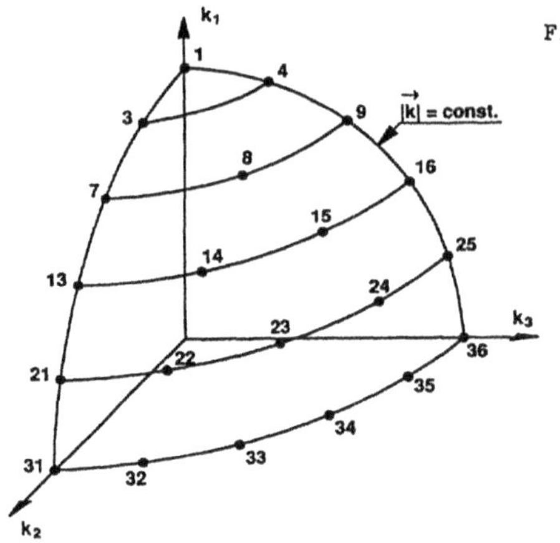

Figure 3. Discretization in spectral space

In most of our calculations, the full transfer is computed at each fourth time step, thus reducing the computer time to about a fourth of what it is when the full 3D-transfer is calculated at each time step.

The computational domain is restricted to the quarter space defined by $k_1 > 0, k_2 > 0$. Taking advantage of symmetry properties, the following flow configurations can be described:

- pure plane strain,
- plane strain with rotation,
- plane shear,
- axisymmetric strain.

The three first configurations correspond to the class of flows discussed previously and represented by eq. (1).

Figure 3 illustrates the type of computational domain discretization adopted here, where, for the sake of simplicity, only the right half of the quarter space $(k_1 > 0, k_2 > 0)$ is presented. The example is for six planes (including the pole) normal to the k_1-direction; the meshing principle consists in adding one additional mesh point on the visible part of the circles when descending from one plane to the next lower one. Thus the total number of grid points on the quarter sphere is equal to the square of the number of planes (here 36 grid points).

The calculations are commonly made with four to six planes beyond which the computer time rapidly increases, without improving the quality of the predictions appreciably. Typical values of the number of spheres are 15 or 20.

As for the initial conditions, the computations have been made with the initial values of the turbulent energy and the dissipation rate taken from our experiment. An energy wave number spectrum is first generated from these values by using the Helland - Rey formulation:

$$E(k) = 1.5\epsilon^{2/3}L^{5/3}\frac{(kL)^m}{[1+(kL)^2]^{\frac{m}{2}+\frac{5}{6}}}e^{-1.5\gamma(k\eta)^{4/3}} \quad \text{with} \quad L = \beta\frac{q^2}{\epsilon}, \eta = (\frac{\nu^3}{\epsilon})^{\frac{1}{4}} \tag{18}$$

β and γ are two constants which are evaluated from the condition that $E(k)$ must restore the given values of the energy and the dissipation rate. Then the initial spectral tensor is determined by using the isotropic relation:

$$\Phi_{ij}(\vec{k}, t=0) = \frac{E(k)}{4\pi k^2}(\delta_{ij} - \frac{k_i k_j}{k^2}) \tag{19}$$

This is a rough approximation of the experimental initial state, which was found to depart significantly from the idealized condition of isotropy.

3. Results and discussion

In the following, some of the more typical results will be discussed, by comparing the experimental results with the EDQNM model predictions. (Note that we use here the notations x, y, z and u, v, w rather than x_1, x_2, x_3 and u_1, u_2, u_3 used previously).

Numerical results obtained from the solution of the linearized equation (7) (i.e. neglecting the transfer term T_{ij}) are also presented for illustration. These solutions (termed hereafter "RDT" (for Rapid Distortion Theory) solutions) have the huge advantage of consuming much less computer time than the solution of the full equations, but on the other hand they cannot claim to simulate closely the evolution of the turbulent field for our experimental conditions. This is why the characteristic "linear" time (τ_L) related to the mean flow distortion (and/or the "residence" time (τ_R) of the flow during submission to the distortion) are approximately of the same order as the characteristic "nonlinear" decay time (τ_{NL}) of the turbulence. Only at the very beginning of the shearing process can the necessary condition for RDT, namely $\tau_R \ll \tau_{NL}$, be satisfied. Nevertheless, the linear solutions of the truncated equation (7) give very useful indications concerning the tendencies of the turbulent field subjected to the distortion, even though they do generally appear to be exaggerated. Therefore they can contribute valuably to our physical understanding of the related mechanisms. The method of numerical resolution of the linear problem was developed at ECL (see [3]).

Figure 4 shows the evolution of the turbulent kinetic energy versus the total shear St. As can be seen from the figure, the presence of shear produces only a slight decrease in the energy decay, compared with the evolution predicted by RDT. This confirms the fact that our problem is strongly dominated by non-linear effects. Both the experiment and the EDQNM prediction show very similar evolutions. The maximum value of St is seen to be not high enough to stop the energy decay.

As for the evolution of the individual normal stresses and the shear stress (not shown here), a good qualitative agreement between the experimental results and the EDQNM predictions is achieved, in spite of the different initial anisotropies. The experimental evolution of the shear stress correlation coefficient corresponds closely to that observed for the longitudinal shear case, with a clear tendency to approach asymptotically a value of the order of 0.5. The EDQNM model was found to underestimate slightly the experimental evolution; the reason for this is not yet completely elucidated.

The different integral length scales are among the most pertinent measurable quantities with regard to the directional anisotropy in spectral space. While all nine integral length scales are readily available from the numerical model, only those relative to the separation in the axial direction are most easily obtainable from experiment. These length scales can indeed be determined from the frequency spectra via a Taylor hypothesis, whereas those with separation in the transverse directions would require more cumbersome two-point measurements.

The evolution of the length scales with separation in the streamwise direction are shown in figure 5. The most spectacular consequence of the shear is seen to be a strong inhibition of the growth of the longitudinal length scale $L_{11,1}$ and a gradual anisotropiza-

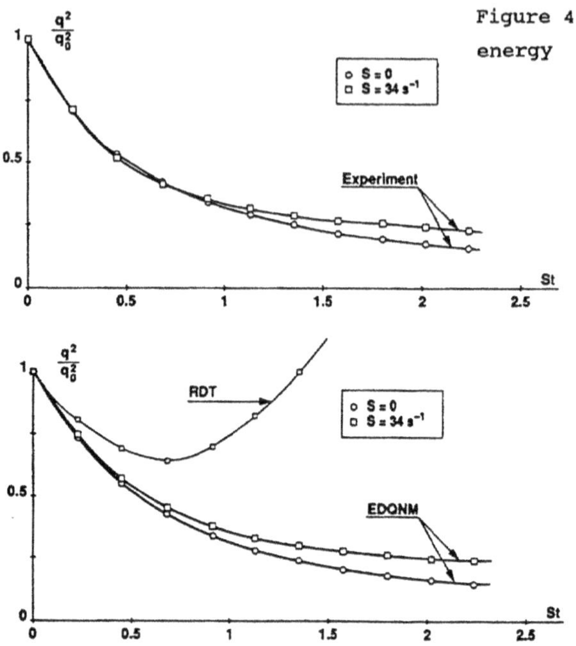

Figure 4. Evolution of the turbulent energy

tion of the lateral ones, with $L_{22,1}$ becoming larger than $L_{33,1}$. The evolutions relative to the absence of shear are indicated by broken lines. The agreement between the experimental evolution and that given by the EDQNM model is very satisfactory, in spite of the different initial states of anisotropy. Here again, the RDT solution clearly indicates the actual tendencies, unavoidably exaggerated for the larger values of St.

A similar good agreement between the experiment and the EDQNM calculation was achieved for the products of the normal stresses u_i^2 with their corresponding length scales $L_{ii,1}$. These quantities represent the energy content of the velocity fluctuations in the equatorial plane $k_1 = 0$, since they are defined from the surface integrals extended over the plane $k_1 = 0$, of the diagonal elements of the spectral tensor:

$$u_i^2 L_{ii,1} = \pi \iint_{k_1=0} \Phi_{ii}(0, k_2, k_3)\, dk_2 dk_3$$

Both, experiment and calculation, provided a distinct increase in the planar energy due to the shear only for the transverse components v and w, whereas the component normal to the distortion plane was found to be scarcely affected.

4. Conclusion

The flow configuration described in the present study belongs to a class of uniformly distorted plane flows, with coupled effects of rotation and strain. Such flows represent suitable test cases for the investigation of homogeneous turbulence. The special case of pure shear is obtained by imposing equal rates of rotation and strain. The experimental facility built according to this principle provides very satisfactory homogeneity conditions, compared to those obtained with more conventional arrangements. The main features of the effect of shear are a gradual reduction in the turbulent energy decay, and

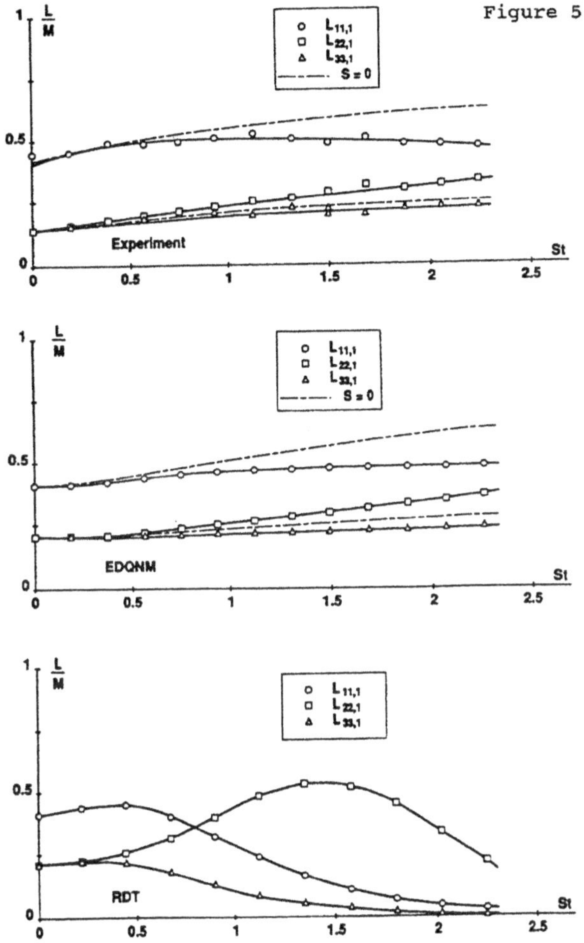

Figure 5. Evolution of the length scales

a strong tendency for the Reynolds stresses and the integral length scales to become more anisotropic. The gradual building up of the shear stress compares well with that observed in previous homogeneous shear flow experiments.

The numerical predictions based on the EDQNM closure model are in good qualitative agreement with the measurements, although in the experiments the initial turbulence is not perfectly isotropic as it is assumed in the calculations. The main features of the action of shear are very satisfactorily recovered. The most interesting numerical result concerns the length scales, as these quantities are hardly available from simple one-point closure models.

Future work will deal with the introduction of anisotropic initial conditions corresponding more closely to the experimental ones. The numerical prediction, as well as the experiment, will be extended to other configurations of distorting mean flows, with different types of rotation-strain coupling.

Acknowledgements

This work has been partly supported by DRET, under grant n° 89 001.

References

[1] V. Moulin, O. Leuchter and P. Geffroy 1989 7th Symp. Turb. Shear Flows, Stanford.

[2] C. Teissedre 1984 Thesis, Lyon.

[3] C. Cambon, C. Teissedre and D. Jeandel 1985 *J.M.T.A.*, **4**, n°5, pp. 629-657.

[4] W.V.R. Malkus and F.A. Waleffe 1990 3rd Eur. Turb. Conf., Stockholm.

[5] J.P. Bertoglio 1981 3rd Symp. Turb. Shear Flows, Davies.

[6] J.P. Bertoglio 1986 Thesis, Lyon.

[7] S. Orszag 1970 *J. Fluid Mech.*, **41**, pp. 363-386.

[8] A. Pouquet, M. Lesieur, J.C. Andre and C. Basdevant 1975 *J. Fluid Mech.* **72**, pp. 305-319.

[9] C. Cambon, D. Jeandel and J. Mathieu 1981 *J. Fluid Mech.*, **104**, pp. 242-262.

New Physics and Models for Compressible Turbulent Flows

O. Zeman and G.A. Blaisdell

Center for Turbulence Research, Stanford University, Stanford CA 94305-3030

This paper reports on recent advances in modeling and direct numerical simulations (DNS) of compressible turbulent flows. In the second-order equations, two principal compressibility terms have been identified: the dilatation dissipation (ϵ_d) and the pressure-dilatation correlation (Π_d). The dilatation dissipation arising from the occurence of shock-like events is treated by the theory of Zeman. A first order approximation to Π_d in homogeneous turbulence is shown to be $\Pi_d \propto -D\overline{p^2}/Dt$. The rate of change of the pressure variance $\overline{p^2}$ is proposed to be governed by nonlinear relaxation (on the acoustic time scale) to its equipartition value. The effect of dilatation dissipation is illustrated by comparing experiments and results of a second-order closure model (SOC) for compressible shear layers. DNS data in decaying, and shear-driven homogeneous turbulence are utilized to illustrate the initial-value dependence of turbulence evolution and to verify the proposed model for pressure-dilatation correlation.

1. Introduction

As the Mach number of a given turbulent flow increases, the density and pressure fluctuations grow and compressibility affects the turbulence structure and dynamics. These changes may be subtle, and in boundary layers are often obscured by rapid variation of mean flow properties near the wall. In boundary-free flows such as wakes and shear (mixing) layers, variations in the mean flow properties are relatively small and compressibility effects on turbulence can be isolated. The most prominent, observed effect of compressibility on turbulence is the inhibition of shear layer growth rate with the (convective) Mach number M_c in experiments of Ikawa and Kubota, Bogdanoff, Papamoschou and Roshko, Samimy and others (for references see Zeman (1990) and Papamoschou and Roshko (1988)).

Homogeneous decaying, and shear-driven turbulent flows are ideally suited for analysing compressibility effects. These can be generated via direct numerical simulations (DNS) of Navier-Stokes equations (Feiereisen et al. 1981, Passot and Pouquet 1987, Passot et al. 1988, Blaisdell et al. 1988, Sarkar et al. 1989, and others). In simulations of 2D turbulence, Passot and Pouquet identified regions of shock-like structures, or shocklets, whose formation was found to depend on a certain combination of initial values of the r.m.s. Mach number (M_t) and the pressure (density) variance $\overline{p^2}$. Similar findings were obtained in the DNS of 3D turbulence (Passot et al. 1988, Sarkar et al. 1989).

In this paper we focus mainly on the physics and analysis of homogeneous compressible turbulence, on modeling the pressure dilatation Π_d and on initial condition

Advances in Turbulence 3
Editors: A.V. Johansson · P.H. Alfredsson

(transient) effects on turbulence evolution. The dilatation dissipation theory and its application to modeling shear layers by Zeman(1990) (hereafter Z90) are reviewed in the following section. The basic equations, and development of the closure for Π_d are in Section 3. Results of DNS and SOC modeling of decaying, and shear-driven turbulence illustrating the effect of initial conditions are presented in Section 4,and 5, followed by Conclusions.

2. Dilatation dissipation in compressible shear layers

Noticable effects of compressibility on the shear layer growth rate are observed to appear when the velocity difference, $\Delta U = U_1 - U_2$, between the fast and slow free streams is about equal to an average sonic speed a_o of the flow, i.e., when the characteristic Mach number $M = \Delta U/a_o \geq 1$. Considering the (baroclinic) effect of unequal free stream densities on the convective velocity (U_c) of large turbulent eddies, Papamoschou and Roshko (1988) defined a more relevant (convective) Mach number M_c. For small baroclinic effects M is twice the value of M_c. Z90 argued that when $M \geq 1$ significant levels of dilatation dissipation can be generated in regions where fluid elements acquire, relative to each other, supersonic speeds and form shock-like structures or shocklets.

In homogeneous turbulence, the dilatation dissipation is formally defined as $\epsilon_d = \frac{4}{3}\nu\overline{(u_{j,j})^2}$. On the basis of a shocklet model, Zeman obtained the following expression for the shocklet contribution to ϵ_d:

$$\epsilon_d \propto \frac{q^3}{L}[(\frac{1+\gamma}{2})^2 \frac{1}{M_t^4} \int_1^\infty \left(\frac{m^2-1}{m}\right)^3 p(m)dm] \propto \frac{q^3}{L} F(M_t, K). \tag{1}$$

Here, $m = (u_j u_j)^{1/2}/a^*$ is the instantaneous Mach number based on the sonic temperature $T^* = (2/(\gamma+1))\widetilde{T}$, $p(m)$ is a probability distribution function (pdf) of m, and L is a turbulence length scale. The function $F(M_t, K)$ represents the expression in square brackets; its values are obtained by numerical integration over the pdf $p(m)$. This pdf is approximated by the Gram-Charlier expansion about a Gaussian distribution with a prescribed kurtosis $K \geq 3$. The r.m.s. Mach number $M_t = q/a(\widetilde{T})$ is the fundamental parameter of compressible turbulence; $q^2 = \widetilde{u_j u_j}$ designates twice the Favre-averaged turbulent kinetic energy and a is the sonic speed at the Favre-averaged temperature $\widetilde{T}$.

In 3D turbulence q^3/L is proportional to the solenoidal dissipation ϵ_s, so the total dissipation rate of turbulence kinetic energy (in 3D) can be expressed as

$$\epsilon_{tot} = \epsilon_s(1 + c_d F(M_t, K)). \tag{2}$$

The constant c_d and the kurtosis K are parameters that are estimated from experiments. In low speed shear layers K varies between 4 and 20 while homogeneous flows are nearly Gaussian with $K \approx 3$. In shear layers, $K = 6$, and $c_d = 0.75$ appear to be about optimum. As evident from (2), ϵ_d augments the solenoidal dissipation, making the total energy dissipation dependent on M_t. As shown in Z90, $F(M_t, K = 6)$ is virtually zero for $M_t < 0.25$ and increases sharply thereafter; hence ϵ_d exerts a strong influence on shear layer turbulence when M_t is sufficiently high. For illustration, Fig. 1 compares the shear layer growth rate $d\delta/dx = \delta'$ inferred from the SOC model simulations (Z90), with experimental data (the growth rates are normalized by the low speed value δ'_o with the same ΔU and U_c). Fig. 1 also includes model results with ϵ_d neglected. It is evident that the new model is qualitatively correct thanks to the inclusion of

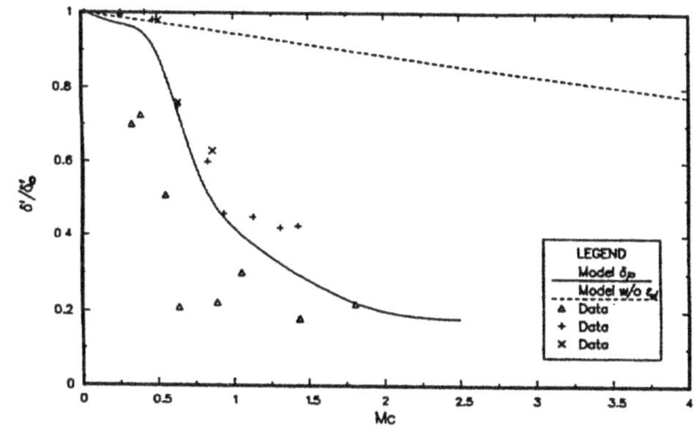

Figure 1. Normalized shear layer growth rate vs. convective Mach number M_c. Solid and dashed lines are Z90 model results (dashed lines with $\epsilon_d = 0$). Experimental data are: (Δ) Papamoschou and Roshko (1988); (+) Bogdanoff (1983); ($\times$) Samimy et al. (1989).

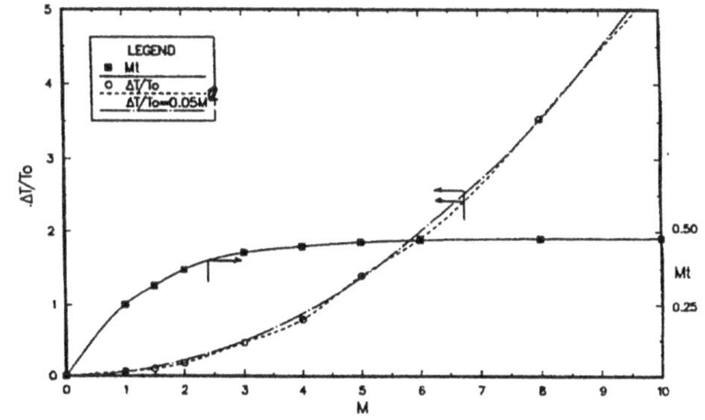

Figure 2. Predicted centerline (maximum) temperature excess (ΔT), and rms Mach number M_t vs. $M = \Delta U/a_o$ ($= 2M_c$), in a shear layer.

the dilatation dissipation through the function $F(M_t)$. An interesting aspect of the dilatation dissipation influence on shear layer turbulence is illustrated in Fig. 2, where the maximum values of M_t are plotted vs. $M = 2M_c$. For $M < 1$, M_t behaves as in a low-speed shear layer with $q \propto \Delta U$. When M exceeds about 3, M_t approaches a (saturation) value $M_{t\infty} \approx 0.5$. This suggests that at sufficiently large M the turbulent fluctuations are determined by the local sonic speed, and not by the (total) shear ΔU. This saturation phenomenon is yet to be verified by experiments.

3. Pressure dilatation correlation in decaying turbulence

Helmholtz' decomposition of the velocity field into the solenoidal (incompressible) and dilatational (compressive) components enables us to cast the equations that govern second-order statistics in compressible homogeneous turbulence in the following form

$$1/2\frac{\partial q^2}{\partial t} = -\nu\overline{(\omega_j)^2} - 4/3\nu\overline{(u_{j,j})^2} + \rho^{-1}\overline{pu_{j,j}} + P_s \quad (3)$$

$$\frac{\partial}{\partial t}(c_p T + 1/2q^2) = P_s \quad (4)$$

$$\overline{p} = \overline{\rho}R\widetilde{T} \quad (5)$$

The viscous term involving the enstrophy $\nu\overline{(\omega_k)^2} = \epsilon_s$ is the familiar solenoidal dissipation which depends only on the solenoidal velocity field. As argued in Section 4, ϵ_s is expected to obey the same model equation as in the incompressible regime. The second viscous term in (3) is the dilatation dissipation (ϵ_d) whose principal contribution is due to shocklets and depends solely on the dilatational velocity field. The third term in (3) is the pressure-dilatation correlation Π_d, which may, in principle, depend on both fields. As we shall see later Π_d mediates the exchange between the kinetic and potential energies. P_s is the rate of turbulence production by mean shear and is discussed in detail in Section 5. In decaying turbulence $P_s = 0$, and the total enthalpy $c_p\widetilde{T} + \frac{1}{2}q^2$ is conserved. In the equation of state (5) we have a choice of fixing either the mean pressure or density; in agreement with the DNS method we set $\overline{\rho} = const.$ In deriving (3), we have neglected fluctuations in kinematic viscosity ν.

The shocklet contribution to ϵ_d is given by (2), and since isotropic turbulence is very nearly Gaussian the function $F(M_t, K)$ is evaluated with K=3. In the present DNS results a nontrivial amount of dilatation dissipation is observed even in the shockless regime (when $M_t < 0.2$). However this contribution depends, apart from M_t, also on the initial conditions and Reynolds number, and is considered negligible. The remaining undetermined term in (3)-(5) is the pressure-dilatation correlation $\overline{pu_{j,j}} = \Pi_d$. In Z90 it was argued that in equilibrium turbulent flows such as spatially evolving layers, Π_d can be neglected, or absorbed in ϵ_d. As shown later, this is not justified in decaying, or rapidly changing turbulence, and a model for Π_d has to be formulated.

Formulation of the pressure-dilatation model: An approximation to Π_d can be readily obtained from first principles. Pressure, density, and entropy $P, \rho, and\ S$ are related by the second law of thermodynamics

$$\frac{1}{\rho}\frac{D\rho}{Dt} = \frac{DP}{Dt}a^{-2} - \frac{DS}{Dt}c_p^{-1}. \quad (6)$$

The variables are decomposed as $P = \overline{p} + p, \rho = \overline{\rho} + \rho\prime, T = \widetilde{T} + T' and\ S = \widetilde{S} + s$. From the mass conservation equation we obtain an expression for fluctuating divergence

$$u_{j,j} = -\frac{1}{\rho}\frac{D\rho\prime}{Dt} - (\rho\prime u_j)_{,j}, \quad (7)$$

and combining (6), (7) and neglecting terms of order $O(\rho\prime/\rho)$ or smaller, we obtain

$$\Pi_d = \overline{pu_{j,j}} = -\frac{1}{2}\frac{D\overline{p^2}}{Dt}(\overline{\rho}a^2)^{-1} + \overline{\frac{Ds}{Dt}\frac{p}{c_p}} - \overline{p_{,j}u_j\rho'}. \quad (8)$$

Let us first consider (8) in the absence of shocks. Since the sources of entropy fluctuations are fluctuating dissipation and temperature variance destruction, the second term in (8) can be neglected for Reynolds numbers $R_t = qL/\nu >> 1$. An order-of-magnitude estimate of the last term in (8) can be made by assuming that large energy-containing eddies conserve total enthalpy. It then follows that $\overline{p_{,j} u_j \rho'} = O(\Pi_d M_t^2(\gamma - 1)/2)$ and the above term is discarded for $M_t \leq 1$. Hence, in this approximation

$$\Pi_d = -\frac{1}{2}\frac{D\overline{p^2}}{Dt}\frac{1}{\overline{\rho}a^2}, \tag{9}$$

as in linear acoustics. The DNS of homogeneous decaying turbulence indicate that the two terms in (9) are virtually identical throughout the tested range $0 < M_t < 0.7$. This suggests that the compressive part of the turbulence obeys linear acoustics for moderate r.m.s. Mach numbers. The physical meaning of (9) is best illustrated by combining (3) and (9) to obtain

$$\frac{D}{Dt}[\frac{1}{2}(q^2 + \overline{p^2}/\overline{\rho}^2 a^2)] = -(\epsilon_s + \epsilon_d). \tag{10}$$

The quantity in square brackets in (10) is the total energy of turbulence; in a purely acoustic mode the total energy is conserved and the kinetic and potential energies are in equipartition. Vigorous exchange of energy between q^2 and p^2 is possible, if the initial state is not in equipartition. Such a "mismatch" can occur in explosions. At larger initial Mach numbers ($M_t > 0.5$) shocks may form in which case the second term on RHS of (6) cannot be neglected. Since Ds/Dt is proportional to fluctuating (dilatation) dissipation, the correlation of Ds/Dt with p/c_p is of order $O(\overline{\rho}\epsilon_d M_t^2(\gamma - 1))$ and is formally absorbed into the dilatation dissipation function $F(M_t)$.

In view of (9), the closure for Π_d has been reduced to determining $D\overline{p^2}/Dt$. Since a formal conservation equation for $\overline{p^2}$ is intractable, it is more preferable to consider a phenomenological model for $\overline{p^2}$. A plausible form of the model was proposed by Zeman (1990a):

$$\frac{D\overline{p^2}}{Dt} = -\frac{\overline{p^2} - p_e^2}{\tau_a}. \tag{11}$$

The above equation states that the pressure variance relaxes to its equilibrium (equipartition) value (p_e^2) on the acoustic time scale $\tau_a \approx L/a$. The equilibrium value p_e^2 is related to the kinetic energy q^2 and M_t by

$$p_e^2/\overline{\rho}^2 q^2 a^2 = \pi(M_t) = \alpha\frac{M_t^4}{1 + \alpha M_t^4}\frac{1}{E}, \tag{12}$$

where $E = p_e^2/\overline{\rho}^2 q_c^2 a^2$ is the partition of potential/kinetic energies in the compressive mode. Equation (12) implies that the ratio of compressive to solenoidal energy is

$$\frac{q_d^2}{q_s^2} = \alpha M_t^4, \tag{13}$$

where the optimal value of $\alpha = 15$. DNS results indicate that for times $t >> L/a$, the partition ratio $E = 1$ (Sarkar et al. 1989). However, in order to match properly the initial value of Π_d between the DNS and the present model it was necessary in some special cases to adopt a (second order) dynamic equation for E (not shown). It should be noted that the initial value theory for $M_t << 1$ (Sarkar et al. 1989) leads to $q_d^2/q_s^2 \propto M_t^2$. However, (13) yielded better agreement with DNS computations, in particular, of shear-driven turbulence. The validity of (13) is questionable in low Mach number decaying turbulence ($M_t^2 << 1$) where the dilatational and solenoidal fields are decoupled, and q_d^2/q_s^2 is determined by initial conditions rather than by M_t.

4. Simulation of decaying turbulence

With the closure equations for Π_d and ϵ_d given by (2), (9) and (11) we need one more closure equation, for the solenoidal dissipation ϵ_s. We use the standard model equation (for incompressible turbulence)

$$\frac{D\epsilon_s}{Dt} = -\beta \frac{\epsilon_s^2}{q^2}, \tag{14}$$

where $\beta = 3.7$ yields the (observed) decay $q^2 \propto t^{-1.2}$. The vortical field and Kolmogorov energy cascade are expected to be only weakly coupled to the dilatational (compressive) field and therefore (14) is kept independent of M_t. Equations (2)-(5) governing the turbulence decay with closure relations (9),(11), and (14) are solved for given initial conditions $q_o^2, \epsilon_{so}, T_o$, and $\overline{p^2}_o$. The model solutions are compared with DNS computations performed by the second author. Briefly, the numerical method used to solve compressible Navier-Stokes equations is a pseudo-spectral Fourier method, with periodic boundary conditions in all three directions. An ideal gas with constant specific heats is assumed and the specific heat ratio is $\gamma = c_p/c_v = 1.4$. Viscosity dependence on temperature follows a power law $\mu/\mu_o = (T/T_o)^{0.67}$. Typical initial Reynolds numbers $Re_t = q^4/\nu\epsilon_s$ are of order $O(100)$.

Fig. 3 compares DNS-model results of the evolution of kinetic energy q^2/q_o^2 for two runs with disparate initial conditions: Run I has a fairly large initial $M_t = 0.7$ while the pressure fluctuation content $P_{to} = [\overline{p^2}/(p\gamma M_t)^2]_o$ is insignificant. Conversely, in Run II, $P_{to} = 4.0$ is large but $M_{to} = 0.05 << 1$. While the q^2 evolution for Run I is comparable with incompressible decay, Run II is markedly different. Here, the relatively large initial pressure fluctuations ($P_{to} > 1$) are converted into kinetic energy within a short time on the order $\tau_a \propto M_t q^2/\epsilon$, the acoustic time scale. The model reproduces the principal features of the behavior of q^2 particularly in higher Mach number Run I. In Run II, the model-predicted levels of kinetic energy are considerably higher than the DNS data. As discussed earlier, a contributing factor to this discrepancy is the failure of the closure relations (12) and (13) when M_t is too low. To illustrate the performance of the pressure-dilatation model, Figs. 4 and 5 compare evolution of Π_d and p^2 for Run I. The model prediction of Π_d is suprisingly good. Considering the complexity of physics involved, the relaxation model (11) for the pressure variance is evidently a good approximation.

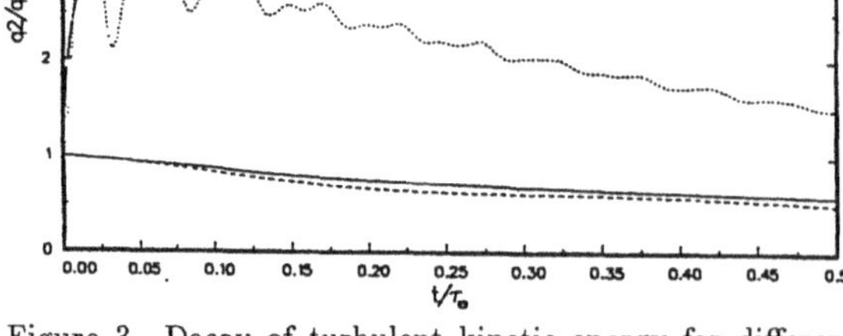

Figure 3. Decay of turbulent kinetic energy for different initial conditions: Comparison between DNS and present model results. Run I: $M_{to} = 0.7$, $P_{to} = 0.045$; Run II: $M_{to} = 0.05$, $P_{to} = 4.0$.

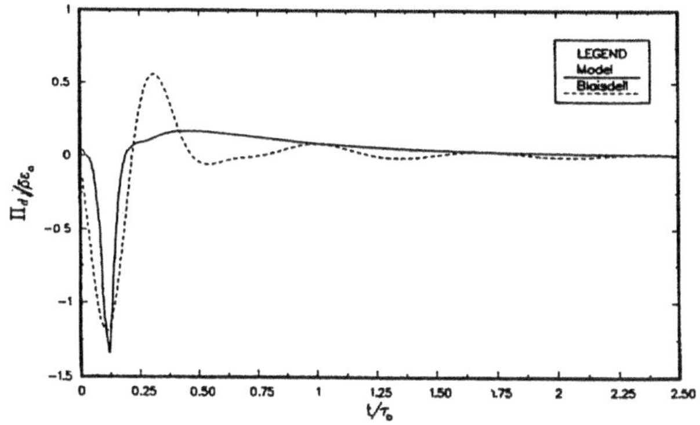

Figure 4. Pressure-dilatation correlation for Run I; otherwise same as Fig. 3.

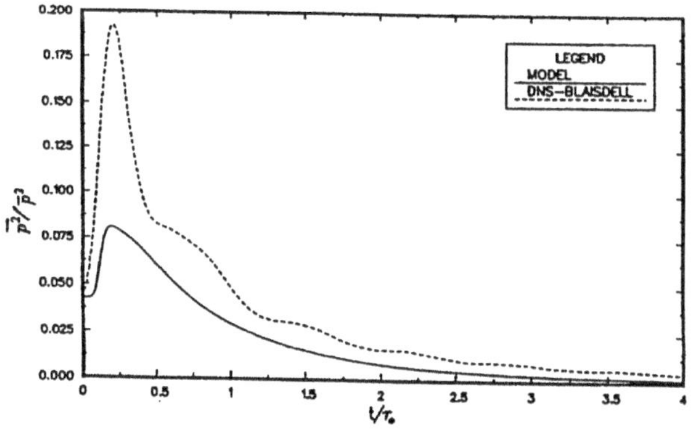

Figure 5. Pressure variance for Run I; otherwise same as Fig. 3

5. Shear-driven homogeneous turbulence

In the presence of mean shear $dU_1/dx_2 = S$, the shear production term in (3) is $P_s = -\widetilde{u_1 u_2} S$ and the kinetic energy equation (3) has to be augmented by two more equations for $\widetilde{u_1 u_2}$, and $\widetilde{u_2^2}$. Using the standard SOC model given in Z90, we obtain the following set (with $r_{ij} = \widetilde{u_i u_j}$)

$$1/2 \frac{\partial q^2}{\partial t} = P_s - (\epsilon_s + \epsilon_d - \Pi_d), \tag{3a}$$

$$\frac{\partial r_{12}}{\partial t} = -0.4 r_{22} S - C_m \frac{r_{12}}{\tau}, \tag{15}$$

$$\frac{\partial r_{22}}{\partial t} = 0.4 P_s - C_m \frac{(r_{22} - 1/3 q^2)}{\tau} - \frac{2}{3}(\epsilon_s + \epsilon_d - \Pi_d), \tag{16}$$

$$\frac{\partial \epsilon_s}{\partial t} = -\beta \frac{(\epsilon_s - 0.75 P_s)}{\tau}. \tag{17}$$

Here, $C_m = 3.25$, is the return to isotropy constant, and the (vortical) turbulence time scale is $\tau = q^2/\epsilon_s$. Otherwise, the closure equations for the dilatation terms Π_d and ϵ_d remain the same as for decaying turbulence dealt with in the previous section.

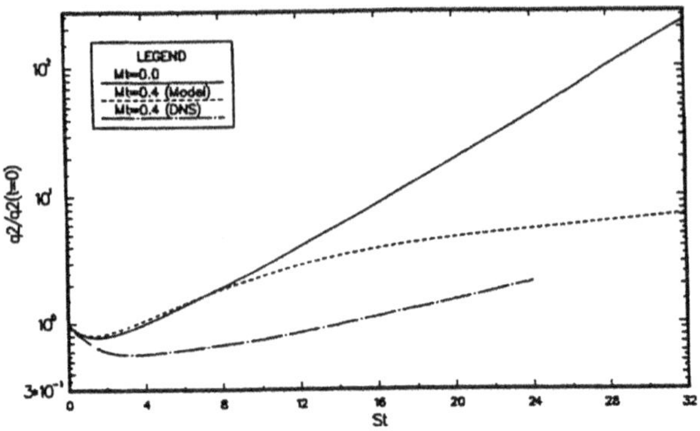

Figure 6. Shear-driven homogeneous turbulence: Evolution of kinetic energy for incompressible regimes (present model and DNS).

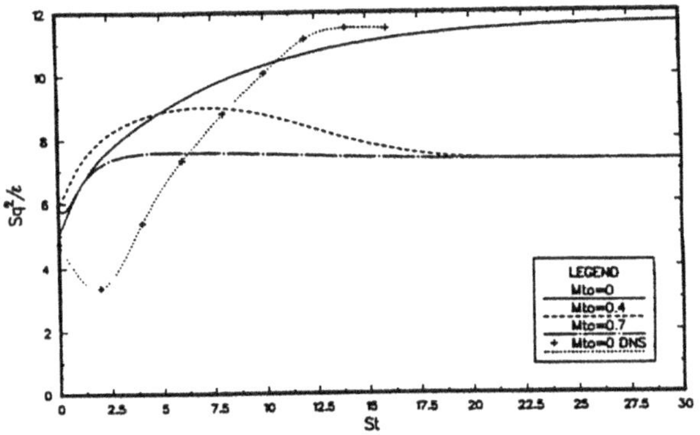

Figure 7. Model-predicted nondimensional turbulent time scale for different initial Mach numbers. Included are the incompressible DNS data of Rogers et al. (1986).

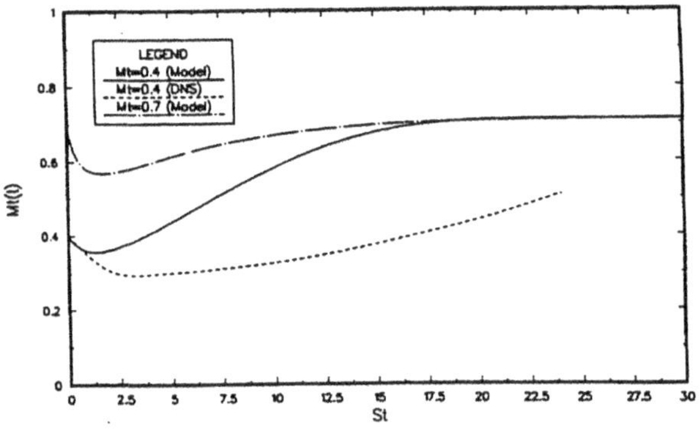

Figure 8. Evolution of rms Mach number (present model and DNS).

Figs. 6, 7, and 8, we present some preliminary results focusing on the influence of compressibility on the long-time behavior of shear-driven turbulence. Fig. 6 shows the evolution of q^2 for initial r.m.s. Mach numbers $M_t = 0.0$, and $M_t = 0.4$. The initial $\overline{p^2}$ is taken to be negligible as in the DNS initial conditions. The model simulations yield the long-time behavior $q^2 \propto exp(\lambda St)$ as expected in the incompressible regime (Tavoularis 1985, Rogers et al. 1986). With $M_t = 0.4$ both the model and DNS suggest that the amplification parameter λ decreases with increasing M_t, although the DNS data exhibit rather different transient behavior. Fig. 7 shows model simulations of nondimensional time scale $S\tau$ for three different initial M_t along with the incompressible DNS data of Rogers et al. (1986). In the incompressible case the model yields the equilibrium value $S\tau \approx 11.7$ in agreement with the DNS value. Interestingly, for finite initial M_t the model simulations asymptote to a state with $S\tau \approx 7.3$. Furthermore, this state has a unique r.m.s. Mach number as indicated in Fig. 8. The DNS data of M_t, also shown in Fig. 8, lag the model simulations and their destiny is not known. We can argue, however, that as in the case of the shear layer (see Fig. 2), homogeneous turbulence subjected to shear ultimately reaches a saturation state where fluctuations are controlled by the sonic speed and excessive values of fluctuating Mach number are cut by the shocklet dissipation.

6. Conclusions

We have proposed a new phenomenological model for the rate of change of pressure variance $D\overline{p^2}/Dt$ and utilized the relationship between the pressure-dilatation correlation and $D\overline{p^2}/Dt$ in order to approximate the pressure dilatation term in the turbulent kinetic energy equation. We have demonstrated that the pressure-dilatation term mediates the exchange between the kinetic and potential energy ($\overline{p^2}$), and that the effect of initial conditions on decaying turbulence and other important physics can be captured by the second-order equations if the dilatation containing terms are considered.
The model simulations of shear-driven turbulence suggest that for finite initial r.m.s. Mach numbers, turbulence bifurcates to a new asymptotic state characterized by a constant M_t and a reduced level of nondimensional turbulent time scale. At the present time, the latter results remain unsupported by DNS (or experiments) because the DNS computations for sufficiently long time, or high M_t, are not yet available.

References

Blaisdell, G. A., Mansour,N.N. & Reynolds, W.C. 1988 *Int. Workshop on Physics of Turbulent Compressible Mixing*, Princeton.

Feiereisen, W. J., Reynolds, W.C. & J. H. Ferziger 1981 *Report No. TF-13*, Stanford University.

Papamoschou, D. & Roshko A. 1988 *J. Fluid Mech.* **197**, 453.

Passot, T. & Pouquet, A. 1987 *J. Fluid Mech.* **181**, 441.

Passot, T., Pouquet, A. & Woodward, P. 1988 *Astron. Astrophys.***197**, 228.

Rogers, M. M., Moin, P. & Reynolds, W.C. 1986 *Rep. No. TF-25*, Stanford University.

Sarkar, S., Erlebacher, G., Hussaini, M.Y. & Kreiss, H. O. 1989 *ICASE Rep. No. 89-79.*

Tavoularis, S. 1985 *Phys. Fluids* **28**, 999.

Zeman, O. 1990 Dilatation dissipation: The concept and application in modeling compressible mixing layers. *Phys. Fluids* **A 2**, 178.

Zeman, O. 1990a On the decay of compressible isotropic turbulence *Rep. No. CTR-115*, Center for Turbulence Research, Stanford University.

Part VI

Drag Reduction and Turbulence Control

Turbulence Control

D. W. Bechert

Abt. Turbulenzforschung, Müller-Breslau-Str. 8 1000 Berlin 12, Germany

Turbulence is quite insensitive to attempts to change or to control it. In this respect, it is very different from laminar or transitional flows. The high resilience of turbulence to perturbations results from the fact that turbulence is the product of strong hydrodynamical instabilities which lead to nonlinear saturation. As in the case of an electronic amplifier in its nonlinear saturation regime, we cannot expect dramatic changes of the output signal caused by small changes of the input signal. Nevertheless, there are several examples where turbulence fluctuation levels and noise radiation can be *enhanced* artifically. On the other hand, it is rather difficult to *reduce* fluctuation levels and related quantities like the turbulent shear stress.

1. Polymer additives in liquids

One example where turbulence and drag can be reduced dramatically is by adding extremely small amounts of polymer chemicals to liquids [1,2]. Obviously, this method has been used for eons in nature in the form of drag reducing mucus of marine animals [3]. As a technological application, polymer additives have become an extremely successful means to reduce pressure losses in pipelines for crude oil [2], for instance in the Alaska pipeline. A typical concentration of about 5 cm^3 polymer additive in 1 m^3 crude oil reduces the pressure loss in a pipeline by abour 30 % [2] and saves energy and installation costs by reducing the number of pumping stations required. In water piping systems polymer additives can reduce the pressure loss by as much as 80 % [2]. The fluid dynamical mechanism of these additives can be roughly described as following: The stretching of the fluid elements is impeded by the resistance to normal stresses of the fluid, which is caused by the long chain molecules of these additives. The issue of polymer additive theory, however, is far from settled [2]. Nevertheless, the maximum attainable drag reduction can be predicted [4]. An interesting feature of flows with polymer additives is that the exchange of matter and momentum perpendicular to the flow velocity, i.e., the v-velocity component of the turbulent motion, is drastically reduced [2].

2. Jets and free shear layers

The investigations of Crow & Champagne [5] have made it evident that acoustically excited axisymmetric turbulent jets behave very similarly to transitional laminar jets. That means that undulatory motion in the jet column eventually develops into ring vortices. The main difference from laminar jets is that there is turbulent jitter in the otherwise deterministic acoustically controlled ring vortex generation. The same has been shown for plane turbulent jets [6] and for turbulent free shear layers [7-9]. The excitation takes place essentially at the nozzle lip. There, under most experimental

Advances in Turbulence 3
Editors: A.V. Johansson · P.H. Alfredsson

conditions, the shear layer thickness is small compared to the vorticity waves being generated. Even an excitation farther away from the nozzle edge leads to dominant jet-edge excited waves [1,11]. Thus, a theory assuming an infinitesimally thin shear layer can be assumed to give reasonable predictions, as it does in the laminar case [10]. It turns out that the fluctuating pressure gradient of the forcing field across the shear layer is the relevant quantity for the excitation [10]. However, the details of the interaction of the artificially generated shear layer waves with the turbulence are less well understood, in particular if shear layer thickness and wavelength become comparable. As expected, the growth of the artificial waves is limited by the increasing stability of the spreading shear layer [2]. In addition, nonlinear effects of the wave itself, i.e., roll-up, and effects due to its interaction with the existing turbulence attenuate the wave growth [7]. On the other hand, at sufficiently high excitation levels of, say, 1 % of the jet velocity and above, the forced wave enhances the fluctuation level of the whole turbulence spectrum [13,14]. As a consequence thereof, the radiated jet noise is also amplified considerably [15,14].

The efforts to reduce mixing and jet noise by acoustical control, however, led to no significant success. At transitional Reynolds numbers and with initially laminar free shear layer, it is possible to reduce shear layer spreading with high frequency excitation [13,16]. A significant reduction of the radiated noise of high Reynolds number jets, however, cannot be achieved in this way. This is only possible by changes in the nozzle exit geometry, which is detrimental for aircraft applications. Only ejector nozzles (with retractable ejectors for flight conditions) maintain the performance and reduce the radiated noise [17]. The change in the boundary conditions for the latter case is, however, quite drastic.

3. Turbulent boundary layers, shear stress reduction

3.1 Manipulators

A number of attempts have been made to reduce turbulent wall shear stress [18], and a few have been demonstrated to be successful. One of the unsuccessful ones is the LEBU (large eddy break up device). It usually consits of two small airfoils in a tandem arrangement, immersed in the turbulent boundary layer. It had been suggested in the beginning [19,20], and, occasionally, this opinion is still repeated [21] that rather large net drag reductions could be achieved with this device. In order to obtain a net drag reduction, the shear stress drag reduction downstream of the airfoils would have to be larger than the drag of the airfoils themselves. However, recent careful measurements with mechanical balances for direct force measurements have shown that *no net drag reduction* can be produced with LEBU's [22-26]. The previous findings, by contrast, were based on the evaluation of local velocity profile measurements and were in fact reflecting the deficiencies of this particular measuring technique.

The finding that LEBU's do not produce net drag reduction could be furnished only by precise measurements, because there is no theory available to reject this idea a priori. In fact, if one considers a flow with a strong and well correlated fluctuation of its flow angle, it is conceivable that an airfoil may be able to reduce drag. That would require that the device drag would be, at least partly, eliminated by the Katzmayr effect [27]. Because it cannot be assumed that this particular effect is well known, a short explanation is given here: The Katzmayr effect can reduce or eliminate the drag of an airfoil if it is exposed to a flow of varying angle of attack. The maximum angle should not be small but should also not exceed the angle at which the flow on the airfoil separates. Under these circumstances the force vector of the airfoil can move into the forward quadrant, i.e., the airfoil can even generate thrust instead of drag. Unfortunately, as it turns out, the turbulence levels and the coherence of the angular fluctuations in a turbulent boundary layer are not sufficient for this effect to occur.

It should not be overlooked, however, that turbulence manipulators, such as airfoils in a turbulent boundary layer, do indeed change the turbulence and do reduce the wall shear stress downstream of the device. In addition, it has been shown clearly that this effect is stronger than just the wake (or drag) effect of the device. To prove this, an experiment with a merely drag producing device has been carried out [26]. The LEBU-airfoils were replaced by wires having identical drag. In order to suppress periodic vortex shedding and wire vibrations, another very fine wire was wound in a spiral around each individual wire. It turned out that in this latter situation the shear stress reduction downstream of the devices is significantly weaker. Thus, the LEBU-airfoils do have a rectifying effect on the turbulent flow, which reduces the momentum exchange in the boundary layer.

Another interesting question is how fast turbulence recovers from such interactions. From LEBU data, this downstream distance is about 80-100 boundary layer thickness for the recovery of the wall shear stress [26,28]. The general observation is that mean flow properties seem to recover earlier than fluctuating quantities, as is also known from other cases, like the perturbation caused by a rapid change in wall curvature [29]. Nevertheless, even the mean flow perturbations persist longer than one would expect, as we have learned recently from our own experiments with adjustable turbulators in a turbulent channel flow. This leads to the question of how well numerical computations, which are unavoidably restricted to a certain computational volume, may be able to emulate equilibrium turbulence conditions. Nevertheless, even with this obvious limitation, these computations have already proved invaluable for the understanding of turbulence [30].

3.2 Nonplanar surfaces

In order to control near-wall turbulence, initial experiments were made at Stanford University [31] with ribs on the wall aligned in the streamwise direction ("riblets"). Michael Walsh then showed with careful measurements that the same concept can indeed be used to reduce drag [32,33]. He carried this idea from laboratory tests to a practical means to reduce surface drag on aircraft [34] and boats. Reif and Dinkelacker [35,36] established the connection to the fine ribs which are found on the skin of fast sharks. In the meantime, a large number of researchers have confirmed Walsh's findings [2,37-41]. Three-dimensional fin-like surface structures have been also tested [42], and they produce a stress reduction similar to that produced by straight ribs, i.e., 6 %. It has been assumed from the beginning that riblets have a straightening effect on the flow and hamper the fluctuating cross flow close to the surface. This assumption is supported by recent quantitative calculations of the flow in the viscous sublayer [43,44].

4. Turbulent separation control by enhanced mixing

Vortex generators are common on wings of commercial airplanes. They have been installed in order to delay separation by enhancing the turbulent mixing. Although they do increase the shear stress, they improve the performance of the wing by delaying separation [45]. Recently, a number of other configurations have been tested, with promising results [46]. In particular, the flow around short axisymmetric bodies can be improved dramatically with rough surfaces like, e.g., coarse transverse grooves, which enhance mixing in the boundary layer [47]. The author expects that the *aerodynamics of cars* could be significantly improved by enhanced turbulent mixing on the rear part of the vehicle.

As an alternative to having rigid surfaces to amplify mixing, oscillating flaps have also been demonstrated to be an efficient means to control separation [48,49]. The limited reliability of oscillating flaps, however, may be a drawback for aircraft applications. Thus, the introduction of oscillations by other means, like fluidic elements [48] or other absolutely unstable flow configurations, like the flow over cavities [47,48] is conceivable.

At the present time, there is very little theoretical background available to support the development of optimal configurations for turbulent separation control by enhanced mixing. Only van den Berg's considerations [50] provide a relation between the pressure gradient in the streamwise direction and a desirable turbulent wall shear stress. Thus, separation control by enhanced turbulent mixing is still a challenge for future research in fluid dynamics.

References

[1] Bermann, N.S.: Drag reduction by polymers. *Ann. Rev. Fluid Mech.* (M. van Dyke, ed.), **10**, 47-64 (1978).

[2] Sellin, R.H.J. & Moses, R.T., eds.: Drag reduction in fluid flows. Proceedings of the 4th. International Conference of Drag Reduction, Davos 31. July - 3. August 1989. Printed by: Ellis Horwood Ltd., Chichester, England.

[3] Hoyt, J.W.; Hydrodynamic drag reduction due to fish slimes. In: *Swimming and flying in nature*, Vol. 2., Wu, T.Y.T. & Brokaw, C.J., Plenum Press, New York 1975.

[4] Virk, P.S., Mickley, H.S., and Smith, K.A.: The ultimate asymptote and mean flow structure in Tom's phenomenon. *J. Appl. Mech., Trans. ASME, Series E*, **37**, 488-493 (1970).

[5] Crow, S.C. & Champagne, F.H.: Orderly structure in jet turbulence. *J. Fluid Mech.* **48**, part 3, 5457-591 (1971).

[6] Bechert, D.W.: Die Steuerung eines ebenen turbulenten Freistrahls durch eine seitliche Wechselströmung, erzeugt in einem Schallfeld. *Z. Flugwiss.* **24**, Heft 1, 25-33 1976.

[7] Oster, D. & Wygnanski, I.: The forced mixing layer between parallel streams. *J. Fluid Mech.* **123**, 91-130 (1982).

[8] Fiedler, H.E. and Mensing, P.: The plane turbulent shear layer with periodic excitation. *J. Fluid Mech.* **150**, 281-309 (1985).

[9] Ho, C.H. & Huerre, P.: Pertubed free shear layers. *Ann. Rev. Fluid Mech.* 1984, 365-424.

[10] Bechert, D.W.: Excitation of instability waves in free shear layers. Part 1, Theory. *J. Fluid Mech.* **186**, 47-62 (1988).

[11] Fourgeuette, D.C. & Long, M.B.: Highly localized pressure perturbations induced by laser absorptive heating in the shear layer of a gas jet. *Optics letters*, **8**, 605-607, (1983).

[12] Michalke, A.: Instabilität eines kompressiblen runden Freistrahls unter Berücksichtigung des Einflusses der Strahlgrenzschichtdicke. *Z. Flugwiss.* **19**, 319-28 (1971).

[13] Vlasow, Y.U. & Ginevski, A.S.: Generation and suppresison of turbulence in an axisymmetric turbulent jet in the presence of an acoustic influence. NASA Tech. Trans. F 15721 (1974).

[14] Moore, C.J.: The role of shear-layer instability waves in jet exhaust noise. *J. Fluid Mech.* **80**, 321-367, (1977).

[15] Bechert, D.W. & Pfizenmaier, E.: On the amplification of broadband jet noise by pure tone excitation. *J. Sound Vib.* **43**, 581-587, (1975).

[16] Zaman, K.B.M.Q. & Hussain, A.K.M.F.: Turbulence suppression in free shear flows by controlled excitation. *J. Fluid Mech.* **103**, 133-159, (1981).

[17] Banerian, G.: Status of some current research in jet noise. *AIAA Journal* **16**, No. 9, 877-888 (1978).

[18] Wilkinson, S.P., Anders, J.B., Lazos, B.S. & Bushnell, D.M.: Turbulent drag reduction research at NASA Langley: progress and plans. *Int. J. Heat and Fluid Flow*, **9**, No. 3, 266-277 (1988).

[19] Corke, T.C.: A new view on origin, role and manipulation of large scales in turbulent boundary layers. *Ph. D. Thesis*, Illinois Institute of Technology, Chicago, III., Dec. 1981.

[20] Plesniak, M.W.: Optimized manipulation of turbulent boundary layers aimed at net drag reduction. *M.S. Thesis*, Illinois Institute of Technology, Chicago III., Dec. 1984.

[21] Guezennec, Y.G. & Nagib, H.M.: Mechanisms leading to net drag reduction in manipulated turbulent boudary layers, *AIAA Journal* **28**, No. 2, 245-252, (1990).

[22] Poll. D.I.A.: A study of LEBU performance by direct total-force measurements. Proc.: Turbulent drag reduction by passive means. The Royal Aeronautical Society, London. 15.-17. Sept. 1987.

[23] Sahlin, A., Alfredsson, P.H. & Johansson. A.W.: Direct drag measurements for a flat plate with passive boundary layer manipulators. Physics of Fluids, **29**, 696-700, (1986).

[24] Sahlin, A., Alfredsson, P.H. & Johansson. A.W.: The possibility of drag reduction by outer layer manipulators in turbulent boundary layers. *Phys. Fluids*, **31**, 2814-20, (1988).

[25] Anders, J.B.: LEBU drag reduction in high Reynolds number boundary layers. AIAA-Paper 89-1011, (1989).

[26] Lynn, T.B., Gerich, D.A. & Bechert, D.W.: LEBU-manipulated flat plate boundary layers: Skin friction and device drag measured directly. Paper presented at the IAHR Drag reduction "89 conference, Davos, Switzerland, July 31-August 3, 1989. And, by the same authors: Direct drag measurements in a LEBU manipulated flat-plate boundary layer. 3rd. European Turbulence Conference, Stockholm 3.- 6. July 1990.

[27] Katzmayr, R.: Über das Verhalten von Flügelflächen bei periodischen Änderungen der Geschwindigkeitsrichtung. *Zeitschrift für Flugtechnik und Motorluftschiffahrt.* **6**. Heft, 13. Jahrg., 80-82, (1922). and: **7**. Heft, 13. Jahrg., 95-101, (1922).

[28] Anders, J.B. & Watson, R.D.: Airfoil large-eddy breakup devices for turbulent drag reduction. AIAA-Paper 85-0520, (1985).

[29] Alving, A.E., Smits, A.J, & Watmuff, J.H.: Turbulent boudary layer relaxation from convex curvature. *J. Fluid Mech.* **2ll**, 529-556, (1990).

[30] Moin, P. & Kim, J.: The structure of the vorticity field in turbulent channel flow. Part 1. Analysis of instantaneous fields and statistical correlations. *J. Fluid Mech.* **155**, 441-464, (1985).

[31] Liu, C.K., Kline, S.J. & Johnston, J.P.: An experimental study of turbulent boundary layer on rough walls. Stanford University, Dept. Mech. Eng., Rept. MD-15, July 1966.

[32] Walsh, M.J.: Drag characteristics of V-Groove and transverse curvature riblets. In: Viscous flow drag reduction (ed. G.R. Hough), *Progress in Astronautics and Aeronautics*, vol. 72., AIAA, (1980).

[33] Walsh, M.J. & Lindemann, A.M.: Optimization and application of riblets for turbulent drag reduction. AIAA-Paper 84-0347 (1984).

[34] Walsh, M.J., Sellers III, W.L. & McGinley, C.B.: Riblet drag at flight conditions. *J. of Aircraft* **26**, No 6, 570-575 (1989).

[35] Reif, W.E. & Dinkelacker, A.: Hydrodynamics of the squamation in fast swimming sharks. Neues Jahrbuch für Geologie und Palaeontologie, Abhandlungen **164**, 184-187, E. Schweizerbart'sche Verlagsbuchhandlung, (1982).

[36] Reif, W.E.: Squamation and ecology of sharks. Courier Forschungsinstitut Senckenberg, Frankfurt/M., No. 78, 255 p.

[37] Nitschke, P.: Experimentelle Untersuchung der turbulenten Strömung in glatten und längsgerillten Rohren. Max-Planck-Institut für Strömungsforschung, Göttingen. Bericht 3 (1983). Transl.: NASA TM 77480 (1984).

[38] Bechert, D.W., Hoppe, G. & Reif, W.E.: On the drag reduction of the shark skin. AIAA-Paper 85-0546 (1985).

[39] Choi, K.S.: Test of drag reducing riblets on a one-third scale racing yacht. Proc.: Turbulent drag reduction by passive means. Royal Aeronautical Society, London, 15.-17. Sept. 1987.

[40] McLean, J.D., George-Falvy, D.N. & Sullivan, P.P.: Flight-test of turbulent skin-friction reduction by riblets. Proc.: Turbulent drag reduction by passive means. Royal Aeronautical Society, London, 15.-17. Sept. 1987.

[41] Sawyer, W.G. & Winter, K.G.: An investigation of the effect on turbulent skin friction of surfaces with streamwise grooves. Proc.: Turbulent drag reduction by passive means. Royal Aeronautical Society, London, 15.-17. Sept. 1987.

[42] Bechert, D.W.: Experiments on three-dimensional riblets. Proc.: Turbulent drag reduction by passive means. Royal Aeronautical Society, London, 15.-17. Sept. 1987.

[43] Bechert, D.W. & Bartenwerfer, M.: The viscous flow on surfaces with longitudinal ribs. *J. Fluid Mech.* **206**, 105-129, (1989).

[44] Bechert, D.W., Bartenwerfer, M. & Hoppe, G.: Turbulent drag reduction by nonplanar surfaces. A survey on the research at TU/DLR Berlin. Proc.: IUTAM-Symposium on structure of turbulence and drag reduction, Zürich, 25.-28. July 1989, Springer (1990).

[45] Chang, P.K.: Control of flow separation: Energy conservation, operational efficiency, and safety. Hemisphere Publishing Corporation, Washington, and McGraw-Hill Book Company, New York (1976).

[46] Lin, J.C. Howard, FD.G. & Selby, G.V.: Turbulent flow separation control through passive techniques. AIAA-Paper 89-0976. Note: further references on vortex generators are given in this paper.

[47] Howard, F.G. & Goodman, W.L.: Axisymmetric bluff-body drag reduction through geometrical modification. *Journal of Aircraft*, **22**, No 6, 516-522, (1985).

[48] Viets, H., Palmer, G.M. & Bethke, R.J.: Potential applications of forced unsteady flows. AFOSR-TR-84-0911 (1984).

[49] Katz, Y., Nishri, B. & Wygnanski, L.: The delay of turbulent boundary layer separation by oscillatory active control. AIIA-Paper 89-0975, (1989).

[50] van der Berg, B.: Drag reduction potentials of turbulence manipulation in adverse pressure gradient flows. National Aerospace Laboratory NLR, the Netherlands, Rept. NLR MP 86060 U, (1986).

Near-Wall Turbulence Structure over Misaligned Riblets

Kwing-So Choi[1] *and Shamim Hamid*[2]

[1]Department of Mechanical Engineering, University of Nottingham, University Park, Nottingham NG7 2RD, UK
[2]BMT Fluid Mechanics, 1 Waldegrave Road, Teddington, Middlesex, TW11 8LZ, UK

1. Introduction

In the early 1980's NASA Langley Research Center conducted a parametric study to show that up to 8 % in turbulent skin-friction reduction can be obtained from riblets on a flat plate with zero pressure gradient, at moderate Reynolds numbers [1]. This extensive work was followed by the studies of turbulent boundary layers with riblets under different flow conditions; on 3-D bodies [2,3,4], with pressure gradients [2,4,5], at high Mach numbers [3,6] and high Reynolds numbers [3,4,6]. Detailed studies on structural changes in the near-wall turbulence were also carried out [7,8,9]. These results indicate, basically, that the drag-reduction performance of riblets under such flow conditions does not differ much from the early NASA results on a flat plate. The recent flight test carried out by Airbus Industries using an A320 aircraft and the field trial by BMT and ARE using a high speed buoyancy propelled vehicle [4] have convincingly demonstrated this point. There are, however, still a few unresolved problems in the practical applications of riblets. The alignment of riblets in a three-dimensional turbulent boundary layer, such as over a swept wing or near the stern of a ship hull, is one of these unsolved problems.

The present paper addresses this important aspect in the use of riblets for turbulent drag reduction. As a first step, we have studied the two-dimensional turbulent boundary layer over misaligned riblets. Mean velocity profiles and turbulence intensity profiles over misaligned riblet surfaces are presented and are compared with the data over a smooth surface and a riblet surface with no misalignment. Turbulence statistics, energy spectra and conditionally sampled burst signatures are also presented here in an effort to explain the effects of misaligned riblets on the near-wall turbulence structure and drag reduction performance.

2. Experimental facility and procedure

The experiments were carried out in the environmental wind tunnel at British Maritime Technology with a test section of width 4.8 m, height 2.4 m and length 15.0 m, under nominally zero pressure gradient condition. A 775 mm diameter turn-table on the tunnel floor 11 m downstream of the beginning of the test section housed the interchangeable smooth and riblet surface test plates, each of width 400 mm and length 600 mm. With the mean freestream velocity of U_∞=3.0 m/s in these experiments the height (1.6 mm) and width (2.5 mm) of the riblets were 12 and 20 wall units respectively.

Mean flow and turbulence data were obtained using DISA 55M constant temperature anemometers with DISA miniature hot-wire probes. TSI hot-film sensors were mounted flush with the wall surface (at the valleys for the riblet plate). In addition to

Advances in Turbulence 3
Editors: A.V. Johansson · P.H. Alfredsson

measurements over the smooth surface, the flow over the riblet surface was investigated for $\alpha = 0°, 10°, 15°, 20°$ and $30°$ by rotating the turn-table. Here the yaw misalignment angle α is the angle between the riblet direction and the freestream direction. All the measurements were made at the centre of the test plate (turn-table axis) where the momentum thickness of the naturally grown turbulent boundary layer was 21.6±1.0 mm over the central 1000 mm of the test section span, with a corresponding Reynolds number of $4.6 \cdot 10^3$. Other details of the experimental set-up are given in Choi [7].

3. Mean velocity and turbulence intensity profiles

The mean velocity profiles for the riblet surface with misalignment angles of $\alpha = 0°, 15°$ and $30°$ are shown in Figure 1 in log-law format. The dashed line in this figure represents the measured data for the smooth surface, while the riblet data were represented by

$$u^+ = 5.5 \log y^+ + B \tag{1}$$

where the intercept B, and hence u^*, was a function of α. The best fit for each misalignment case was obtained in the range $30 < y^+ < 400$ with an appropriate combination of B and u^*. When the riblets were aligned with the streamwise direction an expected upward shift [7] was observed in the log-law profile with B=6.25 and u^*=0.1112 m/s, where the 3.5 % decrease in the friction velocity compared to the smooth surface value (u^*= 0.1155 m/s) corresponds to a 7.0 % reduction in skin-friction drag. This reduction is somewhat larger than the value reported by Choi [7], however both are within the range of values (3-8 %) reported in the literature. The mean velocity profile for the 15° misalignment case is practically the same as that for $\alpha = 0°$ whereas at $\alpha = 30°$ the profile is shifted downwards below the smooth surface data, with $B = 4.18$ and u^*=0.1230 equivalent to a skin-friction drag increase of 13 % relative to the smooth surface.

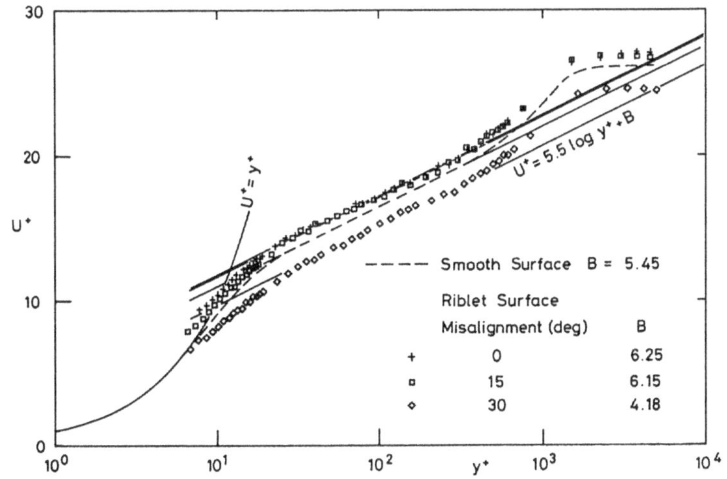

Figure 1. Log-law plot of mean velocity profiles.

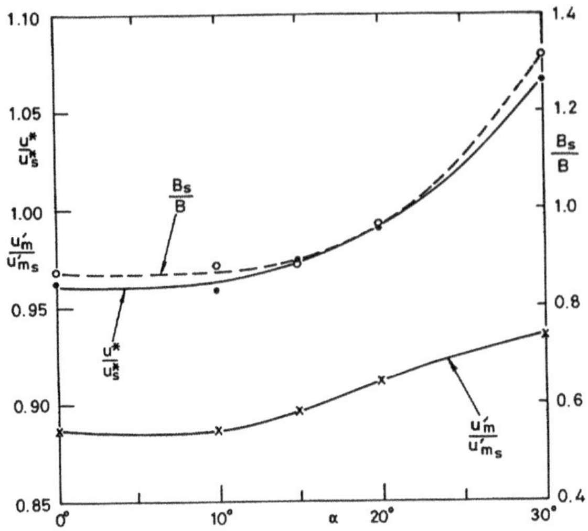

Figure 2. Variation of B_s/B, u^*/u^*_s and u'_m/u'_{ms} with misalignment angle α of the riblet surface.

The variation of these two parameters, B_s/B and u^*/u^*_s, with misalignment angle are summarized in Figure 2, where the suffix s refers to the smooth surface value. From this figure it appears that the skin-friction drag reduction performance of the riblet surface is not impaired for yaw misalignments of up to 5°, with the reduction falling to approximately 2 % at $\alpha = 20°$ and exceeding the smooth surface value beyond $\alpha = 22°$. These trends are consistent with the limited amount of published data on the misalignment effects of riblets [10, 11]. Also plotted in Figure 2 is the peak intensity of turbulence over the riblet surface, u'_m/u'_{ms}, relative to the smooth surface value as a function of misalignment angle α. The shape of this curve is similar to that of u^*/u^*_s and lends further support to the above conclusions since close to the wall surface the turbulence intensity u' is scaled with u^*. Note that within the range of measurements (α up to 30°) the peak turbulence intensity on the riblet surface remained below the smooth surface value, with a reduction of 10 to 11 % for $\alpha \leq 15°$.

Further information about the effect of misalignment can be obtained from the profiles of the turbulence intensity u'/U_∞ shown in Figure 3. Comparing the smooth surface data, dashed line in this figure, it is clear that the turbulence intensity over the riblet surface with α=0 and 15° is reduced in the near-wall region ($y^+ < 70$). There were, however, no noticeable differences between these three profiles above this region. In contrast, the profile at $\alpha = 30°$ displays a different behaviour with u' remaining high relative to the other riblet data and exceeding the smooth surface value at $y^+ \simeq 40$, only approaching it beyond $y^+ = 300$ (not shown). This behaviour is similar to that reported for a flow over a rough surface [12], providing further evidence that the riblet surface changes from a drag reducing to a drag producing surface at large misalignment angles ($\alpha \geq 30°$). Results of some preliminary measurements of the crossflow over the misaligned riblet surface indicated the presence of a measurable change from a two-dimensional to a three-dimensional flow structure up to $y^+ \simeq 50$ for the 30° misalign-

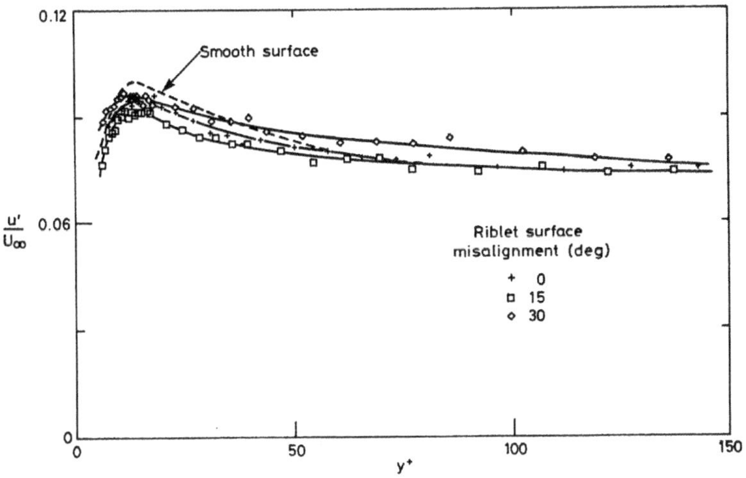

Figure 3. Turbulence intensity profiles.

ment case; the increase in the turbulence intensity seems to be related to this change in the turbulence structure in this region. Note that in all the present cases the maximum turbulence intensity occurred at approximately $y^+ = 15$.

In assessing the overall performance of the riblet surface under different operating conditions it is necessary to consider the total drag. Gaudet [13] has made some direct measurements of drag for a riblet surface for a range of yaw misalignment angles. Unlike the present skin-friction drag data which remained largely unaffected for up to 15°, the results of his investigation show that the total drag increases gradually with increase in up to $\alpha \simeq 20°$ and more markedly for larger misalignment angles, indicating a drag penalty beyond $\alpha = 30°$. This difference seems to be related to the pressure component of the drag, which in turn should be related to the three-dimensionality of the turbulence structure close to the surface which increased gradually with misalignment angles up to 15° and markedly beyond $\alpha = 20°$.

4. Near-wall turbulence structure

Figure 4 shows the probability densities of wall-skin friction fluctuations over the riblet surface with zero, 15 and 30 degrees of misalignment. These probability densities are compared with that of a smooth surface which is similar to the normal distribution [7]. Clearly, the probability in positive tail is greater for the riblet surface than the smooth surface even at the largest misalignment angle of 30°. The frequency of the near-wall burst [7], an event associated with a sharp increase in wall-skin friction due to the downwash of high-momentum fluid between a pair of counter-rotating longitudinal vortices near the viscous sublayer, is therefore greater for the riblet surface even for these misaligned cases. The skewness and kurtosis are, however, peaked at 20° of misalignment (see Figure 5).

Conditionally sampled signatures of the near-wall burst are shown in Figure 6. VITA technique [7] was used to obtain more than one thousand samples of wall-skin

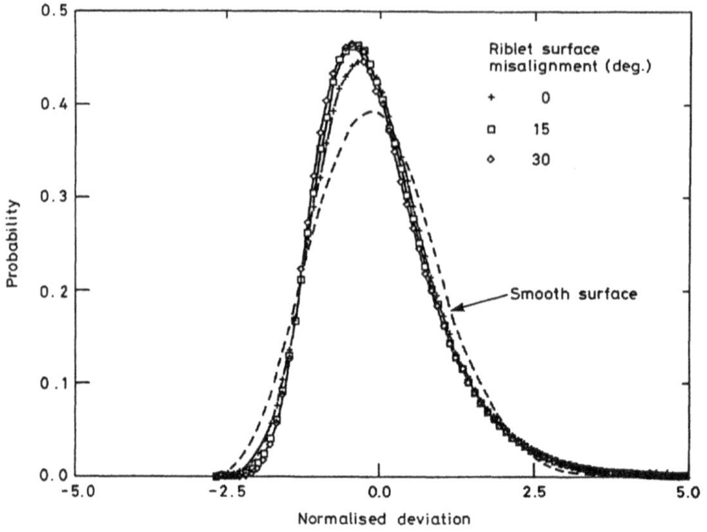

Figure 4. Probability densities of the wall skin-friction signal.

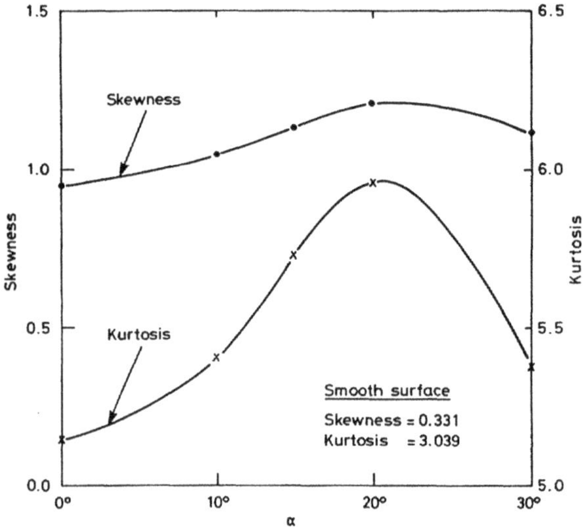

Figure 5. Variation of skewness and kurtosis of wall skin-friction fluctuations with misalignment angle α of the riblet surface.

friction signals during the event which were then ensemble averaged. At zero misalignment, the burst signature had a very sharp rise and fall in wall-skin friction with a noticeable "dip" before the rise. At 15° of misalignment, the overall shape of the burst signature remained very similar to the zero misalignment case with a slight reduction in the rate of rise in wall-skin friction. When the riblets were misaligned to 30°, the rate of rise in wall-skin friction became almost identical to that of the smooth surface. The

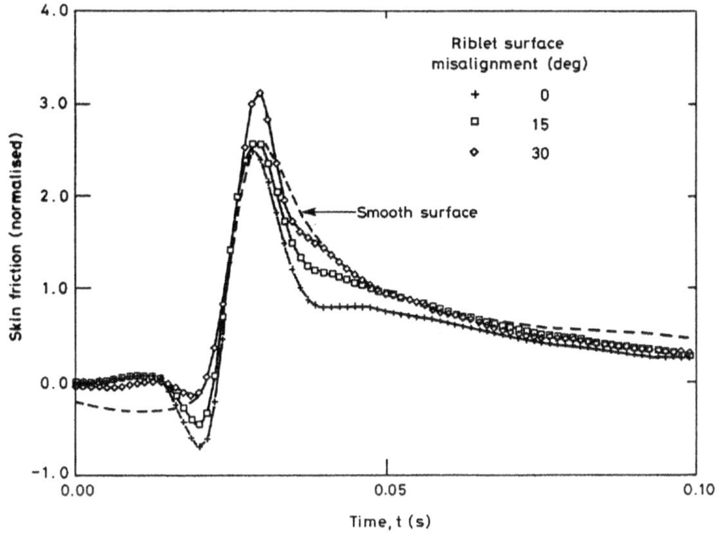

Figure 6. Conditionally sampled burst signatures of wall skin-friction fluctuations.

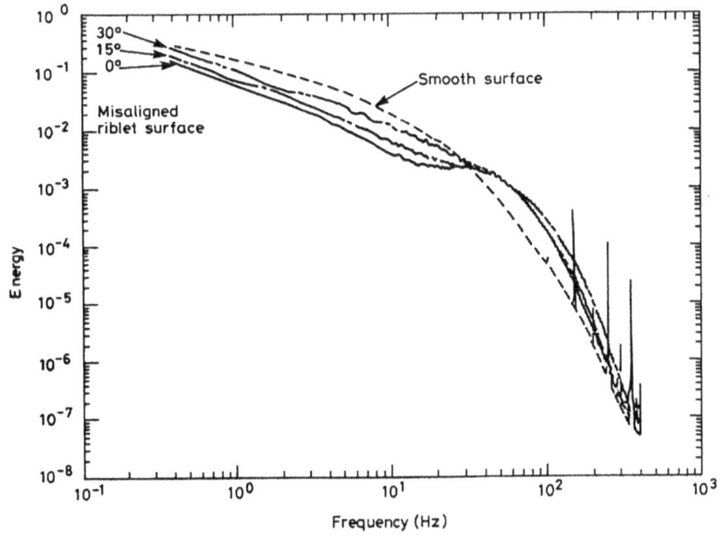

Figure 7. Spectra of wall skin-friction fluctuations.

duration of the burst was steadily increased with the degree of misalignment, nearly doubling the amount within 30 degrees.

The energy spectra of wall-skin friction fluctuations are shown in Figure 7 for three misaligned riblet cases and for the smooth case. Since no efforts were expended to calibrate the hot-film sensors for the measurements the total energy levels in this figure are somewhat arbitrary. However, the energy spectra for the riblet surface at

$\alpha = 0°$ and the smooth surface are in good agreement with the previous data [7], providing confidence in the present result. As indicated previously by Choi [7], the energy spectrum at zero misalignment of riblets shows a noticeable "hump" at higher frequencies centred around 50 Hz which corresponds to the duration of the near-wall burst (see Figure 6) of about 0.02 sec. At 15° of misalignment, the spectrum is still very similar to the zero misalignment case with a slight rise in lower frequency energy. At $\alpha = 30°$, the lower frequency energy is further increased coupled with an increased level of higher frequency energy. The "hump" in the spectrum is nearly vanished at this misalignment angle, although the shape of the spectrum is still quite different from that of the smooth surface.

5. Concluding remarks

Time-averaged quantities, conditionally sampled profiles as well as statistics of turbulence in the near-wall region of turbulent boundary layer over misaligned riblets are presented and the results compared with those over a smooth surface. The results seem to indicate that the turbulence structure over misaligned riblets of up to 15° remains very similar to that of flow over aligned riblets. In fact the turbulence intensity level as well as the friction velocity over a riblet surface at 15° misalignment were almost identical to those at zero degree of misalignment. In other words, the riblets even at 15° misalignment seem to be as effective as zero misaligned riblets in preventing the counter rotating vortices coming very close together during the near-wall burst event [7]. A flow visualised picture given in Figure 8a, which was obtained using a smoke-wire technique with a sheet of laser light parallel to the wall surface, also demonstrates the effectiveness of riblets at 15° of misalignment by showing that the longitudinal vortices close to the wall surface remain aligned with the micro-grooves of riblets. A picture over zero misaligned riblets is given in Figure 8b for comparison.

Although the basic turbulence structure remained unchanged up to 15°, the total drag of riblet surface increases steadily with the misalignment angle [13]. This may be explained by the fact that the pressure component of the drag, which is nearly zero for

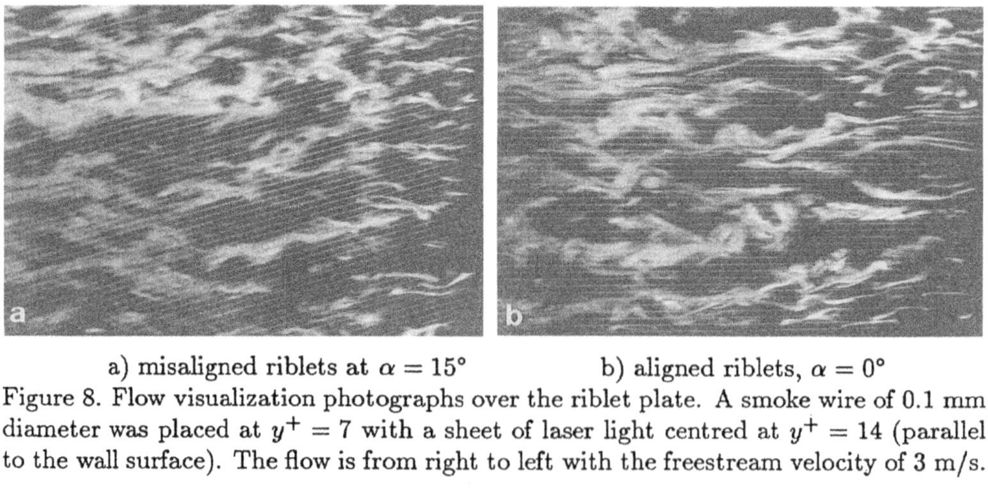

a) misaligned riblets at $\alpha = 15°$ b) aligned riblets, $\alpha = 0°$

Figure 8. Flow visualization photographs over the riblet plate. A smoke wire of 0.1 mm diameter was placed at $y^+ = 7$ with a sheet of laser light centred at $y^+ = 14$ (parallel to the wall surface). The flow is from right to left with the freestream velocity of 3 m/s.

zero misaligned riblets, increases rapidly with an increase in misalignment angle. At 30° of misalignment the total drag of riblet surface becomes very similar to that of a smooth surface [13].

References

1. Wilkinson, S.P., Anders, J.B., Lazos, B.S. and Bushnell, D.M. 1987 Turbulent drag reduction research at NASA Langley - progress and plans. In: *Turbulent drag reduction by passive means.* The Royal Aeronautical Society.

2. Choi, K.-S., Pearcey, H.H. and Savill, A.M. 1987 Test of drag reducing riblets on a one-third scale racing yacht. In: *Turbulent drag reduction by passive means.* The Royal Aeronautical Society.

3. McLean, J.D., George-Falvy, D.N. and Sullivan, P.P. 1987 Flight test of turbulent skin-friction reduction by riblets. In: *Turbulent drag reduction by passive means.* The Royal Aeronautical Society.

4. Choi, K.-S. 1990 Drag-reduction test of riblets using ARE's high speed buoyancy propelled vehicle - MOBY-D. *Aeron J.*, March, 79-85.

5. Choi, K.-S. and Johnson, R. 1989 Pressure gradient effect on the turbulence structure over riblet surface. In: *Advances in Turbulence II.* (ed. H.Fernholz and H.E. Fiedler) Springer-Verlag.

6. Squire, L.C. and Savill, A.M. 1987 Some experiences of riblets at transonic speeds. In: *Turbulent drag reduction by passive means.* The Royal Aeronautical Society.

7. Choi, K.-S. 1989 Near-wall structure of turbulent boundary layer with riblets. *J. Fluid Mech* **208**, 417-458.

8. Hooshmand, D., Youngs, R. and Wallace, J.M. 1983 An experimental study of changes in the structure of a turbulent boundary layer due to surface geometry changes. *AIAA Paper* 83-0230.

9. Bacher, E.V. and Smith, C.R. 1985 A combined visualisation-anemometry study of the turbulent drag reducing mechanisms of triangular micro-groove surface modifications. *AIAA Paper* 85-0548.

10. Coustols, E. 1989 Behaviour of internal manipulators: "riblets" models in subsonic and transonic flows. *AIAA Paper* 89-0963.

11. Walsh, M.J. and Lindemann, A.M. 1984 Optimization and application of riblets for turbulent drag reduction. *AIAA Paper* 84-0347.

12. Pulles, C.J.A, Krishna Prasad, K. and Nieuwstadt, F.T.M. 1989 Turbulence measurements over longitudinal micro-grooved surfaces. *Appl. Sci. Res* **46**, 197-208.

13. Gaudet, L. 1989 Properties of riblets at supersonic speed. *Appl. Sci. Res* **46**, 245-254.

Direct Drag Measurements in a LEBU Manipulated Flat-Plate Boundary Layer

T.B. Lynn, D.A. Gerich and D.W. Bechert*

DLR, Müller-Breslau-Strasse 8, Berlin, Germany
* HFI, TU Berlin, Müller-Breslau-Strasse 8, Berlin, Germany

Direct drag measurements were made in a ribbon-LEBU manipulated flat plate boundary layer using two independent drag balances, to measure C_f drag and device drag directly. No net drag reductions were found for any configuration tested. Contrary to the widely held belief that "tandem" LUBUs perform best, it was found that a single-element LEBU resulted in the least net increase in total drag. A comparison of the available data from the literature indicates that net drag recuction obtained using momentum balance techniques are suspect.

1. Introduction

Although research to date has produced a general consensus regarding the most effective type of LEBU, the optimum configuration, and the effects produced by optimum LEBU manipulators on the structure of the boundary layer, these conclusions are based almost entirely on experiments using indirect methods (i.e., momentum balance methods) to determine "net" drag. The inaccuracies inherent in momentum balance measurements in distorted boundary layer flows have resulted in widely varying drag reduction results obtained for nominally similar LEBU configurations. Figure 1, which shows the local skin friction distributions obtained by different researchers using relatively simple "ribbon" type LEBUs in nominally similar configurations, indicates the degree of scatter in the published data.

One explanation for the large scatter in the published data has been the low chord Reynolds numbers characteristic of these devices. It is well known that low chord

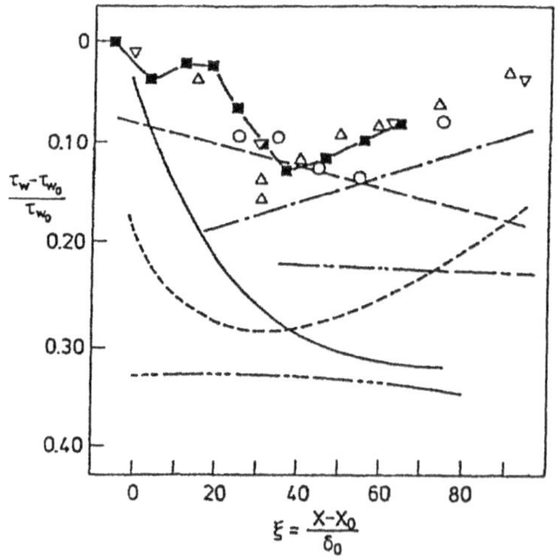

Figure 1: Local skin friction distribution downstream of ribbon-type tandem LEBUs obtained using both direct and indirect methods.
LEBU parameters: $\alpha=0°$, $h/\delta_0\approx0.75$, $R_c\approx30{,}000$, $c/\delta_0\approx1$, $7<s/\delta_0<12$.
Symbols:
Oil-film interferometry:
○ Westphal(1986)
Drag balance:
△ Lemay(1985) ▽ Present data
Sublayer fence: ■ Lynn(1987)
Momentum balance:
—— Bertelrud(1982) — — —Hefner(1983)
-·—·- Anders(1985) - - - - Plesniak(1984)
-·-·- Corke(1981) -·-·—·- Corke(1982)

Advances in Turbulence 3
Editors: A.V. Johansson · P.H. Alfredsson

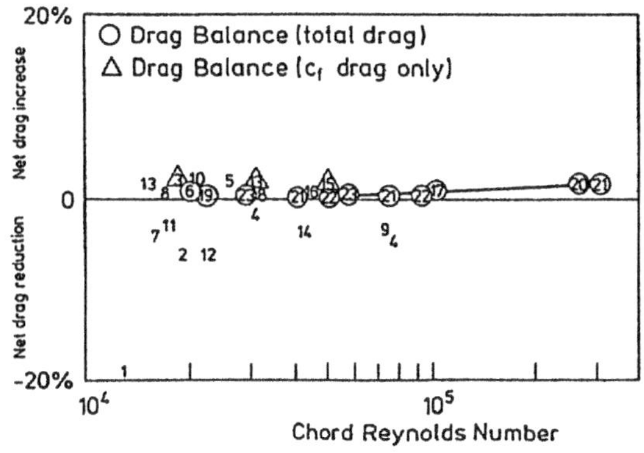

Figure 2: Net drag performance of LEBU manipulators versus chord Reynolds number. For all types of LEBUs.

1 Corke(1981)
2 Plesniak(1984)
3 Nguyen(1984)
4 Anders(1984)
5 Papathanasiou(54)
6 Hefner(1979)
7 Hefner(1983)
8 Bandyopadhyay(86)
9 Anders(1985)
10 Poddar(1985)
11 Coustols(1985)
12 Rashidnia(1986)
13 Mumford(1984)
14 Bertelrud(1982)
15 Lemay(1985)
16 Westphal(1986)
17 Sahlin(1986)
18 Lynn(1987)
19 Poll(1987)
20 Sahlin(1988)
21 Anders(1989)
22 Lynn(1989)
23 Present data

Reynolds numbers airfoils are very sensitive to small spanwise variations affecting laminar separation, resulting in very sporadic behavior reflected in C_d or C_l (Anders 1986). An examination of the effects of chord Reynolds number on the performance of all types of LEBUs, including airfoil-shaped devices, was undertaken. The results are shown in figure 2, where the degree of drag reduction found is plotted against chord Reynolds number for each device. As figure 2 shows, there is indeed a larger scatter at low Reynolds numbers, but only among the indirect-measurement data.

A much stronger correlation is obtained when the method of measuring net drag reduction is considered. Although both net increases and decreases have been found in experiments using indirect methods of measurements (e.g. momentum balance), the experiments using direct methods (i.e., drag balances), indicated by circles in figure 2, have not found a net drag reduction at any chord Reynolds number[1]. Notice also that as shown in figure 1, the variation within the momentum balance data is considerably larger than the variation within the results obtained with other techniques less sensitive to three dimensionalities, although the differences in LEBU geometry are equivalent between the two groups. This also indicates that accurate conclusions can be drawn only from direct drag measurements. Unfortunately, the body of direct drag measurements is relatively small and limited mostly to measurements at higher chord Reynolds numbers. With this in mind, we have undertaken to make direct drag measurements using ribbon type LEBUs at low Reynolds numbers.

2. Experimental set-up

The experiments presented here were conducted in the low speed wind tunnel of the Technische Universität Berlin, with a test section measuring 2m x 1.4m and 10m long. The test section was fitted with a slightly divergent top wall to compensate for the growing boundary layer. This modification resulted in a very nearly zero pressure gradient flow.

[1] Although the data published in Anders (1989) indicate that at the lowest Reynolds number tested a small reduction is possible (approximately 3%), the authors pointed out that due to the large wave drag at these speeds, the error bars were much larger than the drag reduction.

The aluminum flat plate test surface is suspended horizontally along the center line of the test section and can be varied in length from 2m to 10m by adding or removing segments measuring 0.75m in streamwise extent. Each segment is constructed as a 5cm thick hollow "sandwich", with a 6.35mm thick ground ALCOA aluminum plate as the top surface and a 3mm aluminum sheet forming the bottom layer.

The flat plate is equipped with two accurate independent drag balances, one measuring the skin friction drag and the other measuring the device drag. The skin friction drag balance is a floating wall element type measuring 75cm in the streamwise direction and 60cm spanwise. (For a complete description, see Bechert et al. 1985.) The device drag balance consists of a steel frame suspended on spring steel rods. The LEBU tensioning devices, specially designed to allow adjustment of height, angle of attack, and separation, are attached to the steel frame. The entire balance is located below the flat plate, with only the arms of the tensioning devices protruding above the plate. Streamlined fairings enclose the entire balance both above and below the plate so that only the LEBU device is subject to drag forces. (For the complete description, see Lynn et al. 1989.)

The flat plate segments each have a streamwise extent matching that of the skin friction drag balance. The device drag balance is affixed to one such segment, and therefore, by adding additional segments between the device drag balance and the skin friction balance, the complete integrated skin friction drag along the streamwise axis can be measured while simultaneously measuring the device drag directly. See figure 3.

The LEBU devices used for these low speed experiments were of the thin ribbon type, constructed of either 0.15mm x 60mm or 0.20mm x 30mm polished spring steel stock. (The thicker bands were used for the 30mm LEBUs to eliminate flutter problems.) Each LEBU was carefully prepared by hand-grinding the edges to obtain the profiles shown in figure 4, inset. The edges were then inspected under a microscope to ensure that they had the proper shape over the entire span.

Each LEBU was clamped in an end piece specially designed so that the tension distribution could be adjusted fore and aft to ensure that the devices were aeroelastically stable. Strain gauges fitted to each end piece were used to ensure that the devices were not vibrating during a test run. The strain gauges were also used to ensure a proper tensioning force.

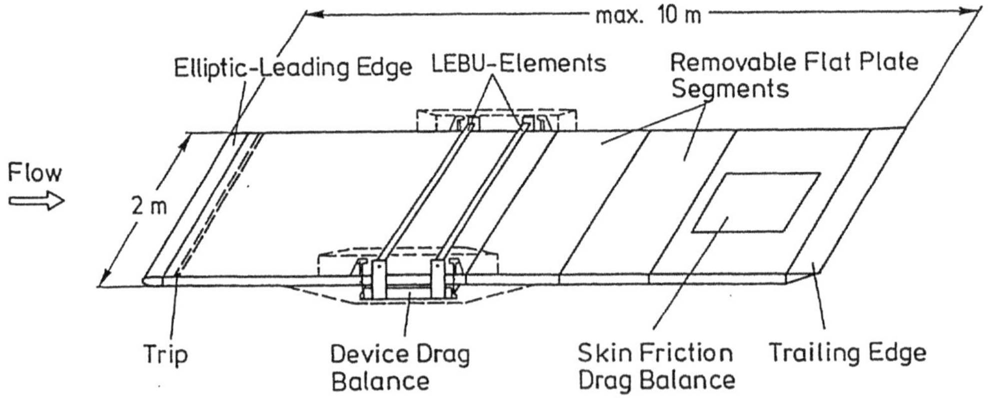

Figure 3: Schematic of plate with device drag balance and floating wall element skin friction balance.

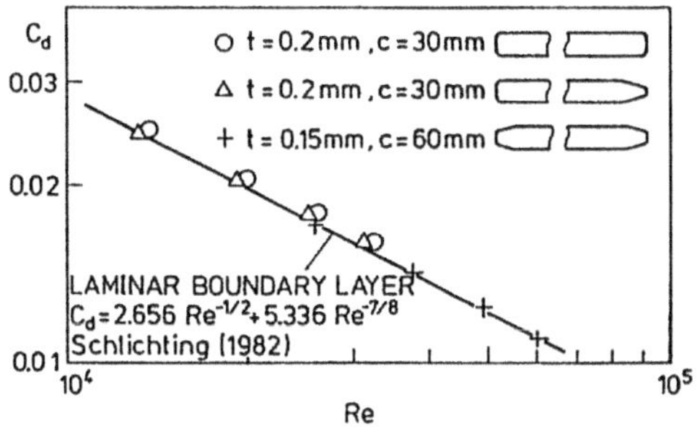

Figure 4: Drag coefficient for ribbon type LEBUs in freestream flow as a function of chord Reynolds number.

For each plate configuration, the skin friction balance, the device drag balance, and the LEBU strain gauges were recalibrated using weights. The skin friction drag and the device drag were then measured for each of the LEBU configurations as well as for the unmanipulated case. To minimize thermal effects due to heating of the wind tunnel, the experiments were run in a timed sequence: ramping the wind tunnel speed from zero to the desired speed, allowing for the drag balances to reach a stable value, noting the balance readings and the flow conditions, and then ramping back to zero. This allowed for zero-drift correction, resulting in an average standard deviation for both force measurements of approximately 0.1% of the measured force. (Each data point is an average over three cycles.)

Reported here are the results from experiments run at a freestream velocity of 15 m/sec ($R_e = 3000$) using a 60mm chord ($c = 2.4\delta_o$) and a 30mm chord ($c = 1.2\delta_o$) in "single" and "tandem" arrangements, each at both $0.5\delta_o$ and $0.75\delta_o$ above the flat plate. These parameters were chosen after the published data was analyzed to determine the characteristics of LEBU devices for which net drag reductions have been reported, and thus those most likely to result in a net drag reduction when measured directly.

3. Results and discussion

Prior to making any skin friction measurements, the prepared LEBUs were tested in the freestream flow above the boundary layer ($h = 2\delta_o$). It was found that the tapered trailing edge LEBUs had a slightly lower C_d than the LEBUs with rounded edges (see figure 4). Although the difference amounted to only 2%, all experiments reported here were run using the tapered trailing edge LEBUs. The drag data for the LEBUs used here lie within 1% of the empirical curve of A.F. Messiter (Schlichting), indicating that the device drag is due primarily to laminar skin friction drag. In this respect, ribbon-type LEBUs perform considerably better than the NACA 0009 profiles used at higher Reynolds numbers (Lynn et al. 1989; Anders 1985).

The skin friction reductions obtained with the six LEBU configurations investigated here are shown in figures 5 a,b. The skin friction distributions show that the manipulated skin friction has returned completely to the undisturbed value by the last measuring station in all cases. The relaxation distances are on the order $200\delta_o$ for both $h = 0.75\delta_o$ and $h = 0.5\delta_o$). The skin friction distribution measured for the $h = 0.75\delta_o$) tandem case agrees well with the other published direct measurements of skin friction distributions, plotted in figure 1. The maximum skin friction reduction was obtained in the tandem

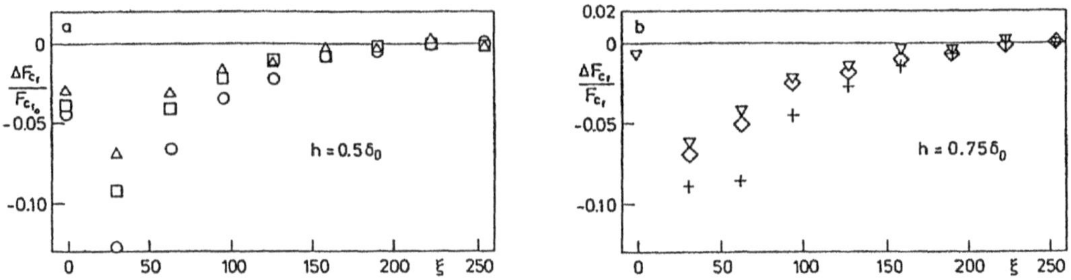

Figure 5 a&b: Skin friction drag savings as a fraction of unmanipulated skin friction drag versus normalized distance downstream.

△ 30mm single, $h=0.5\delta_o$	▽ 30mm single, $h=0.75\delta_o$
□ 60mm single, $h=0.5\delta_o$	◇ 60mm single, $h=0.75\delta_o$
○ 30mm tandem, $h=0.5\delta_o$	+ 30mm tandem, $h=0.75\delta_o$

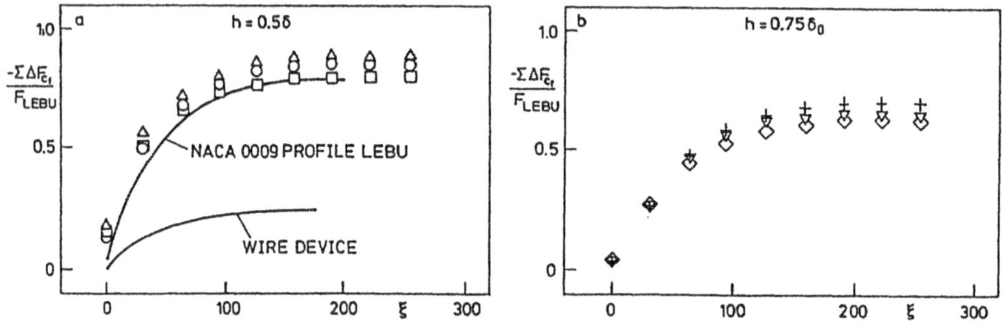

Figure 6 a&b: Cumulative skin friction savings as a fraction of device drag versus downstream distance. Symbols as in figure 5.

$0.5\delta_o$ case, and is on the order of 12-13% (averaged over $30\delta_o$). (Correcting for this averaging effect indicates that the maximum local reduction is approximately 15-16%).

The integral of the skin friction drag savings as a fraction of the device drag, shown in figures 6 a,b, indicates that in spite of its relatively small skin friction reduction, the most efficient LEBU configuration is the 30mm chord single LEBU at $h = 0.5\delta_o$. This configuration, which has the lowest device drag, results in approximately 90% of the device drag being recovered in skin friction savings. This result is contrary to the widely-held belief that a "tandem" configuration performs best. The fact that this is the most efficient configuration is further reflected in the ratios of the total drag on the LEBU system to the drag on just the flat plate. See figures 7 a,b. The resulting best performance, obtained for the 30mm single LEBU $h = 0.5\delta_o$ configuration, is a net increase of 0.25% in total drag. The worst case, a 1.4% increase, was obtained for the tandem configuration (30mm chord) at $h = 0.75\delta_o$. These plots show clearly which type of LEBU most efficiently recovers its own device drag and reveal that varying the distribution of the device chord (i.e., 60mm "single" versus 30mm "tandem") affects the drag recovery distribution very little.

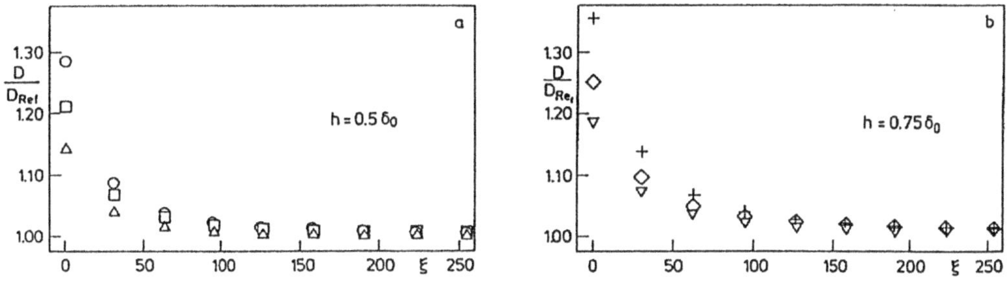

Figure 7 a&b: Ratio of total drag with LEBUs to total drag without LEBUs versus downstream distance. Symbols as in figure 5.

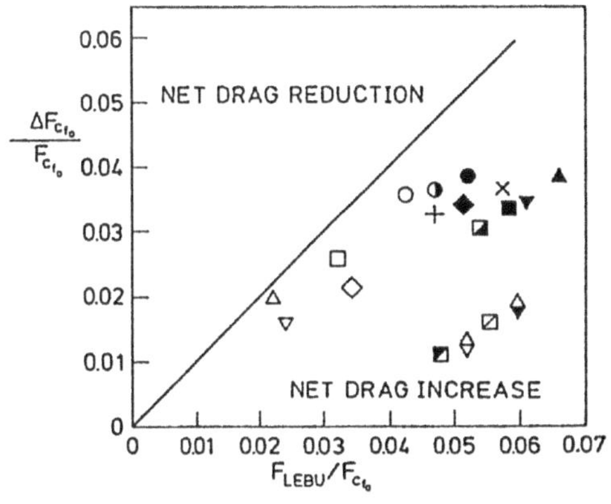

Figure 8: Skin friction savings as a fraction of unmanipulated skin friction drag versus the ratio of device drag to unmanipulated skin friction drag.

Ribbon tandem LEBUs α=0° :
Rc =30,000: ○ h=0.5δo , + h=0.75δo
Ribbon single LEBUs α=0° :
Rc =30,000: △ h=0.5δo , ▽ h=0.75δo
Rc =60,000: □ h=0.5δo , ◇ h=0.75δo
NACA 0009 tandem LEBUs:
Rc =96,000:
h=0.5δo : ◑ α=0°
h=0.8δo : ◪ α=+2° ◆ α=0° ▼ α=-2°
Rc =52,000:
h=0.5δo : ● α=0°
h=0.8δo : ■ α=+2° × α=0° ▲ α=-2°
Wire devices:
Rd =1044: ◩ h=0.5δo , ⊠ h=0.8δo
Rd =1770: ⊖ h=0.5δo , ◆ h=0.8δo

When the integrated skin frition reduction is plotted against the device drag [2] for all cases tested here, some sense can be made of the variation in LEBU performance between the different configurations. Figure 8 reveals that, within a "family"(determined by device height), the skin friction reduction is proportional to the device drag. The data separates then into two groups: The LEBU at $h = 0.5\delta_o$ lie on a line closer to the zero-net-changeline (45°) than do those at $h = 0.75\delta_o$, indicating better performance at $h = 0.5\delta_o$ independent of LEBU configuration. Within each family, increasing the duration of the manipulation (i.e., the length of the device) either by increasing device chord length or increasing the number of devices, results only in additional skin friction savings equal to and thus offset by the additional device drag. Changing the device height shifts the entire curve relative to the 45° line, although it did not shift the curve into the net drag reduction region above the 45° line.

Including data from the NACA 0009 LEBUs shows again that within a height family, the skin friction reduction is directly proportional to the device drag. As expected, the NACA 0009 profile LEBU data is displaced to the right (worse net drag perfor-

[2] The idea that the skin friction reduction varies with device drag was first suggested by Sahlin et al. (1988)

mance), due to the higher device drag (the NACA profiles used had a C_d above the laminar skin friction value, Lynn et al. 1989). The trend is the same; the $h = 0.5\delta_o$ configuration is again better and angles of attack other than zero result in a shift further to the right.The data from wire devices, also shown in figure 8, plot even further to the right. This poor performance is independent of the device height, indicating that the effects of the wake on skin friction drag reduction are constant regardless of location in the boundary layer. The lack of height sensitivity exhibited by the purely drag-producing wire devices also shows that the variation of performance with height for the other types of devices is most likely due to their direct effects on large structures (see Lynn 1989).

4. Conclusion

We have shown that ribbon type LEBUs perform better than NACA 0009 profile LEBUs, due to the ribbon LEBUs' lower device drag. The device drag of ribbon LEBUs is entirely due to laminar skin friction drag, and thus it is not likely to be possible to lower the device drag for a constant chord. A ribbon device with properly sharpened edges is therefore most likely the optimum LEBU design. Both types of devices are far superior to wire type devices, which merely produce drag. Wire devices recover only about 25% of their device drag; NACA 0009 devices recover almost 80% of their device drag; and ribbon devices recover as much as 90% of their device drag. The difference in performance between wire type devices and devices with a chord lenght on the order of δ is most likely due to the effective destruction of large scale turbulent structures by the latter devices. The LEBU manipulators effectively prevent energetic turbulent events from reaching the outer boundary layer interface over a length on the order of a few device chord lengths, thus interrupting the turbulence production cycle (Mumford el al. 1984) The reduction in skin friction due to this decoupling has been shown here to be proportional to the device chord and hence to the device drag (i.e., there is no added benefit to repeated manipulation or increasing the chord length of the device beyond $c \approx \delta$). The better performance of ribbon devices as compared to NACA 0009 devices appears to be due simply to their lower device drag. This improvement, however, merely reduces the amount of net drag increase. Even the most efficient ribbon type LEBU configuration resulted in a 0.25% net increase in total drag. The possibility of sucessfully achieving a net drag reduction is small. All reasonable extrapolations of the data in figure 8 appear to intersect the breakeven line at the origin, indicating that while performance can be improved up to a point by reducing device drag, further reducing the device drag simply reduces the net drag increase proportionally. The best that can be achieved is a zero drag increase.

Unfortunately, the race to produce a net drag reduction has obscured the importance of LEBUs as turbulence manipulators. The effects of ribbon and wire devices on turbulence have been demonstrated by, for example, Lynn (1989), Corke (1981), and Mumford et al. (1984). That the removal of large coherent turbulent structures results in no net change in total drag indicates that the cost of simply destroying energetic turbulent structures present in the boundary layer exactly equals the benefit realized in skin friction savings. The fact that one can, at best, break even as far as drag reduction is concerned is an indication of the relative importance of large scale turbulent structures in the dynamics of the turbulent boundary layer.

Acknowledgement

The authors would like to thank Professor H. Fernholz, Clemens Daase, Wolfram Hage, and Ralf Makris. This work was supported by a grant from the Deutsche Forschungsgemeinschaft.

References

Anders, J.B. 1989 LEBU Drag Reduction i High Reynolds Number Boundary Layers, AIAA-89-1011.

Anders, J.B. 1986 Large Eddy Breakup Devices as Low Reynolds Number Airfoils, NASA SAE Technical Paper 861769.

Anders, J.B., Hefner, J.N. and Bushnell, D.M. 1984 Performance of Large-Eddy Breakup DEvices at Post-Transitional Reynolds Numbers, AIAA-84-0345.

Anders, J.B. and Watson, R.D. 1985 Airfoil Large-Eddy Breakup Devices for Turbulent Drag Reduction, AIAA-85-0520.

Bandyopadhyay, P.R. 1986 Mean Flow in Turbulent Boundary Layers Disturbed to Alter Skin Friction, *J. Fluids Engineering* **108**, 127-40.

Bechert, D.W., Hoppe, G. and Reif, W.E. 1985 On the Drag Reduction of the Shark Skin, AIAA-85-0546.

Bertelrud, A., Truong, T.V. and Avellan, F. 1982 Drag Reduction in Turbulent Boundary Layers Using Ribbons, AIAA-82-1370.

Corke, T.C. 1981 A New View on Origin, Role and Manipulation of Large Scales in Turbulent Boundary Layers , *Ph.D. Thesis*, Illinois Institute of Technology, Chicago, Illinois, December.

Corke, T.C. Nagib, H.M. and Guezennec, Y.G. 1982 A New View on Origin, Role and Manipulation of Large Scales in Turbulent Boundary Layers, NASA CR 165861.

Coustols, E. and Cousteix, J. 1985 Reduction du Frottement Turbulent: Moderateurs de Turbulence 22eme Colloque d'Aerodynamique applique, Lille, France, Nov. 13-15.

Hefner, J.N., Anders, J.B. and Bushnell, D.M. 1983 Alteration of Outer Flow Structures for Turbulent Drag Reduction, AIAA-83-0293.

Hefner, J.N., Weinstein, L.M. and Bushnell, D.M. 1979 Large-Eddy Breakup Scheme for Turbulent Viscous Drag Reduction, AIAA-81-26507.

Lemay, J., Provencal, D., Gourdeau, R., Nguyen, V.D. and Dickinson, J. 1985 More Detailed Measurements Behind Turbulence Manipulators Including Tandem Devices Using Servo-Controlled Balances, AIAA-85-0521.

Lynn, T.B. 1987 Manipulation of the Structure of a Turbulent Boundary Layer, *Ph.D. Thesis*, Yale University, New Haven, CT, May.

Lynn, T.B. 1989 Structures in a Turbulent Boundary Layer, *Advances in Turbulence 2*, Springer-Verlag, Berlin.

Lynn, T.B., Gerich, D.A. and Bechert, D.W. 1989 LEBU-Manipulated Flat Plate Boundary Layers: Skin Friction and Device Drag Measured Directly, IAHR Drag Reduction '89 Conference, Davos, Switzerland, July.

Morse, P.M. and Ingard, K.V. 1968 *Theoretical Acoustics*, New York, McGraw-Hill.

Mumford, J.C. and Savill, A.M. 1984 Parametric Studies of Flat Plate, Turbulence Manipulators Including Direct Drag Results and Laser Flow Visualisation, *ASME Symposium on Turbulent and Laminar Boundary Layers*, New Orleans, Louisiana.

Nguyen, V.D., Dickinson, J., Chalifour, Y., Anderson, J., Haeberle, D. and Larose, G. 1984 Some Experimental Observations of the Law of the Wall Behind Large-Eddy Breakup Devices Using Servo-Controlled Skin Friction Balances, AIAA-84-0346.

Papathanasiou, A.G. and Nagel, R.T. 1954 Boundary Layer Control by Acoustic Excitation, AIAA-86-1954.

Plesniak, M.W. 1984 Optimized Manipulation of Turbulent Boundary Layers Aimed at Net Drag Reduction, *Master's Thesis*, Illinois Institute of Technology, Chicago, Illinois.

Poddar, K. and Van Atta, C.W. 1985 Turbulent Boundary Layer Drag Reduction on an Axisymmetric Body, *5th Symposium on Turbulent Shear Flows*, Cornell University, Aug. 7-9.

Poll, D.I.A. and Westland, P.G. 1987 A Study of LEBU Performance by Direct Total Force Measurements, *Symposium on Turbulent Drag Reduction by Passive Means*, London, Sept.

Rashidnia, N. and Falco, R.E. 1986 Changes in Turbulent Boundary Layer Structure Associated with Net Drag Reduction Layer Manipulators, Michigan State University Turbulent Structures Lab Report TSL-86-1, May.

Sahlin, A., Alfredsson, P.H. and Johansson, A.V. 1986 Direct Drag Measurements for a Flat Plate with Passive Boundary Layer Manipulators, *Phys. Fluids* **29**, 696-700.

Sahlin, A., Johansson, A.V. and Alfredsson, P.H. 1988 The Possibility of Drag Reduction by Outer Layer Manipulators in Turbulent Boundary Layers, *Phys. Fluids* **31**, 2814-20.

Schlichting, H. 1982 *Grenzschicht-Theorie*, Karlsruhe: Verlag G. Braun.

Westphal, R.V. 1986 Skin Friction and Reynolds Stress Measurements for a Turbulent Boundary Layer Following Manipulation Using Flat Plates, AIAA-86-0283.

Part VII

Novel Experimental Techniques

An Analysis of a Natural Turbulent Boundary Layer Through the Processing of Horizontal Hydrogen Bubble Lines

Arno J. Brand and Frans T.M. Nieuwstadt

Lab Aero & Hydrodynamics TU-Delft, Rotterdamseweg 145, 2628 AL Delft, Netherlands

The structure of a natural turbulent boundary layer is described in terms of four length scales that can be identified in instantaneous spanwise profiles of the streamwise velocity.

1. Introduction

The structure of turbulent boundary layer flow is dominated by elongated regions with a lower-than-average velocity (Figure 1a). These so-called low-speed streaks have a mean spanwise spacing $<\lambda_z^+>\sim 100$ (Kline e.a. (1967)) and a mean width $<\lambda_L^+>\sim 20...40$ (Blackwelder (1978), Talmon e.a. (1986)). (Lengths are non-dimensionalized with the viscous length scale $l_* = \nu/u_*$, with ν the viscosity and u_* the viscous velocity.) The streaks can have a very large streamwise extent (Blokland and Krishna Prasad (1984)). The high-speed regions neighboring the streaks are reported to be considerably wider: $<\lambda_H^+>\sim 40...110$ (Talmon e.a. (1986), Robinson e.a. (1988)).

The low-speed streaks give rise to local minima in the spanwise velocity profiles. Associated with these minima are inflection points that mark the position of the shear layers that surround the low-speed streaks. Since these shear layers might be unstable (Rayleigh's and Fjørtoft's theorems, see Drazin and Reid (1981)), the 'bursting of low-speed streaks' (Kline e.a. (1967)) is hypothesized to be the sudden development of hairpin vortices from unstable streaks (Smith (1984)). However, this process is not well understood. To explain some of its dynamics, it was suggested to model a spanwise velocity gradient with a hyperbolic tangent profile (Blackwelder and Swearingen (1988)). This showed that there is a relation between the length scale Δ associated with this gradient and the streamwise wavelength λ_x of the disturbances in the flow. Experimentally it was found that $<\Delta^+>\sim 5...15$ (Blackwelder and Swearingen (1988)).

The quantitative analysis of hydrogen-bubble time-line visualizations gives instantaneous local velocity profiles (Schraub e.a. (1965)). Compared to the use of multiple sensors, this technique has a finer spatial resolution, especially if digital image processing methods are used. For example, in a wind tunnel turbulent boundary layer a typical spatial resolution is $\Delta z^+ \sim 30$ for a rake of hot-wire sensors (Blackwelder and Swearingen (1988), Antonia and Bisset (1990)), whereas in a water channel turbulent boundary layer $\Delta z^+ \sim 1$ can easily be obtained with help of appropriate image processing (Brand and Nieuwstadt (1989)).

Here we report on a study aimed at the description of the spanwise structure of a turbulent boundary layer in terms of four length scales that can be identified in instantaneous spanwise profiles of the streamwise velocity (Figure 1b): (1) the spacing of two low-velocity regions λ_z, (2) the width of a low-velocity region λ_L, (3) the width of a high-

Advances in Turbulence 3
Editors: A.V. Johansson · P.H. Alfredsson

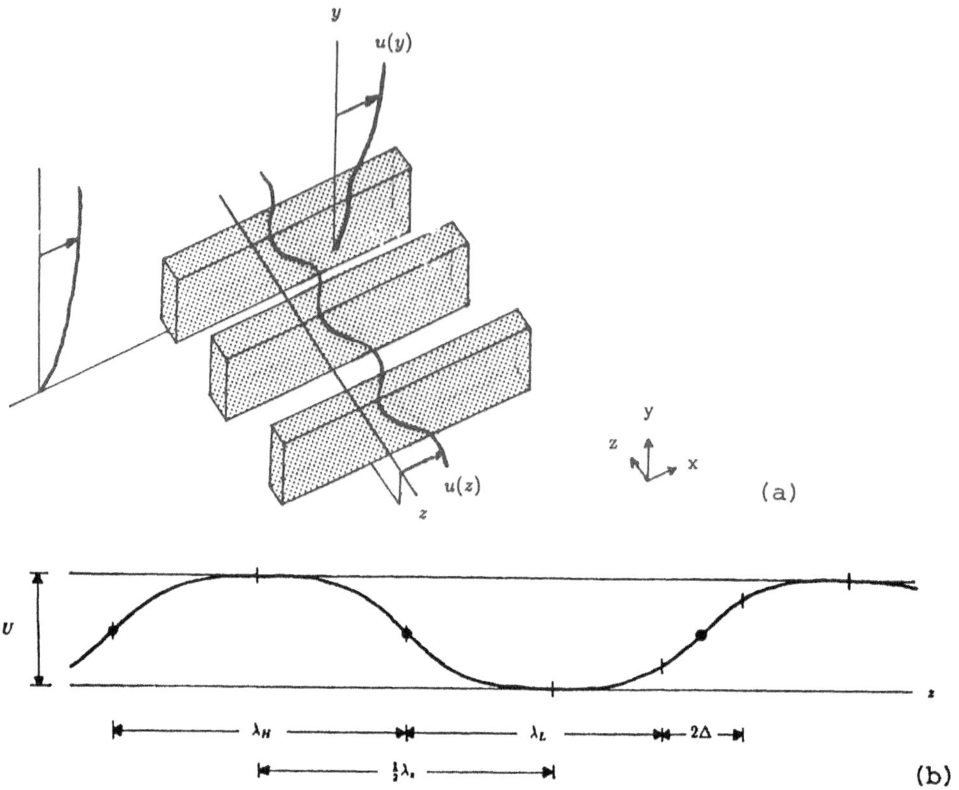

Figure 1: (a) Schematic of the coherent structure of turbulent boundary layer flow. Low-velocity regions (indicated grey) give rise to inflectional intantaneous $u(z)$ and $u(y)$ profiles. (b) A model for an inflectional $u(z)$ profile, with the four associated length scales: the half-spacing of two low-velocity regions ($\frac{1}{2}\lambda_z$), the width of a low and a high-velocity region (λ_L and λ_H), and the length scale of the spanwise velocity gradient (Δ).

velocity region λ_H, and (4) the length scale Δ of a spanwise velocity gradient. First we describe the turbulent boundary layer in which we performed our experiments. Then we address the flow-visualization equipment and the data handling involved in the processing of bubble-line pictures. Finally we present and discuss the results.

2. The turbulent boundary layer

The turbulent boundary layer is generated over a flat plate mounted in an open water channel. This water channel, which was used for combined hydrogen-bubble visualization and laser-doppler anemometry, is described elsewhere (Talmon e.a. (1986)). Modification since then resulted in less vibrations in the test section, and a higher water level. It allows for mean stream velocities up to 19 cm/s with a turbulence intensity of 1.5%.

The flat plate is 175 cm long and 60 cm wide. It has a sharp leading edge, and an adjustible flap at its end in order to control boundary-layer transition and separation. The plate is made from 2 cm thick perspex. The surface of the plate is mounted at a height of 6.5 cm above the bottom of the water channel.

A 7 mm thick tripping wire mounted 25 cm downstream of the leading edge is used to create a turbulent boundary layer. Hot-film measurements showed that $Re_\theta = 978$ (with θ the momentum thickness) at $x = 100$ cm downstream of the leading edge with a mean velocity U_∞ of 18 cm/s. This corresponds to $Re_x = 1.7\ 10^5$. Fitting of the $u(y)$ data to the 'theoretical' profile $u^+ = 2.5 \ln y^+ + 5.0$ yielded $u_* = 8.2$ mm/s and $l_* = 0.13$ mm.

3. Flow-visualization equipment

Figure 2 shows the experimental facility. The hydrogen bubbles are generated at a 20 μm thick and 17.5 cm long platinum wire. The wire holder is traversable in horizontal (x and z) and vertical (y) direction. In the present experiments the wire is oriented perpendicular to the mean flow, and parallel to the surface of the flat plate. The bubbles have a vertical terminal velocity of about 0.5 mm/s. (We assume that the size of the bubbles is of the order of the diameter of the wire.)

The bubbles are illuminated with light from a 5 W Ar-ion laser (Spectra Physics Model 2020). Special fiber-optic cables transport the light to an optical system that produces a thin light sheet. In the test section of the water channel the sheet measures 100x1 mm. (These beam diameters are based on the $1/e^2$-intensity points of the 'gaussian' laser beam.)

The light scattered by the bubbles is recorded with a standard CCD video camera (High Technology Holland MX). An electro-optical shutter (Displaytech Model PV-050C) appropriately synchronizes the light for the camera (i.e. a light pulse of 4 ms duration every 20 ms). The recordings are stored on video tape. In order to optimize the amount of light scattered into the camera, the angle between the camera axis and the axis of the light sheet is about 40°. The viewing field is about 35 mm in streamwise direction and about 85 mm in spanwise direction.

Bubble time lines are created by applying to the bubble wire voltage pulses of 20 ms duration and 30 V amplitude. This results in extremely thin bubble lines: about 2 mm

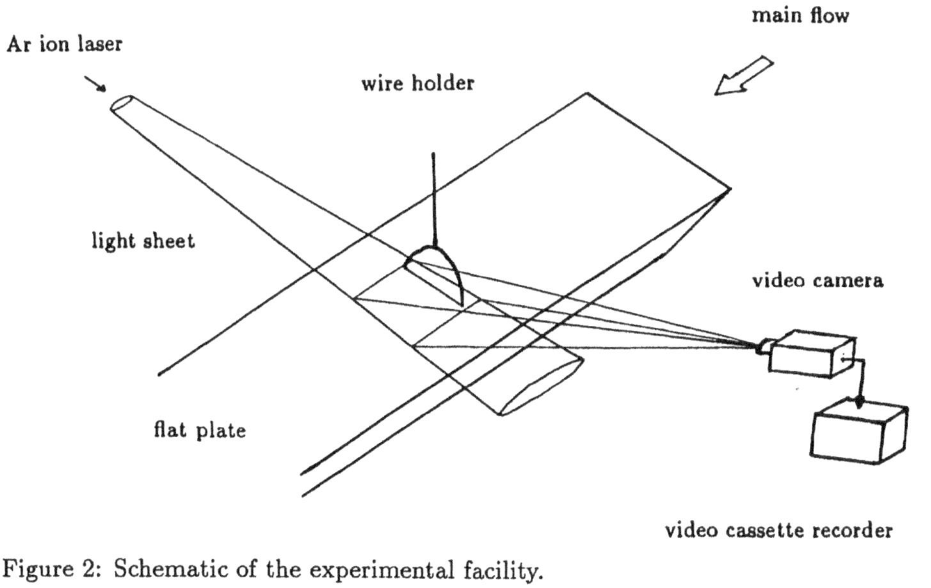

Figure 2: Schematic of the experimental facility.

with a wire at $y^+ = 27$. Single bubble lines are generated at intervals of 1 s because we want a new bubble line to be generated when the previous line is convected out of the viewing field. Tracking of the line then gives a description of the motion of a material line in the turbulent flow (Smith and Paxson (1983)).

4. The data handling

4.1 Pre-processing

A frame grabber (Data Translation DT 2851) mounted in a personal computer (Hewlett Packard Vectra RS/20) is used to digitize the bubble line pictures that are stored on video tape. With help of the 'still/advance' function of the video recorder, the operator grabs a video frame from tape and stores it in a video buffer. The picture is then available in a format of 512x512 non-square pixels and 256 grey levels. Standard and customized commands in an image processing software package (PC Semper) are used for processing of the digitized pictures.

The pictures are re-sampled on a 256x256 grid in order to reduce storage requirements and computation times. This has as additional (and not un-welcome) side effect that image flicker due to the interlaced and non-ideal video signal is removed. Next, a recording of the background is subtracted from the bubble-line pictures. This results in an improved contrast. Uneven illumination is taken care of by applying a maxmin filter to the pictures.

Due to the large angle between the camera axis and the light sheet the viewing field is geometrically distorted. Control grid matching is used to obtain a value for the angle over which the image field must be rotated in order to rectify the images. After this transformation the image plane corresponds to a xz plane in the boundary-layer coordinate system.

Next, the binary versions of the pictures are created. (A binary picture has only two grey values: 0 and 1.) This is possible because the pictures generally contain just one object (the bubble line) and the background. The binary bubble line pictures allow for logical and morphological processing (Brand and Nieuwstadt (1989)). An opening of size 2, followed by a closing of size 2 is used to remove 'salt-and-pepper noise', and to smooth the boundaries of the bubble lines. The thickness of the lines is then reduced to a size of one pixel with the rim function. This gives the position of the bubble line.

After successive processing of the recordings of a bubble line during its voyage through the viewing field, we create the picture that contains the position of the line at intervals of 40 ms. This is done by taking the 'inclusive or' of the thinned bubble-line pictures. Then the u(z) velocity profiles at consecutive time steps can be obtained. For this purpose the distance between neighboring bubble lines is measured at each scan line, and is divided by the value of the time interval. Due to errors introduced at the preceding stages, the uncertainty in the spanwise velocity profiles will be about 10% for u, and about one grid unit for z. Finally, the velocity profiles are stored as data files which are available for post-processing.

4.2 Post-processing

In general, the $u(z)$ profiles suffer from holes which originate from parts of the bubble line that have moved out of the light sheet. The profiles are restorated via cubic spline interpolation. Also outliers are removed. (Outliers are data points u for which $|u - \mu_u| > 2\sigma_u$, with μ_u the mean velocity of the profile and σ_u its the standard deviation.) The cubic splines of the restorated velocity profiles are used to find the position of the velocity minima ($\partial u/\partial z = 0$ and $\partial^2 u/\partial z^2 > 0$), the velocity maxima ($\partial u/\partial z = 0$ and $\partial^2 u/\partial z^2 < 0$), and the inflection points ($\partial^2 u/\partial z^2 = 0$); with the restriction that per grid interval only one minimum, maximum or inflection point is allowed. Inflection points that do not have a minimum and a maximum as neighbors, are not taken into account. So, a spanwise velocity profile is described by alternating minima and maxima, with an inflection point in between two extremes (Figure 1b). This provides a more objective way of defining the position of low-velocity regions than a method that identifies a velocity minimum in a point z if $u(z) - \mu_u < k\sigma_u$, with k a negative threshold level.

The half-spacing of the low-velocity regions $\frac{1}{2}\lambda_z$ is defined as the distance between a minimum and a maximum. The width of the low-velocity regions λ_L is the distance between two inflection points, provided that a minimum lies in between. On the other hand, this distance gives λ_H if a maximum lies in between. Finally, the gradient length scale Δ is found from $2\Delta = U/(\partial u/\partial z)$, in which $U = |u_{low} - u_{high}|$ is the magnitude of the velocity 'jump' (with u_{low} and u_{high} the streamwise velocities in the local minimum and maximum, respectively) and the gradient is evaluated at the inflection point.

5. Results and Discussion

The experiments reported here were performed at $x \sim 100$ cm downstream of the plate's leading edge and with a mean flow velocity $U_\infty \approx 18.0$ cm/s. The wire was mounted at $y^+ = 27$, 20, 12, and 8. At each height bubble lines were recorded during a period of about 5 m. From most of the bubble lines we obtained a spanwise velocity profile. This yielded respectively 254, 300, 281 and 270 profiles, in which a total amount of 1822, 2170, 2169 and 2123 low-velocity regions were detected.

The spatial resolution is $\Delta z^+ \approx 2.6$. The smallest length scale in a turbulent flow is the Kolmogorov scale η. This scale follows from $\eta = (\nu^3/\varepsilon)^{0.25}$, in which the dissipation is given by $\varepsilon = u_\tau^3/\ell$; with characteristic velocity fluctuation $u_\tau \sim 0.2U_\infty$ and characteristic eddy scale $\ell \sim y$. At $y^+ = 27$ this gives $\eta^+ \approx 1$, which means that we have a spatial resolution almost sufficient to resolve the Kolmogorov length scale.

Figure 3 presents the histograms of the four length scales that were determined from the profiles at $y^+ = 20$. It is clear that there is a large variation in individual values for each scale, and that the histograms are skewed towards values lower than the mean. Therefore, we computed three statistical quantities: (1) the mean, (2) the median, and (3) the interval in which 95% of the data lies. Table 1 summarizes these results.

The mean spacing $< \lambda_z^+ > \approx 70$ is low compared to the generally accepted value $<\lambda_z^+> \sim 100$. However, the latter value is based on the spacing of elongated regions in which the bubbles have actually accumulated (Kline e.a. (1967)). Obviously, not all spanwise gradients cause the bubbles to accumulate in streaks. Furthermore, most streak-count schemes (implicitly or explicitly) applied an additional criterion of the form $u_{low}/u_{high} < \alpha$. This has as a result that the streaks originating from small gradients are ignored. Therefore, although the traditional schemes may identify the correct streaks, they underestimate the small local spanwise velocity gradients. From the variation of $<\lambda_z^+>$ with y^+, it is clear that the spacing of the low-velocity regions does not depend on the distance above the wall. This result is in contrast with the observed increase of $<\lambda_z^+>$ with y^+ for $y^+ > 5$. From a comparison of the median and the mean spacings it follows that smaller spacings occur more often than larger spacings. The 95% interval quantifies this observation. What should be surprising about λ_z is not its mean or median, but

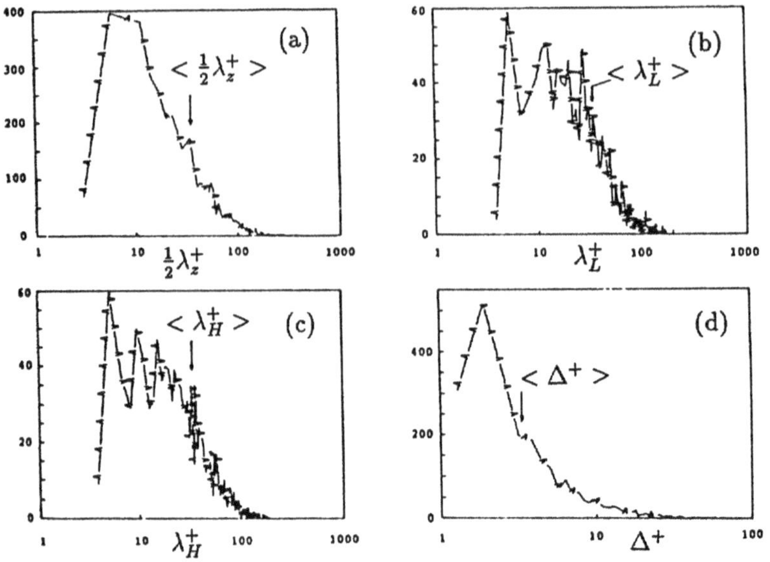

Figure 3: Histograms of the dimensionless length scales at $y^+ = 20$: (a) the half-spacing of low-velocity regions ($\frac{1}{2}\lambda_z^+$), (b) the width of low-velocity regions (λ_L^+), (c) the width of high-velocity regions (λ_H^+), and (d) the length scale of the spanwise velocity gradient (Δ^+). The arrows indicate the mean values.

Table 1: Dimensionless spacing of low-speed regions (λ_z^+), width of low and high-speed regions (λ_L^+ and λ_H^+), and gradient length scale (Δ^+) at different heights above the wall; (a) mean values, (b) median values, and (c) 95% interval.

(a)	y^+	$<\lambda_z^+>$	$<\lambda_L^+>$	$<\lambda_H^+>$	$<\Delta^+>$
	27	71	32	33	6
	20	72	33	34	6
	12	69	31	33	6
	8	68	32	31	6
(b)	y^+	$\overline{\lambda_z^+}$	$\overline{\lambda_L^+}$	$\overline{\lambda_H^+}$	$\overline{\Delta^+}$
	27	48	25	26	4
	20	49	27	27	3
	12	47	26	26	3
	8	47	26	26	3
(c)	y^+	λ_z^+	λ_L^+	λ_H^+	Δ^+
	27	10...267	5...92	5...98	1...26
	20	10...263	5...105	5...102	1...29
	12	10...249	5...92	5...98	1...31
	8	9...249	5...95	5...86	1...28

the range of values that occurs. However, this certainly is not a new observation (Kline e.a. (1967)). Finally, we note that due to our fine spatial resolution, the small scales (in an absolute sense) have a large relative weight in the averaging procedure. Indeed, we found $<\lambda_z^+>\sim 110$ from velocity profiles that were median filtered such that the spatial resolution was reduced to $\Delta z^+ \approx 28$. (Note that we can not decide on the effect of our low Re_θ because we did not vary Re_θ. However, there is some evidence that $<\lambda_z^+>< 100$ for $Re_\theta < 1000$ (Sreenivasan (1988)).)

The mean widths of the low and the high-velocity regions are almost equal: $<\lambda_L^+>\approx <\lambda_H^+>$. This result is considerably different from the generally accepted values $<\lambda_L^+>\sim 20$ (Blackwelder (1978), Talmon e.a. (1986)) and $<\lambda_L^+>/<\lambda_H^+>\sim 2...5$ (Talmon e.a. (1986), Robinson e.a. (1988)). Again this is due to a difference in concept: the latter values are based on the width of the regions in which the bubbles have accumulated, whereas ours are based on the distance between inflection points in the velocity profile. This illustrates that one should be careful in explaining flow phenomena in terms of the behavior of flow markers: the low-speed streaks are regions in which fluid markers have accumulated and need not be elongated and narrow low-velocity regions. The low-speed streaks do exist, but the low-velocity regions are dynamically important.

The value of the mean gradient length scale $<\Delta^+>\approx 6$ is in agreement with the value $<\Delta^+>\approx 10$ that was found with a rake of hot-wire sensors in a wind tunnel turbulent boundary layer (Blackwelder and Swearingen (1988)). It was shown that inflectional $u(z)$ profiles with gradient length scale Δ produce disturbances with streamwise wavelength $\lambda_x = 5\pi\Delta$ (Blackwelder and Swearingen (1988)). If we assume that these disturbances travel with a convection velocity $u_c = 15u_*$ (Robinson e.a. (1988)), then an observer stationary in the boundary layer finds a dimensionless time between two disturbances $T^+ = \lambda_x^+/u_c^+ \sim \Delta^+$. Since this is of the order of the reported time between two turbulent bursts, it explains that there is a large variability in values of the time between bursts (Kim e.a. (1971)). However, it suggests that its distribution should be skewed towards the smaller values.

6. Summary and Conclusions

To get instantaneous spanwise profiles of the streamwise velocity in a natural turbulent boundary layer, we analysed horizontal hydrogen-bubble time-lines with digital image processing techniques. In these profiles we located the position of the extremes and the inflection points. This yielded four length scales: the spacing of the low-velocity regions (λ_z), the width of the low and the high-velocity regions (λ_L and λ_H), and the characteristic scale for the spanwise velocity gradients (Δ).

Irrespective of height above the wall we found $<\lambda_z^+>\approx 70$, $<\lambda_L^+>\approx<\lambda_H^+>$, and $<\Delta^+>\approx 6$. This shows that the traditional schemes to determine the spacing and the width of low-speed streaks, underestimate the significance of small local spanwise velocity gradients. Also it shows that the spatial resolution of the experiment determines the relative weight of the small scales, and that it therefore influences the mean values of the length scales. Finally, it shows that the spanwise structure of a turbulent boundary layer is characterized by a broad range of length scales, rather than by a small interval.

Acknowledgements

The first author acknowledges J.M. Bessem (for developing the light sheet), J. Westerweel (for implementing parts of the image processing software) and H.J.M. Vollebregt (for performing the hot-film measurements and assistence during the visualization experiments.)

References

Antonia R.A. and Bisset D.K., 1990, J. Fluid Mech. **210**, p. 437

Blackwelder R.F., 1978, In: The coherent structure of turbulent boundary layers, Lehigh University

Blackwelder R.F. and Swearingen J.R., 1988, In: Zorac Memorial Sem. on Near-wall Turbulence, Dubrovnik

Blokland R. and Krishna Prasad K., 1984, In: 8th Symp. Turbulence, Rolla

Brand A.J. and Nieuwstadt F.T.M., 1989, In: Open Forum Contr. 7th Turb. Shear Flow Conf., Stanford

Drazin P. and Reid W., 1981, Hydrodynamic Stability, Cambridge University Press

Kim H.T., Kline S.J. and Reynolds W.C., 1971, J. Fluid Mech. **50**, p. 133

Kline S.J., Reynolds W.C., Schraub F.A. and Runstadler P.W., 1967, J. Fluid Mech. **30**, p. 741

Robinson S.K., Kline S.J. and Spalart P.R., 1988, In: Zorac Memorial Sem. on Near-wall Turbulence, Dubrovnik

Schraub F.A., Kline S.J., Henry J., Runstadler P.W. and Littel A., 1965, J. Basic Eng. **87**, p. 429

Smith C.R., 1984, In: 8th Symp. Turbulence, Rolla

Smith C.R. and Metzler S.P., 1983, J. Fluid Mech. **129**, p. 27

Smith C.R. and Paxson R.D., 1983, Exps. Fluids **1**, p. 43

Sreenivasan K.R., 1988, In: Turb. Man. and Rel. IUTAM Symposium, Bangalore

Talmon A.M., Kunen J.M.G. and Ooms G., 1986, J. Fluid Mech. **163**, p. 459

Turbulence Measurements Using Double-Chip Silicon Sensors

L. Löfdahl, G. Stemme**, B. Johansson* and L. Jernqvist**

Dept. of Thermo- and Fluid Dynamics*, Dept. of Solid State Electronics**
Chalmers University of Technology, 412 96 Göteborg, Sweden

A small directional sensitive double-chip silicon based sensor has been designed and fabricated using microelectronic technology. To determine the performance of this sensor the Reynolds stresses in a two-dimensional flat plate boundary layer were determined at a Reynolds number of 4.2 10^6. Comparisons with conventional hot-wire sensors were made and showed that the double-chip sensor was able to determine the turbulent stresses with the same accuracy as a cross-hot-wire.

1. Introduction

For a long time hot-wires have been the dominating sensors in the determination of turbulent stresses, however, the developments in microelectronics have made it possible to use also sensors based on silicon technology for turbulence measurements.

In a recent paper by Löfdahl et al. (1989) the first version of a silicon based sensor for the determination of mean velocity and turbulence intensity was presented. Further developments of the silicon based sensors have been carried out, and a second version of those sensors is described in this paper. The new version was made considerably smaller than the original sensor, and the length to width ratio of the chip was chosen as to increase the sensitivity of the sensor to changes in the flow direction. These sensors could accordingly be used for the determination of turbulent shear stresses as well as normal stresses perpendicular to the main flow direction.

Silicon, silicondioxide and other materials have been used in the fabrication of the sensors. Together with batch fabrication techniques these materials give the possibility to make very small and identical components. New techniques have been developed and added to standard semiconductor processes, for example micromachining of silicon has been used where, by using wet chemical etches, it has been possible to form the silicon into different three-dimensional shapes. Further details of the fabrication processes are available in Stemme (1986) and (1988).

2. Sensor design

The original sensors consisted of a base plate with electric bonding pads and a supporting beam. At the end of the beam a small sensor chip is mechanically attached and electrically connected to the beam. The principle of operation is very similar to those employed in hot-wire technique, however some deviations are important to notice. Instead of using a bridge balance, the operation of the silicon based sensors relies on a voltage difference which is formed between two temperature sensitive diodes. One diode, the "hot" one, is monitoring the temperature of the chip, which is electrically

Advances in Turbulence 3
Editors: A.V. Johansson · P.H. Alfredsson

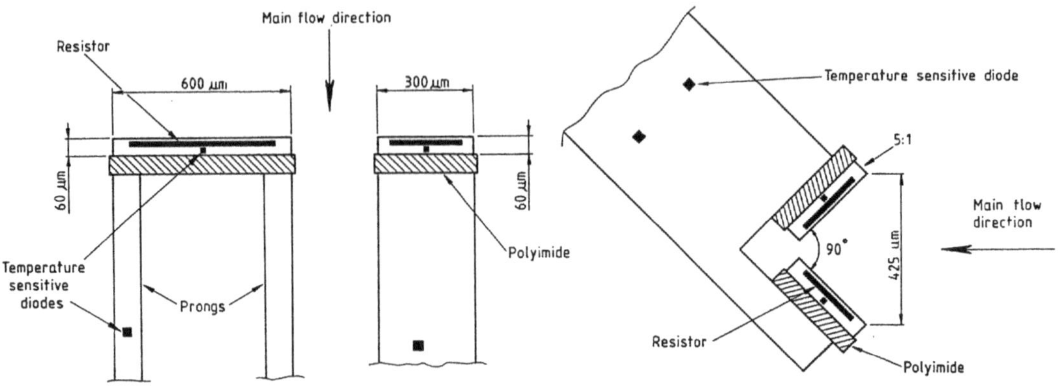

Figure 1. Chips with a length to width ratio of 5/1 and 10/1, respectively, and the double-chip.

heated by an integrated resistor and cooled by the flow. The other diode, the "cold" one, which is positioned on the beam, adjusts the preset overheat ratio to the ambient temperature. A polyimid bridge is carrying the chip, and since poyimid has low thermal conductivity it is also used for the insulation of the "hot" chip. Thus, the power dissipated in the heated resistor is a measure of the instantaneous flow velocity.

Figure 1 shows a principle drawing of the chip, which contains a resistor and a temperature sensitive diode, a polyimid bridge and the part of the supporting beam with the other temperature diode. For further details of the principle of operation and sensor fabrication, see Löfdahl et. al. (1989).

In the second version of the silicon sensors, the overall dimensions of the heated chip were reduced as much as possible due to partly the solidity and partly to the practical handling. Typical dimensions in the plane were 200-300 μm times 40-60 μm. The thickness of all sensors used so far was 30 mm, which presently is the minimum thickness of the fabrication process.

The heated chip of the first version of the silicon sensors had a fairly quadrangular shape, a length to width ratio of 4/3. These sensors were almost unsensitive to changes in the flow direction, and could accordingly only be used for the determination of the mean velocity and turbulence intensity in the main flow direction, see Löfdahl et. al. (1989).

To improve the directional sensitivity of the sensors, the length to width ratios were changed to 5/1 and 10/1, respectively, as is shown in Figure 1. Double chip sensors, where two chips were integrated on the same sensor, were also fabricated with a length to width ratio of 5/1 on each sensor. To enable shear stress measurements the chips in the double sensor were positioned at an angle of 90 degrees to each other and the whole sensor had to be adjusted in the main flow direction as is shown in this figure.

3. Experimental set up

The test measurements were carried out in a low-speed wind tunnel at a constant Reynolds number of 4.2 10^6. All equipment, i.e. wind tunnel, flat plate, hot-wire technique and data acquisition were the same as in previous measurements, so for such details, see Löfdahl et al.(1989).

At the measuring position, 2.0 m downstream the 2.5 m long flat plate, the boundary layer thickness was determined to 37 mm and the friction velocity to 1.18 m/s. The sensors were calibrated in this position, although then positioned well outside the boundary layer. In the calibration model, the silicon sensor was considered as a flat plate with a laminar boundary layer. Then the Reynolds number varies in the interval of 20 through 1200, and according to Schlichting (1978) the heat transfer from the plate is proportional to the square root of the Reynolds number. The same equations, which are employed in hot-wire technique, are thus also applicable for silicon sensors. In the hot-wire reference measurements standard Dantec equipment was used. (Dantec anemometer system No. 5600, boundary layer probe No. 55P05, straight probe No. 55P04 and cross-wire No. 55P51.) In the cross-hot-wire measurements a developed method, very similar to the method presented by Elsenaar and Boelsma (1974), was used.

4. Results and discussion

The directional sensitivity of the silicon sensors is shown in Figure 2, where the mean flow velocity, normalized with the free stream velocity is shown at different flow angles. The flow angle a is defined as the angle between the sensor axis and the main flow direction. As a comparison hot- wire measurements with a straight probe, and a simple cooling law are also included in the figure. From Figure 2 it can be seen that the original silicon sensor, with a length to width ratio of 4/3 was almost insensitive to changes in the flow direction. Of the second versions sensors the 5/1 sensor agrees very well with the hot-wire as well as with the simple cooling law in the angular range of 0 through 30 degrees. Some deviations occur, however, when the flow angle exceeds 30 degrees. If a more sophisticated cooling law is considered, for example the law suggested by Hinze (1959), the yaw correction factor of this law can, for the present case, be computed to 0.4 for a 45 degrees deviation, which is in good agreement with what can be expected of a hot-film probe according to Jörgensen (1971).

Large deviations from the hot-wire data and the simple cooling law may, however, be noticed for the 10/1 sensor. Most probably this is due to the two separate supporting beams (prongs). These beams have a quadrangular cross-section due to the fabrication process which, from a fluid dynamic point of view is not an optimal choice of cross-

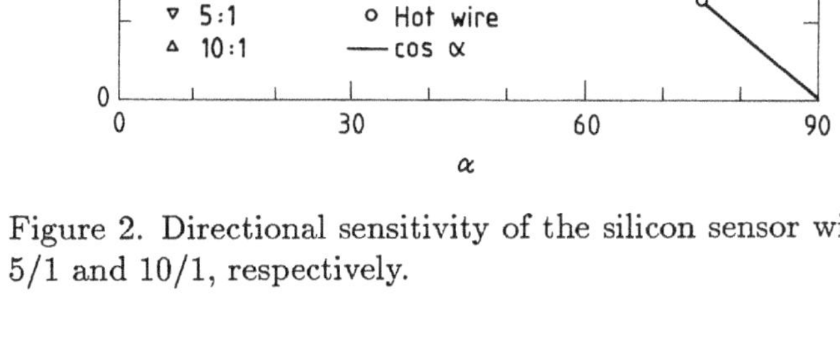

Figure 2. Directional sensitivity of the silicon sensor with a length/width ratio of 4/3, 5/1 and 10/1, respectively.

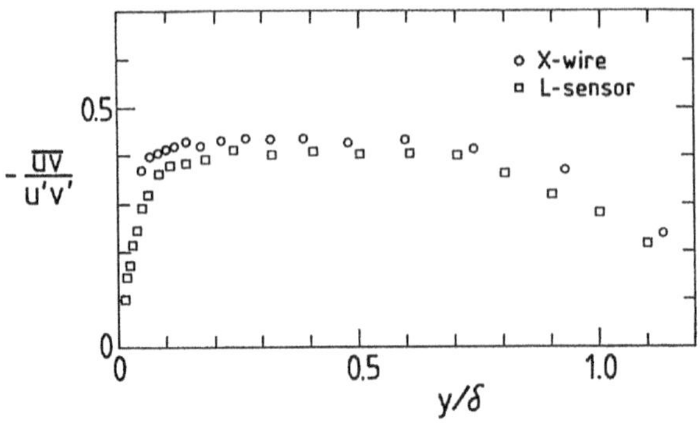

Figure 3. Turbulent shear stresses determined with the double chip sensor.

section, since eddy formation and separated regions may occur close to the heated chip. (The 10/1 sensors have not been used in the turbulence measurements.)

Figure 3 shows the turbulent shear stress, which has been determined with the double chip sensor. Cross-hot-wire measurements at the same position of the flat plate have been included in this figure for reference purpose. Here the shear stress is expressed as a correlation coefficient $\overline{uv}/u'v'$, and as a function of the wall distance normalized with the boundary thickness. The correlation seems to be approximately constant, -0.40, through most parts of the boundary layer, and the present correlation value is approximately the same as the value found by Perry and Abell (1975), who determined the correlation -0.38 in a pipe flow. However, as compared to early measurements of Klebanoff (1955) the present correlation coefficient is slightly low.

The Reynold stresses, which have been determined with the double chip sensor are shown in Figure 4. The profiles are normalized with the friction velocity u_τ, and u_τ^2, respectively, and shown as a function of the wall unit $y^+ = u_\tau y/\nu$. Measurements carried out by Karlsson and Johansson (1988) are used as reference data for the shear stress profile. A comparison between these data and the double chip sensor measurements will

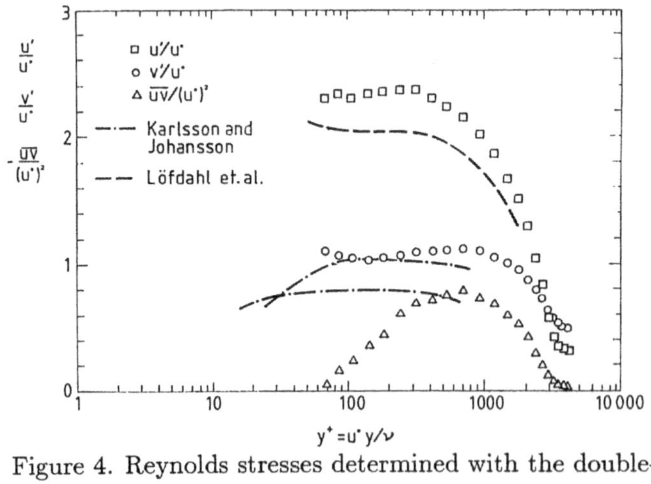

Figure 4. Reynolds stresses determined with the double-chip sensor.

clearly reveal the above mentioned difference in the innermost region of the boundary layer, for y^+ less than 110. This deviation can, however, be explained as follows.

Due to the geometry of the double silicon sensor, the velocity gradient will influence these measurements more than corresponding cross-wire measurements, since the two chips are located on different heights above the wall, while in a cross-hot-wire probe the two wires have their active length at the same wall distance. Furthermore, the double chip sensor has a very small dimension, 450 μm perpendicular to the wall, see Figure 3, its measuring volume is too large for accurate shear stress measurements in this region of the boundary layer. The importance of the spannwise size of the measuring volume has been studied in a recent paper by Johnsson and Barlow (1989), who found that the shear stress profile in the near wall region is extremely sensitive to the size of the measuring volume, and that in the inner part of the boundary layer, for a given y^+, the value of the turbulent shear stress decreases with increasing measuring volume. In the present measurements the characteristic length normal to the wall of the measuring volume is of the order $l^+ = 30$, which is far too large for shear stress determination in the region y^+ less than 200. Thus, fully in accordance with the the conclusions of Johnsson and Barlow (1989), the present measurements will decrease closest to the wall.

In Figure 4, the normal velocity fluctuation, v', is also shown and compared to the measurements of Karlsson and Johansson (1988). The double chip measurements do not reveal such a dramatic dependence of the size of the measuring volume as the shear stress, and a good agreement with the reference data is obtained. This observation is also in accordance with the laser measurements of Johnsson and Barlow (1989), who did not find any direct influence of the size of the measuring volume on this component.

The streamwise velocity fluctuation, u', is also shown in Figure 4 for completeness. A comparison of these data to earlier reference data of Löfdahl et al. (1989) yields, although slightly high, quite good agreement. (Using the double chip sensor, it has not been possible to capture the maximum values of u' in the region $y^+ = 8$ through $y^+ = 25$, since this sensor could not be positioned as close to the wall.)

In the outer portion of the boundary layer, the agreement between the hot-wire measurements, (and also measurements in other works for example Ligrani and Bradshaw (1987)) and the silicon sensor seems to be very good .

5. Conclusions

An investigation has been carried out in which profiles of the Reynolds stresses were measured in a two dimensional turbulent boundary layer, using silicon based sensors as well as hot-wires. Comparisons of the results are made and from these the following conclusions may be drawn.

* The new silicon based sensor with a length to width ratio of 5/1 seems to have the same directional sensitivity as a single standard hot-wire. For larger flow angles more complex cooling laws are necessary also for the silicon sensor.

* Comparisons of the Reynolds stress profiles reveal that the double chip sensor determines these profiles in a two dimensional turbulent boundary layer with the same accuracy as a cross-hot-wire.

The developed silicon based sensors can hardly replace hot-wires in all turbulence measurements, but rather be considered as a good complement in certain flow situations. As compared to hot-wires the silicon sensors have the advantage that other important functions like temperatur diodes, pressure transducers, pre-amplifiers and other control electronics can be integrated in a very small scale on the sensor.

6. Acknowledgements

This work was supported by a grant from the Swedish Board for Technical Development. The authors want to thank Professor Rolf Karlsson of the Swedish State Power Board Älvkarleby Laboratory for valuable and fruitful discussions during the work.

References

Elsenaar, A. and Boelsma, S.H. 1974, Measurements of the Reynolds Stress Tensor in a Three-Dimensional Turbulent Boundary Layer under Infinite Swept Wing Conditions, National Aerospace Laboratory,NLR TR 74095 U.

Hinze, J.O. 1959, *Turbulence*, McGraw-Hill Book Company.

Johnson, P.L. and Barlow, R.S. 1989, Effects of measuring volume length on two-component laser velocimeter measurements in a turbulent boundary layer, *Experiments in Fluids*, **8**, pp 137-144.

Jörgensen, F.E. 1971, Directional Sensitivity of Wire and Fiber-film Probes, *DISA Information*, No.11.

Karlsson, R.I. and Johansson,T.G 1988, LDV measurements of higher order moments of velocity fluctuations in a turbulent boundary layer, Laser Anemometry in Fluid Mechanics, publised by Ladoan-Instituto Superior Tecnico 1096 Lisbon Codex-Portugal.

Klebanoff, P.S. 1955, Characteristics of turbulence in a boundary layer with zero pressure gradient, NACA Report 1247.

Ligrani, P.M. and Bradshaw, P. 1987, Spatial Resolution and Measurements of Turbulence in the Viscous Sublayer Using Subminiature Hot-wire Probes, *Experiments in Fluids*, **5** pp. 407-417.

Löfdahl, L., Stemme, G. and Johansson, B. 1989, A sensor based on silicon technology for turbulence measurements, *J. Phys. E. Sci. Instrum.*, **22**, pp 391-393.

Schlichting, H. 1978, *Boundary Layer Theory*, Seventh Edition, McGraw-Hill Book Company.

Stemme,G. 1986, A Monolithic Gas Flow Sensor with Polyimide as Thermal Insulator, *IEEE Transactions on Electron Devices*, Vol. Ed 33, No.10.

Stemme,G. 1988 A CMOS Silicon Gas Flow Sensor with Puls-Modulated Output, *Sensors and Actuators*, **14**, No.3, pp 293-303.

Perry, A.E., and Abell, C.J. 1975, Scaling laws for pipe flow turbulence, *J. Fluid Mech.*, **67**, 257.

Active Turbulence Generation in a Laboratory Wind Tunnel

H. Makita and K. Sassa

Dept. of Energy Eng., Toyohashi Univ. of Technology, 1-1 Tempaku-cho, Toyohashi 441, Japan

A new type of active turbulence generator was designed. It can induce strong, homogeneous, and quasi-isotropic turbulence with a high turbulence Reynolds number in a laboratory wind tunnel. The resultant turbulence had intensity of more than 30 % and an integral scale exceeding 200 mm just downstream of the generator. The maximum turbulence Reynolds number reached $Re_\lambda \approx 600$, which is about one order of magnitude larger than that of the grid turbulence with the same mesh Reynolds number, Re_M. The energy spectrum of the strong turbulence had a clear inertial subrange spreading for more than two orders of magnitude in wave-number. The features of the spectrum are favorably comparable with those for the grid turbulence of $Re_M \approx 10^6$.

1. Introduction

In most of practical turbulence phenomena, e.g. those in atmospheric and oceanic turbulence, turbulence intensities are much higher and scales of turbulent eddies are much larger than those of wind-tunnel turbulence. Therefore, quite large Reynolds numbers are essentially required to make wind-tunnel simulations on them. Furthermore, statistical theories describing fundamental nature of turbulence assume a priori a satisfactorily large Reynolds number. On the other hand, most of the experimental data ever used to verify the turbulence theories were obtained in facilities of comparatively small size. A typical example is, of course, the theory of isotropic turbulence developed on the basis of grid turbulence experiments. Batchelor (1953) concluded that a turbulence field, in which a turbulence Reynolds number was enough large to observe an inertial subrange in the energy spectrum, cannot be realized in a wind tunnel of conventional size with a static grid. Usually, an enormously large wind tunnel is required to get such a large-scale turbulence field, e.g. the Southern California Co-operative wind-tunnel used by Kistler and Vrebalovich (1966) and ONERA Modane wind-tunnel employed by Gagne (1988).

One of the alternative methods to get such a large-scale turbulence field is to employ some "active" turbulence generator in a wind tunnel of laboratory size. In the present study, we developed a mechanical turbulence generator and then examined the structure of homogeneous, quasi-isotropic and large-scale turbulence induced by it in a laboratory wind tunnel. Two types of active turbulence generators have been derived, i.e. jet grids by Gad-el-Hak & Corrsin (1974) and Tassa & Kamotani (1975), and two-dimensional vibrated grids by Ling & Wan (1972) and Sato & Saito (1974). But, it was difficult for these facilities to control the structure of turbulence, i.e. to attain large turbulence scales, to get smooth wide-band energy spectra, or to keep good homogeneity in the distributions of turbulence quantities. Basically, the present scheme belongs to the latter type. It stirrs up a flow field in a three-dimensional manner by many small randomly-oscillating agitator plates. Good homogeneity and quasi-isotropy were confirmed in the resultant turbulence field and an apparent inertial subrange was observed in its wide-spread spectrum.

Advances in Turbulence 3
Editors: A.V. Johansson · P.H. Alfredsson

2. Turbulence Generator

Figure 1 shows a wind tunnel in Toyohashi Univ. of Tech. (TUT-WT). It has a test section of 0.7×0.7 m^2 in cross section and 6 m in length of which the cross-sectional area is gradually enlarged downstream at a rate of about 0.4 % in equivalent diameter to streamwise length ratio. In the absence of the turbulence generator, the non-uniformity in mean velocity distribution was less than 0.1 % and the background turbulence level was less than 0.04 % of U_0 at 10 m/s in the core of the test section.

The turbulence generator is settled upstream of the test section. It is a sort of active bi-plane grid as shown in Fig. 2. Each plane of the grid placed with 1 mm gap is composed of 15 horizontal or vertical rotatable rods of 6 mm in diameter having many small agitator wings. The rods are independently driven by stepping motors outside of the generator. Each motor rotates

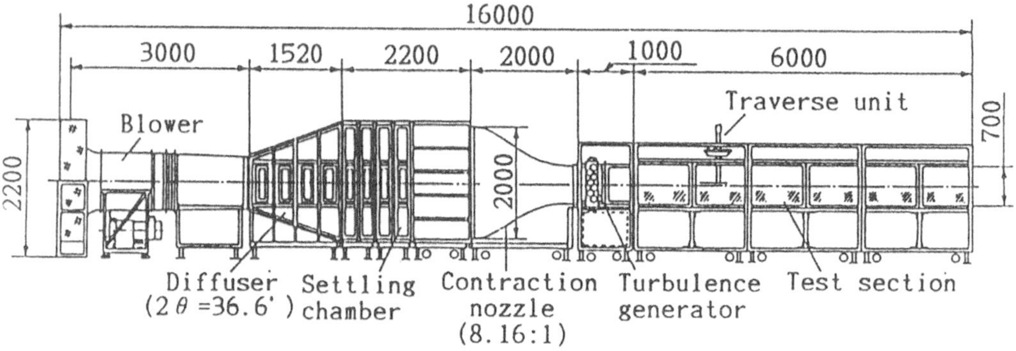

Fig. 1 Wind Tunnel in Toyohashi Univ. of Tech. (TUT-WT) unit:mm

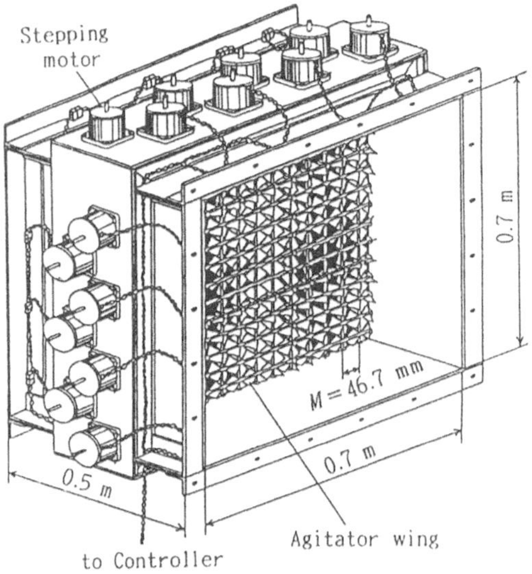

Fig. 2 Active Turbulence Generator

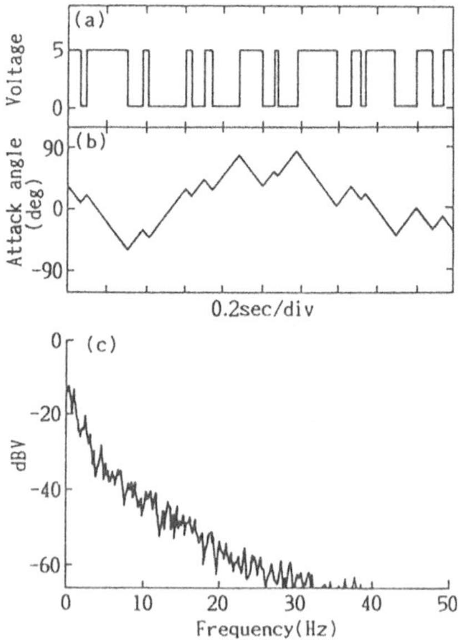

Fig. 3 Turn Pulse (a), Motion of Motor (b) and Spectral Analysis of Motion of Motor (c)

at a constant speed of 2 rps but turns forward and backward randomly when random-frequency turning pulses are fed to the motor's driver-circuit as shown in Figs. 3-a and 3-b. Figure 3-c shows a spectrum of the operating mode of the stepping motor. It tells that the random motion of the agitator wings is apt to enhance spectral components of 0-30 Hz of turbulent fluctuation in the flow passing through the grid surface. Besides the direct agitation by the wings, various sources of turbulence, such as wakes of rods and flow separation at the leading and trailing edges of the wings, will contribute to induce turbulence fluctuations of various frequencies. Hereafter, we call the turbulence field induced by operating the generator the "excited turbulence" and the grid turbulence obtained when the wings are settled at rest parallel to the flow direction the "non-excited turbulence".

3. Experiment

Measurements were made using constant temperature hot-wire anemometers with improved noise level and an X-probe. The X-probe is made of tungsten wires of 2.7 μ m in diameter, and its measuring volume is $0.7\times 0.7\times 0.2$ m^3. The length of the wire was a little larger than Kolmogorov scale, η. The positional arrangement of the wires was made under a microscope. The angle of incidence of the probe was adjusted through a pulley-coil mechanism after the probe was attached to the traversing mechanism in the test section. The errors due to the tangential cooling effect and the effect of the third component of velocity fluctuation were corrected by the schemes of Bruun (1975) and Hinze (1975), respectively. The tangential cooling coefficient was estimated to be $k=0.136$ by measuring the attack angle dependence of the hot-wire output between $\pm 60°$ in a calibration tunnel.

The experiment was conducted for the mean velocity, $U_0 = 1-7$ m/s. Mean velocity, turbulence intensities, length scales, energy spectrum, and anisotropy tensor were obtained from the instantaneous velocity signals recorded in a 12-bit digital recorder by a 16-bit computer, an F.F.T analyzer, an RMS meter, and so on.

4. Results and Discussion

Prior to the measurements, the distributions of mean velocity and turbulence intensities of the excited turbulence were confirmed to be satisfactorily uniform across the test section excepting the near-wall region. Figure 4 shows the decay of the turbulence intensities along the axis of

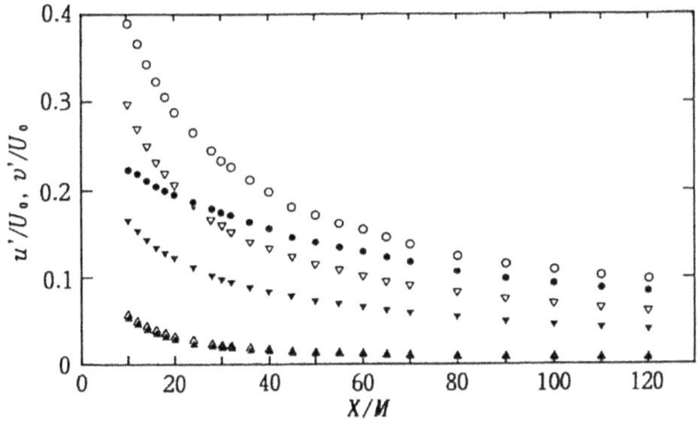

Fig. 4 Streamwise Distributions of Turbulence Intensity.

Excited, U_0=5m/s; ○ : u-component
● : v-component
U_0=1m/s; ▽ : u-component
▼ : v-component
Non-excited U_0=5m/s; △ : u-component
▲ : v-component

Table 1 Comparison of the turbulence characteristics

Experiment	U_0 (m/s)	Re_M (10^3)	u'/U_0 (%)	b_{11} (10^{-2})	L_{ux} (mm)	λ_g (mm)	η (mm)	Re_λ	X/M	n	X_0/M
Present work Excited	5	15.6	17.1	9.5	206	7.1	0.19	387	50	1.43	-12.0
Present work Non-excited	5	15.6	1.5	8.0	23	6.2	0.58	34	50	1.28	7.5
Conventional wind tunnel (a)	10	34.0	2.22	0.0	24	4.9	0.29	72	42	1.25	3.5
Large pressurized tunnel (b)	61	2420.0	1.97	9.8	47.8	2.4	0.047	669	45	1.0	–
Meteorological tunnel (c)	28.9	408.0	2.25	1.3	–	7.0	0.20	280	41	1.0	0
ONERA Modane tunnel (d)	20	–	–	–	–	–	–	2720	–	–	–

(a) Comte-Bellot & Corrsin (1966,1971)
(b) Kistler & Vrebalovich (1966)
(c) Schedvin et al.(1974)
(d) Gagne(1987)

the test section. The vertical and horizontal velocity fluctuations, v' and w', had the same intensity and only u- and v-components are presented in the figure. For the excited turbulence, the value of u'/U_0 at $U_0 = 5$ m/s exceeds 30 % at $X/M = 10$, where X/M is the streamwise distance normalized by the mesh size, $M = 46.7$ mm, of the grid. The turbulence intensity is about one order of magnitude larger than the values obtained for the conventional grid turbulence. Though u'/U_0 decays downstream, it maintains still more than 10 % in intensity at $X/M = 110$.

Just behind the generator, the excited turbulence has an appreciable order of axisymmetric anisotropy. The degree of anisotropy, if expressed by u'/v', reaches about 2.0 there. Toward the streamwise direction, the degree of anisotropy becomes calmed to be less than 1.3 for $X/M > 60$. Such a degree of anisotropy is commonplace for conventional grid turbulence experiments; see Hinze (1975). If the additional small contraction is installed behind the generator, the isotropy will be improved as made by Comte-Bellot and Corrsin (1966).

Reciprocal values of turbulence energy are plotted in logarithmic scales in Fig. 5. The power law, $u'^2 \propto (X/M - X_0/M)^{-n}$, was applied to them. The virtual origin, X_0, and the decay rate, n, depend on the width of X-region to which an iterative least square is applied (Groth & Johansson, 1988). For the excited turbulence, the power law was fitted for $X/M > 60$ where the turbulence field has already attained an quasi-equilibrium state. The values of X_0 are listed in Table 1. The decay rate of the non-excited turbulence, $n = 1.28$, is nearly the same as the value obtained by Comte-Bellot & Corrsin (1966) for their isotropic turbulence; see Table 1. The v-component of the excited turbulence decays somewhat slowly at the initial stage of the decay, because it receives turbulence energy from the u-component during the isotropy is being recovered in the turbulence field. For both of u- and v-components of the excited turbulence, their decay rates reach about $n \approx 1.4$. It seems a little larger than the values of the conventional grid turbulence experiments but is close to the value numerically determined for the isotropic turbulence with $Re \rightarrow \infty$ by Tatsumi and Kida (1980).

Figure 6 shows the streamwise evolution of the longitudinal and lateral integral length scales, L_{ux} and L_{vx}, obtained by integrating auto-correlation coefficients as follows:

$$L_{ux,vx} = U_0 \int_0^{\tau_0} R_{ux,vx} \, d\tau \qquad (1)$$

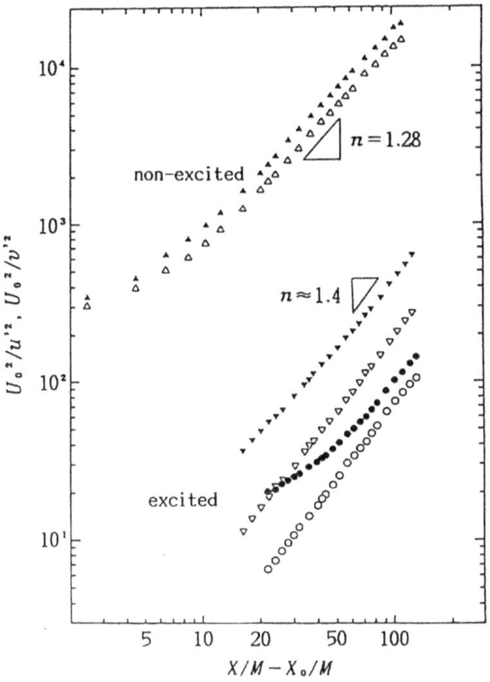

Fig. 5 Streamwise Decay of Turbulence Energy (symbols as in Fig.4)

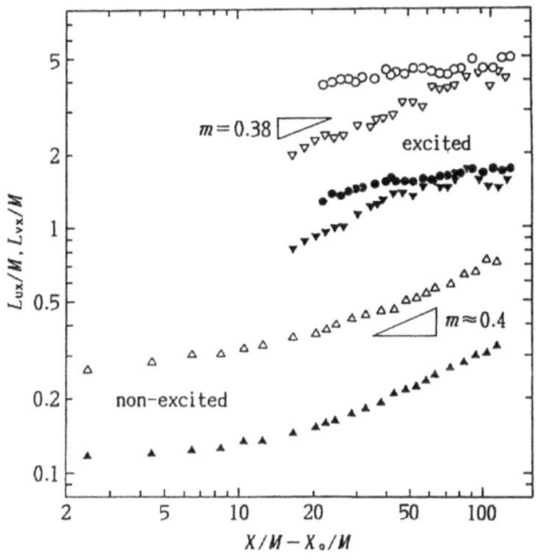

Fig. 6 Increase of Integral Scales with Streamwise Distance. (symbols, as in Fig.4)

where, $R_{ux}=\overline{u(t)u(t+\tau)}/\overline{u^2(t)}$, $R_{vx}=\overline{v(t)v(t+\tau)}/\overline{v^2(t)}$, and τ_0 is the first zero-crossing time-lag in the auto-correlograms. Generally, the grid turbulence could not have the integral scales exceeding the mesh size, M. For the excited turbulence, L_{ux}, attains a value of about 4 times as large as M, being about one order of magnitude larger than that of the non-excited turbulence. The power law, $L_{ux}/M \propto (X/M-X_0/M)^{-m}$, was also applied to the streamwise growth of the integral scales. The growth rate of the integral scale, m, at $U_0=5$ m/s becomes saturated for about $X/M>60$. There, the isotropic relation, $m=(2-n)/2$, given by Batchelor (1953) does not hold. Penven et al. (1985) suggested that a test section is required to have a span of more than 7 times as large as an integral scale to assure the homogeneity of a turbulence field in it. For the present case, the saturation occurs when L_{vx} approaches about one-seventh of the tunnel width. If the test section is enlarged to 1.5 m × 1.5 m in cross section, we will get a more idealized turbulence field where energy-containing eddies can grow up without the wall restriction and we will be able to examine basic hypotheses of the modern turbulence theory more exactly in it.

Representative quantities are compared among the excited turbulence, the non-excited turbulence, and the other typical turbulence fields in Table 1. The turbulence Reynolds number, Re_λ, representing the scale of the turbulence field itself is defined as

$$Re_\lambda=(u_1\cdot\lambda_g)/\nu\ , \tag{2}$$

where ν is the kinetic viscosity, $u_1=\sqrt{q^2/3}=\sqrt{(u^2+v^2+w^2)/3}$, is the characteristic

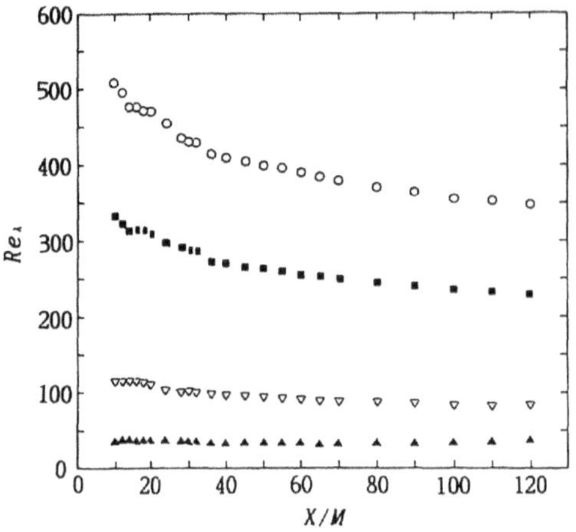

Fig. 7 Streamwise Change of Turbulence Reynolds Number. Excited; ○ : $U_0 = 5$ m/s, ■ : 3 m/s, ▽ : 1 m/s, Non-excited (5 m/s) : △

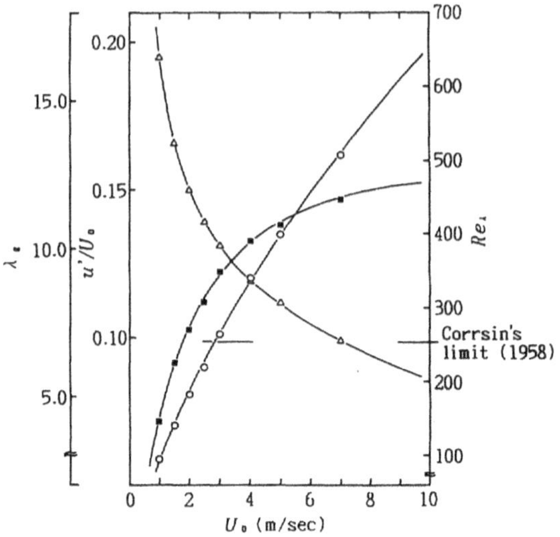

Fig. 8 Taylor's Microscale (△), Turbulence Intensity (■), Turbulence Reynolds Number (○) as a Function of Mean Velocity.

velocity, and λ_g is the Taylor's microscale given as follows:

$$\lambda_g{}^2 = 15 u_1{}^2 \cdot \nu / \varepsilon \tag{3}$$

in which, the energy dissipation, ε, was obtained as a timewise derivative of the turbulence energy, $-(1/2)\cdot dq^2/dt$. The streamwise evolution of Re_λ is illustrated in Fig. 7. Batchelor (1953) predicted that Re_λ is kept constant in a self-preserving turbulence with $n = 1$. For the excited turbulence, Re_λ decreases downstream quite gradually even in the region far from the generator. The excited turbulence, however, seems to be able to offer good experimental fields for the simulations of practical large-scale turbulence phenomena such as the wind environmental problems (see Makita et al. 1989). Figure 8 shows the turbulence intensity, u'/U_0, Taylor's microscale, λ_g, and Re_λ at $X/M = 50$ as a function of U_0. The value of λ_g decreases but u'/U_0 increases with U_0. Consequently, Re_λ grows larger roughly in proportion to U_0. As shown in Table 1, the excited turbulence has a quite large value of $Re_\lambda = 387$ at $U_0 = 5$ m/s, if its mesh Reynolds number, $Re_M = 1.56 \times 10^4$, is taken into account. The values of Re_λ obtained in the present work rank well with those of the grid turbulence with far larger values of $Re_M = 2.42 \times 10^6$ obtained by Kistler and Vrebalovich (1966) and of $Re_M = 4.08 \times 10^5$ done by Schedvin et al. (1974). The value of Re_λ for $U_0 > 3$ m/s exceeds 250; the lowest limit for the existence of an apparent inertial subrange given by Corrsin (1958).

One-dimensional energy spectra measured in various turbulence fields are compared in Fig. 9, Appling the Taylor hypothesis for the excited turbulence with the turbulence intensity of about 17% at $X/M = 50$ does not exert so significant influence on the spectral profile (Champagne, 1978). The non-excited turbulence has $Re_\lambda = 34$ and the inertial subrange obeying $-5/3$ power law of Kolmogorov (1941) is not clearly observed in its spectrum. For the excited turbulence at $U_0 = 5$ m/s, the energy level is enormously enhanced in the low wave-number range and the

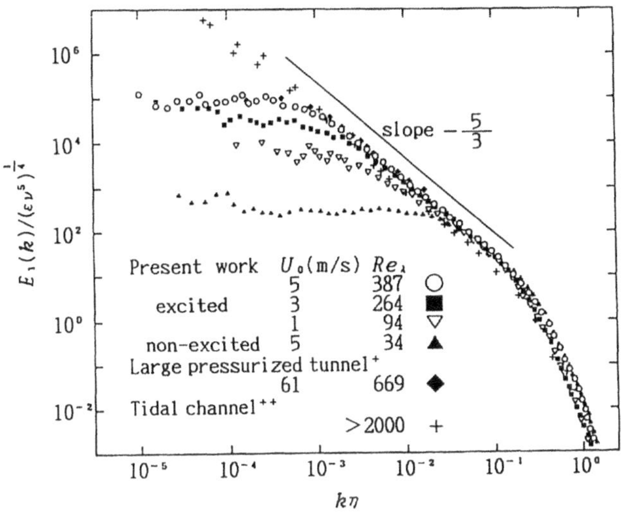

Fig. 9
Comparison of One-Dimensional Energy Spectra
+Kistler & Vrebalovich (1966)
++Grant et al. (1962)

inertial subrange appears for more than two orders of magnitude in wave number. The width of the inertial subrange seems to surpass the result of Kistler and Vrebalovich (1966), though it does not reach that measured in a tidal wave by Grant et al. (1962). More carefully observed, the slope of the inertial subrange is slightly gentler than -5/3 for the present case. A slope of -1.72, being a little steeper than -5/3, was reported by Gagne (1988) for $Re_\lambda \approx 2720$, one of the largest turbulence Reynolds number ever realized in the wind tunnel turbulence.

The value of Re_λ presently obtainable in our wind tunnel is at most 500-640 at $U_0 = 7$ m/s. Practically speaking, getting $Re_\lambda > 1000$ will be required to suffice the condition of $Re_\lambda \approx \infty$. Estimated from Fig. 8, such an ideal condition will be accomplished if the mean velocity can be increased up to 15 m/s in the wind tunnel.

Rotta (1951) introduced the anisotropy tensor, b_{ij}, to evaluate degree of anisotropy of homogeneous turbulence as follows.

$$b_{ij} = \frac{\overline{u_i u_j}}{q^2} - \frac{1}{3}\delta_{ij}, \tag{4}$$

where u_i and u_j are instantaneous fluctuating velocities and δ_{ij} is Kronecker delta. For axisymmetric turbulence, b_{ij} is reduced to only diagonal components b_{11}, b_{22} and b_{33} and the following relation holds.

$$b_{11} = -2b_{22} = -2b_{33} \tag{5}$$

Figure 10 shows streamwise distribution of b_{11}. Even the non-excited turbulence does not attain $b_{ij} = 0$ (complete isotropy) and its b_{11} becomes constant of about 0.06 for $X/M >$ 20. The fact agrees with the other results of the grid turbulence (Comte-Bellot and Corrsin 1966). Batchelor and Stewart (1950) showed that the initial anisotropy of the grid generated turbulence is apt to persist in the final stage of its decay process. The excited turbulence has large b_{ij} just downstream of the generator but its value of b_{ij} decreases downstream. For $X/M > 60$, the b_{11} at $U_0 = 3$ and 5 m/s is reduced to be almost the same level as

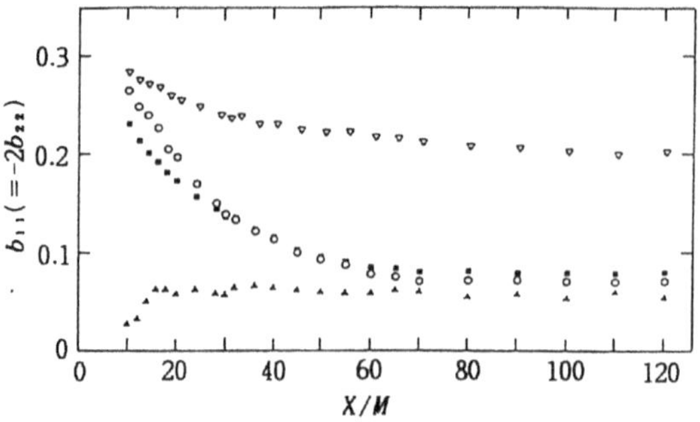

Fig. 10 Streamwise Distribution of Anisotropy Tensor. (symbols as in Fig. 7)

Fig. 11 Turbulence Energy against Invariant II of Anisotropy Tensor. (symbols as in Fig.7)

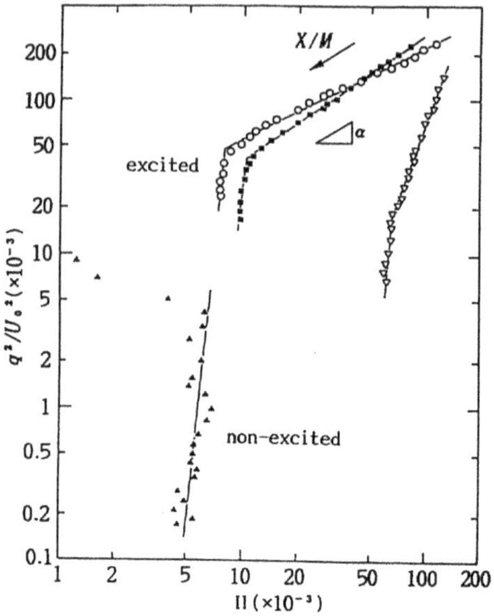

that of the non-excited turbulence. It seems that the turbulence field becomes afforded more excellent ability of recovering isotropy as U_0, and accordingly Re_λ, grows larger.

Figure 11 compares the relations between the rate of returning to isotropy and the decay rate of turbulence energy in logarithmic scales. Here, II is the invariant of anisotropy tensor and defined for the case of the axisymmetric turbulence as:

$$\mathrm{II} = (3/2)b_{11}{}^2. \tag{6}$$

After Penven et al. (1985), the ratio, α, shown as the slope of the plots in the figure denotes the ratio between the time constant of returning to isotropy, τ_R, and that of decay, τ_D, i.e.

$$\alpha = \frac{d(\log q^2)}{d(\log \mathrm{II})} = \frac{(1/q^2)(dq^2/dt)}{(1/\mathrm{II})(d\mathrm{II}/dt)} = \frac{\tau_R}{\tau_D} \tag{7}$$

A horizontal slope means that the turbulence is exclusively recovering isotropy. On the other hand, the vertical one does that the turbulence is merely decaying. For the excited turbulence, the slope is quite gentle in the initial stage of the decay. It means that the excited turbulence returns to isotropy very rapidly in spite of its decay being comparatively slow. For $U_0 = 5$ m/s, the slope suddenly changes at $X/M \approx 70$ to be so steep as observed for the non-excited turbulence. The energy spectra became not to make any essential change in profile for $X/M > 70$, which shows that the turbulence field have already attained some equilibrium steady state (Makita et al. 1987). With increase in U_0, the slope in the returning-to-isotropy stage becomes gentler and the knee point of the slope shifts upstream. These facts suggest that the ability of recovering isotropy of a turbulence field is strengthened with Re_λ as mentioned before. It is more interesting, we think, that a turbulence field is possible to have different nature at the final stage of its decay by the difference in magnitude of its initial turbulence Reynolds number. Judging from the facts mentioned above, the excited turbulence of $Re_\lambda > 250$

can be regarded to have reached an equilibrated steady-state in the region of about $X/M >$ 70 where the quasi-isotropy was already attained and can offer an experimental turbulence field with good utility in treating a variety of turbulence phenomena in it.

5. Concluding Remarks

The newly developed turbulence generator made it possible to get a homogeneous and quasi-isotropic strong turbulence field having turbulence characteristics corresponding to those of grid turbulence with $Re_M \sim 10^6$ in a wind tunnel of laboratory size. By employing the generator, experimental researchers can use high Reynolds number turbulence fields in their wind tunnels for both studying physical structure of turbulence and simulating practical turbulence phenomena in large-scale turbulence fields such as atmospheric turbulence. The present scheme is still required to make further improvements to get more ideal conditions of the turbulence field.

Reference

Batchelor,G.K. (1953) *The theory of homogeneous turbulence*. Cambridge Univ. Press.
Batchelor,G.K. & Stewart,R.W. (1950) *Quart. J. Mech. Appl. Math.* 3, 1.
Bruun,H.H. (1975) *DISA Information*, 18, 5.
Champagne,F.H. (1978) *J. Fluid Mech.* 86, 67.
Comte-Bellot,G & Corrsin,S. (1966) *J. Fluid Mech.* 25, 657.
Comte-Bellot,G & Corrsin,S. (1971) *J. Fluid Mech.* 48, 273.
Corrsin,S. (1958) *NACA RM* 58B11
Gad-el-Hak,M. & Corrsin,S. (1974) *J. Fluid Mech.* 62, 115.
Gagne,Y. (1988) *Thèse de Doctorat*, Institut de Méchanique de Grenoble.
Grant,H.L., Stewart,R.W. & Moilliet,A. (1962) *J. Fluid Mech.* 12, 241.
Groth,J. & Johansson,A.V. (1988) *J. Fluid Mech.* 197, 139.
Hinze,J.O. (1975) *Turbulence* 2nd ed. McGrow-Hill.
Kistler,A.L. & Vrebalovich,T. (1966) *J. Fluid Mech.* 26, 37.
Kolmogorov,A.N. (1941) *C. R. Acad. Sci. U. S.* 30, 301.
Ling,S.C. & Wan,C.A. (1972) *Phys. Fluids* 15, 1363.
Makita,H., Sassa,K. & Iida,A. (1989) *Proc. of 4th Asian Cong. of Fluid Mech.* vol.II, G83.
Makita,H., Sassa,K., Iwasaki,T. & Iida,A. (1987) *Trans. Japan Soc. Mech. Eng.* Ser B. 53, 3180 (in Japanese).
Penven,L.L., Gence,J.N. & Comte-Bellot,G. (1985) *Frontiers in Fluid Mech.* Springer-Verlag, 1.
Rotta,J.C. (1951) *Z. Phys.* 129, 547.
Sato,H. & Saito,H. (1974) *Proc. of 6th turbulence symposium* 103 (in Japanese).
Schedvin,J., Stegen,G.R. & Gibson,C.H. (1974) *J. Fluid Mech.* 65, 561.
Tassa,Y. & Kamotani,Y. (1975) *Phys. Fluids* 18, 411.
Tatsumi,T & Kida,S. (1980) *J. Phys. Soc. Japan* 49, 2014.

Effects of Longitudinal Diffusion in the Fluid and of Heat Conduction to the Substrate on the Response of Wall Hot-Film Gages

S. Tardu, C.T. Pham and G.Binder

Institut de Mécanique de Grenoble, B.P. 53 X 38041, Grenoble-Cedex, FRANCE

1. Introduction

A numerical simulation of the heat transfer from a flush mounted wall hot-film (WHF) used for measurement of the wall shear stress τ is carried out to investigate:

i) its response in reversing flow

ii) the effects of heat conduction to the substrate.

The full problem with the heat transfer equations in the fluid (F) and in the solid (S) coupled at the F-S interface (except on the WHF where the temperature qf is imposed; f:for film) is solved for an oscillating flow: $\tau = \bar{\tau}(1 + a\cos\omega t)$. The solution depends upon the following parameters: $\omega* = \omega l_f^2/\alpha_F, \sigma = Pr(l_f^+)^2, \alpha_F/\alpha_S$ and k_F/k_S (l_f = streamwise length of film, Pr: Prandtl number, α, k = heat diffusivity, conductivity).

The heat transfer law generally used for wall shear stress measurements (Nu = const.$\sigma^{1/3}$) is based on the simplest solution which assumes: $\partial\theta/\partial t = \partial^2\theta/\partial x^2 = 0$ in the fluid (quasi-steadiness and no longitudinal diffusion) and $\alpha_S = 0$. More complete solutions have been considered in the past, for instance with $\partial\theta/\partial t \neq 0$ ((1),(2)) or with $\alpha_S \neq 0$ (in the 1D case with a substrate thickness $= l_f$ (3)). A qualitative analysis (4) has recently implicitly taken into account the ratio k_F/k_S in an attempt to explain the differences measured on $\sqrt{\tau'^2}/\tau$ in air and in water. Yet this alone does not explain some observations of Dengel et al. (5) which show that the measured $\sqrt{\tau'^2}/\tau$ is closer to the true value on a well conducting substrate than on an insulator. The full simulation introduces the additional parameter of the diffusivity ratio α_F/α_S which points to the importance of *how fast* the heat is transfered to the fluid after conduction through the solid in addition to *how much* heat is transferred this way as compared to the direct transfer on the film.

2. Numerical method

The variables in the temperature equation are non-dimensionalized with the following quantities: length: l_f; time: l_f^2/α_F; temperature difference: $\Theta_f - \Theta_\infty$; velocity: u_τ.

The non-dimensional equations in the fluid, in the solid, and at the F-S interface are respectively:

$$F: \qquad \frac{\partial\theta}{\partial t} + u\frac{\partial\theta}{\partial x} = \frac{\partial^2\theta}{\partial x^2} + \frac{\partial^2\theta}{\partial y^2}$$

$$S: \qquad \frac{\alpha_F}{\alpha_S}\frac{\partial\theta}{\partial t} = \frac{\partial^2\theta}{\partial x^2} + \frac{\partial^2\theta}{\partial y^2}$$

Advances in Turbulence 3
Editors: A.V. Johansson · P.H. Alfredsson

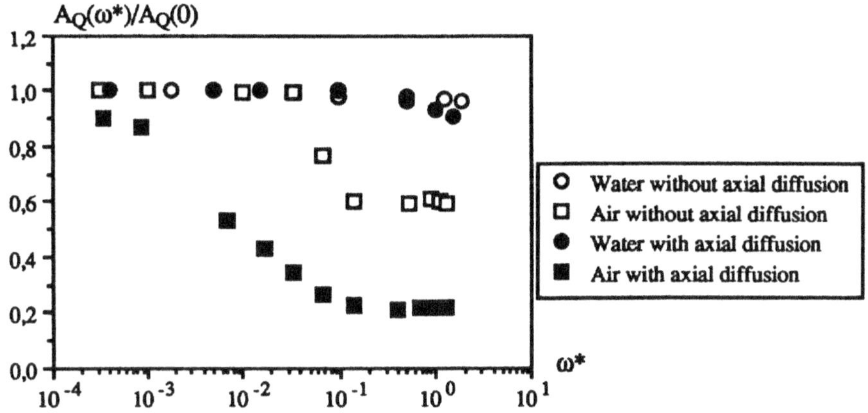

Figure 1 Amplitude ratio of the total flux from the film in three layer model vs non-dimensional frequency

$$F - S: \qquad \frac{k_F}{k_S}\frac{\partial \theta_F}{\partial y} = \frac{\partial \theta_S}{\partial y}$$

where $u = \sigma y(1 + a \sin \omega t)$. In the present computations, the amplitude is throughout: $a = 0.5$.

The boundary conditions are: $\partial\theta/\partial x = 0$ and $\partial\theta/\partial y = 0$ for $x, y \rightarrow \infty$. The size of the computational domain is: in the x direction: 10 to 50 l_f; in the y direction: 2 to 4 l_f in the fluid and 10 to 20 l_f in the solid substrate. Upstream weighted differencing (7) and ADI, combined with stretched coordinates (8)–near the trailing and leading edge of the WHF as well as near the wall–have been used.

The code was tested on the exact 1D time dependent three layer case of Bellhouse and Schultz (3) (F-S-F, with solid thickness $= l_f$ and with the same flow on both sides of the solid slab). Figure 1 shows the relative amplitude of the heat transfer vs. frequency for the case of air and water flow over glass.The reference amplitude is that of the quasi-steady case. The oscillations of the total heat transfer rate are damped with increasing frequency because the part transferred through the solid is filtered. This part is proportionally larger in air than in water flow and is increased by the effect of longitudinal diffusion in the fluid and in the solid (which is, of course, inherently neglected in the 1D solution of Bellhouse and Schultz). Figure 1 shows also that the attenuation of the frequency response of the full simulation which takes axial diffusion into account, starts at a frequency which is about ten times lower than in the 1D exact solution.

3. Results

The unsteady computations were started from the time mean steady field and were carried on until the difference in temperature at any point was less than 0.01% between two successive cycles. A typical example of isotherms is shown on figure 2a for the case of air flow over a glass substrate $k_F/k_S = 25, \sigma = 30$ and a frequency $\omega* = 1.3$. Typical dimensions of commercial films are $l_f = 0.2$ mm, so that $\sigma = 30$ in air corresponds to $u_\tau \approx 0.5$ m/s, i.e. $U_\infty \approx 13.8$ m/s. It is seen on figure 2b that, owing to the

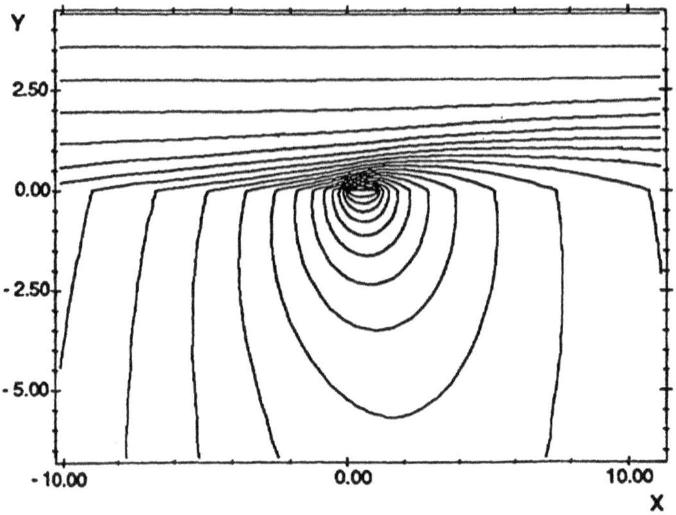

Figure 2a. Isotherms for air flow over a glass substrate. σ=30 ; ω^*=1.3

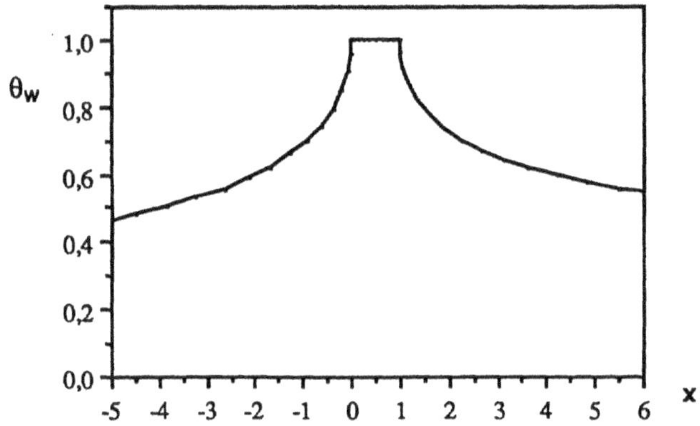

Figure 2b Wall temperature distribution for the combination air/glass. σ=30

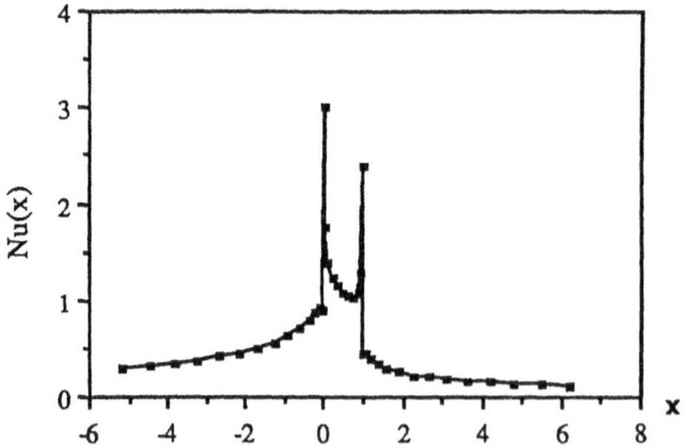

Figure 2c Local Nusselt number for the combination air/glass

high conductivity ratio, the temperature is still equal to half the film temperature at a distance $5l_f$ upstream or downstream. The local Nusselt number too (Fig. 2c) remains large up to distances of $\pm 5l_f$. Even beyond these distances from the film, the contribution to the total heat transfer is not negligible because it occurs over large streamwise extents. The computational domain must consequently be adapted to each flow and fluid-substrate combination. Its streamwise extent was progressively increased until the wall temperature on the upstream boundary was less than 0.1. Figure 2c illustrates also the unsymmetrical heat transfer upstream and downstream of the film: it is smaller downstream because of the hot wake. It also shows that the effects of longitudinal diffusion in the fluid are particularly important at the leading and trailing edges of the film: they reduce the local heat transfer rate at the former which otherwise would be infinite, but they increase it at the latter which explains the maximum reached at this point.

Heat transfer rates in steady flow

The global time-mean heat transfer rates per unit span $\bar{Q}$ for the couple air-glass ($k_{glass}/k_{air} = 25$) are recapitulated in Table 1. The notations are illustrated on the sketch below. In the steady regime, the heat transfered from the film to the sub-

Table 1 Global time mean heat transfer rates per unit span for the couple air/glass

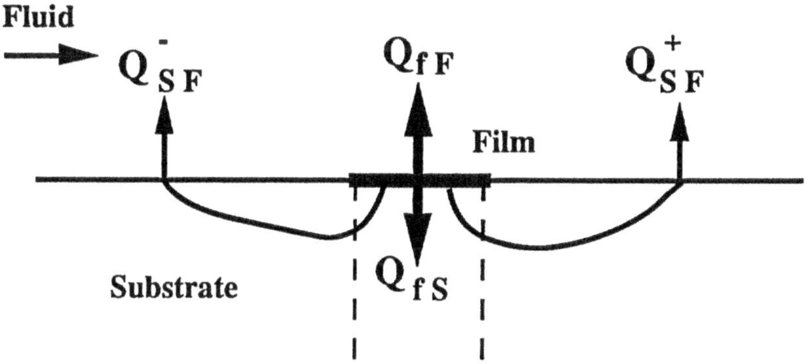

ω^*	$\overline{Q^-_{SF}}$	$\overline{Q_{fF}}$	$\overline{Q^+_{SF}}$	$\overline{Q_{fS}}$
1.7E-4	8.39	1.22	4.85	13.24
2.E-3	8.30	1.22	4.77	13.07
0.30	7.99	1.23	4.50	12.50
0.75	7.73	1.23	4.32	12.05
1.85	7.50	1.24	4.16	11.65
3.00	7.39	1.25	4.12	11.50

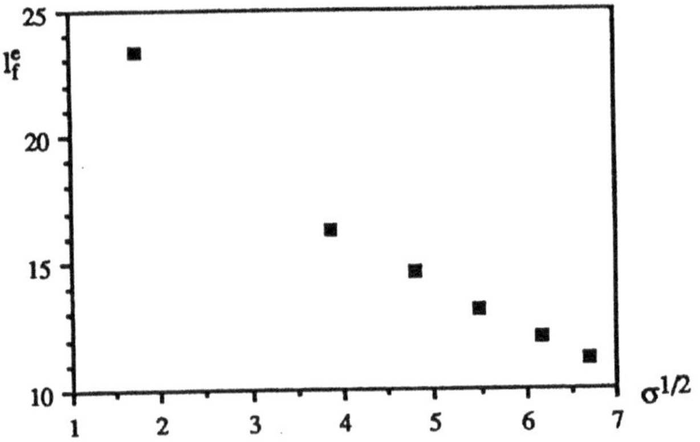

Figure 3 Equivalent film length versus the square root of Péclet number

strate $\bar{Q}_{fS}$ must be balanced by the heat loss from the substrate to the fluid. Thus: $\bar{Q}_{fS} = \bar{Q}_{SF}^{-} + \bar{Q}_{SF}^{+}$. This balance is a good check of computational accuracy.

The dominant fact from Table1 is that for the air/glass combination the heat transfered to the fluid via the substrate $\bar{Q}_{fS}$ is about TEN times larger than the direct transfer from the film to the fluid $\bar{Q}_{fF}$. It is also seen that the upstream transfer through the substrate $\bar{Q}_{SF}^{-}$ is twice as large as the one downstream $\bar{Q}_{SF}^{+}$ and that theses rates decrease slightly with frequency. Note also that the direct transfer from the film to the fluid $\bar{Q}_{fF}$ is only about half the value obtained from the Lévêque formula $\bar{Q}_{fF} = 0.83\sigma^{1/3}$. The reason for this is that the fluid has already been heated before it reaches the film and because of longitudinal diffusion.

The effect of the transfer via the substrate may be characterized by an "equivalent length" l_f^e defined by: $\bar{Q}_{fF} + \bar{Q}_{fS} = 0.83Pr^{1/3}(l_f^e)^{2/3}$. Hence, l_f^e is the length of a hypothetical film on a perfectly non-conducting substrate, obeying the Lévêque law and giving the same total heat transfer as the real film. Figure 3 summarizes the simulations carried out so far. It is seen that the effective length decreases with increasing σ, i.e. with increasing velocity.

A few computations have also been carried out for a three layer configuration where a thin mylar film (thickness $h = 0.5l_f$; conductivity $k_{mylar}/k_F = 5$) is sandwitched between the HWF and the substrate, in order to simulate the response of the DANTEC glue-on probes 55R47. It is seen (Table 2) that the thin mylar film with low conductivity roughly reduces the equivalent length by a factor two. A simulation of the combination

Table 2 Comparison of different substrate/flow conditions for σ=30 ($\bar{Q}=\bar{Q}_{fF}+\bar{Q}_{fS}$)

	Lévêque	Water-Glass	Helium-Glass	Air-Glass	He-Mylar-Glass	Air-Mylar-Glass
$\bar{Q}_{fF}$	2.51	2.31	1.85	1.34	2.35	1.79
$\bar{Q}$	2.51	3.78	6.24	14.00	4.06	8.01
l_f^e	1.00	1.85	3.29	13.18	2.06	5.70

air-mylar-aluminium will be attempted in the future, despite forseable difficulties since $k_{Al}/k_{air} = 6000$! Values for water-glass, helium - glass and helium-mylar-glass are also given in Table 2 and show the expected decrease of the equivalent length with decreasing k_S/k_F ($k_{He} = 5k_{air}$; $k_{water} = 25k_{air}$). The explanation of the decrease of the equivalent length with increasing velocity (or shear) and increasing values of the ratio k_F/k_S of the conduction coefficients lies in the increased capacity of the fluid to carry away the heat conducted through the solid, or in other words, in the capacity of the fluid to cool the solid substrate heated above ambiant temperature in the vicinity of the film.

Response in unsteady flow

The anemometer measures the total heat transfer from the film to the fluid and to the solid: $Q = Q_{fF} + Q_{fS}$. In the unsteady regime, there cannot be an instanteneous balance between Q_{fS} and $(Q^-_{SF} + Q^+_{SF})$ as in the steady regime because the heat flux variations imposed at the solid surface by the velocity changes are filtered to variable degrees by the substrate, depending upon the distance to the film and the frequency. Figure 4 shows the variations of $Q/\bar{Q}$ and $Q_{fS}/\bar{Q}_{fS}$ for a low and a high frequency, i.e. $\omega* = 1.7\ 10^{-4}$ and $\omega* = 3$, respectively. Since $\omega* = \sigma\omega^+$ (where ω^+ is the frequency scaled with inner variables), $\omega* = 3$ corresponds to $\omega^+ = 0.1$, which is the frequency where the energy spectrum of the velocity is maximum near the viscous layer (Compte-Bellot, 1965).

It is seen that for $\omega* = 3$ the transfer Q_{fS} is practically constant and that the variations of Q are considerably reduced. Even at the low frequency $\omega* = 1.7\ 10^{-4}$, is the quasi-steady regime not reached yet. Since the distance to which the heat diffuses during the period of oscillation is $\delta^2 \approx \alpha_S T$, i.e. $\delta^2 \approx (2\pi/\omega*)(\alpha_S/\alpha_F)l_f^2$, the transfer through the substrate can only respond without time lag to velocity variations if $\delta^2 << (l_f^e)^2$ i.e. if $\omega* << 2\pi(l_f^2/l_f^{e2})(\alpha_S/\alpha_F)$. Since for air-glass combination $\alpha_S/\alpha_F = 0.017$ and $l_f^e = 13l_f$ for $\sigma = 30$, this requires $\omega* << 6\ 10^{-4}$, a very small frequency indeed. Clearly the air-glass combination is unfavorable for two reasons: firstly, because of the

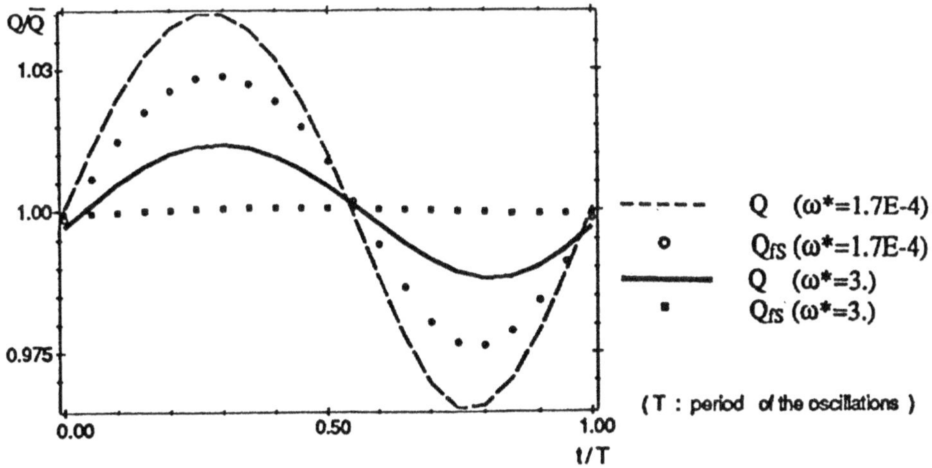

Figure 4 Heat transfer rates from the film versus time; σ=30

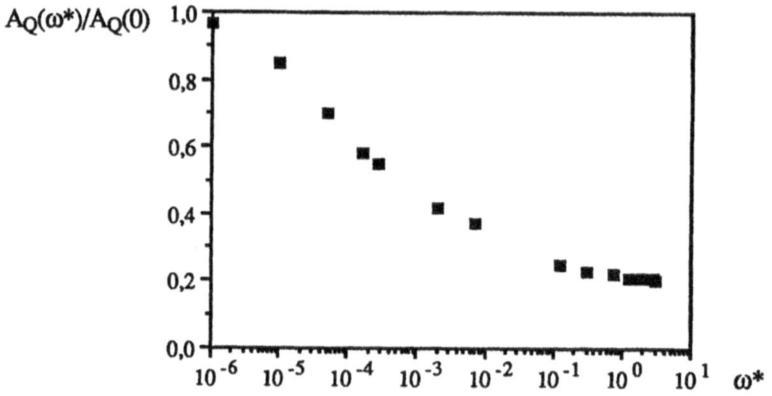

Figure 5 Frequency response of WHF for the combination air/glass. σ=30

larger transfer through the substrate due to the large value of the conductivity ratio k_S/k_F and secondly because $\alpha_S/\alpha_F << 1$ which results in considerable filtering of the variations of upstream and downstream transferts via the substrate.

Some values of the frequency response of the WHF for the air/glass combination are plotted in Fig.5. The limiting value of the attenuation is 1/5 when the mean flow parameter is $\sigma = 30$. This value is in reasonable agreement with the reduced turbulent intensity of the wall shear-stress measured by some authors in air flows as summarized in reference 6. It is seen that the amplitude is already reduced by a factor one half when $\omega* = 3\ 10^{-4}$.

Owing to the finite transit time over the film l_f/u_τ, the thermal boundary layer itself has a finite frequency response. But this frequency limit ($\omega* \approx 1$) is much higher than the limit imposed by the transfer via the substrate.

4. Conclusion

The numerical investigation of the heat transfer of wall mounted hot films shows that heat transfered to the fluid via the substrate has important effects. Thus, for high values of the conductivity ratios k_S/k_F, the transfer from the film to the substrate may be considerably larger than the direct transfer to the fluid. For the air/glass combination, the transfer to the fluid through the substrate is about 10 times larger than the direct transfer. Because of the hot wake, the downstream transfer from the substrate to the fluid is roughly two times smaller than the one upstream.

The importance of the conduction into the substrate increases with decreasing velocity. This effect is attenuated by a thin layer of material with low conductivity inserted between the film and substrate.

The frequency response of the total heat transfer from the film, as measured by an anemometer set, is dominated by the filtering effect of the transient heat transfer via the substrate when $k_F << k_S$ and $\alpha_F >> \alpha_S$. For the air/glass combination, the response is attenuated as soon as $\omega* > 10^{-6}$. This frequency limit is much lower than that of the response of the thermal layer in the fluid itself.

Acknowledgements

This work was supported by the HERMES R&D Program through the Centre National d'Etudes Spatiales and Dassault Aviation.

References

(1) PEDLEY, T.J. 1975 *J.Fluid Mech.* **67**, 209.

(2) FORTUNA, G. and HANRATTY,T.J. 1971 *Int. J. Heat and Mass Transf.* **14**,1499.

(3) BELLHOUSE, B.J. and SCHULTZ, D.L. 1967 *J.Fluid Mech.* **29**, 289.

(4) ALFREDSSON, P.H. and JOHANSSON, A.V. 1989 *Advances in Turbulence 2*, Springer Verlag.

(5) DENGEL, P. , FERNHOLZ, H.H. and HESS, M. 1987 *Advances in Turbulence 1*, p.470, Springer.

(6) ALFREDSON, P.H., JOHANSSON, A.V.,HARITONIDIS J.H. and ECKELMANN H. 1988 *Phys. Fluids* **31**, 1026.

(7) RAITHBY and TORRANCE 1971 *Computers and Fluids*, **2**, 191.

(8) KALNAY DE RIVAS 1972 *J.Computational Physics*, **10**, 202.

Measuring Invariant (Frame Independent) Quantities Composed of Velocity Derivatives in Turbulent Flows

A. Tsinober[1,2], *E. Kit*[1,2] *and T. Dracos*[2]

[1] Institut für Hydromechanik und Wasserwirtschaft, ETH - Honggerberg, CH-8093 Zürich

[2] On summer leave from the Department of Fluid Mechanics, Faculty of Engineering, Tel-Aviv University, Tel-Aviv 69978, Israel

This is a presentation of results of experiments on a turbulent grid flow and few results on measurements in the outer region of a boundary layer over a smooth plate. The air flow measurements included three velocity components and their nine gradients. This was achieved by a 12 hot-wire probe (3 arrays x 4 wires), produced for this purpose using specially made equipment (micromanipulators, etc.), calibration unit and calibration procedure. The probe has no common prongs and the calibration procedure was based on constructing a calibration function for each combination of three wires in each array (total 12) as a three-dimensional Chebishev polynomial of fourth order. A variety of checks were made in order to estimate the reliability of the results. Among the results the most prominent are the experimental confirmation of the strong tendency for alignment between vorticity and the intermediate eigenvector of the rate of strain tensor and the positiveness of the total enstrophy generating term even for rather short events. Emphasis is placed on the necessity to measure invariant quantities, i.e. independent of the choice of the system of reference, as the most appropriate to describe physical processes. From the methodical point of view the main result is that the multi-hotwire technique can be successfully used for measurements of all the nine velocity derivatives in turbulent flows, at least, at moderate Reynolds numbers.

1. Introduction

It is widely accepted in the turbulence research community that experimental information on the field of velocity derivatives is of utmost importance for the progress in understanding of the physical mechanisms in turbulent flows. Due to the three-dimensional nature of turbulent processes it is necessary to measure all the nine velocity derivatives. It is also a consequence of the need to obtain invariant quantities (i.e. independent of the system of reference) which are best suitable to describe physical processes. Information obtained from measuring quantities which are not invariant in the above sense may appear inadequate as it is seen from the three examples below. The first example is one of the results by computations by Rogers and Moin (1987), reported by Narasimha (1990). They clearly demonstrated that the probability density distribution (PDD) of the total dissipation is qualitatively different from that of the square of a single derivative $(\partial u_1/\partial x_1)^2$: while the PDD of the total dissipation is very close to log-normal, it is not the case for the single squared derivative which exhibits a tendency to have a

Advances in Turbulence 3
Editors: A.V. Johansson · P.H. Alfredsson

square root singularity at the origin. The second example is due to Corrsin and Fournier (1982). They showed that the term corresponding to viscous dissipation in the equation of turbulent kinetic energy balance for a single velocity component can have negative regions. The third example is related to enstrophy generation term $\omega_i \omega_j s_{ij}$ (ω_i - are the components of the vorticity vector, s_{ij} - is the rate of strain tensor) - one of the most important quantities in the dynamics of turbulent flows. It was observed in our recent experiments (Dracos et al. 1990) that this term is positive for each event of duration 0.1 sec. (i.e. 70 cm long) manifesting the prevalence of the vortex stretching process. At the same time the term $\omega_1^2(\partial u_1/\partial x_1)$ and similar quantities like $(\partial u_1/\partial x_1)^3$ commonly measured are negative in more than 10% of events.

In this paper we present a brief description of our most important results. More details will be published elsewhere (Tsinober et al. 1990)

2. Facility, Instrumentation, Data Acquisition and Processing

The experiments were performed in the open-circuit wind tunnel of the Institute for Hydromechanics, ETH, Zurich with a rectangular cross section $1.4 \times 1.2 m^2$. The grid was made of wooden rodes 1.5 cm in diameter with square mesh of size 6 cm (solidity of 0.44). Most of the measurements were made at mean velocity of $7m/s$. Some measurements were also made in the outer region of a turbulent boundary layer over a smooth plate at $y/\delta = 0.2$ and 0.8.

We used a multihotwire technique similar to that suggested by Balint et al. (1987) which is based on use of arrays consisting of several hot wires. The voltages obtained from each wire are produced by all the three instantaneous components of velocity. These latter are separated by means of some calibration procedure. This is an intrinsic feature and shortcoming of any method of this kind, since any imperfection in the measuring system (mechanical, electronic, etc.) and/or experimental facility leading to an error in the signals from the hot wires gives rise to errors in all the three velocity components. The important point is that these errors are (spuriously) correlated. Consequently, so will be many quantities obtained from these velocity components. Errors originating in this way may become dominating when measuring quantities like $< u_i u_j >$, $i \neq j$ (in grid turbulence they should vanish), $u_i \omega_i$, etc.

It is therefore of special importance to make any possible improvement of the measuring system, calibration procedure, etc. At the present stage we made the following improvements.

One of the possible error sources of the conventional 9 hot wire probes (Balint et al., 1987) is the presence of a common prong in every array of three hot wires. This leads to electronic cross-talking between these hot wires and consequently to unpredictable errors which may significantly contaminate the data. Thus, we have developed a multihotwire probe without common prongs.

Another problem we encountered is that the single wire (i.e., the one not having a partner) is incapable of catching sufficient information on the corresponding velocity component, say v, in a probe without common prongs. Addition of a fourth wire made the probe symmetrical in obtaining information about the components v and w. We therefore built a probe consisting of 12 hot wires (3 arrays $\times$ 4 wires). A scheme of the tip of one array and of the whole 12-wire probe are shown in Fig. 1a, while in Fig. 1b is shown a close up of an individual array which has also a cold wire (with aspect ratio more than 500) to account for ambient temperature variations and temperature

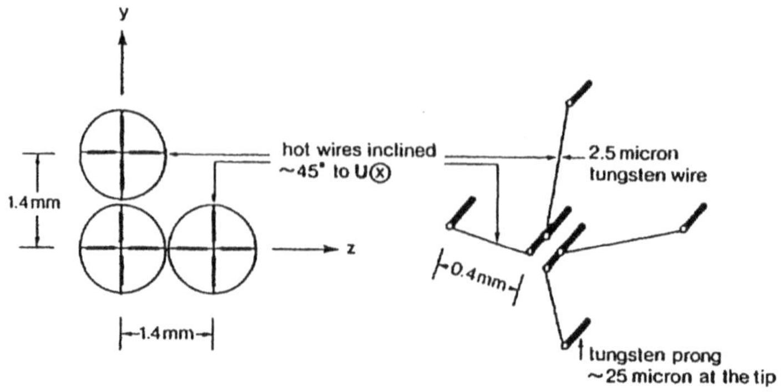

Figure 1. a) Schematic of a 4 wire array and a twelve hot wire probe.

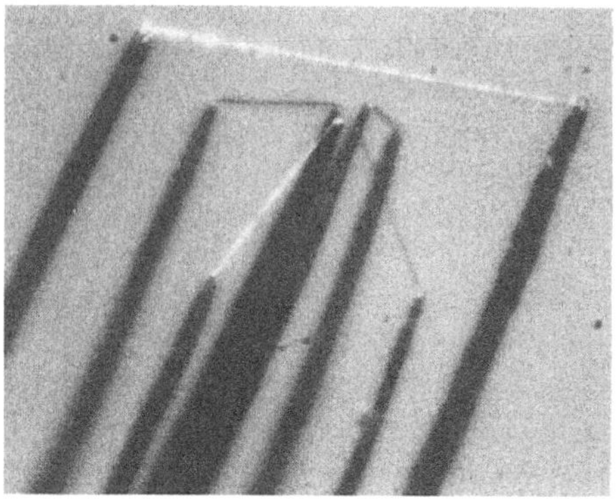

Figure 1. b) Close up of an array.

fluctuations up to 1000 Hz. Details on the probe production technology, etc. are given in Tsinober (1988).

The third problem is that the existing calibration procedure (Balint et al., 1987) had the following shortcomings: i - It employs restrictive assumptions about the symmetry of each array and the flow around the probe, ii - King's law is used for every wire which results in very high requirements on the geometrical precision of the probe and it's alignment in the flow. Imperfections in both lead to uncontrollable errors ("spurrious"), iii - On the basis of *i* and *ii* very limited information is being used to obtain the calibration.

For the above reasons, a new calibration procedure was developed, which is essentially an extention to 3 dimensions of the least-square polynomial approximation used previously in one and two dimensions by several investigators. The described calibration procedure does require neither very precise alignment of the probe with the mean flow, nor extreme precision of it's construction. To implement the above procedure we developed an automatic calibration unit, which reduced the calibration time to 30 minutes only.

The data acquisition was performed by a system based on a MICROVAX computer equipped by an A/D converter (ADF01, DEC) with throughout frequency rate of 350 Khz and a sample and hold feature. The data sampling rate per channel was 10 Khz. The information sampled at each distance from the grid consisted of 50 separate buffers with 1024 points in each buffer. In the following the information contained in a buffer of 1024 points is called an "event".

The above modifications led to a considerable reduction of errors and improvement of results. For example, the isotropy relations between the terms of the matrix $< (\partial u_i/\partial x_j)^2 >$ were satisfied with 15% of error (previously 30% see Kit et al., 1988). The incompressibility - Taylor hypothesis test (correlation coefficient between $\partial u_i/\partial x_1 = -U^{-1}\partial u/\partial t$ and $\partial u_2/\partial x_2 - \partial u_3/\partial x_3$) produced a value of about 0.7 without any smoothing or other similar processing of the raw data (previously 0.35).

Along with the above mentioned improvements some shortcomings still remained. The main of them is that one point correlations $< u_1 u_2 > /u_1' u_2'$ and $< u_1 u_3 > /u_1' u_3'$ where u_i' is RMS value of u_i, were about 0.1 - 0.15, i.e. were not close enough to zero as they should be in ideally isotropic turbulence. This is likely due to intrinsic imperfections of the method leading to errors in u_1, u_2, u_3 which are correlated, and imperfections of the flow.

Finally, the problem of scale resolution should be specially mentioned. This problem is well known both in laboratory and numerical experiments with the obvious difference that in the first case one has the true flow even when some range of small scales is not resolved. In our experiments the overall probe scale was typically 2-3 times smaller than the Taylor microscale and 4-5 times larger than the Kolmgoroff scale, i.e. the scales reasonably resolved where of the order of the Taylor microscale.

3. Results

3.1 Kinematic relations. Along with calculations of the matrix $< (\partial u_i/\partial x_j)^2 >$ and the incompressibility Taylor hypothesis test mentioned above a more delicate test of velocity derivatives measurements was carried out. It is based on the relation imposed by the condition of *homogeneity only* on the one point triple correlations of the velocity derivatives. This condition was obtained by Townsend (1951) and Botchov (1956) and is as follows $< s_{ij}s_{jk}s_{ki} >= -3/4 < \omega_i \omega_j s_{ij} >$. The values of the ratios $2 < \omega^2 > / < s_{ij}s_{ij} >$ and $-3/4 < \omega_i\omega_j s_{ij} > / < s_{ij}s_{jk}s_{ki} >$ obtained from experimental data at different values of x/M are given in Table 1.

Table 1.

Values of $I_i = 2 < \omega^2 > / < s_{ij}s_{ij} >$; $I_2 = -(3/4) < \omega_i\omega_j s_{ij} > / < s_{ij}s_{jk}s_{ki} >$

x/M	8	17	30	38	64	90	B. Layer y/δ	
							0.7	0.2
I_1	0.80	0.82	0.85	0.91	0.84	0.84	0.81	0.78
I_2	0.78	0.85	0.91	0.89	0.93	1.06	0.59	0.32

3.2 Dynamically relevant quantities. First we show some characteristics of dissipation fluctuations $\eta = e- < e >$ (Table 2): $\sqrt{< \eta^2 >}/ < \epsilon >$, skewness S_η and flatness F_η.

All these quantities are essentially constant in gird flow, i.e., independent of the distance from grid. The same is true for the ratio of the spectral densities of ϵ and η. It is an indication that these quantities have to do with universal properties of the grid

Table 2.

Characteristics of dissipation fluctuations								
x/M	8	17	30	38	64	90	B. Layer y/δ	
							0.7	0.2
$\frac{\sqrt{<\eta^2>}}{<\epsilon>}$	1.36	1.36	1.37	1.40	1.36	1.36	1.89	3.65
S_η	2.31	2.31	2.40	2.56	2.38	2.29	13.0	106.5
F_η	8.55	8.44	9.36	10.73	9.29	8.06	624	$1.97 \cdot 10^4$

Table 3.

Values of $s_\alpha =< (\partial_\alpha/\partial x_\alpha)^3 > / < (\partial u_\alpha/\partial x_\alpha)^2 >^{3/2}$ and $s = - < \omega_i \omega_j s_{ij} > / < \omega^2 >< s_{ij} s_{ij} >^{1/2}$								
x/M	8	17	30	38	64	90	B. Layer y/δ	
							0.7	0.2
S_1	0.41	0.46	0.50	0.50	0.50	0.50	0.56	0.35
S_2	0.32	0.41	0.44	0.55	0.40	0.37	0.32	0.045
S_3	0.31	0.34	0.38	0.33	0.20	0.14	0.68	-1.61
S	0.36	0.43	0.52	0.6	0.46	0.52	0.37	0.19

flow in small scales. It is noteworthy that the ratio $\sqrt{< \eta^2 >}/ < \epsilon >$ is a measure of dissipation intermittency (see, e.g. Chen et al., 1989).

The skewnesses for the three velocity derivatives are shown in Table 3 along with the quantity $< \omega_i \omega_j s_{ij} > / < \omega^2 >< s_{ij} s_{ij} >^{1/2}$. It is seen that S_1 agrees well with values known from previous experimental investigations (e.g. Arora and Azad 1980 and references therein). It is also of interest to compare the values of S_i with those obtained by Kerr (1985) in numerical simulations of turbulent flow in a box. In the range of Re_λ relevant for our experiments (60-90) Kerr obtained in both simulations that $S_1 = 0.5$. Values of S_2 and S_3 obtained in our experiments are considerably smaller and are in agreement with preliminary measurements in Kit et al. (1987), though for isotropic flow S_i should be equal. This question requires further investigation.

We would like to emphasize that (see table 3) *measurements of the enstrophy generating term show that its mean is essentially a positive quantity.* (This was first realised by G.I. Taylor 1938).

It was obtained as an average from 200 events each 1024 points long with 10^{-4} sec. interval between two adjacent point (i.e. 0.1 sec duraction or $\equiv$ 70 *cm* length). Moreover, for each such event the enstrophy generating term was positive, while the quantity $< \omega_1^2(\partial u_1/\partial x_1) >$ computed for each event was negative for more then 10% of the events. As mentioned in the introduction this points to the importance of measuring full quantities invariant of choice of the system of reference.

The probability density distribution of ϵ compared with that for $(\partial u_1/\partial x_1)^2$ confirm this point (Fig. 2) and is in agreement with the computations by Rogers and Moin (1987) (see Narasimha 1990). The same point is clearly seen also from comparing the PDD's of $\omega_i \omega_j s_{ij}$ and $(\partial u_1/\partial x_1)^3$ (Fig. 3).

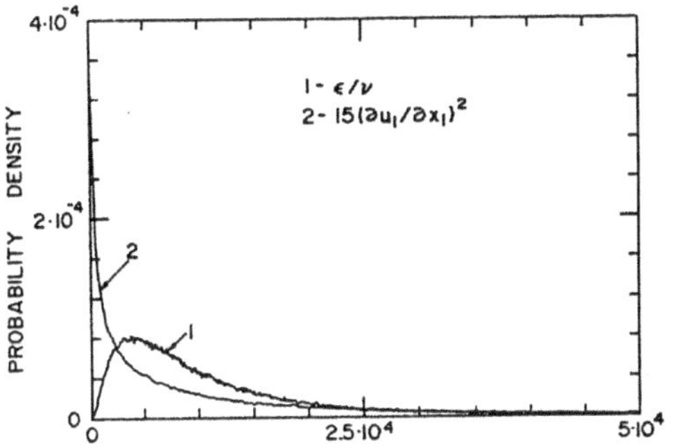

Figure 2. Probability density distributions of dissipation and $(\partial u_1/\partial x_1)^2$; $x/M = 30$.

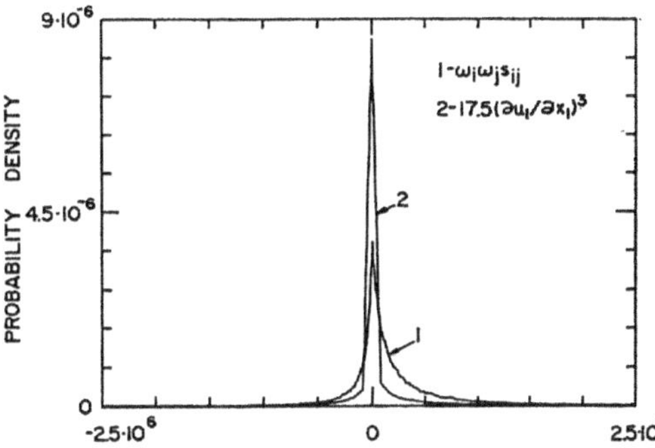

Figure 3. PDD of the enstrophy generation term $\omega_1\omega_j s_{ij}$, and $(\partial u_1/\partial x_1)^3$; $x/M = 30$.

Forth moments of velocity derivatives defined as

$$F_1 = \frac{15}{4}\frac{< s^4 >}{< s^2 >^2},\ F_2 = 3\frac{< \omega^2 s^2 >}{< \omega^2 >< s^2 >},\ F_3 = 3\frac{< \omega_i s_{ij} s_{jk} \omega_k >}{< \omega^2 >< s^2 >}\ and\ F_4 = \frac{9}{5}\frac{< \omega^4 >}{< \omega^2 >^2}, \tag{1}$$

where $s^2 = s_{ij}s_{ij}$, are shown in Table 4 along with their uncorrelated (i.e. Gaussian) values.

Table 4.

Fourth - order moments of velocity derivatives

x/M	8	17	30	38	64	90	B. Layer y/δ 0.7	0.2	Gaussian
F_1	3.97	3.99	4.07	4.27	3.95	3.97	9.09	33.8	3
F_2	4.29	4.42	4.48	4.72	4.62	4.46	11.5	46.4	3
F_3	1.04	0.93	0.88	0.88	0.76	0.82	2.09	3.77	1
F_4	4.77	4.90	5.10	5.30	5.21	4.95	12.3	34.9	3

The values of F_i are in agreement within 20% with the computational results by Siggia (1981) and Kerr (1985). It is noteworthy that as in computations by Kerr (1985) in our experiments $F_4 > F_2 > F_1$, i.e. the qualitative tendencies are the same. Also as in their computations in our experiments F_3 is smaller than 1 implying some alignment in the small-scale turbulent structures. Since F_3 is the normalized variance of the vortex stretching vector $W_i = \omega_i s_{ij}$, the relation $F_3 < 1$ means that there is some reduction of the nonlinear term in the vorticity equation in the sense that $< W^2 >$ is smaller than it's corresponding Gaussian value. Since this reduction is quite moderate it implies that the reduction of the nonlinearity in the Navier-Stokes equation (Kraichnan and Panda 1989, Shtilman and Polifke 1989, Tsinober 1990) is mostly related to the potential part of the nonlinear term and that the solenoidal part of the Lamb vector is reduced quite moderately.

3.3 Some local properties Ashurst et al. (1987) discovered in their computations that the vorticity tends to be aligned with the intermediate eigenvector of the rate of strain tensor and that the strain in this direction is mostly positive (80%). Our experimental data for grid flow clearly show the same tendencies (see Fig. 4), though the alignment is somewhat weaker than in the numerical simulations. Still the effect is very large, e.g. the maximum to minimum ratio in the PDD of the cosine of the angle between ω and α_2 is more than 4. Similar results have been obtained in the outer part of a boundary layer flow. The alignment between α_2 and ω is considerably stronger for $\alpha_2 > 0$ than for $\alpha_2 < 0$. This along with our observations that in 65% of the sample points $\alpha_2 > 0$ is in agreement with Betchov's (1956) conclusions and numerical computations by Ashurst et al (1987). This last result is better seen on Fig. 5 in which the PDD of the normalized intermediate rate of strain $\alpha_2/s\sqrt{6}$ is shown, where $s = \sqrt{\alpha_1^2 + \alpha_2^2 + \alpha_3^2}$. It's most probable value is $\equiv 0.4$. The important point is that the most active part of turbulence is located in neighborhood of this value, which is seen, for example, from the scatter plot of $\alpha_2/s\sqrt{6}$ versus the energy dissipation shown in Fig. 6. Qualitatively similar behavior, i.e. maximum activity around $\alpha_2/s\sqrt{6} = 0.4$, is exhibited in scatter plots of $\alpha_2/s\sqrt{6}$ versus enstrophy generation term. Note that the scatter plot shown in Fig. 6

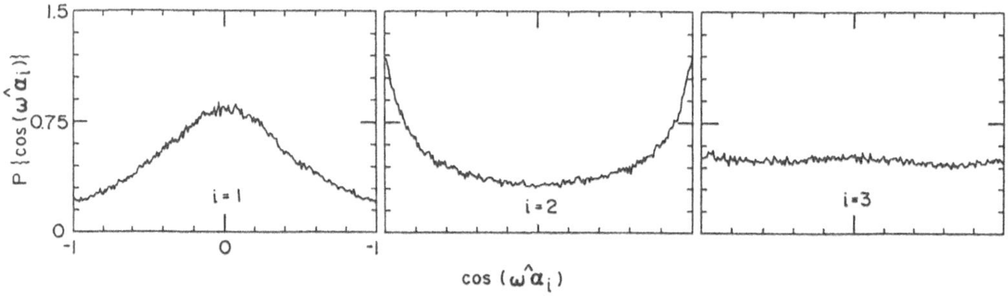

Figure 4. PDD of cosine of the angle between vorticity vector and eigenvectors of the rate of strain tensor. $\alpha_1 < \alpha_2 < \alpha_3$; $x/M = 30$.

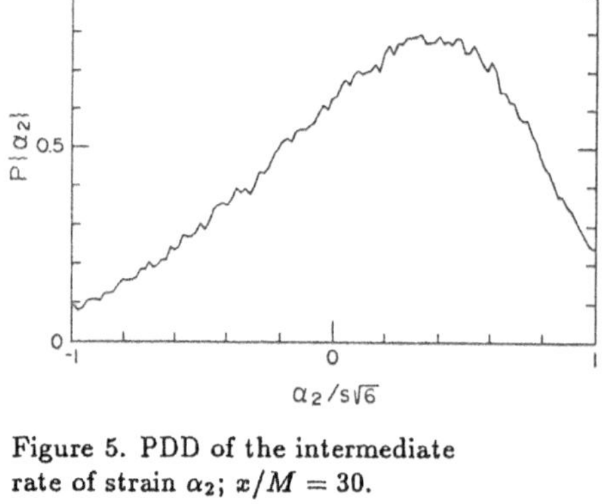

Figure 5. PDD of the intermediate rate of strain α_2; $x/M = 30$.

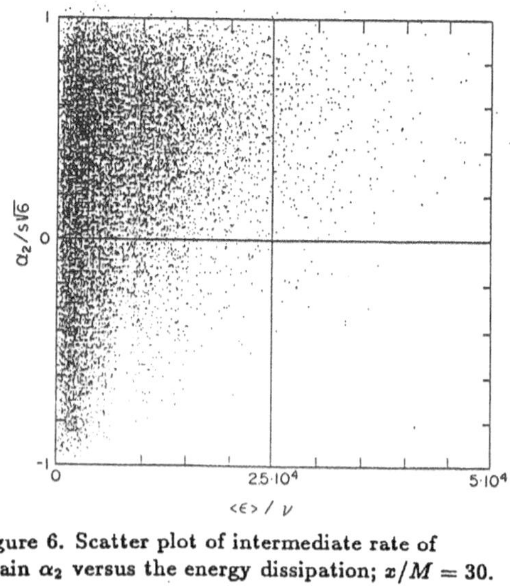

Figure 6. Scatter plot of intermediate rate of strain α_2 versus the energy dissipation; $x/M = 30$.

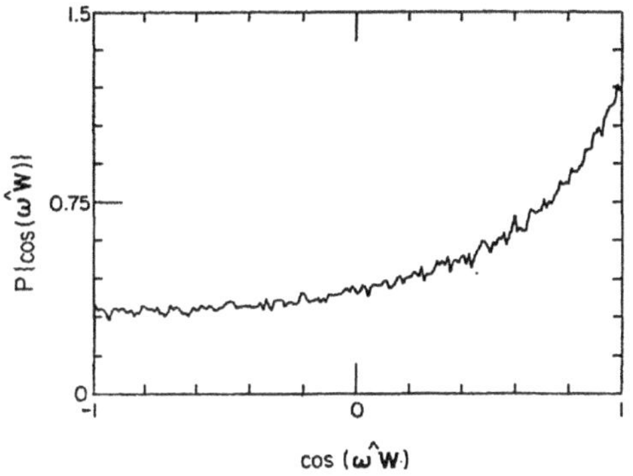

Figure 7. PDD of the cosine of the angle between ω and W; $x/M = 30$.

is very similar to the one obtained by Ashurst et al. (1987) (see their Fig. 1). Also in agreement with their results is the most probable relation between the principal rates of strain, e.g. at $x/M = 30$ this ratio is $3.8 : 1 : 3.1$ while Ashurst et al. obtained $4 : 1 : 3$. Quite reasonable agreement between the experimental results presented here and the numerical ones obtained by Ashurst et al. (1987) is a strong indication of reliability of both, since for such subtle effects it is quite unlikely to be just a coincidence.

We have seen above that the mean of the enstrophy generating term is essentially positive (Table 3) and moreover it was positive for each event 70 cm long. This is the consequence of the prevalence of vortex stretching process over the vortex compressing. The above irreversible tendency can be characterized locally by considering the PDD of the cosine of the angle between the vorticity vector ω and the vortex stretching vector $\mathbf{W}$ introduced earlier (Fig. 7). As expected this probability density distribution is strongly asymmetric and is much larger in the region $\theta = 0$ than in the region $\theta = 180^o$. In other words there is a strong tendency for alignment between ω and W in the strict sense, i,e., tendency for them to be parallel but not antiparallel. This result along with the positiveness of the enstrophy generation term are manifestations of the strong non-gaussianity of the field of velocity derivatives.

Finally, we mention some peculiarities of various quantities during one event (i.e. 70cm long). Perhaps the most spectacular is the behaviour of the enstrophy generating term which exhibit bursts of extremely large amplitude (from 1 to 5 such bursts in one event). This amplitude is up to 50 (!) times larger than the average and more than 60% of contribution to the average of the enstrophy generating term comes from the fluctuations with amplitude larger than three times the average: quite an intermittency. The scale of these extremely violent events (or if one likes very active structures) is $\sim 0.5 \div 1$ cm and in between there exist extremely inactive regions (structures). The scale of the latter is an order of magnitude larger than than of the active events. At this stage we venture to suggest that the enstrophy generating term, being an invariant (i.e. independent of the system of coordinates and in Gallilean sense), seems to be one of the most appropriate quantities to characterize the structural nature of the grid turbulence as well as shear turbulence.

4. Concluding Remarks and Future Work

Our first conclusion is of technical nature. The results obtained and their comparison with those from direct numerical simulations clearly indicate that multihotwire techniques can be successfully used for measurements of all the nine velocity derivatives in turbulent flows at least at moderate Reynolds numbers.

Our second major conclusion is related to the question: does turbulence have a generic, universal structure? We confirmed experimentally the strong tendency for alignment between vorticity and the eigenvector of the intermediate rate of strain tensor (discovered numerically by Ashurst et al. (1987) for quasi - isotropic and shear turbulence), both for turbulent grid flow and the outer part of boundary layer. Our opinion is that this property is one of the important universal features of small - scale turbulent flows. Another universal property of three - dimensional turbulent flows is believed to be the prevalence of the vortex - stretching process over the vortex compressing and closely related to it essential positiveness of the enstrophy generating term. We have demonstrated this to be true via direct measurements of the enstrophy generating term and strong tendency for alignment in the strict sense between vorticity and vortex stretching vectors. The above results became possible due to obtaining quantities like the eigenvectors of the rate of strain tensor quantities, which are invariant of the system of reference. We have demonstrated the necessity of obtaining other invariant quantities like total dissipation and enstrophy generating term. Our suggestion is that quantities of this kind should be among the most appropriate to characterize the structural nature of turbulent flows, e.g. organized motion, etc.

Our main future efforts will be concentrated on improving the technique and it's application to turbulent shear flows.

Acknowledgements

Our special acknowledgement is to Professor J.M. Wallace and Dr. J.L. Balint for their kind hospitality during the summer of 1986 and for introducing two of us (AT and EK) to all the secrets of their multihotwire technology.

This research was supported, in part, by the Eidgenossische Technische Hochschule, Zurich and by grant No. 85-00347 from the U.S. - Israli Binational Scientific Foundation.

References

Ashurst, Wm. T., Kerstein, A.R., Kerr, R.A. & Gibson, C.H. 1987. *Phys. Fluids*, **30**, 2343-2353.

Arora, S.C. & Azad, R.S. 1980. *J. Fluid Mech.*, **97**, 385-404.

Balint, J.L., Vukoslavcevic, P. & Wallace, J.M. 1987. In *Advances in Turbulence* (eds. G. Compte-Bellot and J. Mathieu) (Springer, Berlin), pp. 456-464.

Betchov, R. 1956. *J. Fluid Mech.* **1**, 497-504.

Chen, H., Herring, J.R., Kerr, R.M. & Kraichnan, R.H. 1989. *Phys. Fluids* **A1**, 1844-1854.

Corrsin, S. & Fournier, J.L. 1982. *Phys. Fluids*, **25**, 583-585.

Dracos, T., Kholmyansky, M., Kit, E. & Tsinober A. 1990. In *Proceedings of the IUTAM Symposium on Topological Fluid Mechanics*, Cambrdige, U.K., August 13-18, 1989. (eds. H.K. Moffat and A. Tsinober, Cambr. Univ. Press), pp. 564-584.

Frenkiel, F.N., Klebanoff, P.S. & Huang T.T. 1979. *Phys. Fluids*, **22**, 1606-1617.

Kerr, R.M. 1985. *J. Fluid Mech.*, **153**, 31-58.

Kit, E., Tsinober, A., Teitel, M., Balint, J.L., Wallace, J.M. & Levich, E. 1988. *Fluid Dyn. Res.*, **3** pp. 289-294.

Kraichnan, R. & Panda, R. 1988. *Phys. Fluids* **31**, 2395-2397.

Narasimha, R. 1990. In *Turbulence at the Cross Roads*, (ed. J.L. Lumley), Springer, pp. 13-48.

Siggia, E. 1981. *J. Fluid Mech.* **107**, 375-406.

Shtilman, L. & Polifke, W. 1989. *Phys. Fluids*, **A1**, 778-780.

Taylor, G.I. 1938. *Proc. Roy. Soc.* **A164**, 15-23.

Tennekes, H. & J.L. Lumley, J.L. 1974 A first course in turbulence. *MIT Press*, pp. 87-90.

Townsend, A.H. 1951. *Proc. Roy. Soc. London*, **A208**, 534-542.

Tsinober, A. 1988 Multi-hot-wire probe production for measurement of all nine velocity gradient. *Int. Rep. Fac. Engn., Tel-Aviv Univ.*

Tsinober, A. 1990. *Phys. Fluids*, **2A**, 484-486.

Tsinober, A., Kit, E. & Dracos, T. 1990 Experimental study of velocity derivatives in turbulent grid and boundary layer flows, submitted to the *J. Fluid Mech.*

The Use of Near-Wall Hot-Wire Probes for Time Resolved Skin-Friction Measurements

Peter M. Wagner

Hermann-Föttinger-Institut für Thermo- und Fluiddynamik, Technische Universität Berlin, D-1000 Berlin 12, Germany

Measurements of the mean and the fluctuating wall shear stress in turbulent air flows by means of hot wires mounted at a fixed small distance from the wall were carried out and a calibration procedure was developed. Experimental results in a flat-plate boundary layer are presented and compared to pulsed-wire measurements.

1. Introduction

Measurements of the fluctuating wall shear stress with either hot-wire or hot-film techniques suffer from severe problems caused by the influence of the heat-conducting wall. A good review on the different experimental techniques is given by Alfredsson, Johansson, Haritonidis & Eckelmann (1988).

The values obtained with flush-mounted hot films in wind-tunnel experiments are significantly lower than those obtained with other techniques. Due to the heat transfer from the sensor to the substrate and from the substrate to the flow the dynamic response (to rapid fluctuations) differs from the static response of the probe. As reported by Dengel, Fernholz & Hess (1987) the measured rms level depends strongly on the wall material. The dynamic response can be improved by using wall materials with high thermal conductivity but even with aluminum, the fluctuating wall shear stress will be underestimated by about 25 %.

The values obtained with hot-wires calibrated in the free stream and then used inside the viscous sublayer to measure the local velocity are also affected by the wall proximity. But in this case, the distortion of the temperature field seems to affect only the sensitivity but not the dynamic response of the probe. This problem has been investigated by many authors with different results. Janke (1987) compared several experimental and numerical studies on this subject and presented a relationship between the true velocity and the velocity measured with a constant temperature hot-wire anemometer. This relationship corrects the influence of a wall with high thermal conductivity for both laminar and turbulent flow (see Fig. 2). Since the velocity profile in the viscous sublayer is linear, this formula can be used to determine the skin friction from the velocity, if the wall distance can be measured very accurately.

A more direct approach is to use a hot wire mounted at or at a fixed small distance from the wall to calibrate it directly against the wall shear stress. Such a probe was used by Ajagu, Libby & LaRue (1982), who combined a hot wire near the wall with a flush-mounted hot film, which was used to stabilize the wall-temperature. At small mean skin-friction values their results still deviate from the calibration curve. The hot-wire-on-the-wall technique described by Alfredsson et al. (1988) inspired the present

Advances in Turbulence 3
Editors: A.V. Johansson · P.H. Alfredsson

investigation. Their probe consists of a hot wire mounted only a few wire diameters above the wall. It is calibrated in a turbulent boundary layer against a Preston tube. Due to the low velocities very close to the wall, the sensitivity of such a probe is very low at low Reynolds numbers. Calculations using the heat-loss correction formula given by Janke (1987) show, that the sensitivity can be increased by mounting the wire further away from the wall but still within the viscous sublayer (see Fig. 3).

2. Near-wall hot-wire response

The linear law for the velocity in the near-wall region of both laminar and turbulent boundary layers leads to a linear relationship between the shear stress τ_w at the wall and the velocity u at a distance y from the wall

$$u = \frac{y}{\rho\nu}\tau_w \tag{1}$$

which is valid for nondimensionalized wall distances $y^+ = \frac{yu}{\nu} < 10$ where $u_\tau = \sqrt{\frac{\tau_w}{\rho}}$. Using equ. (1) the power law for the constant temperature anemometer output voltage e can be formulated by means of the wall shear stress as

$$e^2 = A + B\tau_w^\alpha \tag{2}$$

Due to the heat transfer to the wall, the velocity measured with a hot wire calibrated in the free stream does not correspond to the true velocity in the near-wall region. Janke (1987) presents a relationship between the true velocity u and the measured velocity u_m

$$\frac{u_m y}{\nu} = \frac{uy}{\nu} - 0.55\sqrt{\frac{uy}{\nu}} + 3.2 \tag{3}$$

This formula is valid in both laminar and turbulent flow for wall materials with high thermal conductivity (i.e. aluminum or ceramics). Equ. (3) corrects the anemometer reading in the region $y^+ < 5.8$. Above this value, the velocity is slightly underestimated. Using equ. (1) we can determine the relationship between the measured wall shear stress τ_m and the true value τ_w :

$$\tau_m = \tau_w - 0.55\sqrt{\frac{\rho^2\nu}{h^2}\tau_w} + 3.2\frac{\rho^2\nu}{h^2} \tag{4}$$

The value of τ_m can be used in equ. (2) to calibrate the hot wire in the full range of the skin-friction measurements. This leads to a significant deviation from the uncorrected values depending on the wall distance of the sensor (see Fig. 3).

3. Experimental setup and results

The design of the near-wall hot-wire probe is shown in Fig. 1. The sensor is mounted on top of two 0.3 mm prongs in a 3 mm shaft. Probes with prongs 40 to 120 mm long were manufactured. The accurate wall distance which had to be known for the calibration was determined under a precision microscope with an accuracy of approximately 2 μm. The probe dimensions are chosen according to the recommendations of Ligrani & Bradshaw (1987) and Alfredsson et al. (1987). In order to avoid spatial averaging, the active

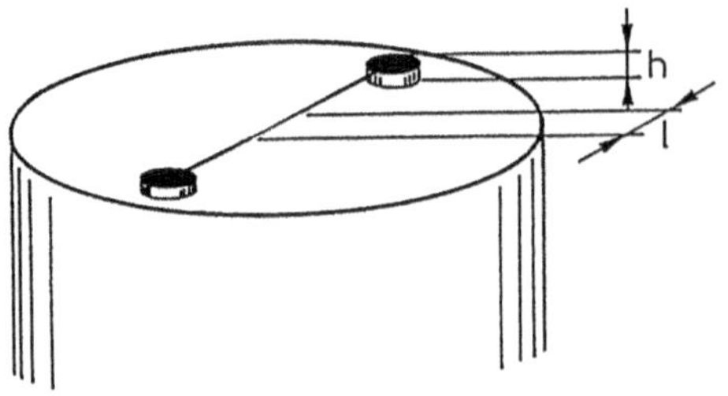

Fig. 1: Design of near-wall hot-wire probes. Sensors with wall distances in the range from 40 to 120 μm have been manufatured and tested.

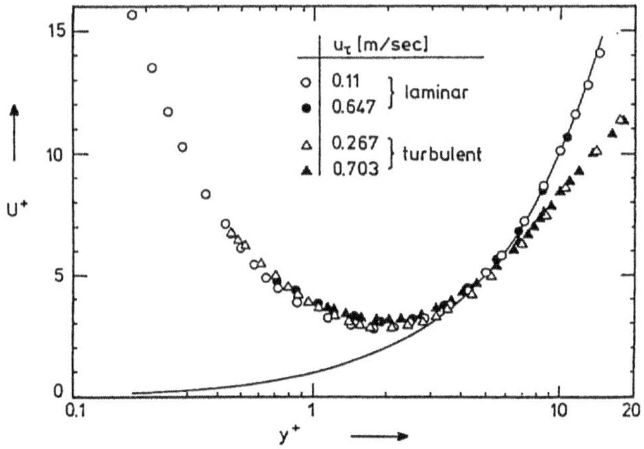

Fig. 2: Hot-wire measurements in the near-wall region of both laminar and turbulent boundary layers (from Janke 1987).

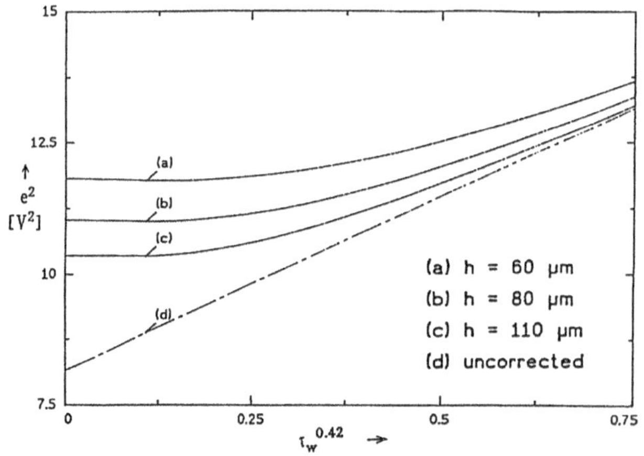

Fig. 3: Calculated deviation of the near-wall hot-wire response to wall shear stress measurements from the uncorrected power law.

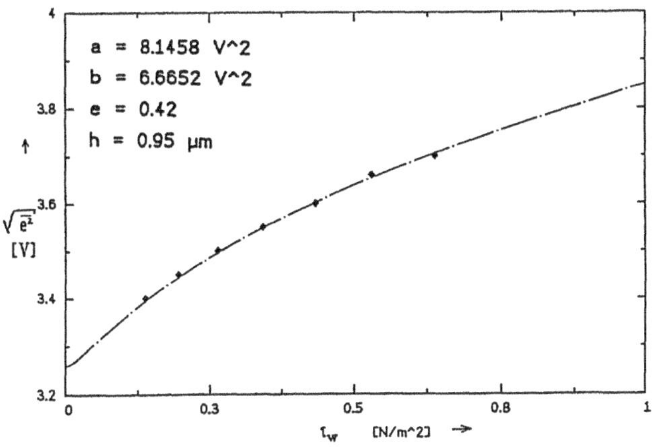

Fig. 4: Typical near-wall hot-wire calibration curve. The calibration points correspond to wall-distances from 2.1 to 4.5 wall units.

length of the wire was reduced by gold-plating to 0.5 mm corresponding to less than 20 viscous length units. In order to obtain a length-to-diameter ratio of 200, 2.5 μm platinum-coated tungsten wires were used.

The probes were connected to a PSI 6110 constant temperature anemometer and operated at an overheat ratio of 35 %. The bridge was read by a 16-Bit microcomputer using a 12-Bit A/D converter. The mean skin friction was measured by a Preston tube. The pressure was determined using a digital micromanometer that could be read directly by the microcomputer. Both the calibration and the measurements were carried out using a computer program. The input values were averaged over at least 60 seconds with sampling rates of 25 Hz for the pressure and 1 kHz for the anemometer signal.

During the calibration both the mean and the fluctuating components of the bridge voltage were evaluated in order to correct for the distortion of the probability density distribution through the nonlinearity of the calibration curve. Furthermore all instantaneous values of the anemometer output were stored. The constants A and B in the power law were calculated by means of a least-square-fit algorithm. The stored time sequences were used to check the calibration curve and the results of this check were used to improve the calibration by means of an error correction procedure. This process could be repeated several times but usually the optimum was reached after one or two iterations with final errors of less than 0.5 % for all reference positions. Fig. 4 shows a typical near-wall hot-wire calibration curve obtained for a probe with a wall distance of 95 μm.

The mean and the rms values and the skewness and flatness of the fluctuating wall shear stress were compared to wall-pulsed-wire measurements, since the wall-pulsed-wire technique developed by Castro, Dianat & Bradbury (1987) does not suffer from most of the limitations of hot-wire anemometry in the viscous sublayer (see Alfredsson et al. (1987) and Dengel et al. (1987)). The main restriction of this technique is that the acquisition rate cannot be increased above about 50 samples per second, which puts some limitation on the evaluation of time-properties of the measured signal. The rms values of the skin fricion obtained with both instruments agree within $\pm 2\%$. In our test flow, which had a slight pressure gradient, it was found to be 0.36 to 0.37 of the

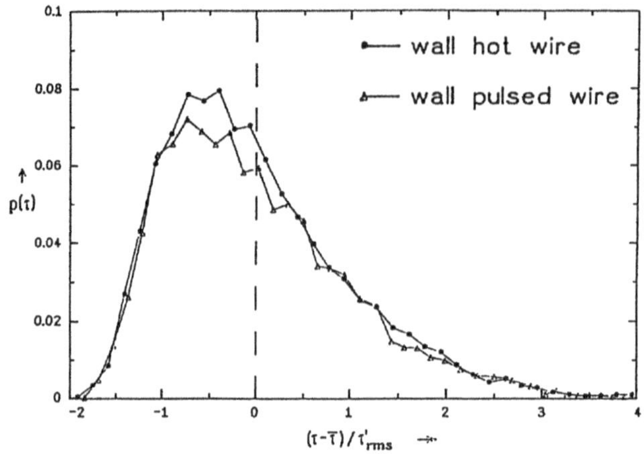

Fig. 5: Comparison of wall shear stress probability density distributions obtained with near-wall hot-wire and wall-pulsed-wire measurements.

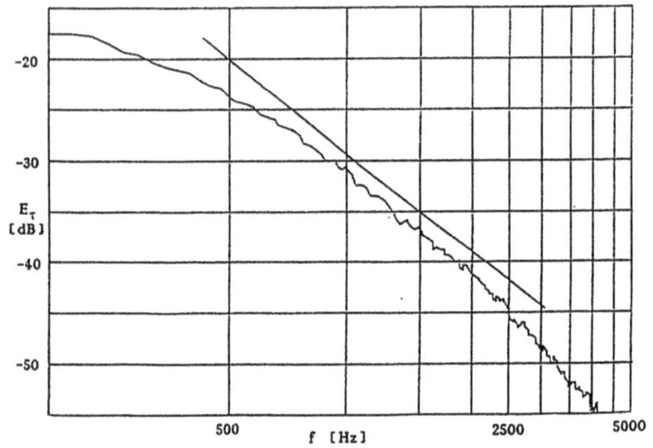

Fig. 6: Power spectral density of the wall shear stress obtained with near-wall hot-wire probes. The straight line indicates a slope of $-\frac{5}{3}$.

mean value. The values of the skewness and the flatness vary between 0.98 and 1.18 and 4.6 and 5.1 respectively. The probability density distributions obtained with both techniques agree fairly well (see Fig. 5). Fig. 6 shows a typical frequency spectrum of the wall shear stress at $u_\tau = 0.56$ m/s obtained with the near-wall hot-wire technique.

4. Conclusions

It was shown, that near-wall hot-wire probes can be used to obtain time-resolved skin-friction measurements in turbulent air flow. The flow disturbance due to the probe holder and the prongs, and the difficulties in the determination of the accurate wall distance of the sensor wire were avoided by mounting the probes at a small fixed wall distance. Comparisions with wall-pulsed-wire measurements show good agreement between both techniques. Due to the low overheat ratio the use of the probe is restricted to experimental setups, where it is possible to stabilize the room temperature within at least ±0.2 K.

The present study was undertaken with financial support from the Deutsche Forschungsgemeinschaft, which we gratefully acknowledge. The author would like to thank A. Dahm, who manufactured the probes, and M. Schober, who developed the microcomputer program.

References

Ajagu, C.O., Libby, P.A. & LaRue, J.C. 1982 Modified gauge for time-resolved skin-friction measurements. *Rev. Sci. Instrum.* **53** (12)

Alfredsson, P.H., Johansson, A.V., Haritonidis, J.H. & Eckelmann, H. 1987 The fluctuating wall-shear stress and the velocity field in the viscous sublayer. *Phys. Fluids* **31** (5)

Castro, I.P., Dianat, M. & Bradbury, L.J.S. 1987 The measurement of fluctuating skin friction with a pulsed wire wall probe. In *Turbulent shear flows 5* (eds. F. Durst, B.E. Launder, F.W. Schmidt and J.H. Whitelaw) Springer Berlin, Heidelberg, New York

Dengel, P., Fernholz, H.H. & Hess, M. 1987 Skin-friction measurements in two- and three-dimensional highly turbulent flows with separation. In *Advances in Turbulence 1* (eds. G. Compte-Bellot and J. Mathieu), Springer Verlag

Janke, G. 1987 Hot wire in wall proximity. In *Advances in Turbulence 1* (eds. G. Compte-Bellot and J. Mathieu), Springer-Verlag

Ligrani, P.M. & Bradshaw, P. 1987 Spatial resolution and measurements of turbulence in the viscous sublayer using subminiature hot-wire probes. *Exp. in Fluids* **5**

List of Participants

A Abdelquari
Ecole Nat Superieure de Mécanique
Lab de Mécanique des Transferts
Turbulents et Diphasiques
1, rue de la Noe
F-44072 Nantes Cedex

A Aharoni
A A Lab Systems Ltd
8 Hakalir Street
Tel- Aviv
Israel

H Alfredsson
Royal Inst of Technology
Dept of Gasdynamics
S-100 44 Stockholm

W Althaus
Univ Dortmund
Abt CT-TV
Postfach 500500
D-4600 Dortmund 50

A Alving
Technische Univ Berlin
HFI
Müller-Breslau Strasse 8
D-1000 Berlin 12

H I Andersson
Div of Applied Mechanics
N-7034 Trondheim-NTH

P Andersson
Royal Inst of Technology
Dept of Hydromechanics
S-100 44 Stockholm

E Aurell
Observatoire de Nice
BP 139
F-06003 Nice Cedex

X-S Bai
Royal Inst of Technology
Dept of Gasdynamics
S-100 44 Stockholm

J-L Balint
The Univ of Maryland
Dept of Mechanical Eng
College Park, MD 20742
USA

D W Bechert
DLR Berlin
Abt Turbulenzforschung
Müller-Breslau Strasse 8
D-1000 Berlin-West 12

S F Belcher
Univ of Cambridge
DAMPT
UK-Cambridge CB3 9EW

A Bertelrud
High Technology Corp
28 Research Drive
Hampton, VA 23666
USA

J-P Bertoglio
Lab de Mécanique des Fluides
Ecole Centrale de Lyon
36, av G de Collongue
F-69130 Ecully

G Binder
Inst de Mecanique Grenoble
BP 53 X
F-38041 Grenoble Cedex

J-P Bonnet
CEAT LEA
43, Route de L'Aerodrome
F-86036 Poitiers Cedex

M S Borgas
CSIRO Div of Atmospheric Research
Private Bag No 1, P O Mordialloc
Australia

A J Brand
TU Delft
Lab for Aero- and Hydrodynamics
Rotterdamseweg 145
NL-2628 AL Delft

J G Brasseur
Pennsylvania State Univ
208 Mechanical Eng Building
University Park, PA 16802
USA

K Breuer
Brown Univ
Center for Fluid Mechanics
Box 1966
Providence RI 02912
USA

J Bruns
RWTH
D-5100 Aachen

M K Bull
Univ of Adelaide
Dept of Mechanical Eng
Adelaide, S A 5000
Australia

A D Burden
Chalmers Univ of Technology
Dept of Theor Physics and Mech
S-412 96 Gothenburg

J L Cano
Dep de Fisica de la Tierra
Astronomia y Astrofisica
Ciudad Universitaria
E-28040 Madrid

K-S Choi
British Maritime Technology
1 Waldegrave Road
UK-Teddington, Middlesex TW11 8LZ

J-P Chollet
Inst de Mécanique de Grenoble
BP 53 X
F-38041 Grenoble Cedex

J Cohen
Technion - Israel Inst of Technology
Haifa 32000
Israel

V Chetkovic
Royal Inst of Technology
Dept of Water Resources Eng
S-100 44 Stockholm

A Dahlkind
Royal Inst of Technology
Dept of Gasdynamics
S-100 44 Stockholm

J Delville
CEAT LEA
43, Route de L'Aerodrome
F-86035 Poitiers Cedex

R Deron
ONERA
BP 72
F-92322 Chatillon Cedex

C Dopazo
Mecanica de Fluidos, E T S I I
M de Luna, 3
E-50015 Zaragoza

T Dracos
Fed Inst of Technology Zürich
Inst of Hydromech and Water Resources Management
CH-8093 Zürich-Hönggerberg

S Drobniak
Technical Univ of Czestochova
Thermal Machinery Inst
Al Zawadzkiego 21
PL 42-200 Czestochova

H Ekander
ABB Lab Research
S-721 78 Västerås

J W Elsner
Technical Univ of Czestochova
Thermal Machinery Inst
Al Zawadzkiego 21
PL 42-200 Czestochova

J Eriksson
The Swedish State Power Board
Älvkarleby Lab
S-81071 Älvkarleby

L-E Eriksson
Volvo Flygmotor AB
S-461 81 Trollhättan

I S Ertesvåg
Div of Thermodynamics
N-7034 NTH-Trondheim

H H Fernholz
HFI/TU Berlin
Strasse des 17 Juni 135
D-1000 Berlin 12

H E Fiedler
HFI/TU Berlin
Strasse des 17 Juni 135
D-1000 Berlin 12

J F Foss
Michigan State Univ
Dept of Mechanical Eng
B100 Research Complex-Eng
East Lansing, MI 48824
USA

U Frisch
Observatoire de Nice
BP No 139
F-06003 Nice Cedex

L Fuchs
Royal Inst of Technology
Inst of Gasdynamics
S-100 44 Stockholm

J C H Fung
Univ of Cambridge
DAMTP
UK-Cambridge CB3 9EW

Y Gagne
Inst de Mécanique de Grenoble
BP 53 X
F-38041 Grenoble Cedex

D Gasser
ETC Centrum
Inst für Fluiddynamik
CH-8092 Zürich

S Gavrilakis
IMHEF-DME-EPFL-Ecublens
CH-1015 Lausanne

W K George
Univ at Buffalo, SUIVY
Turbulence Research Lab
339 Jarvis Hall
Buffalo, N Y 14260
USA

D Gerich
HFI/TU Berlin
Müller-Breslau Strasse 8
D-1000 Berlin 12

S Ghosh
Univ of Cambridge
DAMPT
UK-Cambridge CB3 9EW

A Glezer
Univ of Arizona
Tucson, AZ 85721
USA

M-A Gonze
Inst de Mécanique de Grenoble
BP 53 X
F-38041 Grenoble Cedex

F Grinstein
Code 4410
Naval Research Lab
Washington DC 20375-5000
USA

J Groth
Royal Inst of Technology
Dept of Mechanics
S-100 44 Stockholm

H Gustavsson
Luleå Univ of Technology
Div of Fluid Mechanics
S-95187 Luleå

M R Hajj
The Univ of Texas at Austin
Dept of Civ Eng - WRW 215F
Austin, Texas 7871
USA

M Hallbäck
Royal Inst of Technology
Dept of Mechanics
S-100 44 Stcokholm

P E Hancock
Univ of Surrey
Dept of Mechanical Eng
Guildford
UK-Surrey GU2 5XH

D Henningson
M I T
Dept of Mathematics
Cambridge, MA 02139
USA

M F Hibberd
CSIRO Div of Atmospheric Research
Private Bag No 1
Mordialloc, Vic 3195
Australia

J Hoeven van der
HFI/TU Berlin
Strasse des 17 Juni 135
D-1000 Berlin 31

E Holmberg
Chalmers Univ of Technology
Dept of Theor Phys and Mech
S-412 96 Gothenburg

D A Humphreys
FFA Box 11021
S-161 11 Bromma

J C R Hunt
Univ of Cambridge
DAMPT
UK-Cambridge CB3 9EW

U Högström
Uppsala Univ
Dept of Meteorology
Box 516
S-751 20 Uppsala

G Iuso
Dipartimento di Ingegneria
Aeronautica e Spaziale
Politechnico di Torino
Corso Duce Degli Abruzzi, 24
I-10129 Torino

G H Jansen
EP Dept
Shell Research BVP O Box 1003
NL-1003 AA Amsterdam

J Jiménez
School of Aeronautics
Pl Cardenal Cisneros 3
E-280 40 Madrid

A Johansson
Royal Inst of Technology
Dept of Mechanics
S-100 44 Stockholm

B Johansson
Chalmers Univ of Technology
Dept of Appl Thermodyn and Fluid Mech
S-412 96 Gothenburg

G Johansson
Chalmers Univ of Technology
Dept of Appl Thermodyn and Fluid Mech
S-412 96 Gothenburg

H Joona
Royal Inst of Technology
Dept of Gasdynamics
S-100 44 Stockholm

Y S Kachanov
Inst of Theor and Appl Mech
630090, Novosibirsk
USSR

H J Kaltenbach
Inst für Physik der Atmosphäre
D-8031 Oberhaffenhofen

T Kanzaki
Kyushu Univ
Dept of Chemical Eng
Hakosaki, Fukuoka 812
Japan

A Karlsson
Royal Inst of Technology
Dept of Gasdynamics
S-100 44 Stockholm

R I Karlsson
Royal Inst of Technology
Dept of Gasdynamics
S-100 44 Stockholm

G A Khomenko
Space Research Institute
Profsoyuznaya ul 84/32
Moscow 117810
USSR

M Kiya
Hokkaido Univ
Dept of Mechanical Eng
Sapporo 060
Japan

H-H Kleuker
Univ Dortmund
Abt CT-TV
Postfach 500500
D-4600 Dortmund

B Klingmann
Royal Inst of Technology
Dept of Gasdynamics
S-100 44 Stockholm

J Klingmann
Lund Univ
Dept Heat and Power Eng
PO Box 118
S-221 00 Lund

S Komori
Kyushu Univ
Dept of Chemical Eng
Hakosaki, Fukuoka 812
Japan

G Kowalewski
Univ Erlangen-Nürnburg
Lehrstuhl für Technische Thermodyn
Weichselgarten 9
D-8520 Erlangen

E Krause
Aerodyn Inst der RWTH Aachen
Wüllnerstrasse zw 5 u 7
D-5100 Aachen

R Kristoffersen
Inst for Mekanik
N-7034 Trondheim-NTH

G A Kuzmin
USSR Academy of Sciences
Inst of Thermophysics
Prospekt Lavrentyez 1
Novosibirsk 630090
USSR

O König
HFI/TU Berlin
Sekr HF 1
Strasse des 17 Juni 135
D-1000 Berlin 12

T Lagerstedt
Alfa Laval AB
S-147 80 Tumba

M Landahl
M I T
Dept of Aeronaut and Astronaut
Cambridge, MA 02139
USA

O Leuchter
ONERA
Aerodynamics Dept
29, av de la Division Leclerc
F-92320 Chatillon-sous-Bagneux

M Levenstam
Royal Inst of Technology
Dept of Hydromechanics
S-100 44 Stockholm

Y Li
Royal Inst of Technology
Dept of Gasdynamics
S-100 44 Stockholm

O Likhachev
USSR Academy of Sciences
Inst of Thermophysics
Prospekt Lavrentyez 1
Novosibirsk 630090
USSR

A Lundbladh
Royal Inst of Technology
Dept of Mechanics
S-100 44 Stockholm

T R Lynn
HFI/TU Berlin
Müller-Breslau Strasse 8
D-1000 Berlin 12

L Löfdahl
Chalmers Univ of Technology
Dept of Appl Thermodyn and Fluid Mech
S-412 96 Gothenburg

H Makita
Toyohashi Univ of Technology
Dept of Engineering
1-1 Tempaku-cho, Toyohashi 441
Japan

N A Malik
Univ of Cambridge
DAMPT
UK-Cambridge CB3 9EW

W V R Malkus
M I T
Room 2-369
Cambridge, MA 02139
USA

J-D Marion
Dassault Aviation
78 Quai M Dassault
F-922 14 St Cloud

S Masuda
Keio Univ
Faculty of Science and Technology
3-14-1 Hiyoshi
Kohoku-ku, Yokohama 223
Japan

E Meiburg
Brown Univ
Div of Applied Mathematics
Providence, RI 02912
USA

M Meneguzzi
CERFACS
Ecole Nationale de la Meteorologie
42, av Gustave Coriolis
F-31057 Toulouse Cedex

O Metais
Inst de Mécanique de Grenoble
BP 53X I M Grenoble
F-38041 Grenoble Cedex

U Michel
DLR-Turbulenzforschung
Müller-Breslau Strasse 8
D-1000 Berlin 12

S D Mobbs
Univ of Leeds
Dept of Appl Mathematical Studies
UK-Leeds LS2 9JT

S S Moiseev
USSR Academy of Sciences
Space Research Inst
Profoyuznaya 84/32
Moscow 117810
USSR

H Mårtensson
Volvo Flygmotor AB
S-461 81 Trollhättan

F T M Nieuwstadt
Lab Aero & Hydrodynamics
Rotterdamse Weg 145
NL-2628 Al Delft

L Nycander
Teknikum
Uppsala Univ
Box 534
S-751 21 Uppsala

S G Nychas
Univ of Thessaloniki
Box No 453
GR-540 06 Thessaloniki

K J Nygaard
Univ of Arizona
Dept of Aerospace and Mech Eng
Tuscon, AZ 85721
USA

P Paranthoen
URA 230 CNRS
BP 118
F-76134 Mont Saint Aignan

R J Perkins
Univ of Cambridge
DAMPT
UK-Cambridge CB3 9EW

J C Phillips
Univ of Cambridge
DAMPT
UK-Cambridge CB3 9EW

A Pouquet
Observatoire de Nice
BP 139
F-06003 Nice Cedex

A A Praskovski
Central Aero-Hydrodynamic Inst
Zhukovsky 3
Moscow Region 140160
USSR

C H Priddin
Rolls Royce Pl
P O Box 31
UK-Derby DE2 8JB

S Raghu
HFI/TU Berlin
D-1000 Berlin 12

V I Vasanta Ram
Ruhr Univ Bochum
Inst für Thermo- und Fluidmekanik
D-4630 Bochum 1

J M Redondo
Univ of Cambridge
DAMPT
UK-Cambride CB3 9EW

J M Rees
Univ of Sheffield
Dept of Appl and Comp Mathematics
UK-S10 2TN Sheffield

P Rogberg
Royal Inst of Technology
Dept of Hydraulics Eng
S-100 44 Stockholm

P Rouchon
Centre Reacteurs et Processus
Ecole des Mines de Paris
60 bd St Michel
F-75006 Paris

H Rasmussen
Univ of Cambridge
DAMPT
UK-Cambridge CB3 9EW

I L Ryhming
IMHEF-DME-EPFL-Ecublens
CH-1015 Lausanne

A Sahlin
Royal Inst of Technology
Dept of Mechanics
S-100 44 Stockholm

Y Sakai
c/o Prof J C R Hunt
Univ of Cambridge
DAMPT
UK-Cambridge CB3 9EW

K Sassa
Toyohashi Univ of Technology
Dept of Energy Eng
1-1 Tempaku-Cho, 440 Toyohashi
Japan

P J Schmid
M I T
Dept of Aeronaut and Astronaut
Cambridge, MA 02139
USA

B N Semenov
USSR Academy of Sciences
Inst of Thermophysics
Prospekt Lavrentyev 1
Novosibirsk 630090
USSR

X Shen
Royal Inst of Technology
Dept of Gasdynamics
S-100 44 Stockholm

P Simonsson
Royal Inst of Technology
Dept of Gasdynamics
S-100 44 Stockholm

A-S Smedman
Uppsala Univ
Box 516
S-751 20 Uppsala

H Stapountzis
Univ of Thessaloniki
Dept of Mechanical Eng
Box 443
GR-Thessaloniki 540 06

C Staquet
Ecole Normale Supérieure de Lyon
Lab de Physique
46, allée d'Italie
F-69364 Lyon Cedex 07

L-G Sundström
Royal Inst of Technology
Dept of Mechanics
S-100 44 Stockholm

N Tillmark
Royal Inst of Technology
Dept of Gasdynamics
S-100 44 Stockholm

H Tinoco
The Swedish State Power Board
Älvkarleby Lab
S-810 71 Älvkarleby

R Tryman
FOA 272
Box 551
S-147 25 Tumba

A Tsinober
Tel-Aviv Univ
Faculty Engineering
Dept of Fluid Mech and Heat Transfer
Tel-Aviv 69978
Israel

J Tu
Royal Inst of Technology
Dept of Gasdynamics
S-100 44 Stockholm

R E Uittenbogaard
Delft Hydraulics
PO Box 177
NL-2600 mh Delft

F Unger
Lehrstuhl für Strömungsmechanik
Arcisstrasse 21
D-8000 München 2

L Valino
Menanica de Fluids ETS II Z
M de Luna 3
E-50 015 Zaragoza

J V Vassilicos
Univ of Cambridge
DAMPT
UK-Cambridge CB3 9EW

S I Voropayev
USSR Academy of Sciences
Inst of Oceanology
Krasikova Street 23
Moscow, 117259
USSR

P Wagner
HFI/TU Berlin
Sekr HF 1
Strasse des 17 Juni 135
D-1000 Berlin 12

M Wallace
The Univ of Maryland
Dept of Mechanical Eng
College Park, MD 20742
USA

D Wildermuth
Fed Inst of Technology Zürich
Inst of Hydromech and Water Resources Management
CH-8093 Zürich Hönggerberg

C Wilsby
Alliant Computer Systems AB
Finlandsgatan 52
S-164 75 Kista

C Yagüe
Univ of Cambridge
DAMPT
UK-Cambridge CB3 9EW

S Zahrai
Royal Inst of Technology
Dept of Hydromechanics
S-100 44 Stockholm

G M Zaslavsky
USSR Academy of Sciences
Space Reaearch Inst
Profsoyuznaya 84/32
841 32 Moscow
USSR

V Zeitlin
USSR Academy of Sciences
Inst of Atmospheric Physics
Pyzhevsky 3
109017 Moscow
USSR

O Zeman
Stanford Univ
Center for Turbulence Research
Stanford, CA 94305-3030
USA

Index of Contributors

GPSR Compliance

The European Union's (EU) General Product Safety Regulation (GPSR) is a set of rules that requires consumer products to be safe and our obligations to ensure this.

If you have any concerns about our products, you can contact us on ProductSafety@springernature.com

In case Publisher is established outside the EU, the EU authorized representative is:

Springer Nature Customer Service Center GmbH
Europaplatz 3
69115 Heidelberg, Germany

Batch number: 10761697

Printed by Printforce, the Netherlands